Air Transportation Systems Engineering

Air Transportation Systems Engineering

Edited by
George L. Donohue, Ph.D. (Editor)
George Mason University

Andres G. Zellweger, Ph.D. (Editor)
Embry–Riddle Aeronautical University

Herman Rediess, Ph.D. (Associate Editor)
Federal Aviation Administration

Christian Pusch (Associate Editor)
EUROCONTROL Experimental Center

Volume 193
PROGRESS IN ASTRONAUTICS AND AERONAUTICS

Paul Zarchan, Editor-in-Chief
MIT Lincoln Laboratory
Lexington, Massachusetts

Published by the
American Institute of Aeronautics and Astronautics, Inc.
1801 Alexander Bell Drive, Reston, Virginia 20191-4344

Copyright © 2001 by the American Institute of Aeronautics and Astronautics, Inc. Printed in the United States of America. All rights reserved. Reproduction or translation of any part of this work beyond that permitted by Sections 107 and 108 of the U.S. Copyright Law without the permission of the copyright owner is unlawful. The code following this statement indicates the copyright owner's consent that copies of articles in this volume may be made for personal or internal use, on condition that the copier pay the per-copy fee ($2.00) plus the per-page fee ($0.50) through the Copyright Clearance Center, Inc., 222 Rosewood Drive, Danvers, Massachusetts 01923. This consent does not extend to other kinds of copying, for which permission requests should be addressed to the publisher. Users should employ the following code when reporting copying from this volume to the Copyright Clearance Center:

1-56347-474-3 $2.00 + .50

Data and information appearing in this book are for informational purposes only. AIAA is not responsible for any injury or damage resulting from use or reliance, nor does AIAA warrant that use or reliance will be free from privately owned rights.

ISBN 1-56347-474-3

Progress in Astronautics and Aeronautics

Editor-in-Chief
Paul Zarchan
MIT Lincoln Laboratory

Editorial Board

John Binder
MathWorks, Inc.

Lt. Col. Steven A. Brandt
U.S. Air Force Academy

Fred DeJarnette
North Carolina State University

Leroy S. Fletcher
NASA Ames Research Center

Michael D. Griffin
Orbital Sciences Corporation

Phillip D. Hattis
Charles Stark Draper Laboratory, Inc.

Richard Lind
University of Florida

Richard M. Lloyd
Raytheon Electronics Company

Ahmed K. Noor
NASA Langley Research Center

Albert C. Piccirillo
Institute for Defense Analyses

Vigor Yang
Pennsylvania State University

Ben T. Zinn
Georgia Institute of Technology

Table of Contents

Preface xxi

Chapter 1   Introduction 1

Section I: U.S. and European ATM Systems—Similarities and Differences

Chapter 2   Air Traffic Management Capacity-Driven Operational Concept Through 2015 9
Aslaug Haraldsdottir, Robert W. Schwab, and Monica S. Alcabin, The Boeing Company, Seattle, Washington

Introduction 9
Preliminary Design for the NAS 9
Operational Concept Development 10
Functions, Agents, and Performance 11
ATM System Functional Structure 12
Capacity, Safety, and Separation Assurance 14
Capacity-Driven Operational Concept 17
National Level Flow Management 17
En Route and Outer Terminal Area 19
Approach/Departure Transition 20
Final Approach 22
Surface 23
Efficiency in Low Density En Route Airspace 23
Conclusions 24
References 24

Chapter 3   Comparison of U.S. and European Airports and Airspace to Support Concept Validation 27
Diana Liang, Federal Aviation Administration, Washington, D.C.; William Marnane, EUROCONTROL, Brussels, Belgium; and Steve Bradford Federal Aviation Administration, Washington, D.C.

Introduction 27
Assessment Territory 29
Metrics and Measures 29
Assessment and Findings 29
Conclusion 46
References 47

Chapter 4   Performance Review in Europe..................... 49
Xavier Fron, *EUROCONTROL, Brussels, Belgium*

Introduction 49
Background 49
European Challenge 50
Other Limitations on Growth 57
Conclusions 58

**Chapter 5   United States and European Airport Capacity Assessment
Using the GMU Macroscopic Capacity Model 61**
George L. Donohue and William D. Laska, *George Mason University, Fairfax, Virginia*

Introduction 61
MCM Approach 62
MCM Validation.. 64
MCM Assessment of U.S. and European Airports................... 64
MCM Comparisons.. 70
Conclusions 71
References 72

Section II: Economics of Congestion

**Chapter 6   Forecasting and Economic Analysis for Aviation
Systems Engineering 77**
Peter F. Kostiuk, *Logistics Management Institute, McLean, Virginia;*
and Eric M. Gaier, *Bates White and Ballentine, Washington, D.C.*

Introduction 77
Evaluating National Impacts of ATM Investments................... 79
Generating an Unconstrained Forecast 79
Generating a Constrained Forecast 81
Estimating and Closing the Performance Gap 84
Estimating Airline Benefits from ATM Investments 87
Overview of the Air Carrier Cost-Benefit Model.................... 88
Derivation of the Air Carrier Cost-Benefit Model 90
LVLASO Scenario 97
Conclusions 102
References 102

**Chapter 7   Impact of Air Traffic Management on Airspace
User Economic Performance.............................. 103**
Joseph H. Sinnott and William K. MacReynolds, Ph.D., *MITRE Corporation, McLean, Virginia*

Introduction 103
Airline Cost Drivers and ATM Actions............................ 104
Estimates of System-Wide Excess Cost to Airlines 106
Example of the Impact of ATM Improvements on Long-Term Airline
Costs: Fleet Utilization and ATM Improvements................. 110

The Larger Picture: The Influence of ATM on Demand-Related
Airline Decisions 111

**Chapter 8   Effects of Schedule Disruptions on the Economics
of Airline Operations.................................... 115**
Zalman A. Shavell, *MITRE Corporation, McLean, Virginia*

Introduction 115
Scope of Disruptions.. 117
Alternatives Available to the Airlines for
  Handling Disruptions.. 118
Cost Implications of Disruptions to the Airlines 118
Snowstrom Event at Boston...................................... 119
Aggregated Costs of Disruptive Events........................... 121
Conclusions 125

Chapter 9   Modeling an Airline Operations Control Center 127
Nicolas Pujet and Eric Feron, *Massachusetts Institute of Technology,
Cambridge, Massachusetts*

Introduction 127
Modeling Structure and Hypotheses.............................. 128
Model Identification and Calibration 132
Conclusions 141
References 141

Chapter 10   Pricing Policies for Air Traffic Assignment........... 143
Karine Deschinkel, *ENSAE, Toulouse, France*; Jean-Loup Farges, *ONERA—CERT,
Toulouse, France*; and Daniel Delahaye *LOG, Toulouse, France*

Introduction 143
Model Formulation 144
Identification Problem.. 147
Optimization Problem 148
Principle of Resolution 148
Numerical Experiments... 150
Conclusion and Future Work 156
References 157

Section III: Collaborative Decision Making

**Chapter 11   Improved Information Sharing: A Step Toward the
Realization of Collaborative Decision Making.................. 161**
Peter Martin, *EUROCONTROL Experimental Centre, Brétigny-sur-Orge,
France*; Alison Hudgell, *U.K. Defence Evaluation Research Agency, Great Malvern,
Worcestershire, United Kingdom*; Nicolas Bouge and Sophie Vial, *Aérospatiale,
Les Mureaux, France*

Introduction 161
Collaborative Decision Making 162
Project Overview.. 162

Airline Operational Aspects 162
Airport Operational Aspects.................................. 165
Information Gaps 166
Issues Outstanding 173
Conclusions 175
References 175

**Chapter 12   Air Traffic Control/Air Carrier Collaborative
Arrival Planning 177**
Cheryl Quinn, *NASA Ames Research Center, Moffett Field, California*; and Richard E. Zelenka, *Logicon/Sterling Federal Systems, Herndon, Virginia*

Introduction 177
Current Research: ATC/Air Carrier Information Exchange 180
Future Research 185
Human Factors Issues Associated with ATC-Airline
  Collaborative Tools 187
Conclusions 187
References 188

**Chapter 13   Data Flow Analysis and Optimization Potential
from Gate to Gate 191**
Matthias Poppe, *DFS Deutsche Flugsicherung GmbH, Langen, Germany*; and Georg Bolz, *Lufthansa AG, Frankfurt/Main, Germany*

Introduction 191
Definitions 192
ATM Process Model 192
ATM Process Model Simulations............................... 195
Identification of Potentials.................................... 198
Conclusions 201
References 203

**Chapter 14   Effect of Shared Information on Pilot/Controller
and Controller/Controller Interactions 205**
R. John Hansman and Hayley J. Davison, *Massachusetts Institute of Technology, Cambridge, Massachusetts*

Introduction 205
Why Humans Are Necessary in ATM............................ 206
ATM Interaction Architecture 208
Interaction Assumptions 209
Shared Information in Controller/Pilot Interactions 209
Shared Information in Pilot/Airline Interactions 214
Shared Information in Intrafacility Controller/Controller Interactions..... 215
Shared Information in Cross-Facility Controller/Controller
  Interactions.. 217
Shared Information in Airline/ATM Interactions.................... 221
Flight Information Object 222

Conclusions 223
References 223

Chapter 15 Modeling Distributed Human Decision Making in Traffic Flow Management Operations 227
Keith C. Campbell, Wayne W. Cooper Jr., Daniel P. Greenbaum, and Leonard A. Wojcik, *MITRE Corporation, McLean, Virginia*

Introduction 227
TFM Operations and Implications for Modeling 228
Baseline Schedule Disruption Scenarios Modeled by IMPACT 229
Airline and FAA Agents in IMPACT 231
Basic Analysis of Airline and FAA Decision Making
  with IMPACT 231
Other Analyses with IMPACT 235
Conclusions 236
References 237

Chapter 16 Assessing the Benefits of Collaborative Decision Making in Air Traffic Management 239
Michael O. Ball, *University of Maryland, College Park, Maryland*; Robert L. Hoffman, *Metron Scientific Consulting, Inc., Reston, Virginia*; Dave Knorr and James Wetherly, *Federal Aviation Administration, Washington, D.C.*; and Mike Wambsganss, *Metron Scientific Consulting, Inc., Reston, Virginia*

Introduction 239
Improvements in the Quality of Information and Information
  Distribution 240
System and User Impact 245
Collaborative Routing 249
Conclusions 250
References 250

Section IV: Airport Operations and Constraints

Chapter 17 Fast-Time Study of Airline-Influenced Arrival Sequencing and Scheduling 253
Gregory C. Carr and Heinz Erzberger, *NASA Ames Research Center, Moffett Field, California*; and Frank Neuman, *Raytheon STX Corporation, Moffett Field, California*

Introduction 253
Priority-Scheduling 254
Scope 255
Fast-Time Simulation 255
Order Deviation 261
Simulation Inputs/Outputs 262
Results and Discussion 263
Conclusions 266
References 267

Chapter 18   Capacity-Related Benefits of Proposed Communication, Navigation, Surveillance, and Air Traffic Management Technologies   269
Tara J. Weidner, *Seagull Technology, Inc., Los Gatos, California*

Introduction 269
Assumed Technology Scenarios 269
Capacity-Related Benefits Defined 271
Analysis Methodology Overview 272
Model Assumptions and Results 278
Conclusions 279
References 286

Chapter 19   Collaborative Optimization of Arrival and Departure Traffic Flow Management Strategies at Airports 289
Eugene P. Gilbo, *John A. Volpe National Transportation Systems Center, Cambridge, Massachusetts*; and Kenneth W. Howard, *Arcon Corporation, Waltham, Massachusetts*

Introduction 290
Mathematical Formulation 292
Numerical Examples 295
Conclusions 302
References 303

Chapter 20   Analysis, Modeling, and Control of Ground Operations at Hub Airports 305
Kari Andersson and Francis Carr, *Massachusetts Institute of Technology, Cambridge, Massachusetts*; William D. Hall, *Charles Stark Draper Laboratory, Cambridge, Massachusetts*; and Nicolas Pujet and Eric Feron, *Massachusetts Institute of Technology, Cambridge, Massachusetts*

Introduction 305
Available Data 307
Models 314
Applications 331
Conclusions 339
References 340

Chapter 21   Conceptual Design of a Departure Planner Decision Aid   343
Ioannis Anagnostakis, Husni R. Idris, John-Paul Clarke, Eric Feron, R. John Hansman, Amedeo R. Odoni, and William D. Hall, *Massachusetts Institute of Technology, Cambridge, Massachusetts*

Introduction 343
Departure Process—Results from Field Observations 344
Overview of the Proposed Departure Planner Architecture and
    Operational Context 347
Conclusions 364
References 365

Chapter 22 Modeling Air Traffic Management Automation Metering Conformance Benefits 367
Tara J. Weidner, *Seagull Technology, Inc., Los Gatos, California*; and Steve Green, *NASA Ames Research Center, Moffett Field, California*

Introduction	367
ATM Interruptions Model	368
Illustrative Application	375
Conclusions	381
References	382

Section V: Airspace Operations and Constraints

Chapter 23 Effect of Direct Routing on Air Traffic Control Capacity 385
S. A. N. Magill, *Defence Evaluation and Research Agency, Malvern, Worcestershire, United Kingdom*

Introduction	385
Workload and Capacity	386
Simulation	387
Results	389
Discussion	394
Concluding Remarks	395
References	396

Chapter 24 Performance Measures for Future Architecture 397
Steve Bradford, Dave Knorr, and Diana Liang, *Federal Aviation Administration, Washington, D.C.*

Introduction	397
Architecture	398
Metrics	398
Architecture and Performance	399
Analysis	401
Conclusions	407

Chapter 25 Analytical Identification of Airport and Airspace Capacity Constraints 409
William R. Voss, *Federal Aviation Administration, Washington, D.C.*; and Jonathan Hoffman, *MITRE Corporation, McLean, Virginia*

Introduction	409
Background	410
How to Find Airspace Problems	410
Definition of an Airspace Problem	412
Data Sources	414
Results	414
Conclusions	419
References	419

Chapter 26 Operational Assessment of Free Flight Phase 1 Air Traffic Management Capabilities 421
Dave Knorr, *Federal Aviation Administration, Washington, D.C.*; Joseph Post and Jeff Biros, *CNA Corporation, Alexandria, Virginia*; and Michelle Blucher, *MITRE Corporation, McLean, Virginia*

Introduction 421
System Description 422
Collaborative Approach 424
Metrics Definitions 425
Measurement Process 429
Preliminary pFAST Results 430
Conclusions 434
References 435

Chapter 27 CENA-PHARE Experiment: Requirements for Evaluation of Novel Concepts in Air Traffic Control. 437
Didier Pavet, *CENA, Athis-Mons, France*

Introduction 437
Evaluation Methodology 439
Lessons Learned 441
Discussion: Requirements for Future Developments of Novel Concepts 445
Conclusions 446
References 447

Chapter 28 Restriction Relaxation Experiments Enabled by User Request Evaluation Tool 449
Michael J. Burski, *Federal Aviation Administration, Washington, DC*; and Joseph Celio, *MITRE Corporation, McLean, Virginia*

Introduction 449
URET Utilization 450
URET Benefits 451
Conclusions 460
References 460

Section VI: Safety and Free Flight

Chapter 29 Accident Risk Assessment for Advanced Air Traffic Management 463
H. A. P. Blom, G. J. Bakker, P. J. G. Blanker, J. Daams, M. H. C. Everdij, and M. B. Klompstra, *National Aerospace Laboratory NLR, Amsterdam, The Netherlands*

Introduction 463
Accident Risk Assessment Methodology 467
Mathematical Framework 471
RNP1 in Conventional and Airborne Separation Assurance
  Scenario Examples 474

Concluding Remarks 476
References 477

**Chapter 30   Human Cognition Modelling in Air Traffic
              Management Safety Assessment 481**
  Henk A. P. Blom, Jasper Daams, and Herman B. Nijhuis, *National
  Aerospace Laboratory NLR, Amsterdam, The Netherlands*

Introduction 481
Human Modeling Approaches 483
Modeling for En-Route ATC 488
Reduction of the ATCo Model 497
Example Application 503
Concluding Remarks 507
References 509

**Chapter 31   Probabilistic Wake Vortex Induced Accident
              Risk Assessment 513**
  J. Kos, H. A. P. Blom, L. J. P. Speijker, M. B. Klompstra, and G. J. Bakker,
  National Aerospace Laboratory NLR, Amsterdam, The Netherlands

Introduction 513
Risk Assessment Methodology 514
Wake Vortex Risk Assessment 516
Single Runway Approach 519
Concluding Remarks 524
Appendix: Stochastic Wake Vortex Model 525
References 530

Chapter 32   Free Flight in a Crowded Airspace? 533
  J. M. Hoekstra, R. C. J. Ruigrok, and R. N. H. W. van Gent, *National Aerospace
  Laboratory NLR, Amsterdam, The Netherlands*

Introduction 533
Free Flight 533
Air Traffic Growth 534
NLR Free Flight Study 534
Distrust in Distributed System 537
(Un)Predictability of a Distributed System 538
Complex Geometry Examples 539
Robustness and Redundancy of a Distributed System 542
Effective Conflict Rate for Air and Ground 543
Conclusions 543
References 544

**Chapter 33   Managing Criticality of Airborne Separation Assurance
              Systems Applications 547**
  Andrew D. Zeitlin, *MITRE Corporation, McLean, Virginia*;
  and Béatrice Bonnemaison, *CENA, CS-SI, Toulouse, France*

Introduction 547
Operational Safety Assessment of ASAS 548

Operational Environment of ASAS Applications. 551
Operational Hazards and Mitigating Factors Associated with ASAS 552
Operational Hazard Identification. 553
Allocation of Safety Objectives and Requirements for ASAS
    Applications 556
ASAS Simulations and Trials. 559
Conclusions and Future Work 560
References 560

Chapter 34   Analysis of Aircraft Separation Minima Using a Surveillance State Vector Approach 563
    Tom G. Reynolds and R. John Hansman, *Massachusetts Institute of Technology, Cambridge, Massachusetts*

Introduction 563
Model of a Separation Assurance Budget 564
Need for Surveillance of Intent 566
State Vector Modeling Approach 567
Intent States $I(t)$. 569
State Uncertainty. 571
Relationships Between State Uncertainty and the Current
    Separation Minima 574
Conformance Monitoring. 577
Conclusions 581
References 581

Section VII: Cognitive Workload Analysis and the Changing Role of the Air Traffic Controller

Chapter 35   Passive Final Approach Spacing Tool Human Factors Operational Assessment 585
    Katharine K. Lee, *NASA Ames Research Center, Moffett Field, California*; and Beverly D. Sanford, *Cadence Design Systems, Inc., San Jose, California*

Introduction 585
Methods 586
Results 589
Lessons Learned 593
Concluding Remarks. 596
References 597

Chapter 36   Evaluating Taskload Measures Derived from Routinely Recorded Air Traffic Control Data 599
    Carol A. Manning, *Federal Aviation Administration Civil Aeromedical Institute, Oklahoma City, Oklahoma*; Scott H. Mills, *SBC Technology Resources, Inc. Austin, Texas*; Cynthia M. Fox and Elaine Pfleiderer, *Federal Aviation Administration*

Civil Aeromedical Institute, Oklahoma City, Oklahoma; and Henry Mogilka, *Federal Aviation Administration Training Academy, Oklahoma City, Oklahoma*

Introduction 599
Defining Controller Workload, Taskload, Sector Complexity,
  and Performance 600
Purpose of Study 603
Method 603
Results 606
Conclusions 609
References 613

Chapter 37   Controller Roles—Time to Change   615
   Robert Graham, Alan Marsden, Isabelle Pichancourt, and Franck Dowling,
   EUROCONTROL Experimental Centre, Brétigny-sur-Orge, France

Introduction 615
Controller Tools and Transition Trials—C3T 616
C3T Study Concepts 617
Model-Based Study—Hypothesis 618
Model Based Study Scenarios 618
RAMS Model 620
Model-Based Study Preliminary Results 620
Real-Time Simulation Hypotheses 622
Real-Time Simulation—Preliminary Results 623
Conclusions 625
References 626

**Chapter 38   Trajectory Orientation: Technology-Enabled Concept
   Requiring Shift in Controller Roles and Responsibilities 627**
   Kenneth J. Leiden, *Micro Analysis and Design, Boulder, Colorado*;
   and Steven M. Green, *NASA Ames Research Center, Moffett Field, California*

Introduction 627
Background 628
Trajectory Orientation Concept 629
Research Approach 632
Results and Discussion 633
Conclusions 644
References 644

Section VIII: Emerging Issues in Aircraft Self-Separation

**Chapter 39   Cooperative Optimal Airborne Separation Assurance in Free
   Flight Airspace 649**
   Colin Goodchild, Miguel A. Vilaplana, and Stefano Elefante, *University
   of Glasgow, Glasgow, United Kingdom*

Introduction 649
Operational Methodology 651

Planning Algorithm... 652
Computed Example... 655
Conclusions 660
References 662

Chapter 40 Operational Efficiency of Maneuver Coordination Rules for Airborne Separation Assurance System.................... 665
    R. Schild and J. K. Kuchar, *Massachusetts Institute of Technology, Cambridge, Massachusetts*

Introduction 665
Evaluation of Rule Systems.................................. 666
Rule Design 667
Rule Evaluation Criteria 668
Rule Evaluation.. 670
Human Factors and Rules 674
Conclusions 675
References 676

Chapter 41 Probabilistic Approaches Toward Conflict Prediction.. 677
    G. J. Bakker, H. J. Kremer, and H. A. P. Blom, *National Aerospace Laboratory NLR, Amsterdam, The Netherlands*

Introduction 677
Conflict Prediction Approaches 678
Collision Risk Modeling 680
Comparison of Approaches 681
Discussion of Results 689
Conclusions 692
References 693

Chapter 42 Safe Flight 21: 1999 Operational Evaluation of Automatic Dependent Surveillance Broadcast Applications 695
    James J. Cieplak, Edward Hahn, and Baltazar O. Olmos, *MITRE Corporation, McLean, Virginia*

Introduction 695
Operational Evaluation 1999 696
Method of Test 699
Results 703
Conclusions 711
Selected Bibliography 712

Chapter 43 Conclusions and Observations 713
Introduction 713
U.S. Air Traffic Management System......................... 713
European Air Traffic Management System 714

Public-Private Nature of Air Transportation 714
Safety Is Much Discussed But Little Analyzed 715
Air Traffic Controller—Pilot Cognitive Workload Substitution Function .. 716
Final Comments 716

Index 717

Preface

This book represents a selection of research papers that were presented at two closed forum research meetings held in 1998 and 2000. The United States Federal Aviation Administration and the European EUROCONTROL sponsored these meetings. In December 1995, Jack Fearnsides, Director and General Manager of the Center for Advanced Aviation Systems Development of the MITRE Corporation, and Jean-Marc Garot, Director of the EUROCONTROL Experimental Center, proposed the formation of a joint research seminar to be held approximately every 18 months. These research seminars are designed to share the latest and best of research findings on the complex and emerging field of air traffic management (ATM). These seminars produced formal papers presented in 1997 (Saclay, France), 1998 (Orlando, Florida), and 2000 (Napoli, Italy). The proceedings of these seminars are available on the EUROCONTROL-maintained web site http://www.eurocontrol.fr/atmsem/index.htm. The interested student of air traffic management technology development can find the complete proceedings on this web site.

    The purpose of this book is to select a subset of the papers presented in 1998 and 2000 (Orlando, Florida, 1–4 December 1998 and Napoli, Italy, 13–16 June 2000) and to organize them in a logical sequence. The editors' selections do not necessarily represent our view that these are the only good papers presented in these forums, but represent a collection that best explains a growing understanding of the technical nature of a very complex international air transportation system. To date, there has been no comprehensive collection of system performance data and analysis of this data to identify critical system metrics, a system that is showing the signs of being so successful that its growth is approaching the physical infrastructure capacity limits. Unlike the highway traffic engineering subdiscipline of civil engineering, there is no engineering discipline that deals with the more complex air transportation system. Although there are textbooks on how the current air traffic control system works, there are no books devoted to the underlying theory of the air transportation system. This book is the first that attempts to address the breadth of technical details and the complex factors that drive and limit the air transportation system.

    The editors of this book bring a comprehensive knowledge of the field and the international literature. The principle editor, George L. Donohue, was the Associate Administrator of the U.S. Federal Aviation Administration (FAA) for Research, Engineering and Acquisition from 1994 to 1998. In addition to developing the National Airspace System (NAS) Architecture 4.0, he encouraged Jack Fearnsides and Andres Zellweger (Director of FAA Aviation Research at the time) to initiate the U.S. and European research forums that have produced the papers in this book. Herman Rediess is currently the FAA Director of Aviation Research and was the U.S. cosponsor of the seminars in 1998 and 2000. Christian Pusch was in charge of organizing and sponsoring the 2000 seminar and is currently Head of Research and Development Coordination at the EUROCONTROL Experimental Center, near Paris, France.

To keep these research seminars to a workable size that facilitates a maximum of technical exchange, they have been by invitation only and therefore have provided a limited exposure of the work to a larger audience. In 1997, there were only 60 participants discussing 40 invited papers. In 1998 an international call for papers was issued, and the seminar grew to 110 participants discussing 50 papers selected from a field of 108 papers submitted. In 2000 there were approximately 150 participants discussing 64 papers selected from a field of 127 papers submitted.

The collection of research papers in this book are an initial attempt by the editors to make a comprehensive description of the current state of knowledge available to the interested student of this new emerging subfield of transportation engineering. The chapters are organized into eight sections:

I. U.S. and European ATM Systems—Similarities and Differences

II. Economics of Congestion

III. Collaborative Decision Making

IV. Airport Operations and Constraints

V. Airspace Operations and Constraints

VI. Safety and Free Flight

VII. Cognitive Workload Analysis and the Changing Role of the Air Traffic Controller

VIII. Emerging Issues in Aircraft Self-Separation

Overall, this book includes 43 chapters written by over 85 authors and co-authors.

George L. Donohue
July 2001

Chapter 1

Introduction

AIR transportation refers to the movement of people and material through the third dimension, usually in heavier-than-air vehicles. These vehicles range from 400 lb (182 kg) powered parachutes transporting one person 25 miles (46 km) to 800,000 lb (364,000 kg) jumbo jet aircraft transporting 350 passengers 9000 miles (16,700 km). In fact, jet aircraft have now been designed to the point that they can connect virtually any two points on earth nonstop in less than a day.

The invention and development of the jet aircraft in World War II have led to the use of aircraft as a major mode of both domestic and international transportation. Since 1960, the year that the U.S. Department of Transportation began collecting statistics, the air mode of transportation has grown over six times faster than any other mode of transportation in the United States (that also happens to be over six times faster than the rate of gross domestic product growth). The International Civil Aviation Organization (ICAO) states that more than one third of all international cargo by value was shipped by air in 1998. It should come as little surprise that the technical and physical infrastructure is feeling the strain of this sustained growth rate.

In the United States, there are over 7500 aircraft (over 4500 in Europe) in commercial service at the turn of the century. Roughly 67% of these aircraft are powered by high bypass ratio fanjets, the rest are powered by either gas turbine or piston driven propellers. The fanjet aircraft prefer to fly above 30,000 ft in altitude, whereas the propeller aircraft prefer to fly below 20,000 ft. Aircraft flying above 10,000 ft are usually pressurized because of the lack of adequate oxygen required for passenger comfort and/or survival. The United States operates approximately 40% of the world's commercial air transportation. In addition, the United States has a considerable use of aircraft for private transportation. In contrast, private aircraft play a much less important role in Europe.

There are over 190,000 registered private aircraft in the United States (over 10,000 turboprop or turbojet) with over 600,000 registered pilots. On any given day, there are over 5000 aircraft in the air (between 1000 and 2200 hrs) under positive separation control by the Federal Aviation Administration (FAA) air traffic control (ATC) system. Of this amount, approximately one-third are involved in private transportation. There are also approximately three times this amount of private aircraft in the air that are not under FAA positive control. Europe operates an air transportation system that is approximately 65% the size of the U.S. system, but with very little private air transportation activity. Africa, South America, and Australia operate a considerable amount of private air transportation in addition to commercial air transportation because of the large intercity distances and lack of substantial ground transportation infrastructure.

At the end of World War II, international travel by air became increasingly popular. In 1944, the ICAO was formed as part of the United Nations to regulate international civil aviation. There are approximately 180 member countries at the beginning of the 21st century. Each member country must have a Civil Aviation Authority (CAA) to provide communications, navigation, surveillance, and air traffic management (CNS/ATM) services to internationally accepted standards. For the United States, this agency is the FAA. In addition to the provision of CNS/ATM services, each country must provide aircraft safety oversight for the certification of aircraft airworthiness and aircraft operation. Until recently, these two functions (CNS/ATM and safety oversight) have been supplied by the same government agency. Since 1990, there has been a trend to privatize (through different means ranging from wholly owned government organizations to complete privatization) the provision of CNS/ATM services and retain government safety oversight.

The CNS/ATM function has evolved from the 1920s provision of primitive navigation and communications services to a highly computerized ATM system with central flow control management (CFCM) utilizing space-based communications and navigation equipment. With the advent of radar in World War II, the surveillance function was added to the CAA's provision of services in the late 1950s. The physical limitations of radar at that time set the aircraft separation standards that are still in use today. These separation standards (in conjunction with the number of runways that are available) set the maximum operational capacity that the air transportation system can support. These separation standards are typically 5 n miles in high-altitude airspace (i.e., above 18,000 ft) and 3 n miles within 60 n miles of an airport (typically in low-altitude airspace). Airspace that does not have radar surveillance must maintain procedural separation using aircraft onboard navigation position fixes and ATC communications. These separation standards range from 60 to 100 n miles and are used in oceanic airspace, nonradar airspace, and in undeveloped countries that lack radar services.

The radar physical properties that dictate these standards are beam width and sweep rate. Primary radars have narrow beam widths in azimuth (e.g., 1.4 deg typical) but wide beam widths in the vertical (e.g., in excess of 30 deg). Secondary radars were added to the ATC system to provide cooperative altitude reporting from the aircraft being interrogated using an onboard pressure altimeter and a radar transponder. Over time, this data link added aircraft identification. The aircraft identification allows the ATC computers to correlate each aircraft with its preflight plan and, therefore, display aircraft identification, origin, departure time, destination, estimated time of arrival, altitude, and speed. Today this information is provided to both the CAA-operated ATC centers and also to the airline air operations centers (AOC). This shared information and situational awareness forms the basis for the developing operational procedures known as collaborative decision making (CDM).

In practice, aircraft are routinely maintained at 7–30 miles separation (i.e., in excess of radar limitations) because of air traffic controller cognitive workload limitations. A typical controller can maintain situational awareness on 4–7 aircraft at a time. When airspace sector loading exceeds this amount, controller teams work to maintain aircraft separation. These teams can be as high as three controllers per sector. In the United States, there are over 730 en route sectors and in Europe there are over 460 sectors. The number of sectors that are available to high-density airspace in the United States and Europe is limited by the number of communication channels that are available to the CAA. The number of communications channels available is dictated by the technical efficiency in which the allocated radio spectrum is utilized.

The radio spectrum is allocated and controlled by the International Telecommunications Union (ITU), also a United Nations charter organization. Today most ATC communications are conducted utilizing either 25 kHz or 8.33 kHz double side-band, amplitude modulated VHF frequencies between 108 MHz and 139 MHz. A shift to digital communications began over 20 years ago with ARINC (Aeronautical Radio, Inc.) providing aircraft to AOC digital communications over 25 kHz channels in the 139 MHz frequency range. This data link became known as ACARS (ARINC Communications Addressing and Reporting System) and is a 2400-Bd character oriented link. For the last five years, the FAA has been using this data link to provide predeparture clearances for over 40 high-capacity hub airports in the United States. After 20 years of data communication traffic growth, ARINC is migrating to a carrier sense multiple access (CSMA), fully digital (using D8PSK protocol), 31.5 kbaud data link to accommodate the increasing message traffic. Also, aircraft flying in international oceanic airspace are beginning to utilize the IMARSAT (International Maritime Satellite) data link for Future Air Navigation System (FANS) equipped aircraft to provide position reports, gradually replacing the old HF voice communications system.

The ATC providers are still debating within the ICAO radio navigation forum the exact international standard and implementation timeline for a fully digital ATC communications system. Although there are strong system inter-relationships between the CNS functions and the ATM function, this book is emphasizing only the ATM function. There are several good textbooks on the subject of wireless communications, and this book will not treat this subject any further. A good book that discusses many of the CNS systems commonly used in aviation is *Avionics Navigation Systems*, 2nd edition, by Myron Kayton and Walter Fried.

Most of the world allocates air transportation routes through government agencies. In the United States, prior to 1978, the Civil Aviation Board (CAB) controlled the allocation of routes that commercial air carriers could provide. In 1978, the U.S. government deregulated the air transportation industry and allowed economic forces to shape the air transportation network. This system evolved very quickly to a hub and spoke network. At the beginning of the 21st century, there are approximately 60 hub airports in the United States with a maximum capacity of about 40 million operations per year. Decreasing aircraft separation in the final approach to a runway from an average of 4 n miles between aircraft (the practical limit due to $+/-1$ mile variance of today's system) to 3 n miles could increase this capacity to over 55 million operations per year.

This increased capacity could be achieved by migrating from the use of radar surveillance to the use of aircraft broadcast Global Positioning System (GPS) satellite navigation fixes over a wireless digital data link (this is referred to as automatic dependent surveillance broadcast or ADS-B). This capacity increase cannot be realized, however, without a change in the en route and terminal separation procedures used by air traffic controllers due to human cognitive workload limitations. Today, in both the United States and Europe, approximately 2 min or more of delay per aircraft can be attributed to saturation of the terminal and/or en route sectors. At individual high-density airports, these average delays can be as high as 10 min per aircraft at airport capacity fractions (cf) of over 0.9.

Queuing theory would predict that airport delays will be proportional to $cf/(1-cf)$. Increasingly, a central flow control function is being used to institute ground delay programs to anticipate these delays and hold aircraft on the ground at the point of origin rather than in the air at the point of destination. In the United States, these delays are frequently triggered by a weather event at one or more of the hub airports. It is not clear whether or not a central flow control function can eliminate or even reduce delays at airports with $cf > /0.8$ because of the current first-come-first-served (FCFS) runway assignment protocol and the inherent uncertainties involved with an aircraft flying through a time varying atmosphere.

With this background in mind, one should now realize that there are five main actors that control the utilization of the air transportation system: 1) the CAAs in the provision of regulations and aircraft separation/flow control standards and services; 2) the airlines in their utilization of aircraft enplanement capacity/modern avionics and the utilization of ATM information in their air operations centers; 3) the airport operators in their provision of airport infrastructure; 4) the private aircraft operators in their provision of suitably equipped aircraft and cooperative airspace utilization; and finally, 5) the flying public (who ultimately pay for the services and the infrastructure), who will suffer the consequences of failure to modernize the system. For the capacity and quality of service to increase in the 21st century, each of these players will have to make substantial capital investments in new equipment and/or significantly revise their operational procedures. If these changes are not done, the flying public will pay the additional price in loss of transportation mobility, safety, and economy.

Four chapters are included in the next section to briefly describe the current and envisioned future ATM operational concepts of the United States and of Europe. This section is then followed by five chapters in a section on the economics of the airlines and their linkage to the operations of the ATM system. The next section discusses the emerging practice of CDM as a new paradigm for dynamic optimization of the distributed air transportation command and control structure. The next two sections discuss in some detail the operations in and near airports (six chapters) and en route airspace (six chapters), respectively. These sections not only provide details on operations and current metrics but also discuss early field evaluations of computer embedded decision support systems that are hoped to improve the capacity and productivity of the future ATM system.

It is rare to find a discussion of the relationship between air transportation safety and capacity. It is intuitively obvious, however, that at some separation level the technology will become inadequate to prevent aircraft collisions. This implies that safety and capacity are inversely related in the limit. Figure 1 illustrates a hypothetical inverse relationship between safety and capacity at different levels of technical capability. These curves are analogous to the inter-relationship described in economics theory as labor-capital substitution curves with constant technology isoquants. Over the last 30 years, the world commercial aircraft hull loss rate has been approximately 15 hull losses per year (approximately 20 million departures per year with about 1 million departures per hull loss). The United States operates at about 12 million departures per year with 6 million departures per hull loss and Europe operates at about 7 million departures per year with the same hull loss rate. This has allowed capacity to increase by increasing commercial aircraft safety features such as flight management systems (FMS) and traffic collision Avoidance Systems (TCAS). This continuous improvement in aircraft safety will eventually reach the ATC separation technology limit shown in Fig. 1 and further

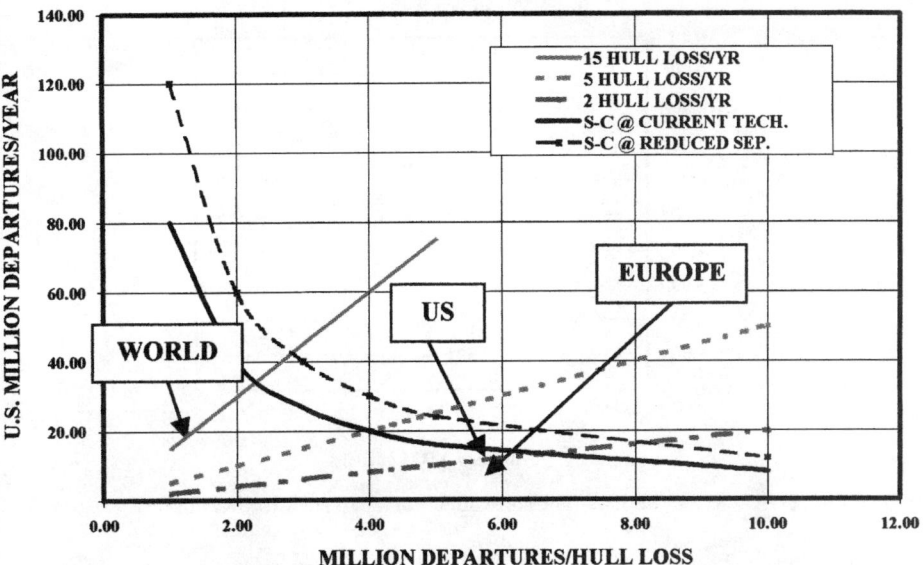

Fig. 1 Hypothetical safety-capacity substitution curves for the U.S. ATC system.

increases in capacity will result in a decrease in overall system safety resulting in increased annual hull loss rates. To avoid this increase in aircraft hull loss rate, a new separation technology must be adopted to move to a new safety-capacity tradeoff curve.

The research communities in both the United States and Europe are converging on the option of transferring aircraft separation authority and responsibility to the aircraft flight deck. The combination of GPS navigation accuracy and wireless digital communication systems is allowing ADS-B systems to provide very precise aircraft location and situational awareness directly to the aircraft flight deck and FMS. This has the potential to decrease the ATM controller workload to allow more aircraft per sector by using the controller to supervise traffic flow control while the pilots are assuring aircraft separation. A hypothetical pilot-controller workload substitution curve is illustrated in Fig. 2. The exact relationship between pilot and controller workload is unknown. If one assumes that it is a simple inverse relationship, even a simple transfer of separation authority could lead to a significant decrease in controller workload. An understanding of this relationship will be required to move to the enhanced technology, higher capacity isoquant shown in Fig. 1.

Sections VI-VIII address the safety issues associated with both the current ATM systems in the United States and Europe and the more important issue of how changes in the system may either increase or decrease safety. Section VI presents five chapters that expressly address the analysis of ATM safety. The next section then presents four chapters on the assessment of ATC controller's cognitive workload and the necessity of changing the role of controllers in the future system. The next section presents four chapters on the emerging trend toward allowing aircraft to provide more self-separation as new technology provides more timely

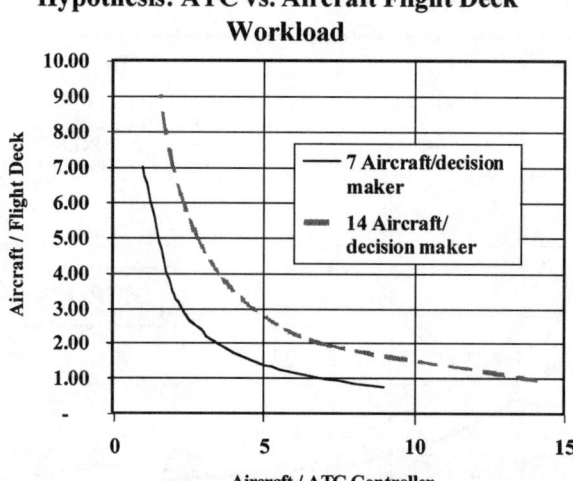

Fig. 2 Hypothetical pilot-controller workload substitution curves.

high-quality information to the aircraft flight deck and the controller's workload is approaching a limit. The final section presents the editors' view of what this all means to future design of the international air transportation system and the technical and operational changes that must occur over the next 5–10 years to keep the system from serious decline as an economical and reliable means of both domestic and international passenger and cargo transportation.

The safety-capacity relationship hypothesized in Fig. 1 was also independently proposed by Dr. Henk Blom in 1993 in an internal, unpublished NLR report.

Section I: U.S. and European ATM Systems—Similarities and Differences

Four chapters are presented in this section that provide an introduction to the two largest and most complex air traffic management (ATM) systems in the world. The chapters in this section present information about the behavior of the U.S. and the European systems—how they operate and how they perform. The latter, of course, is of great interest currently because of the growing delay picture in both Europe and, more recently, the United States. A better understanding of the two systems and their similarities and differences are an important first step in finding solutions to the projected performance problems. It is widely held that these problems must be addressed because the alternative of constrained aviation system activity would have a large negative effect on U.S. and European economic growth.

Perhaps the first step in gaining an understanding of how a system behaves is to develop a concept of operations. This takes the system mission and describes the functions the system has to carry out, the components of the system, and the behavior/performance and interaction of these components. Operations concepts of a future state of a system are valuable because they help guide the researchers and system designers in the development system improvements. Alternative concepts for a future state can provide the basis for comparison and decision about where investments should be made.

Chapter 2 gives an excellent approach to developing an operations concept for the U.S. ATM system for the year 2015. The authors postulate the capacity needs for 2015, develop a concept that will meet those needs, and demonstrate how today's ATM system can transition to the 2015 system. The chapter contains an excellent description of the functional structure of the U.S. ATM system and describes the control loops that make up the system. In Chapter 3, the authors provide operations concepts to show similarities and differences in U.S. and European ATM systems in the 2000–2010 timeframe. The authors present an ATM concept model and uses a network view to explore the potential changes that could bring about increases in capacity.

Chapter 4 contains a discussion of a formal ATM System Performance Review, established in Europe in 1998 with a view toward improving system performance. The performance review group will publish performance indicators for the EUROCONTROL area, propose performance targets, and develop economic regulation guidelines. Performance indicators of interest are access, flexibility, availability, predictability, delay, flight efficiency, and cost effectiveness.

In the last chapter of this section, the authors describe a macroscopic capacity model and, using this model, assess U.S. and European air transportation capacity. The European analysis is based on the 16 top airports in Europe. In the United States the northeast triangle, a region similar in size and also consisting of 16 airports, is analyzed for the comparison. The results give one cause for concern because, particularly in the United States, the system is operating at capacity levels that are in a region where, according to queuing theory, delays increase hyperbolically. The chapter ends with an interesting discussion of where the limiting factors lie in the United States and in Europe.

Chapter 2. Air Traffic Management Capacity-Driven Operational Concept Through 2015, Aslaug Haraldsdottir, Robert W. Schwab, and Monica S. Alcabin, The Boeing Company, 1998.

Chapter 3. Comparison of U.S. and European Airports and Airspace to Support Concept Validation, Diana Liang, Steve Bradford, FAA; and William Marnane, EUROCONTROL, 2000.

Chapter 4. Performance Review in Europe, Xavier Fron, EUROCONTROL, 1998.

Chapter 5. United States and European Airport Capacity Assessment Using the GMU Macroscopic Capacity Model, George L. Donohue and William D. Laska, George Mason University, 2000.

Chapter 2

Air Traffic Management Capacity-Driven Operational Concept Through 2015

Aslaug Haraldsdottir,[*] Robert W. Schwab,[†] and Monica S. Alcabin[‡]

The Boeing Company, Seattle, Washington

I. Introduction

THIS chapter describes an approach to developing an operational concept and architecture trades for the U.S. National Airspace System (NAS). The approach includes an analysis of the system mission, and the functions that the system must perform with the available resources, and presents a logical functional structure for the system that ties together functions, resources, and subsystems. A key aspect of this approach is a flowdown of performance requirements from top-level system performance goals through the concept layers down to the technology level to ensure that design decisions will lead to a system that delivers the desired performance. A preliminary design process for the NAS is proposed, consistent with the systems engineering approach that is used in commercial and military aircraft development programs, adapted to the operational and technical characteristics of the air transportation system.

II. Preliminary Design for the NAS

Figure 1 illustrates a preliminary design process to quantify long-range air transportation system needs, define operational concepts, and translate these needs into subsystem technical performance requirements. The preliminary design will provide trade data relating performance levels, life cycle costs, operational benefits, and technical and functional integration risks.

The process is driven by a set of NAS long-range performance objectives such as capacity, safety, affordability, and environmental impact, obtained by combining market forecasts with public policy. The market forecast is then subjected to a mission analysis that takes into account airport infrastructure development and

---

Copyright © 2001 by The Boeing Company. Published by the American Institute of Aeronautics and Astronautics, Inc. with permission.
[*]Associate Technical Fellow, Air Traffic Management Division. Member AIAA.
[†]Technical Fellow, Retired, Air Traffic Management Division.
[‡]Principal Engineer, Air Traffic Management Division. Associate Member AIAA.

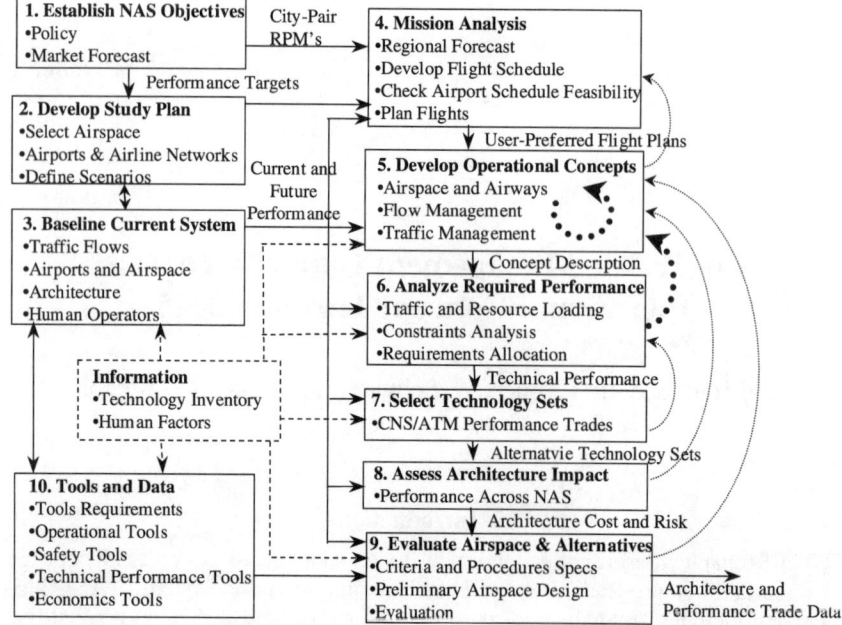

Fig. 1 NAS preliminary design process.

airline networking strategy to develop predictions of fleet mix, desired operator schedules, and an associated traffic demand load on system resources. Along with the mission analysis, an evaluation of the current system operation and performance is done to provide a baseline against which future improvements can be designed and evaluated. The baseline incorporates the current operational concept and existing system architecture for a set of operational scenarios that span a range of states sufficient to define normal, rare-normal, and non-normal behavior to ensure a comprehensive architecture definition.

Using the traffic demand model and the current system baseline, an assessment is made of the performance shortfalls predicted through 2015, and this forms the basis for the synthesis of operational concepts that address the needed improvements. The new concepts take into account anticipated technology available in this time frame, along with the associated human performance characteristics.

The remainder of this chapter discusses the development of operational concepts driven by performance goals, with clear traceability of how performance is allocated to functions, operators, and subsystems. The trade studies required to specify a sound and affordable architecture are described in more detail in (Ref. 1).

III. Operational Concept Development

This section describes a framework for synthesizing air traffic management (ATM) operational concepts, using the foundation developed in (Ref. 2). A number of operational concepts have been developed for the NAS and other parts of the global aviation system[3-8] over the last decade, but it remains unclear how these concepts can be used to drive major architecture decisions and also ensure that system performance objectives are met.

Fundamentally, an operational concept answers the question "what must the system do, how well must it do this, and by whom is this done?" This description captures function, performance levels, and responsibility. Because of the size and complexity of a system such as the NAS, the number of functions, agents, and subsystems involved leads to a large number of possible concept solutions. Considering all the possible solutions, the operational concept could be constructed using strategies such as:

1) Perform all functions desired by all stakeholders and implement all available technology.

2) Randomly search and evaluate all possible concept combinations to find the "best" solution.

3) Insert changes in the current system based on technology opportunity and vocal stakeholder requests.

4) Redesign the operation to solve identified performance shortfalls in such a way that every design decision is guided by performance requirements.

Strategy 1 is very costly and likely to be operationally infeasible; strategy 2 will take too long to converge, and strategy 3 is not likely to deliver the needed long-term performance. This chapter describes strategy 4 and postulates that it is feasible and likely to be the most effective method, as it produces a limited number of concept and architecture choices that can be traded against the overall performance objectives to find the best design.

IV. Functions, Agents, and Performance

Figure 2 illustrates today's ATM work system in terms of top-level functions and agents involved directly in the daily system operation. Not included in the figure are more strategic functions such as airspace management and airline scheduling,

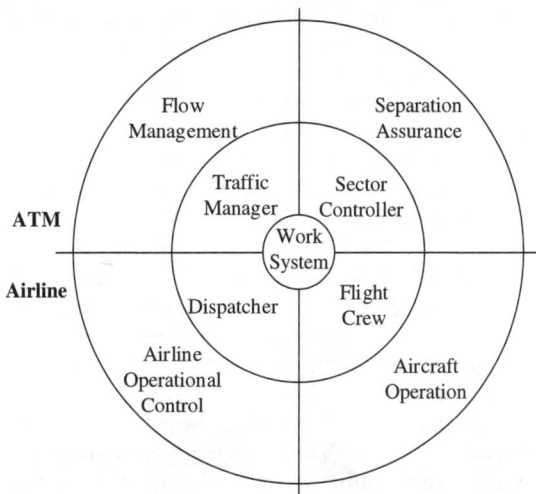

Fig. 2 The air traffic management work system.

some of which may move into the daily operating realm in the future system. The functions can be further divided into subfunctions, and the agents can consist of individuals or teams, supported by technical subsystems, with the human and technology components combining to deliver the required functions with a certain level of performance.

The separation assurance and aircraft operation functions in Fig. 2 are key in realizing the fundamental capacity and safety objectives for the airspace. Nakamura and Schwab[9] proposed to tie the performance of the separation function and its subfunctions and systems to the separation service supported in a given airspace, through a set of required performance indices, as discussed further in Section VI. Key to the successful definition of system performance is the rare- and non-normal performance, which will drive system safety levels and therefore many of the critical architecture decisions. Thus, for communications, navigation, and monitoring, the normal, rare-normal, and non-normal performance (both detected and undetected failure rates) must be specified to insure a system design that will support the future system capacity and safety levels. Additionally, in an environment such as today's radar-based air traffic control (ATC), it is necessary to include the overall impact of human performance, together with decision support and communications, navigation, and surveillance (CNS) elements, on the overall system performance.

V. ATM System Functional Structure

The capacity of the ATM system is fundamentally bounded by the separation standards in effect for the airspace. System throughput is a measure of the realized flow through the system in a given time period and is further constrained by the controller's ability to accommodate traffic demand in the face of operational uncertainty. Periods when demand exceeds capacity in parts of the system can overload the separation assurance agent and thus increase the collision risk, and it is important to include functions in the system that prevent such overload. In the NAS operation this is done through flow planning, where a planning horizon of up to 24 h is appropriate given the daily traffic demand cycle.

The traffic flow planning function is complicated by the fact that the system is subject to a variety of sources of uncertainty. The three most important ones for the daily plan are:

1) Weather prediction uncertainty. Visibility predictions affect primarily the arrival phase, reflected in airport arrival rates. Convective weather predictions can affect any phase of flight, including en route.

2) Aircraft pushback readiness, influenced by a variety of factors in aircraft turnaround at the gate, which affects primarily the departure phase.

3) NAS equipment status, which can affect any phase of flight.

The uncertainty inherent in the daily flow plan often results in situations where the plan is out of phase with the unfolding situation, leading to possible overloads or wasted capacity. To deal with the uncertainty, the system could 1) reduce the uncertainty level (difficult, but some progress is being made), 2) provide plenty of room to safely absorb the uncertainty (procedural control, wasteful), or 3) modify the plan dynamically to manage the situation as it unfolds.

The last option, to modify the plan dynamically, is what the NAS is evolving toward in an effort to achieve an acceptable balance between throughput and

ATM CAPACITY-DRIVEN OPERATIONAL CONCEPT

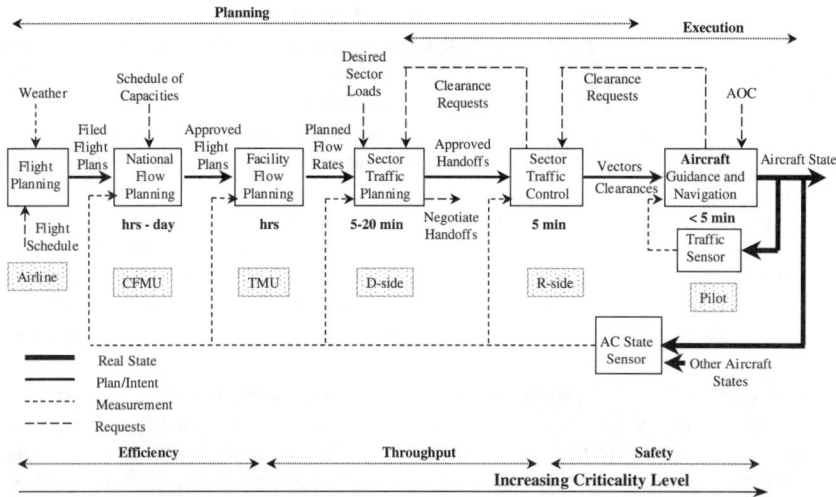

Fig. 3 Air traffic management system functional structure.

safety. Thus the NAS includes several levels of planning: 1) national and regional flow planning, 2) facility-level flow planning, and 3) sector-level flow planning. Each level has a certain planning time horizon and range of possible planning actions. Figure 3 shows the overall structure of the ATM system in terms of functions directly affecting the process that links real-time traffic demand with actual flight through NAS airspace. Figure 3 illustrates the processes and information flow that make up the flight and traffic planning and separation assurance functions of the system. Figure 3 is only one of many possible cross sections through a very large and complex system and hides a considerable amount of detail, but it is useful in tying the operational concept to system safety, capacity, and affordability. It is an idealized representation of a system that is very adaptable because of the presence of human operators, and in which assignment of functions to agents is very dynamic. Figure 3 illustrates the path, starting at the left, from a desired flight schedule and a weather forecast, through filed flight plans, to real aircraft movement on the extreme right. Each block in the diagram indicates a function that is performed in the system today, and the arrows denote either real aircraft state, communication of a plan or intent, measurements, or requests. The functions can be divided roughly into planning and execution, which overlap across the sector control functions.

The diagram indicates the approximate planning time horizons for each function, ranging from a day for national flow planning to minutes or seconds for aircraft guidance and navigation. The actual time horizons employed by system operators vary greatly depending on the airspace and traffic levels, but the numbers in Fig. 3 are reasonable for most components of the NAS. An approximate analogy to the current assignment of functions to agents is shown in the figure through reference to the R-side (radar) and D-side (data) controllers, traffic flow management unit (TMU) positions, and central flow management unit (CFMU). The separation assurance function is here considered to be assigned to the sector controller team, using a radar display and flight plan information, with the aircraft crew as a collision avoidance backup, through visual observation of traffic and/or through use of the Traffic Alert and Collision Avoidance System (TCAS).

The criticality level of the system functions increases from left to right in the diagram. Criticality level is fundamental in all discussions about required performance to support a function and in the level of attention to human factors that a function requires.

Figure 3 illustrates how uncertainty in planning is accommodated through several levels of replanning in the system. Traffic situation data feedback to the planning levels is used for situation assessment, but automation tools to generate replanning options is a weakness in the system today, and therefore it is difficult to update the flow plan comprehensively across facilities or regions as situations change.

To relate back to the system objectives, Fig. 3 illustrates that safety is a primary responsibility of the aircraft, with separation assurance assistance from the sector controller. System throughput is delivered by the execution loop, with overload protection from the planning functions. Efficiency is worked primarily by the flow planning functions, through negotiations of flight plans, with in-flight rerouting performed by the execution loop.

VI. Capacity, Safety, and Separation Assurance

Figure 4 illustrates the separation assurance loop with additional detail showing the primary sub-functions in the loop. The sector planning function's primary objective is to manage the number of potential conflict situations the sector controller may need to process. The set of flight plans inbound and inside the sector can be considered the primary data input to this function, along with the real-time traffic situation as it currently affects the sector controller's workload. The sector planner may also need to assist the controller with clearance requests from

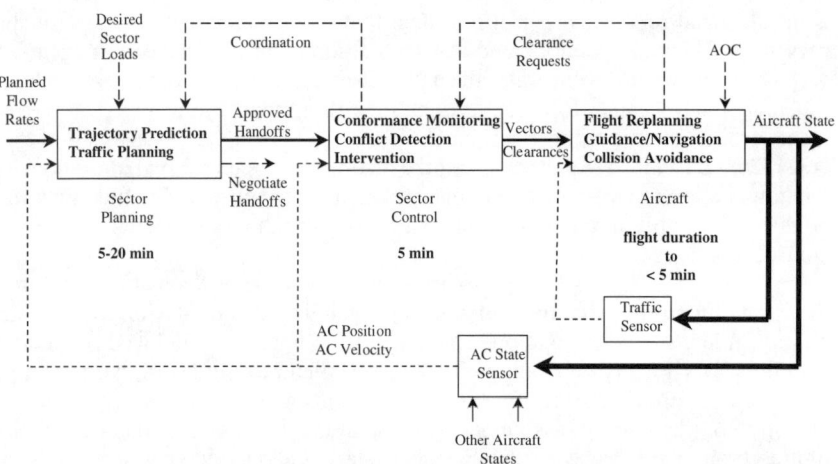

Fig. 4   The separation assurance loop.

aircraft that he/she cannot immediately process. Thus, the sector planner function helps manage the sector controller workload and is therefore the primary agent in managing exposure to collision risk. The sector controller in today's radar control operation is the only traffic management agent that communicates directly with the aircraft. The functions performed by the sector controller are conformance monitoring and short-term conflict detection and intervention, along with receiving, granting, or rejecting route modification requests from the aircraft. Conflict detection performance depends on the accuracy of the aircraft state sensor, the display resolution and update rate, and the controller's ability to predict the aircraft trajectory into the future. Intervention performance involves the decision to act on a potential conflict and the communication of the action to the cockpit crew, which then must intervene and change the flight path. The sector controller is thus a critical element of detection and intervention, and today's system has very limited backup for failures in either the performance of the function or the surveillance and communication systems the function relies on.

The cockpit crew is responsible for guidance and navigation according to an agreed upon flight plan, along with replanning for reasons of safety, efficiency, or passenger comfort. The cockpit crew contributes to the performance of the intervention function through its response to ATC vectors. The crew also has a safety responsibility to monitor and avoid other aircraft in its immediate vicinity, either visually or through TCAS. This is currently a limited safety backup for the sector controller's separation assurance.

Figure 5 is a conceptual representation of effective traffic spacing, depicted as concentric rings, or buffers, with the innermost region (detection) representing the theoretical maximum capacity based on the separation standard. The separation standard in radar-controlled airspace was established primarily as a function of radar surveillance and display system performance. Traffic spacing in busy terminal areas is usually near the separation minimum, achieved by highly

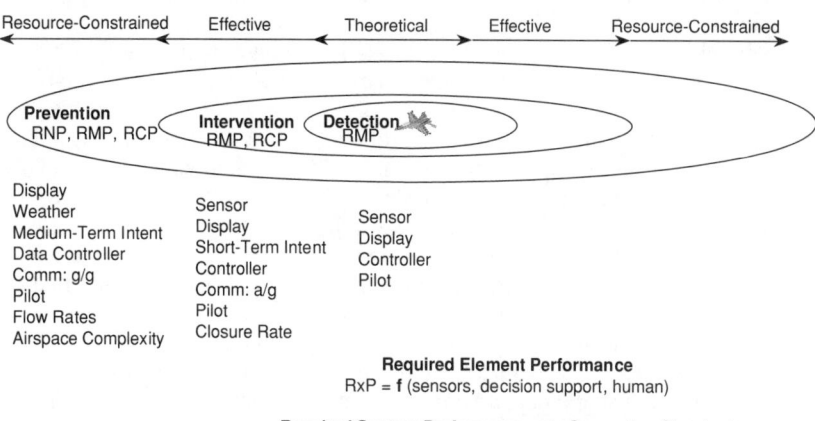

Fig. 5   Traffic spacing and performance factors.

structured airspace design and high controller proficiency. Significant changes to one or more of the communication, surveillance, navigation, and controller/pilot intervention performance are necessary in order to reduce separations beyond the current minimum.

The middle separation ring (intervention) represents the additional separation required to determine that a conflict is imminent and to intervene to resolve it. In a radar-controlled environment, this buffer is a function of the time from detection of intervention need, decision on resolution action, transmission of instruction to the pilot, the pilot's response, and controller verification. The buffer size is also affected by traffic flow patterns (high closure rate encounters require larger control loop margins) and the speed and maneuverability of the aircraft. This buffer can potentially be reduced through the provision of better short-term aircraft intent data, data link communications, tools for conformance monitoring, and short-term conflict alerting.

The outermost ring (prevention) is adjusted by the sector traffic planning function to limit tactical controller workload by preventing too many potential traffic conflicts from developing simultaneously in the sector. As workload increases, the planning controller adopts strategies to prevent the possibility of conflicts. These strategies include level-offs, holding, and rerouting of aircraft away from dense airspace. These strategies must take into account the uncertainty in medium-term (20–30 min) trajectory prediction. The overall effect is to reduce system throughput and to increase aircraft time and fuel burn in the system and is currently evident in the sparse overall traffic spacing patterns in the en route system. This buffer can potentially be reduced through air-to-ground sharing of medium-term intent data through data link, along with automation aids for conflict probing, conflict resolution, and terminal area sequencing and spacing. In a procedural environment, the separation minima are dictated by conflict prevention, with essentially no ability to detect and intervene due to poor performance of the communications and monitoring functions (third-party voice position reports and clearances). This necessitates a large prevention buffer, and thus very low throughput, based purely on navigation performance, to make the likelihood of the need for intervention acceptably small.

Potential improvements in airspace throughput should be proposed in light of Figs. 4 and 5. Specific tools and improvements to CNS/ATM can be targeted to specific factors that influence the size of the individual separation rings. In general, the performance levels required for change increase toward the center of the rings. The inner circles are probably more difficult to affect and thus involve a higher implementation risk. The Future Air Navigation Systems (FANS) 1/A CNS/ATM enhancement primarily addresses the outer ring, as it was targeted at procedurally controlled airspace. Radar-controlled airspace has a much smaller prevention buffer, and significant airspace improvements may demand that the intervention and/or detection buffers be addressed there.

Figure 5 illustrates how prevention, intervention, and detection combine in an overall separation assurance function and lists the performance factors involved in each component. Nakamura and Schwab[9] propose a framework in which the performance of each of these fundamental factors is combined in an overall required system performance parameter, which is then directly related to a minimum allowable separation between aircraft. The navigation function performance has been formalized through the definition of Required Navigation Performance (RNP), as described in (Ref. 10). RNP includes a definition of accuracy, integrity, and availability levels, which are functions of navigation sensors and their sources, cockpit crew interface design, and pilot performance. To compose an overall performance index, Required Total System Performance (RTSP), for the separation assurance function, consideration must be given to required communication performance (RCP) and required surveillance performance (RSP), along with possibly a metric relating to the performance of the traffic planning and intervention functions.

VII. Capacity-Driven Operational Concept

The sequence of transition steps presented here defines one of many possible paths that the system operational concept and architecture could follow through the year 2015. This particular path is constructed with the objective of achieving long-term capacity increases, using the author's best judgment of what system enhancement steps could be taken during this period with available and emerging CNS/ATM technologies. This transition path, and most of the individual steps within it, have not been validated, and thus the overall system capacity impact cannot yet be quantified. The selected technologies will need to be subjected to requirements validation and system tradeoffs. This transition path, however, is a reasonable baseline from which to initiate the trades that must be performed as part of the preliminary design process illustrated in Fig. 1.

VIII. National Level Flow Management

Figure 6 shows the proposed concept transition path for national and local traffic flow management. The diagram shows two parallel paths, one starting at the national level and the other starting at the airport level. The two paths merge in the third transition step into a coordinated traffic flow management system.

A. National Level: Improved Traffic Flow Management

This step involves real-time information exchange between NAS users and central flow management, focused primarily on automatic schedule updates from the airlines and timely notification to airlines of flow management actions.

B. National Level: Collaborative Traffic Management

This operational enhancement includes a collection of changes in flow management aimed at giving users more flexibility in deciding how delay is allocated across their operation. Delay will be allocated to operators according to their published schedule, and the operator in turn allocates the delay to their individual flights. Where arrival airport capacity is the constraint, emphasis will be on arrival

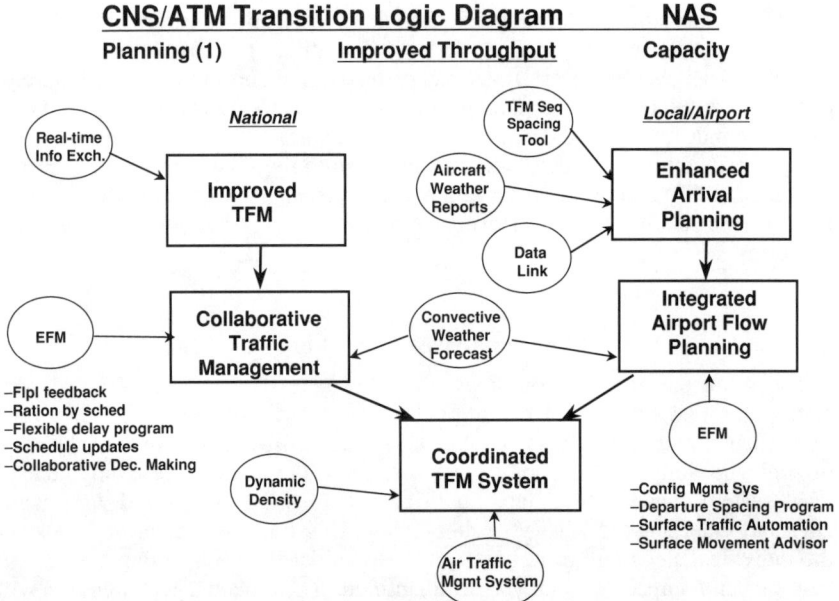

Fig. 6 CNS/ATM transition for flow management.

airport resource management and away from departure gate-hold times. This will allow operators to minimize the overall cost impact of delay on their operation by prioritizing flights according to issues such as passenger and baggage connections.

C. Airport Level: Enhanced Arrival Planning

This enhancement step provides improved terminal area arrival flow planning, including arrival runway load balancing, enhanced arrival sequencing, and improved arrival flow replanning, given a perturbation such as runway change or convective weather.

D. Airport Level: Integrated Airport Flow Planning

This enhancement step involves a group of airport traffic planning initiatives aimed at integrating arrival and departure traffic, along with surface movements, into a coordinated plan. This will include optimal airport balancing of arrival and departure resources and the need for automation to support airport configuration management.

E. Coordinated Traffic Flow Management System

In this step, flow planning at the national, regional, and local level are brought together in a coordinated system. The function allocation strategy to achieve this step and the technologies required are to be determined; the relevant issues are discussed in (Ref. 2).

CNS/ATM Transition Logic Diagram NAS
En Route (6) and TMA (5)   Improved Throughput   Capacity

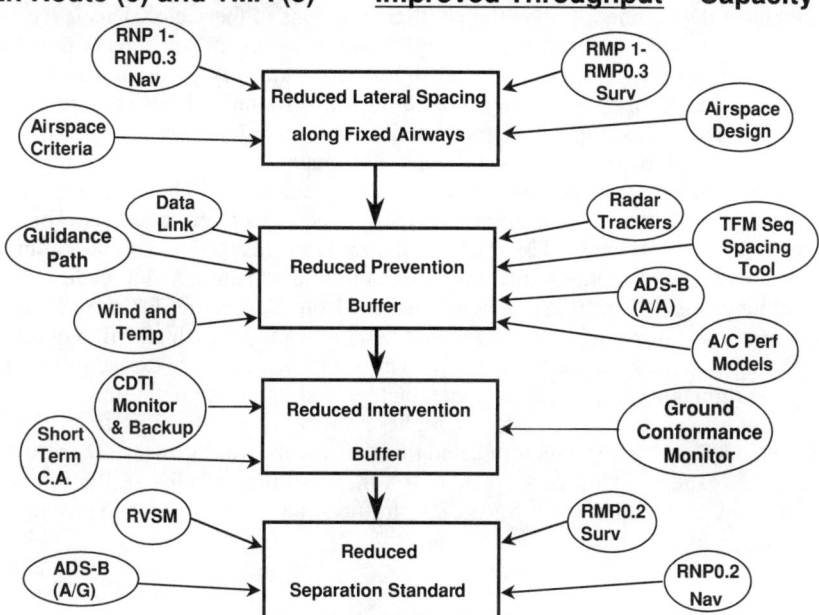

Fig. 7   CNS/ATM transitions for en route and terminal area.

IX.   En Route and Outer Terminal Area

Figure 7 shows the proposed concept transition path to achieve increased capacity in the en route and terminal maneuvering area (TMA) arrival/departure operating phases. The sequence of operational improvement steps represented by the boxes, from top to bottom, addresses a reduction in effective traffic spacing starting with airway spacing criteria, through reduction of prevention and intervention buffers, to the eventual reduction in the separation standard.

A.   Reduced Lateral Spacing for More Arrival and Departure Transition Routes

This enables closely spaced standard arrival and departure routes to fit additional traffic streams within terminal area corridors. This will help avoid congestion over entry points into terminal areas and reduce the need for in-trail traffic spacing that backs up into en route airspace. This enhancement will be most beneficial in terminal areas where airspace is constrained because of proximate airports, special use airspace, or severe weather activity.

Airspace design criteria have to be changed to enable this operational enhancement. Those criteria are likely to be predicated on a level of navigation performance in the range of RNP 1 to RNP 0.3, along with the corresponding surveillance and monitoring performance.

B. Reduced Prevention Buffer

The prevention buffer is the outermost separation ring discussed in Fig. 5, added to reduce the number of potential conflict situations in the sector. This is the role of the sector planning function in Fig. 4, and thus the enhancements proposed here relate to the performance of medium-term trajectory prediction and traffic planning functions. Improvements in both the horizontal and vertical dimensions are included in this step, where the benefits of improved vertical accuracy may be greater but will require more technology investment.

An improvement in medium-term trajectory prediction will be needed to reduce the uncertainty that the controller has today when predicting conflicts. This improvement will be enabled by tracker enhancements that provide higher accuracy and lower latency, better wind and temperature information, and a medium-term conflict probe. The terminal area will benefit from automation for more accurate sequencing and spacing of climbing and descending traffic, which will require accurate aircraft performance models. Data link may be required to exchange weather information, aircraft performance parameters, and trajectory definition.

In addition to the above factors, a higher probability that the aircraft will follow its intended path may be required, and this may involve four-dimensional terminal area navigation capability. Depending on the level of criticality of the function, there may be a requirement for cockpit traffic situation awareness to provide redundancy of function.

C. Reduced Intervention Buffer

To reduce the intervention buffer, it is postulated that data link may improve the delivery time and integrity of communications from controller to pilot. A ground-based conformance monitor is assumed that alerts the controller to aircraft deviations from intended trajectory, and a short-term conflict alert function is also assumed. Criticality level is expected to be high, which may require an independent airborne monitoring function.

D. Reduced Separation Standards

This refers to both vertical and horizontal separation. Reduced vertical separation minima (RVSM) in domestic airspace would likely be predicated on vertical path following performance similar to what is required in the North Atlantic. Horizontal separation is likely to require improvements in the surveillance sensors both for en route and terminal areas, and better navigation performance. The detailed requirements will have to be worked out, starting with the development of a risk evaluation methodology to determine the influence of technology and human factors on collision risk in positively controlled airspace.

X. Approach/Departure Transition

Figure 8 shows the proposed concept transition path for increased capacity in the arrival and departure transition phases. The sequence of operational improvement steps represented by the boxes, from top to bottom, addresses a reduction in effective traffic spacing starting with route spacing, intervention buffers, through reduction in the basic separation standard.

CNS/ATM Transition Logic Diagram   NAS
Arr/Dep Trans (4)   Improved Throughput   Capacity

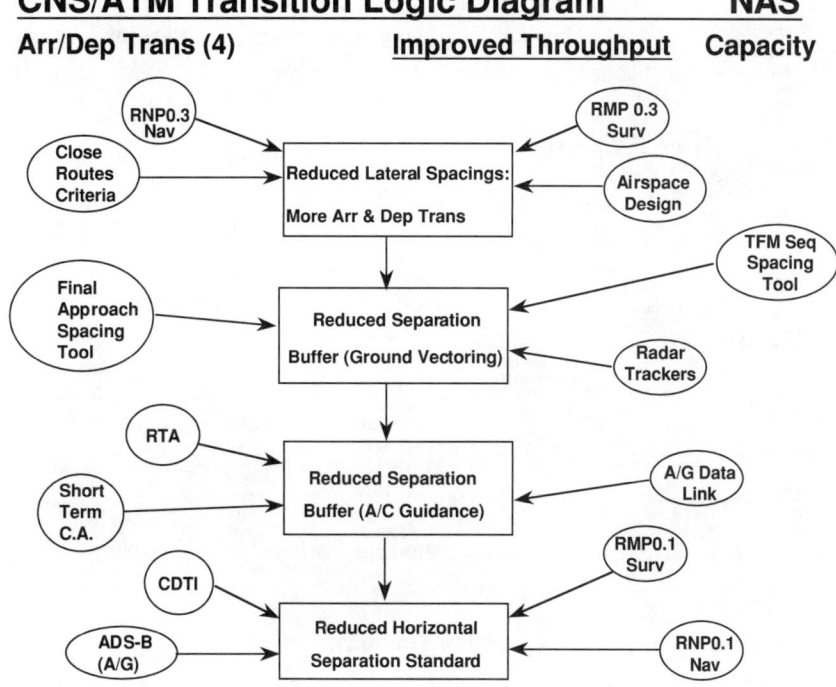

Fig. 8   CNS/ATM transition for arrival and departure.

A. Reduced Lateral Spacing for More Arrival and Departure Transitions

This enables closely spaced arrival and departure routes to fit additional traffic streams within terminal area corridors. Airspace design criteria have to be changed to enable this operational enhancement. They are likely to assume RNP 0.3 along with corresponding surveillance performance.

B. Reduced Separation Buffer (Ground Vectoring)

This enhancement involves more accurate timing of aircraft delivery to the final approach fix through more effective ATC vectors. The improvement will be enabled by better trackers for trajectory prediction, automation for accurate traffic sequencing and spacing, and support to generate accurate ATC vectors for final approach spacing.

C. Reduced Separation Buffer (Aircraft Guidance)

The component of the spacing buffer at the final approach fix that is contributed by the aircraft guidance and navigation performance will be improved in this step. This will involve the use of required time of arrival functionality, and data link to deliver clearances with accurate timing information. Short-term conflict alert functionality may be needed to improve conformance monitoring.

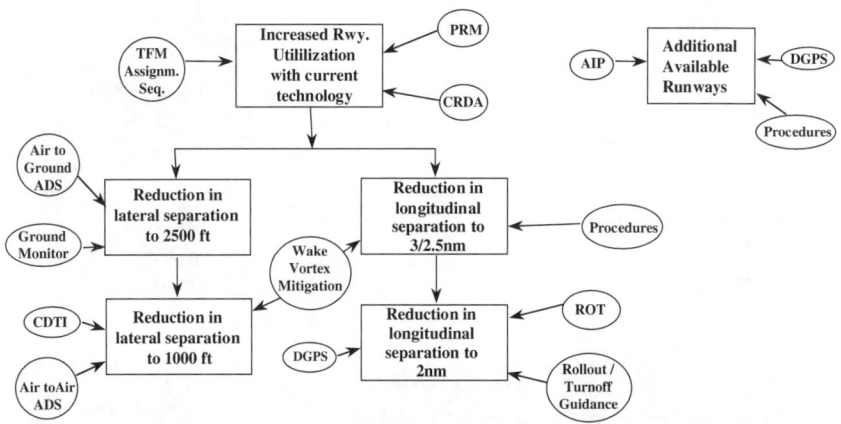

Fig. 9 CNS/ATM transition for final approach and initial departure.

D. Reduced Horizontal Separation Standard

In this operating phase it is normally spacing on final approach that determines the separations applied. As seen in Fig. 9, the concept includes a plan to reduce spacing on final approach, and thus the approach transition phase may need corresponding separation reductions. The improvement and enablers would be analogous to the last box in Fig. 7.

XI. Final Approach

Figure 9 shows the proposed concept transition path to achieve increased capacity in the final approach and initial departure operating phases. The chart shows two independent enhancement paths, the one on the right centered on additional runways, the one on the left centered on increased runway utilization.

A. Additional Available Runways

This improvement involves new runways being built and existing runways being made more available through development of instrument approaches. The Federal Aviation Administration's (FAA) Airport Improvement Program (AIP) is the enabler for new runway construction, which also may rely on new approach and procedure design to address airport noise concerns. Flight management system (FMS) capabilities can be utilized to reduce both the spread and the severity of noise impact through tailored approach and departure procedures. Instrument approaches to a larger number of runways in the continental U.S. (CONUS) will be enabled by differential Global Positioning System (GPS) down to category (CAT) III minima.

B. Increased Runway Utilization, Current Technology

This improvement step involves the installation of existing technology where needed to increase throughput of closely spaced parallel and converging runways in instrument meteorological conditions (IMC). To fully take advantage of the precision runway monitor and converging runway display aid technologies, it may be necessary to include arrival and departure sequencing and spacing automation.

C. Reduction in Lateral Separation to 2500 Feet

This reduces further the minimum lateral separation between parallel runways for independent operations. To assist with blunder detection, automatic dependent surveillance (ADS) event-based position reporting and improved monitoring on the ground will be needed. Precision missed approach guidance may be needed.

D. Reduction in Lateral Separation to 1000 Feet

The reduction below 2500 ft between independent parallel runways in IMC is currently being discussed in the context of airborne separation assurance through cockpit display of traffic information (CDTI). Wake vortex is an issue here. This is an ambitious step, and the exact requirements will have to be worked out carefully.

E. Reduction in Longitudinal Separation to 3 or 2.5 Nautical Miles

In IMC the longitudinal separation on final approach is currently set by wake vortex considerations, and therefore this enhancement step must address wake detection and avoidance through wake prediction/detection technology and new procedures.

F. Reduction in Longitudinal Separation to 2 Nautical Miles

Further reduction in longitudinal spacing on final approach would address runway occupancy and the need to ensure rapid braking and turnoff performance of the aircraft. In low visibility this may require improved rollout and turnoff guidance, perhaps based on differential GPS. Included here might be the possibility of allowing two aircraft on the runway at the same time, assuming the required braking performance to stop short.

XII. Surface

The proposed concept transition path to achieve increased capacity on the airport surface is described in (Ref. 2) with two independent enhancement paths, one centered on low visibility operations, the other on good visual conditions.

XIII. Efficiency in Low Density En Route Airspace

The affordability objective can be further addressed through accommodation of user preferences in airspace where capacity is not constrained. This consideration could be added to the set of transitions presented in this paper, along with the required enhancements in airspace management, including dynamic airspace allocation and resectorization. Whether the performance gain is sufficient to justify the cost must be evaluated along with other proposed improvements, along with the operational and technical feasibility.

XIV. Conclusions

This chapter presents an operational concept for the NAS in 2015, assuming that increased system capacity is the primary modernization driver. The concept is presented in the form of a transition path from today's operation to 2015, describing a series of operational improvement steps along with suggested technology enablers. The concept is presented in the context of an overall NAS preliminary design process that emphasizes system performance and trade studies to arrive at a high-performance architecture. Key to the successful application of the process is a careful flowdown of performance requirements from the overall mission to functions, operators, and systems, which guides the system design to deliver the desired performance outcome. The NAS preliminary design process will need to be supported by an analysis toolset to produce quantifiable performance predictions. The development of this toolset is the current focus of the authors.

Acknowledgments

This work was supported in part by NASA's Advanced Air Transportation Technologies program, under subcontract with the National Center of Excellence for Aviation Operations Research. The authors are grateful for the support of NASA, the FAA, and The Boeing Company for the work presented in this paper. The reader is referred to (Ref. 2) for a more complete list of contributors to this effort, to whom the authors are indebted.

References

[1] Schwab, R. W., Haraldsdottir, A., and Warren, A. W., "A Requirements-Based CNS/ATM Architecture," *World Aviation Congress*, SAE/AIAA, Anaheim, CA, Sept. 1998.

[2] Haraldsdottir, A., Alcabin, M. S., Burgemeister, A. H., Lindsey, C. G., Makins, N. J., Schwab, R. S., Shakarian, A., Shontz, W. D., Singleton, M. K., van Tulder, P. A., and Warren, A. W., "Air Traffic Management Concept Baseline Definition," Boeing Commercial Airplane Group, Seattle, WA, 31 Oct. 1997.

[3] "Air Traffic Services Concept of Operations for the National Airspace System in 2005," Federal Aviation Administration, 30 Sept. 1997.

[4] "A Joint Government-Industry Operational Concept for the Evolution of Free Flight," RTCA, Washington, DC, Aug. 1997.

[5] "EATMS Operational Concept Document," Eurocontrol, Brussels, Belgium, Jan. 1997.

[6] "Operational Concept for Air Traffic Management (ATM)—Aeronautical Operational Control (AOC) Ground-Ground Information Exchange," RTCA, Washington, DC, 29 April 1997.

[7] "Final Report of RTCA Task Force 3, Free Flight Implementation," RTCA, Washington, DC, 26 Oct. 1995.

[8] "Flight Management System—Air Traffic Management Next Generation (FANG) Operational Concept," Federal Aviation Administration, July 1997.

[9]Nakamura, D. A., and Schwab, R. W., "Development of Minimum Aviation System Performance Standards for RNP for Area Navigation," ICAO Review of General Concept of Separation Panel, Working Group A, Montreal, Canada, 1996.

[10]"Minimum Aviation System Performance Standards: Required Navigation Performance for Area Navigation," RTCA, Doc. DO-236, Washington, DC, 27 Jan. 1997.

Chapter 3

Comparison of U.S. and European Airports and Airspace to Support Concept Validation

Diana Liang*
Federal Aviation Administration, Washington, D.C.

William Marnane[†]
EUROCONTROL, Brussels, Belgium
and
Steve Bradford[‡]
Federal Aviation Administration, Washington, D.C.

I. Introduction

THE Federal Aviation Administration (FAA) and EUROCONTROL are engaging in efforts to modernize their respective airspace and are developing operational concepts for their own regions. Both recognize that a harmonized path toward modernization would greatly benefit both the users and service providers. To this end, a joint activity for concept development and validation has been proposed, accepted, and sponsored by the FAA and EUROCONTROL Research and Development (R&D) Steering Committee. It was hoped that better understanding of similarities and differences would support modernization efforts and enable future joint participation on concept development, validation, and implementation activities. The first activity undertaken was a high-level comparison of the two operational concepts that found many similarities and agreements. A key concern coming out of that comparison was that the level of analysis did not provide clear assurance that the described environments were actually comparable. To allay concerns, a follow-on activity was proposed to compare in more detail the operations of U.S. and European airports and airspace.

Core airspace of the U.S. East Coast and Europe was selected for comparison and analysis. Both areas are characterized by high-density en route traffic and

---
This material is a work of the U.S. Government and is not subject to copyright protection in the United States.
*OPS Research Analyst, NAS Concept Development Branch
[†] ATM Expert, Strategy, Concept and System Unit
[‡] Chief Scientist, NAS Concept Development Branch

contain the busiest airports. The challenge has been, because of differences in history, culture, and social sensitivity, to assemble directly comparable data and to make explicit the reasons why the current situation is as it is.

Concerning airport capacity, the commonly held belief that there is an "Inclement Meteorological Conditions (IMC) vs Visual Meteorological Conditions (VMC)" operational paradigm difference between Europe and the United States was examined. Putting aside the observation that U.S. airports generally have a larger number of runways available for simultaneous use, the comparison study could not find major systemic differences in the manner in which airports and runways are managed. There is a difference in the coordination of airport slots vs the market-driven schedule for airports. The coordinated schedule is a function of politics, perceived capacity, and acceptable delay to the users. It provides the appearance of a more uniform demand in Europe vs peak operations in the United States, but overall it leads to similar actual situations at airports working close to their capacity limits. The appearance of uniformity also exists in the United States at coordinated airports such as Chicago O'Hare.

This led to a realization that the standard method of examining performance is not adequate for overall comparison. The number of operations scheduled and the manner in which the schedule is set (negotiated vs market driven) become less significant as sliding peak hours are examined, that is, when the peak consecutive 60-min performance is compared vs the hourly count (or declared capacity).

We do see that the performance of airports at true peak is comparable. The hourly count approach most commonly used masks the maximum achievable performance at airports such as Charles de Gaulle and Atlanta Hartsfield, and it is clear that this will be the case in the examination of most airports. This is an important point highlighted by the comparison and should be considered further in the joint development of measures of operational performance. It further highlights the necessity to examine actual real-time operations and procedures rather than only statistical data.

In the case of en route airspace, the similarities seem to outweigh the differences, although the study did not looked at staffing aspects. There is no great difference in throughput or procedures for sectors where it seems that "best current practices" are fully used in both Europe and the United States. When coupled with the airport results, this indicates that both systems are not diametrically different. It is only when traffic flow and its management and specific, specialized sector/practices are considered that conclusions can be drawn. It seems that in the United States there is a larger recourse to specialized control techniques, such as miles in trail, to help in managing extreme cases rather than preflight resource assignment. Another example is one-way sectors with highly structured traffic flows in the United States that have no real equivalent in Europe, although flight-level allocation techniques have been used for years to strategically deconflict flows.

Differences may be best explained by how air traffic control (ATC) developed and adapted to traffic conditions and external events. In particular, it was important to consider why at the strategic flow level the Europeans opted for the central flow management unit (CFMU) vs the apparently more ad hoc system used in the United States. This led to examining the traffic flow management objectives and the processes employed to meet these objectives.

When the impact of the overall traffic flow was considered, it became clear why some decisions are made over others. It is the flow options available to meet demand that need investigation. The scope of information and span of influence of the individual air traffic management (ATM) components begin to provide insight into existing and future concepts. To this end, the team identified the traffic management phases of flights as a means to reflect that a particular flight may need to be considered simultaneously involved in an arrival sequence and en route deconfliction while finishing its departure phase and being processed for the planning of the next flight. It is recognized that in the gate-to-gate perspective the network effect is linking all decisions from departure to destination and the attempts to choreograph the mass of flights into an efficient aggregate.

The comparison has contributed to an original understanding of concept, services, and performance, as well as their inter–relationship. Although it has not addressed all issues in all detail, it also provides the context for further mutually defining the future global air navigation systems (ANS) concept and overcoming the present shortcomings.

II. Assessment Territory

Core airspace areas (see Fig. 1) have been selected to represent the most typical problems of the highest traffic density on both sides of the Atlantic. It is in those environments that we can best understand how current concepts can deliver today's performance and the challenges the expected traffic growth will present.

The busy traffic areas are generally of the same size, though the concentrations of major airports appear denser in Europe. The similarities extend to raw traffic statistics. At the 1997 ATM 2000 conference in Naples, Italy, the European members reported that average instrument flight rules (IFR) flight length within the European Civil Aviation Conference (ECAC) countries was 470 n miles and 1 h and 20 min in duration. The U.S. participants reported similar numbers for the United States 470 n miles and 1 h and 23 min in duration. It was the appearance of similarity and not difference that helped initiate this comparison.

III. Metrics and Measures

The measures and metrics developed for this activity describe operations and not the more customary values of delay. The goal is to match like with like as much as possible in terms of traffic needs and problems to be solved. Once this is achieved, the next step is to look at procedures and performance. To this end, the assessment looks at characteristics such as fleet mix, operations per day, peak operations, individual user penetration (percentage of flights), average transit time, average sector size, average staffing, etc.

IV. Assessment and Findings

A. Airports Assessment

Several airports within the core airspace were selected for the comparison exercise. Understanding today's limitations, operations, and business paradigms for these airports will enable service providers and users to plan for transition and implementation of future operational concepts for national and global airspace systems.

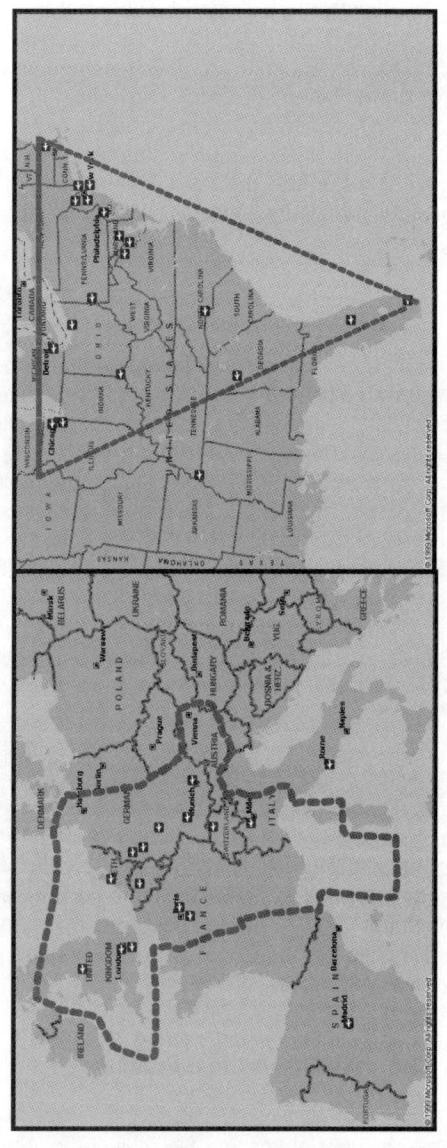

Fig. 1 European and U.S. core airspace.

1. Airports Activities and Operators

In the United States the aviation system is a hybrid system of hub-and-spoke and point-to-point (direct) services. The major carriers in the United States all operate large hubs. The general characteristics of hub-and-spoke services are 1) one or two carriers are the dominant operators at a major airport, and 2) observation of airport traffic shows a bank of aircraft arriving within a time period, followed by a bank of aircraft departing at 45–60 min after the arrival bank.

In Europe the hub-and-spoke business paradigm is not yet as prevalent. Even so, the data reflect airline domination (percentage of activities) at the major European airports. This is a legacy of regulation and has not changed significantly. For example, in the United Kingdom, the major operator at the two primary airports (London Heathrow and London Gatwick) is British Airways (BAW). Similarly, for Schiphol Airport in the Netherlands, Koninklijke Luchtvaart Maatschappij (KLM) is the dominant operator. In France, the major air carrier at Charles de Gaulle (CDG) is Air France (AFR). Thus, the national operator for the country is still the major operator at the country's major airport(s). Those airlines developing hubs do it first at these historical bases.

Table 1 reflects major operators at selected European and U.S. airports. At some of the top U.S. airports, for example, at the hub airports such as Chicago O'Hare (ORD), Atlanta Hartfield (ATL), and New York Newark (EWR), over 50% of the traffic is due to one or two carriers.

The data in Table 2 compare traffic at U.S. and European airports. The percentages of heavy aircraft reflect the transatlantic/long-haul activities, and the medium aircraft type reflect flights within the European or U.S. continent. In both areas there is a high proportion of medium aircraft. This may indicate the aircraft type is based on the length of the city pairs. There is no evidence of using heavy aircraft to increase the passenger carrying capacity of each slot.

2. Airport Capacity, Configuration, and Movements

In Europe and the United States, each airport publishes a declared capacity based on airport infrastructure (e.g., runways), supporting airspace, and other variables (e.g., staffing and negotiation). The shared understanding is that declared capacity for the airport in Europe is based on IMC operation, whereas the declared capacity for the United States is based on VMC operation.

Consider first the relationship between capacity and delay. Delay is a decision variable in scheduling in both the United States and Europe. In Europe, unlike most airports in the United States, the number of slots provided at an airport is the subject of negotiations between the airport and carriers, and it is based on the acceptable level of delay, political, and environmental considerations. Airports such as Heathrow will have higher "planned" delays by providing a larger number of slots. Carriers accept the delay performance as a tradeoff to access. The sense that capacity (slots) is both a technical and political consideration extends to the United States as we note the current political struggle over slots at Washington's Ronald Reagan National Airport (DCA), where the number of combined general aviation (GA) and commercial slots often exceeds the capacity provided on a scheduled basis.

In the United States, other than the four slot-controlled airports (Chicago's ORD, Washington's DCA, and New York's LGA and JFK airports), the users' business

Table 1 Major operators at top European and U.S. airports

Airport	% of Total
Atlanta—Harts Field International (ATL)	
Operator	
Delta Air Lines Inc (DAL)	62.6
Atlantic Southeast Airline (ASE)	11.9
Valujet Airlines (VJA)	8.2
Continental Airlines (COA)	2.5
American Airlines Inc (AAL)	2.3
Total	87.46
Paris—Charles de Gualle (CGD)	
Operator	
Air France (AFR)	49.68
Deutsche Lufthansa German Airlines (DLH)	5.42
British Airways (BAW)	4.39
Aitalia (AZA)	3.04
(ARP)	2.19
Total	64.72
London—Heathrow (LHR)	
Operator	
British Airways (BAW)	31.06
British Midland (BMA)	13.63
British Airways shuttle ( )	5.90
Lufthansa	4.00
Aer Lingus (EIN)	3.23
Total	57.82
Newark—Newark International (EWR)	
Operator	
Continental Airlines (COA)	57.3
United Air Lines Inc. (UAL)	6.5
US Airways (USA)	5.8
American Airlines Inc (AAL)	5.6
Delta Air Lines Inc (DAL)	3.7
Total	78.88
Chicago—O'Hare International (ORD)	
Operator	
United Air Lines Inc. (UAI)	42.9
American Airlines Inc (AAL)	39.7
Delta Air Lines Inc (DAL)	2.4
Northwest Orient Airlines (NWA)	2.3
US Airways (USA)	1.8
Total	89.04
Amsterdam—Schiphol (AMS)	
Operator	
Koninklijke Luchtvaart Maatschappij (KLM)	37.99
Air UK Ltd. (KM UK)	11.23
Eurowings AG (EWG)	6.55
Transavia Holland B.V. (TRA)	4.33
Martinair Holland B.V. (MPH)	2.24
Total	62.34

Table 2 Summary of aircraft categories by wake vortex

Airport	Peak daily movements	Total annual movements	Heavy, %	Medium, %	Light, %
New York— John F. Kenedy (JFK)	982	342,814	38.99	57.98	3.69
London— Heathrow (LHR)	1279	451,073	33.55	66.54	0.89
Germany— Frankfurt (FRA)	Not available	415,688	33.00	66.00	1.00
London— Gatwick (LGW)	783	251,291	20.95	75.00	4.25
Amsterdam— Schiphol (AMS)	1032	376,810	18.87	77.80	3.33
Paris— Charles de Gualle (CDG)	1200	367,222[a]	18.17	81.48	0.24
Atlanta— Hartsfield (ATL)	2298	817,492	14.89	83.34	1.78
Washington— Dulles (IAD)	1124	379,621	14.54	77.80	7.67
New York— Newark (EWR)	1264	444,370	12.20	79.73	8.07
Chicago— O'Hare (ORD)	2536	887,551	11.16	76.16	12.68
New York— La Guardia (LGA)	1009	361,135	5.66	74.56	19.77
Memphis (MEM)	717	252,941	1.80	87.72	10.48

[a] 1998 data.

cases and their relationship to delay are major contributors to scheduling but not to the declared capacity. The users adjust their schedules to delay performance. Increased delays can lead to both changes in block times as well as changes in number of operations scheduled. It is clear that the marketplace provides schedules that are more susceptible to disruption of service because of weather, but in the end the schedule is again a function of trading access with delay. The explicit negotiations in the European process may result in choosing points lower on the knee of the delay curve.

The methods for measurement of capacity declaration also need to be examined. For instance, as related by Frederic Rico, Director of Air Traffic Operations, Aéroports de Paris, Charles de Gaulle Airport's declared hourly capacity in 1996 was 84, while the peak hourly rate was actually as high as 106.[1] This is the difference in measuring the schedule from n : 00 to n+1 : 00 and capturing the peak sliding 60-min throughput. In addition, Rico makes a point that, while the declared capacity is set by several factors, the capacity the airport operators are working to achieve through both airspace and airport initiatives is the fair weather capacity. This allows the flexibility associated with the "sliding window" operations.

A related factor is environment. Are other airports in close proximity and are there dependencies between these airports? For instance, Ref. 1 shows that the Charles de Gaulle Airport's declared capacity went from 76 in 1993 to 84 in 1996. The number of runways stayed the same, but changes at nearby Le Bourget Airport and other airspace infrastructure changes allowed independent operations between the two airports providing the capacity difference. Similar considerations occur in the United States for areas such as the Washington terminal radar approach control (TRACON) and the New York TRACON.

Airport configuration also provides insight into airport usage. For example, a runway serving a mixed population of aircraft will have different results from a runway serving a homogenous category of aircraft. At airports with multiple runways, are the runways operated dependently or independently? Independent operations may allow far greater services, whereas dependent operations may mean only one runway is being used at a time. Dependent and independent operations are reliant on the runway layout (separation, crossings, etc.). Are runways operating in mixed mode (arrival and departure) or single mode (arrival only or departure only)?

Breakdown of airport movements shows the ebb and flow of activities across a day. Little activity occurs in the early morning and late evening hours. This could be explained by 1) certain airports are under noise restrictions, and therefore the number of operations at this time are reduced due to airport curfew; and 2) passengers prefer not to travel at these hours. Furthermore, airport movements are also dependent on passenger demand. Passenger demand can vary with seasons, with an increase in movements at certain times because of vacation travel.

a. Coordinated, partially coordinated, and noncoordinated airports. European and U.S. operations are both similar and different for airports. With respect to European airports, they are:

1) Fully coordinated airports in which the slots allocated to aircraft operators as a result of the deliberations of the airport scheduling committee are coordinated with CFMU with respect to departures. These slots are taken into account when the CFMU is in the strategic planning phase.

2) Partially coordinated airports in which the slots agreed by the airport scheduling committee are not considered by CFMU. Requests for clearance and CFMU regulation, if any, are on an "as required" basis.

3) Noncoordinated airports in which traffic density requires neither an airport scheduling committee nor special handling by CFMU.

Most of the large European airports are coordinated. This is ruled by European Commission Directive 95/93 of 18 January 1993, which in essence stipulates that, because of demand and the increasing number of congested airports in the community, allocation of slots is necessary. The allocation of slots at congested airports should be based on neutral, transparent, and nondiscriminatory rules, and it is considered that the requirement of neutrality is best guaranteed when the decision to coordinate an airport is taken by the member state responsible for that airport on the basis of objective criteria.

The majority of U.S. airports are noncoordinated or nonslot controlled airports with the exception of the four identified previously. These four airports are located in high-density areas and fall within the boundaries of the U.S. core airspace. Some airlines, airports, and other advocates are working to increase flights operating at these airports. Several reasons are given: 1) opening up the coordinated airports will encourage competition; 2) coordinated control at these airports is an outdated concept (with the advance of decision tools and automation, these airports are capable of supporting more operations); and 3) access to other airports may be limited because of operational caps related to night operations and local noise agreements.

b. Amsterdam Schiphol. Airport characteristics (see Fig. 2) are:

1) There are five runways, capacity 100 movements per hour and one terminal, 96 gates, 33 stands, and an additional 19 stands for cargo operations.

2) Runways are 1L/19R (length 3300 m/10,827 ft and width 45 m/148 ft), 1R/19L (length 3400 m/11,155 ft and width 45 m/148 ft), 6/24 (length 3490 m/11,450 ft and width 45 m/148 ft), 9/27 (length 3450 m/11,319 ft and width 45 m/148 ft), and 4/22 (length 2018 m/6621 ft and width 45 m/148 ft).

3) Annual movement is 376,810 (1998).

4) Average daily movement is 1032 (1998).

5) Declared capacity is 100 movements per hour.

6) Environmental constraints will ultimately limit the capacity of the airport. Special standard instrument departures (SIDs) are used between 2300 hrs and 0700 hrs. There are limited landings on R/W22 and on converging runways; there are limitations imposed in certain visibility and cloud-base conditions.

c. New York, John F. Kennedy International Airport. Airport characteristics (see Fig. 3) are:

1) There are four runways.

2) Runway names and size are 4 L/22 R (length 11,351 ft/3460 m and width 150 ft/46 m), 4 R/22L (length 8400 ft/2560 m and width 150 ft/46 m), 13L/31R (length 10,000 ft/3048 m and width 150 ft/46 m), 13 R/31 L (length 14,572 ft/4442 m and width 150 ft/46 m), and 14/32 (length 2560 ft/780 m and width 150 ft/46 m).

3) Annual movement is 352,305 (1997).

4) Average daily movement is 982.

5) Declared capacity is 92 movements per hour.

6) JFK launches the bulk of departures between 1600 hrs local and 1900 hrs local.

7) When weather and winds force JFK to utilize the ILS RWY13L approach, LGA must change to an ILS RWY13 approach. In this runway configuration, because of wake turbulence, LGA must use extra spacing between arrivals to allow for LGA heavy jet departures. This will decrease the arrival acceptance rate at LGA. If traffic demand is light, the runway change will take less than 5 min. If the traffic demand is heavy, the runway change will take between 15 and 20 min to allow N90 to clear the airspace prior to conducting operations for the ILS RWY13L approach at JFK. During this transition one should expect holding delays for JFK and LGA.

3. Airport Conclusion

The comparison study cannot find major systemic differences in the manner in which airports and runways are managed (see Table 3). The negotiation of slots vs the market-driven schedule provides the appearance of uniform demand vs peak operations. The number of operations scheduled and manner in which the schedule

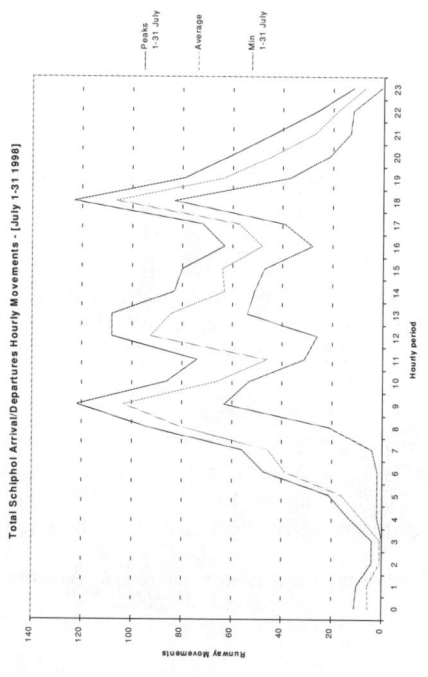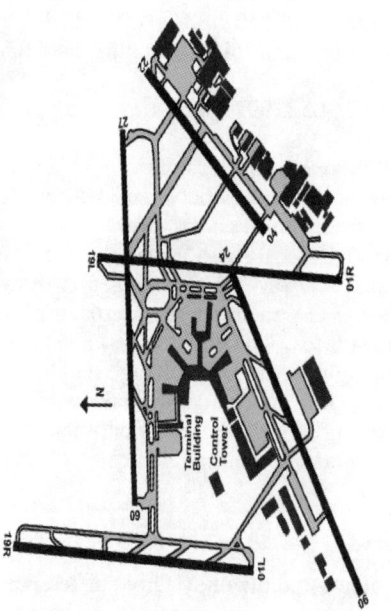

Fig. 2 Amsterdam Schiphol.

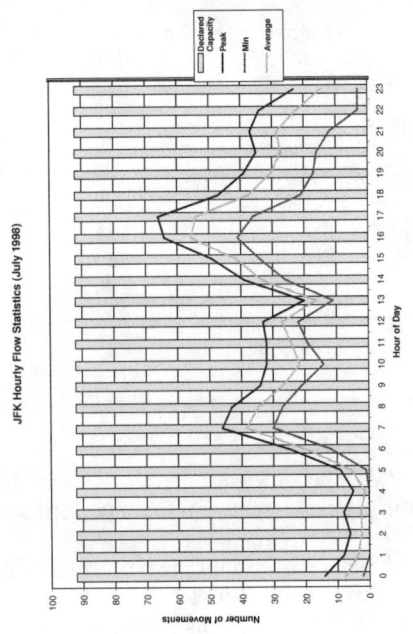

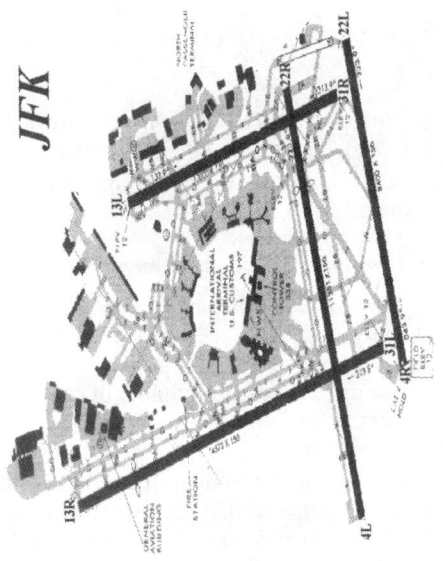

Fig. 3 New York, John F. Kennedy International Airport.

Table 3 Comparison of European and U.S. airports

Airport	Number of runways	Runway capacity (1996)	Annual movements (1996)
Gatwick	1	42	394,104
LGA	2	76	342,618
CDG	0.02	84	367,222
IAD	3	120	330,439
Schiphol	5	131	303,000
DTW	5	150	531,098
EWR	3	100	443,431
PIT	4	162	447,436
Averages European	2.0066667	86	354,775
U.S.	3.4	101	419,004

is said to be set (IMC vs VMC) becomes less significant as sliding peak hours are examined, that is, when the peak consecutive 60-min performance is compared vs the hourly count.

We see that the standard method of examining performance is not adequate for overall comparison. The hourly count approach most commonly used masks the maximum performance, and it is clear that this will be the case in the examination of most airports. This is an important point, which is highlighted by this comparison and should be considered further in the joint development of measures of operational performance.

B. Airspace

Moving away from the runway, the comparison turns to the structure of and operations in the airspace. This comparison looks at volumes of airspace and associated organizational units, staffing, flight planning and clearance delivery, procedures, and traffic volumes.

1. Airspace Volumes and Organizational Assignments

To compare centers, we have included statistics related to any center intersected by the boundaries of the study airspace (see Table 4). Thus, the centers chosen provide a "covering" of the study airspace. In comparing the core airspace of the United States and Europe, one difference is the number of centers. As shown, there are nearly twice as many centers in Europe. This is a reflection of the effect of national boundaries on center definition.

Another subject of comparison is the volume of an air traffic control center (ATCC). The volume used for comparison in this study has been calculated by multiplying the area of the ATCC, expressed in miles,[2] by the number of available flight levels within the ATCC. This removes the inherent difference when comparing the cubic volume of airspace with 2000- vs 1000-ft vertical separations. This also means that the volumes are not fixed. For instance, the introduction of reduced vertical separation minima (RVSM) will increase the number of available flight

Table 4 European and U.S. centers

Centers	Volume, $nm^2 \times ft$	Number of sectors
European centers		
Amsterdam	671,106	6
Dusseldorf	304,574	11
Geneva	307,234	6
London	2,144,174	29
Manchester	349,097	3
Munich	655,639	11
Reims	670,821	11
Zurich	438,687	7
Brussels	243,327	6
Frankfurt	593,032	18
Karlsruhe	445,373	17
Maastricht	1,112,142	10
Marseilles (Aix)	2,829,51	22
Paris	1,681,904	19
Vienna	2,807,077	14
U.S. centers[a]		
Atlanta (ATL)	2,466,960	45
Chicago (ZAU)	2,044,424	46
Indianapolis (ZID)	1,903,233	35
Miami (ZMA)	9,618,567	30
Washington (ZDC)	3,226,551	43
Boston (ZBW)	3,063,002	30
Cleveland (ZOB)	1,817,732	43
Jacksonville (ZJX)	3,905,773	37
New York (ZNY)	654,440	31

[a] up to 60,000 ft for U.S. centers.

levels within upper airspace (above FL245) and will have the effect of increasing the volume of some ATCCs.

The total volume of airspace under investigation shows an inverse to that relationship with the combined U.S. centers having nearly twice the volume as the European centers. The median center in Europe, Munich, has a volume equivalent to the smallest U.S. center, New York. Only two European centers, Vienna and Marseilles (Aix), have volumes larger than the median value of the U.S. centers. Only one U.S. center has a volume that is less than the European average.

Data for individual sector volumes were not easily accessible. However, average values can be compared. The 15 centers in Europe have 190 sectors as compared to the 340 sectors in the nine U.S. centers. This works out to an average of 13 sectors per center in Europe and 38 per center in the U.S. The average volume of the sectors is nearly equal.

To use this as a major point of similarity is an oversimplification, because sector volume is based on design characteristics to reduce complexity for the controller and to keep traffic assigned at moderate levels. As an example, within the U.K. FIR Scottish ATCC has a volume of 5,521,018 with 10 sectors, whereas London

Table 5 Total number of flights for September 1996

Centers	Avg. transit time (per ACC), min	Total movements for Sept. 1996
European centers		
Amsterdam	8.6	35,553
Dusseldorf	19.1	41,007
Geneva	11.3	43,574
London	20.1	113,203
Manchester	12.3	28,779
Munich	14.6	62,547
Reims	19.4	55,047
Zurich	12.1	51,966
Brussels	10.5	42,631
Frankfurt	11.7	62,278
Karlsruhe	18.9	60,129
Maastricht	20.1	80,057
Marseilles	27	61,705
Paris	14.3	90,850
Vienna	19.6	43,402
U.S. centers		
ZAU (Chicago)	29.7	382,320
ZDC (Wash. DC)	32	216,360
ZJX (Jacksonville FL)	36.2	236,640
ZNY (New York)	22.9	128,010
ZTL (Atlanta)	30	340,770
ZBW (Boston)	31.3	105,540
ZID (Indianapolis)	28.6	367,230
ZMA (Miami)	40.6	124,530
ZOB (Cleveland)	26.8	342,000

has a volume of 2,144,174 with 29 sectors, less than half the volume with three times as many sectors.

Total number of flights is available for each center along with the average transit time for the center (see Table 5). As shown, the transit times for European centers are smaller than the transit times for U.S. centers. London center in Europe is similar to New York center in the United States in terms of traffic, type of traffic, and average transit time and number of sectors.

To put the number of centers, size of centers, and transit times into perspective, values from an early assessment are brought into the mix. That assessment of aircraft activities showed the average flight time in the United States is 1 h and 23 min and the average flight length is 470 nm. Average IFR flight length within the ECAC countries was 470 n miles and 1 h and 20 min in duration. The average flight in the United States crosses three center boundaries vs five center boundaries in Europe.

This is a point of significance because our future concepts imply and in some cases explicitly state that interfacility coordination is a source of inefficiency. The concepts call for communication and coordination between facilities to be similar to those practices that occur intrafacility. This may be a key factor and warrants further examination into the degree to which interfacility coordination acts as a barrier to efficiency, and if significant, what is the compounding effect that increasing numbers of coordination have on individual flights and on the effective management of flow.

a. Declared sector capacity. An important factor for the declared capacity of the sector is the staffing in the sector. Qualifications of controllers and the progress of trainees can have a significant effect on traffic throughput. Because working practices, especially in the manning configuration, vary not only from country to country but perhaps also from unit to unit, assessment of workload for comparative purposes is very difficult. Controller workload is also closely tied to staffing and controller pay, making any examination into workload sensitive. In addition, confidentiality and other sensitive issues relative to local work practices and procedures can cause difficulties in compiling meaningful data. Because of the sensitivity of these issues, limited data were made available for this analysis.

b. Comparative statistics. A detailed assessment at the sector levels, including the analysis of declared sector capacity in Europe and traffic count and sector loading, was performed on the basis of the following parameters that could be assembled for a number of sectors:

1) Instantaneous count is the maximum number of flights observed within the sector at any time during a 15-min period.

2) Daily entry rate (DER) is in both ACCs and sectors. (Note that there is no difference in DER between requested and regulated flight plan data.)

3) Maximum instantaneous count in each sector is based on requested and regulated flight plan data. The requested flight plan data are not available in the United States.

4) Graphical presentation of declared sector capacity, instantaneous count, and sliding hourly entry rate (SHER). SHER is the number of flights entering the sector within one hour. This figure is calculated every 15 min. It is calculated by summing the number of aircraft entering 45 min before and 15 min after the reference time. An example of a graphical presentation is given in Figs. 4–6.

5) Traffic is divided into seven different classes in Europe. These clearances are also indicative of U.S. operations but with a different nomenclature. These classes are cruise, climb, descend, climb-cruise, cruise-descend, climb-descend, and climb-cruise-descend.

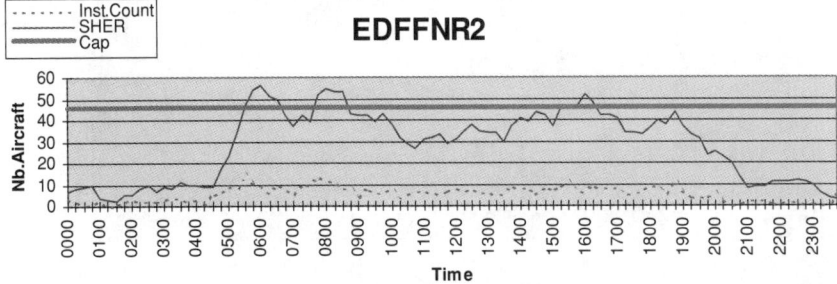

Fig. 4 Sector NRe (Frankfurt).

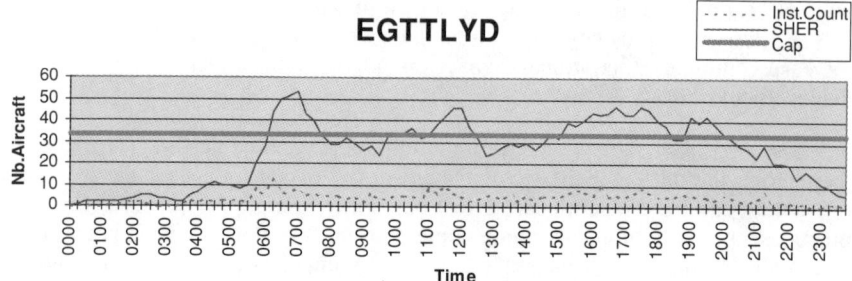

Fig. 5 Lydd sector, requested and regulated.

The assessment addressed the magnitude of, and relations between, the sector parameters for a typical busy day. For the reasons indicated the assessment considered sectors as black boxes, and the report highlights the overall characteristics for some examples only.

Frankfurt sector NR2 has a declared capacity of 46 aircraft. At 0600 hrs the SHER is 56, which means that 56 aircraft entered the sector during the period 0515 hrs to 0615 hrs. At 0530 hrs the maximum instantaneous count of 15 flights can be identified within the sector at the same time.

The second example shows that the Lydd sector at London exceeded declared capacity (33) several times. Indeed, from 0600 hrs to 2030 hrs the sector was close to or exceeding capacity. The DER was 624 aircraft. The instantaneous count was reasonably steady and below 10 for most of the busy period. It is interesting to note that requested flights and regulated flights are exactly the same. This could be that,

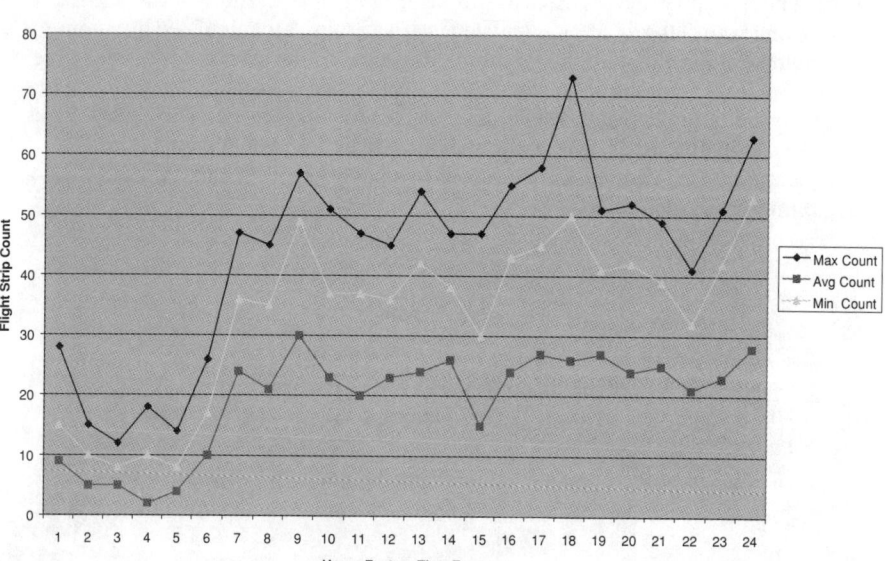

Fig. 6 New York sector A.

while actual traffic was exceeding capacity, the number of aircraft in the sector at any one time (instantaneous count) was within the capacity of the controllers. This could indicate that declared sector capacity is set at a level that guarantees that traffic will be manageable in all circumstances in spite of uncertainties in the system that may result in actual traffic being greater than that regulated. Only 13% of the 624 aircraft were in the cruise. The remaining 87% were evolving traffic with 23% climbing and 42% descending.

For comparison purposes sectors from New York center were chosen (Fig. 6). Statistics that match the European formulation were not available for these sectors. However, hourly flight strip counts that are an indicator of throughput were available. These numbers are based on Tuesday for a six-month period from October through March. When one compares these sectors (and others not shown), there are similar throughput rates.

In the United States, sector 48 is an en route sector that has an instantaneous sector capacity of 18. An hourly sector capacity number for each sector is not available. Instead, a negotiated monitor alert threshold (MAT) value is used. The MAT number for each sector is based on the sector size, the type of traffic, and the complexity of the traffic. The MAT number is an indication of how much traffic a controller can safely work at any one time. We have computed the instantaneous capacity as MAT + 20%.

The trend shows traffic exceeds capacity in the latter part of the day. Approximately 43% of the flights in sector 48 are cruising, 34% are climbing, and 23% are descending. Sector 48 has a minimum throughput of 40 per hour and a maximum of 106 with the SHER (the sliding hourly count). This high throughput rate was achieved by establishing a very fixed structure to traffic and airspace to allow for an efficient, one-way flow of the traffic-only corridor.

2. Airspace Conclusion

Within the individual flight or sector statistics there are no striking differences. There is no great difference in throughput or staffing procedures. This is good, for when coupled with the airport results, this says that we can be reasonably assured that we are describing systems that are not diametrically different from an infrastructure point of view. Nevertheless, it seems that in the United States there is larger recourse to specialized control techniques to help in managing extreme cases. The example of sector 48 has no real equivalent in Europe, although flight-level allocation techniques have been used for years to strategically deconflict flows. An explanation could be found in how ATC developed and adapted to traffic conditions and external events and was able to deal with management of traffic as a whole.

However, there are the appearances of significant differences in the manner that flow is managed and how facility boundaries may affect flow. There are two issues that deserve more detailed examination. If coordination across facility boundaries can be disruptive to flow, then the significantly higher number of facilities faced by an average flight in Europe vs the United States is significant. The second issue involves the role of traffic flow and the traffic flow position in European facilities and the United States and the techniques they apply. In the United States this position takes an active role in smoothing flow internally with adjacent facilities nationally, and it has a variety of flow techniques.

This examination of flow leads one to consider and, if possible, provide explanations of why at the strategic flow level the Europeans opted for the CFMU vs the apparently more ad hoc system used in the United States, or the U.S. use of single direction sectors. This also leads into the next section, where we examine the idea flow objectives and the processes employed to meet these objectives.

C. Traffic Management

1. Traffic Management Phases

The focus of the airspace comparison section was the examination of the management of individual flights in the National Airspace System (NAS). Although the statistics are aggregate, the snapshots into airspace focus on the aircraft and the individual flight as it moves through the system—the transit time of the flight, the phase of flight the aircraft is in, the instantaneous count of flights, etc. The values measured can give indications of general health from one day to the next just as body temperature does in humans. The values do not provide insight into the underlying objectives that the ATC system had for these flights. The problem with extrapolating from these statistics to a definitive comparison occurs whether one is comparing one sector to the next in the same center as well as sectors from opposite sides of the Atlantic.

Consequently, in the course of conducting the comparison, the team had to go from measurement to hypothesis in order to try to use the statistics as a means of comparison. In developing the hypotheses, it became clear that an aircraft is subject not only to the individual phases of flights, departure, cruise, arrival, etc., but also to a series of traffic management phases. The phases, shown in Fig. 7, are moving from left to right in the figure: 1) Ramp Management moves the aircraft in and out of the gates, 2) Airport Surface Management takes aircraft in departure from the ramp to the departure queue, 3) Departure Management manages the departure queue to launch aircraft from the queue(s) into the airspace, 4) Dispersion Management, as its name implies, has the objective to get flight up and out of the terminal into the en route structure, 5) En route Management shows the aircraft at altitude and moving towards their destinations, but it is not yet subject to actions associated with their arrival, 6) Collection Management sequences and spaces aircraft to bring

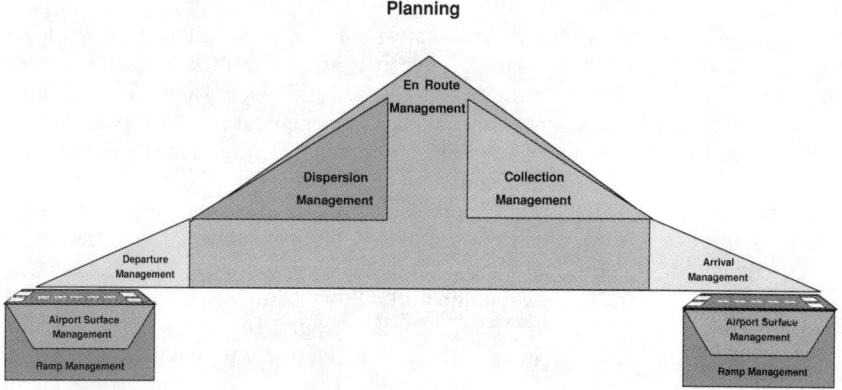

Fig. 7 Traffic management phases.

them into the terminal area, 7) Arrival Management assigns aircraft to runways and gets them onto the surface, 8) Airport Surface Management gets the aircraft off the runways and moves the aircraft to the ramp, and 9) Ramp Management works to get them into gates.

An assessment can be made of the ATM system and the related traffic management phases:

1) Gate-to-gate is not equivalent to dropping an aircraft in at one end, turning the crank, and popping it out at the other. Traffic is mixed in size and direction, and so the phases need to be scaled to manage the uncertainty. The chokepoints in the system flow should only be the natural physical points, i.e., the concrete infrastructure runways, taxiways, and gates.

2) The phases are not disjointed. In the Newark arrival case, aircraft departing Washington will be in both the departure management phase and the collection management phase at the same time. In fact, that is probably the most common occurrence. A flight that is required to fit into a specific slot en route may be in the airport surface management, the departure management, and the dispersion phases simultaneously.

3) The phases need to be managed to ensure that a preceding phase does not overload a succeeding phase.

4) Aircraft are not segregated by traffic management phases. Some aircraft in cruise might be subject only to en route management, while others in the same sector might already be part of a sequence and undergoing spacing for the collection management phase. These characteristics are considered by the controller as a conflict, and probing and resolution planning are conducted. It is also a level of complexity currently not considered by most tools.

5) The technique used in the Newark arrival example is the imposition of miles in trial on this arrival route to ensure that the load over the arrival fixes will not exceed the terminal and airport rates. A specific form of demand capacity balancing invariant process is taking place.

Herein may lie the differences that are seen in what is characterized as the ad hoc U.S. system vs the more structured CFMU control flow in Europe. There has always been the characterization that the bottleneck in Europe is en route, and the bottleneck in the United States is the airports. However, the large recourse to schedule coordination at European airports might create a false impression here. Characterizing the flow based on the traffic management phases provides insight into what that may actually mean. Comparing things airport by airport, sector by sector, did not show great differences in throughput. The airport-by-airport comparison showed mainly differences in scheduling philosophies and infrastructure size.

In neither environment is there a problem in filling up the airspace, especially close to the airports (with close being a subjective term for which a value of 200 n miles is not unreasonable). The differences seem to be in the techniques available for managing flow. The techniques have a relationship to the scope of information and the span of influence available in the technique. In the United States requiring miles-in-trail (MIT) restrictions on internal center sectors as well and especially those in the previous center can achieve the management of flow. The same influence was not possible in Europe, in particular because different centers often depend on different providers or states, and therefore do not share the same culture and are less concerned by problems in the second or third center downstream theirs. It is interesting to note that for flows to Paris, the French ACCs feeding the Paris ACC provide help to their colleagues by tactically prearranging the flows using some form of MIT, therefore directly contributing to a simplification of the traffic complexity in the en route and terminal maneuvering area (TMA) sectors.

The inability to extend influence back into the en route sector results in an imbalance of demand and capacity close to the airport and will require large amounts of holding. Considering the example from Ref. 1, this is equivalent to free flowing a rate of 90 aircraft up to the terminal boundary and then trying to manage it with only a 60-aircraft outflow. It is clear that any airspace buffer will soon be exhausted. Without the ability to manage the flow in the en route sector through controller imposed restrictions, the alternative is to manage the flow at the source, hence the CFMU.

The restriction method is not without problems. First, as can be seen in the Newark example, the technique requires a volume of airspace where aircraft can be staged through vectoring and other techniques to space and sequence the flow. As traffic volumes grow in general, the buffer airspace may become overloaded with other traffic. The en route traffic problems experienced in the United States in the Cleveland and Indianapolis centers may be related to this and is a subject of follow-on analysis.

Second, the method of metering to fixes and not managing flow to the airport results in inefficiencies and underutilization of runways. This is the subject of much of the research represented in NASA's Traffic Management Advisor. A major component of its utility will be extending accurate arrival trajectory modeling across center boundaries and providing individual flight strategies to upstream sectors regardless of facility.

Finally, there is always the problem of the close-in flight. Both the restriction method and the improved Traffic Management Advisor method work best when the flow consists mainly of aircraft that are airborne at the time of first management. When the flow has a large component of short flights, as may be represented in the core of Europe or in the Washington to New York flows, the techniques become more difficult and less efficient. This is made even worse with small centers (in geographical terms) and with very interacting flows such as in the core area of Europe: it is located in the center of the continent and fed from all directions, whereas the busy U.S. areas mainly extend at both extremities of the continent.

V. Conclusion

The comparison exercise has led to increased understanding on both sides of the ocean into the vagaries associated with shared knowledge. While experts in each nuance may say, "Of course, how could you have thought differently?", too often the shared knowledge carries the day in discussions and decisions. We investigated the IMC vs VMC operations in Europe and the United States. The common knowledge provided the starting point, but the subtleties of access, negotiations, and politics are missing in that shared knowledge. When these components were added, no great difference on a runway-by-runway basis could be derived.

In the airspace, when staffing practices and like sectors are compared, once again the similarities outweigh differences. It is only when we began to consider flow that we began to consider very specialized sector/practices and some specific conclusion could be drawn. In addition, it is also when we considered the impact of the overall traffic flow that we began to ascertain why some decisions vs others are made. It is the tool options available to meet aggregate goals of the air navigation systems that need the investigation. It is the scope of information and span of influence that begin to provide insight into our existing and future concepts.

To this end, the team identified the traffic management phases of flights. When we say gate-to-gate, we often think of individual flights and the individual flight phases. However, on reflection gate-to-gate is not about individual flights but flow. It is a concept that recognizes the network effect linking all decisions from departure to destination and attempts to choreograph the mass of flights into an efficient aggregate.

We have only begun to explore the richness that this view into the concept provides. This framework improves our ability to tie together concepts, services, and performance. It also provides the context in which we can further mutually define the ANS global concept.

Reference

[1] Rico, F., "Air Traffic Management Constraints," *Proceedings, ECAC/EU Dialogue with the European Air Transport Industry, Airport Capacity—Challenges for the Future*, European Civil Aviation Conference, April 1999, p. 41.

Chapter 4

Performance Review in Europe

Xavier Fron*
EUROCONTROL, Brussels, Belgium

I. Introduction

THE purpose of this paper is to examine the challenges facing the EUROCONTROL organization in improving the performance of air traffic management (ATM) in Europe. It mentions in particular the new performance review system established in 1998, whose role is to publish ATM performance indicators for the EUROCONTROL area, to propose performance targets, and to develop economic regulation guidelines.

II. Background

The EUROCONTROL organization is presently composed of 28 member states in Europe and of its permanent agency. It is responsible for the provision of air traffic services (ATS) in its area. A number of initiatives were taken to meet the challenges associated with growing traffic as follows: 1) 1988, central flow management unit (CFMU); 2) 1990, European Civil Aviation Conference (ECAC) Strategy for the 90s, European Air Traffic Control Harmonisation and Integrated Program (EATCHIP); 3) 1992, European Air Traffic Management System (EATMS) Airport/Air Traffic System Interface (APATSI) (airports); 4) 1994, prepare institutional arrangements; 5) 1997, new institutional strategy EUROCONTROL revised convention; and 6) 1999, strategy for the years 2000+ (ATM 2000+).

Two main features of the new strategies are particularly worth mentioning here. First, they are performance-oriented rather than solution-oriented, as was the ECAC Strategy for the 1990s. Second, a compromise was adopted between two extreme institutional options, i.e., leaving each individual state responsible for its own ATM and having a single European ATM body. The provision of ATM remains decentralized, but a "strong, independent and transparent" authority, namely the Performance Review Commission (PRC), monitors ATM performance and sets targets. This ATM system should be as efficient as if it were run centrally. The PRC reports at the level of Directors General of Civil Aviation, as shown in Fig. 1.

Copyright © 2001 by the American Institute of Aeronautics and Astronautics, Inc. All rights reserved.
*Head, Performance Review Unit.

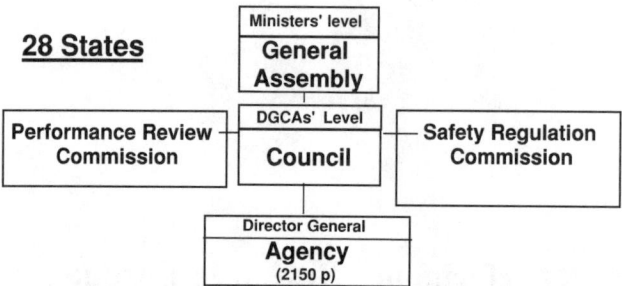

Fig. 1   EUROCONTROL institutional structure (simplified).

The PRC's task is to define and monitor performance indicators, to propose targets to be met by ATS providers, and to publish guidelines for economic regulations, which could include incentives to meet those targets. Figure 2 shows the key performance areas (KPA) as tentatively defined.

In three key performance areas, called conditions, minimum levels of performance decided outside the PRC have to be met (safety, environment, and equity). The ATM system performance is then to be optimized for the other KPAs from an airspace user's perspective. The first step is to measure the present performance and to understand where major deficiencies are. The next step is to understand tradeoffs between performance items before targets are proposed. A further step will be to develop incentives for all players (users and providers) to behave in such a way that the system optimum is approached. The relevance of this paper to the ATM 1998 seminar should be clearer at this stage.

III.   European Challenge

A.   Traffic Growth

While the volume of controlled civil air traffic in Europe remained virtually static between 1975 and 1985, there was sustained growth of between 5 and 12% per annum from 1985 onward. (Note that this date coincides with the signature of the Single Act introducing a single market in 1993, the impact of which was anticipated by users.) Despite competition from alternative means of transport and telecoms, experts widely predict sustained air traffic growth in Europe to accompany further

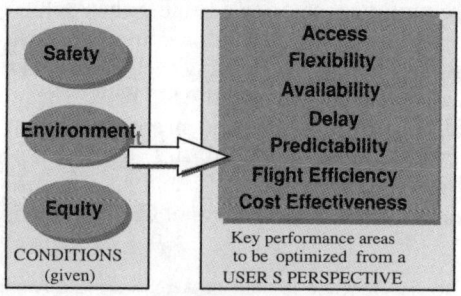

Fig. 2   Key performance areas.

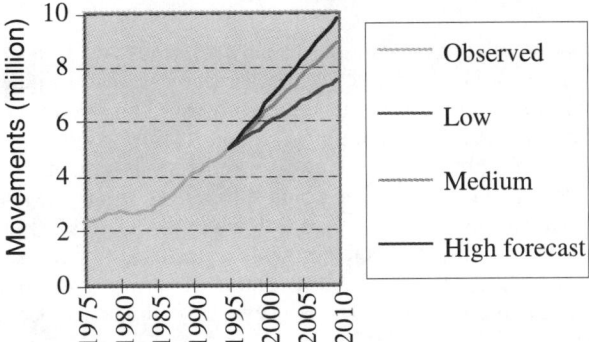

Fig. 3 IFR movements (Euro 88 area[1]). The "Euro 88" area comprises 11 states of western Europe, i.e., approximately 75% of traffic in the ECAC area (most of the European states to the west of the CIS).

European integration and the process of globalization. In fact, more than half of the 200 main city pairs in Europe are still within the same country, and intercountry city-pairs are expected to develop (see Fig. 3).

B. ATM Key Performance Areas

The performance of ATM must be improved, in particular with regard to safety, delays, cost efficiency, and other key performance areas (predictability, environmental sustainability, etc.), in order to respond to traffic growth and users' concerns. Corresponding potential benefits and savings amount to hundreds of millions of Euro each year. (One Euro is worth about 1.1 U.S. dollar.) The performance of ATM in Europe must be addressed at the continental level, because domestic airspaces are extremely interdependent, and the effectiveness of the network is largely dependent on its weakest link. [Ninety-eight percent of European traffic departs from or lands in Europe, 74% of European air traffic is internal to the region, whereas a fraction remains within a single country (30% in the case of France, for example).]

C. Safety

Risk increases approximately as the square of traffic. If traffic doubles, the risk per flight hour has to be divided by four to keep the same absolute level of risk. In this context it is essential to increase safety in order to guarantee public confidence in air transport. Safety does, however, come at a price, and a balance must be struck between safety and other forms of performance, such as the cost of the ATM system. Deciding between safety and performance does not fall within the remit of the PRC and should be carried out at the political level (Council, Permanent Commission, European Union, etc.). The Safety Regulation Commission (SRC) and the PRC are deliberately separate. Performance review must take safety into account, but it has no authority with regard to the regulation thereof. A balance must therefore be found between both commissions with regard to safety matters.

D. Capacity and Delays

The ATM system may be regarded as a network whose every node and arc has a limited capacity. When demand exceeds capacity in certain places, the most immediate solution is to manage the situation by holding aircraft on the ground. This reduces risk, cost (holding in the air is around 3.5 times more expensive than on the ground), and pollution, but does, however, generate air traffic flow management (ATFM) delays which we have all no doubt suffered. (It should be noted that there are many other reasons for delays, such as loading passengers/luggage, technical problems with the aircraft, insufficient runway capacity, unfavorable weather, etc.) Flow management is therefore one of the solutions that was developed in response to the crisis at the end of the eighties. However, any imbalance between capacity and demand materializes in ATFM delays.

Since 1996, flow management has been run centrally by EUROCONTROL's CFMU on behalf of all the participating states (33 in 1997). A takeoff slot is allocated to all flights passing through an overloaded area on a "first-planned, first-served" basis. Experience has shown that centralized, individual management of queues on the ground was more effective than decentralized management on the basis of origin-destination flows, as was the case previously. Furthermore, the high level of protection centralized management provides against random traffic peaks means that the acceptable traffic threshold can be raised in control centers. Delays on the ground imposed by the CFMU are the result of insufficient capacity at certain control centers and airports. In 1996, 15% of control centers were responsible for 90% of ATM-related delays. Furthermore, of some 800 airports capable of receiving commercial traffic in Europe, only a few of them (Heathrow, Milan, Athens in the summer, etc.) were responsible for the majority of delays attributable to insufficient airport capacity.

As with many queuing systems, the cumulated delays in the area have an explosive behavior as soon as the demand/capacity ratio approaches one. Network capacity is difficult to determine with certainty, because there are alternative routes and the traffic structure changes, in particular between weekdays and weekends. System capacity is therefore shown as the relationship between demand and delays. An elasticity of the order of 5 has been measured between delays and demand, i.e., 1% extra demand increases delays by 5% with constant capacity and, inversely, 1% extra capacity reduces delays by approximately 5% with constant demand. An increase in capacity can be measured as that traffic increase that can be handled with delay remaining constant (+6.2% between 1996 and 1997). This is illustrated in Fig. 4.

Accumulated delays associated with insufficient ATM capacity in the CFMU area totaled 20.6 million min in 1996 for approximately 7 million flights, i.e., an average of a 3-min delay per flight. Approximately 20% of delays are ATM related, and 7% exceed 15 min. The causes of the various delays are analyzed inter alia by the CFMU and EUROCONTROL's Central Office for Delay Analysis (CODA). The cost to operators of ATM-related holding is estimated at Euro 400 million (the average cost of 1-min holding on the ground is Euro 18.5, according to IATA) in 1996 (this figure does not take account of inconvenience to passengers). There are, however, less reliable figures for analyzing the other causes of delays (due to technical problems with aircraft, loading of luggage and passengers, etc.). The late arrival of aircraft flying the previous flight segment, which is itself partly caused by ATM, is the main known cause of delays (around 40%). Information regarding

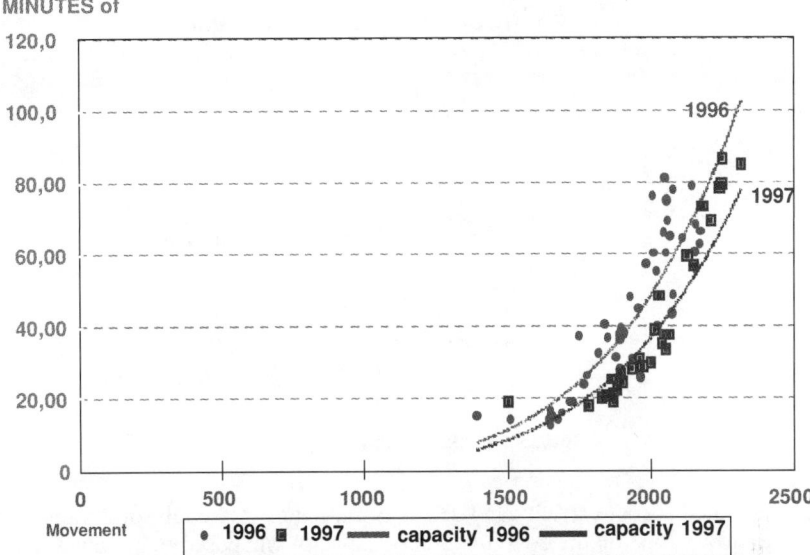

Fig. 4 ATM measured as the relationship between demand and delays.

in-flight delays is still less widespread, but they also contribute to late arrivals and penalties.

Air traffic control (ATC) capacity thresholds are largely dictated by the limited workload that can be absorbed by controllers in each sector of airspace. (A control sector is the geographic unit or organization for control tasks, under the responsibility of a team of controllers.) The traditional method of increasing capacity by subdividing airspace into increasingly smaller sectors is anticipated to reach its limits around 2005 in areas with the highest traffic density. This is the basis for the notion of the capacity wall. The capacity of increasingly larger portions of airspace will be limited by this capacity wall unless radically different ATM methods are introduced. The introduction of new communications, navigation, and surveillance (CNS) technologies, such as satellites, is useful only insofar as it promotes the use of more efficient procedures or reduces the cost of the infrastructure, but it does not provide a solution per se.

E. Cost Effectiveness

The cost of the ATM service in Europe, an amount of Euro 3500 million in 1997, is recovered entirely through the route charges paid by airspace users (air transport, military, and general aviation). Route charges account for some 4% of the operating costs of the major airlines. On top of these come charges for terminal and airport services. Total charges are of the same order as fuel costs (10% of operating costs).

There is great pressure from airspace users to reduce the costs of the ATM service. However, the objective should not be to minimize route charges at any price. There are tradeoffs between a number of performance items such as costs and delays. Figure 5 shows that there is an optimum capacity/demand ratio where

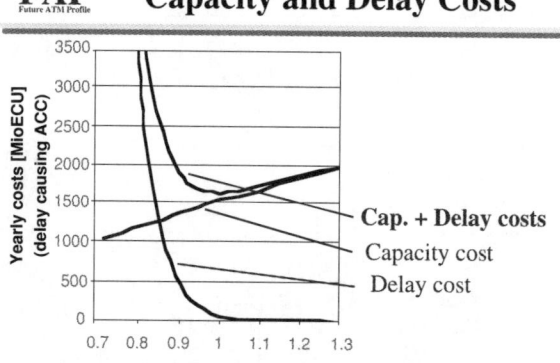

Fig. 5 Capacity and delay costs.

the cumulated cost of capacity and delays is minimum. Although some validation remains to be done on the cost function of capacity (supposed to be linear here), the result remains.

F. Other Performance Areas

A reduction in the operational penalties associated with nondirect routes in upper airspace could save hundreds of millions of Euro per annum. Other performance areas, such as predictability, are considered extremely important in the United States. Indeed, accurate forecasts of traffic conditions a number of months in advance would enable reliable timetables to be fixed and observed, which would be extremely beneficial not only to airlines but also to passengers.

Last, insufficient ATM service quality also has a cost. The PRC should therefore act to identify the optimal balance between the various performance factors (cost, capacity, etc.) which are themselves linked to production factors (staff, investments, etc.) and work toward achieving this balance.

G. Economic and Financial Stakes

Table 1 summarizes orders of magnitude for the economic and financial stakes associated with ATM in Europe. Being based on rough estimates, conclusions there have to be taken with caution. The table tends to indicate that a 10% productivity gain in ATM services, a stabilization of delays, or a 50% reduction in the lengthening of routes would each bring savings in the order of hundreds of million of Euro per annum as compared with natural trends. These objectives are ambitious but not unattainable.

The main value added by ATM could be to enable further air transport growth. A doubling of traffic would increase the annual revenue of air carriers in Europe by approximately Euro 40,000 million per annum, with direct and indirect benefits for employment and economic growth.

Several factors could, however, restrict such growth, in particular the capacity of the major airports, environmental pollution, and ATM capacity. All of these restrictions must be overcome if this added value is to be created.

Table 1 Potential savings

Forecast/actual value, Euro million		Date	Assumed improvement, %	Annual savings, Euro million
Route charges	3500	1996		
	5200	2005	−10	+500
	7000	2010–15	−30	+2100
Ground delays	400	1996		
caused by ATM	800	2000	−50	+400
	2000[a]	2005	−80	+1600
Ground delays				
caused by	500	1996		
insufficient airport	1000[a]	2000	−50	+500
capacity	2500[a]	2005	−80	+2000
In-flight delays			not available	
Lengthening of	600	1996		
routes	900	2005	−50	+450

[a] Extreme scenario in which no further capacity is created (do nothing).

H. Cost and Financing of ATM

Air traffic control has traditionally been financed from state budgets because it is the responsibility of states pursuant to the Chicago Convention signed in 1944. This is still the case in the United States, where the federal government levies a tax on tickets for commercial flights and on fuel for private flights. In Europe receipts are being assigned to ATC service providers, who increasingly have the status of private companies. Users now fully finance the ATM service in Europe through charges. The recent International Civil Aviation Organization (ICAO) conference in Rio highlighted the importance of associated financing and control mechanisms for the implementation of ATM/CNS systems. The EUROCONTROL organization has a powerful and effective means of financing in the form of its route charges mechanism.

EUROCONTROL's Central Route Charges Office (CRCO) is responsible for the recovery of charges on behalf of the participating states. These totaled Euro 3500 million for 27 European states in 1997. In Europe route charges are proportional to the distance flown and the square root of the aircraft's weight. This formula enables costs to be apportioned according to contributing capacity, without excessively penalizing operators of the heaviest aircraft. It is accepted as fair and equitable by all parties. The service unit is based on control for an aircraft weighing 50 tons over a distance of 100 km. The unit rates are determined by the enlarged Committee for Route Charges in such a way that 100% of the costs are recovered.

Since January 1998, flight distance has been calculated on the basis of the route length per state overflown (calculation method known as RSO), as opposed to that of the most frequently flown route, and the charges are paid to the state actually overflown. Choice of route will henceforth introduce an element of competition between ATM service providers, particularly in Europe, where national airspaces are rather small. Nevertheless, the principle of recovery of costs still applies today, i.e., a state may in principle bill users for the entire cost incurred,

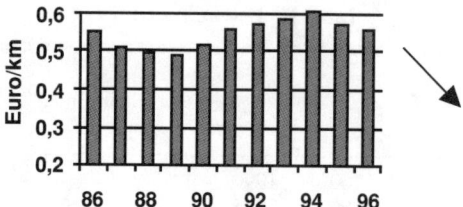

Fig. 6   Average rate of route charges in Europe (deflated).

irrespective of the amount. Apart from having an inflationary effect, this formula offers no encouragement to increase capacity or the quality of the service provided, because the costs are recovered regardless of the service offered. The RSO system could therefore achieve full efficiency only if all the charges paid were linked to the service actually provided, for example by means of a maximum price mechanism.

Although the users, who are also the payers, have no direct authority over the setting of the unit rates, they nevertheless exert an influence, and average unit rates have remained relatively stable over a long period (see Fig. 3). As from 1998, they are entitled to attend all meetings of EUROCONTROL's decision-making bodies with observer status. The stagnation of traffic at the beginning of the 1980s led to a substantial reduction in ATM investments and caused capacity to stagnate. The considerable traffic growth at the end of this period brought a substantial increase in the delays caused by ATM, and ultimately led to measures being taken by ECAC Transport Ministers as described at the beginning of this chapter. The situation was redressed between 1990 and 1994 at the expense of an increase in the unit rates, despite the fact that all other air transport costs were falling. The average unit rate then fell from 1995 onward because of pressure from users.

Figure 6 also shows that it has hitherto been possible to increase capacity at a relatively stable long-term unit cost. This in itself is remarkable, because the traditional method of creating capacity involves dividing airspace into ever smaller control sectors, a process with diminishing return. Thus, increasing the number of sectors by 40% only increases capacity by approximately 20%. Productivity gains have therefore in the past compensated for the lower efficiency caused by subdividing airspace. Users are now calling not only for increased capacity and reduced penalties but also for lower unit costs. This adds an extra dimension to the challenge faced by ATM services.

I.   Penalties Associated with Air Traffic Control

ATM generates direct costs associated with route and airport charges, and also penalties due to delays, imposition of longer flight paths or flight times, nonoptimal use of aircraft in the vertical plane, etc.

Multisensor and/or satellite surface navigation systems now enable aircraft to follow an optimal route between two distant points without overflying radio beacons. Nevertheless, controlled aircraft still often follow a network of published routes, because this facilitates control (low rates of closure at the same altitude, finite number of crossing points to monitor, more orderly and predictable traffic). Average route lengthening is in the order of 10% (45% in certain cases). It would

Table 2 ATM performance improvement targets

	Natural trend from 1995 to 2010/2015	Assumed target
Traffic	×2	×2
Accidents/incidents	×4	×1
Airport/ATM delays	×10–20	×0.5
Charging rate	×1	<1
Flight penalties	×2	×1

be possible to reduce this by approximately half by means of direct routes, because departure and arrival paths around airports must continue to be highly organized. The financial benefits would be in the order of Euro 450 million per annum in 2005, and emission pollution would be reduced by 5%.

J. Summary of Expected Improvements in Performance

The challenge facing ATM to deal with a doubling of air traffic in the years to come may be summarized as follows:

1) The capacity wall must be breached around 2005, otherwise ATM delays will mushroom to such an extent that aviation development will be jeopardized.

2) The rate of incidents/accidents per flight hour must be reduced to a quarter of the current present figure.

3) Unit costs must be reduced.

4) Service quality must be improved and needless penalties reduced. This is summarized in Table 2.

The EUROCONTROL organization's objective could therefore be to manage ATM and the corresponding CNS infrastructure in such a way that traffic demand is met, that agreed minimum levels of safety, environmental sustainability, and equity performance are met, and that the performance of the ATM system is optimized for specified key performance areas. The Performance Review Commission could play a decisive role here.

It is reassuring that the research and development work carried out in the last 10 years, in Europe in particular, enables these challenges to be faced with reasonable confidence. However, the benefits will materialize only if the other limiting factors such as airport capacity and environmental sustainability are tackled concomitantly.

IV. Other Limitations on Growth

A. Airport Capacity

As we have seen, only about 20 of approximately 800 airports capable of receiving commercial traffic in Europe are responsible for the majority of airport delays. Airport delays are the direct consequence of mismatching capacity and demand at the major airports, even where demand is limited by scheduling committees, which do not allow the number of scheduled flights to exceed the number of slots available. Much is now known about airport-related delays owing to the work carried out by EUROCONTROL's CODA. These analyses are, however, based on data that are still insufficient or incomplete.

Many factors combine to increase and concentrate demand on the most overloaded airports, in particular the formation of worldwide alliances between the major airlines, built around "hubs." This causes traffic concentrations at certain airports at certain times, with waves of arrivals, followed by waves of departures. To reduce airport constraints, some airlines establish their hubs at secondary airports, which are reserved almost entirely for themselves, like Delta Airlines at Cincinnati, Ohio. Such developments are dictated by market opportunities and are therefore difficult to predict.

On the other hand, increasing airport capacity is a complex matter entailing political considerations (traffic rights, slot allocation, environmental constraints, intermodal transport policy). It is difficult to build new runways, and landing frequency is limited by physical phenomena (wake turbulence, braking time, runway clearance time). ATM can only optimize the use of existing capacity.

The distribution of traffic between airports will thus have to change in line with the market, competition between means of transport and with means of telecommunication, environmental pressures, market trends, etc. ATM will have to be able to adapt to the new traffic flows. In conclusion, airport capacity will be a decisive factor influencing air transport growth in Europe in the years to come, as is already the case in the United States. Initially, we need to get a better idea of the current causes of airport delays, and it is even more essential to predict how they will develop. To this end there is a need for reliable, consistent sources of information on all major European airports and for demand forecasts that are as accurate as possible. An initial step in this direction might be to encourage OOOI (OOOI: Out, Off, On, In, i.e., off-block, take-off, and landing times) reports to be submitted for scheduled flights, as is already the case in the United States.

B. Environment

Preservation of the environment might be a major constraint on the development of air transport, in particular the control of noise pollution. Certain airports (such as Amsterdam and Orly) are already subject to major environmental restrictions. ATM can influence the impact of aviation on the environment only marginally by reducing flight distances (around 5% of emissions) and through less noisy takeoff and landing procedures. Progress will essentially have to come from engine improvements and regulations intended to eliminate the most noisy aircraft.

V. Conclusions

Air transport has a good potential for growth and for creation of highly qualified jobs in a context of European integration and globalization. Nevertheless, the associated benefits will materialize only if the limiting factors, in particular airport capacity and environmental pollution, are addressed with the necessary vigor.

As far as ATM is concerned, major changes in systems and methods of flow control, airspace management, and aircraft separation will be needed to breach the capacity wall predicted around 2005 in areas of high-density traffic.

The initiatives under way, in particular within the framework of the revised EUROCONTROL Convention, leave little doubt that solutions are being found, and will continue to be found, to enable air traffic control in Europe to satisfy users' requirements in terms of quantity and quality in the decade to come. Airspace users, who bear the entire cost of the ATM service, will undoubtedly be exerting pressure to ensure that these promises are kept. Swift and effective action to improve the performance of the ATM system in Europe should be a decisive factor in this process.

Chapter 5

United States and European Airport Capacity Assessment Using the GMU Macroscopic Capacity Model

George L. Donohue* and William D. Laska[†]
George Mason University, Fairfax, Virginia

I. Introduction

INTERNATIONAL air transport is the fastest growing segment of transportation. This can be partially attributed to a combination of market trends and institutional reforms combined with rising incomes and increased leisure time. Air passenger traffic since 1960 has grown worldwide at an average yearly rate of 9% with freight and mail traffic growing at 11 and 7%, respectively. In 1995 approximately 1.3 billion passengers were carried by the world's airlines. For European air passenger traffic growth, the projected annual growth rate is 5.2% between 1993 and 2000, 4.2% between 2000 and 2005, and 3.8% between 2005 and 2010.[1] This growth is graphically displayed for the top 16 European airports analyzed in this study (representing 78% of total European passenger transport) in Fig. 1.

In addition to passenger transport, air transportation has become an important form of freight transport. Freight traffic is predicted to grow at 30% over the same time. The International Civil Aviation Organization (ICAO) estimates that over 30% of world trade by value is transported by air with forecasts of it rising another 400% by 2015.

This continued rise in air transport activity has placed enormous pressure on the finite capacity of the air transportation system. In particular, the effect of reaching capacity limits has caused the number and length of air transportation delays to increase. This could affect future economic growth, especially for export-oriented economies such as Germany.[2] Therefore, understanding delays and their relationship to capacity becomes very important.[3]

There are a number of high resolution models available (LMINET, DPAT, TAAM, and RAMS models in particular) that will assist in the understanding

---

Copyright © 2001 by the American Institute of Aeronautics and Astronautics, Inc. All rights reserved.
*Professor of Systems Engineering and Operations Research. Fellow AIAA.
[†]Ph.D. Candidate, School of Public Policy.

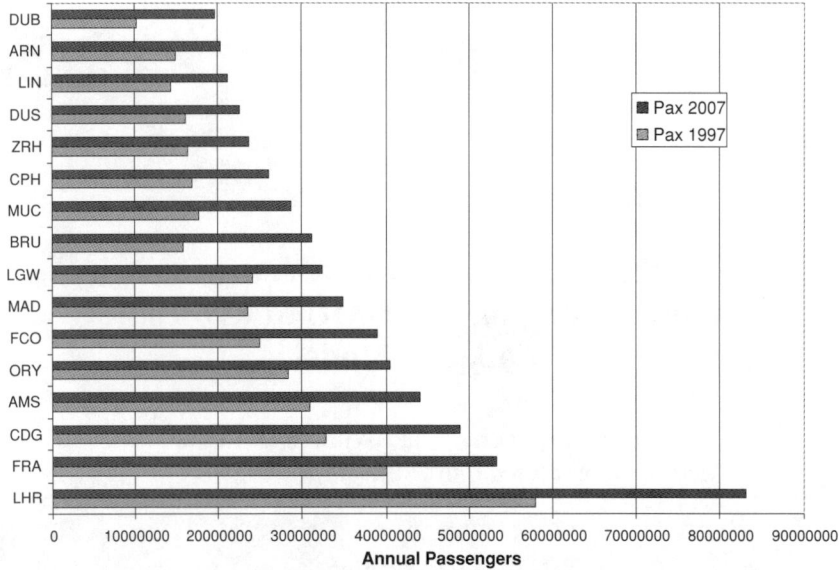

Fig. 1 Current and projected enplanement at the 16 European airports used in this study.

of the factors involved with capacity and delays. These models provide a detailed analysis but require significant amounts of data that are sometimes difficult to obtain. Learning to use these models takes considerable time and effort, limiting their use to specialized individuals.

This chapter will focus on the Macroscopic Capacity Model (MCM)[4] developed at George Mason University, which takes a simpler approach to analyze the expected capacity limitations of airports while using aggregate data. Having a reduced data-dependent model will allow greater access for initial capacity/delay analysis. The work being done on this model will be of benefit to policy makers who are responsible for making national and/or local air transportation systems decisions and for those responsible for measuring the air transportation systems operational capacity.

The remainder of this chapter will briefly discuss the methodology used by the MCM. Thirty-two high-density U.S. and European airports will be analyzed and compared using the MCM. The final section concludes with a utility analysis of the European air transportation system compared to the U.S. air transportation system.

II. MCM Approach

The MCM is designed to estimate the maximum capacity of national air transportation regions. The model is based on empirically observed aircraft arrival rates, government published annual operational rate statistics, and analytically derived airport arrival and departure functions. This is a simple model based on aggregating the effects modeled in a number of airport capacity and delay simulation models.[5]

The MCM assumes that the air transportation system can be modeled as a network of queues in equilibrium whose maximum capacity is the sum of twice the maximum airport arrival rates, less airspace human factors limitations. (In equilibrium the total mix of operations is 2 × arrivals. Arrival rates are typically less than departure rates.) This underlying theory of modeling the air transportation system as a queuing problem has been incorporated into both the Logistics Management Institute (LMI) model[6] and the MITRE Decision Policy Analysis Tool (DPAT) model.[7]

The MCM model approach is much like highway engineering and communications theory.[8] Knowing the maximum capacity has much to offer in gaining theoretical insight into the air transportation problem. The MCM model for system maximum capacity can be expressed as:

$$C_{\max} = 2 \times C_{\text{AR max}} \, S\Sigma_i \, (XGR)_i - C_{\text{AS max}} \Sigma_K A_K$$

where,

$$A_K = (A/C_{\text{request}} - A/C_{\text{accept}})/C_{\text{AS max}}$$

The factor R_i is the number of runways at the ith airport, X_i is an airport runway design efficiency factor, and G_i is the gate utilization factor. The variable S is an aircraft separation factor, largely independent of airport. ($S = 0.5$ at 4.0 n mile separation, roughly current technology limit.)

The second term in this equation represents human factor limitations in the en route airspace. It will be seen that this term can be significant in high capacity airspace such as Europe and the northeast triangle of the United States. The factor A_K is the receptivity factor for the Kth sector and ideally should be minimized.

The term A/C_{request} is the number of aircraft requesting sector entry, and A/C_{accept} is the number of aircraft granted sector entry (i.e., not restricted via a ground delay program or holding pattern).

The factor $C_{\text{AR max}}$ is estimated to be 64 aircraft arrivals/hour, as shown in Fig. 2. (This represents approximately 2 n miles seperation on final approach at 130 kn.) The factor $C_{\text{AS max}}$ represents the maximum number of aircraft per hour in a sector

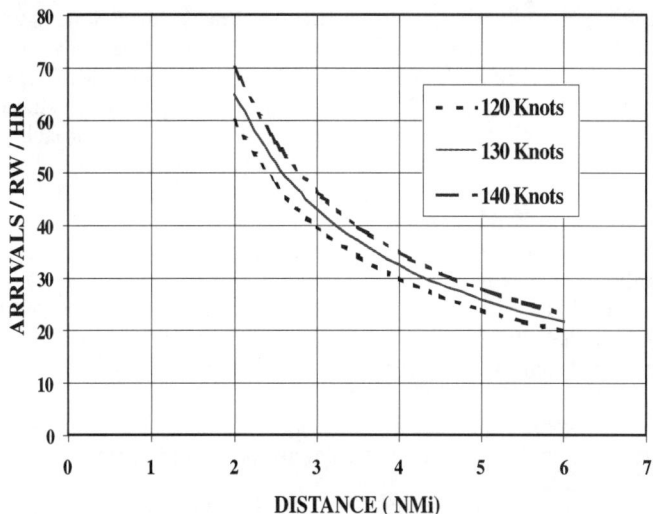

Fig. 2 Runway arrival rate as a function of aircraft spacing.

(approximately 120). (This assumes a 15-s entry and exit acknowledgment without data link.)

III. MCM Validation

It is difficult to validate any model. One way of validation is to see how predictions of a model correlate with empirically derived dependent variables and theoretical correlations. To evaluate whether or not this macrocapacity model formulation is reasonable, we examine how well delay is correlated with airport capacity ratio predictions from established models.

Comparing the MCM results to that of the DPAT model has provided a preliminary evaluation of the MCM.[9] Although different parameters are computed, the results were consistent in predicting delay growth, suggesting that the MCM captures the essence of the ATM system behavior as represented in a more highly detailed model.

IV. MCM Assessment of U.S. and European Airports

An analysis of the air transportation capacity in the United States and Europe using the MCM has been conducted to compare relative operational efficiency using a common macrocapacity model. The data used for the United States came from the Airport Capacity Enhancement (ACE) plan and 1997 Consolidated Operations and Delay Analysis System (CODAS) data. From this data, a sample of 16 airports was selected representing 7.6 million operations per year in an airspace volume similar to that of Europe.

European data came from Ref. 1. An equal sample of 16 airports was used representing 4.3 million operations per year (over 78% of total European air transportation operations). Annual operational rates were divided by 350 days and 16 h/day to determine peak hourly operational rates for both U.S. and European airports.[10]

Based on this analysis, MCM estimated that the northeast triangle of the United States (a 1000 n mile triangle from Boston, Massachusetts, to Minneapolis, Minnesota, to Tallahassee, Florida[4]) is currently operating at 74% of maximum capacity, as shown in Table 1, and will be at 89% capacity by 2012 (Table 2).

Using this same definition of capacity, European capacity is estimated to be currently at 40% of maximum capacity if the average final approach spacing is assumed to be 4 n miles (4 n mile spacing equates to approximately 32 operations per runway per hour) (Table 3). Estimating capacity fraction for 2007 resulted in 55% capacity fraction, respectively (Table 4).

Figures 3 and 4 illustrate the relative efficiency of these airports as a function of the number of runways (normalized for runway and gate efficiency). Note that all of these airports are operating at less than maximum capacity. Queuing theory would suggest that the maximum capacity will never be reached for a random access queue but that mean delays will increase hyperbolically above a capacity fraction of 50%.

If the air transportation system can be approximately modeled as the sum of airport queues, then delay at any particular airport would be predicted to grow as:

$$\text{Delay} \approx 4[\%AP_{\text{cap}}/(1 - \%AP_{\text{cap}})]$$

Table 1 Capacity and delay parameters for 16 U.S. airports in the northeast triangle (MCM analysis, 4 n mile separation, 1997)

Airport	#R/W	Gates	G	X	Total OPS/yr FAA1997	MCM max EST. at 4 n mile	Peak OPS/hr FAA 1997	Capacity Fraction	Predicted delay	Reported delay
Chicago (ORD)	7	173	0.97	0.5	892,665	217	159	0.73	11	24
Atlanta (ATL)	4	180	1	0.7	785,854	179	140	0.78	14	32
Detroit (DTW)	5	99	0.99	0.4	547,350	127	98	0.77	13	8
St. Louis (STL)	5	86	0.92	0.4	528,746	118	94	0.80	16	31
Minneapolis (MSP)	3	73	0.9	0.55	496,091	95	89	0.93	55	7
Charlotte (CLT)	3	62	0.86	0.7	473,800	116	85	0.73	11	6
Boston (BOS)	5	88	0.95	0.36	473,127	109	84	0.77	14	25
Newark (EWR)	3	92	0.97	0.47	461,500	88	82	0.94	64	58
Pittsburgh (PIT)	4	122	1	0.6	454,259	154	81	0.53	4	3
Philadelphia (PHL)	3	63	0.86	0.5	422,493	83	75	0.91	42	16
Cincinnati (CVG)	3	120	1	0.7	413,579	134	74	0.55	5	12
New York (JFK)	4	180	1	0.36	362,305	92	65	0.70	9	18
La Guardia (LGA)	2	60	0.92	0.55	348,854	65	62	0.96	101	49
Washington Dulles (IAD)	3	78	1	0.67	337,383	129	60	0.47	4	6
Washington Reagan (DCA)	3	48	0.93	0.47	311,105	84	56	0.66	8	4
Cleveland (CLE)	4	50	0.92	0.36	300,620	85	54	0.63	7	6
Total	61	1574			7,609,731	1873	1359			
Average				0.52				0.74	24	19
Median								0.75	12	14
Standard Deviation								0.15	28	17

Table 2 Predicted delay increases for 16 U.S. airports using FAA forecast for 2012 (MCM analysis, 4 n mile separation). Some airports are predicted to be unable to meet forecast demand. The overall capacity fraction for this region is predicted to increase to 89% +/− 13%.

Airport	#R/W	Gates	G	X	Total OPS/yr FAA 2012	MCM max EST. at 4 n mile	Peak OPS/hr FAA 2012	Capacity Fraction	Predicted delay	Reported delay
Chicago (ORD)	7	173	0.97	0.5	1,110,000	217	198	0.91	42	
Atlanta (ATL)	4	180	1	0.7	985,000	179	176	0.98	213	
Detroit (DTW)	5	99	0.99	0.4	700,000	127	125	0.99	291	
St. Louis (STL)	5	86	0.92	0.4	655,000	118	117	0.99	588	
Minneapolis (MSP)	3	73	0.9	0.55	530,000	95	95	1.00	953	
Charlotte (CLT)	3	62	0.86	0.7	613,000	116	109	0.95	72	
Boston (BOS)	5	88	0.95	0.36	527,000	109	94	0.86	25	
Newark (EWR)	3	92	0.97	0.47	485,000	88	87	0.99	374	
Pittsburgh (PIT)	4	122	1	0.6	590,000	154	105	0.69	9	
Philadelphia (PHL)	3	63	0.86	0.5	460,000	83	82	0.99	788	
Cincinnati (CVG)	3	120	1	0.7	716,000	134	128	0.95	78	
New York (JFK)	4	180	1	0.36	415,000	92	74	0.80	16	
La Guardia (LGA)	2	60	0.92	0.55	360,000	65	64	0.99	533	
Washington Dulles (IAD)	3	78	1	0.67	437,000	129	78	0.61	6	
Washington Reagan (DCA)	3	48	0.93	0.47	329,000	84	59	0.70	9	
Cleveland (CLE)	4	50	0.92	0.36	415,000	85	74	0.87	28	
Total	61	1574			9,327,000	1873	1666			
Average				0.52				0.89	252	
Median								0.95	75	
Standard Deviation								0.13	309	
Official FAA OPS/yr reduced to not exceed predicted maximum										

Table 3 Capacity and delay parameters for 16 European airports at 4 n mi separation (MCM analysis, 4 and 6 n mile, 1997)

Airport	R/W	Gates	S 4 n mile	G	X	Projected OPS/yr, IATA 1997	MCM max EST. at 4 n mile	Peak OPS/hr 1997	Capacity fraction 4 n mile	Predicted delay 4 n mile	Reported delay ADM[a]
Heathrow	3	172	0.5	1.00	0.71	428,600	136	77	56%	1	3
Rheim/Main	3	145	0.5	1.00	0.76	389,600	146	70	48%	1	4
Charles de Gaulle	2	193	0.5	1.00	0.77	374,998	99	67	68%	2	4
Schiphol	5	144	0.5	1.00	0.80	353,000	256	63	25%	0	7
Orly	3	103	0.5	1.00	0.77	250,000	148	45	30%	0	3
Leonaro Da Vinci	3	72	0.5	1.00	0.77	245,757	148	44	30%	0	4
Barajas	2	93	0.5	1.00	0.55	252,400	70	45	64%	2	5
Gatwick	2	90	0.5	1.00	0.71	207,679	91	37	41%	1	4
Brussels National	3	107	0.5	1.00	0.66	277,006	127	49	39%	1	6
Munich	2	83	0.5	1.00	0.77	255,948	99	46	46%	1	4
Copenhagen	3	128	0.5	1.00	0.63	280,800	121	50	41%	1	2
Zurich	3	60	0.5	1.00	0.50	268,352	96	48	50%	1	5
Dusseldorf	3	67	0.5	1.00	0.65	187,549	125	33	27%	0	6
Linate	2	35	0.5	1.00	1.00	165,283	128	30	23%	0	5
Arlanda	2	264	0.5	1.00	1.00	255,000	128	46	36%	1	2
Dublin	3	95	0.5	1.00	0.55	134,300	106	24	23%	0	2
Total	44	1851				4,326,272	2022	773			
Average					0.73				40%	1	4.10
Median									40%	1	3.91
Standard deviation									0.14	0.5	1.3

[a] Average delay per movement data from Eurocontrol Annual Report, 1998.

Table 4 Predicted delay increases for 16 European airports using International Air Transport Association (IATA) forecast for 2007 (MCM analysis, 4 and 6 n mile, 2007)

Airport	R/W	Gates	S 4 n mile	G	X	Predicted total OPS/yr, 2007	MCM max EST. at 4 n mile	Peak OPS/hr, 2007	Capacity fraction 4 n mile	Predicted delay 4 n mile	Projected delay (ADM)[a]
Heathrow	3	172	0.5	1.00	0.71	634,433	136	113	83%	5	4.62
Rhein/Main	3	145	0.5	1.00	0.76	461,100	146	82	56%	1	5.48
Charles de Gaulle	2	193	0.5	1.00	0.77	533,739	99	95	97%	29	6.38
Schiphol	5	144	0.5	1.00	0.80	520,056	256	93	36%	1	6.57
Orly	3	103	0.5	1.00	0.77	250,000	148	45	30%	0	4.59
Leonaro Da Vinci	3	72	0.5	1.00	0.77	351,529	148	63	42%	1	5.64
Barajas	2	93	0.5	1.00	0.55	392,487	70	70	100%		8.03
Gatwick	2	90	0.5	1.00	0.71	243,360	91	43	48%	1	6.05
Brussels National	3	107	0.5	1.00	0.66	363,000	127	65	51%	1	8.43
Munich	2	83	0.5	1.00	0.77	357,000	99	64	65%	2	5.69
Copenhagen	3	128	0.5	1.00	0.63	415,653	121	74	61%	2	3.53
Zurich	3	60	0.5	1.00	0.50	397,227	96	71	74%	3	7.43
Dusseldorf	3	67	0.5	1.00	0.65	238,925	125	43	34%	1	8.84
Linate	2	35	0.5	1.00	1.00	244,659	128	44	34%	1	7.67
Arlanda	2	264	0.5	1.00	1.00	298,000	128	53	42%	1	3.57
Dublin	3	95	0.5	1.00	0.55	168,524	106	30	28%	0	3.17
Total	44	1851				5,869,692	2022	1048			
Average					0.73				55%	3	5.98
Median									49%	1	5.87
Standard deviation									0.23	7.3	1.8

[a] Average delay per movement data from Eurocontrol Annual Report, 1998.

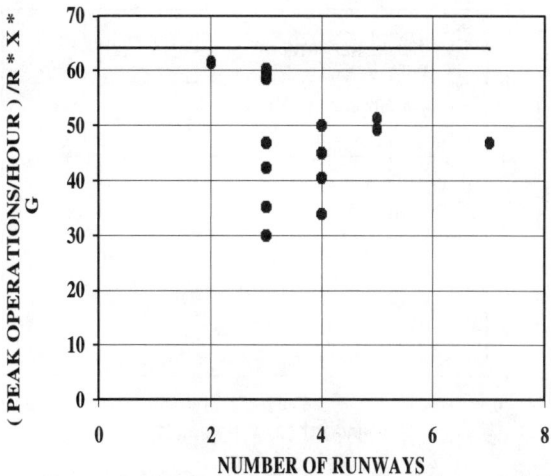

Fig. 3 Comparison of 16 U.S. airports in the utilization of runway capacity (OPS/RW/HR) normalized for runway and gate efficiency.

where $\%AP_{cap}$ is the predicted airport capacity fraction and 4 is observed to be approximately equal to the mean number of U.S. delays greater than 15 min per 1000 operations at 50% capacity fraction. Both the airline reported delays and the model predictions are listed in Table 1 and shown in Fig. 5.

Similar data are not available for the European airports considered but are available in delay minutes per movement.[11,12] For that reason U.S. delays in minutes were plotted using CODAS data for 1997 and shown in Fig. 6. Since

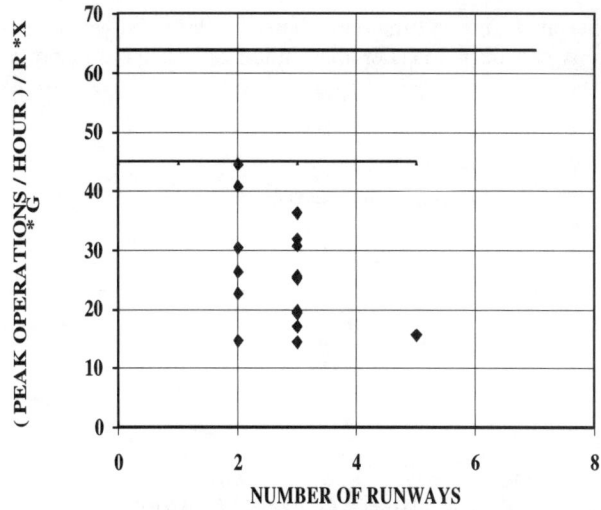

Fig. 4 Comparison of 16 European airports utilization of runway capacity OPS/RW/HR normalized for runway and gate efficiency.

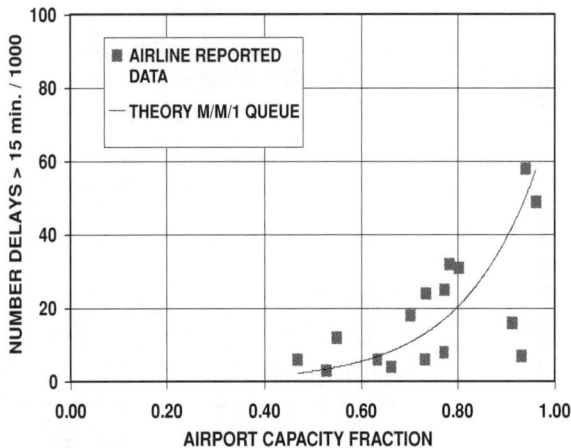

Fig. 5 Comparison of airline reported delays at 16 major U.S. airports compared to M/M/1 queuing theory prediction.

the units of delay are now minutes of delay per movement, the equation becomes:

$$\text{Delay} \approx 1 \text{ min/mv}[\%AP_{cap}/(1\%AP_{cap})]$$

Both the EUROCONTROL reported delays and the model predictions are listed in Table 3 and illustrated in Fig. 7. Comparing the reported delay to airport M/M/1 queuing predictions shows a different outcome than in the United States.

V. MCM Comparisons

From earlier analysis[9] and the data shown in the previous section, a number of observations can be made. For the United States, air transportation system

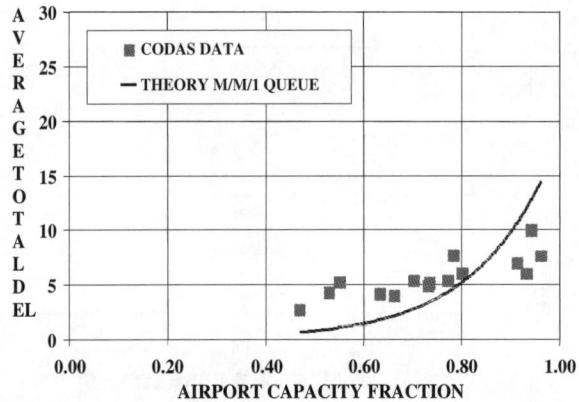

Fig. 6 Comparison of CODAS reported delays at 16 major U.S. airports compared to M/M/1 queuing theory prediction.

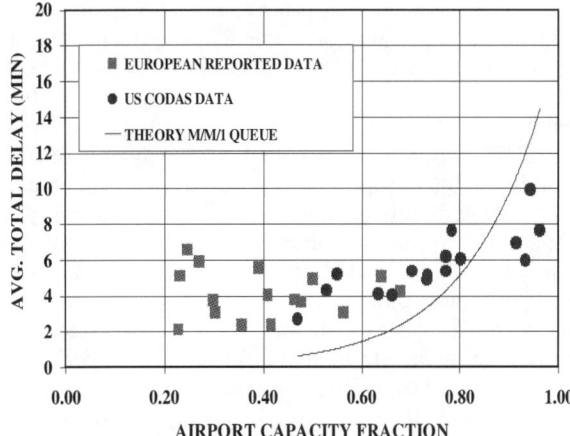

Fig. 7 Comparison of reported delays at 16 major European airports compared to U.S. CODAS data and queuing theory prediction at 4 n mile separation.

capacity seems to be largely airport queue limited. The 16 airports selected out of the northeast triangle also indicate that delays are higher at moderate capacity fractions, indicating that congested airspace is also leading to delays in this region of the United States.

Results from the European airports analyzed indicate a different situation. Capacity does not follow the M/M/1 predicted levels. Figure 7 illustrates that delays are significant even at airports that are operating at low capacity fractions. This would suggest that externalities outside of the airport environment are causing delays. To determine with greater precision the factors causing this would require further analysis.

A macroscopic observation may be suggested, however. Because of the historically based random access/random service, i.e., first-in-first-out (FIFO), nature of air traffic control, delays are very sensitive to demand/capacity ratios above 50%. For Europe, with airport capacity remaining constant, it has been observed that a 1% increase in demand generates approximately a 6% increase in air traffic management (ATM) delays.[12] Fron[13] has postulated that a system relationship exists between demand and delays for Europe that show an elasticity of five between demand and delays. Queuing theory would imply that overall Europe is operating at 60% (d(delay)/d(%Capacity) $= 0.064$ at %capacity $= 60\%$) of system capacity, compared to 40% as predicted by the model considering airport capacity alone. This would imply that 20% of the potential capacity of the European air transportation system is lost to en route inefficiencies.

VI. Conclusions

This chapter has focused on the MCM and its use as a reliable macrocapacity model to be used as an initial level of air transportation capacity analysis. In the first section of the chapter, the MCM was described and evaluated as a model able to predict capacity levels using aggregate data. The next section assessed the MCM's capability by using data from 32 high-density airports in the United States and Europe.

It is observed that the United States is achieving a higher utilization of existing runway infrastructure and possibly en route airspace than is Europe (i.e., up to 30 aircraft per runway per hour vs 20 aircraft per runway per hour). Also, it is observed that the United States has constructed more runways at its major airports but has invested in far less gate and terminal infrastructure, which is becoming a limitation at some U.S. airports (i.e., U.S. gate-to-runway ratio of 26 vs a European ratio of 42). In addition, the European airport designs are in general more efficient that the older U.S. designs found in the northeast triangle (i.e., U.S. average X factor equals 0.52, whereas the equivalent European X factor equals 0.73).

For the European airports selected, the MCM indicated an adequate capacity margin at the airports even though overall delays are increasing.[11] This result may not be due to the inadequacy of the MCM, but to factors external to the airports themselves. These factors are represented in the model's second term that represents en route airspace capacity degradation.

Hüttig et al.[2] have stated, "The inconsistent development of air traffic and airport capacity (in Europe) is the reason for the looming capacity crisis..." (p.). This is supported in that delays on the ground imposed by the CFMU result from insufficient sector capacity at certain control centers. In 1996, 15% of control centers were responsible for 90% of ATM related delays.[13] EUROCONTROL has stated that a small number of sectors (i.e., 15 sectors, roughly 3%) caused about 45% of the ATM delays during the summer of 1998.[11] In 1999, 44% of ATM delays originated from a demand/capacity mismatch in 30 sectors out of 468 sectors.[12] This may be one cause for reactionary delays, i.e., late departures due to late arrivals, that are the largest single category of departure delay causes.[12]

Airport capacity in Europe may be determined more by operational norms than by physical airport runway capacity. These operational norms appear to be more conservative than those used in the United States. Overall capacity may possibly be set artificially lower than can actually be safely achieved.

Acknowledgment

This research was funded by financial support from the Federal Aviation Administration and George Mason University. The opinions expressed in this study represent those of the authors alone and not the sponsoring institutions.

References

[1] Air Transport Action Group, *European Traffic Forecasts 1980–2010* [electronic], http://www.atag.org/ETF/Index.htm [cited 18 Aug. 1999].

[2] Hüttig, G., Busch, W., and Gronak, N., "Growing Demand and Capacity of Airports," *Transportation Research*, Vol. 28A, No. 6, 1994, pp. 501–509.

[3] Reynolds-Feighan, A. J., and Button, K., "An Assessment of the Capacity and Congestion Levels at European Airports," *Journal of Air Transport Management*, Vol. 5, 1999, pp. 113–134.

[4] Donohue, G. L., "A Simplified Air Transportation System Capacity Model," *Journal of Air Traffic Control*, April-June 1999, pp. 8–15.

[5] Odoni, A., Bowman, J., Delahaye, D., Deyst, J., Feron, E., Hansman, R. J., Khan, K., Kuchar, J. K., Pujet, N., and Simpson, R. W., *Existing and Required Modeling Capabilities*

for Evaluating ATM Systems and Concepts, NAG2-997, International Center for Air Transportation, Massachusetts Inst. of Technology, Cambridge, MA, 1997.

[6]Long, D., Lee, D., Johnson, J., Gaier, E., and Kostiuk P., *Modeling Air Traffic Management Technologies with a Queuing Network Model of the National Airspace System,* NASA/CR-1999-208988, 1999.

[7]Wieland, F., "The Threshold of Event Simultaneity," *The Society for Computer Simulation International,* Vol. 16, No. 1, 1999, pp. 23–31.

[8]Sheffi, Y., *URBAN TRANSPORTATION NETWORKS: Equilibrium Analysis with Mathematica Programming Methods,* Prentice-Hall, Upper Saddle River, NJ, 1985.

[9]Donohue, G. L., and Shaver, R., "United States Air Transportation Capacity: Limits to Growth Parts I and II." Papers 00-582, 00-0583, Transportation Research Board 79th Annual Meeting, Washington, DC, Jan. 9–13, 2000.

[10]Donohue, G. L., "A Macroscopic Air Transportation Capacity Model: Metrics and Delay Correlation," Advanced Technologies and Their Impact on Air Traffic Management in the 21st Century, Capri, Italy, Sept. 1999.

[11]EUROCONTROL, *Performance Review Commission First Performance Review Report,* Rept. 1, EUROCONTROL, Brussels, 1999.

[12]EUROCONTROL, *Performance Review Commission Special Performance Review Report on Delays (January–September 1999),* Rept. 2, EUROCONTROL, Brussels,1999.

[13]Fron, X., *ATM Performance Review in Europe,* 2nd USA/Europe Air Traffic Management R&D Seminar, Orlando, FL, 1–4 Dec. 1998.

Section II: Economics of Congestion

FIVE chapters are presented in this section that examine the air transporation system from an airline service provider prespective. The first two chapters are concerned with the generic approach to understanding the economic impacts of air traffic management (ATM) system changes and of operation distruptions. Chapter 8 presents a very informative description of how schedule disruptions of can propagate through the transportation system. Chapter 9 is the first formal description of the airline air operations centers (AOC). The AOC is becoming as important to the operation of the commercial public transportation system as the Federal Aviation Administration (FAA) air traffic control (ATC) center.

The final chapter in this section, Chapter 10, explores the concept of using pricing of services instead of technical and operational solutions for solving the congestion problem. Economic techniques of differential pricing, slot auctions, and route or bank auctions will increasingly be examined as alternative means of dealing with the increasing levels of delay and cancellations associated with the decreasing usable capacity margins of the system.

Chapter 6. Forecasting and Economic Analysis for Aviation Systems Engineering, Peter F. Kostiuk, Logistics Management Institute; and Eric M. Gaier, Bates White and Ballentine, 1998 and 2000 papers combined.

Chapter 7. Impact of Air Traffic Management on Airspace User Economic Performance, Joseph H. Sinnott and William K. MacReynolds, Ph.D., MITRE Corporation, 1998.

Chapter 8. Effects of Schedule Disruptions on the Economics of Airline Operations, Zalman A. Shavell, MITRE Corporation, 2000

Chapter 9. Modeling an Airline Operations Contol Center, Nicolas Pujet and Eric Feron, MIT, 1998.

Chapter 10. Pricing Policies for Air Traffic Assignment, Karine Deschinkel, ENSAE; Jean-Loup Farges, ONERA-CERT; and Daniel Delahaye, LOG, 2000.

Chapter 6

Forecasting and Economic Analysis for Aviation Systems Engineering

Peter F. Kostiuk*
Logistics Management Institute, McLean, Virginia
and
Eric M. Gaier[†]
Bates White and Ballentine, Washington, D.C.

I. Introduction

WITH romantic visions of soaring aircraft and exotic technologies that make the air transportation system function, it is sometimes easy for decision-makers in the aviation community to forget that air transportation is an *economic* activity. Few passengers sitting back in coach are there to enjoy the flight; they want to get somewhere, and the more quickly and cheaply the better. Consequently, a crucial component of aviation systems engineering involves forecasting future demands for air travel and analyzing the economic implications of different alternatives. Selecting the best alternatives also requires an understanding of the economic motivations of major stakeholders and their likely responses to changes in the aviation system infrastructure.

The joint public-private nature of investments in air traffic management (ATM) systems presents unique challenges to decision-makers and analysts. Public authorities, at least in the United States, provide most of the infrastructure, such as airports and air traffic management services. Private companies, such as airlines, purchase equipment and other resources that work with the publicly provided infrastructure. Consequently, for any significant change two benefits criteria must be met:

1) Do the economic and other benefits to the traveling public and national economy exceed the cost and justify public expenditures?

2) Does the airline investment generate sufficient increased revenue or cost savings to meet minimum profit objectives?

When the need for significant change and investment is as acute as it is today, the necessity for public-private agreement on a course of action is heightened. The

---

Copyright © 2001 by the American Institute of Aeronautics and Astronautics, Inc. All rights reserved.
*Program Director
[†]Manager

remainder of this chapter describes methods developed to support the economic analysis component of aviation systems engineering. The approach emphasizes two required analyses. First, we need a national or system-level analysis that forecasts the demand for air travel and the ability of the current airport and airspace system to meet those demands. This analysis generates estimates of the performance gap between the underlying demand and what is feasible with current systems and operational practices. By translating that demand gap into economic measures, we quantify the potential value of system improvements. A key part of the national analysis is the identification and assessment of potential behavioral responses to constraints. That analysis acknowledges that airlines, passengers, and airport operators will not persist in unproductive strategies if the required level of air traffic management services is not available to support demand growth.

The second economic analysis provides preliminary cost-benefit estimates for specific airline investments in communications, navigation, and surveillance/air traffic management (CNS/ATM) systems. This part of the analysis is crucial, because benefits that may be positive to the overall economy may not justify private investments, either because the airline cannot recoup all of the benefits or because they do not meet the higher threshold for return on investment required by commercial companies.

Although we distinguish between the national and airline benefits analyses, they maintain common components and approaches. As shown in Fig. 1, the starting point for any analysis is a baseline demand prediction. This forecast provides input to an operational model that incorporates system capacity. Through queuing models, the interaction between capacity and demand generates estimates of delay or block times. For an analysis at the National Airspace System (NAS) level, the model should provide sufficient coverage of total system operations with adequate detail to support policy analyses. For an airline cost-benefit analysis, the analyst substitutes a limited network model that captures key features of the airline operation, or focuses on a specific airport that is crucial to the airline's strategic objectives.

In both analyses, the analyst then sends output from the operational impact analysis to the economic models. In the simplest and most common case, analysts translate delay estimates into costs by applying a cost per flying hour. This

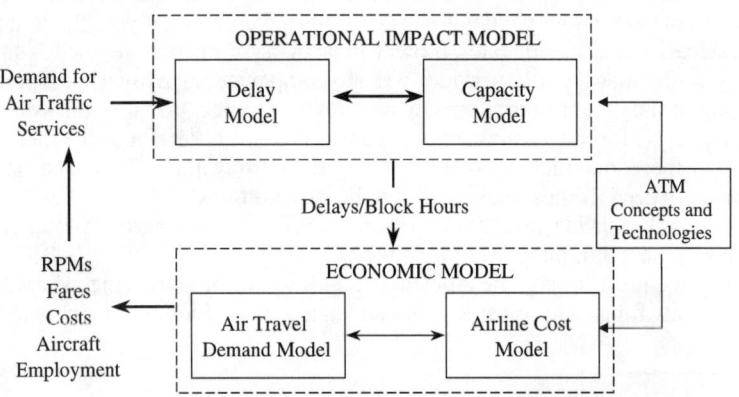

Fig. 1 Integrated operational and economic analysis.

approach adequately captures the impact of small changes in system performance, but neglects the potentially more significant effects on airline revenue. This issue is discussed more fully later in this chapter.

Crucial to the success of this approach is the focus on airline and passenger behavior, which traditionally are ignored in benefit analyses. This chapter describes the first successful effort to develop a methodology for closed-loop modeling of air traffic and aircraft flight delays. It therefore provides a more realistic framework for forecasting future demands on the ATM system and for evaluating the possible impacts of different strategies in response to system capacity constraints.

II. Evaluating National Impacts of ATM Investments

The primary objectives of national forecasting and economic analysis in aviation systems engineering are to 1) provide credible travel demand forecasts to estimate the required level of air traffic services and provide scenarios for estimating benefits; 2) quantify the operational and economic impacts of proposed technologies on capacity constraints and system performance; and 3) evaluate the impacts of alternative public and private policies in response to system constraints. Meeting these objectives requires an integrated analysis that incorporates air traffic operational models with economic models that forecast airline costs and fares. In a series of projects conducted for NASA over the past several years, Logistics Management Institute (LMI) developed such a suite of models and devised methods for using them to generate closed-loop forecasts of air travel in the presence of congestion.[1,2] The approach provides a specific implementation of the general approach of Fig. 1. It accommodates analysis of the impact of technology insertion and other public and private policies, such as airport expansion and schedule smoothing.

III. Generating an Unconstrained Forecast

The first step in the analysis is to generate an unconstrained baseline forecast calibrated to the Federal Aviation Administration's (FAA) terminal area forecast (TAF). We describe the forecast as unconstrained because the FAA intentionally does not include the impact of airport and airspace capacity constraints. That forecast can be viewed as an estimate of the underlying trend demand for air travel, based on assumptions about fuel costs, airline productivity, economic growth, and other factors.

To build the unconstrained forecast, we minimize departures from current airline scheduling practices as much as possible. We start with the current Official Airline Guide (OAG) schedule, grow it over time in a manner consistent with the TAF airport operations growth rates, and construct a future OAG schedule that retains current operational practices. Changes from that schedule occur only in response to capacity constraints at the airports when we construct the constrained forecast. Figure 2 describes the steps in constructing the unconstrained forecast.[3]

The objectives of the demand forecast are to provide quantitative information on the volume of future air traffic, along with information on its temporal and geographic distribution. Because forecasting is inherently uncertain, we seek to minimize the degree of uncertainty by employing an evolutionary approach that

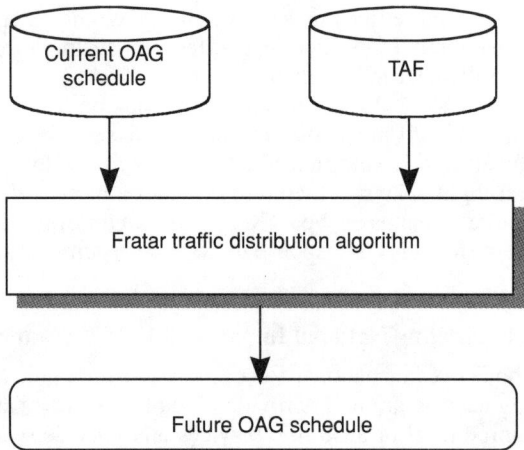

Fig. 2 Creating the unconstrained forecast.

starts with a characterization of current demand and grows it over time with a minimum of changes in current demand patterns.

The FAA's TAF provides a forecast over an extended time period of the FAA's prediction of enplanements and aircraft operations at hundreds of airports. By design, these forecasts represent unconstrained estimates of air travel based on demographic and economic factors, without regard to potential system capacity constraints. Although the FAA forecast has limitations, it is the most detailed forecast available, and it is widely publicized and vetted throughout the industry. The FAA also regularly brings in outside experts to improve its forecasting methods. For the initial baseline forecast, we recommend using the TAF for growth rates in enplanements and operations.

The FAA makes forecast operations in the TAF in the following way:

1) It forecasts the enplanements based on outputs of socioeconomic models, such as gross domestic product (GDP) and demographic growth rates, with due consideration of originating traffic and connection traffic. Each major airport has its own specific models.

2) It forecasts the load factors to and from each airport based on the demand, fare yield, and airlines cost.

3) It forecasts the average number of seats per aircraft for arrivals and departures at the airport.

4) It divides the forecast enplanement by the forecast load factor and by the forecast average number of seats per aircraft to get forecast operations.

Although airline schedules will obviously change significantly by 2010, our approach relies on the strong tendencies of the airline industry since 1978 for major carriers to construct tightly coupled hub and spoke schedules to obtain economic benefits. These schedules provide opportunities for other airlines in many markets, but the economic benefits from current practices are so great that it is unlikely that significant changes will occur unless operational and capacity constraints become intolerable. Because one of the objectives of forecasting in aviation systems engineering is to identify system performance requirements, the baseline forecast should closely reflect current operational strategies.

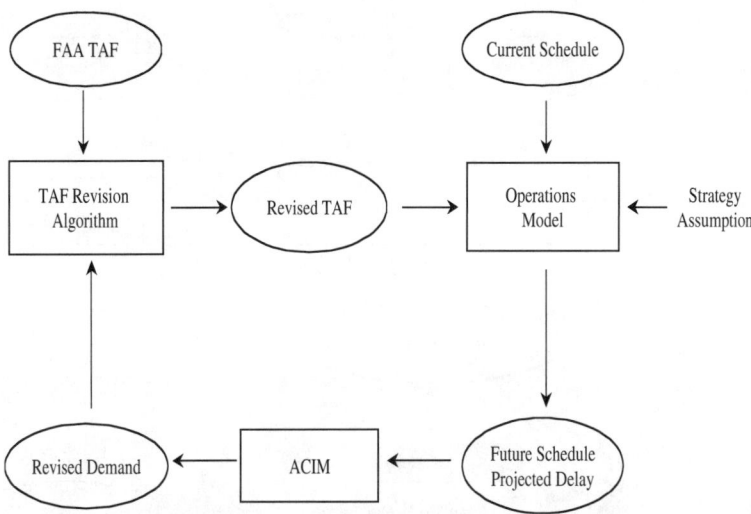

Fig. 3 Constrained forecast methodology.

Based on the TAF, the baseline unconstrained forecast for 2010 is 1,061 billion revenue passenger miles (RPM), which is about two-thirds higher than now. Considering the level of stress the current system experiences, and the lack of any plans to increase capacity anywhere near two-thirds greater than today, one might hypothesize that the unconstrained forecast is infeasible. The next section describes how we test that hypothesis.

IV. Generating a Constrained Forecast

As shown in Fig. 3, the basic approach links delay forecasts from the operations model, which are driven by traffic projections at the airports, with industry-level supply and demand characteristics embedded in the air carrier investment model (ACIM). The impact of various policies to alleviate congestion and its resultant airline delays can be analyzed by modifying parameters of the operations model. The result is a revised forecast that can be compared with the unconstrained forecast to measure the success of the proposed strategy in accommodating air travel demand.

The operations model depicted in Fig. 4 takes projected airport operations growth rates and combines them with the current traffic schedule and possible policy options to generate the future traffic schedule and delay estimates. Because the operations model computes airport delays according to the traffic schedule, delays are amplified by a set of multipliers to account for the rippling delays for the rest of the day. The delay multipliers are derived by American Airlines using operational data. In general, the earlier in the day the delay occurs and the longer the delay lasts, the larger the delay multiplier. Next, the total delays computed from the operations model are used to generate the block time changes. Increases in block times increase airline costs, which generates a revised industry forecast.

The revised RPM forecast starts with the unconstrained operations growth rates from FAA's TAF. Our approach modifies the airport operations growth rates

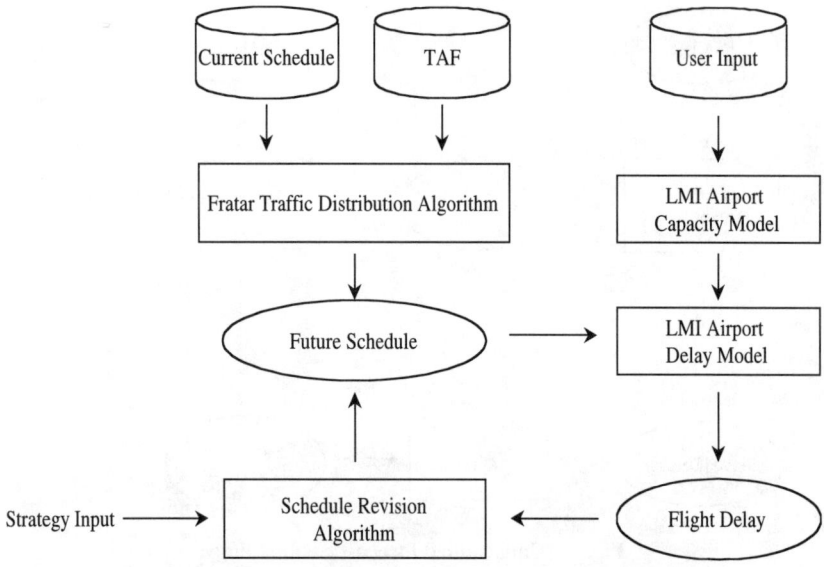

Fig. 4 Operations model schematic.

based on the revised RPM forecasts from the ACIM, which is accomplished through the TAF revision algorithm. The system converges when total system traffic (operations or RPMs) from both the operations model and the ACIM agree.

The operations model also includes parametric capacity models for the 64 airports included in the network. These capacity model parameters can be modified to capture the impact of changes in minimum separation standards, position reporting accuracy, communications delay, aircraft performance, aircraft mix, and other factors that affect airport capacity. We have used the model to estimate the impact of several NASA ATM technologies.

The value of the integrated approach shown on Fig. 3 is that it generates a more realistic forecast that directly accounts for airport and airspace capacity limitations. We generate that constrained baseline forecast in two ways. First, we can model airport capacity constraints directly by imposing a limit on the increase in average flight delay (or block time). Once that limit is met, no further operations growth is allowed at that airport, regardless of the TAF forecast or underlying demands. This approach is fairly stringent, because it implies a hard physical limitation on airport capacity. Nonetheless, it provides a good estimate of the limitations of current capacity and operational practices.

As an alternative to this rigid constraint, we developed another approach that uses the ACIM to ration the limited capacity through higher fares. The premise of using the ACIM to evaluate the impact of delay on air travel demand is that increases in air carrier operating costs are passed along to consumers in the form of higher fare yields, which further slow the growth rate of demand. Thus, an equilibrium is achieved in which the costs of delay are balanced by the passenger's willingness to pay for additional travel. Under this approach, the ACIM produces an estimate

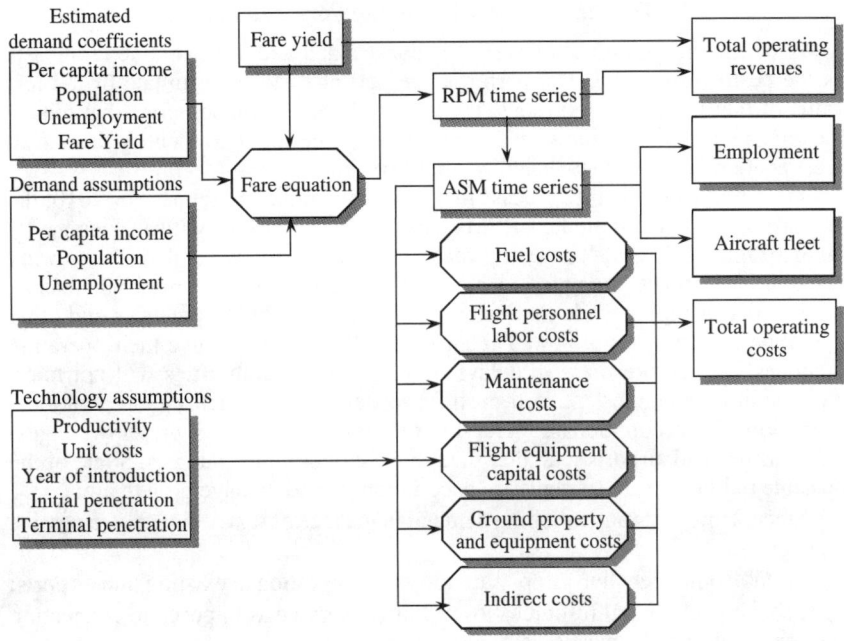

Fig. 5 Air carrier investment model.

of the reduction in aggregate air travel demand due to the increased costs of congestion. Note that this approach understates the impact of the delay, because it only accounts for the direct delay effect on airline operating costs. Operational issues, such as whether it is feasible for an airline to operate with much lengthier delays, even if embedded into the schedule, are not captured. These effects are likely to be much greater than the direct cost and fare impacts.

Figure 5 describes the structure and key components of the ACIM. The key features for the forecasting analysis are the fare yield effects from the air travel demand model and the cost effects of delay in the airline cost model. Cost changes are ultimately reflected in fare changes through the target profit margin, which may be modified by users of the model. Extension modules translate the aggregate fleet requirements into seat-size categories and estimate the impact on U.S. aircraft manufacturing and related employment.

In the ACIM operating costs are calculated for six functional cost categories as functions of input prices, input factor productivities, and total output. As shown in Fig. 5, the cost categories consist of fuel, flight personnel labor, maintenance, aircraft capital, ground property and equipment capital, and a residual category termed other indirect. The measure of output that drives the cost calculations is available seat miles (ASM). The ACIM solves for industry equilibrium by iterating fare yields until the specified profit constraints are satisfied. Thus, any changes in airline operating costs are passed on to the traveling public in the form of changes in fares. Implicitly, such analysis assumes that the commercial air travel industry will remain price competitive.

V. Estimating and Closing the Performance Gap

We define the difference between the constrained and unconstrained forecasts as the performance gap. The performance gap provides an estimate of the lost value of that travel and represents the potential gains from solutions that reduce the gap. The challenge for aviation systems engineering is to identify the safest, most economical solutions that close that gap.

Let us first examine the forecast for the unconstrained baseline. For 2010, the unconstrained forecast predicts RPMs will be at 1,061 billion, with an average delay per flight of 78 min. Those huge delays will not actually happen. They *would* occur if air carriers attempted to meet all future demand by simply increasing the number of flights while maintaining the same scheduling practices and other operating methods in use now. The airlines will certainly change their operating practices long before massive delays develop. Public authorities will not tolerate that situation either and will push for system improvements and operational limitations. To accommodate increasing demand in the face of growing congestion, airlines and airports will alter their current operating strategies. Some of the possible public-private strategies to accommodate and manage growth are:

1) increasing fares and rationing demand in the face of scarce capacity (a passive strategy);

2) establishing new hub airports to mitigate congestion at existing hub airports;

3) shifting additional resources toward direct service as opposed to connecting service to mitigate congestion at major hub airports;

4) smoothing the peaks and valleys of typical bank operations to mitigate the growth of delay at major hub airports;

5) shifting additional resources toward nighttime operations;

6) employing larger aircraft, as opposed to growth in frequency; or

7) combining the five active strategies.

We analyzed the impact of these strategies on two major outcomes, average flight delay and total system throughput as measured by RPMs. We did not include the cost of implementing the strategies in the analysis. Table 1 shows the effect on RPMs and the amount of predicted growth that the strategy enables. An LMI report describes the details of how we analyzed each strategy.[4]

Figure 6 summarizes the impacts of these strategies on average flight delay. To a first approximation, system-wide delay is a function of the number of operations at the various LMINET airports. All other things being equal, more operations will tend to increase the total amount of delay. Other factors include whether additional operations occur at already busy airports or during congested times of the day.

Table 2 and Fig. 6 show average delays per operation according to the potential strategies and the unconstrained forecast. For each strategy, Table 2 shows a low and high estimate of its impact. The worst system-wide delay is observed for the higher fares scenario in which the airlines simply continue their current method of conducting business despite a large increase in operations demanded by forecast RPM growth. Delay costs are passed along to the traveling public in the form of higher fares. At the opposite end of the spectrum, delay is the least when the airlines follow a combination of active strategies. This is closely followed by the stand-alone strategy of incremental growth in average aircraft size (above that already incorporated into the FAA forecast). Night operations also are effective as a stand-alone strategy because airline departures and arrivals are moved from heavily congested times to the less busy evening period.

Table 1 Impact of strategy alternatives on RPMs

	Strategy							
	High fare	New hubs	Direct service	Schedule smoothing	Night operations	Larger aircraft	Combined	Unconstrained
RPMs delivered in 2015	1,233.5	1,239.6	1,242.0	1,245.6	1,257.9	1,277.6	1,300.5	1,331.4
Level of satisfied RPM growth from 1997 to 2015	86.8%	87.7%	88.0%	88.5%	90.1%	92.8%	95.8%	100.0%

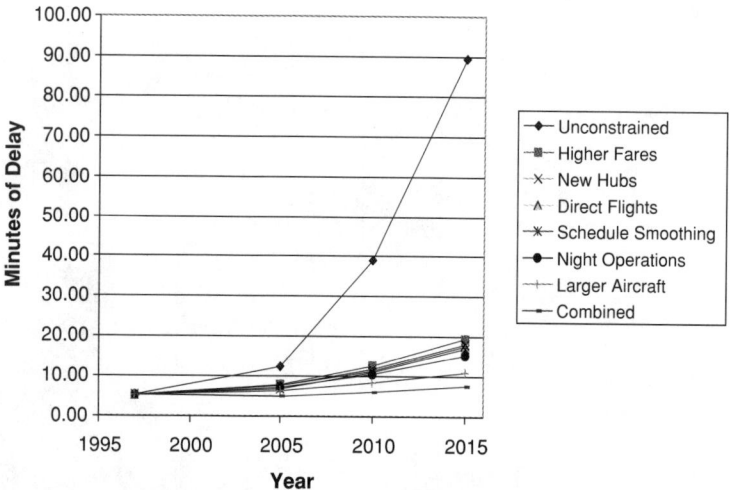

Fig. 6 Average minutes of delay per operation.

The effects of the remaining three strategies on delay are very close. Schedule smoothing has a slight effect because the hours immediately surrounding peak arrival and departure times also tend to be very busy. Direct flights reduce average delays because operations at hub airports are somewhat reduced, although there is a slight net increase in system wide operations (2.5% in 2015). "New hubs" appears to be the least effective strategy because the system wide number of operations remain virtually unchanged, although operations are shifted from busy hub airports to less busy potential hub airports (but these airports may face their own capacity constraints).

Note that the average flight delay under all of these forecasts is very high and probably infeasible. The forecasts are based solely on feedback through cost and fare increases, without addressing the operational feasibility issue. In addition, we have not yet included into the cost and fare analysis the costs of implementing the

Table 2 Average minutes of delay per operation

Scenarios	1997	2005		2010		2015	
		Low	High	Low	High	Low	High
Unconstrained	5.3		12.5		39.1		89.4
Higher fares	5.3		8.0		12.5		19.4
New hubs	5.3	7.8	7.7	12.2	11.9	18.8	18.2
Direct flights	5.3	7.7	7.5	11.9	11.4	18.5	17.8
Schedule smoothing	5.3	7.2	6.8	11.5	11.0	18.0	17.1
Night operations	5.3	7.2	7.0	11.0	10.4	16.5	15.2
Larger aircraft	5.3	7.1	6.3	10.3	8.4	14.7	11.0
Combined	5.3	6.3	4.9	8.3	6.1	11.4	7.8

Table 3 Delay tolerances and total system operations in 2007

Departure tolerance, min	Arrival tolerance, min	Total daily operations	Attained traffic growth, %
3.0	2.0	59,106	73.9
3.83	2.71	60,120	81.3
5.0	4.0	60,849	86.7

strategies. Those costs, with the exception of the high fare strategy, will be quite substantial, and in some cases probably too high for the strategy to be feasible. If these omitted factors were included in the analysis, we expect that average fares will be significantly higher, with an accompanying reduction in average delay and industry output.

With an alternative approach, we can analyze the impact of imposing a hard delay constraint. In this policy, we set a maximum limit on the increase in average delay or block time. Once an airport meets that limit, no further operations are allowed. The results of this analysis are shown in Table 3. If we believe that the system can tolerate only limited increases in average delay, such as 3.0 min. on departure and 2.0 min on arrivals, we will lose over one-quarter of the projected growth in operations from 1997 to 2007. The impact on RPMs and enplanements will be even greater, because operations growth rates are predicted to be smaller.

Much of this discussion has revolved around forecasts of delay and industry output, as measured by RPMs or operations. The more critical question for aviation planners, however, is the value of the performance gap and what can we do to overcome that gap. For the lost operations listed in Table 3, for example, the difference between enabling 74% of operations growth vs 87% is about $3.5 billion in annual revenue in 2007. Furthermore, the direct cost of the extra 4 min of delay is about $1.3 billion, when costed at $30 per minute of delay.

Similarly, we can provide an estimate of lost RPMs by valuing them at a yield of 13 cents per RPM. For the difference between the constrained and unconstrained baselines in Fig. 6 for the year 2010, the lost RPMs are worth $5.7 billion. Despite that lost traffic, average flight delays still exceed current levels by about 15 min.

VI. Estimating Airline Benefits from ATM Investments

The previous section described the substantial shortage of capacity that will inflict serious economic damage to the air transportation industry unless solutions are found very soon. Most of the solutions proposed require joint investments by public authorities and airlines. Depending on how those costs are allocated and who reaps the benefits, improvements will or will not occur. To assess those issues requires a cost-benefit analysis of proposed solution from the airline perspective.

This section describes an air carrier cost-benefit model (CBM) that meets these requirements. The model is part of the Aviation System Analysis Capability (ASAC), an integrated suite of models and databases developed by LMI for NASA. ASAC provides a mechanism to assess the operational, economic, and safety impacts of aviation technologies. The CBM is distinguished from many of the existing aviation cost-benefit models by its focus exclusively on commercial air carriers. The model considers such benefit categories as time and fuel savings, utilization opportunities, reliability enhancements, safety and security improvements, and capacity enhancements. A distinction is made between benefits that are predictable and those that occur randomly. Such a distinction captures that ability of air carriers to reoptimize scheduling and crew assignment decisions in the face of predictable benefits. With regard to the costs of new technologies, the model incorporates a life-cycle cost module that applies nonrecurring acquisition, recurring maintenance and operation, and training costs to each aircraft equipment type independently.

The core operating cost calculations of the CBM follow an activity-based cost approach first developed for the ACIM. This approach estimates operating costs in six cost categories as a function of output, input prices, and input productivities. The default price and productivity parameters of the model are populated with publicly available data from the largest three U.S. carriers. Thus, the default model is developed for a representative airline that facilitates its use to build consensus regarding aviation investments. In addition, the model incorporates a database of alternate parameters that allows the user to customize analysis for specific air carriers or groups of air carriers.

The basic outputs of the model include net present value (NPV) and duration calculations. We supplemented these basic outputs with a sensitivity analysis and simulation module that allows the user to select variables for sensitivity analysis and input data ranges. The sensitivity analysis algorithm produces a tornado diagram that summarizes the sensitivity of the results to independent variations in selected variables. The simulation algorithm uses Monte Carlo simulation to produce a distribution for the basic outputs as a function of the simultaneous variation in the selected variables.

Finally, this section illustrates the use of the model in conjunction with other operational models to evaluate the projected costs and benefits of a NASA research program called Low Visibility Landing and Surface Operations (LVLASO). This program seeks to develop a set of innovations that reduce runway occupancy time and approach separation standards in poor weather and low visibility conditions. We find that the technologies of the LVLASO program will demonstrate net benefits to the representative air carrier but will contain substantial risk. The model identifies the variables that contribute to the range of uncertainty.

VII. Overview of the Air Carrier Cost-Benefit Model

In creating the CBM, we had some specific goals in mind. A primary objective was to create a flexible financial analysis tool to support credible estimates of benefits to airline operators from proposed technical and procedural innovations. Underlying this objective was a realization that future technologies must demonstrate net benefits to the user community. In addition, we recognized that existing aggregate level cost-benefit methodologies, which consider a much broader scope of benefits than those affecting only commercial air carriers, often lack sufficient operational complexity to establish credibility with airline operators. Therefore, we chose to focus exclusively on commercial air carriers for this model.

We envisioned the capability to evaluate the financial impact to airlines under a variety of user-defined technology scenarios. Because investments in new technology are subject to a great deal of uncertainty, we determined early on that a sensitivity analysis capability was essential. In addition, we envisioned the capability to input costs, benefits, and penetration assumptions differentially by aircraft equipment type. Finally, we envisioned the capability to customize analysis to represent specific air carriers or groups of air carriers.

To define the required functionality of the model, we reviewed a total of nine aviation cost-benefit models and methodologies. To assist in organizing the materials, we developed a two-dimensional classification system. The first dimension was the scope of the costs and benefits considered by the model. The scope of the models ranged from extremely narrow, in which the costs and benefits were limited to a single equipment type, to extremely broad, in which the benefits to the aviation community, flying public, and general society were considered. The second criterion was the level of detail of the modeling approach. Methods ranged from highly detailed bottom-up approaches, in which the operating costs were calculated differentially by phase of flight and equipment type, to aggregate-level top-down approaches, in which industry averages were applied uniformly to all equipment types and carriers. As expected, there was a high degree of correlation between these dimensions.

We found that, other than airline proprietary analysis, no general cost-benefit models exist that focus exclusively on the air carriers and model operating costs at an appropriate level of detail. This finding echoes concerns we heard during our visits with industry representatives. Therefore, we concluded that many of the existing models either did not provide enough modeling detail, or attempted to provide more detail than could credibly be modeled in a financial analysis framework. An example of the former is that most models did not distinguish operating costs by aircraft type. An example of the latter is that several of the models differentiated fuel burn by phase of flight through the use of differential thrust settings. Although important to consider, our contention is that an analysis of such topics is more appropriately conducted with an operational model than with a financial analysis model. Therefore, we envisioned a cost-benefit model that recognized the important distinction between operational issues and financial analysis issues. The CBM takes a bottom-up approach in which operating costs are estimated at the aircraft equipment level and aggregated to obtain airline costs. Thus, the parameters that determine aircraft direct operating costs, such as crew labor rates, are different for each type of equipment. However, some parameters, such as those that determine revenue and indirect operating costs, are available only at the airline level of aggregation. The default parameters of the model are derived from the most recent Department of Transportation (DOT) Form 41 Reports for the largest three U.S. carriers American, Delta, and United. Thus, the parameters of the model represent a hypothetical airline composed of a weighted average of these carriers. Therefore, financial analysis using the default parameters of the model is representative of a large major carrier.

In addition to the default parameters of the model, we have also developed a database of alternate parameters for each carrier or carrier group, such as small majors, or nationals. This database allows analysis to be tailored to a particular set of carriers. Like the default parameters, the alternate parameters are drawn from publicly available DOT Form 41 Reports. Further detail regarding the database of alternate parameters is provided in a later section of this chapter.

VIII. Derivation of the Air Carrier Cost-Benefit Model

This section describes the derivation of the CBM. We begin with a high-level discussion of the model's structure. We then discuss the types of benefits that can be addressed by the model. This discussion is followed with a description of the life-cycle cost module that is used to estimate any costs streams associated with new technology. We then discuss the model's core operating cost calculations that employ a variant of the activity based cost approach developed for the ACIM. Finally, we discuss the output of the model.

A. Structure of the Model

The CBM measures the impact of technological change against a clearly defined baseline. Analysis, therefore, requires the specification of two distinct scenarios, a baseline scenario and a revised scenario. The baseline scenario is intended to capture the most likely future set of outcomes in the absence of new technology (other than innovations explicitly treated in the forecast). We have provided a set of default assumptions that we believe accurately reflect the future expectations. However, we have also provided the capability to modify all of the baseline assumptions so that a user may specify a customized baseline. Conversely, the revised scenario is intended to capture the most likely set of outcomes in the presence of additional new technology. Thus, any differences between the revised scenario and the baseline scenario, with regard to the financial status of the carrier, are attributed to the incremental new technology. Figure 7 illustrates this concept.

As shown in Fig. 7, the primary inputs to the model consist of a baseline scenario and a set of revised assumptions that capture the impact of technology. This set includes parameters related to air travel demand, airline cost and productivity, life-cycle costs for new equipment and training, and the timing and penetration of the technological impact. The main outputs of the model are NPV and duration calculations. In addition, the user may access a set of additional outputs such as annual cash flows, operating costs, and operating revenue by equipment type or aggregated at the airline level.

B. Benefits Assessed by the Model

From our review of existing cost-benefit models, we identified a set of standard benefit categories for inclusion in the model. Although any variable in the model may be modified to assess the benefits of technology, these categories represent the

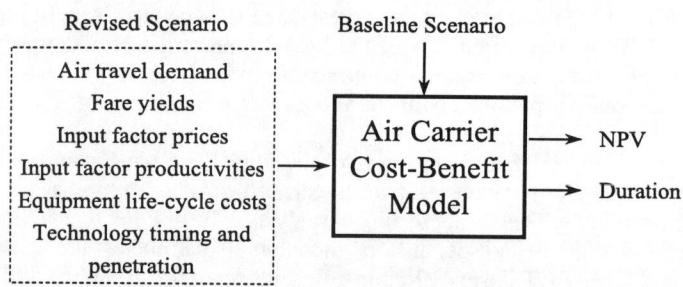

Fig. 7 Air carrier cost-benefit model schematic.

FORECASTING AND ECONOMIC ANALYSIS

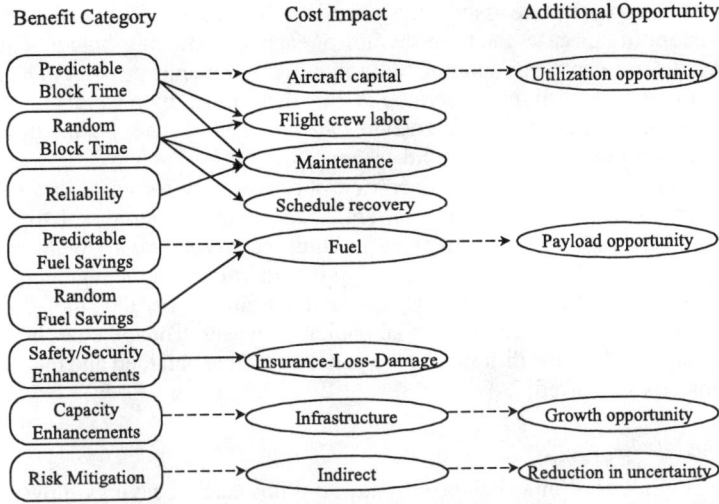

Fig. 8 Benefit categories.

most likely drivers of future benefits. In several cases these categories represent predefined links between the primary impact of an innovation on cost, and subsequent secondary impacts such as revenue enhancement. The main types of benefits that are addressed by the model are shown in the first column of Fig. 8. Each benefit category has a primary impact on costs as shown in the second column. Some categories lead to further impacts by enabling additional benefit opportunities. For example, in the case of predictable fuel savings, additional payload opportunities arise for flights that are currently payload or range constrained. Benefit categories that enable additional opportunities are denoted in Fig. 8 with dashed lines.

We make a distinction between predictable and random time and fuel savings. Generally, predictable savings are more valuable than random savings because predictable savings allow the airline to reoptimize the scheduling and fuel load calculations. This is reflected in Fig. 8, with predictable time and fuel savings leading to additional opportunities while random savings do not. In actuality, the value of predictable savings also depends upon the time horizon.[5]

1. Utilization Opportunity

When predictable time savings are realized, it may be possible for an aircraft to obtain an additional flight segment at the end of a schedule day. To determine whether predictable time savings are sufficiently large, we compare the predicted time savings with a critical value that depends upon the flexibility of the airline's aircraft and crew scheduling decisions. The basic question we are addressing is what magnitude of savings is required to generate additional flight segments at the end of a schedule day. On one extreme we assume that there is no flexibility in the scheduling decision. In that case, each aircraft in the fleet must generate enough time savings itself to allow an additional flight. For example, if a particular aircraft flies 5 flight segments per day at an average block time of 2 h per flight, then—abstracting from the possibility of increasing the number of daily block hours—a total savings of 20 min per flight is required to generate one additional flight.

At the other extreme we assume that there is unlimited flexibility in the scheduling decision. In this case, the time savings of each aircraft contribute to a general pool that determines the number of additional flight segments possible. The actual number of additional flights generated is then determined by a weighted average of the low and high estimates. The weights are adjusted by the schedule flexibility parameter that ranges between 0 and 1.

The analysis described above is carried out separately for each aircraft type. We assume that the length, duration, and load factor for additional flights are equal to the average value for the relevant equipment type. We apply the average passenger yield to the traffic generated by the additional flight segments. Also, because aircraft capital expenses are assessed per aircraft per day, the additional flight segments do not incur additional capital expenses. Thus, the net benefit of an additional flight is the difference between the revenue obtained and the variable operating costs incurred.

2. Schedule Recovery

When flights are running behind schedule, unforeseen time savings allow the carrier to recoup a portion of the costs associated with the delay. These schedule recovery benefits are in addition to any variable operating cost savings and capture the value from reducing the occurrence of passenger and baggage misconnect, crew and aircraft reassignment, and loss of customer goodwill. Because schedule recovery opportunities exist only when a flight is behind schedule, the magnitude of the benefits depends upon the proportion of flights expected to be behind schedule.

3. Payload Opportunity

For flights that are payload constrained, predictable fuel savings also enable additional payload opportunities. The basic premise is that, because the weight of the fuel load is reduced, the revenue payload can be increased. We assume that additional payload is exclusively in the form of cargo as opposed to passengers.

C. Life-Cycle Cost Module

To address the cost streams associated with innovation, we developed a life-cycle cost module for the CBM. As shown in Fig. 9, the life-cycle cost module consists of two primary components: aircraft-related expenditures, and nonaircraft and infrastructure expenditures. Although infrastructure expenditures operate at the airline level of aggregation only, aircraft expenditures can be input globally or differentially by equipment type. Aircraft-related expenditures consist of 1) acquisition and installation costs that are input on an aircraft basis; 2) nonrecurring training costs that are input on a flight crew basis; 3) recurring operation and maintenance costs that are input on a block hour basis; and 4) recurring training costs that are input on a flight crew basis.

In addition to the cost items, the model requires certain equipage timing and penetration assumptions. These consist of an initial equipage year, an initial proportion of the fleet, and a terminal proportion of the fleet. As shown in Fig. 9, the life-cycle cost algorithm automatically draws input from the revised scenario to determine the number of aircraft, flight crews, and block hours impacted by aircraft related expenditures. The result is an estimate of the annual life-cycle cost stream for each equipment type affected. These are subsequently aggregated to determine the total impact to the airline.

FORECASTING AND ECONOMIC ANALYSIS

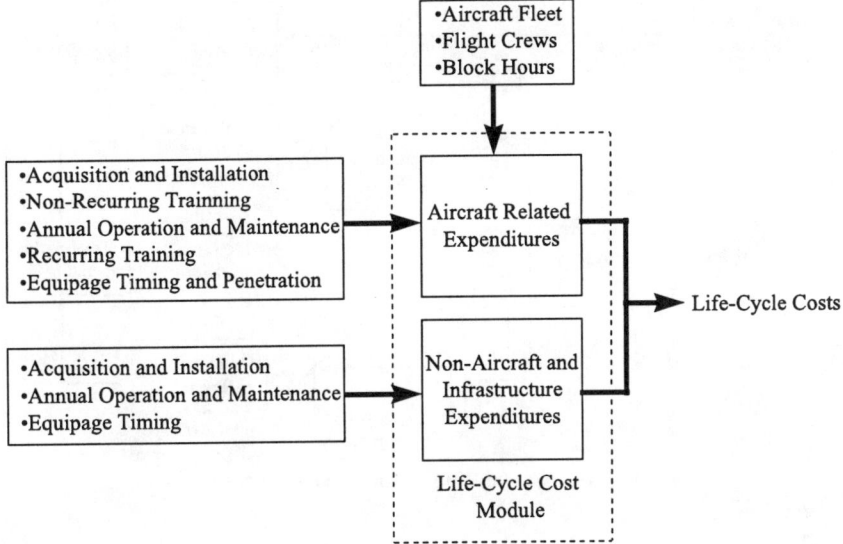

Fig. 9   Life-cycle cost module schematic.

The infrastructure expenditures consist of nonrecurring acquisition and installation costs, annual operation and maintenance costs, and an initial equipage year. The infrastructure costs are input as expenditures for the entire airline and are not dependent upon other inputs from the revised scenario. The infrastructure expenditures are subsequently combined with the aircraft-related expenditures to produce an estimate of total annual life-cycle costs. Finally, the life-cycle expenditures are combined with operating cost and revenue projections to estimate airline profit. The following section discusses the calculation of direct operating costs.

D.   Calculating Air Carrier Operating Costs

To estimate direct operating costs, the CBM follows an activity-based cost approach originally developed for the ACIM.[6] The approach explicitly calculates operating costs in each of six categories as a function of total output, input factor productivities, and per-unit input prices. The cost analysis is based upon observations from DOT Form 41 data in conjunction with detailed aircraft fleet inventories from AvSoft's Airborne Collision Avoidance System (ACAS) Fleet Information System and airline cost of capital information from Ibbotson Associates.[7] The cost data follow each air carrier with annual observations from 1985 through 1995.

Whereas the ACIM focuses on 26 air carriers and calculates operating costs at the airline level of aggregation, the CBM focuses on a single carrier and calculates operating costs at the aircraft equipment level. Figure 10 illustrates this concept. The more finely detailed approach of the CBM allows the user to evaluate the impact of technology differentially by equipment type. The model has the capability to consider up to 23 different equipment types. This set includes the 18 equipment types in use at year end 1996 by the largest three carriers, an additional

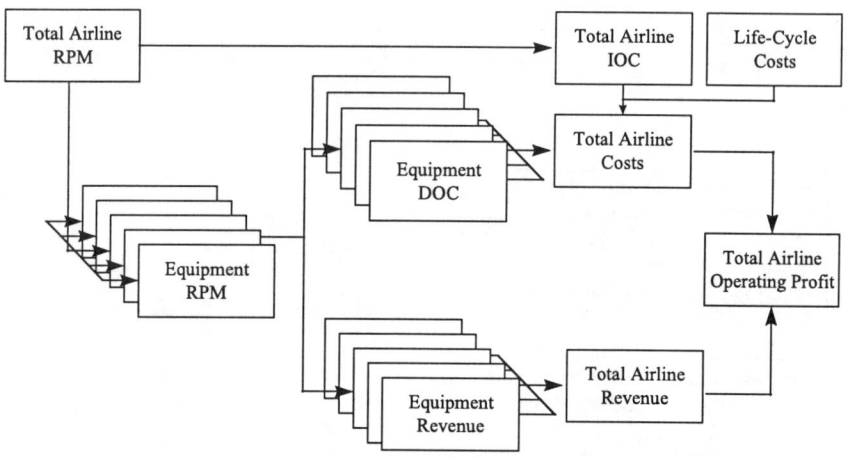

Fig. 10 Calculating airline operating costs.

4 equipment types in use by the alternate carriers, and a vacant equipment type for use in evaluating future aircraft models. To facilitate various types of analysis, the model accepts input parameters at the equipment level of detail, by groupings of equipment types, or globally. The predefined groupings capture such characteristics as single-aisle aircraft, multi-aisle aircraft, Boeing aircraft, and Airbus aircraft.

As shown in Fig. 10, the algorithm begins with the projected RPM for the entire airline. This aggregate traffic forecast is then allocated to each of the equipment types in accordance with certain RPM share assumptions specified by the user. These assumptions allow the user the flexibility to phase out older equipment types, grow existing equipment types, and add new equipment types. Passenger traffic at the equipment level, as measured by RPM, subsequently drives the calculation of direct operating costs and revenue. Next, the equipment-level direct operating costs and revenue calculations are aggregated at the airline level. Estimates of indirect operating costs, derived from the airline-level traffic, are combined with cost estimates from the life-cycle cost module to obtain total airline costs. Finally, total operating expenses are compared with total operating revenues to determine operating profits.

To estimate equipment-level operating costs from equipment-level traffic projections requires several intermediate steps. As shown in Fig. 11, the equipment-level RPM forecast is first converted to available seat miles using a set of equipment-specific load factor assumptions. From ASM, we obtain the required aircraft miles using the seating configuration employed by the carrier. Using a set of equipment-specific assumptions regarding block speed, we obtain the number of block hours flown from the number of aircraft miles. Finally, we obtain the aircraft fleet requirements from the number of block hours using a set of equipment-specific utilization assumptions.

As shown in Fig. 11, the majority of the operating costs are derived from the block hour projections. These consist of fuel; flight personnel labor; maintenance; insurance, loss, and damage; and a residual category termed other direct expenses. Aircraft capital costs, however, are driven by the number of aircraft

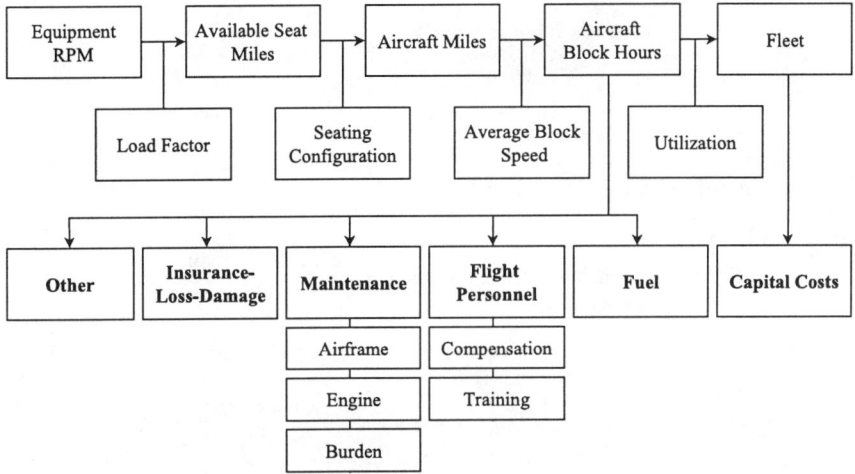

Fig. 11   Calculating equipment-level direct operating costs.

in the fleet as opposed to the number of block hours flown. This distinction allows the airline to take full advantage of any additional aircraft utilization benefits without incurring additional capital charges. Some cost categories contain more than one cost item. Maintenance costs, for example, are composed of aircraft and engine subcategories in addition to overhead, or burden. Maintenance burden is driven by the sum of airframe and engine maintenance costs, as opposed to block hours.

Within each cost category the operating expenses are determined by the interaction of one or more productivity parameters and a per-unit input cost parameter. For example, in the case of fuel expenses, total costs are the product of total block hours flown (output), fuel consumption per block hour (productivity), and fuel price per gallon (input price). Figure 12 illustrates the calculations used by the model for each cost category.

With the exception of aircraft capital charges, each parameter is derived from the equipment-specific base year DOT Form 41 observations. Thus, for each equipment type the base year cost estimates exactly match the carrier's Form 41 filing. To the extent that the parameters follow predictable trends, the cost estimates remain accurate over the forecast horizon.

Flight equipment capital costs were estimated in an especially detailed manner. We began with the 1996 inventory of aircraft from the AvSoft fleet database. This database provides detailed information on the age of each aircraft in a carrier's fleet. Using model-specific resale price information from Airclaims' International Aircraft Price Guide, we estimated the value of each aircraft as a function of its age.[8] Summing over all of the aircraft in a carrier's fleet gives a measure of the total value of the flight equipment.

Next, we applied depreciation and cost of capital charges to the value of the flight equipment. The parameter for depreciation charges is 3.3%, which results form the standard straight-line approach with a useful life of 30 years and no residual value. The parameter for cost of capital charges is 9.8%, which was derived by aggregating carrier-specific cost of capital charges published by Ibbotson Associates. Thus,

$$\text{Fuel costs} = \text{block hours} \times \frac{\text{fuel price}}{\text{gallon}} \times \frac{\text{gallons}}{\text{block hour}}$$

$$\text{Flight personnel compensatio} = \text{block hours} \times \text{labor rate (burdened)}$$

$$\text{Engine maintenance} = \text{block hours} \times \frac{(\text{maint. labor} + \text{maint. mat})}{\text{block hour}}$$

$$\text{Airframe maintenance} = \text{block hours} \times \frac{(\text{maint. labor} + \text{maint. mat.})}{\text{block hour}}$$

$$\text{Maintenance burden} = \text{burden rate} \times (\text{airframe} + \text{engine maint.})$$

$$\text{Flight equipment capital costs} = \text{aircraft} \times \frac{\text{capital charges}}{\text{aircraft}}$$

$$\text{Insurance} \cdot \text{loss} \cdot \text{damage costs} = \text{block hours} \times \text{insurance} \cdot \text{loss} \cdot \text{damage rate}$$

$$\text{Other DOC} = \text{block hours} \times \text{other DOC rate}$$

Fig. 12 Operating cost calculations.

the flight equipment capital costs were calculated as 13.1% of the carrier's aircraft inventory value. Like all parameters in the CBM, the cost of capital parameter represents a constant dollar value.

The advantage of this approach is that the resulting measure of capital cost includes the opportunity cost of the carrier's investment in equipment, whereas depreciation charges taken directly from Form 41 reports do not. Thus we take an economic approach to determine the costs of capital as opposed to a less desirable accounting approach. Nevertheless, the impact of this economic approach must be considered when interpreting the operating profits output by the model. As in the ACIM, there is a discrepancy between the operating profits determined by the model and those reported in Form 41 caused by the opportunity cost of flight equipment capital. We call the profits measured by our approach adjusted operating profit.

With regard to indirect operating costs, we distinguish three cost categories. These consist of landing fees, air traffic control charges, and a residual category termed other indirect charges. Although landing fees are incurred system wide, air traffic control charges are currently incurred only during international operations. An exception would be a flight between U.S. domestic locations that passes under the jurisdiction of a foreign air traffic control authority such as Nav Canada. Indirect charges are calculated using the same activity base cost approach as for direct charges. The cost driver for landing fees is the number of operations, whereas the driver for other indirect charges is ASM. Similarly, air traffic control charges are a function of the block hour rate and the percentage of block hours subject to charges. We approximate this percentage by the proportion of block hours incurred in international service.

E. Model Output

In addition to the sensitivity analysis capability, the model has several basic outputs. The first is a calculation of the net present value of the technology investment under consideration. The second is a calculation of duration, which measures the time dimension of the cash flows. In addition, the model provides access to many of the underlying calculations, such as the discounted and nondiscounted cash flows, total airline revenues and expenses under the baseline and revised scenarios, and equipment-specific cost calculations under the baseline and revised scenarios.

IX. LVLASO Scenario

To illustrate the use of the CBM in the context of other economic analysis to support aviation systems engineering, this section evaluates the benefits of a set of air traffic management technologies under the LVLASO program. LVLASO, part of the NASA Terminal Area Productivity (TAP) Program, seeks to augment existing airport capacity by reducing aircraft runway occupancy time (ROT), separation requirements, and taxi times in low visibility conditions.

We begin our analysis with the ASAC airport capacity model.[9] We model the impact of the new technologies on airport capacity independently for each of five major airports Atlanta Hartsfield (ATL), Dallas Ft.Worth (DFW), Detroit (DTW), Los Angeles (LAX), and NewYork-LaGuardia (LGA). Airport capacity is a function of wind and weather conditions, airport configuration, and a set of technology related parameters such as ROT and arrival separation. Output from the airport capacity model is subsequently passed to the ASAC airport delay model, which projects arrival and departure delay as a function of hourly demand and airport capacity. For each airport, we estimate delay with and without the capacity enhancing technologies. The projected difference between the two scenarios becomes input for the CBM.

A. LVLASO Technologies

The TAP program aims to safely achieve visible meteorological conditions airport operating capacity during instrument meteorological conditions. The TAP program includes three technology elements: reduced spacing operations (RSO), LVLASO, and ATM. Sub-elements of LVLASO include high-speed roll-out and turn-off (ROTO); taxi, navigation, and situation awareness (T-NASA); and dynamic runway occupancy measurement (DROM). These subelements are designed to cut delays on the runways and taxiways during periods of poor visibility.

In addition to ground equipment requirements, the subelements of LVLASO require a set of upgrades to existing cockpit avionics capabilities. Table 4 lists the avionics requirements. The components included are based upon discussions with NASA personnel and on NASA briefings.

B. Deriving the Cost-Benefit Model Inputs

We model the impact of the LVLASO technologies on airport capacity by modifying the poor visibility [instrument meteorological conditions (IMC)], arrival ROT, and separation standards to equal the good visibility [visual meteorological

Table 4 LVLASO avionics requirements

Technology subelement	Requirement
T-NASA	Differential global positioning system (DGPS)
	Head-up guidance system (HGS)
	Cockpit display of traffic information (CDTI)
	Automatic dependent surveillance-broadcast (ADS-B)
	Controller to pilot data link communications (CPDL)
	Taxi map display hardware/data
DROM	Audio card
ROTO	DGPS
	Stereo headsets
	ADS-B
	DGPS
	Taxi map display hardware/data
	HGS

conditions (VMC)] values for each aircraft class. The result is a revised capacity for poor weather conditions for each airport configuration that approximates good weather capacity.

Our technology scenario assumes that the benefits of the new technologies will be realized beginning in the year 2005. Accordingly, we specify projected traffic demand patterns for 2005 at each airport in the airport delay model. The model uses a queuing engine to calculate the average arrival and departure delay on an hourly basis for each airport. For this analysis, we exercised the airport delay model over an entire year of actual meteorological conditions for each airport. We then aggregated the hourly and daily results to obtain average delay statistics for arriving and departing flights on an annual basis. This analysis was performed for both a baseline and improved technology scenarios. The results from the airport capacity and delay models are summarized in Table 5.

Because the CBM requires input in the form of changes in block time, the next step was to convert the figures from Table 5 to percent changes in block time. This requires an assumption regarding the average block time for departing and arriving flights at each airport. We used the 1995 DOT T-100 Reports to define the current average block time for each. These were subsequently adjusted by the projected increase in delay from 1995 to 2005 to determine the projected average block times for 2005. As described in an earlier section, the default parameters and assumptions of the CBM represent a large major carrier. Therefore, we used the T-100 Reports for the largest three carriers only to project average block time. The result was a projected change in arrival and departure average block times from the baseline scenario to the revised scenario for each airport.

To aggregate the impact of the technologies across all five airports, we constructed weights according to the number of operations at each airport by the largest three carriers. The result is a weighted average change in block time that

Table 5 Projected 2005 delay statistics

Airport	Scenario	Average arrival delay, (min)	Average departure delay, (min)
ATL	Baseline	59.8	29.4
	Technology	55.5	25.9
DFW	Baseline	16.2	15.8
	Technology	15.9	16.0
DTW	Baseline	15.6	—[a]
	Technology	12.7	—[a]
LAX	Baseline	24.3	20.6
	Technology	23.9	20.4
LGA	Baseline	22.0	20.7
	Technology	19.71	18.60

[a]The web version of the DTW Airport delay model does not calculate departure delay.

will be used to extrapolate to the system-wide impact. Table 6 illustrates this methodology.

The final step in deriving the CBM inputs is to project the proportion of air traffic that will benefit from the new technology. Our LVLASO scenario assumes that these technologies will be in place at 10 major airports by 2005. In addition, we assume that the technologies will be installed incrementally at the next largest 10 airports over the remainder of the forecast horizon. To determine a benefit penetration curve for our representative air carrier, we further examined 1995 T-100 Reports. For each flight segment in the Department of Transportation T-100 Report, one of four possibilities must be realized. These possibilities are 1) the flight segment both departs and arrives at airports with the new technologies; 2) the flight segment departs at an airport with the new technologies, but arrives at one without; 3) the flight segment departs at an airport without the new technologies, but arrives at one with; or 4) the flight segment both departs and arrives at airports without the new technologies.

Table 6 Deriving cost-benefit model input

Airport	Annual operations[a]	Change in arrival block time, %	Change in departure block time, %
ATL	199,073	−2.7118	−2.6593
DFW	246,276	−0.2014	0.1385
DTW	12,476	−2.3990	−2.3990[b]
LAX	98,331	−0.1863	−0.1110
LGA	52,147	−1.6803	−1.3773
Weighted average		−1.1924	−0.9997

[a]1995 operations for American, Delta, and United.
[b]In the absence of departure delay information for DTW, we assume that departure delay equals arrival delay.

Table 7 Penetration assumptions

Departure airport	Arrival airport	Operations 2005, %	Operations 2016, %	Change in block time, %
New technology	New technology	14.9	31.3	−2.1921
New technology	Baseline	31.6	28.7	−0.9997
Baseline	New technology	31.6	28.7	−1.1924
Baseline	Baseline	21.9	11.3	0.0000
2005 Weighted average				−1.3049

Categorizing each flight segment according to the criteria above yields estimates of the proportion of flights benefiting from the new technology. We exercised these criteria separately for 2005, with the 10 TAP airports, and 2016 for the top 20 airports. However, since the CBM can incorporate only a single parameter for change in block time, it was necessary to construct a weighted average across these categories to represent the benefit penetration. This methodology is illustrated in Table 7.

Thus we adopt an initial benefit penetration of 78.1% with an initial reduction of 1.3049% in block time. Over the forecast period, the penetration grows to 88.7% although the impact remains constant. This assumption does not account for the fact that the block time impact itself is growing over time as more and more flights both depart from and arrive to airports with the new technology. For this reason, our estimates of the benefits of the LVLASO technologies should be viewed as conservative.

To evaluate the net benefits of the LVLASO scenario, it is necessary to employ a set of life-cycle cost assumptions regarding the requirements for new avionics equipment. Table 8 summarizes the results of a previous study LMI conducted for NASA.

Aggregating the results from Table 8, we arrive at the following life-cycle cost assumptions:
1) $355,200 per aircraft for acquisition and installation of new cockpit avionics;
2) $2,500 per flight crew as initial training expense;
3) $1.15 per block hour as operation and maintenance expense; and
4) $500 per flight crew as recurring annual training expense.

Finally, our equipage penetration assumption is that all aircraft will be equipped during 2005 to take advantage of the block time benefits.

C. Scenario Results

We exercised the model under the assumptions discussed above. The main result shows that the LVLASO technologies do provide modest benefits to the representative carrier. We estimate the NPV of this investment to be $141 million at a discount rate of 8%. In addition, the investment has a large duration of 25.3 years, which correctly indicates that the stream of benefits is far into the future.

To analyze the sensitivity of the main results to variation in the input data, we exercised the sensitivity analysis module for several key variables. As shown in Fig. 13, these include change in block time, discount rate, penetration assumptions,

FORECASTING AND ECONOMIC ANALYSIS

Table 8 Life-cycle cost assumptions

Equipment item	Life-cycle cost category	Expense
VHF digital radio	Acquisition	$30,000 per radio
	Installation	$1,000 + 4 hours out of service
	Recurring maintenance and operation	$0.10 per block hour
HGS	Acquisition	$219,000 per HGS
	Installation	$50,000 + 2 days out of service
	Initial training	$2500 per flight crew
	Recurring training	$500 per flight crew
	Recurring maintenance and operation	$1.00 per block hour
Taxi map display hardware/data	Hardware acquisition	$0.00 (currently in place)
	Data acquisition	$10,000 annually per fleet
Three-dimensional audio warning system	Headset and sound card acquisition	$2,000 per aircraft
	Installation	$200 + 2 hours out of service
	Recurring maintenance and operation	$0.05 per block hour
CDTI software	Acquisition	$1000.00 per aircraft

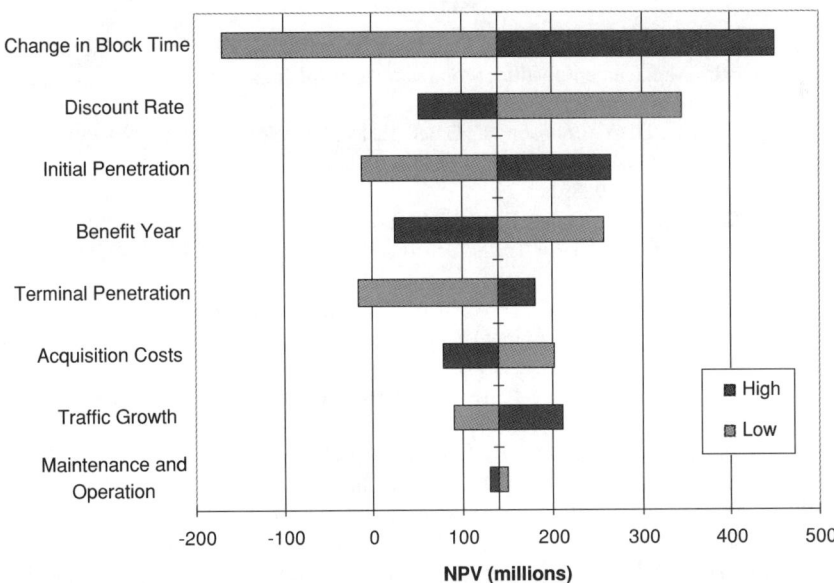

Fig. 13 Sensitivity results.

life-cycle costs, and traffic demand growth. In exercising the sensitivity analysis module, we made a simple assumption that the low and high values were 50 and 150% of the middle values, respectively.

Under these assumptions, it is clear that the LVLASO technologies contain several risks that threaten the projected benefits. The most substantial risk is due to uncertainty in the magnitude of the block time savings. This issue might be particularly risky if the magnitude of the time savings depended upon the equipage of other carriers' aircraft. This dependence is likely when considering air traffic management technologies that impact variables such as separation standards. Other important risks are due to the timing and penetration assumptions. It is clear that, if the technology benefits are slipped relative to the year of equipage, the benefits will be eroded quickly. Thus, the analysis indicates several variables that decision makers would need to investigate further before committing valuable resources.

X. Conclusions

The analysis presented in this chapter predicts a serious capacity shortfall within the next several years. This mismatch between the demand for air travel and the airport and airspace system will result in significant increases in delays, travel times, and average fares. Valuable economic resources will be wasted and there will be large losses of potential output and revenue.

This chapter also demonstrates the usefulness of an integrated suite of operational and economic models in generating more credible air travel forecasts. These revised forecasts include airline and passenger feedback and provide a basis for quantifying the economic value of capacity enhancement.

References

[1] Kostiuk, P. F., Gaier, E., and Long, D., "The Economic Impacts of Air Traffic Congestion," Logistics Management Institute, *Air Traffic Control Quarterly*, Vol. 7, No. 2, Spring 1999, pp. 123–145.

[2] Long, D., Lee, D. A., Johnson, J., Gaier, E. M., and Kostiuk, P. F., "Modeling Air Traffic Management Technologies with a Queuing Network Model of the National Airspace System," NASA CR-208988, Jan. 1999.

[3] Long, D., Lee, D. A., Gaier, E. M., Johnson, J., and Kostiuk, P. F., "A Method for Forecasting the Commercial Air Traffic Schedule in the Future," NASA CR-208987, Jan. 1999.

[4] Long, D., Wingrove, E., Lee, D., Gribko, J., Hemm, R., and Kostiuk, P., "A Method for Evaluating Air Carrier Operational Strategies and Forecasting Air Traffic with Flight Delay," LMI Report N590251, Oct. 1999.

[5] Chew, R. G., "Free Flight Preserving Airline Opportunity," American Airlines, Ft. Worth, TX, Sept. 1997.

[6] Wingrove, E. R., Gaier, E. M., and Santmire, T., "The Aviation System Analysis Capability Air Carrier Investment Model (Third Generation)," NASA CR-207656, April 1998.

[7] "Industry Cost of Capital," Ibbotson Associates, Chicago, IL, 1998.

[8] "International Aircraft Price Guide," Airclaims Ltd., London, Winter 1996.

[9] Lee, D. A., Nelson, C., and Shapiro, G., "The Aviation System Analysis Capability Airport Capacity and Delay Models," NASA CR-207659, April 1998.

Chapter 7

Impact of Air Traffic Management on Airspace User Economic Performance

Joseph H. Sinnott* and William K. MacReynolds, Ph.D.[†]
MITRE Corporation, McLean, Virginia

I. Introduction

THE MITRE Corporation's Center for Advanced Aviation Systems Development (CAASD) has worked with the Federal Aviation Administration (FAA) since 1958 providing system research and development expertise. The FAA-sponsored CAASD programs and projects serve to foster collaboration between government and industry regarding communication, navigation, and surveillance technology and in understanding and improving the design and operation of the air traffic management (ATM) system to meet user needs. The FAA and CAASD have worked together to develop new concepts and prototypes for industry to produce and deliver to the aviation community.

The FAA has made great progress in the last several years in understanding user concerns and their relationship to ATM performance. The FAA has developed and applied metrics to assess ATM performance and to manage ATM system evolution. The next logical step is to understand the impact of ATM on air carrier economic performance. This chapter reports the work CAASD has done with the active participation of the FAA's Office of System Capacity. This body of work addresses the issue of what changes in ATM performance would generate value for users and how proposed ATM changes should be prioritized.

One goal of this research is to increase FAA effectiveness by understanding what performance really means to users and to aid in identifying, developing, and prioritizing ATM improvements. Research reported in this chapter does not seek to replace operational ATM metrics nor to determine for users what their needs should be. Rather, this work explores the relationship between ATM actions and user decisions, especially those of air carriers. Key relationships are shown in Fig. 1. Solid lines indicate reasonably well understood relationships; shaded lines indicate areas where some relationships are understood,

---

Copyright © 2001 by the MITRE Corporation. Published by the American Institute of Aeronautics and Astronautics, Inc., with permission.
*Senior Principal Staff.
[†] Senior Staff.

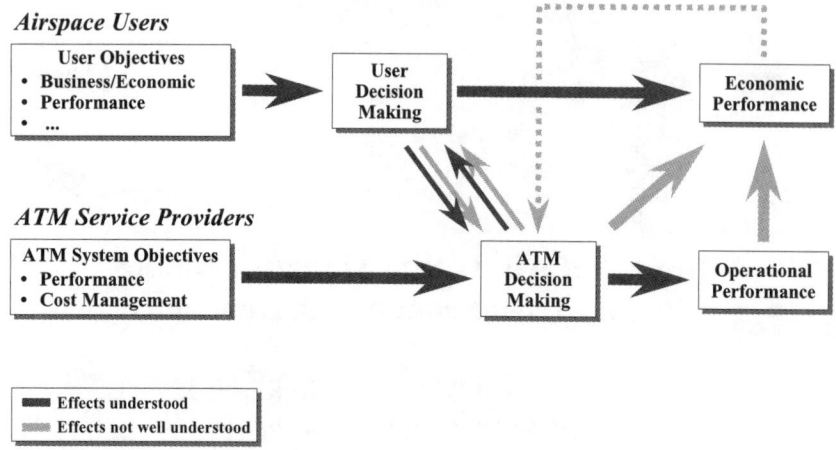

Fig. 1 ATM performance: operational and economic perspectives.

but where there is much to learn; the dashed lines indicate poorly understood relationships.

This chapter summarizes recent CAASD research into the relationship between ATM actions and airline decisions and economic performance. There are broader concerns involved in deciding appropriate future ATM changes than is involved in this research. The purpose of this body of work is to illuminate important but poorly understood relationships between ATM and air carriers so that both FAA and airlines better understand the impact they have on each other.

II. Airline Cost Drivers and ATM Actions

CAASD has examined the available public airline cost reporting systems, corporate annual reports, and Department of Transportation (DOT) airline financial data. Each of these sources was developed for other purposes. CAASD developed a cost structure tailored to illuminate our understanding of cost impacts of ATM decisions and actions. This new structure is appropriate for understanding the effects of ATM actions on airline costs (defined by functional activity and category) and places all operating costs in a consistent context that mitigate differences between direct and indirect operating costs.

Using these cost categories, industry-wide cost drivers were identified. Costs included under the functional activity Flying Operations encompass 36% of total airline operating costs (Table 1). The largest costs in this category are pilot salaries and benefits, fuel, and passenger service expenses. Costs included under the functional activity Prepare To Fly are 17% of total operating costs. This category includes expenses for servicing the aircraft and handling baggage and passengers prior to flight. Marketing/Scheduling costs constitute 17% of total operating costs. Capital Assets and Direct Maintenance costs represent 11 and 9% of total operating costs, respectively.

Airlines can be categorized by the type of service offered, routes served (short or long haul, including international flights), hubbing activity, and fleet mix; and

Table 1 CAASD functional cost categories for selected air carriers (1996 data)

Cost categories	Percent of total operating expenses
Flying Operations (pilot and copilot salary and benefits, fuel, passenger service)	36%
Prepare to Fly (traffic and aircraft servicing)	17%
Direct Maintenance (flight equipment maintenance)	9%
Capital Assets (flight equipment depreciation, aircraft rentals)	11%
Marketing and Scheduling (reservations and sales, advertising, and promotions)	17%
Other (G&A, ground property and equipment maintenance, transport related expenses)	8%

cost differences can be identified for different types of airlines. A low-cost carrier that provides mainly point-to-point service is likely to spend a higher proportion of its total operating expenses on aircraft, maintenance, and fuel than the industry-wide average, a reflection of shorter stage length and an emphasis on greater aircraft utilization. Its expenditures, as a percent of total operating costs, on ground activities and passenger service costs are necessarily lower than the industry-wide average. Major international hubbing airlines spend a higher portion of operating costs on passenger service than the industry-wide average. Cash-strapped airlines spend a smaller portion of their operating costs on crew salaries and benefits than the industry-wide average.

The costs that are related to ATM action were evaluated by dividing costs into fixed and variable components. Whether a cost is constant (fixed) or variable depends on the time horizon. In the short term most costs are fixed. In the long term all costs are considered variable. Airline operations were partitioned into short-, medium-, and long-term decisions. In the short run, airlines are focused on flight operations and costs incurred in flight. Only a few costs are variable; therefore, only a limited number of cost categories can change in the short run. In the medium term, an airline has more time to modify schedules, assess routes, and mitigate persistent disruptions to their networks. In the long term, an airline has the flexibility to respond to ATM system performance changes by changing its fleet mix or moving a hub.

This analysis is to judge the impact of ATM system performance on fuel cost, crew cost, reserve crew requirements, passenger ill-will, crew utilization, aircraft utilization, hub scheduling, route selection, and block times, among other things. The key question is how does ATM performance affect airline cost and service? An airline's schedule is driven by revenue concerns. Fulfilling the schedule involves costs. Increasing ATM efficiency allows shorter scheduled block and ground turn times leading to increased aircraft productivity. Improved aircraft utilization and productivity translates into fewer aircraft or reduced costs to fly a given set of routes or could, of course, lead to additional flights per aircraft and increased revenue.

Recommended improvements to the ATM system have focused on adding capacity, flexibility, and access to reduce system-wide delay and to increase schedule predictability. Based upon statements by carriers, their major problem with ATM is unpredictable delay.

III. Estimates of System-Wide Excess Cost to Airlines

Estimates of system-wide excess costs were made for both short- and long-term costs. The former reflect excess operating costs for fuel and crew. The latter include all excess operating costs. Excess costs come from increased operating costs due to operational delay—the difference between actual time and a feasible minimum time.

Operational delay was measured in different flight phases: ground turnaround, taxi-out, airborne, and taxi-in. Airborne and ground turnaround operational delay was calculated from samples, extrapolated to all flights. Actual times for each flight phase were found in the DOT Airline Service Quality Performance (ASQP) database.[1] The ASQP database contains all the domestic flights of the 10 largest U.S. air carriers. The ASQP carriers are Alaska, America West, American, Continental, Delta, Northwest, Southwest, TWA, United, and US Airways.

Table 2 shows operational delay in millions of minutes per year by phase of flight. It contains the estimate bounded by lows and highs that capture most of the variation in annual delay due to sampling error. Table 2 also shows extrapolated values of operational delay for non-ASQP air carriers. Non-ASQP carriers are assumed to have the same distribution of delay in the sampled phases as ASQP carriers. Together, operational delay for non-ASQP and ASQP carrier's domestic flights comprise a National Airspace System (NAS)-wide operational delay in this chapter.

Table 3 shows operational delay per flight for the roughly 8.4 million domestic flights in 1997. Over all phases, the estimate is 32.0 min per flight. The airborne phase is estimated to have 8.2 min of operational delay per flight, or 25% of the total. The taxi phases combined have 16% of operational delay. Almost 60% or 18.9 min per flight of operational delay occurred at ground turn.

Each phase of flight has different direct operating costs with a short- and long-term aspect. The short-term aspect is a period when most costs are fixed by prior

Table 2 Operational delay by phase of flight for ASQP and non-ASQP carriers in millions of minutes per year (lower bound/estimate/upper bound)

Air carrier/phase of flight	ASQP	Non-ASQP	Total[a]
Ground turnaround	98/100/102	60/61/63	158/162/165
Taxi out	20.2	12.3	32.5
Airborne	34/44/53	18/25/32	52/69/85
Taxi in	6.5	4	10.5
Total[a]	95/103/111	159/170/182	254/273/293

[a]Discrepancies are due to rounding.

Table 3 Operational delay per flight by phase of flight for all carriers in minutes per flight (lower bound/estimate/upper bound)

Air carrier/phase of flight	All carriers
Ground turnaround	18.5/18.9/19.3
Taxi out	3.9
Airborne	6.2/8.2/10.1
Taxi in	1.3
Total	29.7/32.0/34.3

decisions, such as the number and mix of aircraft, routes flown, gates available, maintenance schedules, etc. In the short term, decisions made by airlines concern how to fly given routes and whether certain flights should be canceled because of weather problems or schedule integrity issues. These decisions have impact on only a limited number of cost categories. Operational delay in the short term causes extra expenditures for fuel and crew above what could potentially be attained if scheduled block time were at a feasible minimum.

Table 4 shows the short-term weighted averages of operating cost per minute for all carriers. The data come from 1997 financial filings from all commercial air carriers to the DOT in what is called the Form 41 database.[2] Non-ASQP carriers (cargo, regional, commuters, air taxi, and charter) employ a wide variety of aircraft, as do ASQP carriers. However, on average, non-ASQP carriers utilize aircraft that have lower fuel and crew costs during each flight phase in the short term and lower indirect costs in the long term (see Table 5 for short-term operating costs and Table 6 for long-term operating costs). Although non-ASQP carriers are assumed to have the same distribution of delay per flight as ASQP carriers, unit costs are quite different.

The method of analysis to determine excess cost is to multiply annual operational delay for each phase of flight by the cost per unit of time for that phase and carrier type. For short-term excess cost, this method yielded the results in Table 5. Table 5 shows that the estimate of excess airline operating cost in the short term is $2.3 billion or roughly 10% of total fuel and crew expenses for all carriers. The estimate is derived, in part, from samples in the ground turnaround and airborne phases. Ranges depicting 95% confidence intervals for each sample are reported beginning in Table 2. Based on an extrapolation to non-ASQP carriers from ASQP

Table 4 Short-term unit cost by phase of flight for ASQP and non-ASQP air carriers

Air carrier/phase of flight	ASQP, $/min	Non-ASQP, $/min
Ground turnaround	$14.28	$7.55
Taxi out	$16.29	$8.87
Airborne	$26.34	$15.47
Taxi in	$16.29	$8.87

Table 5 Short-term system-wide excess airline operating costs by phase of flight for ASQP and non-ASQP carriers in billions of dollars (lower bound/estimate/upper bound)

Air carrier/phase of flight	ASQP	Non-ASQP	Total
Ground turnaround	0.13/0.13/0.14	0.04/0.04/0.04	0.2/0.2/0.2
Taxi out	0.3	0.1	0.4
Airborne	0.9/1.1/1.4	0.3/0.4/0.5	1.2/1.5/1.9
Taxi in	0.1	0.03	0.1
Total	1.5/1.7/2.0	0.5/0.6/0.7	1.9/2.3/2.7

airborne and turnaround time samples, it is estimated that short-term excess costs are in the range of $1.8 billion to $2.7 billion.

In Table 5, note that the highest excess cost (about two-thirds of the total) occurs during the airborne phase. Airborne operational delay amounts to 8.2 min per flight, and it costs over $26 per min for ASQP carriers and over $15 per min for non-ASQP carriers (as shown in Table 3). Excess cost for turnaround time is quite low. The reason is that short-term ground turnaround unit costs are, for all practical purposes, zero for all minutes up to scheduled departure time. The cost per minute accrues only when the flight pushes back late. The study estimated that just 9.3% of the operational delay minutes at ground turnaround occurred beyond scheduled departure time.

In the long term all operating costs are variable. Given sufficient time, airlines can alter their schedules, change routes, consolidate gate operations, and alter their fleet mix. Long-term airline operating cost per minute estimates require careful judgments. For example, some costs are closely related to phases of flight. During airborne time, fuel is being burned rapidly, about six times faster than on the ground during taxi. Maintenance costs are heavily influenced by time spent airborne. Taxi times have a different fuel burn rate, but the same crew costs as in the air.

Turnaround time costs involve crew expenses only when the aircraft stays at the gate beyond its scheduled departure time. In addition, there are some airline cost categories that support all phases of flight, called indirect operating expenses. Indirect costs are apportioned to each phase of flight proportionate to the actual amount of time spent in each phase. Cost per minute in the long term (Table 6) is

Table 6 Long-term unit cost by phase of flight for ASQP and non-ASQP air carriers

Air carrier/phase of flight	ASQP, $/min	Non-ASQP, $/min
Ground turnaround	$59.30	$25.67
Taxi out	$61.42	$26.99
Airborne	$78.17	$45.20
Taxi in	$61.42	$26.99

Table 7 Long-term system-wide excess airline operating costs by phase of flight for ASQP and non-ASQP carriers in billions of dollars (lower bound/estimate/upper bound)

Air carrier/phase of flight	ASQP	Non-ASQP	Total
Ground turnaround	4.6/4.6/4.7	1.1/1.2/1.2	5.7/5.8/5.9
Taxi out	1.3	0.3	1.6
Airborne	2.7/3.4/4.1	0.8/1.1/1.5	3.5/4.5/5.6
Taxi in	0.4	0.1	0.5
Total	8.9/9.7/10.5	2.4/2.7/3.1	11.3/12.4/13.6

consequently higher than short-term cost per minute because more categories of operating costs are included.

Long-term excess costs are estimated to be $12.4 billion, as shown in Table 7, or approximately 15% of all carriers' operating costs. Based on extrapolation from a sampling of ASQP carriers to non-ASQP carriers and 95% confidence intervals for airborne and turnaround samples, long-term excess costs are estimated to be in the range of $11.3 billion to $13.6 billion.

The airborne phase (Table 7) does not contribute the most to excess costs even though the largest portion of indirect costs are placed in the airborne phase (because it typically entails more time than other phases). In the long term excess cost during the turnaround phase is greater than for the airborne phase because operational delay minutes are the highest for the turnaround phase (more than twice the amount of operational delay during the airborne phase), as shown in Table 2, and indirect costs are applied to all those minutes.

The airborne origin-destination (OD) sample was stratified by length of flight and flight frequency. Sample results show that the longer the flight, the greater the airborne delay and, hence, the higher the excess cost. This suggests that there is some operational delay experienced in the en route portion of the airborne phase. The airborne sample was also stratified by frequency of OD routes, low (below 722 flights per year), medium (between 722 and 1730 flights per year), and high (over 1730 flights per year). None of these strata exhibited operational delay per flight averages significantly different from the average for all of the flights. That is, interestingly, there is no difference in airborne delay and excess cost per flight between less and more frequently traveled routes.

It is important to note as part of this discussion of excess costs that not all, or even possibly a significant percentage, of these excess costs may be recoverable in practical terms. Excess costs are related to a variety of causes: airport and ATM system capacity and operating practices, adverse weather both at airports and in the en route system, traffic congestion, and airline operations and scheduling. For example, most of the nation's major airlines operate hubs to provide extensive route networks and cost-effective service. However, the cyclical operation of arrival and departure banks often results in congestion-related delays, especially in bad weather, and necessitates longer aircraft turnaround times to permit passengers to make connections. Thus, some excess costs are an integral and necessary part of providing cost-effective network service. Some excess costs, however, clearly are recoverable through investments or procedural changes to increase the efficiency and capacity of the nation's aviation system.

IV. Example of the Impact of ATM Improvements on Long-Term Airline Costs: Fleet Utilization and ATM Improvements

The magnitude of airline fleet capital costs makes fleet impacts a key category when evaluating the economic impact of ATM initiatives. The CAASD fleet requirements model estimates fleet requirements among scheduled passenger airlines, establishing a chain of causality from ATM improvements to financial performance, as shown in Fig. 2. The chain begins in Phase I with a change in ATM performance. Performance changes typically stem from an initiative like Free Flight Phase 1, a new tool like passive final approach spacing tool (pFAST), or a new procedure like closely spaced parallel approaches. Phase I quantifies the improvement using ATM performance measures, whether in terms of increased maximum airport arrival rates, reduced taxi delays, etc.

Phase II translates ATM impacts to airline operating measures such as actual flight times. In Phase III operating efficiency improvements drive changes in scheduled activity times such as scheduled airborne time and scheduled ground turnaround time. Because most of an airline's assets, including the fleet, are allocated to produce the schedule, scheduled times provide a better basis for estimating fleet requirements than daily operating times. A particular benefit of using scheduled times is that they reflect not just average operating times but also variation in times, establishing a tie to operational predictability.

In Phase IV airlines re-allocate their fleets to exploit time savings by saving aircraft. This is an exceedingly complicated process because many airline schedules must mesh together. Crew, aircraft, ground resources, and maintenance schedules must all be carefully managed. Even with careful management, there is likely to be considerable slack in any schedule. Some slack (in aircraft and crew utilization and maintenance deadlines) is useful to respond to schedule disruptions. Phase IV estimates potential aircraft saving assuming that some slack is useful and that crew and maintenance schedules can adapt to new fleet assignments. Any redundant

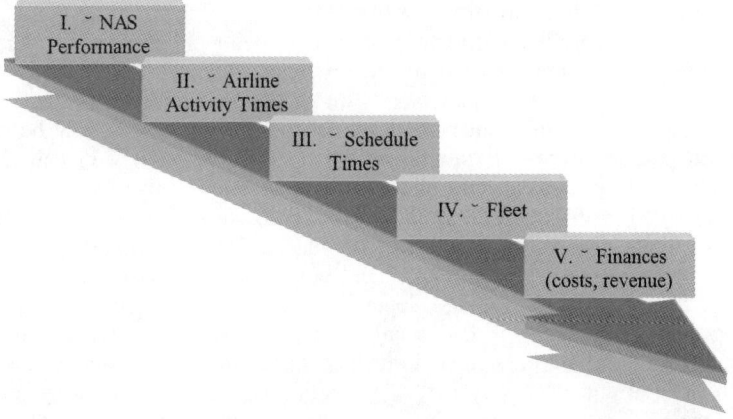

Fig. 2 Chain of impact (phases) from ATM change to fleet finances.

Table 8 Estimated fleet impact from eliminating excess airborne time

	Base fleet size	4,429 aircraft
Phase II	Reduction in airborne time	1.0%
Phase IV	Estimated reduction in fleet[a]	19 aircraft
Phase IV	Percentage reduction in fleet[a]	0.5%
Phase V	Approximate one-time savings (replacement)	$1.1 billion
Phase V	Estimated revenue increase	0.5%

U.S. Scheduled Passenger Airlines, domestic service, 1997

[a] Assumed base fleet of 4,429 aircraft.

aircraft identified in Phase IV can be removed from service or redeployed onto new flights. Phase V converts these potential fleet changes into financial measures like cost and revenue.

Suppose all excess airborne time related to ATM performance is eliminated. Assume that this reduction amounts to 1.0% of actual airborne time. The CAASD fleet utilization model initially assumes that airlines adjust their fleets but maintain their existing network structure. In that case, how much of their current fleet could be saved?

Table 8 summarizes the results of the analysis conducted to answer this question. Estimated fleet savings are 19 of the 4,429 aircraft in the airlines' 1997 domestic fleet, or 0.5% of the total. If airlines eliminate the excess aircraft, the one-time replacement value of the 19 aircraft is approximately $1.1 billion. Alternatively, airlines could use the extra aircraft to add new service. The estimated increase in passenger revenue is about 0.5% of base revenue.

ATM improvements in the airborne phase can occur in the terminal area or in the en route portion of flight. The distribution of Phase IV and V benefits among airlines depends on whether time savings accrue in the terminal area (affecting arrival and departure airspace) or whether they occur en route. Figure 3 illustrates the estimated percentage fleet reduction by airline for the two cases. We assume that terminal area time savings tend to save a fixed length of time per flight. Airlines whose aircraft operate many flights per day benefit most from terminal area savings (Fig. 3a). We assume that en route time savings are approximately proportional to the average flight duration. Airlines whose aircraft spend the most time per day airborne benefit most from en route savings (Fig. 3b). A comparison of Figs. 3a and 3b shows that percentage decreases in airborne times produce a more even distribution of benefits than fixed time reductions, which vary widely among airlines.

V. The Larger Picture: The Influence of ATM on Demand-Related Airline Decisions

ATM improvements can definitely save short-term excess costs of fuel and crew and probably provide aircraft fleet savings in the long term. The next question is what impact might ATM improvements have on airline strategic planning? Airline planning for routes determines what aircraft airlines need and their potential revenue. To secure the revenue, airlines invest in aircraft, ground equipment, hubs

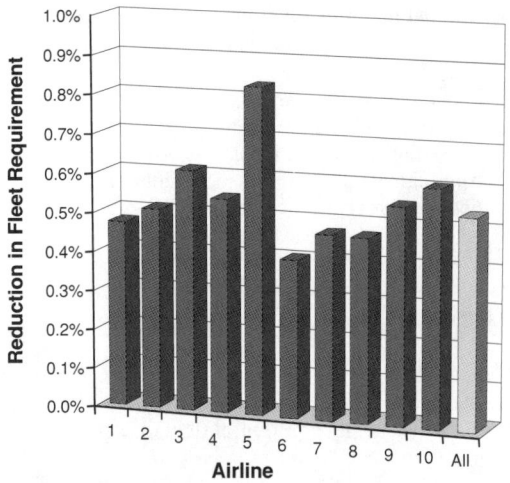

a) 1.2 minutes per departure

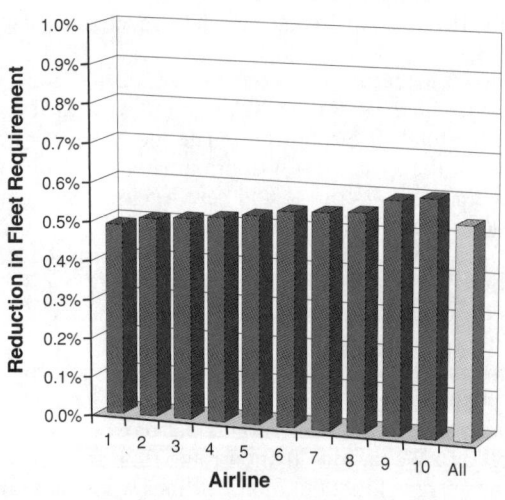

b) 1.0% of airborne time

Fig. 3   Reduction in fleet requirement.

(if necessary), and reservation or ticketing systems. ATM, through the services it provides, the restrictions it imposes, and its role in congestion, can influence scheduling, route planning or development, and fleet planning. To the extent that ATM impacts these decisions, ATM can indirectly influence decisions on ground equipment investments, hub development, and reservation systems. The issue is the extent to which ATM affects airline product decisions with respect to both the importance to any one decision and the prevalence throughout decision-making.

An air carrier's chief business interest is revenue generation. Key decisions involve route development, scheduling, and fleet planning. As Fig. 4 indicates,

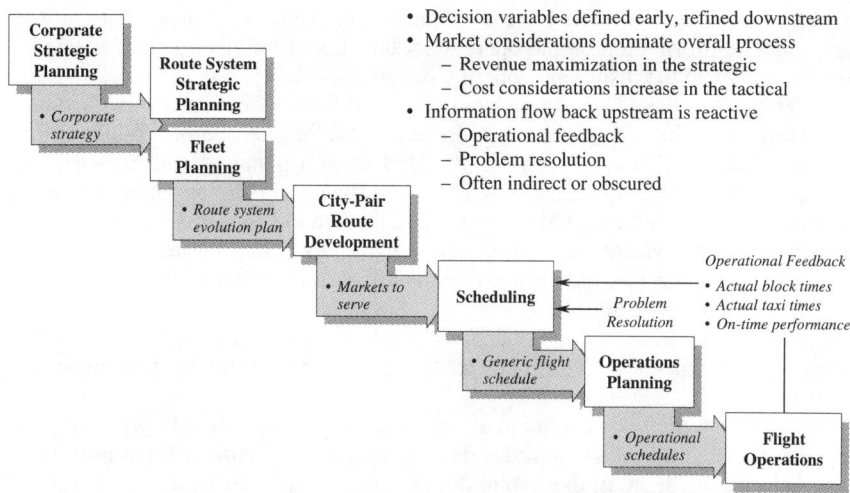

Fig. 4 Airline demand-related decision-making processes.

decision-making starts with strategic planning. Route planning and selection lead to fleet requirements. As specific schedules are made regarding city-pair routes, the chosen fleet can be assigned an itinerary. Operational activities provide feedback that may serve to alter a schedule.

Scheduling relies heavily on historical operational data, especially block times. Block times can be influenced by ATM, and airlines are sensitive to changes in block times. ATM affects block times through the direct services it offers, with ATM restrictions and in the role ATM plays in contributing to congestion. To obtain their desired on-time performance, airlines will add padding into a schedule to reflect an amount above average block times to allow for delay and seasonally experienced variation in block times.

Revenue is the main driver of route development decisions, although congested routes are sometimes a concern. The issue is the extent to which ATM influences are recognized by airlines at this stage. Logically, if ATM improvements can lead to more economical routes, it can add to revenue and return to investors.

The ATM impact on fleet planning is mostly indirect. ATM can impact block times, and block times can affect fleet assignment and fleet utilization. ATM can also influence congestion. Congestion indirectly impacts fleet planning by raising concerns about aircraft size. Thus, ATM influences on airline costs and revenues may be significant. Longer-term consequences of ATM changes and their influence on airline planning must be more clearly understood in order to assess more fully the benefits of ATM improvements.

VI. Conclusions

The research conducted to date confirmed that the aviation community as a whole has a limited understanding of the nature and extent of the impact of the ATM system on airspace user economic performance. Most of what is understood focuses on direct operating cost impacts associated with a specific operational problem or a specific ATM system enhancement. Moreover, there is often little quantitative information on the operational benefits of improvements that can be used to estimate the resulting economic benefits.

It appears that the ATM system does have broader effects beyond short-run direct operating costs, including impacts on revenue and long-term costs. While airline revenue is not a first order concern of ATM system planning and management per se, revenue is a significant concern to airlines and drives many decisions related to their use of the ATM system. As such, revenue concerns do affect, albeit indirectly, ATM system demand, system performance, and future requirements. Revenue, of course, also clearly and directly affects airline economic performance. It is a key factor that must be understood. Some of the broader effects on airline economics are understood in concept by some and not at all by others, including some in the airline industry; and few have systematic methods for estimating many of them.

Quantitative methods can and have been developed to estimate these broader effects. Much of the data required to develop and apply first order versions of such methods are available in the public domain. However, such methods need to be developed and validated collaboratively with one or more large airlines to draw upon their broad, deep, and specific airline management and operations knowledge and data.

References

[1] *Airline Service Quality Performance (ASQP) Reports of 14 CFR Part 234 (On-Time) Data,* Office of Airline Information, Bureau of Transportation Statistics, U.S. Dept. of Transportation, Washington, DC.

[2] *Air Carrier Financial Statistics—Form 41 Financial Schedules,* Office of Airline Information, Bureau of Transportation Statistics, U.S. Dept. of Transportation, Washington, DC.

Chapter 8

Effects of Schedule Disruptions on the Economics of Airline Operations

Zalman A. Shavell*
MITRE Corporation, McLean, Virginia

I. Introduction

SEVERE disruptions in the National Airspace System (NAS) significantly affect the economics of airline operations. Thunderstorms, blizzards, equipment outages, and other unscheduled, erratically occurring disruptions wreak havoc on air carriers' ability to fly published schedules, causing them to lose passenger revenues and incur additional costs. Because hubs function effectively only when passengers are able to make scheduled connections and are often located at congested airports, the effects of disruptions are magnified when hubs are affected. These effects become most visible when the disruptions occur during peak arrival or departure pushes. Delays, cancellations, and diversions are the most visible evidence of the effects of these disruptions on the airlines. Generally, each of these results in aircraft and crews being out of position relative to planned itineraries. Passengers are inconvenienced as arrivals are delayed and scheduled connections missed. As a result, an airline may become responsible for the cost of alternative transportation, lodging, food, and, if the delay is sufficiently long, a cash payment to compensate the traveler for any inconvenience.

This chapter has resulted from work performed by the MITRE Corporation's Center for Advanced Aviation System Development (CAASD) for the Federal Aviation Administration (FAA). It is but one element of a multiyear effort that is focused on the impact of air traffic management on the economics and performance of airlines.

The research reported in the first section of this chapter provides an overview of the types of disruptions that affect air carriers and the challenges these disruptions pose for them. Two case studies follow. The first case study estimates the cost to the affected airlines of a specific event in the northeast United States in the fall of 1998. (In the context of this chapter, the term "event" generally refers to weather, mechanical failure, and other causes of disruptions to

---

Copyright © 2001 by the MITRE Corporation. Published by the American Institute of Aeronautics and Astronautics, Inc., with permission.
*Senior Business Analyst. Senior Member AIAA.

airline schedules.) Estimates include the primary and secondary effects of delays, diversions, and cancellations on airline costs. The primary effects of a delay include its direct consequences on the affected flight (e.g., crew costs, tickets for delayed passengers on competitors' flights, etc.). The secondary effects of a delay reflect the consequences of the aircraft not being able to meet the requirements of its schedule (e.g., flights on additional legs scheduled for the aircraft are delayed, crew is not able to connect with other aircraft as scheduled, passengers miss connections, etc.).

The second case study aggregates the costs of disruptions for the domestic operations of the 10 largest U.S. airlines, those with at least 1% of total domestic scheduled- service passenger revenues, for 1998. The results of that study include an estimate of the magnitude of the effects of severe disruptions above those of the more minor disruptions that are a part of normal day-to-day operations.

The fundamentals of airline operations are no different from those of any other firm operating in an open market: control costs to earn a profit within the price structure imposed by the market place. On the cost side of the equation, each firm faces two categories of cost, fixed and variable. Fixed costs are those that managers cannot change in time to affect the outcome of events. On the other hand, management is able to control variable costs to shape events to their liking.

Manufacturing firms and some service providers, when confronted with disruptive events, are able to react to preserve the uninterrupted flow of product and service to their customers. The customers of airlines, on the other hand, are either trying to board or are already on the airplanes when flights are canceled, delayed, or diverted. There is nothing airlines can do to avoid inconveniencing their customers; their only option is to mitigate, as best they can, the resulting adverse consequences. Regardless of what actions they may take, the consequences of these events are such that the airlines can do little that will correct all damage to their competitive positions that results from the customers' dissatisfaction. Confronting largely uncontrollable events that impose additional costs while, at the same time, directly affecting service quality and competitive market position is a problem that is particularly applicable to airline operations.

The airline industry sells their ability to transport passengers and freight safely to their intended destinations. Their schedule embodies their service. Events of varying severity compromise their ability to fly their schedules. At a basic level, airplanes break, crews fail to report, caterers fail to keep their delivery schedules, and the National Airspace System becomes congested, all of which compromise the ability of airlines to meet their schedule commitments in the short term.

Relatively small numbers of flights are delayed or canceled for the majority of these events. Inconvenience to airline customers is minimal. Most importantly, the airline industry recognizes that these occurrences, at a minimal level, are inherent to airline operations.

Some disruptions are so severe that they cause flight operations in an entire region of the country to be curtailed. Weather is the primary cause of these severe disruptions, although air traffic control outages, labor actions by unions, the closing of runways, and airport construction can also severely affect the flow of air traffic over a wide area. When faced with these severe disruptions, even if only one airport is affected directly, the effect ripples and relatively large numbers of flights may be delayed, canceled, or diverted. The integrity of bank operations at hub airports is disrupted, and passengers and freight miss connections. When such severe disruptions occur, they have major economic implications for the affected airlines. Estimating the costs of severe disruptions is the focus of this chapter.

II. Scope of Disruptions

The range in the magnitude of the disruptions that can affect airline operations is considerable. A single flight may be delayed a few minutes to correct a false warning in the cockpit, or numerous flights from all over the country may be canceled because a storm has closed a major hub in the northeast. Events such as a severe line of thunderstorms may curtail traffic at a single airport, or a major snowstorm can shut down traffic at major airports in an area that encompasses several states. A strike can effectively shut down an airline; and, if it is one of the major carriers, the entire national air transportation system can feel the effects. On relatively rare occasions, an airline will hold one or more flights because a particularly critical connecting flight has been delayed. Accommodation of large groups traveling together or a desire to avoid repeating a missed connection that has recently occurred is among the factors that could motivate such a self-imposed delay.

For example, if an airport is severely impacted by a storm, the FAA's Air Traffic Control System Command Center (ATCSCC) may institute ground stops on traffic heading toward that particular airport, disrupting the flow of traffic over a wide geographic area. The result can be an overcrowding of gates and parking spaces at the airports not directly affected by the storm but serving as points of departure for flights bound for the weather-impacted airport. When ground stops are lifted, if the aircraft are released at a time that conflicts with a later departure bank, they can cause queues to build up for departure runways. The arrival volume of released flights at the airport that necessitated the ground stop could, in turn, challenge its capacity.

Delays and cancellations are the most frequent results of disruptive events. For the airlines these are the least damaging in terms of creating schedule disruption and customer dissatisfaction. Aircraft are at least located at a point that was included in their itineraries, and the crews assigned to them are likely to be available for service. If the delayed or canceled flight is at a hub, the airline is likely to have alternatives on its own system for accommodating affected passengers and freight.

On the other hand, when a flight is diverted, the affected passengers and crew are likely to find themselves at an airport where the airline has substantially fewer rebooking or substitution options. First, the airline may have few, if any, operations in and out of the airport to which the flight was diverted. Passengers either have to be transferred to another carrier or be accommodated until their flight is able to proceed to its original destination. Second, the airline may have neither gates nor service facilities at the alternate airport. Consequently, it may find itself paying higher than usual prices for fuel, catering services, and access to gates at which it can off-load passengers rather than keeping them confined within the aircraft.

III. Alternatives Available to the Airlines for Handling Disruptions

When an event has not been forecast, affected airlines have no choice but to react to the event. However, when some notice of an impending event is available, airlines are able take preemptive actions designed to mitigate the effects of the disruption. For example, they may choose to cancel flights in the face of a snowstorm and move aircraft from the affected airports in order to have them available to serve stations that will not be affected directly by the adverse weather.

Generally, airlines have demonstrated a marked preference for delays and cancellations over diversions. Diversions are the most costly of the three alternatives in terms of both direct cost and the creation of ill will among the passengers. However, assuming that the FAA has not imposed a ground stop, some airlines will choose to launch flights in the face of significant forecast delay in the hope that the weather will have lifted by the estimated time of arrival and that the aircraft will be permitted to land. If the aircraft is unable to land at its destination, it must be diverted to an alternate airport, causing the airline to incur the costs of a diversion.

Airlines also develop strategies for dealing with an actual or anticipated ground delay program. The ATCSCC clears flights for specific takeoff times, but the airlines retain the option of substituting another flight for one originally scheduled for each slot. Using this flexibility, they will use a variety of business and operating factors to choose to operate those flights that the airline perceives to be in its best interest.

Within limits, airlines also have the ability to modify the itineraries of the aircraft they have available. In many cases, the aircraft available are sufficiently similar to those scheduled for a flight and one can be substituted easily for another. When this is true, flights may be canceled so that the aircraft assigned to them may be used for other operations the airline deems to be operationally advantageous. Some flights, as a result of an airline's analysis of its options, may be intentionally delayed rather than canceled in the belief that an aircraft to operate them will be available from the pool of delayed incoming flights.

The options available to the airlines for dealing with disruptive events are constrained by crew availability and aircraft maintenance schedules. Crews approaching their limits with respect to duty time may not be able to continue with a flight that has been delayed or diverted. In such a case a "fresh" crew may have to be called upon to operate the diverted aircraft. Similarly, a delay or diversion can take an airframe out of service if it is operating close to its limits in terms of required scheduled maintenance. If an aircraft due for maintenance is diverted, it may have to "dead head" to the maintenance facility rather than reaching it at the end of a scheduled revenue operation. Crew time and the maintenance schedule play a vital role in the ultimate delay/cancellation decision.

IV. Cost Implications of Disruptions to the Airlines

Disruptions to schedules affect airline costs through three distinct paths. First are the direct costs such as those that are incurred for additional fuel, crew time, and maintenance. Second are passenger-related costs that include such varied items as meals and lodging for individuals subject to delay or payments to other airlines and fare revenue lost when passengers switch to competitors. Third are secondary costs such as the ill will created in passengers that are subject to delays and the lost revenue from trips that are canceled.

Data needed to estimate some of these cost items are available and data for others are not. Secondary costs are difficult to measure; and, therefore, not all airlines attempt to estimate these types of costs when determining the cost impact of delays or diversions. Placing a value on ill will presents problems for cost estimation that are particularly difficult to resolve, whereas estimating the revenues lost to other airlines by passengers transferring to them is not as odious.

Each event that disrupts airline schedules, regardless of its magnitude, is unique. Costs to the airlines of disruptive events vary widely. Even within a class of events, for example, snowstorms, the costs will vary greatly depending on the specifics of such items as the airports involved, the airlines and the type of aircraft they use, and the magnitude and duration of the storm.

However, although each event is unique, case studies can provide insights that are useful. Although the results of case studies are not scientifically rigorous, they do help identify variables and relationships among them to be considered, creating a higher level of understanding.

Disruptive events influence airline economics at two levels, the individual airline and the industry. At the level of individual airlines, losses to one may be gains for another, and the loss of revenues to the industry as a whole is likely to be minimal. For example, when an airline needs to cancel a flight, passengers may be ticketed on competitors' flights so that only the smallest inconvenience is endured by each. On the other hand, losses due to interruptions to flight operations computed at the industry level are not likely to be recoverable. Trips will be canceled rather than rescheduled, and fixed costs incurred will not be balanced through generation of additional revenues that could cover them.

Therefore, research reported here was conducted at two levels. One, a particularly severe snowstorm event that affected operations in the northeast, was analyzed to investigate the pattern by which an irregular operations event evolves and the economic effects that it has on the affected airlines. The second example analyzes the magnitude of the economic effects of disruptions to airline operations when they are aggregated to the national level. The analysis shows that there is a minimal, inescapable level of disruption that airlines, in the aggregate, view as an ongoing fixed cost of operations.

V. Snowstorm Event at Boston

The disruptions to airline operations that centered on Boston, Massachusetts, on 8–10 October 1998 were chosen as the first case study. Over a period of three days, the area affected expanded from the immediate area of Boston to cover a substantial portion of the country. Before it was over, ground delays had been instituted for a sizable portion of the United States. Arrival data for these three days clearly show the pattern of delays that emerged as the event progressed, and that of the recovery that took place on the third day. These data also show how the lack of scheduled activity during the night provided a respite during which the airlines were able to partially recover their schedules.

Figure 1 shows variations in delays, cancellations, and diversions at Boston on 8–10 October 1998 as a severe weather pattern that affected flight operations first developed and then dissipated. They began during the afternoon and evening of 8 October, peaked during the afternoon of 9 October, and dropped off during 10 October. The vertical bars in Fig. 1 show the number of flights that were unable

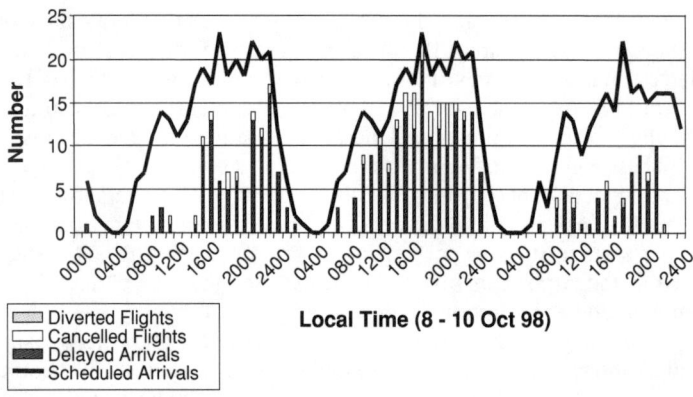

Fig. 1 Delays, cancellations, and diversions for Boston.

to meet their scheduled arrival time during each hour of the three days analyzed. For example, at the height of the disruption during the early evening of 9 October, virtually all incoming flights were delayed.

Three aspects of this chart are of particular interest. First, note that no flights were diverted until late (one each in the hours beginning at 1600, 1700, and 2100 hrs) on 10 October. This reflects a definite hesitancy on the part of airlines to accept diversions as they are the most costly of the options for dealing with the disruptive events. Second, delays were by far the most common effect of the disruptive events. Third, the absence of scheduled flights during the night provided the airlines a chance to recover from the effects of disruptions that occurred the previous day. As the curve shows, large numbers of flights scheduled to arrive in Boston during the day were delayed. Many of these were able to land in the evening during the lull (not shown) when few flights were scheduled and any queues built up during the day could be dissipated.

Note also the cumulative pattern of delays, cancellations, and diversions shown in Fig. 2. Of particular interest is the significant slope of the curve on 9 October (A) showing the increase in delays as the intensity of the weather activity grows. Conversely, the comparatively flat segment of this curve on 10 October (B) demonstrates the decreasing frequency of delayed flights as the recovery from irregular operations evolved.

As noted, the effects of the disruptive activity were widespread, covering a substantial portion of the northeast coast. Although this case study focuses on the events at Boston and the costs the airlines operating there incurred, other major airports were involved. For example, two of the airports affected by the same weather event were Philadelphia, Pennsylvania, and Newark, New Jersey. Although not as great as the cost burden imposed on airline operations at Boston, it was substantial at each of the other two airports, demonstrating how substantial and widespread the effects of a large weather system can become.

Table 1 shows the cost categories for the losses incurred at Newark and Philadelphia in addition to those experienced at the focal point of the disruption, Boston. In this calculation, only crew costs are attributed to the delay. Data for each of the three days of the event are shown. Clearly, the costs incurred on 9 October at Boston are the largest for one day at any one of the three locations for which data

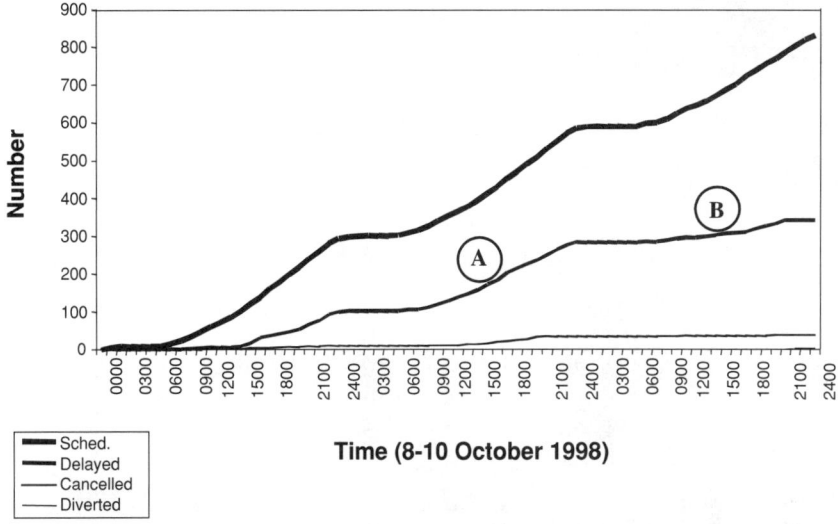

Fig. 2 Cumulative cancellations, delays, and diversions.

are available. Again, the data show the preferences of the airlines for choosing delays and cancellations rather than diversions.

A more refined computation of delay costs, which includes costs such as fuel in addition to crew costs and estimates the revenue lost due to a delay, raises the estimated total for Boston alone from just over $1 million U.S. dollars (USD) to about $2.3 million USD. This is a more realistic measure of the cost of disruption and reflects more of the true magnitude of the costs that irregular operations can impose on the airlines operating from an affected airport.

Because of the severity of the implications of irregular operations, airlines and the FAA take proactive steps whenever possible to mitigate the losses suffered. In doing so, they take as much of their destiny into their own hands as is possible and often are able to mitigate the overall effects of disruptions on operating costs. On the other hand, if the airlines are in a position where they can only react to events as they occur, their options are much reduced, and, therefore, the costs of disruptions are likely to be higher than they could have been otherwise.

VI. Aggregated Costs of Disruptive Events

For the year 1998 the total estimated direct costs to the airlines of irregular operations incurred by the 10 U.S. airlines that report Airline Service Quality Performance (ASQP) data to the Department of Transportation were $1.826 billion USD. Cancellations and delays dominated with total costs of $858 million USD and $909 million USD, respectively, and diversions imposed an additional $59 million USD in costs.

Calculation of costs of delays, cancellations, and diversions for the nation as a whole during 1998 was somewhat easier than for the events at a single airport. For example, no allowance for lost revenue was needed as it was assumed that individuals who were not able to complete their travel plans on time on the airline originally

Table 1 Initial estimated irregular operations losses (USD)

Boston		Newark		Philadelphia	
	Costs, K		Costs, K		Costs, K
8 October 1998		8 October 1998		8 October 1998	
10 cancellations	$116	10 cancellations	$116	6 cancellations	$70
No diversions	$0	2 diversions	$45	1 diversion	$22
9514 min of arrival delay	$127	10337 min of arrival delay	$138	8462 min of arrival delay	$113
	$243		$299		$205
9 October 1998		9 October 1998		9 October 1998	
25 cancellations	$290	9 cancellations	$105	11 cancellations	$128
No diversions	$0	1 diversion	$22	3 diversions	$67
23852 min of arrival delay	$318	15187 min of arrival delay	$203	14380 min of arrival delay	$192
	$608		$330		$387
10 October 1998		10 October 1998		10 October 1998	
4 cancellations	$46	3 cancellations	$35	6 cancellations	$70
3 diversions	$67	1 diversion	$22	No diversions	$0
4015 min of arrival delay	$54	3212 min of arrival delay	$43	2188 min of arrival delay	$29
	$167		$100		$99
Estimated total cost	$1,018		$729		$691

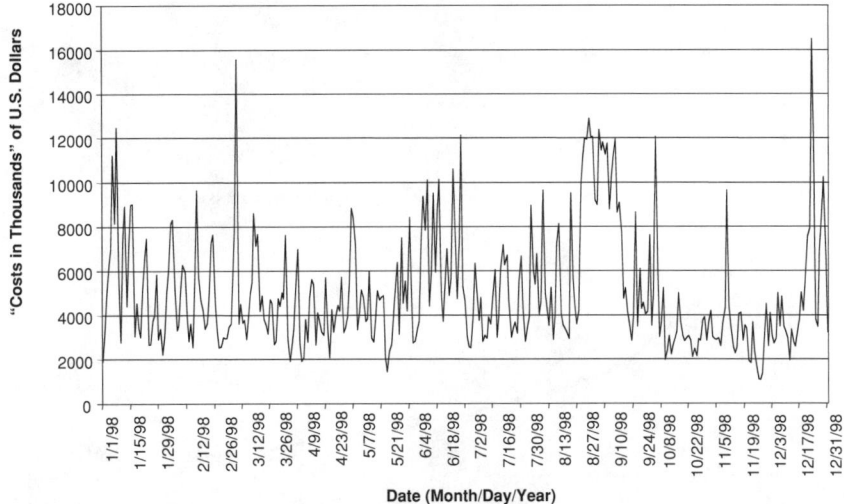

Fig. 3 ASQP cost by day.

chosen would make the trip on another. The national data when examined for the year also provided insights that were not apparent from the Boston case study.

The plot of total costs due to disruptions incurred by the airlines reporting ASQP data is shown in Fig. 3. It shows that disruptive events were spread throughout 1998. Two periods during which the costs of disruption were relatively high and sustained can be noted: mid-June to early July and late August to mid-September. The latter is the more severe of the two and was largely attributable to a labor action by Northwest Airlines' pilots. Sharp peaks are also visible in February and December. However, this curve in itself does not imply that these periods are indicative of a systemic propensity of disruptive events to cluster during specific periods of the year.

The data also show that, in general, days that experienced particularly severe disruptive events are followed by ones on which the airlines are able to deliver on their schedules. Rarely are two or more disrupted days grouped together. This suggests that, when faced by severe disruptions, the airlines tend to forgo short-term solutions aimed at fixing the immediate problems and focus, instead, on positioning resources to provide as complete a schedule as possible on the following day. On the other hand, when disruptions are less severe, there is reason to believe that the airlines will attempt to recover the current day's schedule rather than deferring its recovery to the following day.

When, as shown in Fig. 4, costs are plotted by day in decreasing order of magnitude, the picture becomes somewhat clearer. The slope of the curve decreases shortly after crossing the 90th percentile line, pointing to the fact that relatively few events account for a disproportionate share of the total costs attributable to disruptive events. The leftmost portion of the curve clearly suggests that a relatively small number of days were subject to disruptions that were substantially more costly than most. In fact, the 20% of the days with the highest cost from irregular operations in 1998 averaged about 27% above the average for the year. Further,

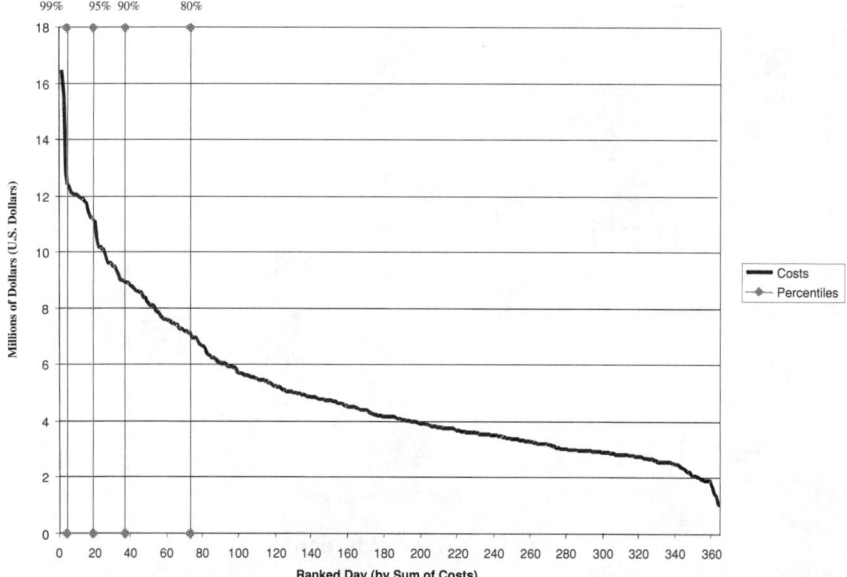

Fig. 4 Rank order of the sum of ASQP carrier costs (1998).

the worst 5% of the days (19 days) accounted for 12.8% of the total annual cost of disruptions to the ASQP airlines in 1998.

The right side of the curve also suggests the existence of a baseline of low-level disruptive events that constitute an inescapable element of airline operations. Included in this baseline are routine mechanical failures, in-flight medical emergencies that result in diversions, unanticipated congestion at airports and en route, and failure of aircrew and other personnel to report as scheduled. Specifically, the curve suggests that 20% of the days (73 days) can be taken as forming the baseline of the cost of disruptions to the ASQP airlines. About $3 million USD per day comprise the unavoidable cost of delays, diversions, and cancellations inherent in the system.

Although the data that have been presented are for 1998 only, Fig. 5 shows that 1998 was not particularly unusual in terms of the events that disrupt airline

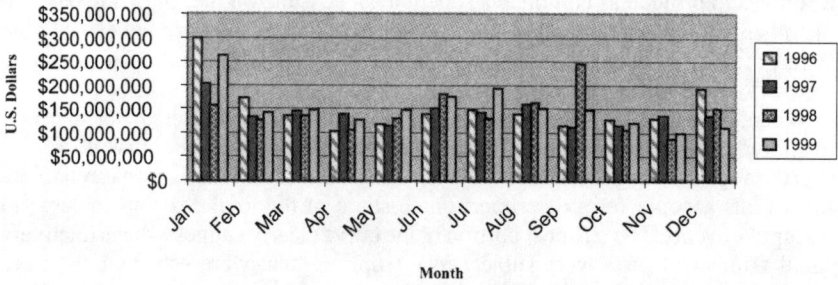

Fig. 5 Total disruption costs (major airlines).

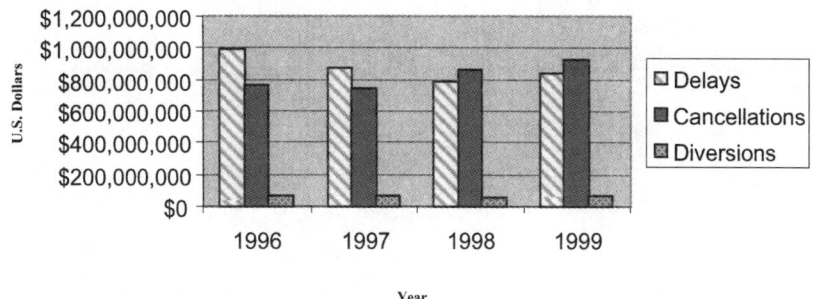

Fig. 6 Annual disruption costs by source (major airlines).

operations and the costs for the airlines that result. January in each of the four years for which data are plotted is the worst month of the year. Again, the large cost shown for September 1998 is attributable to the labor problems of Northwest Airlines. More importantly, the consistency of the values shown for each of the months supports the conclusion that the findings of this study are representative of the costs airlines incur when faced with disruptive events.

In addition, as shown in Fig. 6, the effects of disruptive events on airline operations have also been consistent across at least four years. The distribution of delays, cancellations, and diversions is constant for the years 1996 through 1999, further supporting the conclusion that the results presented here have broad applicability.

VII. Conclusions

Airlines, more than most businesses, are subject to external events that affect both their costs and the attractiveness of their product, their schedule, in the market place. These costs are substantial and difficult to estimate.

The unpredictability and uncontrollable nature of most disruptions make it difficult to mitigate their effects. Variations in cost and revenue structures among air carriers make it difficult to estimate their economic impact. However, as shown in this chapter, the economic impact of severe disruptions is very significant. Therefore, air carriers, the FAA, and other aviation organizations should continue developing means for providing more advance notice of pending disruptions, sharing information about existing disruptions, and collaboratively making decisions regarding air traffic affected by disruptions.

Chapter 9

Modeling an Airline Operations Control Center

Nicolas Pujet* and Eric Feron[†]
Massachusetts Institute of Technology, Cambridge, Massachusetts

I. Introduction

A KEY challenge for major U.S. airlines is to achieve efficient information management to alleviate the impact of unforeseen schedule disruptions. In addition to planning up to 2500 flights per day, the operators in the airline operations center (AOC) of a major airline adjust in real time the movements of the hundreds of aircraft and thousands of crewmembers of the airline to minimize costly delays and cancellations, while complying with complex contractual and maintenance routing constraints.[1] The benefits of the evolution of the National Airspace System (NAS) toward more user operational flexibility (as embodied in the free flight concept) will depend in part on the ability of the airlines to make operational decisions quickly.[2] Thus the economic impact and real-time nature of AOC decisions motivate an in-depth analysis of the performance and dynamic characteristics of the AOC. In particular, such an analysis would help evaluate the performance limits of the current AOC and predict the impact of future AOC decision aids and processes (such as collaborative decision making).

This chapter presents an approach to modeling the AOC as a network in which each operator is a queueing server. Queueing network models were initially developed for telecommunication systems[3] and have been analyzed and applied extensively in academic research[4] and industry, in particular in the context of multiprocessor computers[5] and flexible manufacturing systems.[6] However, to our knowledge they have not been applied to modeling a manned real-time command and control center.

Section II of this chapter introduces the real-world workings of the AOC and discusses the corresponding model structure and hypotheses. Section III details how the model parameters are identified using a combination of on-site observations, archived operational data, and computer transactional data.

Copyright © 2000 Air Traffic Control Association Institute, Inc. Reprinted from the Air Traffic Control Quarterly, Vol. 7, No. 4, by permission of the Air Traffic Control Association Institute, Inc.

*Formerly with the International Center for Air Transportation, Department of Aeronautics and Astronautics.

[†]Associate Professor, Laboratory for Information and Decision Systems, Department of Aeronautics and Astronautics. Senior Member AIAA.

II. Modeling Structure and Hypotheses

The AOC of a major airline is composed of 50–100 operators working in three shifts around the clock. These operators are responsible for dispatching flights (i.e., preparing and filing flight plans and following flights from departure to destination), and for adjusting the airline schedule (including the flight schedule, departure slot assignments, aircraft assignments, and crew assignments) in response to external perturbations such as thunderstorms, airport capacity restrictions, equipment failures, etc. Most of the real-time activity in the AOC involves the following classes of operators:

1) System operations controllers (SOCs) oversee the operations of the airline and are responsible for the major adjustments to the schedule, such as cancellations.

2) Aircraft routers are specialized in monitoring and adjusting the routing of aircraft through the network of flights while complying with aircraft maintenance routing constraints.

3) Crew schedulers monitor and adjust the assignment of crew members (pilots and flight attendants) while complying with contractual constraints.

4) Dispatchers are responsible for preparing flight plans (i.e., deciding on fuel loads, alternate destinations, etc.), releasing flights, and following en route flights.

5) Air traffic control (ATC) coordinators interface between the AOC and the Federal Aviation Administration (FAA), particularly during ground delay programs.

The control authority of the AOC ranges from a few hours before scheduled flight departure to one hour before departure (see Fig. 1). Within one hour of departure, most of this control is handled locally by the airline's station control center (SCC) at the departure airport.

For modeling purposes, it is hypothesized that each operator is working on only one task at any given time, and therefore each individual operator acts as a single multiclass G/G/1 queueing server (see Fig. 2).[7] The "customers" of this queueing server are tasks waiting to be worked on by the operator (e.g., writing a flight plan, finding a replacement for an aircraft that has broken down, etc.). Observations made in the AOC of a major airline have confirmed this hypothesis. The identification of the queueing characteristics of each operator is addressed in Section III. Because each operator in the AOC can be modeled as a queueing server, the AOC can be modeled as a network of queueing servers.

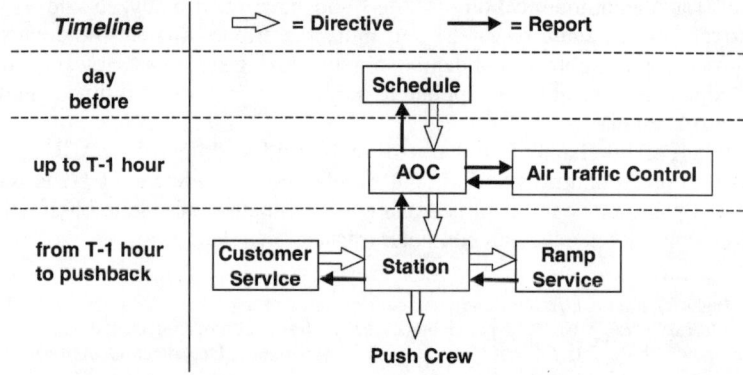

Fig. 1 Airline operations timeline.

MODELING AN AIRLINE OPERATIONS CONTROL CENTER 129

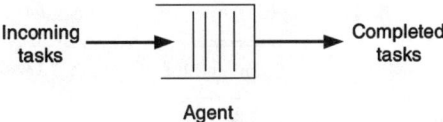

Fig. 2   Modeling an operator in the AOC.

The order in which various tasks are performed by the operators is quite constant, so that accurate process maps can be charted to represent the order in which these tasks occur. Figure 3 shows an example of such a process map. Observations in the AOC and extensive interviews conducted with the operators have also confirmed that a small set of tasks and processes is sufficient to give an almost exhaustive description of AOC activities. Each process is triggered by an external event such as mechanical failure of an aircraft, announcement of a ground delay program by the FAA, etc. Note that some queueing occurs any time a task is given from one operator to another. Table 1 presents the processes that occur under irregular operations, and Table 2 lists the background processes that take place on a more routine basis (regardless of weather and traffic conditions).

It is important to note that a tractable and useful model of the AOC can be obtained by focusing on the states and transitions of the queueing servers representing the AOC operators. The model does not need to include the position, status, and scheduled assignments of the hundreds of aircraft and thousands of crewmembers, nor the hundreds of operational constraints and decision variables that must be considered by the AOC operators. This simplification occurs because the model does not account for the slow AOC feedback loops linking current decisions to future inputs (in the context of this model, "slow" means taking more than a few hours). Instead, the AOC inputs are modeled by deterministic or stochastic arrival processes whose parameters are chosen to match statistical measurements of the real system (see Fig. 4 and Section III). This simplification is based on the observation that operators do not typically consider potential long-term effects of their decisions (e.g., the evening effects of a morning adjustment to several aircraft and

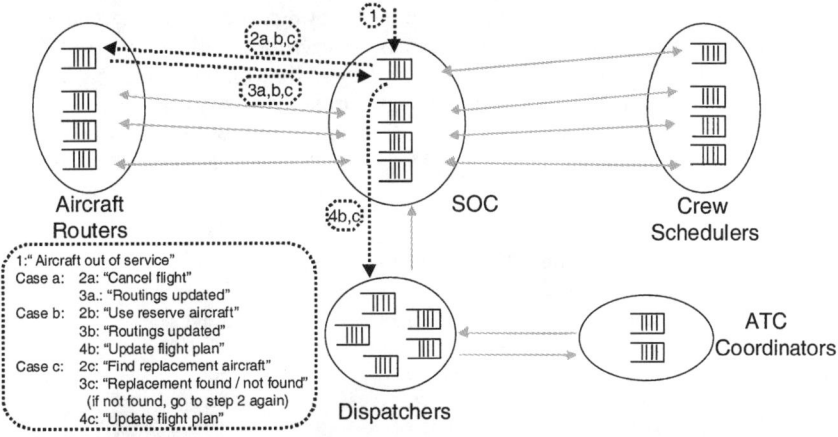

Fig. 3   An irregular operations process: recovery from mechanical failure.

Table 1   Irregular operations processes

Process	Launched by	Operators
Ground delay program	Message from ATC	SOCs Aircraft routers Crew schedulers Dispatchers ATC coordinators
Cancellation plan	Severe weather forecast	SOCs Aircraft routers Crew schedulers
Recovery from long delay	Expected delay > 30 min	SOCs Aircraft routers Crew schedulers
Recovery from mechanical failure	Aircraft breakdown	SOCs Aircraft routers
Crew problem	Scheduled crew unavailable	Crew schedulers
Flight plan revision	Change of aircraft or crew	Dispatchers
Severe weather pilot report	Call from pilot	Dispatchers
Airborne holds and diversions	Pilot reporting airborne hold/diversion or airport closure due to bad weather	Dispatchers ATC coordinators
Assist flight	Flight lost contact with ATC	ATC coordinators

crew routings), because these effects will be influenced and thus rendered unpredictable by many highly uncertain variables such as future decisions and weather events. For example, the model does not represent how a change in a particular aircraft routing affects the frequency of aircraft routing checks later in the day. Instead, the statistics of aircraft routing checks are collected in the real-world AOC and used to build a stochastic model, thus embedding the slow feedback from current decisions to future taskload as a source of stochastic noise.

The queueing network model of the AOC is parameterized as follows:

1) Routing topology: the routes and sequences that different tasks take among the various operators.

Table 2   Background processes

Process	Launched by	Operators
Revise flight plan	Schedule	Dispatchers
Check weather conditions	Weather forecast updates	Dispatchers
Check aircraft routing	Schedule	Aircraft routers

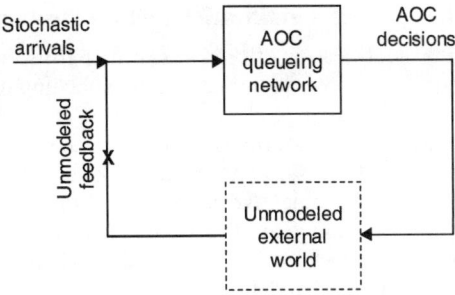

Fig. 4 Slow feedback replaced with stochastic arrivals.

2) Customer arrival process: the frequency of occurrence of background and irregular operations processes.

3) Task priorities and service times: how different tasks will be prioritized by each operator, and the length of time required to complete a given task.

4) Branching probabilities: the probability that a process will follow one of several alternative task sequences.

Example 1 illustrates the role of these parameters for a specific process.

Example 1: Model Parameters for Mechanical Failure Process: Consider the AOC process initiated by the mechanical failure of an aircraft, as modeled in Fig. 5 with standard queueing network notations.[8] The occurrence of mechanical failures and other processes is assumed to occur according to independent Poisson processes with the following parameters:

λ_{MF}: Poisson rate for aircraft mechanical failures
λ_S: Poisson rate for other background SOC tasks
λ_R: Poisson rate for other background aircraft router tasks

The branching probabilities are denoted as follows:

p: probability that the SOC calls on the aircraft router to help solve the problem
q: probability of a new iteration (if a proposed plan is infeasible)

The task service times are denoted by the random variables S_1 through S_5. If the SOC or the aircraft router typically set aside pending background tasks to address the mechanical failure, then the preemptive service discipline of each operator is also important.

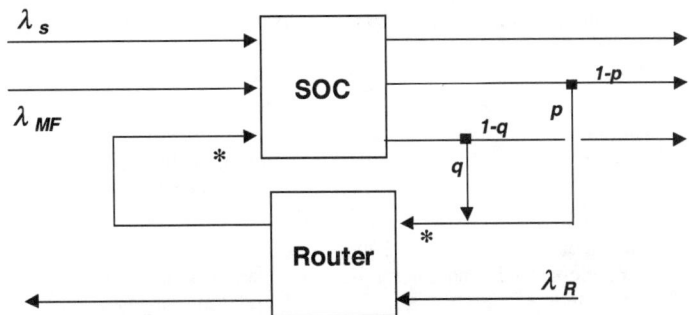

Fig. 5 Queueing network model for mechanical failure recovery process.

III. Model Identification and Calibration

The objective of the identification effort was to obtain from the real AOC system some statistics on corresponding parameters of the queueing network model. An initial intensive on-site study was conducted, including interviews with operators and manual measurements of service times, in order to build a baseline quantitative model of the current AOC processes and dynamics (Section III.A). However, it became clear that a systematic, objective calibration effort was needed in order to obtain reliable and useful results from the model. This led to the collection and analysis of operational and transactional data from the main AOC computer system (Section III.B). These transactional data were also combined with Airline Service Quality Performance (ASQP) data to examine the relations between scheduled flight operations, AOC processes, and actual flight operations (Section III.C).

A. Identification Through On-Site Manual Measurements

The most straightforward approach to obtain identification data is to observe the operators on-site, conduct focused interviews, and make manual measurements. Table 3 illustrates this approach by showing some of the process parameters of the airborne holds and diversions process. These values were averaged over a small number of samples or were estimated by the dispatchers during focused interviews. Similarly, a first estimate of service times and queueing disciplines was obtained

Table 3   Airborne holds/diversions process parameters

Parameter	Value	Comments
Time to read/hear pilot report of first holding pilot	30 s 2 min	If pilot sends electronic message If pilot calls
Time to build list of future affected flights	60 s	Need only one command to print all flights
Minimum interarrival time of affected flights	3 min	Typical of busy hub airport at rush hours (e.g., San Francisco)
Maximum interarrival time of affected flights	10 min	
Maximum airborne holding time of each flight	20 min	Typical fuel reserve allocated by pilot/dispatcher to airborne holding
Minimum duration of event	10 min	Shortest time to get airport arrival process back to normal
Time to call an individual flight and give update	3 min	About 1 min to establish radio patch and 2 min to give update

by manual observations and interviews with AOC operators. Some standardized tasks were found to have a fairly constant service time (e.g., preparing a flight plan), whereas some other tasks were found to have a widely varying service time (e.g., finding a replacement aircraft after a mechanical failure). Queueing discipline rules were also inferred from direct observations and interviews. The priority of a task was found to depend on the estimated time until departure, and the type of market and route. International flights are generally given the highest priority, followed by hub-to-hub flights, with simple out-and-back flights given the lowest priority.

Note that this data collection methodology presents many shortcomings. The results of focused interviews may be biased by the perception that the operators have, or feel they should have, of their functions. Rare events such as major irregular operations, which could reveal much about the performance limits of the AOC, are rarely witnessed, and their size and complexity make it very difficult for a single on-site observer to capture significant data. The number of parameters required for a realistic simulation of the AOC is quite large; even without accounting for different weather situations or differences between individual operators of the same class, more than 120 parameters must be identified to analyze the processes in Tables 1 and 2. A rigorous identification of these parameters would require a large number of samples in many different representative conditions (e.g., regular day, bad weather day, etc.), and gathering sufficient samples by hand is largely intractable.

B. Identification Using Archived Historical Data

The difficulties involved with on-site manual data collection lead to the use of archived historical data. Data sources include operations data archived by the FAA and by the airlines, and operator transaction logs from the AOC computers.

Both the airlines and the FAA archive irregular operations data, e.g., data on cancellations or ground delay programs (GDPs). The statistics of occurrence of several types of irregular operations processes from Table 1 were estimated using these historical data. For example, the occurrence of GDPs is accurately modeled as a Poisson process. The parameter of this Poisson process, the duration of the delay program, and the magnitude of the delays were estimated for each hub of the airline from the list of all GDPs experienced in a given month of interest.

While operations data are quite useful, they only show the end results of AOC interventions. However, most of the AOC activity is computer based, and operator terminal transactions are continuously recorded in the AOC for auditing purposes. Until now, these logs have apparently never been used to study the dynamic aspects of information flows and decision making in the AOC. These transactional data contain the times at which each operator started or finished work on each task (but do not contain the time at which each task arrived in the operator's queue). These logs can be used to estimate process occurrences and priorities, task service times, and branching probabilities, as illustrated in the following example.

Example 2: Model Calibration Using Transactional Data: Transactional data were used to study the work patterns and service times of four dispatchers during the same period of time, an afternoon shift in March 1998. Dispatchers 1 and 2

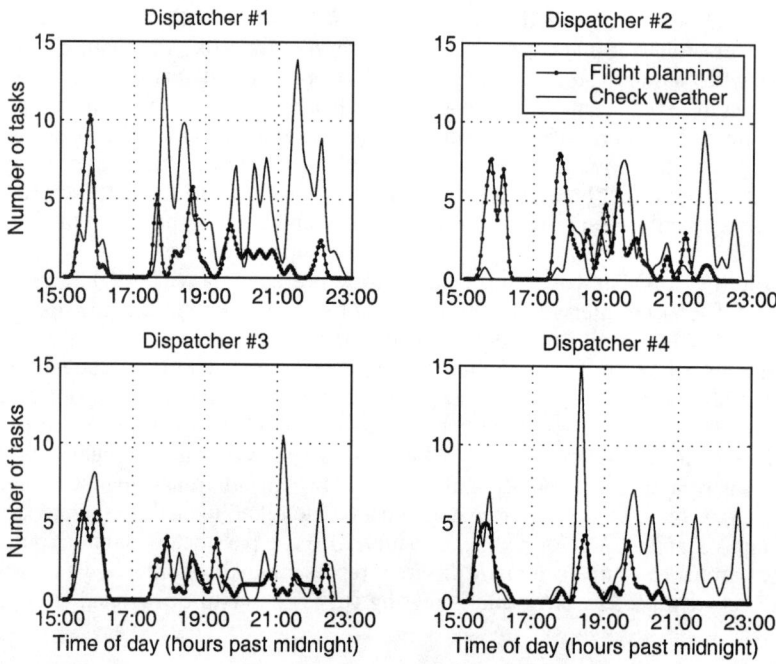

Fig. 6 Distribution of tasks over time for four dispatchers (no data available from 1625 to 1730 hrs).

were working on sectors in the western United States, while dispatchers 3 and 4 were working on sectors in the eastern United States (none of these dispatchers was working on high-frequency shuttle-type operations, which are organized differently).

The distribution of two dispatcher tasks over time is shown in Fig. 6. These charts display the distribution of the flight plan service start times over the course of the shift along with the distribution of weather checks. There appears to be a high frequency of weather checks and flight plans at the beginning of the shift (around 1530 hrs), as the dispatchers are trying to get a picture of the situation in their sectors. Other peaks in weather check frequency appear as dispatchers check newly available forecasts. The precise time of these peaks varies somewhat between dispatchers. These plots also show that the number of flight plans that are started during the clusters of weather checks (typically just after a new weather forecast is issued) is usually small, which indicates that the two activities are usually not mixed but rather are worked on in batches. This fact has been observed on site, but only qualitatively. Because flight plan tasks arrive in these dispatchers' queues regularly throughout the shift (following the airline schedule), it can thus be inferred that weather checks have priority over flight plan tasks. It appears that dispatchers 1 and 2, whose sectors in the western United States were affected by severe weather disruptions, spent more time checking the current state of flight plans in the beginning of their shift than dispatchers 3 and 4, and performed more frequent weather checks in the second half of the shift, alternating with batches of flight plan updates.

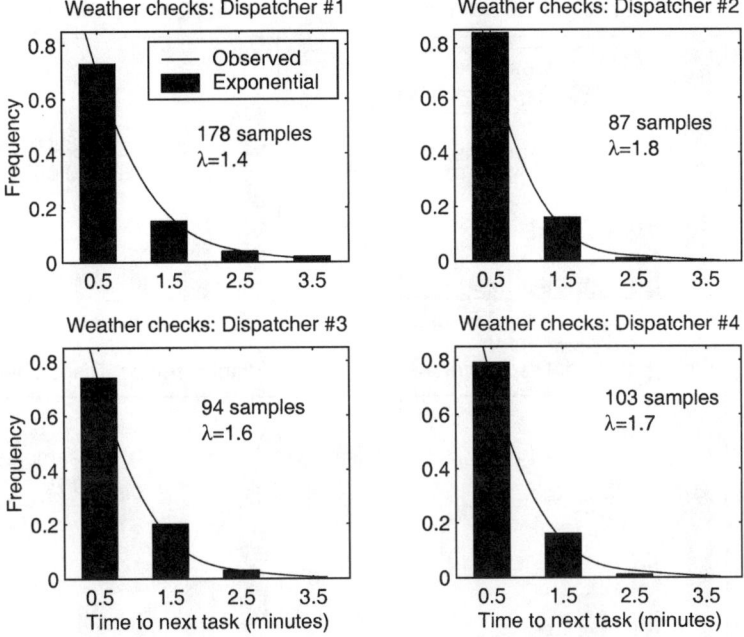

Fig. 7 Distribution of estimated weather-check service time.

In our current model, weather checks and flight plan tasks arrive according to a deterministic process; the rate of arrival depends on the time of day and the type of operations (severe weather day or routine day). Note that the charts give the times at which tasks are started (service commencement epochs), but not the rates at which these tasks come due (arrival rate). These rates could conceivably be obtained from other sources [respectively, the airline's meteorology department and the Official Airline Guide (OAG)], and then these charts could be used to estimate the dispatchers' workloads. This workload estimate could be used to estimate their availability to take over new tasks, and thus the potential benefits of decision aids and/or improved communications.

The computer transactional data can also be used to evaluate task service times under certain assumptions (namely, that the operator idle time is small). Figure 7 shows an estimate of the service times of the weather checks for the same four dispatchers. The observed distributions are approximated well by exponential probability distributions. The spread in each distribution could reflect differences in forecast complexity or criticality to the current operations. The parameters of the exponential distributions vary between 1.4 and 1.8. These variations could reflect different work habits among these four dispatchers. Note that dispatcher 1, who appears to have the fastest service rate, also performed twice as many weather checks during his/her shift than the other dispatchers did. This could be a result of the severe weather situation in his/her sector and/or of personal work habits. In general, the data set indicates that a good model of the service time for weather checks would be an exponential distribution with parameters between 1.4 and 1.8, depending on the weather conditions and the individual operators.

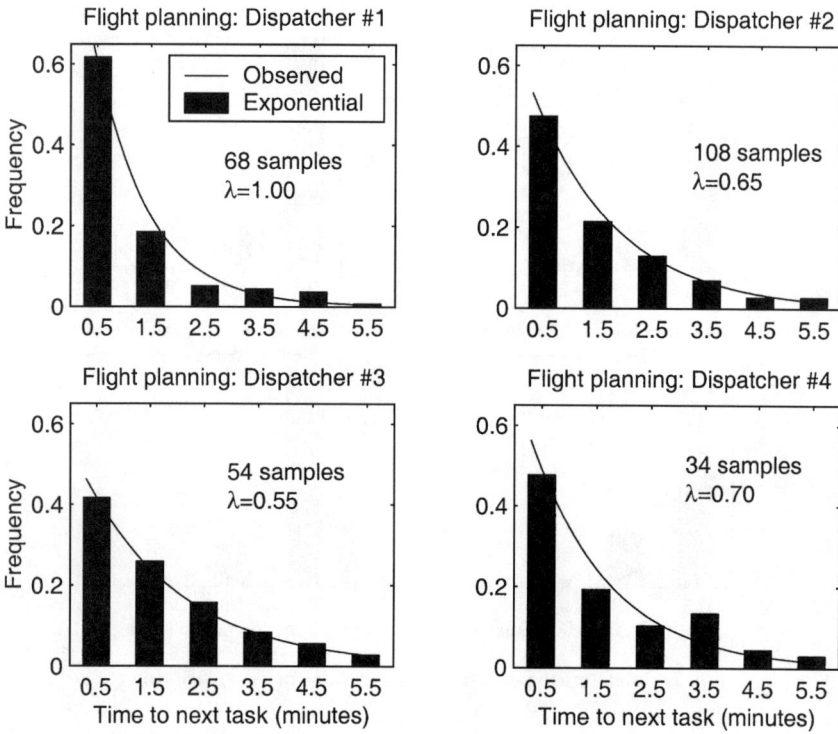

Fig. 8 Distribution of estimated flight-planning service time.

Figure 8 shows the distribution of an estimate of the service time for flight planning that was observed for the same four dispatchers. These observed distributions are also approximated well by exponential distributions, with parameters that vary between 0.55 and 1. Dispatcher 1 again demonstrates the fastest service rate, but dispatcher 2 processed more flight plans at a slower rate. Again, this could be a result of the weather conditions and/or of personal work habits. In general, the data set indicates that a good model of the service time for flight planning would be an exponential distribution with parameters between 0.55 and 1, depending on weather conditions and the individual operators.

The analysis given in this example was used to compile similar statistics for all the other AOC operators and tasks to obtain an extensive numerical calibration of the model.

C. Combining Transactional and ASQP Data

The transactional data were also combined with the ASQP database. This database consists of ACARS reports from the scheduled domestic flights of the 10 major U.S. carriers (except Southwest Airlines, which collects these data manually) and is publicly available.[9]

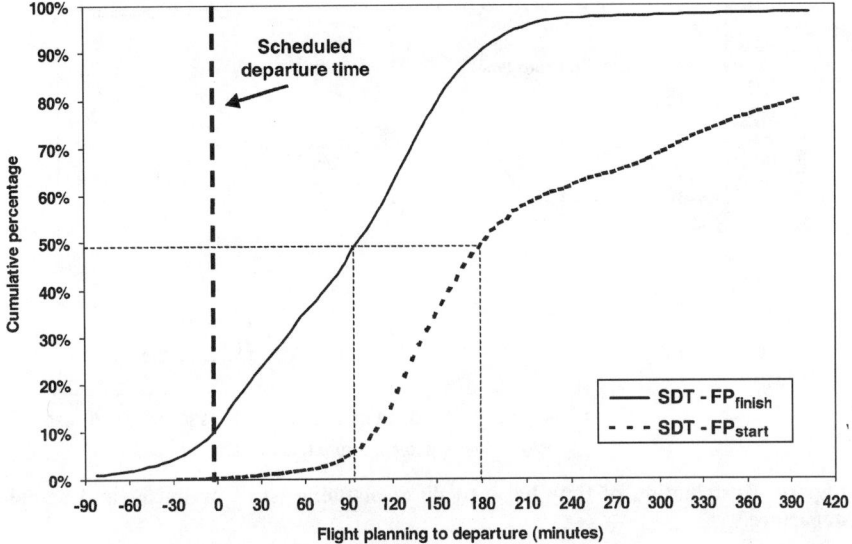

Fig. 9 Distribution of time between flight planning (start or finish) and scheduled departure.

The two following examples demonstrate that combining these two data sources provides insight in the relation between the timing of tasks within the AOC and the timing of events outside of the AOC. This in turn can give indications of the level of performance of the AOC, specifically how well the AOC keeps up with the airline schedule.

Example 3: Timing of Flight Planning: For a given flight, define

FP_{start}: time at which dispatcher started work on the flight plan

FP_{finish}: time at which dispatcher entered the last flight planning command into the computer system

SDT: scheduled block departure time of the flight

ADT: actual block departure time of the flight

Figure 9 shows the observed distribution functions of $(SDT - FP_{start})$ and $(SDT - FP_{finish})$ over 4180 domestic flights of a major U.S. airline on 7 and 8 March 1998. Note the following: for 50% of the flights, the flight plan was started less than 180 min before scheduled departure; about 60% of the flight plans were started between 1.5 and 4 h of scheduled departure; for 50% of the flights, the flight plan was finished less than 90 min before scheduled departure; for 10% of the flights, the flight plan was finished after scheduled departure.

Note that if a flight is subject to changes in crew and/or aircraft assignments, the flight plan has to be revised, which increases the dispatcher's workload. Hence, knowing when a flight plan is typically started enables one to predict the increase in workload that aircraft/crew changes would bring to the dispatcher. For instance, note that for about 35% of the flights, the flight plan was started less than 2.5 h before the scheduled departure time; hence, if the aircraft/crew changes are made 2.5 h before the scheduled departure time, one could estimate a 35% chance that the dispatcher has not started working on the flight plan, and thus that the changes

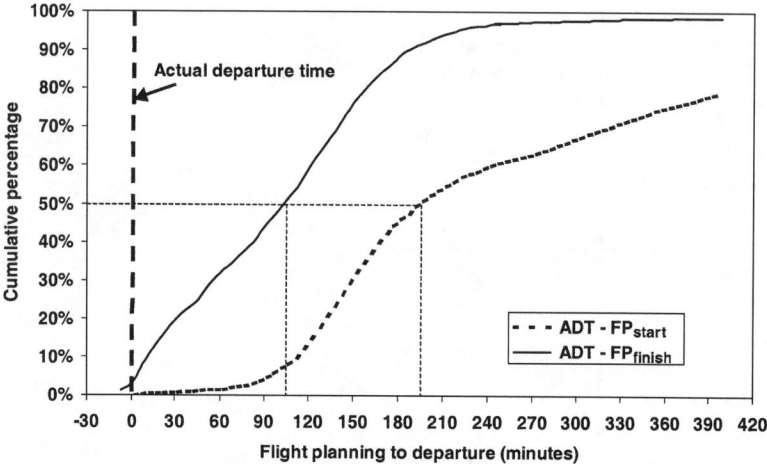

Fig. 10 Distribution of time between flight planning (start or finish) and actual departure.

will not increase his/her workload. Conversely, one could estimate a 65% risk that the dispatcher has already worked on the flight plan, and thus would have to go back to this flight plan and revise it. If the dispatcher is expected to experience a high level of workload in the next 2 h (e.g., due to airborne holding and diversions, other flight plan revisions, or many airborne flights to follow), a decision could be made to assign the flight plan revision to another dispatcher.

Figure 10 shows the observed distribution functions of $(ADT - FP_{start})$ and $(ADT - FP_{finish})$ over the same set of flights used in Fig. 9. Note the following: for 50% of the flights, the flight plan was started less than 195 min before actual departure; for 50% of the flights, the flight plan was finished less than 105 min before actual departure. An interesting feature of Fig. 10 is that many flight plans are worked on almost until the aircraft leaves; about 17% of the flight plans are finished less than 25 min before actual departure (i.e., while the crew is already on board and the bags and passengers are being loaded into the aircraft). We hypothesize that dispatchers tend to work closer to actual departure on delayed flights. The rationale is that a delay at the departure gate introduces more uncertainty about passenger and cargo loads, because the load planner and customer service agents may take advantage of the delay to add last-minute passengers and cargo loads onto the aircraft. If these weight additions exceed a given threshold, they have to be reviewed and approved by the dispatcher, which would explain the last-minute flight planning tasks. The data help confirm this hypothesis. Figure 11 shows the distribution of $(ADT - FP_{finish})$ separately for flights with less than 5 min of delay and for flights with more than 5 min of delay. Note that dispatchers do work closer to actual departure on delayed flights. For instance, 24% of the delayed flights (compared to only 14% of the on-time flights) were worked on in the last 25 min before actual departure. Some delayed flights apparently suffered from a vicious circle in which an initial delay enticed AOC and airport employees to increase the aircraft load, which resulted in additional load planning and eventually a few minutes of additional delay. A possible way to break this circle would be to freeze

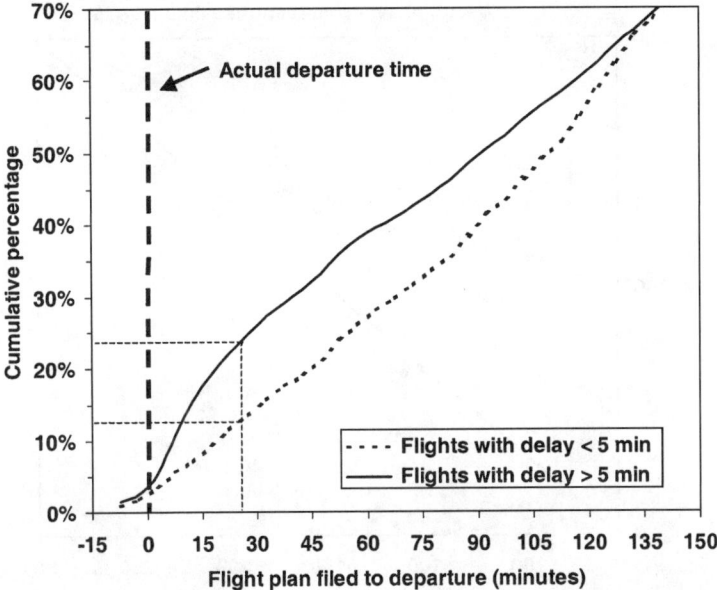

Fig. 11   Comparison of flight planning time for delayed and on-time flights.

the aircraft loads roughly 10 min before the desired departure time. The benefit would be a reduction in last-minute replanning delays. The drawback would be the loss of some passengers and cargo load (some of which could be recovered on later flights). An analysis is under way to study correlations between last-minute load planning and delays.

Example 4: Dispatcher Reaction to Aircraft Change: If the aircraft assigned to a flight is changed (e.g., because the aircraft that was originally assigned is suffering a mechanical failure), the dispatcher has to revise the flight plan. Let D_S be the time interval between the new aircraft assignment decision and the scheduled departure of the flight; this is the time available for the dispatcher to respond. Let R be the time that the dispatcher takes to finish the flight plan revision after the new aircraft assignment decision; this is the dispatcher's actual response time. Figure 12 shows the relationship between R and D_S. Each data point represents one irregular flight.

The closer a data point is to the horizontal axis, the faster the dispatcher has finished responding to the new aircraft assignment. The closer a data point is to the diagonal, the later the dispatcher keeps updating the flight plan. A data point on the diagonal means that the last flight plan task was done at the time of scheduled departure (a data point above the diagonal would mean that the dispatcher was still looking at the flight plan after scheduled departure time—these flights are not shown here). Figure 12 suggests that the dispatcher has a different strategy depending on D_S. If D_S is smaller than 100 min, the dispatcher tends to revise the flight plan very quickly; if D_S is larger than 100 min, the dispatcher is in no hurry to revise the flight plan since he/she knows that he/she will have to look at the flight later anyway.

A more detailed demonstration of these statements is shown in Fig. 13, where the ratio R/D_S is plotted against D_S. This ratio represents the fraction of the available response time that the dispatcher actually used. The actual response time of the

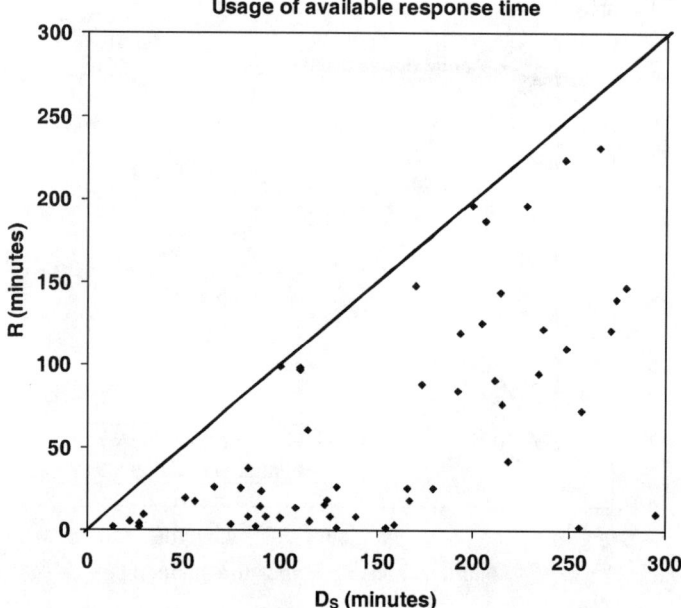

Fig. 12   Dispatcher reaction to new aircraft assignments.

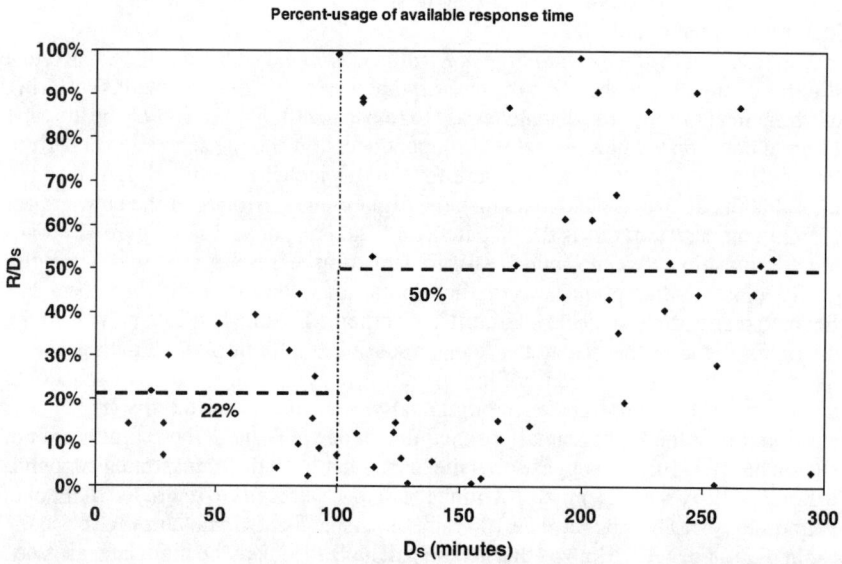

Fig. 13   Dispatcher reaction to new aircraft assignments (normalised).

dispatcher averages 22% of the available response time when D_S is under 100 min, but averages 50% of the available response time when D_S exceeds 100 min. This analysis shows that in the AOC model, a new aircraft assignment should result in a high-priority flight plan revision task in the queue of the dispatcher if the flight is scheduled to depart in less than 100 min, but should result in only a low-priority task if the flight is more than 100 min away.

IV. Conclusions

The economic benefits of ensuring timely decision making in the AOC motivate an in-depth study of its dynamics. To this end, a modeling approach has been proposed that represents the operators of the AOC as queueing servers and charts the flow of tasks in this network as fixed process maps. A preliminary calibration effort to estimate the parameters of this model has been conducted using focused interviews and direct observations in the AOC. A more complete identification was conducted using archived operational data and computer transactional data. A computer simulation and a graphical user interface have been built to further investigate the dynamic behavior of the model.

Acknowledgments

This chapter was reprinted from the *Air Traffic Control Quarterly*, Volume 7, Number 4, by permission of the Air Traffic Control Association Institute, Inc. We would like to thank the airline that provided the data, as well as Honeywell and NASA Ames Research Center for their support in this research.

References

[1]*Airlines Operational Control Overview,* Draft 1.4, Airline Dispatchers Federation and Seagull Technology, Inc., 1995.

[2]Delcaire, B., and Feron, E., "Dealing with Airport Congestion: Development of Tactical Tools for the Departure Flows from a Large Airport," International Center for Air Transportation, Rept. 93-3, Massachusetts Inst. of Technology, Cambridge, MA, 1998.

[3]Gelenbe, E., and Pujolle, G., *Introduction to Queueing Networks,* Wiley, New York, 1987.

[4]Kleinrock, L., *Queueing Systems Theory,* Vol. 1, Wiley, New York, 1975.

[5]Robertazzi, T. G., *Computer Networks and Systems: Queueing Theory and Performance Evaluation,* Springer-Verlag, New York, 1990.

[6]Gershwin, S. B., *Manufacturing Systems Engineering,* Prentice-Hall, Upper Saddle River, NJ, 1994.

[7]Schmidt, D. K., "A Queueing Analysis of the Air Traffic Controller's Work Load," *IEEE Transactions on Systems, Man, and Cybernetics,* Vol. SMC-8, No. 6, 1978.

[8]Morrison, J. R., and Kumar, P. R., "New Linear Program Performance Bounds for Queueing Networks," *Journal of Optimization Theory and Applications,* Vol. 100, No. 3, March 1999, pp. 575–597.

[9]Pujet, N., and Feron, E., "Flight Plan Optimization in Flexible Air Traffic Environments," *AIAA Guidance, Navigation and Control Conference,* San Diego, July 1996.

Chapter 10

Pricing Policies for Air Traffic Assignment

Karine Deschinkel*
ENSAE, Toulouse, France

Jean-Loup Farges[†]
ONERA – CERT, Toulouse, France
and
Daniel Delahaye[‡]
LOG, Toulouse, France

I. Introduction

DURING the past 10 years, air traffic has grown in a constant way, and forecasts indicate that this tendency will continue. In Europe as in the United States, the significant growth of air traffic (approximately 8% per year) is at the origin of the congestion observed not only in large airports but also in airspace. Indeed, airspace in which the density of air traffic becomes significant has to be controlled to ensure the safety of flights. The airspace under control is divided into sectors. A sector is a volume of the space defined by a floor and a ceiling, and it is crossed by air routes. A sector is assigned to a controller who has a global view of the current traffic distribution in the airspace and can give orders to the pilots in order to avoid collisions. If the number of planes, the number of conflicts, and the input flow in a sector are too high, the workload of the controller increases, and he is not able to ensure his work in optimal conditions of security. A sector in this situation is congested, and reactive procedures have to be applied to ensure safety. Congestion can be reduced by modifying the structure of the airspace (increasing the number of runways, increasing the number of sectors by reducing their size).

An alternative way to reduce congestion is to perform flow control. Ground delay programs consist of controlling the flow at its origin. Several previous studies have focused on slot allocation to reduce congestion. As shown in Ref. 1, this problem consists of finding departure slots for all flights so that en route capacities

---

Copyright © 2001 by the American Institute of Aeronautics and Astronautics, Inc. All rights reserved.
* Ph.D. Student.
[†] Research Scientist.
[‡] Research Scientist.

and declared airport capacities are respected and that the total ground delay is minimized. In Ref. 2 this problem is extended to include route allocation. In this approach, the modification of flight plans of aircraft, by changing their slot of departure and their route, reduces congestion of sectors. This optimization of new slots and routes for each aircraft is performed with a genetic algorithm and significantly reduces the peak workload in the most congested sectors and in the most congested airports.

However, the application of flight plan modifications can be a source of problems. Indeed, the allocation optimized for the system does not respect equity between users. For instance, for the same origin-destination pair, two users can be affected on routes having very different costs. This is why routes and slots cannot be directly imposed to the companies. In the traffic theory, the difference between a system approach and a user approach has been highlighted by Wardrop.[3] In the system approach, a route and a departure time is assigned to each user by a central organism. In the second approach, users are free to choose their route and their departure time.

The impact of travel time information on users' choices has been modeled and associated traffic assignment techniques have been developed by Cascetta and Cantarella.[4] The evolution of users' choices is modeled by introducing a learning mechanism based on generalized transportation costs, which spreads the traffic demand in time and in spatial dimension. Congestion is expected to be reduced by moving the time of departure and by changing the current path. Ben-Akiva et al.[5] study the impact of a toll on users' choices in term of route and departure time. They show that there exists a toll that brings the congestion to a minimum.

In the problem addressed by this chapter, the route-slot allocation that provides the minimum of congestion from the point of view of the system is known. It is supposed to be computed by an existing method such as the one developed by Oussedik and Delahaye.[2] An economical strategy to reach this allocation is searched. The purpose of this chapter is to find a mechanism for pricing so that the choices of companies lead to the target allocation.

This chapter has been organized as follows. In the next section a mathematical formulation of the model is given. Two sections are devoted to the formulation of parameter identification and price optimization problems, respectively. Then, the principle of resolution of the optimization problem is explained. Finally, some numerical examples of traffic are presented.

II. Model Formulation

A. Structure of the Model

Figure 1 presents the relation between the different variables of the model. This model is used to estimate the average number of flights expected on each option for given values of sector prices. To perform this task, reliable information about scheduled flights, sequences statistics, and airline companies' costs must be available.

B. Basic Notations and General Assumptions

For a given day and for each origin-destination pair, it is assumed that the requested time of each flight is known. To fill out flight plans, airline companies

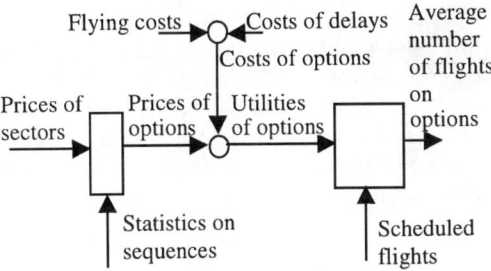

Fig. 1 Schematic representation of the model.

have to choose among different options of routes and departure times by taking into account different indicators (delay, travel time). It is assumed that the introduction of a price modifies the choices of the companies.

Some of the variables used throughout this chapter are given as follows:

W = set of origin-destination (OD) pairs
ω = an element of set W
R_ω = set of alternative routes for OD ω
T = number of alternative time periods in the horizon
Δ = length of a time period
S = number of sectors
(i, j) = option of route i and takeoff period j
$NVP^\omega(j)$ = number of planned flights on period j
$ND^\omega(i, j)$ = desired number of flights on option (i, j)
$NE^\omega(i, j)$ = average number of flights on option (i, j)
$P^\omega_{(i,j)}$ = price of option (i, j)
$x_{k,n}$ = price of sector k during the period n
X = vector of sector prices for all sectors and all periods
$C^\omega_u(i, j)$ = cost of option (i, j) for a flight initially planned on period u
$c^\omega(i, j)$ = cost of option (i, j) associated with the flight itself. This cost depends mainly on the choice of the route i but may also depend on the departure period j because time varying meteorological conditions can have an impact on aircraft fuel consumption.
$r(u, j)$ = cost of delay for a flight initially planned on period u and taking off on period j. This cost covers several aspects, such as impact on the reputation of the airline and problems induced by the perturbation of the planning of aircraft and crew for next flights.

The cost $C^\omega_u(i, j)$ can be expressed by the summation of two terms, the cost associated with the flight itself and the cost associated with the disrespect of the schedule:

$$C^\omega_u(i, j) = c^\omega(i, j) + r(u, j) \tag{1}$$

C. Price for an Option

The strategy of pricing could be to consider each origin-destination pair separately. In this case, a price for each option (i, j) and for each OD should be calculated. However, with this strategy flights of different OD that travel through the same sector do not pay the same tax. This is why it is preferred to associate a tariff with each sector and over each period. The price of option (i, j), $i \in R_\omega$, is a combination of sector prices:

$$P_{(i,j)}^\omega = \sum_{k=1}^{k=S} \sum_{n=1}^{n=T} a_{(i,j)}^{(k,n)} x_{k,n} \qquad (2)$$

where $a_{(i,j)}^{(k,n)}$ represents the probability, for an aircraft taking option (i, j), to enter the sector k during the period n.

This probability is computed taking into account some variability because the sequence of sectors crossed by an aircraft depends not only on the route but also on the type of the aircraft. Indeed aircraft present different rates of climb, and flights levels are chosen in function of aircraft characteristics. It is assumed that for each route a statistic of sequences of sectors and associated entry times are known. For each route, the probability q_i^ℓ is associated with the ℓth sequence of sectors, seq_i^ℓ. For each sector k in seq_i^ℓ, the interval of time between the takeoff and the entry time in the sector, $t_{i,k}^\ell$, is given.

Let:

$b_{(i,j)}^{(k,n,\ell)}$ = the probability that an aircraft on option (i, j) and traveling the ℓth sequence of sectors enters the sector k during the period n
$E(z)$ = integer part of real z
$D(z)$ = decimal part of real z

then $b_{(i,j)}^{(k,n,\ell)}$ is given by:

$$b_{(i,j)}^{(k,n,\ell)} = \begin{cases} 1 - D\left(t_{i,k}^\ell/\Delta\right) & \text{if } k \in seq_i^\ell \text{ and } n = j + E\left(t_{i,k}^\ell/\Delta\right) \\ D\left(t_{i,k}^\ell/\Delta\right) & \text{if } k \in seq_i^{\ell l} \text{ and } n = j + E\left(t_{i,k}^\ell/\Delta\right) + 1 \\ 0 & \text{otherwise} \end{cases} \qquad (3)$$

Finally, $a_{(i,j)}^{(k,n)}$ is given by the average value over the sequences:

$$a_{(i,j)}^{(k,n)} = \sum_\ell b_{(i,j)}^{(k,n,\ell)} q_i^\ell \qquad (4)$$

D. Model of Choice Behavior

Travel decisions entail choices among discrete set of alternatives, such as departure time, destinations, and routes. The outcomes of such choices are provided by a class of mathematical models called probabilistic discrete choice models by Horowitz.[6] These models give the probability that a traveler will be attracted to a given alternative among the set of alternatives available to him. The multinomial Logit and multinomial Probit are two well-known examples of probabilistic discrete choice models (see Refs. 7–9). The Probit model is more complex, and the Logit model is particularly applicable to travel demand. Therefore, the Logit model is adopted in this chapter to predict the decisions of airline companies for their flights. Their decisions depend on the utility associated with each option. The utility of a trip via route i departing at period j for a flight initially planned on period u is assumed for simplicity to have the linear form:

$$V_u^\omega(i, j) = C_u^\omega(i, j) + P_{(i,j)}^\omega \tag{5}$$

The probability that a flight planned on slot u is assigned to option (i, j) is expressed as follows:

$$PR_u^\omega(i, j) = \frac{\exp[-\alpha V_u^\omega(i, j)]}{\sum_{r \in R_\omega} \sum_{s=u-J_{\min}}^{s=u+J_{\max}} \exp[-\alpha V_u^\omega(r, s)]} \tag{6}$$

In this formula, α is a positive scale parameter. The assumption that the moving of the slot of departure must be done in a limited domain $(J_{\min}, J_{\max})$ is made. These predictions give the average number of flights assigned to option (i, j):

$$NE^\omega(i, j) = \sum_{u=j-J_{\max}}^{u=j+J_{\min}} NVP^\omega(u) PR_u^\omega(i, j) \tag{7}$$

with $NVP^\omega(u) = 0$ for $u < 1$ and $u > T$.

III. Identification Problem

Assuming that for each period j and each route i the number of flights taking off, $NO^\omega(i, j)$, has been observed, some parameters of the model can be estimated. The problem consists of minimizing the square differences between the observed number of flights and the average number of flights given by the model:

$$\min \sum_{\omega \in W} \sum_{i \in R_\omega} \sum_{j=1}^{j=T} [NO^\omega(i, j) - NE^\omega(i, j)]^2 \tag{8}$$

The costs perceived by the companies are usually not well known. For that reason the minimization has to be performed with respect to the variables related to costs: $c^\omega(i, j)$ and $r(u, j)$. Identification of the α parameter of the Logit model is not relevant. Indeed, a modification of this parameter is equivalent to a change of unit for the expression of the costs and prices. To have a ratio number of measurements/number of identified parameters high enough, it may be necessary to make additional assumptions for the cost structure. Two reasonable assumptions are:

1) The meteorological conditions prediction is so uncertain that the predictable part of the cost associated to the flight itself depends mainly on the route:

$$c^\omega(i, j) = c^\omega(i) \tag{9}$$

2) The cost induced by a delay is proportional to the absolute value of the deviation with respect to the schedule:

$$r(u, j) = \beta |j - u| \tag{10}$$

With such assumptions, the error criterion has to be minimized with respect to β and with respect to the different $c^\omega(i)$.

IV. Optimization Problem

It is assumed that an optimal route and slot allocation has been found. The problem consists of finding prices that minimize the difference between the average number of flights on each option of each OD couple and the desired number of flights:

$$\min_X \sum_{\omega \in W} \sum_{i \in R_\omega} \sum_{j=1}^{j=T} [ND^\omega(i, j) - NE^\omega(i, j)]^2 \tag{11}$$

The average number of flights is the result of the choices of companies. The desired number of flights on each option is the result of a slot and route allocation that can be obtained by the method presented in Ref. 2.

The sector prices are positive variables and should be small with respect to the costs of options. To compute only realistic prices, the following constraint should be used:

$$0 \leq x_{k,n} \leq \bar{x} \quad \text{for } k = 1, S \quad \text{and} \quad n = 1, T \tag{12}$$

where $\bar{x}$ is the maximum allowable price for a sector.

The similarity between the identification problem and the price optimization problem allows the use of the same kind of optimization methods for the resolution of both problems.

V. Principle of Resolution

As the criterion of the price optimization problem is sometimes nonconvex, it is interesting to assess the resolution of the problem by local and global optimization methods. A gradient algorithm and a simulated annealing algorithm are proposed for obtaining local and global minima, respectively.

A. Gradient Algorithm

The gradient algorithm modifies the prices $x_{k,n}$ iteratively (see Ref. 10). At each iteration q the new prices $x_{k,n}^{q+1}$ are computed from the current prices $x_{k,n}^q$ by performing the following steps:

1) Computation of the sensitivity $\nabla f_{k,n}^q$ of the criterion with respect to each price by:

$$\nabla f_{k,n}^q = \left. \frac{\partial F(X)}{\partial x_{k,n}} \right|_{X=X^q} \tag{13}$$

2) Optimization in the direction defined by $\nabla F(X)$ taking into account the constraints:

$$x_{k,n}^{q+1} = \min[\max(x_{k,n}^q - \lambda^q \nabla f_{k,n}^q, 0), \bar{x}] \quad (14)$$

3) At each step $F(X^{q+1}) \leq F(X^q)$

The algorithm stops either when the criterion reaches a given value or when the number of iterations is too large. At each iteration λ^q is obtained by minimizing the criterion with respect to this parameter using a golden search method that requires 10 computations of the criterion. An analytical expression is used for $\nabla f_{k,n}^q$:

$$-2 \sum_{\omega} \sum_{i \in R_\omega} \sum_{j=1}^{j=T} \{ND^\omega(i, j) - NE^\omega(i, j)[X]\} \nabla nve_{k,n}^\omega(i, j) \quad (15)$$

with

$$\nabla nve_{k,n}^\omega(i, j) = \sum_{u=j-J_{\max}}^{u=j+J_{\min}} NVP^\omega(u) PR_u^\omega(i, j) \nabla pr_{k,n}^{\omega,u}(i, j) \quad (16)$$

and

$$\nabla pr_{k,n}^{\omega,u}(i, j) = -\alpha \left\{ a_{(i,j)}^{(k,n)} + \frac{\sum_{r \in R_\omega} \sum_{s=u-J_{\min}}^{s=u+J_{\max}} a_{(r,s)}^{(k,n)} \exp[-\alpha V_u^\omega(r, s)]}{\sum_{r \in R_\omega} \sum_{s=u-J_{\min}}^{s=u+J_{\max}} \exp[-\alpha V_u^\omega(r, s)]} \right\} \quad (17)$$

B. Simulated Annealing Algorithm

The simulated annealing algorithm works with two values for the prices: $\tilde{X}^q$ is the best value obtained until iteration q and X^q is the value for iteration q. At each iteration, the steps of the algorithm are:

Sort price Y^q in the neighborhood of X^q

If $F(X^q) < F(Y^q)$

- $X^{q+1} = Y^q$

  - if $F(Y^q) < F(\tilde{X}^q)$

    $\tilde{X}^{q+1} = Y^q$

  else

    $\tilde{X}^{q+1} = \tilde{X}^q$

Else

- $\tilde{X}^{q+1} = \tilde{X}^q$

- if $random[0, 1] < \exp(F(X^q) - F(Y^q)/T^q)$

$$\tilde{X}^{q+1} = Y^q$$

else

$$\tilde{X}^{q+1} = \tilde{X}^q$$

The algorithm stops either when the criterion reaches a given value or when the number of iterations is too large. Three versions of the algorithm exist. Those versions differ in the neighborhood definition. For the first version, each component $y_{k,n}^q$ of Y^q is sorted in a uniform distribution over the interval $[x_{k,n}^q - \delta, x_{k,n}^q + \delta]$. The two other versions of the algorithm use a sector load factor given by:

$$L_{k,n}[X] = \sum_\omega \sum_{i \in R_\omega} \sum_{j=1}^{j=T} a_{(i,j)}^{(k,n)} \{ND^\omega(i,j) - NE^\omega(i,j)[X]\} \quad (18)$$

For the second version, if $L_{k,n}[X] > 0$, $y_{k,n}^q$ is sorted in a uniform distribution over the interval: $[x_{k,n}^q - (\delta/2), x_{k,n}^q + (3\delta/2)]$. Otherwise it is sorted in: $[x_{k,n}^q - (3\delta/2), x_{k,n}^q + (\delta/2)]$. For the third version $y_{k,n}^q = x_{k,n}^q + \text{random}[0, \delta]L_{k,n}[X^q]$. For all versions, when the sorted value is outside the interval $[0, \bar{x}]$, it is projected on one of the interval boundaries. The variable T^q decreases by steps as q increases:

$$T^{q+1} = \begin{cases} \lambda T^q & \text{if } q \equiv 0[m] \\ T^q & \text{otherwise} \end{cases} \quad (19)$$

where $0 < \lambda < 1$, m is the step length, and T^1 is given.

VI. Numerical Experiments

A. Example 1

The network used is presented in Fig. 2. It includes two OD couples and three routes for each OD. The control is performed through 13 sectors. The capacity of sectors 0 to 8 is assumed to be one aircraft per sample time, and the capacity of sectors 9 to 12 is assumed to be three aircraft per sample time. For each route there is only one sequence of sectors, and the time to travel any sector is equal to one sample time.

The time horizon considered is equal to eight sample times. The demand for each OD consists of three aircraft scheduled at the first sample time. The maximum ground delay that airlines integrate in their choice is equal to three sampling periods. The α parameter is set to 0.1.

It is assumed that in the observed situation each aircraft takes off on time and flies on the shortest path (sector sequences {9, 3, 4, 5, 10} for OD 0 and {11, 1, 4, 7, 12} for OD 1). This situation induces large overload of sectors culminating with a load of six aircraft per sample time in sector 4 at the third sample time.

Some remarks about this identification problem must be made:
1) Some possible options are not used. In theory, they correspond to infinite costs. To obtain finite costs, the observed value for unused options is set to 0.1, and consequently the number of aircraft taking off on time on the shortest path is set to 2.89 instead of 3.

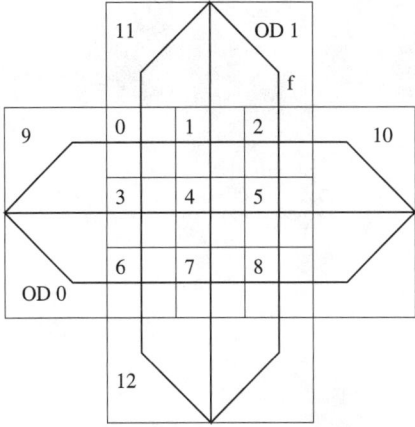

Fig. 2 Network for the academic example.

2) If, for an OD, a constant is added to the utility of all options, the result of the formula giving the probability of an aircraft choosing an option is not changed. Thus, it is not possible to identify costs directly, but only the difference between the costs of different options.

3) The problem presents symmetries; there are no differences between unused routes.

4) Demands and observed situations are equal for the two ODs.

As a consequence, the identification problem consists of finding only two parameters: β and the cost difference between the unused routes and the shortest path routes.

This identification was performed with a simulated annealing algorithm and leads to a criterion of 0.007102. The optimal values are $\beta = 45$ and a cost difference of 46.3. The computation of the cost of the shortest path routes is performed assuming that the cost of flying is equal to three times the cost of a ground delay. Considering that the flying time is five sample periods, this leads to a value of 675.

The target was constructed manually in order to respect the sector capacity constraints. For OD 0 aircraft must be on time and each aircraft must take one of the three routes. For OD 1 one aircraft must be delayed for one period and take route $\{11, 0, 3, 6, 12\}$, another aircraft must be delayed for two periods and take route $\{11, 1, 4, 7, 12\}$, and the last aircraft must be delayed for three periods and take route $\{11, 2, 5, 8, 12\}$.

The bound on sector prices was computed assuming that the impact on total company cost should remain reasonable. As each route crosses five sectors, the cost of the shortest path route is divided per five, and the maximum price of a sector is obtained by taking a proportion (between 0.05 and 0.95) of the result. The value obtained for $\bar{x}$ varies between 6.75 and 128.25.

Prices were optimized using the gradient algorithm and the third simulated annealing algorithm. The gradient algorithm does not reach the global minimum when the maximum price is high. Figure 3 presents the results obtained with the simulated annealing algorithm. Those results indicate that:

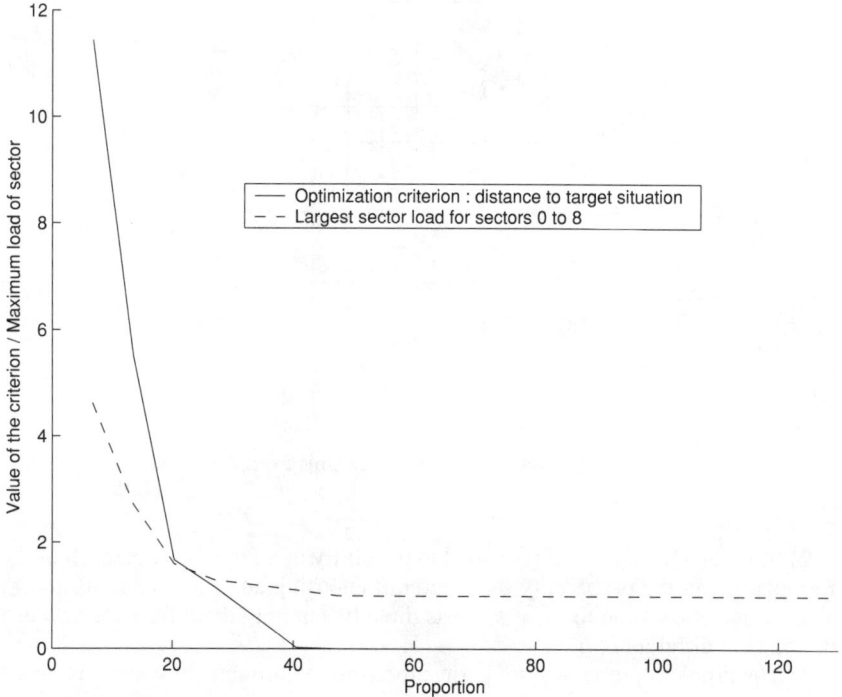

Fig. 3 Relation between the bound on sector prices (given as a proportion of the cost of the shortest route) and the optimization results.

1) High prices (about 35% of the cost of the shortest path route) are needed to reach the target situation.

2) For smaller prices the target situation is not reached, but the increase of the price decreases continuously the load of the sector with highest flow.

This academic example shows that a pricing policy could be an adequate tool in the treatment of a saturation reduction problem.

B. Example 2

1. Network

The algorithms previously described are tested on a part of the French airspace, which includes 23 sectors. Four origin-destination pairs are considered. One pair presents four routes and the others present three routes.

2. Demand

The demand corresponds to the traffic measured on these OD with a sample time of 15 min. The α parameter of the Logit model is set to 0.115. Two demands are used; their optimization horizons are the morning peak (0500–0800 hrs) and all day (0000–2400 hrs), respectively. The constraints on maximum prices are not activated.

3. Target

For both optimization horizons, two target traffic assignments are used. For the morning peak, the first target corresponds to a uniform distribution over time periods and routes, and the second one to integer numbers of flights randomly allocated on routes and periods. For the all-day demand, the first target corresponds to a uniform distribution over the routes, and the second one to a uniform distribution over time periods and routes.

4. Size of the Problem

For the morning peak the number of optimization variables $S \times T$ is $23 \times 12 = 276$. For the all-day demand, the number of optimization variables $S \times T$ is $23 \times 96 = 2208$.

5. Elementary Computation Time

On a SUN-ULTRA 10 with a memory of 64 megabytes, time for computing one value of the criterion, one value of the gradient, and one value of the load factor for the morning peak are $T_c = 0.02$, $T_g = 0.54$, and $T_l = 0.03$, respectively. For the all-day demand, the values are $T_c = 0.06$, $T_g = 10.6$, and $T_l = 0.36$. The computation times associated with tests and random generation are negligible. The gradient algorithm performs one gradient computation and about 10 criterion evaluations per iteration, whereas the simulated annealing algorithm performs only one criterion evaluation per iteration. Thus, a fair comparison of the algorithms implies a number of iterations $[(T_g + 10T_c)/T_c]$ higher for the first version of the simulated annealing algorithm than for the gradient. The values of this ratio are 37 and 186 for the morning peak and the all-day demand, respectively. For the two other versions of the simulated annealing algorithm, the number of iterations should be $[(T_g + 10T_c)/(T_c + T_l)]$ higher than for the gradient. The values of this ratio are 14.8 and 26 for the morning peak and the all-day demand, respectively.

6. Gradient

Figures 4 and 5 depict the evolution of the criterion during the optimization for the gradient algorithm. The distance to the target decreases significantly in the 100 first iterations, and further criterion improvement is obtained continuously. Table 1 presents the values of the criterion before and after optimization.

7. Simulated Annealing

The criteria given by the second and third versions of the simulated annealing algorithm for the second target of the morning peak are between 14 and 16 for δ and T^1 varying on [0.1-10] and [0-0.5], respectively. For the same problem, the first version of the algorithm gives values between 12 and 14. For this version, values of $\delta = 0.1$ and $T^1 = 0.02$ allow good solutions for the morning peaks' problem: the criteria obtained after 74,000 iterations are between 0.072 and 0.095, respectively, (depending of the initial value of the random generator) for the first target and 12 for the second target.

Figure 6 presents the evolution of the criterion for the first target and two different initial values of the random generator. The evolution of the distance to the second target during the optimization with the simulated annealing algorithm is shown in Fig. 7. The criterion decreases from 19 to 12 in 74,000 iterations.

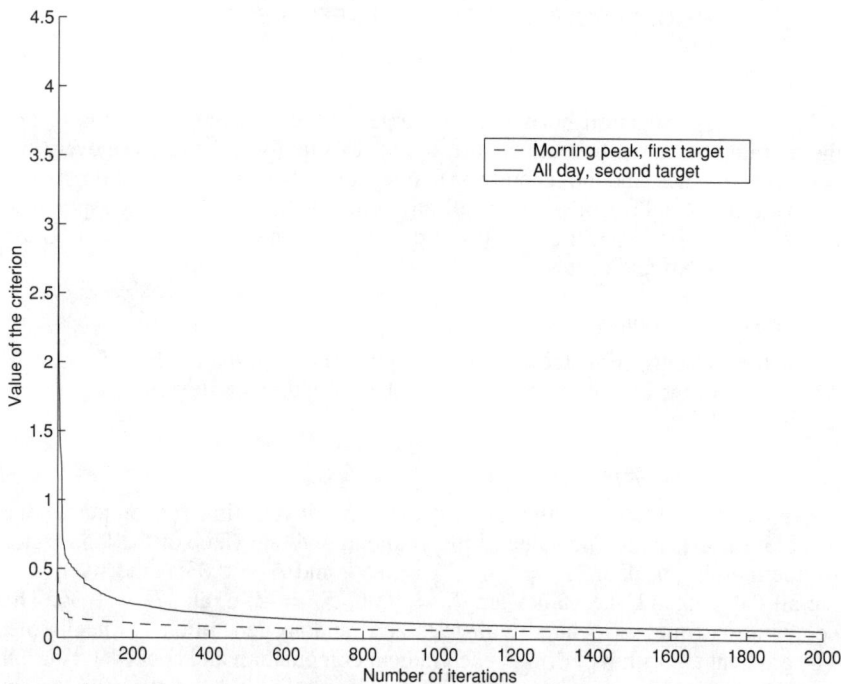

Fig. 4 Evolution of the criterion with gradient algorithm.

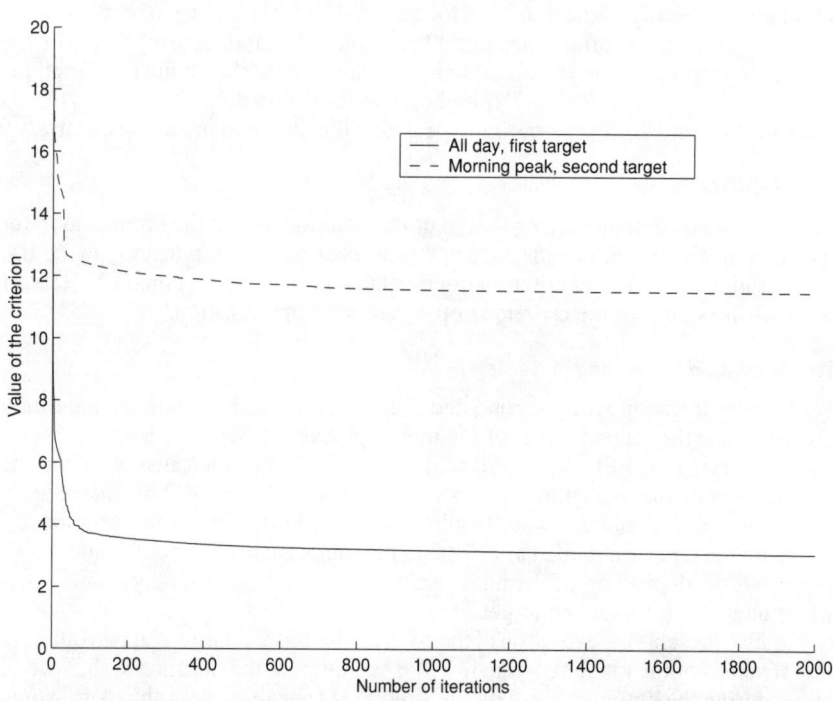

Fig. 5 Evolution of the criterion with gradient algorithm.

Table 1 Criterion after 2000 iterations (number of flights^2) for an optimization with the gradient algorithm

Demand	Morning peak		Allday	
Target	1	2	1	2
Null prices	0.62	19	13	4.3
Result	0.036	12	3.1	0.069

8. Comments

The results obtained lead to the following comments:

1) The optimization by the two algorithms shows that a large decrease of the initial value of the distance to the target is obtained. This indicates the feasibility of the flow control by prices.

2) The gradient algorithm is more efficient than the simulated annealing algorithm. For the first target and a similar computation time, the simulated annealing algorithm finds a solution with a value between 0.072 and 0.095, whereas the gradient algorithm produces a solution with a criterion of 0.036. Moreover, the decrease of the criterion in the first iteration is significantly higher for the gradient

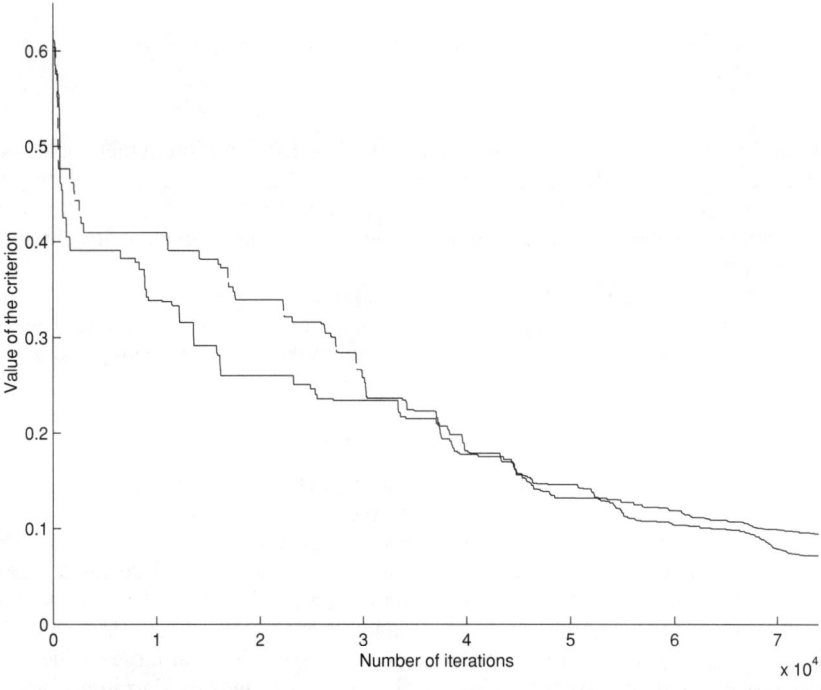

Fig. 6 Evolution of the criterion with simulated annealing algorithm, morning peak first target (two different initial values of the random generator).

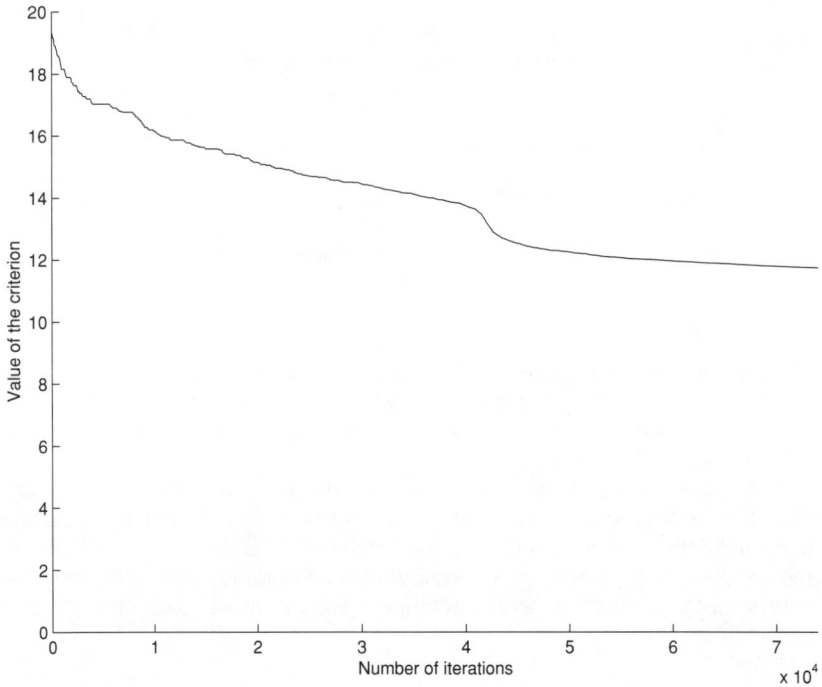

Fig. 7 Evolution of the criterion with simulated annealing algorithm, morning peak second target.

algorithm. It seems that for this problem there is no local minimum that could give an advantage to the simulated annealing.

3) The convergence of the gradient algorithm is very slow. The criterion is significantly reduced in the first iteration but further improvements are obtained asymptotically.

4) For the gradient method, a significant number of prices are equal to zero, indicating that prices are only raised locally to avoid congestion. This could be the reason for the convergence to a local minimum in the academic example.

VII. Conclusion and Future Work

A simple assignment model of the relationship between the toll applied to aircraft flying a given sector at a given instant and the choice made by airline companies in terms of takeoff time and route has been proposed. For a given demand, the model is able to provide the flow at the origin for each route all during the day. An identification procedure for estimation of some parameters of the model has been proposed and demonstrated in an academic example. An optimization problem that corresponds to the objective of setting those flows at target values has been formulated. The solution of the problem by two kinds of algorithms, gradient and simulated annealing, has been studied. The results obtained by the two methods with actual data are similar, indicating that the criterion does not present local optima. However, for the academic example with high bounds on prices, the gradient algorithm converges to a solution that is significantly worse than the solution of the simulated annealing algorithm. This indicates the presence of a local minimum. The conditions for the existence of a local minimum remain to be clarified. The target values are not necessarily reached, indicating that the system optimum may not always be obtained through a sector pricing policy.

Research on pricing policies for the reduction of congestion in the area of air transportation is a new domain and many options can be studied. Concerning the specific approach proposed in this chapter, further research will address direct optimization of congestion and modeling of traffic that does not follow timetables.

References

[1] Maugis, L., "Mathematical Programming for the Air Traffic Flow Management Problem with en-route capacities," *Proceedings of the 14th Triennial World Conference of the International Federation of Operational Research Societies*, 1996.

[2] Oussedik, S., and Delahaye, D., "Dynamic Air Traffic Planning by Genetic Algorithms," *Proceedings of the Congress of Evolutionary Computation*, IEEE, 1999, pp. 1110–1116.

[3] Wardrop, J. G., "Some Theoretical Aspects of Road Traffic Research," *Proceedings of the Institute of Civil Engineers*, Vol. 2, No. 1, 1952, pp. 325–378.

[4] Cascetta, E., and Cantarella, G. E., "A Day-to-Day and Within-Day Dynamic Stochastic Assignment Model," *Transportation Research*, Vol. 25, No. 5, 1991, pp. 277–307.

[5] Ben-Akiva, M., de Palma, A., and Kanaroglou, P., "Dynamic Model of Peak Period Traffic Congestion with Elastic Arrival Rates," *Transportation Science*, Vol. 20, No. 3, 1986, pp. 164–181.

[6] Horowitz, J. L., "Statistical Comparison of Non-nested Probabilistic Discrete Choice Models," *Transportation Science*, Vol. 17, No. 3, 1983, pp. 319–350.

[7] Dial, R. B., "A Probabilistic Multipath Traffic Assignment Model Which Obviates Path Enumeration," *Transportation Research*, Vol. 5, No. 2, 1971, pp. 83–111.

[8] Yang, H., "Multiple Equilibrium Behaviors and Advanced Traveler Information Systems with Endogenous Market Penetration," *Transportation Research*, Vol. 32, No. 3, 1998, pp. 205–218.

[9] Akamatsu, T., "Decomposition of Path Choice Entropy in General Transport Network," *Transportation Research*, Vol. 31, No. 4, 1998, pp. 349–362.

[10] Minoux, M., "Optimisation non linéaire sans contrainte," *Programmation Mathématique*, Tome 1, Dunod, Paris, 1983, Chap. 4.

Section III: Collaborative Decision Making

Six chapters discuss the developing concepts and theory behind the new distributed optimization strategy of information sharing and collaborative decision making. The first three chapters in this section are taken from the 1998 conference and postulate the potential benefits of collaborative decision making between the air traffic control/central flow control authority and the airline air operations center. Chapter 14 describes a different sort of collaboration, the relationship between the pilot and air traffic controller.

Chapters 15 and 16 are more recent studies that begin to investigate how CDM behaves in actual practice. Chapter 15 is an innovative and pioneering model of the ATM system as a Complex Adaptive System (CAS). CAS systems are nonlinear systems that have multiple actors optimizing different object functions. Chapter 15 describes a modeling approach that attempts to understand the complex game playing that is observed to be occuring in the U.S. CDM system. Chapter 16 is an excellent description of the initial U.S. CDM system and an evaluation of its performance using new metrics of air traffic management (ATM) performance.

Chapter 11. Improved Information Sharing: A Step Toward the Realization of Collaborative Decision Making, Peter Martin., EUROCONTROL; Alison Hudgell, DERA and; Nicolas Bouge and Sophie Vial, Aérospatiale, ATM 98.

Chapter 12. Air Traffic Control/Air Carrier Collaborative Arrival Planning Cheryl Quinn, NASA Ames Research Center; and Richard E. Zelenka, Logicon/Sterling Federal Systems, ATM 98.

Chapter 13. Data Flow Analysis and Optimization Potential from Gate to Gate, Matthias Poppe, DFS; and Georg Bolz, Lufthansa AG, 1998.

Chapter 14. Effect of Shared Information on Pilot/Controller and Controller/Controller Interactions, R. J. Hansman and Hayley J. Davison, MIT, ATM 2000 paper 19.

Chapter 15. Modeling Distributed Human Decision Making in Traffic Flow Management Operations, Keith C. Campbell, Wayne W. Cooper, Jr., Daniel P. Greenbaum, and Leonard A. Wojcik, MITRE Corporation, 2000.

Chapter 16. Assessing the Benefits of Collaborative Decision Making in Air Traffic Management, Michael O. Ball, University of Maryland; Robert L. Hoffman and Mike Wambsganss, Metron Scientific Consulting, Inc.; and Dave Knorr and James Wetherly, FAA, 2000.

Chapter 11

Improved Information Sharing: A Step Toward the Realization of Collaborative Decision Making

Peter Martin[*]
EUROCONTROL Experimental Centre, Brétigny-sur-Orge, France

Alison Hudgell[†]
U.K. Defence Evaluation Research Agency, Great Malvern, Worcestershire, United Kingdom

Nicolas Bouge[‡] and Sophie Vial[§]
Aérospatiale, Les Mureaux, France

I. Introduction

THIS chapter reports on an investigation and analysis of the requirements of European aircraft operators and airport organizations for the improvement of information distribution among air traffic management (ATM) service providers and the user community. Such improvements in information management are crucial to the realization of the concept of collaborative decision making (CDM) in Europe.

The work resulted in the identification of a number of information gaps that exist between the actors concerned with ATM and operations. These demonstrated the range of needs that must to be taken into account in development of an information management solution. Furthermore, it was demonstrated that the organizational complexities of the different actors (airlines, airports, and ATM) must be considered when developing system-wide information management and distribution solutions.

Copyright © 2001 by EUROCONTROL. Published by the American Institute of Aeronautics and Astronautics, Inc., with permission.
[*]Project Manager, Collaborative Decision Making.
[†]Senior ATM R and D Engineer.
[‡]Project Manager.
[§]ATM R and D Engineer.

II. Collaborative Decision Making

CDM is a concept that has gained widespread currency in ATM research and development.[1,2] It is recognized as an important approach in trying to make best use of scarce resources such as airport runways, airport terminal gates, and air traffic flow management takeoff slots. A number of different levels of collaboration can be identified in CDM scenarios: 1) an enhanced distribution of information to ensure that each user has a good a picture as possible of the situation, 2) active cooperation to improve planning estimates, 3) a consideration of additional actors' priorities in an actor's own planning processes, and 4) redistribution of decision-making responsibility, for example a delegation to users of responsibility for managing the allocation of a scarce resource.

The project team carried out a study to identify what information users require to make best use of their resources. This should provide a basis from which operational procedures and supporting tools may be developed.

III. Project Overview

Initially an object model of the information distribution processes was developed using the ROSE4 Object Modeling tool with the purpose of developing a baseline understanding. A set of questionnaires was then developed targeted at identifying what information on scarce resources is lacking and hence the information that is needed to make better use of these resources. Two separate questionnaires were developed, one for airlines and the other for airports and air traffic control (ATC) authorities.

Thirteen airlines agreed to participate in the study (Air Liberté/TAT, Alitalia, Britannia, British Airways, Cargolux, Easyjet, Magec Aviation, Monarch, Olympic Airways, Regional Airlines, Swissair, Virgin Atlantic, and Virgin Express). Similarly, seven airport authorities and corresponding ATC organizations agreed to contribute (Aéroports de Paris, Regie der Luchtwegen-Régie des Voies Aériennes (RLW-RVA) Brussels Airport, U.K. National Air Traffic Services Ltd. and Heathrow Airport Ltd., LVB and Schiphol Airport, Swisscontrol and Zurich airport authority, Hellenic Civil Aviation Authority, and Nice Airport).

Face-to-face interviews were conducted, typically taking up a whole day. This gave the opportunity for a full discussion of the constraints and operational problems. In addition, some ad-hoc discussions were carried out with domain experts to gain supplementary information and views.

Following each interview a summary was prepared, and the study was completed by a full analysis of these summaries. The analysis was published as a EUROCONTROL report.[3] The authors would like to stress that the conclusions reached are theirs alone and do not necessarily represent the views of the participating airlines, airports, and ATC service providers.

IV. Airline Operational Aspects

A. Organizational Background and Implications

The airline companies interviewed in the course of the project represented a wide sample of the different types of organizations and methods that the future European ATM System (EATMS) will have to accommodate. Most of the companies interviewed operate mainly within Europe: short-to-medium-haul flights within countries bordering the European Civil Aviation Conference (ECAC) represent 70% of their business. The sample included a cargo carrier, a specialized business jet provider, and regional, charter, and low-cost operators, as well as scheduled service providers.

Some of the companies operate primarily in the hub-and-spoke model, whereas others operate point-to-point or shuttle flights. All airlines use at least one center from which the airline planning, operations, and commercial activities are managed. Many had complex associations of alliances and subsidiaries, leading to code sharing arrangements and the potential for coordinated scheduling and operations. This is not yet widely developed, but it is likely to be extended significantly in the future. Furthermore, while many of the airlines operated the majority of their flights from coordinated airports, several did not and were thus not subject to significant levels of airport slot control.

The functions of flight planning, fleet management, and dispatch were often carried out by dedicated airline operations centers (AOCs), particularly in the larger companies. Several of these had sophisticated computer support systems including worldwide communications links with their aircraft via ACARS.

For the small- and medium-sized companies, functions were frequently outsourced to service providers, such as SITA, handling agents, and airline reporting offices (AROs) provided by airports. For example, slot management may be dealt with by other airlines, airports (e.g., Aéroports de Paris), or specific service companies (e.g., Transair). Similarly, they often relied on an outside supplier for flight plan preparation and submission service for which SITA and Jeppesen were providers. Handling agents are often used at remote stations (i.e., away from the airline's main operating base) to assure flight preparation, boarding, and dispatch.

Thus some airline companies have only limited or partial links with other actors in the ATM system. Given the requirement for equitable treatment for all EATMS system users, any solutions for improved information distribution must be easily accessible to all users whatever their operating arrangements.

B. Operational Issues

1. Airline Operating Concept

All airlines do not use the same operating concept. For example, many European airlines are developing hubbed modes of operation. This approach is not yet widespread except among the major airlines, and it is difficult to identify an airport hub in the U.S. sense with operations dominated by a single company's operations. Such an operating mode is convenient for connecting passengers and switching aircraft and crews, but it imposes certain constraints:

1) For the airline, a hub is very delay sensitive. Feeder flights must not be delayed, otherwise transit passengers miss their connection, and if there are few feeder flights, the passengers can experience long delays.

2) For the airport, transit passengers and their baggage must be transferred from one aircraft to another in a very short time, necessitating sharp peaks in activity in the airport facilities.

3) For ATC a hub imposes a greater load than point-to-point operations because arrivals and departures are bunched instead of being spread out in time.

All of these place a premium on having good information flow within the airline, for example, warning of late departure. In the U.S. model such information exchange can be assured by the dispatcher-pilot information exchange. In airlines equipped with a datalink such as ACARS, pilot-AOC messages can fill the gap. However, in other cases efficiency must rely on prompt communication by the handling agent at the departure airport, and this cannot always be assured.

2. Turnaround Management

Airframe turnaround times range from 20 min to 1.5 h for passenger carriers. Typically turnaround can be considered to cover the period from on-blocks to pushback, including disembarkation and boarding by passengers, baggage handling, refueling, and safety checks on the aircraft. Turnaround times depend on:

1) Company operating strategy: some airlines plan a greater margin for turnarounds in their schedule to help manage the effects of delays.

2) The aircraft type: bigger aircraft require longer turnarounds and some types are easier to load and unload by virtue of location of baggage doors on the airplane. For example, the minimum turnaround time for a B747 is 1.5 h.

3) Passenger connection times if the airline operates a hub: sharp peaks in activity are necessitated.

4) Airport: turnaround times are often longer in international airports.

5) Whether the flight is short haul or long haul: short-haul flights are operated with higher frequency than long-haul flights.

3. Management of Delays and Disruption

Between 30 and 100% of the flights of each of the airlines participating in the study are regulated. (A regulation is applied by the central flow management unit (CFMU) to limit traffic flows to safe levels when demand exceeds capacity.) This result is not surprising; the major activity of the airlines was operating short-haul flights inside Europe where the sky is the most congested. Thus these flights are subject to chronic delays resulting from insufficient capacity in the airspace and airports concerned.

The key impact of delays is to disrupt an airline's planned flying schedule. To maximize the return on their assets, airlines try to increase the proportion of time spent flying passengers. However, this means that schedules become tighter and more prone to disruption.

Thresholds quoted for delays that disrupt the company schedule range from 0 to 30 min. If the delay of a flight is greater than this time, the airline cannot absorb it during subsequent flight or turnaround, and the remaining schedule is affected (knock-on delays). Parameters taken into account in determining this threshold are usually the forward schedule and crew working hours, and the parameters may vary from flight to flight.

In practice, airlines prefer to deal with chronic delays by accepting the disruption and continuing the schedule dealing with the knock-on delays as best they can. This strategy is usually preferable to cancellation. In fact, they seldom cancel because they do not want their customers to switch to a competing airline's flight on the same route. Instead airlines prefer to fly half-empty.

However, a separate and arguably more significant form of delay arises in disruption situations such as when fog, snow, or an incident such as a strike severely reduces the capacity either locally or throughout a region. When this occurs, airlines are very badly affected because the disruption often catastrophically upsets the planned schedule and imposes very high costs because of aircraft and crews being in the wrong locations.

On these occasions it is difficult for an airline, even with the appropriate tools, to consider different operational scenarios and their consequences in response to delay disruption. Looking more than one flight ahead is not always rewarding, as so much can happen in the meantime to make plans obsolete. Hence all the airlines interviewed emphasized the critical importance of providing them with more information, more quickly, and more accurately to enable better responses to disruption situations.

V. Airport Operational Aspects

It is very difficult to generalize when discussing airport operations because every airport is different. Within Europe one can identify major international airports, regional airports, hub airports, and airports that are not hubs. There are many different sizes of airports, in terms of surface area, number of runways, number of stands, number of terminals, and so on. The level of sophistication of automation and information systems varies, as does the range of facilities available to passengers and airlines. Each airport operates under different constraints, such as environmental, political, and commercial constraints; and there are a variety of problems, for example unfavorable weather conditions, to contend with. For all of these reasons, the operational priorities of different airports can be quite varied. Furthermore, the organization and division of responsibilities vary significantly. These variations mean that information distribution needs to be flexible enough to accommodate the demands of the different organizations and requirements.

Airports are very complex enterprises. In general, a number of different organizations are involved in the operation of an airport. The precise boundaries of responsibility for each organization vary from country to country and from airport to airport, as do the relationships between the different organizations. Each airport is organized as appropriate for that airport, for commercial, historical, and political reasons.

For the purposes of this study, the roles of the airport authority and ATC provider were defined in terms of their main responsibilities. It should be noted that, in practice, more than one organization may contribute to a given role, and/or a single organization may fulfill (part of) several roles.

The role of the airport authority is operation of the airport, including 1) terminal management, check-in counters, departure lounges, baggage belts and reclaims; 2) provision and allocation of stands and gates; 3) guidance and control of vehicles and aircraft on the apron (apron control), including provision and operation of follow-me cars and marshals where required; 4) provision and allocation of buses to transfer passengers to remote stands; 5) towing operations; and 6) provision of de-icing facilities. In addition, in the context of a multimodal traffic system, the airport authority may be considered responsible for providing information and connections to local and regional transport services.

The role of the ATC provider is provision of air traffic services, including 1) control of taxiing aircraft on taxiways and runways; 2) control of aircraft approaching and taking off from the airport; 3) ATC in terminal area and en route airspace; and 4) liaison with the CFMU. To demonstrate the lack of a consistently identifiable organization, apron control is the responsibility of an ATC organization at some of the airports, whereas at others that responsibility is assigned to an airport management company (although startup clearance must always be coordinated with ATC).

Similarly, responsibility for bus transfers from gate to aircraft, including provision and operation of the buses, may rest with airlines or handling agents. However, at some airports buses are a resource owned and managed by the airport authority.

In some cases the airport authority and ATC provider roles are carried out by different parts of the same organization. In other cases they are performed by completely separate organizations operating in very different ways. Often, the airport authority is a commercial company, whereas the ATC provider is government-owned. This can result in a difference in culture and response time between the two organizations. For example, an airport authority may work to encourage a rapid growth in traffic at an airport, while the ATC service cannot respond sufficiently quickly to accommodate the increase in traffic. The result will be delays at the busiest times of day.

Some examples of the actual splits of airport authority and ATC provider roles at airports are:

1) At Brussels airport RLW-RVA performs both airport authority and ATC provider roles. This has led to a very close cooperation between the two operations. Furthermore, the terminal-operating company and the airport side of RLW-RVA have recently formed a joint company with airport authority responsibilities, resulting in an organization for Brussels Airport close to that defined at the start of Section V.

2) At Athens the ATC provider and airport authority roles are fulfilled by two separate, independently operating branches of the Greek Civil Aviation Authority (GCAA).

3) The major home-base airline operates its own terminal and apron facilities at Athens (Olympic Airlines at West Terminal, for Olympic aircraft only) and London Heathrow (BA at Terminals 1 and 4, for BA and alliance carriers' aircraft). The airport authority operates the other terminal(s).

4) In Paris the airport authority Aéroports de Paris employs air traffic controllers to provide tower ATC, and also provides aircraft handling.

A given item of information will be held by different parties at different airports and will be managed and used differently depending on the culture of the party concerned. Thus a given organization, for instance the airport authority, will see a different "part of the picture," leading it to deal with the situation in different ways. Achieving a consistent picture is essential for good decision making.

VI. Information Gaps

The analysis of the returns from the interviewees led to the identification of information gaps or links between entities that appear to be missing. That is not to say, however, that information exchange should be seen as a one-to-one process. Historically there has been a good deal of centralization in ATM systems, but we must be aware that modern technology is moving the trend toward a more decentralized world view with users being able to seek information over multiple channels.

The information gaps identified by airlines, airports, and ATC providers are described in the following subsections. Figure 1 summarizes these information gaps.

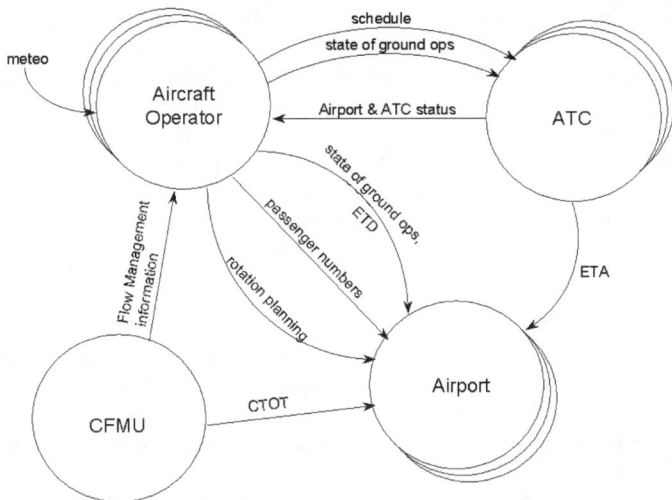

Fig. 1 Summary of information gaps.

A. Airline Information Gaps

The airlines identified a number of areas in which information would be of particular assistance to them, as described below.

1. Airport and ATC Status Information

Airlines noted that it would also be useful to have a variety of additional information related to airport and ATC status. Examples of information items that were mentioned include 1) pretactical data and live updates on airport capacity, 2) information on airport gates and aircraft parking, and also for alternate airports, 3) information on terminal, and local and regional transport system problems, 4) coordinated airport and flow management slots, and 5) expected holding times in stacks from ATC to better organize turnarounds.

The need for more information from airports was particularly strongly felt. For many companies information is readily available concerning the airport where they have their main operating base. However, information concerning other airports served by the company is much more difficult to obtain easily and cheaply.

One example of this lack of information concerns airport capacity. Companies often lack basic information on capacity in various situations. It may be useful to standardize airport capacity declarations because at present some airports "oversell" their capacity, whereas others allow in extra flights without slots. This information should be supplemented by updates to reflect the real situation (e.g., changes in wind direction) that would provide a basis for the airline to respond by modifying its operations.

Because ECAC airports are increasingly congested, if flights do not arrive or depart on time, the disruption of gate and parking allocation planning results in additional ground delays for the airlines and information. One major carrier noted that it has experimented with having aircraft approaching its main operating airport slow down to avoid an irritating wait on the tarmac for its passengers if the allocated gate is not clear. Ideally it would like to do this for other airports it serves.

Information on terminals and local transport is also important for companies. For example, early warning of rail or road disruption can enable an airline to modify its operating schedule.

Regarding airspace congestion in and around airports, some companies observed that there is a mismatch of capacity and airport slot allocation, and that it could be helpful to improve coordination between the flow management slot and the airport slot.

2. Flow Management Information

Airlines identified a need for more accessible flow management information, such as 1) pretactical (i.e., a few hours before the flight takes off) forecasting of constrained sectors, 2) pretactical forecasting of likely average delays on particular routes, 3) tactical (i.e., very shortly before the flight takes off) information on approximate foreseen delays, 4) tactical information on possible alternate routes with approximate delay indications, 5) highlight the timing of regulations in comparison with the flight plan in question, and 6) more information on the reasons behind the delays, such as in which Area Control Centers (ACCs) the capacity/demand balance is creating a bottleneck. In general, information should be presented indicating what is available rather than what is forbidden. For example, it would be efficient for users if displays could be organized to indicate available slots as opposed to a map of constraints.

Some of the information proposed (constrained sectors, delays) is already available through Air Traffic Flow Management (ATFM) Notification Messages (ANMs). This introduces another important consideration in improved information exchange. Shortage or cost of manpower and effort required to supply or extract information effectively means that it is important to consider manpower issues when developing new applications. If not, these concerns will prevent the best being made of the available information.

As an example, ANMs give codes for constrained sectors, but the keys to localize them must be looked up separately. The time required to find the locations means that, except for frequently occurring problems, it is not worth the operators investing the time required to find alternate routes to avoid the restrictions.

Visual map-based displays are an important consideration in improving accessibility. For example, information on capacity, constrained sectors could be displayed with different colors depending on their load levels. Meteorological data could be overlaid, and routes affected by routing schemes such as the traffic orientation scheme (TOS) or conditional routes (CDRs) could be highlighted. In addition, a flight plan could be superposed and manipulated by the user (or flow manager). Customized filters could be used to display selected layers of the airspace (lower sectors included), selected routes, cities, airports, etc. Finally, customized alert systems (e.g., by flagging changed information on the map display) and data discrimination filters could be introduced.

Given that such facilities could require considerable investment if provided centrally by ATM service providers [as is the case with the current remote terminal access/remote client access (RTA/RCA)] an alternative approach would be for a data stream to be provided, which aircraft operators or other users could then integrate within their own operational systems.

B. Airport Authority Information Gaps

This section describes information that airport organizations noted was unavailable or could be improved. As discussed earlier, it should be remembered that airports are typically composed of multiple different organizations that are integrated to varying levels of effectiveness.

1. Earlier Information on Planned Rotations

Currently the stand allocation process is necessarily largely tactical. Last-minute delays and rotation changes are inevitable and mean that stand allocation cannot be completely planned in advance. Although recognizing this fact, a number of airport authorities would like to be able to plan stand allocations further in advance and more proactively than at present. To enable this, earlier and more accurate planning information from aircraft operators would be necessary. For each rotation at its airport, the airport authority needs information on approximate on-blocks time, approximate off-blocks time, aircraft type, and estimated passenger and cargo load.

Any strategic planning of stand capacity for the new season's traffic is typically carried out using extrapolated traffic from the previous year. By the time airline schedules giving details of rotations are available to airport authorities, it is often too late to make any major changes to the stand allocation plan in time for the start of the season. Instead these have to be made as the season progresses. Earlier schedule and aircraft type information from the airlines could be merged with traffic history to give a more reliable basis for strategic stand allocation planning.

Better quality information from the airlines would benefit the airlines themselves because the airports would be in a better position to provide the services demanded by the airlines. Rotation information is generally provided by airlines a few days before the start of the season, in time for the airport authority to begin pretactical planning. Some airlines already provide this information very accurately and very promptly, but there are significant gaps. Therefore, even at this later stage, the information held by individual airlines is generally better than that received by airport authorities, and airport authorities would benefit from improved accuracy and coverage.

2. Rotation Planning Updates

Allied to the requirement for better baseline information on planned rotations, the airport authority needs to receive updates as the airlines' planned rotations change to keep its stand allocation plan in line. Better advance information allows them to plan more proactively.

It is not unusual for planned rotations to change a number of times before the flights actually take place. The type of aircraft performing a flight may vary from day to day, as the number of passengers booked on the flight varies. Changes may be made with a very short notice, particularly in the case of home-base airlines or those having more than about 10 aircraft at the airport.

Some airlines already send rotation planning updates, but often these are not reliable. Some send few or no planning updates, and so a change may not be apparent to the airport authority until the aircraft arrives at the airport. Sometimes the airport authority may have had no advance notification of, for example, aircraft type, and so it will not know what kind of stand a flight requires until they actually see it (or are informed verbally by ATC).

The effectiveness of airport authorities' stand allocation planning is reduced by the fact that planning information is not complete. Out-of-date or missing information from some airlines reduces the benefit of high-quality information from others. An on-time flight that behaves exactly as the airport authority expects fits smoothly into the plan, whereas an arrival for which the airport authority has incorrect or no information causes a lot more work. Airport authorities would therefore like 1) advance information on the aircraft type and expected length of stay of all arrivals and 2) reliable updates on all airlines' rotation planning.

Airlines are currently not obliged to send the required information and updates, and may see little direct benefit, particularly from sending all updates as they occur. Thus many airlines will not bother to send updates when they are busy, and some will never consider it worth the manpower and communications cost. The effort required would be reduced by electronic links, so that when an airline updates its own fleet planning (or flight planning) system, updates to rotation plans are sent automatically to the airport authorities concerned. For airlines with manual planning systems, effort required is likely to remain an issue with provision of full planning updates. However, the possibility to provide such automatic links could be considered a requirement for any new systems being procured.

3. Passenger Numbers

For the allocation of terminal resources and/or transfer buses, the airport authority requires the number of passengers expected on a flight. However, passenger load often remains unknown, especially for arriving flights. This can lead to waste of the airport's passenger capacity, or to inadequate facilities being provided for the customers (of the airport and the airline). Passenger numbers for departing flights are more readily available to the airport authority via the handler at the airport.

As a minimum, airport authorities would like to know final passenger load before a flight arrives. This could be provided by the check-in handler at the departure airport, or by the airline (these may in practice be the same organization). Passenger load is currently already provided to the airport authority for billing purposes in the postflight phase; all that is required is earlier transmission of this information.

To go further, early notification of expected passenger numbers (preferably with updates) would aid stand/gate allocation planning. However, the airlines may consider this commercially sensitive information.

4. Estimated Time of Arrival

Another key component of the information about rotations that the airport authority needs is the estimated time of arrival (ETA) of a flight. More accurate updates can allow stand allocation to be more proactive and more efficient. Every airport authority interviewed identified a requirement for improved ETAs, although the details of what was required varied from airport to airport depending on what was already available.

Some airport authorities [e.g., Nice, Heathrow Airport Ltd.] receive automatic notification when a flight joins the stack. This notification would be of more use to them if they also knew how long it was expected to *stay* in the stack, thus enabling them to derive an ETA. Accurate predictions of arrival taxi times could improve the accuracy of estimated time of arrival at the stand, where an accurate estimate of landing time is already available (for example, from the ATC system).

Where the airport authority has access to flight plans, notification to them of an arriving flight's actual time of departure (ATD) could be used to update the flight plan information. This would provide a reliable ETA at the earliest possible opportunity. The airport authority could use information on predicted or actual departure delays to update expected arrival times. One airport authority commented that it had considered giving priority to on-time flights, to encourage airlines to send accurate planning information and updates, especially of ETA.

ETA updates could be provided from a number of sources:

1) The CFMU has filed flight plans. ETA from the flight plan is not always accurate, but it represents an update to an airport authority that is working from only airline schedules.

2) AOCs may have updates of ETA, derived from pilot reports or ACARS communications. Many airlines already provide their handlers (and sometimes airport authorities) with updates of ETA for long-haul flights. However, to send ETA updates for all flights could imply a lot of extra effort and communications costs, which airlines are unlikely to be prepared to meet without justification (preferably in terms of direct benefit to the airline).

3) The local ATC system is likely to have accurate ETAs once the flight is in the local flight information region (FIR). These could be linked directly into the airport authority's system. (Systems are already linked at some airports; some others plan links in the future.) Accurate ETAs would be available earlier if neighboring ATC centers' systems were also linked in.

4) Information about expected delays and ATD from which ETA might be extrapolated could be sent from the airport of departure.

5) The EUROCONTROL Enhanced Tactical Flow Management System (ETFMS) will provide a centralized source of ETAs. A number of airport authorities noted that the ETFMS would be most useful if its information were available as an electronic data feed, so that ETAs could be fed directly into their own systems.

Different sources might be most appropriate for different airport authorities, or at different times in advance of a flight arriving at its stand. ETA is clearly of great interest to many different parties; a consistent, regularly updated estimate available to all would be widely valued.

5. Estimated Time of Departure

As with ETA, estimated time of departure (ETD) is a key piece of information defining a rotation. To be able to predict pushback times more accurately would be a major contribution to airport authorities' allocation planning capability. Many airport authorities expressed a requirement simply to have a 10 min warning of expected push back.

Again, a number of different aspects were identified by the airport authorities interviewed. The two major items identified were information on departure delays and information on the progress of ground handling operations.

ETD is of interest to many different parties, and a consistent, regularly updated estimate available to all would be widely valued. It appears that the information from which ETD could be determined is scattered across a larger number of sources, and hence an accurate ETD is less likely to be held in electronic form. These information gaps are discussed in the following two sections.

a. Information on departure delays. Departure delays may arise because of handling or technical problems, or as a result of ATC delays, either from CFMU slots or delays in receiving startup clearance from the tower. Notification of problems encountered and expected delays would allow the airport authority to keep ETD updated.

Currently, information on departure delays most often comes by telephone coordination between airport authority operations and the handler or AOC. Aircraft operators often avoid informing airport authority operations of problems or expected delays, in case they are asked to move the aircraft to a remote stand for the duration of the delay. (This would allow the airport authority to make better use of its available pier service, but it is inconvenient for the airline that has to move.) Moreover, given the present slot allocation system, airlines may be penalized if they declare any delays: if they cannot meet the slot and are not granted a slot extension, they will be put back at the end of the slot allocations queue and incur greater delays. The result is "wait and see" behavior with the hope that the problem will be solved before request for clearance time.

A number of airport authorities would like to receive flow management slots, to warn of expected departure delays and as an indication of ETD. However, they do not generally consider it necessary to prioritize apron operations in favor of regulated aircraft, and so knowledge of flow management slots is not required for that purpose.

b. Progress of ground handling operations. In general the airport authority is not aware of the state of airline ground handling operations for a particular flight. Such information would be useful as an aid to prediction of departure time. Confirmation that each part of the operation (cleaning, catering, baggage, boarding of passengers) had been completed would give an indication of whether the flight was ready to depart on schedule. An indication of any problem encountered (lost baggage, lost passenger, late catering with an estimate of the arrival time of the caterers, etc.) would help further.

Much of this information is already passed verbally, for example, between the handlers and the pilot, but it is not made available to all at the airport who could profit from it. However, it is worth noting again the variation in operations between different airports and between different airlines. Heathrow Airport Ltd. receives updates on the progress of handling operations for some airlines via ACARS messages; other airlines rely on them to pass estimated or actual time of departure to the AOC.

Aéroports de Paris was unusual among the airport authorities interviewed in that it is also responsible for handling and airport ATC, and so the different aspects of airport operations are more closely linked than at many other airports. It already passes information on handling delays to the tower to improve their estimates of departure times, and it noted that it could also provide this information to CFMU if necessary.

A number of airport authorities are addressing their requirement for more information about handling operations. For example, Zürich Airport Authority is investigating methods for tight monitoring of the movement of passengers and bags around the airport.

C. ATC Provider Information Gaps

This section describes information requirements noted by ATC providers at airports, particularly tower control.

1. Estimated Time of Arrival and/or Actual Time of Departure

At present most ATC providers at airports have little direct information on the ETA or ATD. An estimate can be obtained from, for example, filed flight plans (FPLs) or CFMU slot allocations, but this is subject to a high degree of inaccuracy. Better quality information would be a significant step to allowing ATC to improve arrivals sequencing and stack management.

2. Airline Schedule Information

ATC believed it needed better information from airlines on schedules. This would help them improve planning for optimizing both arrival and departure schedules. For example, the Dutch LVB noted that they employ different runway combinations for inbound and outbound traffic peaks, and airline schedule information is needed to plan the timings of switches. Nice ATC noted that they receive no feedback from the IATA slot conference because it is not a coordinated airport. Frequently, schedules are available only at the last minute, and often later than the dates when they are supposed to be provided. As a result, they proposed that early publication of schedules should be mandatory because this would help airport organizations take more efficient decisions and consequently also help the users.

Charter flight information was found to be particularly variable. Special events such as football matches could introduce a high level of uncertainty into planning.

3. State of Airline Ground Operations

It was noted by several ATC [e.g., GCAA, U.K. National Air Traffic Services (NATS), Swisscontrol] that they have no information on the state of airline ground operations. Improving this would bring capacity benefits by allowing ATC to make early planning of taxiing and departure sequence, give time to negotiate slot extensions, and help with arrivals planning (because it would be known better when a gate would be free).

Several suggestions for improving the current situation were proposed. These included:

1) A 10-min advance warning of the aircraft's call for startup would be of significant help.

2) Several ATC authorities said that that they would like to have general information from airlines on what is going on at the gates.

3) Airlines should be responsible for sending accurate information on ground operations progress and delays to the tower (and CFMU).

4) Useful information could also be provided by apron control. For example, it would be useful to know if a pushback tractor is available, if a pushback tractor is in place, and if the baggage is loaded. One authority noted that, in the future development of their system, they will send the planned departure sequence to the airlines, who would have the responsibility to update this planning to reflect unforeseen events during their ground operations.

VII. Issues Outstanding

If the first step in information management is just to provide the required information, the associated step to consider in collaborative decision making is that of data management. The following subsections address particular aspects of data management.

A. Cost of Providing Information

An important concern raised by many of those interviewed was that the overall business case must be justified. Information may well provide a benefit, but it is not always easy to identify exactly what that benefit might be in cost terms.

Companies are concerned that the cost of providing or exploiting the information is properly considered. Costs of provision include data gathering, new information systems, and communications costs. In particular, costs of information use should not ignore the manpower effort required to extract the key information from what is provided, and this may be significant if the information is poorly presented or has to be located among a large volume of other less relevant information. To better bear the cost of provision of information implementation, it is necessary to find routes that emphasize win-win situations between the different actors.

B. Safety and Liability

As for all developments of the ATM system, safety must be of paramount concern in collaborative decision-making applications. New developments should ideally enhance prevailing safety levels, and in any case must not reduce them. There are a number of important safety-related concerns, such as information may not be correct and updates may not be made in a timely way.

This will impose costs on implementations. At the least, information must have associated timestamps and checking mechanisms. At the extreme of safety-critical information, it will be important to look closely at the solutions implemented.

This introduces the issue of liability. If information is used for operational purposes, users need to know who is responsible if something goes wrong.

C. Confidentiality

Closely related to safety and liability is the issue of information confidentiality. Although transparency of process and open access to information are essential elements of CDM, the confidentiality of certain information will need to be assured and information will need to be protected from unwanted interference. This includes preventing unauthorized reading or malicious writing, modification, or deletion of information.

For example, information may be confidential to aircraft operators for commercial reasons, and information on certain categories of flight may be confidential to individual states for reasons of national or military security. Information must also be safeguarded against misuse by terrorists, etc.

These concerns imply the need for a proper consideration of security issues to protect information. It may include adoption of encryption, use of limited access networks, and other equivalent methods.

D. Standardization

Standardization is a prerequisite for efficient information flows. The diversity of the information processing systems and information usage within the ATM actors is such that the data available from the different information sources exist with different formats or material supports. Gradual convergence of these, steered by aviation authorities, will stem from consensus of the actors.

VIII. Conclusions

The initiative described here was a starting point for the development of CDM applications in the European ATM environment. The starting point is well recognized: the improved distribution of information. However, in this study the analysis has served to demonstrate the complexity of the organizational structures that will have to be supported in European CDM applications.

Airports must been seen as key players in the information exchange process because they are the focus of both the start and the end of the flight operations processes. They are complex enterprises that differ greatly among themselves; information is available from them more or less easily and with different degrees of feasibility. Also, they currently have many information distribution solutions developed locally that must be taken into consideration in development of system-wide applications. However, there are significant gains to be made from introduction of new sources of information to aid their decision makers.

Airlines and ATC also have very significant benefits to gain from new sources of information. Once again, the focus for both is on getting airport-related information, and this is therefore the area in which future work should be focused.

Finally, those interviewed identified the risk of duplication of work on improving information sharing. To avoid this, proper coordination of efforts needs to be encouraged and developed at a European-wide (or Worldwide) level.

Acknowledgments

During the project, information was gathered with the active participation of 13 aircraft operators, seven airports, and nine supporting ATC authorities. The authors of this paper appreciate greatly the contributions made by the representatives of each of these organizations to the success of the study. In addition, the authors would like to stress that the conclusions reached are theirs alone and do not necessarily represent the views of the participating airlines, airports, and ATC service providers.

References

[1] Kollman, K., Kupper, K., Wambsganss, M., and Wetherley, J., "Decision Making in Aviation Transportation: Improving Decision Making Through Shared Information," FAA ATM IPT R&D, Oct. 1997.

[2] Miaillier, B., "ATM Strategy for 2000+," EUROCONTROL, Rept. FCO.ET1.ST07.DEL02, Issue 2.0, Brussels, May 1998.

[3] Martin, P., Hudgell, A., Bouge, N., and Vial, S., "FASTER (Future ATFM-AO-Airport Synergies Towards Enhanced opeRations)," EUROCONTROL Experimental Centre, Rept. 332, Paris, Aug. 1998.

Chapter 12

Air Traffic Control/Air Carrier Collaborative Arrival Planning

Cheryl Quinn*
NASA Ames Research Center, Moffett Field, California
and
Richard E. Zelenka[†]
Logicon/Sterling Federal Systems, Herndon, Virginia

I. Introduction

BECAUSE of the continued increase in air travel and a shift toward hub-and-spoke operations by major air carriers in the United States, the National Airspace System (NAS) may be reaching its effective capacity. Daily peaks in the volume of air traffic can result in in-flight delays, holding, or ground delays. Events such as severe weather or airport closures may result in aircraft diversions and flight cancellations that can impact hub-and-spoke operations for days. To address the capacity problems of the NAS, a NASA research initiative, the Advanced Air Transportation Technologies (AATT) Program, is focused on developing operational concepts and their associated decision support tools, procedures, and hardware systems to maximize safety, efficiency, and flexibility of operations in the NAS.[1] The program is comprised of ground-based air traffic management tools and some flight deck-based tools to support a move toward free flight. Free flight involves the removal of air traffic control restrictions and increased flight deck capabilities so that operators have the freedom to select their path and speed in real time.[2]

One ground-based system under development since the 1980s is the Center-Terminal Radar Approach Control (TRACON) Automation System (CTAS), a suite of decision support tools designed to effectively manage arrival traffic into a terminal area. CTAS (Build 2) is comprised of two major components, the Traffic Management Advisor (TMA) and the Passive Final Approach Spacing Tool (pFAST). The TMA provides air route traffic control center (ARTCC) traffic

---

Copyright © 2001 by the American Institute of Aeronautics and Astronautics, Inc. No copyright is asserted in the United States under Title 17, U.S. Code. The U.S. Government has a royalty-free license to exercise all rights under the copyright claimed herein for Governmental purposes. All other rights are reserved by the copyright owner.
*Human Factors Engineer, Terminal Area ATM Research Branch.
[†]Director, Business Development. Senior Member AIAA.

management coordinators (TMCs) with time-based scheduling information to deliver a flow of arrival traffic into a terminal area that matches the current capacity of the facility.[3] TMA can also provide sector controllers with advisories indicating the amount of delay that each aircraft must absorb in order to achieve the desired flow rate. Passive FAST increases airport throughput by providing TRACON controllers with balanced runway assignments and landing sequence advisories.[4] The basic component of these tools is a set of four-dimensional trajectory synthesis algorithms that use aircraft performance models and wind models to provide very accurate predictions of aircraft position over time.[5] TMA and pFAST have been extensively evaluated in Fort Worth ARTCC and the Dallas/Fort Worth (DFW) TRACON, respectively, and have shown significant reduction in arrival delays and increases in throughput without adversely affecting controller workload.[4,6,7] TMA remains a primary decision support tool for daily traffic management at Fort Worth ARTCC.

The CTAS TMA consists of two primary displays: the Timeline Graphical User Interface (T-GUI) and the Planview Graphical User Interface (P-GUI). The T-GUI displays arrival aircraft on a series of timelines referenced to arrival fixes into the TRACON or airport runway thresholds. For each aircraft, TMA can display estimated time of arrival (ETA), CTAS scheduled time of arrival (STA), per aircraft delay, and runway assignment (Fig. 1). The P-GUI provides a planview display of

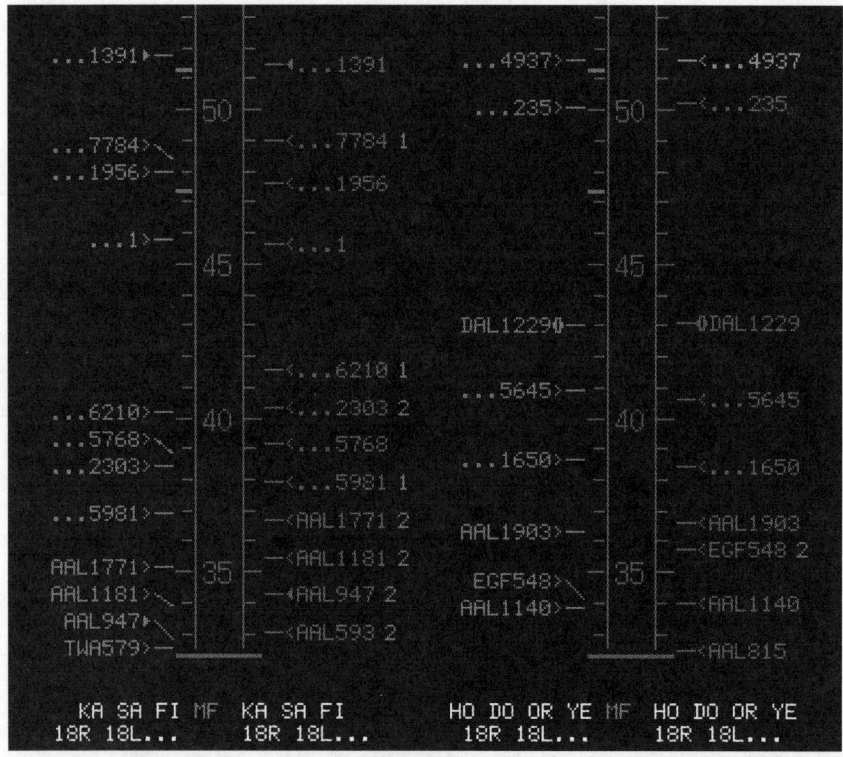

Fig. 1   CTAS TMA Timeline Graphical User Interface.

AIR TRAFFIC CONTROL/AIR CARRIER COLLABORATIVE 179

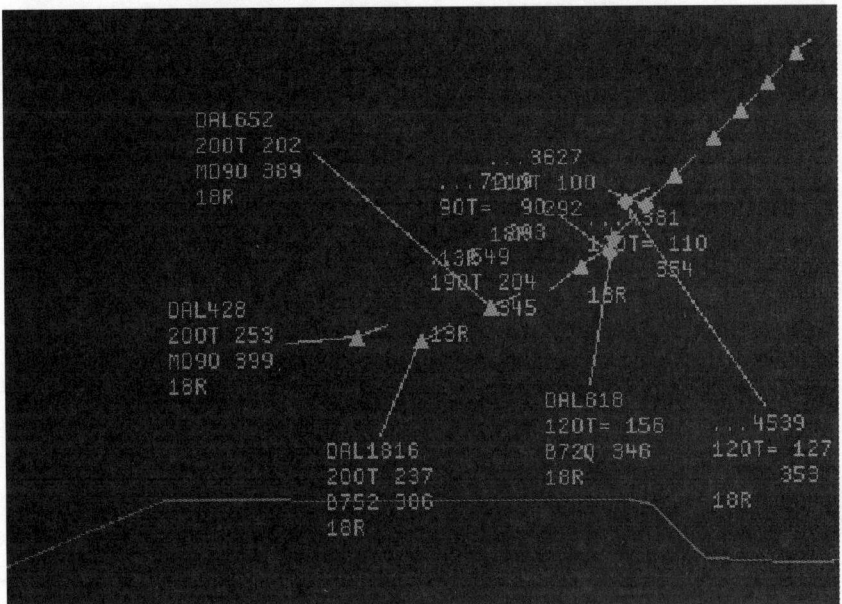

Fig. 2 CTAS TMA Planview Graphical User Interface.

arriving aircraft. The view is similar to a controller's radar display with aircraft data tags presented on a map of the airspace (Fig. 2). These displays are updated with every 12-s ARTCC Host radar update. The functionality of T-GUI and P-GUI has been developed with the participation of a variety of FAA controller personnel to ensure interface acceptability and usability.[4,6]

The Collaborative Arrival Planning (CAP) Project, an initiative within the NASA AATT program, is built on the concept that a great deal can be gained through collaboration between air traffic management and the users of the airspace. By providing air carriers with more accurate scheduling information and allowing them to have more input into flight path selection, airlines and cargo operators can more effectively manage both airborne and ground operations. The CAP Project investigates the concept of sharing information between air traffic control (ATC) and air carrier facilities, including airline operational control (AOC) centers and their airport ramp towers. Increased exchange of data between ATC and the air carrier is expected to improve air traffic management decision making while providing improved efficiency of air carrier operations and greater scheduling flexibility.[8,9]

The first phase of the CAP Project provides one-way flow of filtered CTAS TMA scheduling information to the AOC. The airspace user can then plan and make decisions based on more reliable estimates of arrival times. The next phase of the project involves transfer of air carrier information to CTAS, including aircraft departure times and weight information, which may help to improve the CTAS scheduling predictions. Later phases of the project will investigate two-way data exchange for assisting with individual aircraft arrival preference requests and incorporating fleet-wide user preferences into CTAS scheduling algorithms. This chapter provides a summary of the results of the fielding of a CTAS TMA system at an airline AOC facility, and a description of future research efforts in two areas: the real-time exchange of information between air carriers and ATC and the incorporation of user-preference information into ATC scheduling.

II. Current Research: ATC/Air Carrier Information Exchange

A first step toward greater air carrier and air traffic management collaboration is information sharing. ATC data on airport configuration, arrival scheduling, and runway assignments are expected to be of high value to air carrier operations. Such information can be provided by CTAS. Similarly, air carrier operational data, such as actual per-aircraft weight, accurate gate departure and takeoff times, and aircraft-sensed wind information, can be of benefit to ATC decision support tools. These types of data exchange between the air carriers and the ATC provider are being addressed and advanced through the CAP Project.

A. ATC to Airline Information Sharing

The following sections focus on the results of a field evaluation of airline AOC use of CTAS-generated ATC scheduling information. The CTAS CAP Display System, a filtered repeater of the CTAS TMA system used by ATC, was installed at the American Airlines System Operations Control (AALSOC) Center in Fort Worth, Texas. Because CTAS Build 2 TMA is currently operational at Fort Worth ARTCC and because American Airlines has a major AOC facility in the Fort Worth area, the AALSOC was selected for the initial field evaluation. At the AALSOC, the majority of requests and communication with air traffic control facilities are handled by an ATC coordinator, and so it was expected that the CAP Display System would be of the most value if installed at this position (Fig. 3).

Primary implementation objectives for the experimental airline repeater included ease of implementation, accurate repeating of the operational CTAS TMA at Fort Worth ARTCC, and the guarantee of noninterference between the airline repeater and the operational CTAS TMA. To meet these objectives, a separate, complete CTAS TMA system was installed parallel to the operational system at Fort Worth ARTCC. One dual monitor workstation was provided to AAL's SOC facility to run GUI processes that display the CTAS information. This workstation was connected to the NASA North Texas CTAS field site network via four 128 kbps ISDN lines, to provide the one-way communication necessary for the AAL-resident displays. This one-way slave-shadow configuration does not allow any interruption or manipulation in the airline repeater system to propagate back to the operational system.

A Memorandum of Agreement (MOA) between the Federal Aviation Administration (FAA), NASA, and AAL enumerated the restrictions and airline operating rules for the CTAS CAP Display System. Requirements imposed included the scrubbing of nonhost airline identifiers on the aircraft tags and data blocks, and disabling the airline's ability to enter airport configuration changes. The digital data recording capability of the TMA system was disabled, and all data were provided only as a display through the P-GUI and T-GUI displays. In addition, the host airline was also restricted from using CTAS to question the real-time decisions and operations of Fort Worth ARTCC.

Fig. 3   ATC coordinator and CAP Display System at AALSOC.

Human factors engineers worked with software engineering staff to ensure that the CAP Display System complied with all the requirements of the FAA MOA. The CTAS user interface was modified so that certain features were removed, or it was clearly indicated that features were disabled. Specialized training materials were prepared to help AALSOC personnel better understand the CTAS tools and how they assist with air traffic management at Fort Worth ARTCC. Rather than use existing CTAS documentation, quick-reference guides for T-GUI and P-GUI were prepared to correspond to the modified functionality. AALSOC ATC coordinators and dispatchers were also given a quick-reference card that provided step-by-step instructions for the most frequently used P-GUI and T-GUI functions.

Thirty-two ATC coordinators and other AALSOC personnel were trained on CTAS and the use of the CAP Display System and were given a detailed, hands-on orientation. During an initial preoperational evaluation period lasting three weeks, NASA personnel collected observations of system use and were available to answer questions. During this time, NASA observers were able to modify the T-GUI and P-GUI displays to help the ATC coordinators and dispatchers access CTAS scheduling information more easily. Four timelines were referenced to each of the four meter gates, and two timelines were referenced to the four arrival runways (one runway per side of each timeline). A load graph was set up to show graphs of the anticipated demand into DFW, the CTAS scheduled traffic into DFW, and the average per-aircraft delay. This display configuration allowed the ATC coordinators to locate a specific aircraft to determine its STA, quickly assess the average delay into DFW, or determine whether aircraft were in holding outside a meter gate. Data were collected on how CTAS information is used by

AALSOC personnel and what information format is most useful for AOC tasks. Data were also collected to generally assess the benefits to the AALSOC of shared CTAS scheduling information. These data provide a better understanding of how CTAS scheduling information influences AOC information flow and its impact on operational decisions concerning holding and diversions.

Based on the observations conducted during the initial deployment and observations conducted after the repeater had been operational for about two months, CTAS TMA scheduling information was found to be beneficial to the AOC environment. AAL ATC coordinators found that they made fewer phone calls to Fort Worth ARTCC to check on delays into DFW. Dispatchers concerned with individual flights were able to use the system to find where an aircraft was and what its scheduled arrival time was likely to be. The more accurate arrival time predictions provided by the CAP Display System enabled AALSOC personnel to make better scheduling and fleet-planning decisions concerning flights arriving into their DFW hub. Results also show that more accurate arrival time estimates assist the airline in avoiding flight diversions to alternate airports, benefit airline strategic planning, and reduce the number of status phone calls from the airline to the FAA service provider.[10]

B. Time of Arrival Prediction Accuracy

Major airlines estimate the time of arrival for each of their flights by tracking time deviations from nominal flight plans. The flight plan, once updated to reflect the actual takeoff time, is generally a good approximation of the en route portion of a flight. However, any unexpected rerouting or delays can significantly alter the flight's landing or on-time arrival. Even if no en route deviations occur to alter the nominal flight plan navigation fix times, a flight is typically subject to terminal-area delays (e.g., speed reductions, vectoring, holding) that significantly alter its landing, or on time. Such terminal-area ATC delays differ from day to day for each arrival rush and are difficult for airlines to predict accurately.

Many airlines, including AAL, attempt to minimize the inaccuracies in their estimated on times (E ON) by having their flight crews update their on time estimates during the flight, typically 30 min before landing. While helpful, such updating does not account for terminal-area ATC delays at the destination. At Fort Worth ARTCC, the terminal-area delays are being calculated and assigned to *each* aircraft by CTAS TMA. TMA delays are assigned to match throughput to the current airport capacity, and because the controllers are vectoring aircraft to meet those times, the TMA scheduled arrival times are more accurate than the airline estimates.

Histograms of AAL time of arrival prediction accuracy vs those provided by CTAS TMA are given in Figs. 4a and 4b. The prediction horizon is 60 min from actual time of arrival, using the crew-updated E ON prediction, if available (Fig. 4a). CTAS TMA scheduled time of arrival (Fig. 4b) is used for the same 63 aircraft at the same prediction horizon of 60 min.

The improved time of arrival prediction accuracy of CTAS TMA over typical flight plan/flight crew-adjusted airline estimates is shown in Figs. 4a and 4b. The values from the TMA-derived data are more closely centered around the mean (Fig. 4b), and those provided by AAL are more variable, resulting in a wider, more flattened distribution (Fig. 4a). The arrival time estimates from AAL have

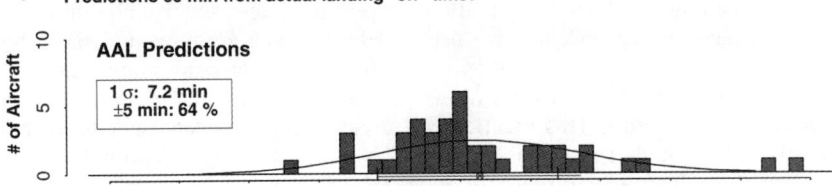

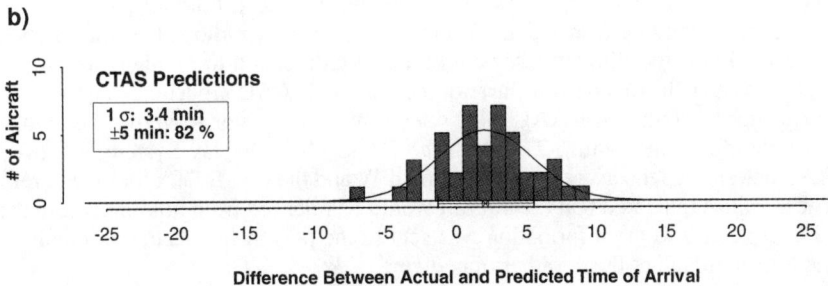

Fig. 4 Accuracy of a) American Airlines predicted arrival times vs b) CTAS predicted arrival times.

a 1-sigma accuracy within 7.2 min of the actual arrival times, whereas the TMA-derived arrival time estimates have a 1-sigma accuracy within 3.4 min. These comparisons do not illustrate the impact that improved time of arrival predictions (and delay information) can have on individual flights. Improved time of arrival estimates, and associated terminal-area per-aircraft delay information, assist AOCs in the management of potential flight diversions when confronted with otherwise uncertain ATC terminal-area delays.

C. Flight Diversions

Arrival aircraft approaching terminal areas are routinely delayed for separation or airport ATC constraints. Delays can be manifested as speed reductions, vectoring, or airborne holding. Aircraft are legally required to be dispatched with enough fuel to allow 45 min of airborne holding. Dispatchers typically request additional fuel based on their historical knowledge of en route and terminal-area delays for each particular flight. As a cost-saving measure, extra fuel beyond the 45-min hold is not commonly carried unless severe weather or abnormal ATC delays are expected at the destination city. As such, arriving aircraft encountering unexpected, severe terminal-area delays must sometimes divert to an alternate airport.

When an ARTCC cannot absorb all of the required delay for a particular aircraft through speed reductions or vectoring, that aircraft is placed into a holding pattern. Generally, the first aircraft to arrive in the holding area is placed at the lowest holding altitude, and subsequent aircraft are placed at increasing altitudes.

Each of these aircraft is then issued an expected further clearance (EFC) time, which is the time at which the aircraft can expect to be released from holding and sent to a metering fix and into the airport. EFCs, which are typically issued by ATC during very dynamic, high-workload periods, are often conservative, inflated times, so that the controller does not to have to recompute the times or reissue them for each aircraft. This practice is difficult for airline fleet operations, especially when an aircraft that is in holding is issued an EFC that would require the aircraft to divert due to fuel limitations (if the aircraft were actually held that long).

On several occasions the CTAS CAP Display System was used by AAL to assist in preventing diversions for flights into their DFW hub. These aircraft encountered unexpected terminal-area delays and were placed into a holding pattern. ATC issued EFCs ranging from 20 to 45 min; the flight crew radioed their dispatcher, concerned that fuel limitations would require a diversion to an alternate airport if they were held for the full duration of the EFC. AAL procedure requires the dispatcher to contact the AAL ATC operations coordinator to verify the actual, likely holding times with ATC. With the CTAS CAP Display System, however, AAL determined the average delays into DFW and the CTAS STA for the aircraft. These values indicated that the aircraft would be released from holding before the EFC expired, and the information was sent to the pilot, who elected to remain in the hold and land at DFW rather than divert.

AALSOC personnel also used the CAP display to proactively divert an aircraft before it would have been forced to divert, thereby saving fuel. The aircraft in question had just been rerouted by ATC to an alternate arrival corner. According to the CTAS P-GUI, the magnitude of the holding over the alternate arrival corner was such that the aircraft would exceed its fuel reserves. Given that the flight was already low on holding fuel, AAL diverted the aircraft to an alternate airport immediately rather than expend extra fuel in transit to another holding location, which would likely have led to a diversion.

In general, the cost of an aircraft diversion is a function of many variables, including equipment type, degree of coupling or connectivity between an airline's hub and feeder spoke flights, and how early in the daily flight schedule the diversion occurs. Costs associated with flight diversions described above include the direct operating costs of flying the aircraft to its alternate airport and back, those associated with holding aircraft on the ground, and lost future revenue from passenger ill will caused by the excessive delays.[11] Such costs can be quite high, with estimates between $20,000 and $100,000 per diversion.

D. Strategic Fleet Planning

The CTAS CAP Display System has been found useful in airline strategic fleet planning, specifically in the areas of aircraft equipment move-ups and the previously described airline-initiated rerouting of aircraft through alternate arrival fixes. Aircraft equipment move-ups refer to aircraft substitutions as a result of the originally scheduled aircraft becoming unavailable. Circumstances forcing equipment move-ups include unexpected maintenance and aircraft diversions. The AAL operations coordinators attempt to shuffle available aircraft to minimize the impact of these delayed aircraft on their operations. A critical item required for such shuffling is the knowledge of what aircraft equipment is or will be available to build an alternate, smaller departing bank of aircraft. With the CAP display, the airline was able to locate aircraft and determine their arrival times to select potential candidates for schedule swaps. The CTAS information provided a new level of stability and efficiency to equipment move-up planning, and facilitated the recovery from the off-schedule operation.

E. TMA Repeater Impact on Airline Ramp Operations

In addition to proven benefit in the AOC environment, TMA scheduling information may also benefit airline ramp operations. Fueling trucks, baggage handlers, caterers, and gate personnel are assigned to each gate based on the airline's estimated arrival times of each flight. These resources could be managed more effectively if ramp management personnel had more detailed, up-to-date schedule information and the expected landing runway. Both of these could be derived from a CAP Display System updated by a TRACON radar source.

An experimental fielding of a TMA repeater system at Delta Air Lines (DAL) Airport Coordination Center (ramp tower) at DFW was conducted in 1999 and provided insight into how CTAS scheduling information could benefit ramp operations, specifically the impact of CTAS scheduling information on airline allocation of gate resources. Both DAL and AAL have suggested that CTAS-updated arrival information would be especially useful if it were integrated into their existing, highly specialized scheduling and ramp management software tools.

III. Future Research

A. Airline to ATC Information Sharing

The exchange of air carrier operational data to CTAS is planned and is currently being developed in a laboratory simulation environment. Actual aircraft weights, accurate gate departure and takeoff times, and aircraft-sensed wind information can assist ATC automation decision support tools such as CTAS. Airline preferences, such as descent profiles/speeds and per-aircraft cost index (ratio of time and fuel costs), are either generically modeled (in the case of descent speeds) or not considered (such as cost index).

Current CTAS algorithms assume aircraft parameters such as weight, engine performance, and aircraft performance based on generic models of a limited number of aircraft types, e.g., Boeing 757-200. Although these assumptions are adequate for arrival trajectory synthesis, more accurate knowledge of aircraft weight and performance is essential for climb modeling. Providing airline-derived data to CTAS has the potential to improve the accuracy of the CTAS trajectory calculations, which could affect CTAS arrival scheduling. Improvements in arrival scheduling could benefit overall capacity by increasing arrival throughput. CTAS scheduling that takes into account airline cost index ratios for individual aircraft would likely yield airline operational efficiencies.

Once data are supplied by the airlines, the CAP Project objectives will be to evaluate the impact of these airline-supplied data on TMA and pFAST, as well as developmental CTAS tools in a laboratory setting. Should these data provide significant improvements in CTAS performance, more extensive data sharing efforts with other carriers would be explored.

B. Collaborative ATC/Air Carrier Tools

CTAS TMA scheduling is accomplished on a first-come-first-served (FCFS) basis. Delays are distributed among aircraft in such a way as to minimize delays to the system as a whole. This method of assigning delays to aircraft is objective and equitable; however, there are times when the air carriers have distinct preferences regarding which of their aircraft receive delays. A given air carrier may have certain flights that are considered more important than others. For example, an international flight or a flight that carries a large number of connecting passengers would have a greater economic impact on the air carrier if the flight were seriously delayed. In certain cases, an air carrier can request that ATC do whatever possible to expedite a certain aircraft. In general, ATC will attempt to expedite the aircraft and may delay another of the same airline's aircraft to do so. Under no circumstances other than emergencies would ATC delay a second airline's aircraft to accommodate the first airline's preference. A long-range goal of the CAP Project is to be able to incorporate airline preferences into ATC scheduling. There are two efforts under way in the CAP Project to facilitate airline arrival preferences: the distribution of CTAS digital data and the strategic CAP tool.

C. Distribution of CTAS Digital Data

Because it is impractical for NASA to deploy CAP Display Systems to a large number of airlines, NASA and the FAA are exploring the distribution of CTAS digital data to airlines. Agreements with the FAA would enable a filtered stream of the Fort Worth Center CTAS data to be distributed to airline clients via the Volpe National Transportation Center. This program will serve as a proof of concept for making CTAS digital data available to airlines at other Free Flight Phase One CTAS deployment sites. By having the CTAS data available in digital form, the airlines can then make use of the more accurate arrival time estimates within their own scheduling and ramp management software tools.

D. Strategic CAP Decision Support Tool

The second effort, known as strategic collaborative arrival planning (S-CAP), investigates the possibility of developing new CTAS sequencing and scheduling algorithms that take into account airline preferences. Recent research in this area explores the feasibility of exchanging CTAS scheduling delays between pairs of aircraft as a means of accommodating an airline request for an earlier arrival. Carr et al.[12] uses fast-time simulation to demonstrate that it is possible to schedule an earlier arrival time for an aircraft by assigning an equivalent amount of delay to another aircraft from the same carrier. Results indicate that an aircraft could be scheduled up to five minutes earlier than the original ETA without impacting the scheduled delays of aircraft outside of the delay exchange pair.[12]

Another study currently under way focuses on a method of scheduling a bank of arrival aircraft according to a preferred order of arrival. The current CTAS scheduling algorithm sequences aircraft using a FCFS method based on ETA at the runway threshold. The CTAS scheduling algorithm generates STA at the runway that may delay aircraft in order to meet sequencing, separation, and airport capacity constraints. In fast-time simulation, an alternative method of scheduling was used to schedule a bank of aircraft in a preferred arrival order. Results show that when compared with FCFS scheduling, the alternative method is often successful in reducing deviations from the preferred arrival order while resulting in little or no increase in scheduled delays.[13]

Future work will focus on improving the success rate of the new scheduling algorithms and gaining a better understanding of the effectiveness of the algorithms under various traffic conditions. Another element to this research effort is to explore the costs of non-FCFS scheduling in terms of overall system delay increases and impact of delay exchange on controller workload.

IV. Human Factors Issues Associated with ATC-Airline Collaborative Tools

CTAS tools have been developed and tested with a great deal of controller involvement. Future CAP tools will follow the same user-centered approach but for the first time will include air carriers as system users. The CAP Display System leveraged the existing TMA interface, which was modified to accommodate FAA MOA requirements. Thus, the system is intended for ATC, and it displays information in a manner that may not be suited to the airline's needs. For the initial evaluation, it was determined that a specialized airline interface for CTAS scheduling information was not necessary. The focus of the research was to determine whether the CTAS scheduling information was useful for AOC tasks, not to evaluate the existing interface. In fact, the airline would prefer that CTAS scheduling information be used to update their existing displays, which were designed specifically for the AOC environment.

The test deployment of the TMA repeater system has shown that the AOC environment can benefit from CTAS-derived scheduling information. It is also evident from observations that ATC and AOC have different goals when it comes to the routing of aircraft. The AOC has a more global view of each specific flight and makes decisions for a flight from point A to point Z based on what is most beneficial for maintaining the air carrier's fleet-wide schedule. Air traffic, on the other hand, makes decisions based on what is best for all of the air traffic within the airspace on a sector-by-sector basis from A to B, B to C, and so on. The sharing of ATC information with airline AOC is the first step toward achieving a common understanding of each group's goals and intentions, which will lay the groundwork for more extensive collaborative decision making efforts in the future.

Throughout the tool development, the design tradeoffs between improving air carrier business choices and impacting controller workload will be carefully considered. As the capability for handling user preferences is incorporated into CTAS scheduling algorithms, it will be important to ensure that the controllers who are attempting to accommodate the requests are not overloaded. Efforts will be made to determine the impact of non-FCFS scheduling on controller workload, and, if too extreme, further research will examine ways of providing advisories to the controllers to help achieve the preferred arrival schedule.

V. Conclusions

Initial observations show that access to ATC scheduling information can be beneficial to the AOC by providing a more global picture of the air traffic coming into the airport. It provides the air carrier with information on expected delays and holding that might otherwise require phone calls to the ARTCC facility. By being able to see the amount of delay to be incurred by each aircraft, the air carrier can make informed decisions about when to divert aircraft during extended holding operations.

Based on discussions with Fort Worth ARTCC and DFW TRACON personnel, there has been no adverse impact on FAA ATC operations as a result of the experimental fielding of the CTAS TMA Repeater at American Airlines SOC. In fact, results indicate a decrease in the number of telephone inquiries to FAA facilities regarding specific aircraft in holding or concerned with excessive metering delays. As discussed, on the occasions when AAL used the CAP display to locate aircraft concerned with, or issued, long EFCs, AAL did not contact an FAA facility. Such a reduction in telephone inquiries, especially those that occur during the high workload periods of arrival rushes or severe weather, is viewed positively by FAA facilities.

Additional deployment efforts investigated the impact of precise estimates of arrival time on ramp and hub management operations. Precise arrival time estimates allowed air carriers to improve gate management operations through more efficient allocation of ground resources and personnel. The eventual goal of the CAP program is to incorporate user preferences into the CTAS scheduling algorithms. An air carrier can make business decisions about which flights are most important, and these preferences can be accommodated if there are opportunities within the scheduling process. To this end, the CAP program will study the current methods for handling air carrier preference requests and develop ways of automating the requests. The next step will be the investigation of the feasibility of performing non-FCFS CTAS arrival scheduling through S-CAP.

References

[1]Ballin, M. G., Coppenbarger, R. A., Schleicher, D. R., and Johnson, S. C., "Summary Overview and Status of Advance Air Transportation Technologies Program Development Activities," Flight Management and Human Factors Division, NASA Ames Research Center, Moffett Field, CA, May 1997.

[2]RTCA, Inc., "Final Report of RTCA Task Force 3: Free Flight Implementation," RTCA, Inc., Washington, DC, Oct. 1995.

[3]Denery, D., and Erzberger, H., "The Center/TRACON Automation System: Simulation and Field Testing," *Proceedings of the Advanced Workshop on Air Traffic Management*, Capri, Italy, Oct. 1995.

[4]Davis, T. J., Krzeczowski, K. J., and Bergh, C., "The Final Approach Spacing Tool," *Proceedings of the 13th IFAC Symposium on Automatic Control in Aerospace*, edited by Schaecter, D. B., and K. R. Lorell, Lockheed Missiles and Space Company, Inc., Sunnyvale, CA, 1994, pp. 70–76.

[5]Erzberger, H., and Tobias, L., "A Time-Based Concept for Terminal-Area Traffic Management," *Proceedings of the 1986 AGARD Conference No. 410 on Efficient Conduct of Individual Flights and Air Traffic*, Brussels, AGARD-CP 410, Advisory Group for Aerospace Research and Development, Neuilly-Sur-Seine, France, 1986, pp. 52-1—52-14.

[6]Swenson, H. N., Hoang, T., Engelland, S., Vincent, D., Sanders, T., Sanford, B., and Heere, K., "Design and Operational Evaluation of the Traffic Management Advisor at the Fort Worth Air Route Traffic Control Center," *Proceedings of the 1st USA/Europe*

Air Traffic Management Research and Development Seminar, Saclay, France, 17–19 June 1997.

[7]Hoang, T., and Swenson, H., "The Challenges of Field Testing the Traffic Management Advisor in an Operational Air Traffic Control Facility," *Proceedings of the AIAA Guidance, Navigation, and Control Conference*, AIAA-97-3734 AIAA, Reston, VA, Aug. 1997, pp. 1409–1417.

[8]Perry, T., "In Search of the Future of Air Traffic Control," *IEEE Spectrum*, Vol. 23, No. 8, Inst. of Electrical and Electronics Engineers, New York, Aug. 1997, pp. 18–35.

[9]Beatty, R., "Cooperative Problem Solving Between Airline Operations Control and ATC Traffic Flow Management," *Proceedings of the Airline Dispatcher's Federation Conference*, Airline Dispatchers Federation, Washington, DC, 1996, pp. 204–208.

[10]Zelenka, R., Quinn, C., Beatty, R., and Heere, K., "The Impact of CTAS Information on Airline Operational Control," *Air Traffic Control Quarterly,* Vol. 8, No. 1, Air Traffic Control Association Institute, Inc., Arlington, VA, 2000, pp. 33–62.

[11]Irrgang, M., "Airline Irregular Operation," *Handbook of Airline Operations*, McGraw-Hill, New York, 1995, pp. 349–366.

[12]Carr, G. C., Erzberger, H., and Neuman, F., "Delay Exchanges in Arrival Sequencing and Scheduling," *Journal of Aircraft,* Vol. 36, No. 5, AIAA, Reston, VA, 1999, pp. 785–791.

[13]Carr, G. C., Erzberger, H., and Neuman, F., "Fast Time Study of Airline Influenced Arrival Sequencing and Scheduling," *Journal of Guidance, Control, and Dynamics,* Vol. 23, No. 3, AIAA, Reston, VA, 2000, pp. 526–531.

Chapter 13

Data Flow Analysis and Optimization Potential from Gate to Gate

Matthias Poppe*
DFS Deutsche Flugsicherung GmbH, Langen, Germany
and
Georg Bolz[†]
Lufthansa AG, Frankfurt/Main, Germany

I. Introduction

THE main objective of the Joint Air Navigation Experiments (JANE) is the evaluation and demonstration of a future air traffic management (ATM) system and required components to provide support for decisions on new ground systems and avionics. Guidelines are the ideas established in the framework of the development of the European Air Traffic Management System (EATMS), the gate-to-gate concept, collaborative decision making (CDM), and free flight.

The program was initiated in 1997 in a joint effort by Deutsche Lufthansa AG (DLH) and Deutsche Flugsicherung GmbH (DFS). The motive for JANE is the assessment of the potential for capacity improvement and benefit aspects by defining and carrying out studies, simulations, and field trials with all of the involved parties, i.e., air traffic control (ATC), airspace users, and airports.

The process modeling and simulation contributes in particular to the objectives of JANE by

1) assessment of new concepts: from a business point of view all relevant ATM processes are analyzed and optimization potential could be assessed independent of the organizations and current roles of the partners; the business case is the airline request for passenger and/or cargo transportation;

2) proposal of scenarios for implementation: as a result of this business modeling and simulation, the need and benefit for a new local decision support system (LDSS) has been clearly identified, and the data interfaces and underlying methodology (implemented in the algorithms and decision rules) were deduced.

Copyright © 2001 by DFS. Published by the American Institute of Aeronautics and Astronautics, Inc., with permission.
*Project Manager Research and Development.
[†]Director IT-Strategy.

II. Definitions

A. Business Process

A business process is an activity that is relevant for achieving the business goals. Here, the business goal is the air transport of passengers, cargo, or mail.

The process (In a hierarchical view, a process can consist of a group of other processes that can be described in subsequent refinement steps.) is initiated by an event. The completion of the process itself is an event that triggers one or more succeeding processes, thus forming a continuous sequence of process steps. A process usually requires time and resources to be performed.

B. Types of Processes

An (organizational) procedure is a sequence of atomic operations performed by one operative person. The workflow is a network of procedures at the group level, thus involving the work of one or more operative persons. The business process is a network of workflows and procedures at the level of a single organizational unit. The term business process is also used in order to refer to an enterprise-wide or an enterprise-overlapping network of business processes, which involves one or more organizational units working together on a common task.

C. Functional, Information, and Organization View

The method of business process modeling allows the creation of an organizational model that integrates three views of model descriptions: 1) the function view, which describes the sequence of activities that are necessary to process a job; 2) the information view, which describes the information interchange among activities and between activities and information stores; and 3) the organization view, which shows the responsible organizational unit.

The process model reflects the dynamic behavior of the organizational system. This provides an easy-to-understand overview of interactions within complex organizational systems.

III. ATM Process Model

The scope of this model with its main process groups is explained. The focus has been shifted from the gate-to-gate view to a new process-oriented view. While preserving the central idea of the gate-to-gate approach, this eases the identification of possible bottlenecks of the whole ATM system.

A. Scope of the Model

1. Overview

The model describes presently existing processes of ATM in the approach and departure sector, ground control and handling procedures at the airport, and other air transport activities between approach and departure phase of a flight. Organizations involved are airlines, airports, and air traffic control.

The processes are related to a single flight. Multiple instances of this abstract flight interact through usage of resources. The full scenario is triggered by external events representing the entrance and exit of flights.

The model covers normal operational procedure. Parameters allow to representation of a wide range of normal operating conditions; these can be covered in simulations. Process performance can be investigated by various measures. However, it does not include specific descriptions for deviations from procedures under exceptional conditions.

Simulation in the sense of this model can be efficiently applied to validate consistency, to quantify resource usage, to identify bottlenecks, and to analyze measures of the overall process performance under different scenarios. Therefore, it could be a useful instrument in a preparatory analysis phase for fast-time or real-time simulations.

Simulation as used here represents a typical flight event. It may not be employed to analyze effects depending on details of the properties of a particular flight or the correlation between consecutive flights. It is especially unsuited to predict the singular features of a specific series of events on a particular day of operations.

The model is generic and can in principle be applied to other airports. The actual situation of a specific airport has to be provided for simulations. Different scenarios can easily be incorporated on the specific process levels. These scenarios can be described by varying parameters inside the existing specific model or by modifying the process organization.

2. ATM Context Level

The context level is the abstract description of the entire ATM processes. It consists of three main process groups:

1) Dispatch and Consolidate Flightplan. This service process is a necessary precondition for each single flight. It contains dispatch of an operational flight plan by the airline or the handling agent, check and distribution of the flight plan by ATC, air traffic flow management (ATFM) measures.

2) Handle, Support, and Operate Flight. This is the core process group of the ATM processes. It contains the ground handling part, the ground movement part, and the airborne part.

3) Optimize Usage of Air Traffic System. This support process consists of measures for global optimizing procedures in order to achieve collaborative decision making.

The external entry points are "Scheduled (Planned) Flights Departure," "Actual Arrivals" (from the en route phase), and "Time-Event" (time-triggered process). These are used to create the number of events with respect to the simulation.

The external exit point "En Route" describes the end of the ATM process with respect to the focus chosen for this project. It can be used as an interface to other process models describing the en route phase in detail.

B. Gate-to-Gate View and Process View

Tasks were identified together with operational experts.[1] The tasks have been selected and grouped according to objectives and methodology of the project. Accordingly, the tasks highlighted in gray in Fig. 1 were selected to fall within the scope of the ATM process model. Reasons for this choice were to focus on

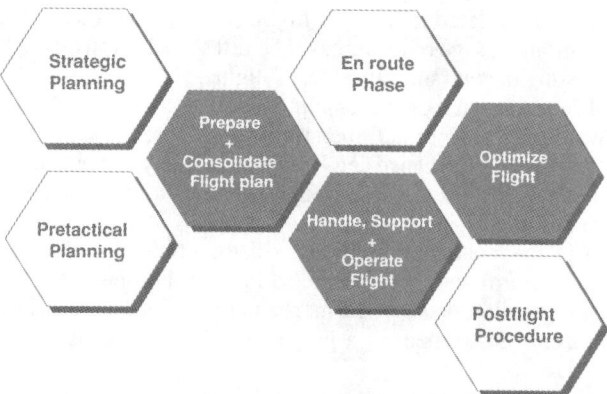

Fig. 1 Task areas within the JANE reference scnario.

operational procedures (the preoperational planning phases generate state parameters for the external inputs inside the ATM model), to aim at preparation for a collaborative decision-making process during the phases where *all* partners are involved, and to avoid complexity beyond the objectives of the JANE process modeling.

The objectives of the project necessitate a deviation from the conventional view of the flight process from the gate-to-gate view. The following reasons demand a view of the flight process from en route to airport to en route (refer also to Fig. 2):

1) All participants are involved during the arrival, ground handling, and departure phase.

2) The creation of a process model with the focus on the en route phase requires an investigation of at least two airports. This would require a higher effort without increasing general information content.

3) The goal of the process modeling project is the creation of a generic model. Therefore, it is not necessary to distinguish between different airport infrastructures and airport-specific variables.

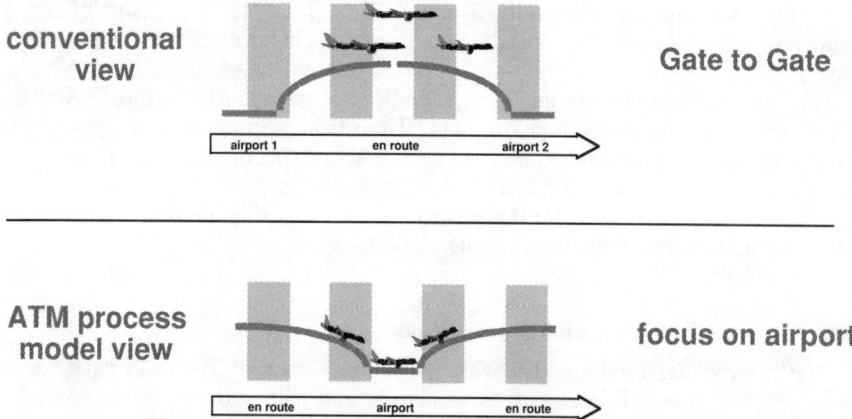

Fig. 2 Different views of the ATM process.

4) The interaction of departing and arriving flights at the airport is an important criterion for identification of potentials.

5) En route and ATFM processes are dealt with but not modeled per se.

Although the en route phase has not been modeled per se, the process interfaces are available. A detailed en route investigation is more suited to fast-time and/or real-time simulations, which, on the other hand, provide limited capabilities in terms of airport and ground process activities including ATFM.

IV. ATM Process Model Simulations

An overview is provided of the simulation capabilities and the benefits and limitations of a process simulation.

A. Simulation Concept

Process simulation is a part of process optimization. Simulation is the transition from the single process analysis to the multiple-event analysis. The realization of a process is called a process instance. In the ATM process model, each process instance reflects a single flight event. Every process instance is created at an external input (Scheduled (Planned) Flights Departure and Actual Arrivals in the ATM process model). A flight event created at an entry point runs through the whole process and its refinements. The duration between entry point and exit point is the sum of total handling times, total waiting times, total transport times, and total setup times. Waiting times occur when physical resources, human resources, or information required are unavailable.

An analysis based on a simulation run can reveal: 1) resource interdependencies (resources cannot be used independently) and relations among the partners ATC, airports, and airlines; 2) deficiencies caused by process synchronization (required information was not provided on time by activities performed simultaneously); 3) delays caused by high transport times of information; and 4) bottlenecks caused by limited resources (several activities use one resource). Note that the process instances interact exclusively via resources, a fact that is central to the understanding of multiple-event analysis.

B. Dynamic Process Analysis

The dynamic process analysis consists of three major steps:

1) In the first step several simulation runs were performed for calibration of the process model by varying process parameters, such as distribution of processing times. Simulation data were compared to real-world data to find the optimal adjustment.

2) In the second step the calibrated model was analyzed with respect to detecting capacity bottlenecks of resources (e.g., load of the runways during one day of operation); detecting dependencies between resource capacities and control procedures (e.g., different procedures for the coordination of arrivals depending on runway capacity); detecting delays due to resource unavailability (e.g., physical resources such as the taxiway or human resources such as handling personnel); and detecting the upper capacity limit of the analyzed ATM environment (e.g., maximum number of arrivals and departures).

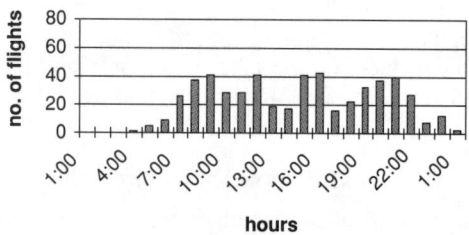

Fig. 3 Distribution of actual arrivals on a typical day of operation (538 arrivals in total). Each column represents the number of arrivals within an interval of 1 h.

3) After the identification of weaknesses, what-if analyses were used to create alternatives to find an improved solution. Examples of what-if analysis are variation of certain process scenarios (within the scope of this study this was used for testing the model's consistency); variation of input data (Section C); and variation of process parameters (e.g., aircraft separation or runway capacity).

C. Example: Variation of Input Data

This is an example of a what-if analysis by variation of input data. The aim is to analyze the dependency between distribution of arrivals and occurrence of waiting times caused by sequencing and required separation standards.[2]

Figure 3 shows the frequency distribution of actual arrivals per hour based on real-world data from the DFS Statistical Analysis System (STANLY) at Frankfurt airport (EDDF). The diagram reveals that the maximum number of flights per hour does not exceed 42. Four peaks can be identified.

Figure 4 shows the distribution of process waiting times (delays) based on the distribution of arrivals shown in Fig. 3. The diagram reveals that for 348 of the

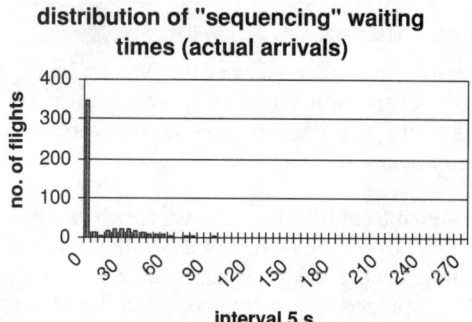

Fig. 4 Distribution of process waiting times caused by sequencing of aircraft. Each column represents the number of flights within an interval of 5 s (first column: no waiting times).

distribution of test arrivals per hour

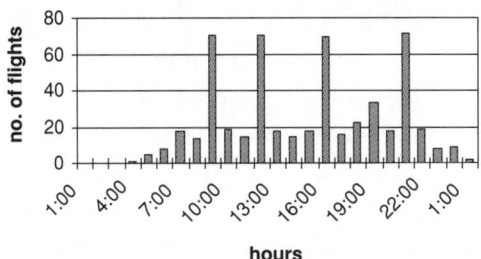

Fig. 5 Modified distribution of arrivals per hour (538 arrivals in total). Each column represents the number of arrivals within an interval of 1 h.

538 flights, no waiting times exist (refer to first column), and for the most part the waiting times for the remaining fights do not exceed 60 s. This indicates that the traffic situation on this particular day was very stable.

Figure 5 shows a test scenario in which the number of flights within the four peaks were artificially increased to approximately 70 flights per hour while the sum over the entire day is kept unchanged to demonstrate the possible effects of bulk arrivals.

The new distribution of process waiting times (shown in Fig. 6) was calculated with the simulation results using the distribution of the test scenario as input. The number of aircraft delayed is greater than in Fig. 4. The diagram reveals that for 255 of the 538 flights, no waiting times exist (refer to first column), and for most of the remaining flights the waiting times do not exceed 80 s.

A comparison of Fig. 4 and Fig. 6 shows that the artificial redistribution of flights with a significant increase of flights within the peaks led to an increase in process waiting times.

This variation of input data shows that bulk arrivals do not cause a serious problem for the sequencing processes. However, possible disturbing effects for

distribution of "sequencing" waiting times (test arrivals)

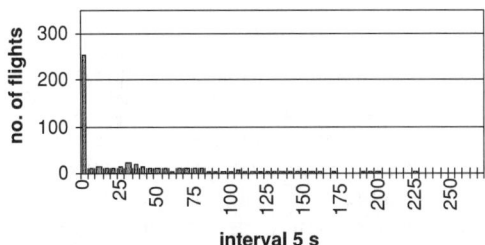

Fig. 6 Distribution of process waiting times caused by sequencing of aircraft in the test scenario. Each column represents the number of flights within an interval of 5 s (first column: no waiting times).

airport and/or airline operation could be revealed by extending this exercise to other processes (e.g., effects on taxiway utilization or gate allocation).

V. Identification of Potentials

Operational areas with potential for short-term and long-term improvements are identified from a process-oriented view by taking into account the interdependencies which other processes from other partners. They depend in their concrete form on the specific situation at Frankfurt airport, which is regarded as rather typical.

They serve as examples for how the process modeling led to identification of areas for improvement. They constitute candidates for further investigations into specification of process modifications, which could furnish business benefits for the overall ATM process at the airport.

A. Processes with Short-Term Improvement Potential

1. Start-Up Given and Push-Back Requested

The sequence of process steps around "Start-Up Given" shows that the main purpose here is a coordination between flight crew and tower and a consolidation of flight-related information to adjust eventual remaining discrepancies. This could well be automated.

The issuing of start-up and push-back clearances by the apron controller depends on the traffic situation around the aircraft position. A first-come-first served (FCFS) principle prevails. The ordering of push-back clearances issued does not reflect the most effective sequencing appropriate for aircraft to arrive for takeoff at the runway. Furthermore, the apron controller lacks information on short-term slot alterations or on priority situations within the queue of flights ready for departure.

For the timely availability of push-back vehicles at the gate, it would be beneficial to have an advance indication about 10 min ahead of the push-back request. This is an example of usefulness of proactive control.

Again, as long as no agreements have been reached on how to guarantee fairness among airlines, at least within those geographic areas of the airport used primarily by one airline and among the flights of one airline, the concerned airline should be contributing to an optimized sequencing by providing a list of priority rights.

The necessary information to arrive at an optimized sequencing is available in principle. It needs to be delivered to decision makers together with optimization criteria provided by collaborating participants.

2. Actual Slot Usage

The actual usage of central flow management unit (CFMU) slots as opposed to planned slot usage leaves slots unutilized due to extensive preplanning times and the lack of replanning. This occurs when delayed aircraft cannot use their planned slots. No regulated process of requeuing these unused slots exists here. A kind of short-term trading would remedy this situation. Also, prioritization according to importance of afflicted flights, in particular their impact on hub connectivity, should be included. Although from the ATC point of view an equal treatment of all airlines has to be guaranteed, this prioritization could, without taking further measures, at least be performed among flights of one airline with involvement of this airline's operations control center.

B. Processes with Long-Term Improvement Potential

1. Actual Position and Gate Allocation

At the planning level, the allocation of positions and gates is coordinated. However, when deviations occur in actual operations, there is only ad-hoc distribution of available resources. The information necessary for prioritization is not available for decision making at this time.

By providing this information and by implementing coordination, improvement could be achieved here. The optimization goal for gate allocation is close positioning of flights with high flow of transfer passengers.

2. Arrival and Departure Management

The present ground handling processes are orientated toward maintaining departure punctuality. However, this is not a primary business goal from the point of view of customer services an airline provides. What is called for is on-time arrival for the passenger reaching the final destination and maintaining connectivity for the connecting passenger.

Therefore, an integrated and coordinated arrival and departure management that guarantees arrival punctuality taking into account flow management, departure and arrival control, ground movement, and ground handling aspects while being flexible with respect to departure time could help to limit the impact of delays on other flights and to secure hub connectivity.

The necessary information to arrive at an optimized arrival and departure management is available in principle. It needs to be delivered to decision makers together with optimization criteria.

C. Need for Change in Process Understanding

It is a general feature of the processes analyzed that they lack communication, cooperation, and coordination between process instances and among actors, while being regulated and interrelated solely by the usage of central resources. Insights gained on the basis of the process model underline the necessity of a fundamental change in the process understanding. Guiding principles are collaborative decision making and a better coupling between the planning and control phases (replanning).

1. Need for Process Improvements

Process improvements in particular are required because of insufficient coordination. The missing procedural interdependencies inside the control phase and between the control and planning phase lead to local optimization decisions that might have global ramifications of a disturbing nature. To overcome this global suboptimality, appropriate coordination and information exchange steps need to be implemented.

This situation is aggravated by the fact that different phases from different process instances overlap in time, i.e., planning phases for later flights are executed at the same time as control steps for earlier flights. Disturbances in the actual operations caused by deviations from the planned course of events influence the leeway for future planning steps due to tight resource constraints and complicated time sequencing patterns. The process execution and the information of these time interleaved planning and control phases stay separated, whereas the target of the operational planning is changing due to impact from the actual course of events.

The information base for decision making in the planning phase is not updated by actual operational data. Short-term planning often occurs in a situational unawareness from the actual operational status.

2. Limitations of Traditional Process Organization

Several instances of the process flight as understood here in the ATM process model are solely coupled via resources. They interact indirectly by making use of the same resource (the notion of resource is to be understood here in an abstract way in the wide sense, e.g., the tractor or the gate position is a resource as well as the tower controller communicating with the cockpit). This is not a limitation of the methodology employed or the tools used but a feature of the real processes as they are designed and/or have evolved.

Deficiencies in actual process organization occur mainly in coordinating activities in which the coordinating agent is unaware of the goals and optimization criteria of other parties affected. As an example, consider the process step "Arrival Sequencing" carried out by the ATC controller in the approach phase, which lacks prioritization criteria according to urgency and importance even though this aspect could be critical for hub connectivity. Note that present International Civil Aviation Organization (ICAO) regulations mandate a FCFS principle here.

A particular aspect is the lack of well-defined process termination triggers. It occurred as a problem to describe interdependencies between process steps involving several participants. The process cannot be terminated before all subprocess threads involved are finished.

The situation emerged that the preceding process step lacks a clear termination signal and an indicator for when the termination signal is to be expected. The next process step with its waiting period may be on a time-critical path of the overall process. This leads to intentional deviations from the designated course of events in order to speed up the overall process. This happens to ensure the timely entrance into succeeding process steps but from a local perspective with limited information. For example, with a push-back request; the request is typically issued by the pilot even before all operational conditions are met because he knows from his experience that it takes a few minutes until the push-back car will arrive. With this look-ahead action, he may gain valuable time for the next activities. However, this is not a systematical and predictable behavior.

What is required is a look-ahead in time based on an expected remaining duration of the prior process step. A well-specified process could be set up ensuring an optimum time efficiency for all push-back requests if 1) an improved interconnection of process steps would exist together with the required information exchange and 2) the necessary look-ahead in time by early indicators could be made available.

3. Guiding Principles for Process Improvements

Three principles are identified that will be useful to conceptualize the approach toward process redesign in the area of ATM.

a. Collaborative decision making. CDM is a paradigm for process organization in a context where several organizational entities with different goals and decision horizons work together on a common task. Traditionally, responsibility for each subtask and subgoal is delegated to one of the organizational units, which decides according to its own priorities, based on its limited view of the process. The concept of CDM constitutes the principle of sharing (internal state) information relevant to the common process and coordinating decisions based on a balance of interests in a collaborative way, reconciling local restricted views and global common goals.[3]

Each participant still has a limited information horizon, whereas the joint decision must emerge from a consideration of all information relevant to the process. However, the possibility of readjustments and, for security reasons, override possibilities must be foreseen. It should be decided case by case to what extent CDM will be implemented as a rule-driven process between autonomous actors or as coordinated by a central instance composed of all participants.

b. Replanning and proactive control. In air transport operations there is a continuous interleaving in time of planning and control activities.[4] Operational planning is an ongoing activity with a typical time horizon of 2 h (in case of airport operations). The necessary information is available in due time as a planning information base. However, modifications of the planning assumptions due to deviations from plan during actual operations are not available.

No activities for readjusting planning assumptions and for performing a replanning could be identified. Actual status information about flight operations do not enter into the planning process for later flights despite the fact that they are available and used in other activities. If made available, they would constitute valuable information to achieve a more realistic and up-to-date planning.

A crucial element to achieve the look-ahead in time (forecast of events) is the usage and availability of accurate and up-to-date estimated times for the occurrence of certain process states as input data for proactive control. While these exist in various information systems they are not formally collected for other processes.

The combination of both elements will allow to implement the principle of Replanning and Proactive Control (RPC).

c. Cooperating intelligent agents (CIA). The idea of active objects CIA is applicable on several levels of the ATM and ground handling process. They should be equipped with the capability of targeted information transmission, "autonomous intelligence," and local decision-making capability for mutual coordination and partial planning. As an example, the whole concept of free flight considers aircraft equipped with advanced flight management systems and data link as active objects, deciding autonomously on their preferred trajectories, actively engaging in communication with a ground control station and negotiating their assigned trajectories in a collaborative way.

As a process design paradigm, CIA will be helpful to conceive the processes to be designed under the new procedural concept. As a software design paradigm, CIA will constitute an element of interconnected decision support systems to implement the type of business processes envisaged for CDM. CIA could supplement and support the two other paradigms, CDM and RPC. It may be successfully applied in long-term reorganization as well as in short-term adjustment.

VI. Conclusions

The main achievements with the present model are summarized. The possibilities of usage of this ATM process model are outlined, and a look into the future shows a first concrete use by supporting the development of the local decision support system.

What has been achieved with the present model for ATM and ground handling processes at the airport can be summarized as follows.

1) The approach chosen, i.e., to analyze the process flight centered at the airport, and the methodology employed, i.e., to base the analysis in a coherent way on a business process modeling methodology concentrating on processes independent of organizational structures, have been successfully proven to be applicable to modeling ATM and ground handling procedures. All of the required information to optimize the overall ATM is available in principle and has been identified, but problems are caused by the different systems, different interpretation of data, and different time horizons. The approach chosen and the methodology employed have also been used to identify areas for process improvements with high business value, both for short-term improvements through process adjustments and for long-term optimization through process reorganization.

2) With this tool it can be shown how singular (process) improvements of one partner impact the operations of the other partners, e.g., whether there is an overall benefit if ATC increases the arrival/departure capacity, or if other bottlenecks arise that had not been taken into consideration before.

3) The process model with its simulation capabilities can further be used to support fast-time and real-time simulations by what-if-analyses to show interdependencies of the whole ATM environment, including airport and airline operations.

Business process modeling has been applied in a coherent, rigorous, and comprehensive way to ATM processes. The consistent application of the business process modeling methodology and the unified view of the complete process is the key to identifying optimization potentials in the overall process beyond organizational boundaries and a prerequisite for future work on process reorganization.

For future use of the model, the model interfaces with the outside world, i.e., with surrounding processes not described explicitly in the model, through well-defined entry and exit points. It can be extended by replacing these with connections to other newly described processes. Typical extensions could be to include the en route phase explicitly and to form a complete ATM scenario by connecting different airports. However, both from a practical point of view (to keep complexity of the model at a manageable level) as well as from a conceptual point of view, it is desirable to extend the model not simply by enlarging it but also by consolidating it on a subprocess level.

To arrive at detailed analyses of particular subprocesses or specific aspects, the level of detail present in the model needs to be refined. This would apply in particular when more information on subprocesses is to be gained through simulation, e.g., to evaluate scenarios for process improvement in particular areas, or if information flows between various information technology (IT) systems involved were of interest.

The following topics have been identified as most promising to approach next:

1) What-if analysis: investigate how the distribution of peaks in the arrival frequency affects runway usage. This will be a preparatory analysis for fast-time simulations.

2) Feasibility study: extend the model to cope with different runway configurations, standard arrival and departure routes, and integration of detailed ground handling activities.

The abstract concepts presented in Section V.C (CDM, RPC, CIA) will be shown to be applicable and useful in the context of JANE through fast-time and real-time simulations. The LDSS closes the gap between planning and control by supporting the following objectives: 1) optimize local airport throughput, 2) enable collaborative decisions by providing common situational awareness, 3) increase ATC capacity for approach and departure, and 4) support gate-to-gate operations. The LDSS uses a rather simple methodology to reach these objectives. All inbound and outbound flights are supervised and managed by time control related to the local airport runway capacity. Main parameters are the required time of arrival and the required takeoff time. The LDSS rules and internal calculations will take into account all relevant data, such as the runway configuration, CFMU slots, weather, and a number of flight-related information. Algorithms and optimal strategies can be chosen based on the results of theoretical studies and of the what-if analyses mentioned previously.

A prototype of the LDSS is currently being operated and is under evaluation at Frankfurt airport. It is an excellent example of the transition of research and development results to a pre-operational system.

References

[1] Lauf, Wolfgang (ed.), *JANE Model of Air Traffic Management Processes*. Model Documentation. DFS, Langen, Germany, Sept. 1998.

[2] de Vèrdiere, D. C., *Adaptation of Demand to Available Runway and Airspace Capacity*. Version 1.0, ECTF/IP/016, EATMS Concept Task Force EUROCONTROL, Brussels, 1996.

[3] *Potential Applications of Collaborative Planning and Decision Making*. Final Rept. EATCHIP Task CSD-4-01, EUROCONTROL, Brussels, 1998.

[4] *Future ATFM-AO-Airport Synergies Towards Enhanced Operations*. Executive Summary, EEC Rept. No. 332, EATCHIP Task CSD-4-01, EUROCONTROL, Bretigny, France, Aug. 1998.

Chapter 14

Effect of Shared Information on Pilot/Controller and Controller/Controller Interactions

R. John Hansman* and Hayley J. Davison[†]

Massachusetts Institute of Technology, Cambridge, Massachusetts

I. Introduction

TO RESPOND to the increasing demand on limited airspace system resources, a number of applications of information technology have been proposed, or are under investigation, to improve the efficiency, capacity, and reliability of air traffic management (ATM) operations. Much of the attention in advanced ATM technology has focused on advanced automation systems or decision aiding systems to improve the performance of individual pilots or controllers. However, the most significant overall potential for information technology appears to be in increasing the shared information between human agents such as pilots and controllers or between interacting controllers or traffic flow managers. Examples of proposed shared information systems in the United States include controller pilot datalink communication (CPDLC), traffic management advisor (TMA), automatic dependant surveillance (ADS), collaborative decision making (CDM), and National Airspace System (NAS) level common information exchange.[1]

ATM is fundamentally a human-centered process consisting of the negotiation, execution, and monitoring of contracts between human agents for the allocation of limited airspace, runway, and airport surface resources. The decision processes within ATM tend to be semistructured. Many of the routine elements in ATM decision making on the part of the controllers or pilots are well structured and can be represented by well-defined rules or procedures. However, in disrupted conditions, the ATM decision processes are often unstructured and cannot be

---

Copyright © 2001 by the Massachusetts Institute of Technology. Published by the American Institute of Aeronautics and Astronautics, Inc., with permission.

*Professor, Department of Aeronautics and Astronautics, MIT International Center for Air Transportation. Associate Fellow of AIAA.

[†]Graduate Research Assistant, Department of Aeronautics and Astronautics, MIT International Center for Air Transportation.

reduced to a set of discrete rules. As a consequence, the ability to automate ATM processes will be limited, and ATM will continue to be a human-centered process in which the responsibility and the authority for the negotiation will continue to rest with human controllers and pilots. As the ATM system leans more toward distributed authority to improve efficiency and safety, the use of information technology to support the human decision process will therefore become an important aspect.

The premise of many of the proposed shared information systems is that the performance of ATM operations will improve with an increase in shared situation awareness between agents (pilots, controller, dispatchers). This will allow better informed control decisions and an improved ability to negotiate between agents. A common information basis may reduce communication load and increase the level of collaboration in the decision process.

In general, information sharing is expected to have advantages for all agents within the system. However, there are important questions that remain to be addressed. For example, what shared information is most important for developing effective shared situation awareness? Are there issues of information saturation? Does information parity create ambiguity in control authority? Will information sharing induce undesirable or unstable gaming behavior between agents?

This chapter will explore the effect of current and proposed information sharing between different ATM agents. The chapter will primarily concentrate on bilateral tactical interactions between specific agents (pilot/controller, controller/controller, pilot/dispatcher, controller/dispatcher). However, it will also briefly discuss multilateral interaction and more strategic interactions. The information about the interactions was based on field observations in air traffic control facilities, flight decks, and dispatch units as well as a series of surveys inquiring about the status of information sharing in the NAS today.

II. Why Humans Are Necessary in ATM

The need for human decision makers within ATM can be demonstrated from an analysis of ATM decision processes based on recent decision theory models.[2] Semistructured decision theory asserts that humans add value to unstructured decision processes. A semistructured decision process consists of structured and unstructured subprocesses. A structured decision process is defined as a process that can be reduced to a well-defined set of rules and therefore could be reliably automated. Unstructured processes are those for which a reliable set of rules governing the decision process cannot be defined a priori. There are many reasons why it may be inappropriate to structure a decision process, including insufficient understanding of the process due to complexity, ambiguity of goals, insufficient data on which to base a structured decision, uncertainty, and changing environments, as well as humanistic elements such as creativity or subjective and moral judgment.

Within ATM examples of structured decisions are those completely defined by standard operating procedures, routine processes or currently automated processes such as radar tracking algorithms. However, many important decisions within ATM are unstructured and therefore difficult to reliably automate. Even relatively simple decision problems such as conflict resolution are difficult to automate using classical optimization techniques. More complex problems such as dealing with multiaircraft conflicts, irregular operations, and ambiguous situations tend to be even less structured and clearly require unstructured solution strategies. It should be noted that decision processes are more likely to become unstructured as the state of the system moves away from nominal operations. Therefore, information systems that support the human in unstructured decisions are most important in non-nominal situations due to irregular operations, weather, unanticipated failures, or emergency situations.

In unstructured decision theory, it is not possible to define the optimal decision a priori, but it is sometimes possible to identify well-informed or ill-informed decision situations. In many cases, situations are unstructured simply because they lack sufficient data to execute a well-structured decision rule. Often, in these situations, the human is called upon to diagnose the situation and to infer the missing data or to project a probable future state of the system.

In general, the more informed the human agent, the better the decision process. However, if too much information is presented in time-critical situations, the human can become saturated and may not be able to integrate important information due to distraction, fixation, or simple overload. Expert operators (pilots and controllers) demonstrate the ability to filter out noncritical information and have a higher tolerance for information saturation; however, even this filtering process appears to be unstructured.

In processes involving multiple humans, the ability to share information generally makes each agent more informed and can result in improved overall negotiations and decisions, particularly if the shared information supports a well-structured decision process with a clearly optimal course of action. One aspect that complicates multiagent decision systems is resolving goal differences between the agents. In most unstructured decisions, humans exhibit extremely complex goal sets. Although agents may share a partial set of goals, there are elements in their goal sets that diverge. For example, in a comparison of controller and pilot goal sets[3–5] found that controllers and pilots share the goal of maintaining safety of flight, but they have differing goals with regard to optimizing the performance of the system. Controllers tend to have the goal of maintaining the flow within the sector, whereas the pilots have the objective of optimizing the performance of their specific flight.

In cases of weakly differing goals, the decision/negotiation process may become difficult because of ambiguity in which it may not be clear which goal set has the higher priority. In time-constrained decisions such as tactical ATM, it is therefore necessary to have a clearly defined hierarchy to resolve ambiguity. In more strongly differing goal sets, such as direct competition for resources, agents may exhibit gaming behavior in which their actions may be strongly coupled to other agents in the system. Potential issues of undesirable gaming behavior must be considered.

Based on the semistructured nature of decision processes within ATM, as well as issues of responsibility and societal trust, it is clear that human pilots, controllers, and dispatchers will continue to have supervisory decision-making responsibility in future ATM systems.[6,7] In developing decision support and information systems, considerations on how the proposed systems will impact the interaction need to be addressed.

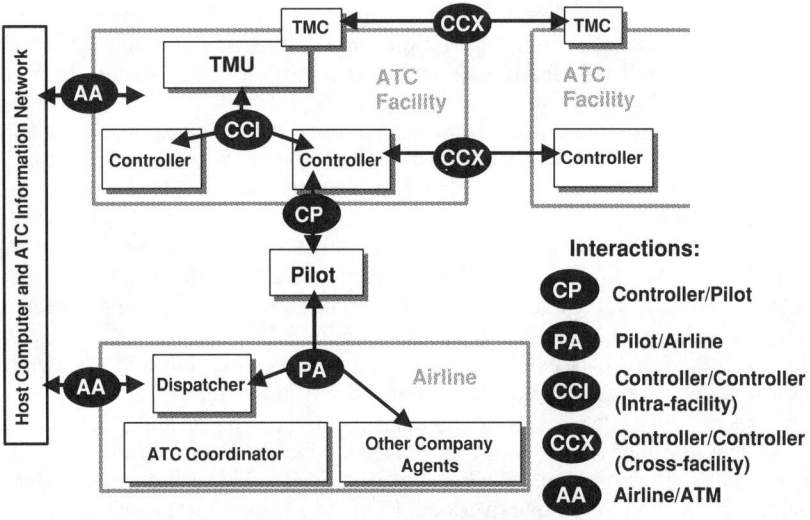

Fig. 1 ATM tactical interaction architecture.

III. ATM Interaction Architecture

To discuss the impact of shared information, it is useful to define the interaction architecture. The interactions between agents within ATM can be generally separated into tactical and strategic. The tactical level is focused on the management of individual flights as they propagate through the airspace systems. The key tactical interactions for a particular flight are shown schematically in Fig. 1. The central interaction is between the pilot and the current tactical controller who is managing that flight, and this is shown as the CP interaction in Fig. 1. However, the pilot (for airline operations) also interacts with other company representatives including the dispatcher (PA interaction) who provides technical and decision support services (e.g., flight planning, weights, weather), and other company agents (e.g., station, ramp, etc.) who interact with the flight. The tactical controller also interacts with other controllers who have current or future interest in the flight. In terms of the information architecture, it is important to distinguish which interactions occur within the same facility (interfacility, CCI interaction) from those interactions that occur between facilities (cross-facility, CCX interaction). Within each facility there is a traffic management coordinator (TMC) as a part of the traffic management unit (TMU) who interacts with the tactical controllers and is the principal external contact point for both tactical and strategic interactions with the facility. Performing a similar function for the airlines, the ATC coordinator is responsible for the tactical and strategic interactions with ATM (AA interaction), such as receiving information about NAS delays and conveying the delays to the dispatchers as well as submitting and making changes to canned flight plans.

At the strategic level, the interactions, shown schematically in Fig. 2, focus principally on the management of traffic flows and airspace system resources. In the United States, the central elements of strategic interaction are performed by the air traffic control system command center (ATCSCC). These elements interact with

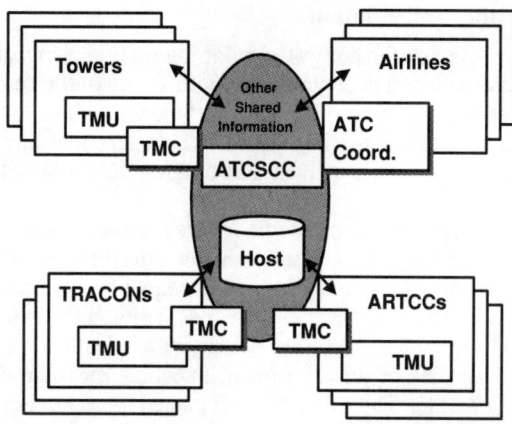

Fig. 2 ATM strategic interaction architecture.

each of the control facilities [air route traffic control center (ARTCCs), specified terminal radar approach control (TRACONs), and towers], as well as with the airlines generically represented as airline operations centers (AOC).

IV. Interaction Assumptions

The interactions between human agents within the ATM system are based on an assumed set of common rules, information, procedures, background, language, and culture. The common rules are the operating regulations such as the Federal Aviation Regulations. Common information includes published navigational, airway facility, and Notices to Airmen (NOTAM) information. Common procedures include published flight procedures such as instrument approach procedures and standard terminal arrival routes, as well as established operational procedures such as airborne holding procedures. The common language of ATM is English with International Civil Aviation Organization (ICAO) standard phraseology. The common background and culture has evolved over decades of continuous safety-critical interaction. In general it is a culture of professionalism and shared respect between agents. The culture is transmitted through the apprentice training that both pilots and controllers experience.

V. Shared Information in Controller/Pilot Interactions

The interaction between pilot and controller is focused on the management of the specific flight for which they have common responsibility. The principal interaction element is the assigned clearance that constitutes the contractual agreement between the pilot and controller for the airspace, runway, or airport surface resources assigned to the flight. The need for shared information increases when the clearance must be amended due to some conflict (e.g., weather, traffic, airspace). In an analysis of information required to support shared controller/pilot situation awareness, several key areas have been identified, including weather, traffic, intent, and affective states such as workload, urgency, or stress.[4,5] Each will be discussed briefly.

A. Shared Weather Information

A primary causal factor for controller/pilot interactions is weather, which can make planned trajectories unacceptable for both safety and ride quality reasons. In en route operation, convective weather, turbulence, winds, and icing are the major weather factors that must be considered. In terminal and surface operations, visibility, ceiling, surface winds, and braking action are also key weather factors.

In current en route operations, most pilots have access to airborne weather radar and can visually observe the en route weather, whereas the controllers have limited weather presentation on their displays. As a consequence, controllers rely on pilot reports of convective weather, turbulence, winds, icing, and deviation requests to build a mental representation of the spatial extent of weather within their sector. Figure 3 shows examples of an experiment in which en route controllers were asked to draw their mental map of weather-impacted areas after controlling traffic around convective weather for approximately 20 min. In most cases the controllers' mental maps captured the major convective areas, but there were some gaps and the process is not reliable.[4]

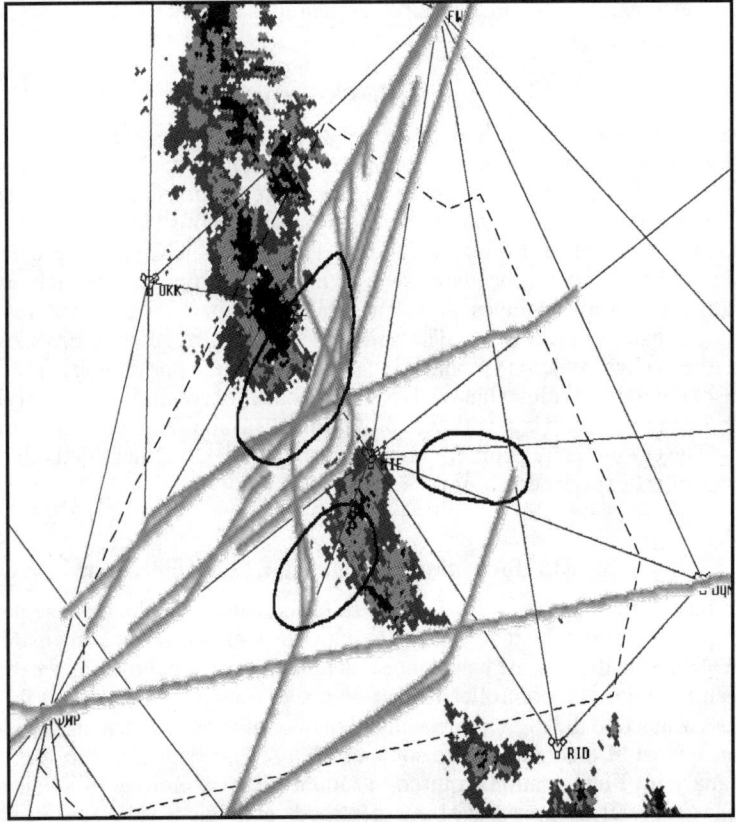

Fig. 3 Example of controller's inferred weather location (black), aircraft tracks (gray).[14]

In field observations, controllers are observed to build similar mental maps for turbulence, icing, and winds. Clear air turbulence (CAT) is not observable to the pilots and is a key consideration in route/altitude deviation requests. As a consequence, a collective mental representation of the turbulent regions is built up through informal pilots' reports (PIREPs). Access to turbulence information is generally through the controller's mental representation but is also sometimes obtained through "party-line" communications.[8,9] The principal difficulty in informal PIREP-based weather information is the inconsistency in coverage and age of the information. Often reports are not available, particularly for late night or early morning operations or during the early part of a controller's shift. Shared representations of wind and icing information are generated by similar processes but are not as common as turbulence reports.

To explore the effect of shared weather information on pilot and controller interactions, an integrated simulation study was conducted for six pilot/controller teams in a sector of the Indianapolis Center airspace. In the weather study, active en route controllers worked high-density traffic flows through broken lines of convective weather. In the shared information condition, the controllers' display included a Next Generation Radar (NEXRAD) weather radar overlay that replicated the weather radar information displayed on the pilot's electronic horizontal situation indicator (EHSI). The simulated weather scenarios included testable response probes[10] in order to evaluate the controllers' and pilots' weather situation awareness. In the case in which the controllers did not have convective weather information displayed, the testable response metric indicated that they were aware of the weather in only 50% of the cases. In the shared information condition, the results indicate that they were aware in 94% of the test cases. In 6% of the cases the results were ambiguous, but there were no cases in which the controllers were clearly unaware of the weather situation.[4] In addition to the improved weather situation awareness enabled by the shared information, an improvement in controller performance was observed in the shared information case. In the 36 test cases, five loss of separation events (5 miles, 1000 ft) were observed, all of which occurred in the condition without the shared awareness. The increased performance appears to be due to an improved ability to infer pilot intent in maneuvering around convective weather (two cases) and a reduction in workload enabled by the enhanced weather situation awareness (three cases).[4,5]

B. Shared Traffic Information

Traffic is also a key factor necessitating interactions between pilots and controllers. In the analysis of goals and information requirements, it was found that pilots and controllers have related but differing goals with regard to traffic. The pilots are observed to have an aircraft-centered view and are primarily concerned with traffic that will impact their current or planned trajectories. Conversely, the controllers have a more system-centered, "big-picture," view and are concerned with how the trajectories and overall flows will interact.[4,5]

In the current radar-controlled environment, controllers have access to current state information (position, altitude, velocity) of all radar-observed aircraft as well as flight plan information. Unless equipped with the Traffic Alert and Collision Avoidance System (TCAS), the pilot's knowledge of the traffic situation is entirely dependent on being informed by the controller or on inferred mental representations developed by monitoring the party line. Because of the clear information superiority of the controller, pilots will generally defer to the controller's requests with regard to traffic. Controllers will often point out proximate traffic both to increase the pilot's level of traffic situation awareness and to distribute responsibility for traffic separation. If the pilot acknowledges visually acquiring the traffic, separation responsibility can be transferred to the pilot and reduced separation criteria can sometimes be used.

When equipped with TCAS, pilots have limited information on proximate traffic (typically within $+/-$ 2000 ft in altitude and 40 miles), and the potential for shared situation information is increased. As TCAS has become more common and the pilots and controllers have gained confidence in the system, the use of TCAS to support shared traffic situation awareness has increased. Although there are indications that TCAS provides a basis for shared information, the procedures for TCAS-based interactions have not been fully developed or exploited. Only limited applications of TCAS-based interactions, such as passing on oceanic tracks, have been approved.

C. Shared Intent Information

One of the most useful elements of shared information between pilots and controllers is shared information on intent. Past researchers have advocated the value of intent information in achieving the objective of distributed air-ground traffic management without defining what they consider intent information to be.[11] The current mechanisms for shared intent information are flight plans and published operating procedures. The flight plan is the basic mode of shared information on the goal and intent of the pilot. The elements of the standard flight plan were developed in the period of (low bandwidth) teletype communication and are quite sparse. The flight plan elements relating to aircraft intent include: 1) aircraft identification, 2) aircraft type and equipment, 3) departure point, 4) departure time, 5) initial cruising altitude, 6) route of flight, 7) airspeed, 8) destination, 9) alternate airports, 10) estimated time en route, and 11) fuel on board.[12,13]

It is interesting to note that the airway/waypoint structure of the airspace system is designed to efficiently communicate the route of flight by both voice and low bandwidth datalink communications. By designating the route clearance in terms of predefined airways, waypoints, or procedures (e.g., arrival procedures), both the pilot and controller have a shared representation of the clearance. The current airway structure was developed based on the location of ground-based navaids and was efficient when air navigation was exclusively based on these navaids. With the proliferation of satellite and inertially based area navigation, there is the desire to fly more efficient trajectories. However, to be able to communicate the intended trajectory and to coordinate with other controllers often result in the use of less efficient procedures. There is the need to update the airway system and airspace procedures to take advantage of improvements in navigation and communication capability. However, any future system must be capable of operating in degraded modes, and the ability to efficiently articulate the intended routing through voice must be maintained.

At the more tactical level, there is very little direct information on the pilot's immediate intentions. The controller has only the information on the pilot's radar-measured states (which exhibit significant lag) and what can be inferred from the flight plan and voice communications. As a consequence, the controller must allow a significant separation buffer to allow for uncertainty in pilot intent. In cases in which the pilot's intent is well known (e.g., final approach), it is sometimes possible to reduce the separation criteria. One of the potential advantages of ADS will be a reduction in sampling delay and the possible incorporation of feed forward states indicating aircraft intention such as heading, turn rate, or even the programmed trajectory within the aircraft's flight management system (FMS).

One study that addressed this issue tested airline pilots' performance using three traffic displays with varying levels of intent against the basic TCAS display. The first intent display provided only the rate at which the intruder was approaching, shown with an arrow, as well as a conflict probe and a profile view. The second display provided both the commanded altitude and heading of the intruder, and the last display included FMS trajectory including lateral navigation (LNAV) and vertical navigation (VNAV) paths. Various performance measures were taken while using each display, and it was found that during conflict situations the pilots began avoidance maneuvers later using TCAS than with both the command and rate displays, and that the pilots maneuvered later using the commanded or rate displays than when using the FMS trajectory display. However, the pilots seemed to find the FMS display the most useful in only the most complicated of situations, such as when there was a conflict with two or more aircraft at once. In simpler scenarios, it appeared that the high clutter factor in the FMS display outweighed the benefits of intent information, and the pilots began their avoidance maneuvers later. Thus the study seems to have found that a tradeoff between the amount of intent information presented in a display and the clutter it creates has a significant effect on performance.[14]

A final issue in shared intent information is the ability to communicate goals or rationale behind a flight plan request (by the pilot) or a route change or vector (by the controller). In current voice communications a pilot or controller may provide information supporting the cause of the request. In currently planned CPDLC datalink systems, only a limited syntax is expected to be available to provide rationale behind commands or requests.

D. Shared Affective Information

An important category of shared information between pilots and controllers is affective information such as emotional state, workload, urgency, or capability.[15] In current systems, affective information is inferred from the prosodic content of voice communications.[8] Controllers can clearly communicate urgency or their level of workload by the inflection and rate of their communications. Controllers also routinely assess competency, attentiveness, and capability by the speed of response and the prosodic content of pilot voice transmissions. The importance of shared affective information should be considered in future datalink environments as well as issues regarding the loss of party-line information (PLI). Simulator studies of party-line information usage indicate that PLI is an important but unreliable mode of information transfer. Mechanisms to transmit the important PLI elements should be included in future datalink communication systems such as CPDLC.[8]

VI. Shared Information in Pilot/Airline Interactions

During in-flight operations, the pilot has access to a number of company representatives through voice or datalink (ACARS) communications. In areas of flight planning, the principle interaction between the pilot and the airline is through the dispatcher, who provides flight planning services and supports the pilots in the event of any disruptions to the planned flight (e.g., weather, delays, diversions). In most airline operations the dispatcher develops and files the initial flight plan and fuel loading plan, which is subsequently reviewed and coauthorized by the pilot. Pilots frequently amend the fuel load but rarely amend the filed flight plan prior to departure. The flight plan provided to the pilot is typically a detailed plan that expands on the basic flight plan that is the basis of the pilot/controller interaction. The expanded flight plan includes planned altitude changes (step climbs), recommended speeds, as well as weather data. However, at most airlines the rationales behind route planning decisions (made by the dispatcher) are not communicated to the pilot directly.

One example of the lack of a shared information basis occurred during the initial phases of the National Route Program (NRP) in the United States. The NRP increased the flexibility in high-altitude routings for long-distance flights and allowed airlines to specify desired routings. Dispatchers typically used the flexibility of the NRP to file "wind optimal" or otherwise optimized routes. Because the rationale was not provided and because the routings used standard waypoints to define the routing, pilots often requested direct routings that negated the efforts of the dispatchers and resulted in suboptimal routing. Several airlines issued advisory instructions to their pilots to avoid direct routings when flying NRP flight plans.

The increased use of the ACARS datalink can be seen in Fig. 4. As datalink has become more popular among airlines, the frequency of information exchange between pilot and dispatcher has increased. In the 1980s, the average aircraft sent and received approximately 1700 messages per month. More recently in the 1990s, that number jumped to 2700.[16] When pilot/dispatcher interactions were primarily conducted through VHF voice channels, the pilots and dispatchers would communicate

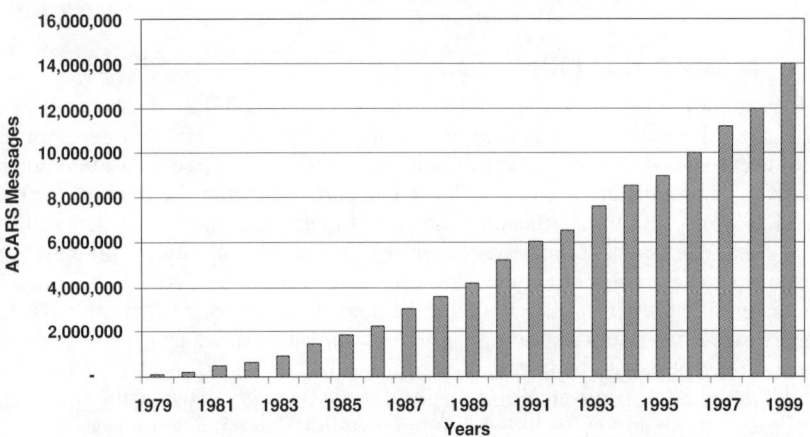

Fig. 4 Monthly datalink messages as a function of year.[16]

only if there was a problem. The interaction between pilot and controller tended, therefore, to be reactive, and during disrupted operations (e.g., hub airport closure), the dispatcher and communication channel was often overloaded. In the ACARS datalink environment, the interaction between pilots and dispatchers appears to have become more proactive and cooperative. Many pilots and dispatchers use ACARS as a low bandwidth e-mail link. It is more common for pilots to inform dispatchers of anticipated problems and for dispatchers to make suggestions or provide advance information. In large airlines the dispatchers and pilots often sign their ACARS messages, creating a personal relationship that was not present before.

The use of ACARS has also influenced pilot/airline (PA) interactions between other airline agents such as maintenance control or the local station. Pilots can communicate with maintenance for decision support or to alert maintenance of required maintenance prior to landing. One shared information set between the pilots and maintenance for most airlines is a set of maintenance codes that allow efficient transmission of these requests. Local station interactions include gate assignments, in-range alerting, and requests for special services.

VII. Shared Information in Intrafacility Controller/Controller Interactions

The interaction between controllers within a single facility tends to focus on coordinating the control of individual flights as well as regulating the flow of aircraft within the facility. Most facilities are arranged so that the primary traffic flow is between adjacent controllers in order to facilitate intercontroller communication and coordination. This is shown in Fig. 5, which schematically represents the flow of aircraft and control responsibility within a tower facility as well as the flow of flight progress strips that are surrogates for the movement of aircraft through the facility's airspace.[17] The flight progress strip contains the flight plan information and is annotated to record changes in flight status and other control information. The strips provide a communication and information sharing mechanism between

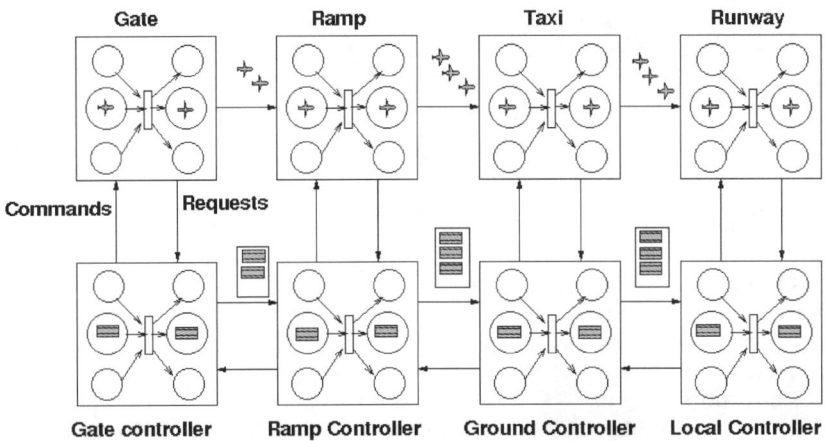

Fig. 5 Schematic representation of the flow of aircraft and flight strips through a tower facility.[17]

the radar (R) and data (D) side controllers as well as between different sector controllers and relief controllers.[18]

The transfer of responsibility of aircraft between controllers is unambiguously represented by physical "hand-off" of the flight progress strip. The order of the strips is often a representation of the controller's planned sequence and intent. In some cases special positioning of the flight progress strip within the strip rack indicates a requirement for special handling. As we discuss increasing the amount of shared information through updating dynamic information in the host computer, one must also keep in mind the practical concern of the time constraints imposed on the controller. Even if the information is updated at a reasonable rate, workload induced task-shedding prevents controllers from reloading the strip information from the host computer at the same rate. This is an issue with the conflict predictor User Request Evaluation Tool (URET), which integrates flight plan information into its algorithms.

One aspect of controller/controller interaction that is often neglected is the affective state of the controller. Because of their physical proximity, affective state is often communicated by posture, gesture, or voice. For example, urgency may be transmitted directly by shouting or by a sudden change in posture. Because of safety and working relationship issues, the workload of an adjacent controller is an important element of shared information. This may be communicated directly or indirectly. For example, in tower operations upstream controllers may reduce traffic flow based on the accumulation of strips.[17]

To improve the interaction between controllers and underlying automation systems, there is a trend to shift from paper flight progress strips to electronic emulation. One advantage of these systems is the potential for intercontroller communication through the electronic flight progress strip. However, it is important to consider the role of the shared information that is currently communicated in the physical movement of paper strips. In addition, electronic flight progress strips may not be appropriate for some tower facility control positions that require visual attention to the external scene.

One area in which automation has successfully improved controller performance is in integrating the tasks of two interdependent local controllers. For example, the Converging Runway Display Aid electronically integrates the aircraft from the two runways on each controller's runway display through "ghost planes" from the other runway to ease the aircraft converging process. Using this aid, the controller does not need to acquire the position of the other controllers' aircraft and mentally integrate the position information with the information with his or her own aircraft. This electronic aid has been successfully integrated into several airports, including Boston, Newark, and St. Louis, which all use converging runways.[7]

Because most aircraft operations and ATM control actions are routine, an important role of shared information between controllers is to communicate flight status and intent in irregular cases such as nonstandard routings or emergencies. If the controllers are in close proximity, this is often accomplished by voice communication (sometimes shouting) or gesture. If the controllers are separated, communication may be through telephone or through an intermediary such as the D-side controller.

In some high-traffic TRACON facilities, the flow patterns are sufficiently rigid that the simple location of an aircraft along with its radar data block defines its flight plan. In these cases flight progress strips are not used and the shared information is the underlying procedure and flow pattern that provides a shared-control strategy. This system is efficient in minimizing the need for intrafacility communication but makes it very difficult to modify the underlying procedures at the facility level.

An additional important area of shared information is the current and projected state of the facility, particularly with regard to traffic and weather. A number of tools have been developed to help supervisors monitor traffic and weather [weather radar displays, enhanced traffic management system (ETMS), TMA]. These tools have become widely used by controllers to develop a shared situation awareness at the facility level. A simple method of providing facility-wide awareness of current traffic restrictions is used by the New York TRACON. In the TRACON TMU an erasable white board is used to update the traffic management coordinators and other controllers of traffic restrictions that are in effect for the TRACON at that time. This method is quick, clear, and reaches the controllers, who can retrieve the most current information whenever they need it.

VIII. Shared Information in Cross-Facility Controller/Controller Interactions

The restricted ability of controllers to interact and share information across facility boundaries limits the flexibility and efficiency of ATM operations. The principle interactions between controllers, at the single aircraft level, involve coordinating aircraft trajectories, hand-offs, and special situations. The key mechanisms for information sharing are the host computer flight information (at the time the flight progress strip is printed), Standard Operating Procedures (SOP) or Letters of Agreement (LOA) between facilities, flow control programs such as miles in trail (MIT) restrictions, as well as direct telephone voice communication between sector controllers.

Because most current cross-facility interaction is bilateral, it is difficult to coordinate across multiple sectors, and there is a restricted ability to accommodate pilot requests for nonstandard routing. Controllers have significant flexibility within their sector, but this decreases as the number of involved sectors increases. It is important to consider that the sector issues are three-dimensional, and so the degree of coordination can increase significantly when altitude changes are involved. Even when multisector rerouting clearances are given, workload considerations often make it difficult for route amendment to be entered into the host computer. As a consequence, it is not uncommon for discrepancies to exist between downstream controllers' and the pilots' clearances.

There is a clear need for future systems that support shared information and multilateral controller interactions to allow more sophisticated use of the airspace in the future. Planned information architectures recognize the need for sharing of flight information across the airspace system.[19] However, there are fundamental issues that must be addressed, ranging from how rationale and intent are communicated to how to input workload and how consensus is achieved in multilateral electronically mediated negotiations.

An additional area of cross-facility interaction is the coordination operation at the facility level through the TMCs in each facility. Information sharing at this strategic level is extremely important because of the significance of negotiation in the decision-making process. Information sharing also becomes more critical as the facility becomes busier. In normal operations, the TMCs may have time to communicate with other facilities to give information and seek information. In periods of high traffic or interrupted operations, the cross-facility communication time must be streamlined, making the information sharing tools a necessary element in achieving this goal.

There are several categories of critical information needed to make traffic management decisions in every facility. TMCs need to know the traffic demand expected to enter the facility, the weather to expect, how the facility can adapt to the traffic situation (performance abilities of tactical controllers), the status of the airports in the region, and the current restrictions on the system.

A. Shared Traffic Demand Information

The primary tool for shared traffic demand information is the ETMS. The ETMS consists of a traffic situation display (TSD), a flight schedule monitor (FSM), and several other tools that provide data on all airborne traffic and projections for traffic load. An example of the TSD with the sector monitor/alert function overlay can be seen in Fig. 6. TMCs may also use radar to determine the short-term traffic demand from a sector adjacent to the facility that is feeding aircraft into the facility. In the New York facilities, the departure spacing planner (DSP) is used to display the current departure queue for each individual airport that is updated by the towers, the TRACON, and New York Center. The DSP's initial function was to reduce the need for the tower controllers to call for departure in restricted conditions. However, it has also provided a means to indicate the departure demand—a critical task for TRACON and center TMCs.

Another example of a tool used for information between facilities is the Center-TRACON Automation System (CTAS) TMA, which is used for the coordination

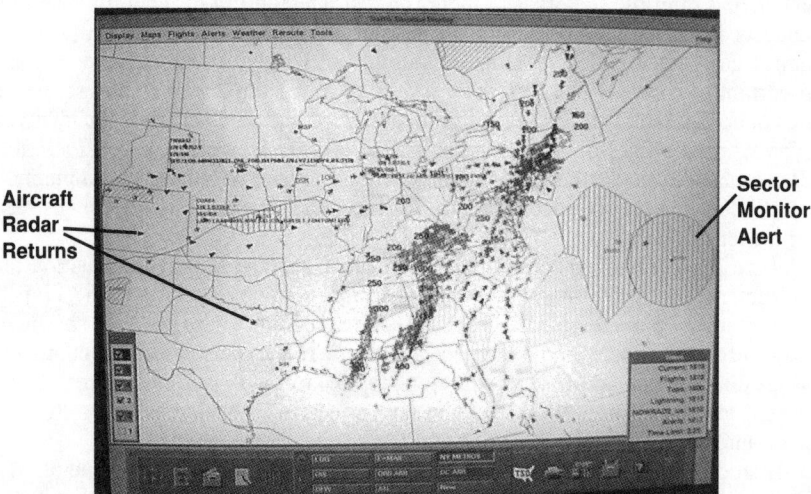

Fig. 6 Traffic situation display depicting traffic demand and weather with the sector monitor/alert function overlay.

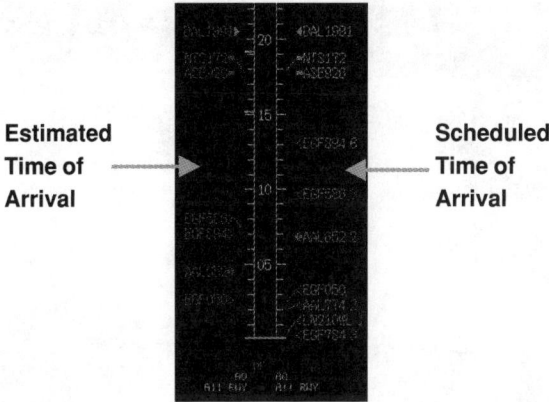

Fig. 7 Example of a traffic management advisor (TMA) for use between en route center and TRACON.

of arrival traffic between center and TRACON airspace. An example of the TMA timeline is shown in Fig. 7. Both en route and TRACON controllers have access to the scheduled and estimated times of arrival of aircraft to the feeder fix so that the en route controllers can provide information to the TRACON about what kind of a flow to anticipate as well as information about canceled or delayed flights. While NASA designed TMA to perform as a decision aid, optimally sequencing aircraft at the TRACON boundary, a key reported benefit of TMA is improved interaction enabled by the shared information content of TMA.[7] Currently TMA supports only bi-level interaction between a single center and TRACON. However, development is under way for a multicenter TMA that will support multilateral interactions. The CTAS technologies have been tested in both Dallas/Fort Worth and Denver airports.

B. Shared Weather Information

TMCs have multiple sources of weather information (e.g., National Weather Service forecasts, facility meteorologists, independent weather contractors), but often these sources do not agree on weather forecasts. Each facility is most likely to trust its own meteorologists. When two ATC facilities disagree on traffic initiatives due to different forecasts, the command center (ATCSCC) must often be the "tiebreaker" in deciding which initiative to pursue (using its own meteorologists or a historically validated procedure).

C. Shared Controller Workload Information

This category of information is difficult to convey to a TMC at another facility. Individual tactical controller abilities and workload can be conveyed within the facility through affective states (speech, posture, etc.). Although TMCs at the facility may be able to witness signs of overload as they unfold, TMCs at another facility often must wait for a telephone call from that facility before they are able to take action to remedy the situation.

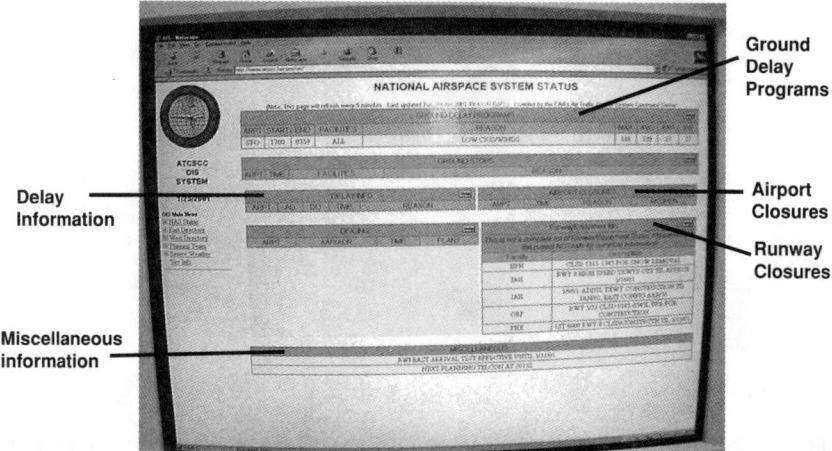

Fig. 8 Command center (ATCSCC) operational information system website is a way to share nationwide restrictions and airport information.

D. Shared Traffic Restriction Information

Once a traffic initiative has been implemented at the command center (ATCSCC), it is important to communicate the initiative in a timely manner to each affected tactical controller. One way that this information is communicated is through the command center's operational information system (OIS) website, available even to the public through the Internet. Figure 8 depicts the information found on the OIS website. The Internet has proven to be a rapid and accurate way to communicate information to interested controllers, and it is an area that should be more fully exploited in future air traffic management information sharing tools.

Other information on restrictions is communicated to the center (ARTCC) TMCs, who then input the information into a general information (GI) message system that is sent out electronically to all facilities within the jurisdiction of the center. The problem with the GI message system is that it is dependent on the center TMC to quickly input messages and on the TMCs at smaller facilities to then distribute the information to the tactical controllers. By inserting more of the human element into the communication process, the information is less likely to find the tactical controller in a timely and accurate fashion.

As part of the CDM initiative, the command center (ATCSCC) standardized many of the historically validated restrictions and common traffic flow reroutes into a "National Playbook." These standardized procedures are presented in the metaphor of a football playbook because of the expectation that when a "play" is called by the command center, each center (ARTCC) will execute measures consistent with the national plan, cutting down on interfacility communications.

During the summertime when severe thunderstorms are common across the NAS, a hotline is established between the command center (ATCSCC), the ATC facilities in the east, and the airlines. This severe weather hotline provides a means by which the traffic managers coordinate aircraft releases and tactical reroutes around the dynamic weather.

E. Shared Airport Status Information

Airport runway configuration, arrival and departure rate, and current maintenance status are all important factors considered in strategic traffic management decisions. To quickly acquire this information, TMCs may consult the command center (ATCSCC) website for the most current update on the airport. Often, though, terminal TMCs may not update the airport status information, and so other facilities' TMCs still rely on the most reliable method of retrieval—the telephone.

IX. Shared Information in Airline/ATM Interactions

The principle basis of shared information between airlines and ATM is the airline published schedules (e.g., Official Airline Guide) and filed flight plans. Because actual operations often differ from the planned schedule due to weather, equipment, or traffic factors, the controllers often have a poor understanding of the actual plan or intent of the airline. Controllers do not have direct access to the current schedule and gate information available to the public in the terminal, and often must guess if a flight is canceled or delayed if a flight strip is not activated. This is particularly acute during interrupted operations.

There are several efforts under way to improve the level of shared information between airlines and ATM. The most recent effort was an act by the Federal Aviation Administration (FAA) to consolidate weather information between the airlines and the air traffic control facilities through a joint FAA/airline severe weather website. This was a first step in recognizing the importance of shared weather information, however, the specific elements of shared weather information have yet to be determined. As further experience in incorporating shared information is gained, the utility of providing specific information will be further established.

At the strategic level, the CDM and the severe weather action plan (SWAP) provide mechanisms to support information sharing in response to schedule interruptions. The ground delay program also improves overall airline efficiency by allowing the carrier to "swap" controlled time of arrival (CTA) slots with another of that airline's flights. Other strides implementing cooperative problem solving are the Pacific Track Advisory Program, which combines the efforts of the Oakland, California, en route center and the airline dispatchers, and the NRP discussed previously.[20]

One tool used at en route air traffic control facilities is the FSM, shown in Fig. 9, in which the ATC and the separate AOCs share information about flights throughout the day that have been canceled or delayed to both predict the traffic flow and become prepared for heavy flow periods at a specific airport. The FSM appears to be highly valued by the personnel in the ATC TMU. In a survey conducted by the authors on current information sharing practices, one TMC commented that, "Recent developments and access to new technology has made the job of Traffic Management Coordinator more functional. Implementation of the FSM has provided a real-time data tool pertaining to actual aircraft operations, flight cancellations, and arrival slot re-allocations."

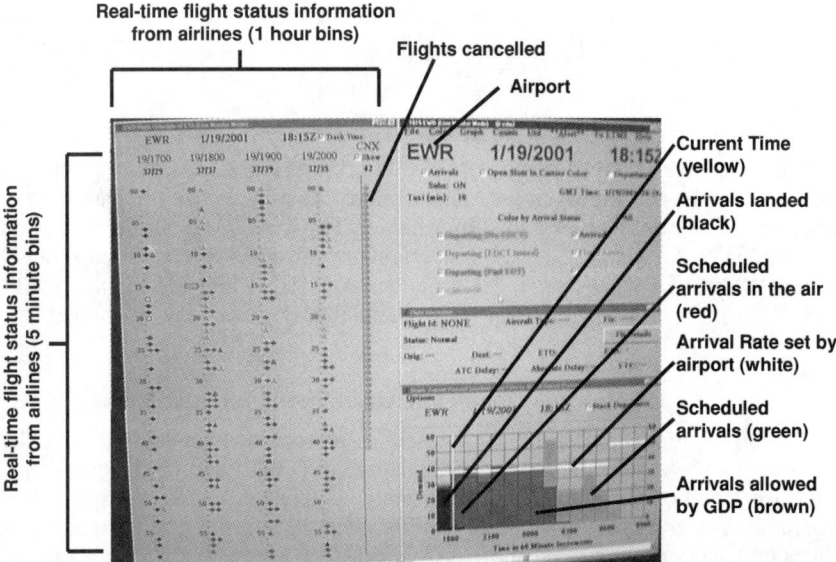

Fig. 9 Flight schedule monitor predicts traffic flow by using shared information between ATM and AOC.

At the tactical level, the surface movement advisor (SMA) or the surface movement system (SMS) provide shared information on schedule and gate changes as well as aircraft in the vicinity of the airport and planned landing runways to coordinate operations on the airport surface.

The airport resource management tool (ARMT) is another research effort in which Delta Airlines and Atlanta Hartsfield International Airport (ATL) collaborated to share information about traffic in the terminal environment. Another collaboration, dynamic aircraft route planning (DARP), recruits the pilot and air traffic managers in addition to the controller and dispatcher to allow aircraft flight plan changes after the aircraft is en route. DARP is intended to be used on trans-Pacific routes to increase fuel efficiency and decrease flight time.[20]

X. Flight Information Object

One example of the direction that the future of information sharing is taking is the flight information object (FIO). The FIO is a shared set of information between the controller, the pilot, and other agents regarding a particular flight, as seen in Fig. 10a. The current manifestation of the FIO is the flight plan data in the host computer. Initial FIO enhancements would be based on data available to the host computer from planned information tools (e.g., URET, TMA, etc.) and may include additional aircraft data (such as weights, fuel state, etc.).[21]

The FIO concept can be expanded to be accessed by additional agents who have direct interest in the flight, such as the airline station, passenger service providers (e.g., hotels, rental cars, limousines), airline maintenance, etc., as seen in Fig. 10b. The FIO may expand from the host computer to link additional databases and

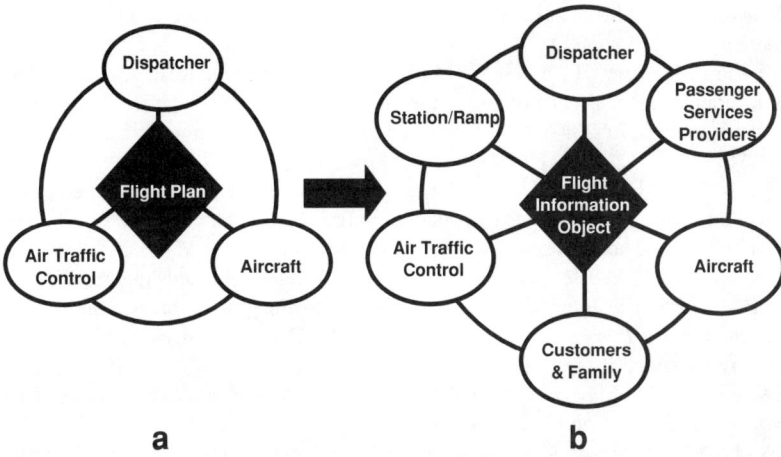

Fig. 10 Flight information object as a tool for information sharing.

information sources. At the tactical level, the advanced FIO may include intent and state information linked down from the aircraft autoflight systems by Automatic Dependent Surveillance-Broadcast (ADS-B) or CPDLC. At the airline level, the advanced FIO may include passenger itinerary, cargo manifest information, as well as aircraft maintenance status or other operational information.

XI. Conclusions

Because of complexity, multiple objectives, and the need to be robust to unexpected conditions, ATM will continue to be a distributed multi-agent process with humans executing semistructured decision processes. The key role of advanced information technology in future ATM systems will be to provide mechanisms for information sharing between the human agents, which can be expected to improve the collective decision processes within the system.

Acknowledgments

This work was supported by the NASA Ames Research Center under grants NAG 2-716 and NAG 2-1128 and by TASC as part of the FAA Center of Excellence in Operations Research. The authors acknowledge the contributions of Todd Farley, Mica Endsley, Husni Idris, Ioannis Anagnostakis, Amedeo Odoni, Alan Midkiff, Eric Feron, John-Paul Clarke, Amy Pritchett, and Tom Reynolds.

References

[1] *National Airspace System Architecture Version 4.0*, Department of Transportation, Federal Aviation Administration, Washington, DC, Jan. 1999.

[2] Kaliardos, W., "Semi-Structured Decision Processes—A Conceptual Framework for Understanding Human-Automation Systems," Ph.D. Dissertation, Dept. of Aeronautics

and Astronautics, Massachusetts Inst. of Technology, Draper Laboratory Rept. T-1360, Cambridge, MA, May 1999.

[3]Endsley, M., Farley, T., Jones, W., Midkiff, A., and Hansman, R. J., "Situation Awareness Information Requirements for Commercial Airline Pilots," International Center for Air Transportation, Rept. ICAT-98-1, Massachusetts Inst. of Technology, Cambridge, MA, Jan. 1998.

[4]Farley, T., and Hansman, R. J., "An Experimental Study of the Effect of Shared Information on Pilot/Controller Re-Route Negotiation," International Center for Air Transportation, Rept. ICAT-99-1, Massachusetts Inst. of Technology, Cambridge, MA, Jan. 1999.

[5]Hansman, R. J., Endsley, M., Farley, T., Vigeant-Langlois, L., and Amonlirdviman, K., "The Effect of Shared Information on Pilot/Controller Situation Awareness and Re-Route Negotiation," *FAA/Eurocontrol 2nd International Air Traffic Management R&D Seminar (ATM 98)*, Orlando, FL, Dec. 1998.

[6]Wickens, C. D., Mavor, A. S., and McGee, J. P. (eds.), *Flight to the Future, Human Factors in Air Traffic Control*, National Academy Press, Washington, DC. 1997.

[7]Wickens, C. D., Mavor, A. S., Parasuraman, R., and McGee, J. P. (eds.), *The Future of Air Traffic Control, Human Operators and Automation*, National Academy Press, Washington, DC, 1998.

[8]Midkiff, A., and Hansman, R. J., "Identification of Important 'Party Line' Information Elements and Implications for Situational Awareness in the Datalink Environment," *Air Traffic Control Quarterly*, Vol. 1, No. 1, 1993, pp. 5–30.

[9]Pritchett, A., and Hansman, R. J., "Variations Among Pilots from Different Flight Operations in Party Line Information Requirements for Situation Awareness," *Air Traffic Control Quarterly*, Vol. 4, No. 1, 1997, pp. 29–50.

[10]Pritchett, A., Hansman, R. J., and Johnson, E., "Use of Testable Responses for Performance-Based Measurement of Situation Awareness," *International Conference on Experimental Analysis and Measurement of Situation Awareness*, Nov. 1995.

[11]Smith, P. J., Billings, C., Woods, D., McCoy, E., Sarter, N., Denning, R., and Dekker, S., "Can Automation Enable a Cooperative Future ATM System?" *Proceedings of the 1997 Aviation Psychology Symposium*, 1997, pp. 1481–1485.

[12]*Air Traffic Control Overview: Kansas City ARTCC*, MIT Lincoln Laboratory Group 41, Air Traffic Automation, Federal Aviation Administration, Lexington, MA, 1998.

[13]*Air Traffic Controllers Handbook-Air Traffic Control Directive # 7110.65L*, Federal Aviation Administration, Washington, DC, 1998.

[14]Barhydt, R., and Hansman, R. J., "Experimental Studies of the Effect of Intent Information on Cockpit Traffic Displays," *Journal of Guidance, Control, and Dynamics*, Vol. 22, No. 4, July/Aug. 1999, pp. 520–527.

[15]Picard, R., *Affective Computing*, MIT Press, Cambridge, MA, 1997.

[16]"GLOBALink/VHF," ARINC Co., Phoenix, AZ, 1999.

[17]Idris, H., Delcaire, B., Hall, W., Anagnostakis, I., Hansman, R. J., Feron, E., and Odoni, A., "Observations of Departure Planning Processes at Logan Airport to Support the Development of Departure Planning Tools," *FAA/Eurocontrol 2nd International Air Traffic Management R&D Seminar (ATM 98)*, Orlando, FL, Dec. 1998.

[18]Mackay, W., Fayard, A., Frobertam, L., and Medini, L., "Reinventing the Familiar: Exploring an Augmented Reality Design Space for Air Traffic Control," *Proceedings of the ACM CHI '98*, 1998.

[19]*A Concept of Operations for the National Airspace System in 2005*, Rev. 1.3, Federal Aviation Administration, Washington, DC, 27 June 1997.

[20]*Airline Operational Control Overview,* Airline Dispatchers Federation and Seagull Technology, Inc., Cooperative Research and Development Agreement 93-CRDA-0034, Sept. 1995.

[21]Taber, N. J., and Viets, K., "Flight Object Operational Concept" (draft version), The MITRE Corp., McLean, VA, March 2000.

Chapter 15

Modeling Distributed Human Decision Making in Traffic Flow Management Operations

Keith C. Campbell,* Wayne W. Cooper Jr.,† Daniel P. Greenbaum,‡ and Leonard A. Wojcik§

MITRE Corporation, McLean, Virginia

I. Introduction

THIS chapter describes results from a state-of-the-art computer simulation model of distributed human decision making in traffic flow management (TFM) operations when weather disrupts airline schedules. The computer model, called Intelligent Agent-Based Model for Policy Analysis of Collaborative TFM (IMPACT), is believed to be the world's first model to capture the behavioral complexity of human decision making in TFM operations.

TFM is a process in which the economic stakes of the airlines are high, time is precious, the number of possible actions is large, and the interests of decision makers often conflict. Complexity arises from the way in which prior actions affect later decisions as decision makers struggle to adapt to a changing environment. IMPACT has been applied to help understand the value of increased information and collaboration between airlines and the air traffic management (ATM) authority in complex, dynamic TFM scenarios involving many airlines. The results show how collaborative decision making can produce clear economic gains for all airlines and for the system as a whole. IMPACT shows how overall system performance is influenced by airlines whose motivation is to improve their individual performance, and how system behavior depends on the characteristics or "personalities" of the individual airlines. IMPACT also has been used to search for new ways for the whole TFM process to evolve toward better economic performance for airlines and improved service for the flying public.

Copyright © 2001 by the MITRE Corporation. Published by the American Institute of Aeronautics and Astronautics, Inc., with permission.
*Lead Software Applications Development Engineer.
†Lead Simulation Modeling Engineer.
‡Senior Multi-Disciplinary System Development Engineer.
§Project Team Manager.

IMPACT uses agent-based modeling technology, in which both individual airlines and the ATM authority are represented as self-interested, idiosyncratic agents, each with its own volition and ability to make decisions and take actions. Agent-based modeling can represent the evolution of conflict resolution and collaboration by multiple stakeholders in a dynamically changing environment. It also can be used to model possible "gaming" among airlines, as they try to exploit the collaborative system in ways that were not originally intended. Agent-based modeling, as implemented in IMPACT, is believed to be the best approach to date for modeling the complexity of decision making in TFM operations and for generating and exploring new TFM policies based on information sharing and collaboration.

II. TFM Operations and Implications for Modeling

The execution of the TFM function is complicated by the many schedule uncertainties that arise during the course of any day, especially those due to weather. Airline schedules in the United States, for example, are typically designed for good weather days. When bad weather limits the capacity of one or more airports, U.S. Federal Aviation Administration (FAA) TFM specialists (subsequently referred to simply as FAA) must institute TFM procedures to delay takeoffs of flights to the destination airports with reduced capacity. The delays given to flights before they take off are called ground delays.

The economic consequences of delay for an airline vary considerably among the airline's flights, especially when the complex economics of hub operations is considered. One flight, for example, may have many connecting passengers to international flights, whereas another flight may have relatively few such passengers. Therefore, when the FAA assigns ground delays to particular flights, the FAA is implicitly making significant economic decisions for the airlines. From the standpoint of the FAA's conduct of the ATM system, this poses a serious problem because the FAA does not have purview into real airline costs. Only the airlines themselves are in position to know the economic consequences of delays to their flights. This has led to the concept and development of collaborative decision making (CDM) in the execution of TFM. Under CDM, the FAA controls the arrival rate at a reduced-capacity airport, but FAA collaborates with airline flight planners and dispatchers to determine *which* of the airlines' scheduled arrivals should be given priority.

The full potential of CDM on the economic performance of the ATM system is unknown and transcends the arrival substitution question just described. Perhaps most interesting is the question of how overall system performance can be enhanced if driven by decisions made to enhance individual airline performance, and relatedly, how these individual decisions would change with more information sharing among the players, who may be highly competitive with one another.

Over the years, various attempts have been made to model the complex dynamics of the ATM system, including stochastic optimization models and discrete-event dynamic simulations.[1-3] There has been a previous attempt to simulate multiple-agent decision making in TFM operations based on a statistical multiple regression model; however, this attempt used a very simplified representation of airline schedules, and results were reported only for scenarios in which there were just two airlines.[4] By comparison, ground traffic behavior in congested cities has been simulated at various levels of detail and geographical area, including a detailed model of vehicle behavior for an entire metropolitan area based on a cellular automaton (CA) approach.[5] We saw a need to look beyond these previous approaches to better represent the behavioral complexity of the system in order to address the questions posed.

The approach presented here builds on pioneering complex-systems research of the Santa Fe Institute (SFI). SFI is the foremost research center for the science of complexity, a field that studies how aggregate phenomena emerge from underlying patterns. Complexity science has been applied in many disciplines, including physics, sociology, anthropology, biology, and economics, and researchers at SFI study the factors that such problems have in common. In the vernacular of complexity science practiced at SFI, the U.S. operational TFM system, including both FAA elements and airline elements, is a complex adaptive system (CAS). It is complex both in the combinatorial sense that there are many possible decisions, and in the behavioral sense that decision making is distributed among many players who may have very different goals. The TFM system is adaptive, because the airlines and the FAA constantly receive feedback from the system and have the opportunity to change what they do.

A powerful means of studying a CAS like the TFM system in the context of its operations is through the use of agent-based modeling (ABM). The ABM approach typically is most successful when the simulated agents are kept as simple as possible within the constraints of the situation to be simulated.[6] This facilitates the possibility of emergent behavior, which is aggregate behavior by the system as a whole that would be extremely difficult to predict from the attributes of the individual agents, which are individually relatively simple to understand.

IMPACT, the agent-based computer simulation model of TFM interactions developed by MITRE Center for Advanced Aviation System Development (CAASD), represents individual airlines and the FAA as self-interested, idiosyncratic software agents, each with its own volition and ability to make decisions to take actions. Agents have either economic or policy motivations and have information about their environment and other agents. Once a scenario is populated with agents and a randomly generated system event is introduced, agents are permitted to act according to the rules of the system, and the scenario evolves spontaneously. This is described in more detail in the following sections.

III. Baseline Schedule Disruption Scenarios Modeled by IMPACT

Three kinds of baseline scenarios have been modeled by IMPACT to represent past and present-day TFM operations in the United States. These are called no-action, ground delay program (GDP), and CDM. The three kinds of baseline scenarios are simplified representations of approaches that have been used in actual operations to respond to schedule disruptions caused by capacity-reducing weather at airports. The three baseline scenario types function as starting points from which to create new scenarios needed for IMPACT analyses of various kinds.

In all of the three scenario types, weather conditions are expected to cause a capacity reduction at a single airport for a portion of a single day, and airlines and the FAA respond (within the rules defined by the scenario type) to the resulting schedule disruption as the day's events unfold. To simplify the simulation, only arrivals to the affected airport are modeled. In a typical scenario, approximately a thousand flights from the Official Airline Guide (OAG) are simulated. About 10 to 15 airlines are represented as decision-making agents, and the FAA is represented as a single agent. In the scenarios presented in this chapter, weather information is assumed to be perfect.

No-action is the simplest of the three scenario types. In no-action scenarios, the airlines simply send their scheduled flights to the affected airport with no change in departure times. Similarly, the FAA takes no action to respond in advance to the weather. Thus, in no-action, agents make no decisions, which may result in a large amount of airborne holding. No-action is an exaggerated representation of operations before the early 1980s, when responses to anticipated weather problems were not as well orchestrated as they became after the institution of ground delay programs into standard TFM practice. No-action also represents what could happen in the present system without any response to anticipated or unanticipated future weather conditions.

The GDP scenario represents decision making after ground delay programs were introduced into standard TFM practice in the early 1980s. In a GDP scenario, the FAA responds to a future anticipated capacity reduction at the affected airport by declaring a ground delay program. The airlines respond by canceling, substituting, exchanging, and delaying flights within their individual schedules. Without collaborative decision making in TFM operations, information exchange among the participants in ground delay programs is very limited, and the model represents this (in an exaggerated way) as no information available to agents about when other airlines' flights will approach the airport. In this chapter, two examples of GDP scenarios are described, one in which the airlines take no action following the GDP, and another in which they take action based on static information about other airlines' flights.

In CDM scenarios, the FAA also responds to a future anticipated capacity reduction by declaring a ground delay program, but it has improved information that allows the FAA to compress the arrival schedule by eliminating canceled flights from the program. The model can be configured to represent different policies with respect to compression and other parameters of collaborative decision making. Airlines can cancel, substitute, exchange, or delay their own flights, but unlike the GDP scenarios discussed previously, they have information about each other's intended arrival schedule. This feature represents information sharing of the current CDM program in the United States. In this program, airlines and the FAA exchange flight intent and other information using a tool called the flight schedule monitor (FSM). FSM removes airline-specific information so that users are given expected flight arrival times without airline identifiers and flight numbers. Agents modeled in IMPACT make decisions in light of what other airlines as well as the FAA have already decided, just as real airlines involved in the CDM program make decisions in light of information about what other airlines and the FAA have already decided, using FSM. Thus, a comparison of a CDM scenario against GDP scenarios under otherwise identical conditions can be used to estimate the value of this kind of information sharing.

To represent actual operational practice in CDM scenarios, the CDM simulation software was developed to include various features that are used in real CDM operations. For example, the FAA may extend the GDP beyond the period of reduced arrival capacity as a means to manage demand. In this chapter we compare a case in which the FAA does not extend the GDP to another case in which the FAA does extend the GDP. Also, simulated CDM scenarios include a variable window of time (e.g., 20 min) in which airlines are permitted to send flights in advance of the departure time needed to arrive at the capacity-limited airport at the assigned arrival time. This feature gives simulated airlines greater flexibility than they would have without such a window.

IV. Airline and FAA Agents in IMPACT

Airline agents in IMPACT make decisions based on the expected cost to themselves of alternative actions. Modeled costs are based on available industry data and include direct costs like fuel and crew costs, as well as other less tangible indirect costs from expected future lost revenue caused by passenger ill will.[7] High incremental costs are assumed for canceled flights, and very high costs are assumed for diverted flights. These costs are estimates and have not been validated with the airlines.

When an airline agent makes a decision, the model computes the expected cost to the airline of a limited set of possible alternative decisions. In keeping with the spirit of ABM, the agents do not attempt to optimize their decisions (such an optimization would be impossible in any case because they do not know what other agents will do), but they make incremental changes to attempt to improve their situation. Agent attributes can be changed to represent different airline strategies, and airline agents can learn over many simulated events to improve their performance. In the results presented here, the airline agents make decisions by evaluating the approximate cost impact of a limited number of alternative possible decisions. However, the structure of IMPACT also facilitates experimentation with a variety of approaches to airline decision making, including those based upon powerful heuristic algorithms such as simulated annealing and genetic algorithms. Experimentation with these algorithms is a possible research direction for the future.

The FAA agent in IMPACT makes decisions based on such policy motivations as keeping system demand within the capacity of the relevant resources and promoting system efficiency and equity between airlines. In scenarios described here, the FAA agent monitors the capacity/demand balance at the weather-influenced airport, and responds within the rules of the scenario being modeled to keep demand within system capacity. In many simulations, we have assumed that the FAA schedules the ground delay program to end as soon as the expected capacity reduction is over, and does not attempt to account for the "bow wave" of pent-up demand that typically follows a restriction on demand. The bow wave is observed in real aviation operations. As with the airline agents, the model facilitates experimentation with a variety of approaches to FAA decision making.

V. Basic Analysis of Airline and FAA Decision Making with IMPACT

Figure 1 shows an arrival traffic summary for an IMPACT simulation of a no-action scenario for a sample capacity-reducing weather event. Figures 2 and 3 show

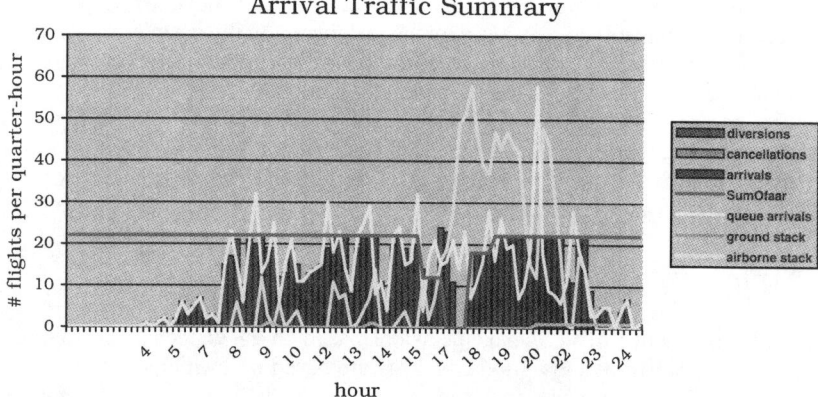

Fig. 1 IMPACT output for no-action scenario.

arrival traffic summaries in two types of GDP scenarios for the same weather event. In Fig. 2, the FAA declares a GDP whose arrival rate exactly matches the storm, and airlines take no further actions to modify their original schedules, which results in a large bow wave following the GDP. Figure 3 shows the outcome when the same GDP is declared by the FAA, but airlines respond based on a static assumption about other airlines' flights. Each airline modifies its schedule following declaration of the GDP but, in the absence of information about what other airlines are doing, simply assumes the portion of the bow wave created by other airlines does not change. Figure 4 shows the arrival traffic summary for a CDM scenario with the same weather event and ground delay program. Finally, Fig. 5 shows the arrival traffic summary for a CDM scenario with the GDP extended to the end of the day. The same original arrival schedule applies across all five scenarios. Table 1 compares key parameters for the five scenarios in which delay and cost values are averaged across all flights for all airlines during the entire day. The total number of flights scheduled to arrive during the day is 1117.

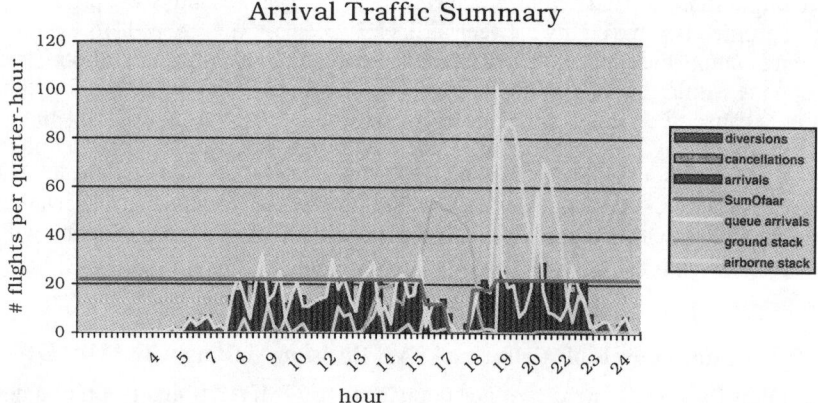

Fig. 2 IMPACT output for GDP-only scenario.

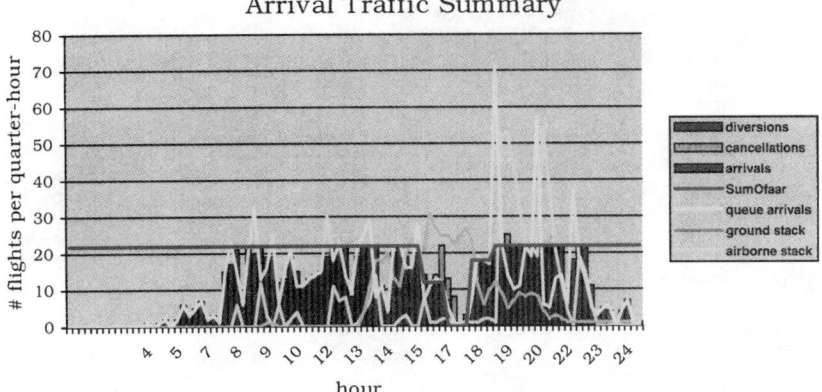

Fig. 3 IMPACT output for GDP scenario with airline response to static information.

Figures 1–5 show the capacity profile [denoted as Sum Of aar (airport arrival acceptance rate)] for the weather-influenced airport as a function of time of day. Capacities are in number of flights per quarter hour. The capacity profile is identical for all scenario types. For the GDP and CDM scenarios, the ground delay program declared by the FAA agent exactly matches the actual capacity profile. In these simulations, agents have perfect knowledge of the future weather.

Figures 1–5 show the numbers of diversions, cancellations, and arrivals at the airport as the day unfolds. Diversions and cancellations are shown at the original scheduled times or arrival. Arrivals are plotted at actual time of arrival; note that the number of arrivals per quarter hour never exceeds the capacity of the airport. The curve labeled queue arrivals shows the number of flights per quarter-hour interval entering the arrival queue. The curve labeled ground stack shows the number of flights held on the ground as a function of time. The airborne stack curve shows the number of flights in airbone holding as a function of time. (In real operations, not all flights would have to be held around the airport; they could be held en route as well.)

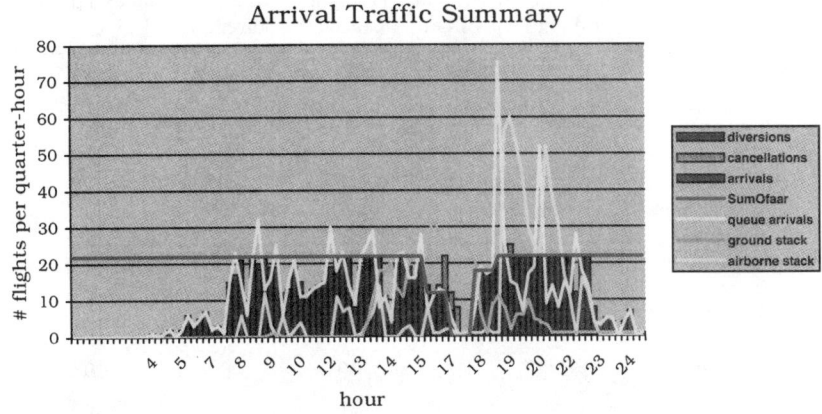

Fig. 4 IMPACT output for CDM scenario.

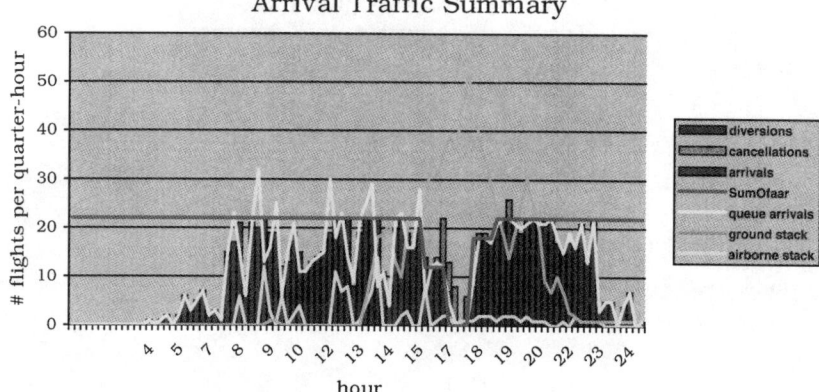

Fig. 5 IMPACT output for CDM and extended GDP scenario.

Comparison of Figs. 1 and 2 shows that the GDP eliminates the large airborne stack in the no-action scenario during the storm but creates an even larger stack after the storm. This large bow wave after the GDP occurs because of heavy arrival demand after the storm. In the GDP scenario with airline response to static information (Fig. 3) and in the CDM scenario (Fig. 4), the airlines reduce the size of the bow wave after the GDP by taking actions to cancel flights, but the airlines do not completely eliminate the bow wave. The sizes of the bow waves in Figs. 3 and 4 are similar, and Table 1 shows that the costs per flight for these cases are less than for no-action and GDP-only. (The fact that the cost is slightly less in the

Table 1 Comparison of the four baseline scenarios

	Average airborne holding time, min per flight	Average ground holding time, min per flight	Number of flights canceled	Number of flights diverted	Average cost per flight, thousands of U.S. dollars
No-action scenario	12.5	0	0	39	4.5
GDP-only scenario	11.4	8.1	0	25	4.4
GDP scenario with airline response to static information	6.9	6.4	33	0	2.2
CDM scenario	7.7	6.1	30	0	2.3
CDM with extended GDP	1.8	11.2	40	0	1.9

GDP scenario with airline response to static information, compared to the CDM scenario, is not believed to be significant. In other weather and traffic scenarios, the opposite relationship has been observed.)

It is possible that different airline decision algorithms would permit airlines to reduce the bow wave further than shown in Figs. 3 and 4, but there are limits on what airlines can accomplish as long as they are behaving independently of one another. As Figs. 1 and 2 show, the original storm and the GDP produce periods of rapidly increasing airborne holding. For a particular airline's flight originally scheduled during this period of increasing airborne holding, there is incentive for the airline to have the flight arrive as early as possible, rather than later, because early arrival would cause the flight to arrive when the queue is smaller. This applies to *all* flights arriving during the rising portion of the airborne queue, and so there is an incentive for airlines to either send flights as soon as possible or cancel them. If all flights were owned by a single airline, that airline would be able to delay its flights on the ground to prevent a large airborne stack. However, with decision making distributed among a number of airlines, the incentive to send flights early drives the system to maintain a bow wave, even though it is costly to every airline. This situation is analogous to the well-known "tragedy of the commons," in which agents acting in their own self-interest do not necessarily promote the interest of the system as a whole.[8]

Figure 5 shows the case in which CDM is used but the ATM authority extends the GDP well beyond the period of reduced capacity, which eliminates the bow wave. Table 1 shows that the cost in this scenario is lowest among the five scenarios. The secnarios illustrate a basic theorem in TFM, namely that the best result occurs when both airlines and the ATM authority respond to a schedule disruption. The ATM authority by itself cannot know the priorities of the airlines, but self-interested, noncolluding airlines cannot resolve excess congestion by themselves. The best result is achieved when airlines are able to adapt their schedules to changing conditions, while the ATM authority manages overall demand.

VI. Other Analyses with IMPACT

Analyses have been performed with IMPACT to address a variety of issues related to TFM operations, including the influence of airline characteristics or personalities on system economic performance, and the effects of possible changes in rules for collaborative decision making. Other analyses have focused on decision making by the ATM authority, for example, to look at the effect of changing its planning horizon when weather forecasts are noisy. IMPACT provides a useful platform for investigating decision making with such noisy information because it can be run repeatedly to show the range of possible outcomes in scenarios with stochastic elements. Work has also begun to integrate IMPACT with decision analysis to show the effects of changes in decision-making criteria used by the ATM authority.

In one set of experiments, IMPACT was used as an open-ended search tool to find ways to achieve better aggregate economic performance. Although this application of IMPACT is in a very preliminary stage, the approach could be used to suggest future enhancements to the CDM process to include more extensive forms of collaboration. The open-ended search was accomplished by configuring IMPACT to simulate a weather-induced schedule disruption in which the entire set of airlines behaves as if it were a single composite airline (the composite-airline scenario). In the composite-airline scenario, airlines effectively behave in ways to improve overall system performance, rather than necessarily improve their own performance. Although this kind of behavior is not realistic in current operations, the composite-airline scenario shows how economic improvement might be possible if mechanisms could be produced to ensure that overall system gains are distributed fairly among all participating airlines. These mechanisms could include, but are not restricted to, slot trading and purchasing among airlines within the ground delay program.[9]

Although the preliminary results suggest that additional improvement might be possible through more extensive forms of collaboration, these benefits may be reduced when imperfect weather predictions are introduced. Also, airline flight banking, which is not included in the present results, may influence the outcome. Finally, experiments need to be performed with different approaches to agent decision making, such as those based on powerful heuristics like simulated annealing and genetic algorithms, to attempt to improve the performance of the airline agents.

VII. Conclusions

The agent-based IMPACT model has been used to model information sharing and collaboration in weather-induced schedule disruptions. IMPACT is uniquely suited to understanding how overall system performance (and the public good) can be improved in a system in which self-interested individual agents each attempt to improve the outcome for themselves. IMPACT permits experimentation and analysis in ways that would be infeasible to try with the real system or with conventional simulation methods. An example of this is in the area of decision making when there is uncertainty about future weather. IMPACT can be applied to show the range of possible outcomes that could result when an uncertain weather prediction is made, and IMPACT can show how the decision-making strategies of individual airlines and the FAA play out in such scenarios. IMPACT also can be used to perform experiments in which airlines or groups of airlines attempt to game the system to take advantage of information sharing and collaborative opportunities. These experiments may perform a valuable function in developing confidence among airlines and government regulators in further information sharing and collaboration possibilities for the future. There are also questions regarding the influence of banking operations, and expected or unexpected events in other parts of the system, such as reduced capacity at other airports. Another key area for investigation with IMPACT is how to help the FAA to continue to improve the quality of its decision making in TFM operations, including guidance on how to evaluate past decisions in light of the uncertainties that existed at the time the decisions were made. By shedding light on these questions, IMPACT can help us understand what is the best path for TFM system evolution towards better economic performance through improved information sharing and collaboration.

Acknowledgment

The IMPACT model was developed by MITRE CAASD as part of a mission-oriented investigation and experimentation project sponsored by the FAA. The contents of this document reflect the views of the author and The MITRE Corporation and do not necessarily reflect the views of the FAA or the DOT. Neither the Federal Aviation Administration nor the Department of Transportation makes any warranty or guarantee, expressed or implied, concerning the content or accuracy of these views.

References

[1] Richetta, O., and Odoni, O., "Solving Optimally the Static Ground Holding Problem in Air Traffic Control," *Transportation Science*, Vol. 27, No. 3, 1993, pp. 228–238.

[2] Lindsay, K. S., Boyd, E. A., and Burlingame, R., "Traffic Flow Management Modeling with the Time Assignment Model," *Air Traffic Control Quarterly*, Vol. 1, No. 3, 1999, pp. 255–276.

[3] DeArmon, J. S., and Lacher, A. R., "Aggregate Flow Directives as a Ground Delay Strategy: Concept Analysis Using Discrete Event Simulation," *Air Traffic Control Quarterly*, Vol. 4, No. 4, 1996, pp. 307–323.

[4] Klein, G. L., and Anton, P. S., "A Simulation Study of Adaptation in Traffic Management Decision Making Under Free Scheduling Flight Operations," *Air Traffic Control Quarterly*, Vol. 7, No. 2, 1999, pp. 77–108.

[5] Ricker, M., and Nagel, K., "Experiences with a Simplified Microsimulation for the Dallas/Fort Worth Area," *International Journal of Modern Physics C*, Vol. 8, No. 3, 1997, pp. 483–503.

[6] Epstein, J. M., and Axtell, R., *Growing Artificial Societies: Social Science from the Bottom Up*, MIT Press, Cambridge, MA, 1996.

[7] Irrgang, M., "Airline Irregular Operations," *Handbook of Airline Economics*, edited by D. Jenkins, McGraw-Hill, New York, 1997, pp. 349–365.

[8] Hardin, G., "The Tragedy of the Commons," *Science*, Vol. 162, 13 December 1968, pp. 1243–1248.

[9] Adams, M., Kolitz, S., Milner, J., and Odoni, O., "Evolutionary Concepts for Decentralized Air Traffic Flow Management," *Air Traffic Control Quarterly*, Vol. 4, No. 4, 1996, pp. 281–306.

Chapter 16

Assessing the Benefits of Collaborative Decision Making in Air Traffic Management

Michael O. Ball[*]
University of Maryland, College Park, Maryland

Robert L. Hoffman
Metron Scientific Consulting, Inc., Reston, Virginia

Dave Knorr and James Wetherly
Federal Aviation Administration, Washington, D.C.
and
Mike Wambsganss
Metron Scientific Consulting, Inc., Reston, Virginia

I. Introduction

COLLABORATIVE decision making (CDM) was conceived within the Federal Aviation Administration's (FAA) airline data exchange (FADE) experiments that began in 1993. These experiments proved that having airlines submit real-time operational information to the FAA could improve air traffic management decision making. CDM is an effort to improve air traffic management through information exchange, procedural improvements, tool development, and common situational awareness.

The initial focus of CDM, known as ground delay program enhancements (GDP-E), began its prototype operations at San Francisco (SFO) and Newark (EWR) airports in January 1998. Under GDP-E, participating airlines send operational schedules and changes to schedules to the air traffic control systems command center (ATCSCC) on a continual basis. This schedule information includes, but is not limited to, flight delay information, cancellations, and newly created flights. Through the use of the flight schedule monitor (FSM), the ATCSCC uses this information to monitor airport arrival demand and to conduct ground delay

---

Copyright © 2001 by the American Institute of Aeronautics and Astronautics, Inc. All rights reserved.
[*]Robert H. Smith School of Business.

programs (GDPs). The airlines are also able to monitor arrival demands and model ground delay programs via FSM but do not have the capability to alter or implement ground delay programs.

In addition to improving the execution of GDPs, CDM has been found to have application to other air traffic management problems, such as airspace congestion due to heavy traffic or severe en route weather avoidance. The collaborative routing effort is intended to improve handling of potential flow problems that are likely to require rerouting or other flow management actions. The National Airspace System Status Information (NASSI) activities are aimed at employing CDM technology and concepts to share critical safety and efficiency data among NAS users.

Although one can point to a variety of concepts and technologies that are fundamental to CDM's success, probably the most vital underlying element has been a strong and continuous interaction among all important players, including the FAA, the airline industry, other NAS users, and members of the research and development community. Regular, e.g., monthly, meetings of the larger CDM group as well as smaller subgroups have been held throughout the life of the CDM project. Through these interactions a new GDP paradigm was developed and agreed upon by all the players. Essential to this paradigm is the implicit definition of new fair allocation principles that are embodied in the ration-by-schedule and compression algorithms. These algorithms are key components of FSM, the CDM decision support tool. Another fundamental CDM technology is the CDMNet, a private extranet that connects the ATCSCC, the participating airline operational control centers (AOCs), the hub site of the enhanced traffic management system (ETMS), as well as certain other parties. More detailed descriptions of CDM technologies and concepts can be found in Refs. 1 and 2.

A final key aspect of the CDM effort is its reliance on data analysis and objective critique. In support of this, an analysis subgroup has worked closely with the CDM community to highlight its accomplishments and to point out those areas in which it needs to be more effective. With regard to the benefits of CDM, there have been two major reviews of CDM activities: one in the spring of 1998 by NEXTOR, the National Center of Excellence for Aviation Operations Research,[3] and one in December 1999 by the FAA's Free Flight Phase I Office.[4]

Because CDM has had the most impact to date on GDPs, this will be the primary focus of our discussion of benefits assessments. Section II of this chapter contains the results of analysis on the impact of CDM on information quality. Section III presents results related to the system and user impacts of CDM. Section IV summarizes ongoing activities and challenges in the area of collaborative routing.

II. Improvements in the Quality of Information and Information Distribution

The goal of the initial implementation of CDM was to support GDP planning at SFO and EWR airports. The infrastructure put in place to achieve this goal involved broad information collection and distribution mechanisms. In this section, we organize the presentation of our analysis by first considering the quality of the new information infrastructure, then the impact on SFO and EWR GDPs, and, finally, the impact on more general decision making.

ASSESSING THE BENEFITS 241

The use of CDM has produced new information by combining FAA and airline data sources. All CDM airline participants, including American Airlines, Continental Airlines, Delta Airlines, Northwest Airlines, Southwest Airlines, Trans World Airlines, United Airlines, and US Airways, have implemented data feeds from their operations systems into the CDMNet. Using these data feeds, the airlines provide information on flight cancellations, mechanical delays, and other events that impact the demand on the NAS. This information is merged with FAA-generated information by systems at the Volpe Center in Cambridge, Massachusetts into a real-time data feed, known as the CDM string.

Through the CDMNet, the CDM-enhanced information has been distributed in an unprecedented fashion. In fact, probably the most significant aspect of the new CDM information infrastructure is that the airline operations centers receive the same information as the FAA ATCSCC specialists. Such information is critical in enabling airline operations specialists to plan responses to changing conditions and possible FAA control actions. Previously, such information was not available to airline operations planners or was only available after the fact, when it could no longer be used to influence decision making.

Our analyses have found that the information flowing over the CDM string is of higher quality. Moreover, we have found that the improvements are most dramatic when the system is under stress and the information is most critically needed. Next, we summarize the results by data type: flight departure prediction and cancellation data.

A. Predictive Accuracy of Flight Departures: Integrated Predictive Error Metric

With CDM, a concerted effort has been made to improve the accuracy of flight departure predictions. Participating air carriers have voluntarily augmented ETMS flight data with their own departure predictions. The premise is that each airline has the most complete picture of its operations (delays due to connectivity, gates, etc.), thus enabling it to make more accurate predictions of its departure times than ETMS.

We used the integrated predictive error (IPE) metric to monitor long-term trends in flight departure predictive accuracy. IPE is a weighted average of the errors in a stream of predictions made over time for a single event. This number is called the IPE of the event. A numerical suffix indicates the number of hours over which the metric was tracked. IPE units are normalized by the tracking time and can be thought of as an average error (usually given in minutes). For instance, an IPE-6 value of 10 min would be obtained by making a steady stream of predictions over 6 h, each of which is off from the actual departure time by 10 min.

For each day between January 1997 and June 1999, we computed the average IPE-6 departure error over all flights bound for SFO or EWR airports. This process was repeated for the metrics IPE-5, IPE-4, ... , IPE-1, to arrive at six averages for each airport, for each day in the 30-month period. The results were then stratified into GDP days and non-GDP days, and averaged over the month in which the day occurred. To detect long-term trends, we smoothed the natural variance in the monthly IPE values by plotting a cumulative average (over all prior months) for each of the six metrics. The results for SFO appear in Fig. 1.

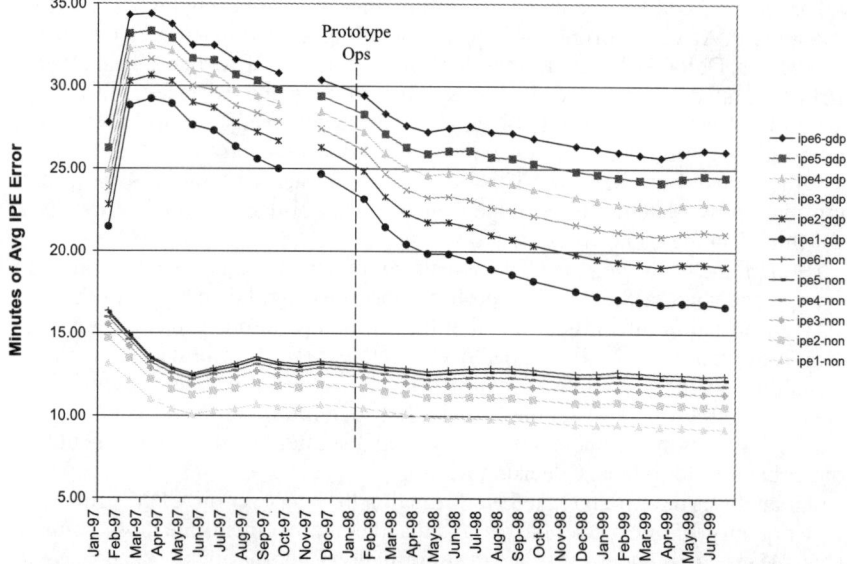

Fig. 1 IPE results for SFO, January 1997 through June 1999.

The monthly IPE averages for GDP days are substantially higher than for non-GDP days. This is to be expected because the reassignment of arrival times by the FAA in a GDP leads to unpredictable variance in departure times and, subsequently, higher IPE values. Also, a GDP throws air carriers into a state of irregular operations that can have other adverse consequences affecting departure prediction.

The most notable feature of the results is that on GDP days, when accurate flight data is most crucial, the average error in departure prediction dropped over most of the 30-month study, indicating an improvement in the accuracy of flight data. For instance, between January 1998 (the inception of CDM) and June 1999, the average IPE-6 departure error for GDP days at SFO dropped by 3.40 min per flight (5.58 min per flight for GDP days at EWR). [These are changes in the *cumulative* average since January 1997. The noncumulative average IPE-6 values for GDP days at SFO in January 1998 and June 1999 were 28.82 and 26.02 min, respectively (25.46 and 22.33 min, respectively, for EWR)]. Average IPE values for non-GDP days at SFO and EWR have dropped as well. In fact, for both airports, the departure prediction error has been pushed below 15 min, the industry-wide standard for an on-time event.

These positive (downward) trends actually began before the inception of CDM prototype operations at SFO. However, we note that the CDM participants began submitting flight data several months before the prototype operations period, which is characterized by the use of the FSM resource allocation tools.

All the metrics IPE-6, IPE-5, ..., IPE-1, exhibit a common pattern for both GDP and non-GDP days, but the lower the tracking period, the lower the IPE value. For instance, the IPE-5 curve is essentially a downward shift of the IPE-6 curve, the IPE-4 curve is a downward shift of the IPE-5 curve, and so on. This indicates that, on average, departure prediction accuracy increases (has less error) as the departure of a flight approaches.

It is unlikely that departure prediction error for GDP days will ever be reduced to the levels obtained on non-GDP days; both FAA and air carrier manipulations of departure schedules during a GDP will always introduce unpredictability in departure time predictions. Also, it is possible that improvements in departure time accuracy are approaching a limit due to the fact that, in the aggregate, there will always be an inherent amount of uncertainty in prediction of aviation events.

B. Timeliness of Flight Cancellation Messages

We analyzed the impact of CDM on the notification of flight cancellations. Prior to CDM, there was no mechanism by which the airlines could notify the ATCSCC of a flight cancellation other than a telephone call or a flight substitution message. The ATCSCC relied solely on NAS-generated cancellation messages. Via the CDMNet, CDM has enabled the airlines to directly submit a flight cancellation message independently of all other ETMS cancellation fields. This is intended to generate a more accurate picture of demand for NAS resources, especially during a GDP.

To measure the timeliness of cancellation messages received, we chose the original estimated time of departure (OETD) of a flight as the base time against which to measure the amount of notice given for a flight cancellation. For all effective purposes, this is the time of departure listed in the Official Airline Guide (OAG). More strictly speaking, we used the first estimated time of departure for each flight listed in our database, which is based on aggregate demand lists (ADL). Usually, the OETD appeared in our database 12 h prior to the OETD. Although there are strong arguments for using an arrival-oriented metric as opposed to a departure-oriented metric, it seems reasonable that a cancellation notice should be submitted prior to the scheduled departure time of a flight.

To compare the ETMS and CDM systems, two groups of cancellation fields were established: group $G5^+$, which models ETMS cancellation messages, and group G6, which models the CDM cancellation messages. $G5^+$ is based on five out of the six ETMS cancellation fields. (The diversion field, dv, was excluded from the study due to lack of usage.) The + indicates that we have added a logic that mimics the ETMS time-out cancellation field; this models what would have happened in the ETMS system if an airline cancellation message (fx) had not been received (fx messages suppress the activation of the ETMS time-out field). G6 is based on the fields in group $G5^+$ plus the CDM-provided airline cancellation field, fx.

Because $G5^+$ is contained in G6 and because $G5^+$ contains a time-out logic that eventually records flights canceled by an fx message, the groups share the same knowledge of cancellations. For each group, we defined the cancellation notice time of a flight f to be the amount of time *before* the OETD of f that the earliest cancellation notice was received in any of the fields in that group. Thus, the cancellation notice time is the earliest time that the cancellation group had knowledge of the cancellation, and the analysis is reduced to comparing the cancellation notice times of each group for each canceled flight. A flight is considered to be canceled

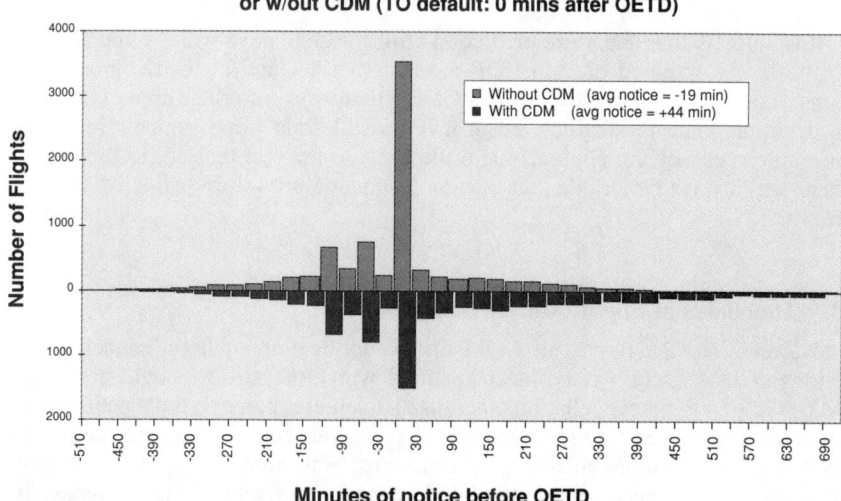

Fig. 2 Average cancellation notice times for SFO with default timeout of 0 min.

if and only if at least one of its cancellation fields turned positive and the flight showed no activity (i.e., no departure, no arrival, etc.).

We partitioned flight cancellation times that occurred at SFO between 1 January 1998, and 31 May 1999, into the two groups, $G5^+$ and G6, and averaged the results to form two distributions of cancellation times: one with CDM messages (G6) and one without CDM messages ($G5^+$). Figure 2 is a violin plot of the two distributions; the upper distribution is without CDM and the lower distribution is with CDM. Note the shift in the distribution from left to right after CDM is added, which indicates that cancellations are known earlier after CDM cancellation data are added. In fact, without CDM, the average cancellation notice was received 19 min *after* the OETD, but with CDM, it was received 44 min *before* the OETD for an improvement of 63 min.

The 63-min performance gap at SFO is a very conservative figure, given that the default time-out mechanism used to model ETMS cancellation notice times was set at 0 min. This assumes that ETMS would have canceled each nonactive flight immediately after its OETD. Experts at the Volpe National Transportation Systems Center claim that a more realistic default time-out would be 120 min after OETD.

Figure 3 gives the distributions of cancellation notice times at SFO for a time-out default time of 120 min. Again, we see a shift in the distribution toward earlier cancellation notice times. This time, the average cancellation notice time was 49 min *after* the OETD, whereas with CDM it was received 44 min *before* the OETD. That is, the average notice was received 93 min earlier under CDM. Similar results have been obtained for EWR. Because most traffic flow initiatives are made on a planning horizon of a few hours, this is certainly enough of an improvement to have a positive impact on traffic flow management decisions.

Further analyses have shown that after May 1998 the gap between CDM performance and ETMS performance continued to widen, especially on GDP days when

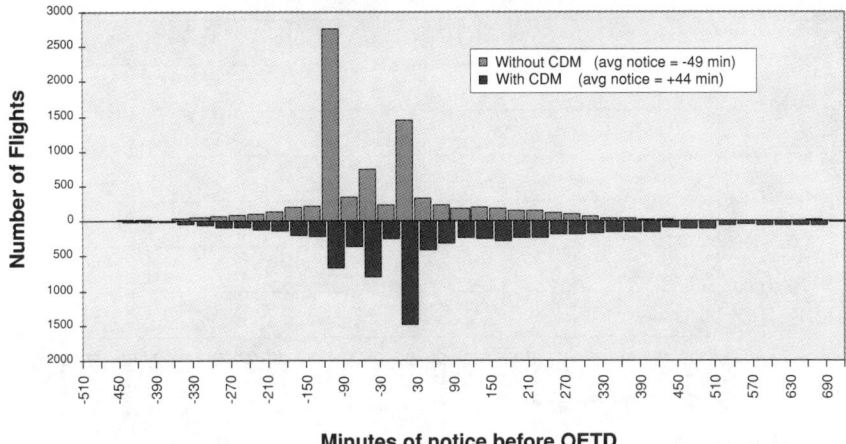

Fig. 3 Average cancellation notice times for SFO with default timeout of 120 min.

timeliness of flight cancellation information is most crucial. Informal examination of the average cancellation times for individual cancellation fields has revealed that the superior performance of the CDM flight cancellation mechanism is attributable to the direct submission of flight cancellation messages by CDM participating air carriers via the CDM-provided fx field. These messages were often received several hours in advance of an OETD, hence they drive up the average cancellation notice time considerably.

III. System and User Impact

Based on a series of interviews, we observed a consensus among ATCSCC specialists that CDM procedures do yield more effective GDPs. It is difficult, if not impossible, to measure the total impact of CDM on GDP planning, because the improved information quality and FSM decision support features can influence GDP planning in subtle and varied ways. The measure of success of a GDP depends on several factors, such as the throughput achievable for the existing weather conditions, the amount of assigned ground delay, and the airborne delay encountered. In this section, we isolate some of the CDM benefits.

A. Compression Benefits

The compression algorithm is a procedure unique to CDM. It eliminates vacant slots and reduces overall delays by altering the assignment of slots to airlines in a way that treats all airlines fairly. Between 23 January 1998, and 15 July 1999, the ATCSCC executed 1385 cycles of compression over a total of 21 airports and 539 GDPs. The percent of planned (ATCSCC assigned) delay reduction (at airports with 10 or more compression cycles) ranges from 7.5% at Atlanta's Hartsfield Airport (ATL) to 18.2% at Boston's Logan Airport (BOS). The percent savings in

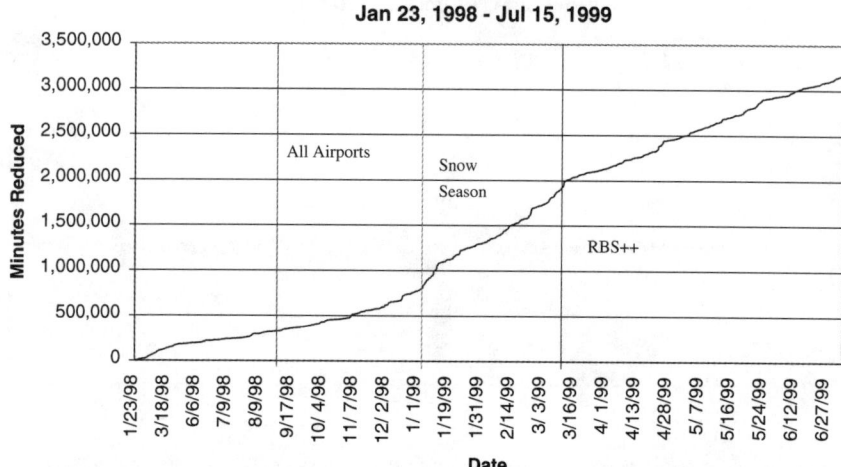

Fig. 4 Cumulative planned delay savings from the compression algorithm.

planned delay at EWR and SFO, respectively, were 13.0 and 9.7%. The average over all of the GDP airports was 12.7%.

Figure 4 shows the rise in cumulative delay savings over the period from 20 January 1998, to 15 July 1999. The time horizon is marked by three epoch periods of CDM history. The first is the time at which GDP prototype operations went into effect at all airports. The second is the snow season of 1999, which led to notoriously bad weather conditions. Because this season induced many more GDPs, there were many more executions of compression, and the rate of growth of compression savings began to rise noticeably. In the third period, the ration-by-schedule (RBS) and compression algorithms were run in a combined fashion known as RBS++. By the end of the overall time period 15 July 1999, the total compression savings had climbed to 3,165,925 min.

We estimate that roughly one-half of the 3,165,925 min in savings (approximately 1,582,962 min) could be obtained by intra-airline substitution processes that existed prior to CDM. The other half could only be obtained through the inter-airline slot swapping mechanism provided by the compression algorithm particular to CDM. To put these savings in perspective, at a conservative industry standard of $25.00 per min, this 1,582,962 min of planned delay savings due to compression represents a savings of (approximately) $39.574 million for an average of (approximately) $28,574 per compression cycle.

We note that these are savings in *assigned* ground delay. Further analysis is required to determine the degree to which these savings could be mitigated by corresponding changes in airborne delays.

B. User Reported Benefits

A major component of CDM benefits for GDPs is intended to lie in providing each airline with greater ability to control the allocation of unavoidable delay among their flights. CDM allows an airline to reduce the impact of delays on passengers and airline operations by shifting delay away from flights in which the delay has the most detrimental (and costly) impact. As an indication of the benefits in this area, we note that United Air Lines has reported that it has derived significant delay reduction from the use of CDM-based GDPs at SFO and EWR and also from the use of FSM to plan its responses to GDPs at Chicago O'Hare (ORD) airport. United Air Lines estimated that the total savings over the initial 1.5 months of CDM prototype operations was 11,000 min with a value of $3–4 million.

C. New Measure of GDP Performance: Rate Control Index

GDP enhancements have brought with them the need for metrics that evaluate the performance of a GDP as a whole, rather than just a single component of a GDP. One such metric is the rate control index (RCI). RCI measures the flow of air traffic into an airport and compares it to the targeted flow that was set by the traffic flow managers at the ATCSCC during a ground delay program. A single index, or percentage, is reported for the entire performance of a GDP on a single day. A higher score (e.g., 95%) corresponds to better performance, meaning the flow of traffic into the airport closely matched the targeted pattern of traffic, both in quantity and in distribution. See Ref. 5 for more details on the metric.

Though still in development, RCI has been applied to traffic flow into the terminal space of SFO. Figure 5 shows the trend of the RCI metric for SFO over the 30-month period from January 1997 to March 1999. We smoothed out the variance of the monthly points by computing a moving average over four months (see the lines with the square icons). Further smoothing was obtained by computing a cumulative average over all months since January 1997 (see the lines with the triangular icons). Three checkpoints are worth noting: the first month for which a

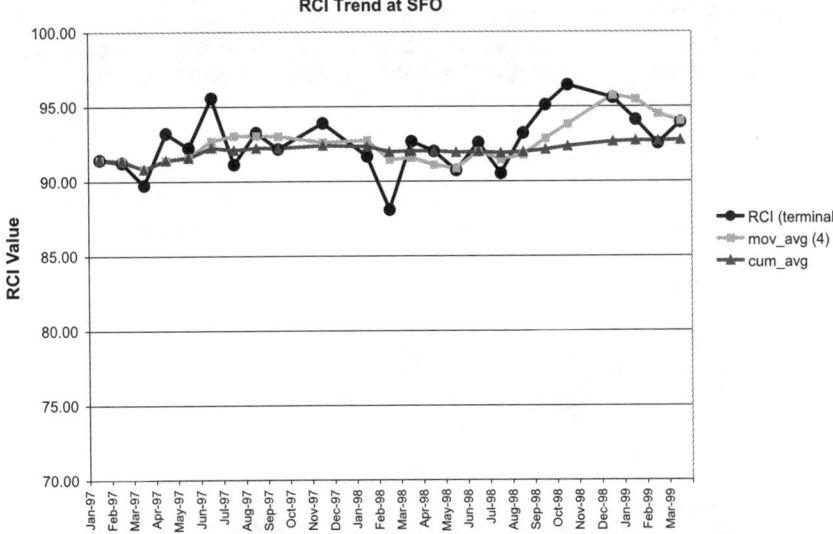

Fig. 5 RCI trend at SFO, January 1997 through March 1999.

four-month cumulative average had been computed (April 1997), the start of CDM (January 1998), and the last point (March 1999). For SFO, these points were 91.42, 92.33, and 92.75. This indicates a slight rise in the (cumulative average) RCI value for SFO since the start of CDM.* However, this trend is probably not statistically significant. We feel that a longer history is required to judge the significance of CDM's impact.

Preliminary applications of the metric to EWR show lower index rates (poorer performance) than at SFO and little change in the long-term trend. Lower index values at EWR are to be expected because of the complexity of its terminal space (bordering on different traffic centers) and the less predictable nature of east coast traffic. Also, we caution that the results at EWR are less conclusive than at SFO because the computation of this metric is dependent on the modeling of airborne holding, which is more difficult at EWR than at SFO.

D. Added Benefits

1. Revisions

Prior to CDM, the ATCSCC did not have the capability to revise a program once it was in effect, meaning they were not able to modify ground delay program parameters such as the airport acceptance rate (AAR) and the scope or duration of the program. Although they did have the ability to affect GDP-controlled traffic flow by means such as blanket delays (adding a fixed number of minutes of delay to all flights), the methods for program modification were cruder and less effective than the revision capability now provided by CDM.

One of the most powerful revisions that can be made to a GDP is to extend the length of a program. This allows the ATCSCC to control later-arriving traffic when adverse weather effects last longer than expected and to smooth out pent-up demand (a stack) that may accumulate toward the end of a program. This tool has been used frequently since the inception of CDM and has proven to be highly effective for controlling traffic flow. Log entries made by the ATCSCC attest to the effectiveness of revisions to smooth out the traffic (and reduce) departure delays. The flexibility of this tool has resulted in the avoidance of underutilized capacity and excessive airborne holding.

2. Near-Term GDP Cancellations

A near-term cancellation of a GDP is when a GDP is aborted within 30 min of its planned start time. Because ground delay is served by flights prior to takeoff, many flights will have absorbed delays well in advance of the start time of the GDP. Thus, all assigned ground delays absorbed prior to the start of the canceled GDP are (in hindsight) unnecessary. For this reason, near-term cancellations of GDPs are considered undesirable.

We tracked the number of instances of near-term GDP cancellations both pre- and post-CDM at six major airports. We conjectured that the combination of

---

*Because the number of ground delay programs varies with the month, these results do not give equal weight to ground delay programs. Different results are obtained when equal weight is given to each program, i.e, averaging over all prior ground delay programs. The method we have adopted screens out some of the seasonal effects of weather. The legitimacy of this method is confirmed by averaging over each season of the year (results similar to those above are obtained).

improved demand information and the power run feature of FSM that allows ATCSCC personnel to delay the implementation of a GDP to the last possible minute should decrease the number of near-term cancellations. Some airports showed improvement, others showed none at all. However, there has been a remarkable improvement (decrease) at Lambert-St. Louis (STL) airport in the percentage of near-term GDP cancellations. We believe that this is the result of superior data quality of the two major airlines that dominate the airport. This caliber of data quality is attributable to the use of daily download, the replacement of potentially obsolete OAG information with fresh airline operational data at the start of each day.

3. Impact on Overall Airline Decision Making

Information provided by airlines indicates that the CDM-enhanced information has been used to improve their decision making beyond the realm of GDPs at SFO and EWR. On at least two separate occasions, United Air Lines used CDM information to reduce the number of flights canceled in anticipation of a GDP by 25% over the number that would have normally been canceled; the estimated total cost savings was $1.5 million. We have also identified instances in which airlines have solved capacity-demand imbalances by reducing demand in response to CDM information, thereby eliminating the need for an FAA traffic management action such as a GDP.

Somewhat surprisingly, we have also found that the airlines have used CDM-supplied information for a variety of purposes totally outside the intended application domain of GDP planning. Specifically, airline operations managers have used this information to support fuel planning, diversion decisions, and management of flow into hubs. Delta Airlines reports that the more accurate information provided by FSM has allowed them to preserve the destination of flights that normally would have been diverted to other airports.

IV. Collaborative Routing

Early on it was recognized that CDM had applicability to en route airspace management, and a subgroup has been actively pursuing this area. A set of initial collaborative routing (ICR) tools and procedures were prototyped and tested during the 1999 severe weather season. These included a national collaborative routing coordination tool (CRCT), a concept prototype tool developed by the MITRE Corporation, which provides the FAA traffic flow management specialist with automated features that support the identification of flights impacted by congestion and aid in the development of alternative routes; a collaborative convective forecast product (CCFP), a national convective weather forecast, which represents a consensus based on inputs from AOC and ARTCC weather units; use of information and application distribution products (PictureTel and World Wide Web) to support collaborative routing decision making; low altitude arrival and departure routes (LAADR), which embodies a set of procedures for allowing the use of low altitude routes to avoid congested airspace; coded departure routes (CDR), which involves a set of procedures and a database for the creation, storage, and dissemination of alternate routes to be used to avoid airspace blocked by severe weather.

In anticipation of the 2000 severe weather season in the United States, a number of collaborative routing tools and procedures are now being tested and put in place.

These include deployment and use of more powerful versions of the CDR, CCFP, and LAADR capabilities from ICR. A national "playbook" database containing standard rerouting strategies to be used to avoid closed or impeded airways has been created with access provided to both the FAA and NAS users. A new strategic planning team has been established at the ATCSCC to design and implement national route planning (NRP) strategies, which in many cases will be based on the playbook.

Longer term efforts are ongoing to develop new decision support tools for en route airspace management. The fundamental challenge for these efforts is to adapt the resource allocation principles developed for GDP-E to the en route setting. Another challenge of similar difficulty is to analyze the performance and assess the impact of such collaborative routing tools and procedures. Although GDP planning decomposes on an airport-by-airport basis, the typical en route planning problem must take into account a wider set of flights and airspace components. This difference implies that en route problems have a much larger systems nature, and this leads to substantial challenges for both tool development and analysis.

V. Conclusions

CDM provides a new framework and philosophical approach to air traffic flow management. To date, it has been applied principally to the planning and control of ground delay programs. In this chapter, we have reported on an analysis that investigated the impact of CDM. That analysis demonstrated improvements in information quality, system efficiency and user operations. Work is ongoing to apply CDM concepts to the management of the en route airspace. We feel the CDM philosophy is applicable in a very wide range of air traffic management contexts and that CDM represents rich research area.[6]

References

[1]Hoffman, R., Ball, M., Odoni, A., Hall, W., and Wambsganss, M., "Collaborative Decision Making in Air Traffic Flow Management," NEXTOR TR, RR-99-2, Inst. for Transportation Studies, Univ. of California, Berkeley.

[2]Wambsganss, M., "Collaborative Decision Making Through Dynamic Information Transfer," *Air Traffic Control Quarterly*, Vol. 4, 1997, pp. 107–123.

[3]Ball, M., Hoffman, R., Hall, W., and Muharremoglu, A., "Collaborative Decision Making in Air Traffic Management: A Preliminary Assessment," NEXTOR TR, RR-99-3, Inst. for Transportation Studies, Univ. of California, Berkeley, 1999.

[4]"An Operational Assessment of Collaborative Decision Making in Air Traffic Management," Doc. Control No. R90145-01, Free Flight Phase I Office, Federal Aviation Administration, Washington, DC, 1999.

[5]Hoffman, R., and Ball, M., "Measuring Ground Delay Program Effectiveness Using the Route Control Index," *Journal of Air Traffic Control*, Vol. 42, 2000, pp. 19–23.

[6]Ball, M. O., Hoffman, R., Chen, C., and Vossen, T., "Collaborative Decision Making in Air Traffic Management: Current and Future Research Directions," edited by G. Galati and A. Odoni, Proceedings of Capri ATM-99 Workshop, sponsored by IEEE Joint Chapter—Signal Processing and Aerospace and Electronic Systems, Central and South Italy section, 1999, pp. 267–276.

Section IV. Airport Operations and Constraints

Six chapters are presented in this section with an emphasis on the airport as a limiting factor in total system capacity. Chapter 17 describes the pioneering work of Heinz Erzberger at NASA Ames Research Center on attempting to increase National Airspace System (NAS) system capacity by optimizing the arrival stream into an airport. This is a principle determinant of overall system capacity. Chapter 18 uses initial data and modeling of Center-Terminal Radar Approach Control (TRACON) Automation System (CTAS) tools to estimate the benefits of adopting the arrival sequencing and separation decision support tools described in Chapter 17.

Chapter 19 is an excellent description of the airport arrival-departure curves that are used to characterize nodal service rates in network queuing models of the NAS system. Chapter 19 also describes how these curves can be used to optimize the arrival and departure sequencing of any airport.

Chapters 20 and 21 describe the often overlooked complexity of airport ground movement operations and how inattention to this phase of flight can decrease overall airport departure rates. Reduced ground movement and restricted departure rates can adversely affect overall NAS capacity by decreasing the networks nodal service rates.

Chapter 22 is a more recent modeling evaluation of the terminal airspace decision support system tools being developed and evaluated at the NASA Ames Research Center.

Chapter 17. Fast-Time Study of Airline-Influenced Arrival Sequencing and Scheduling, Gregory C. Carr and Heinz Erzberger, NASA Ames Research Center; and Frank Neuman, Raytheon STX Corporation, 1998.

Chapter 18. Capacity-Related Benefits of Proposed CNS/ATM Technologies, Tara J. Weidner, Seagull Technology, Inc., 1998.

Chapter 19. Collaborative Optimization of Arrival and Departure Traffic Flow Management Strategies at Airports, Eugene P. Gilbo, VOLPE National Transportation System Center; and Kenneth W. Howard, Arcon Corporation, 2000.

Chapter 20. Analysis, Modeling, and Control of Ground Operations at Hub Airports, Kari Andersson, Francis Carr, Eric Feron, and Nicholas Pujet, MIT; and William D. Hall, Charles Stark Draper Laboratory, 2000.

Chapter 21. Conceptual Design of a Departure Planner Decision Aid, Ioannis Anagnostakis, Husni R. Idris, John-Paul Clarke, Eric Feron, R. John Hansman, Amedeo R. Odoni, and William D. Hall, MIT, 2000.

Chapter 22. Modeling of ATM Automation CD & R and Metering Conformance Benefits, Tara J. Weidner, Seagull Technology, Inc.; and Steve Green, NASA Ames Research Center, 2000.

Chapter 17

Fast-Time Study of Airline-Influenced Arrival Sequencing and Scheduling

Gregory C. Carr* and Heinz Erzberger[†]
NASA Ames Research Center, Moffett Field, California
and
Frank Neuman[‡]
Raytheon STX Corporation, Moffett Field, California

I. Introduction

THE continued growth of air traffic within the United States, combined with the use of hub and spoke operations by air carriers, has led to increased congestion and delays in the terminal airspace surrounding the nation's busier airports. The problem of congestion is exacerbated at hub airports, where air carriers schedule large numbers of flights to arrive and depart within a short time period. To air carriers, hubbing makes good economic and competitive sense.[1] At the same time, however, hubbing operations often lead to overcapacity periods and precipitate delays that can directly impact the economic efficiency of an air carrier's flight operations. To ensure that the safe capacity of the terminal area is not exceeded, air traffic management (ATM) often places restrictions on arriving flights that are transitioning from en route airspace to terminal airspace. The constraint of arrival traffic is commonly referred to as arrival flow management and includes techniques such as time-based metering, vectoring, and the imposition of ground delays or miles-in-trail restrictions. Arrival flow management is typically performed without consideration for the relative priority that airlines may place on individual flights, based on factors such as crew criticality, passenger connectivity, critical turnaround times, gate availability, on-time performance, fuel status, or runway preference. The development of new arrival flow management techniques that consider priorities expressed by air carriers will allow airlines to have greater control over individual arrival banks. This will lead to increased airline

---

Copyright © 2001 by the American Institute of Aeronautics and Astronautics, Inc. No copyright is asserted in the United States under Title 17, U.S. Code. The U.S. Government has a royalty-free license to exercise all rights under the copyright claimed herein for Governmental purposes. All other rights are reserved by the copyright owner.
*Research Engineer. Member AIAA.
[†]Senior Scientist. Fellow AIAA.
[‡]Research Engineer.

economic efficiency and reduce the economic impact of ATM restrictions on the airlines.

Air traffic control automation tools are being used in arrival flow management to enable collaboration between air carriers and ATM and to assist controllers in efficiently matching traffic demand and airport capacity. The self-managed arrival resequencing tool (SMART) allows air carriers to affect arrival demand through self-imposed ground delays or through company speed control.[2] The collaborative decision making program's flight schedule monitor allows air carriers and ATM to use more efficiently available arrival slots during Federal Aviation Administration- (FAA) imposed ground delay programs.[3] NASA and the FAA have designed and developed a suite of software decision support tools (DSTs) to improve the efficiency of high-density airspace.[4] Collectively known as the Center-TRACON Automation System (CTAS), these tools use sequencing and scheduling algorithms to automatically plan the most efficient landing order and landing times for arriving aircraft.[5] Operational evaluation of the CTAS tools has shown them to be effective in improving airport throughput and reducing delays while maintaining controller workload at a reasonable level.[6]

One of the CTAS tools, the traffic management advisor (TMA), is currently being used at the Fort Worth Air Route Traffic Control Center to perform arrival flow management of traffic into the Dallas/Fort Worth airport (DFW). The TMA is a time-based planning tool that assists traffic management coordinators and Center controllers in efficiently balancing arrival demand with airport capacity.[6] The primary algorithm in the TMA is a real-time scheduler that generates efficient landing sequences and landing times for arrivals within about 200 n mile from touchdown.[7] Aircraft are scheduled so that they arrive in a first-come-first-served (FCFS) order based on an estimated time of arrival (ETA) at the runway. Although FCFS scheduling establishes a fair order based on estimated times of arrival, it does not take into account individual airline priorities among incoming flights. As part of its collaborative arrival planning research and development program, NASA Ames Research Center is exploring the possibility of allowing airlines to express relative arrival priorities to ATM through the development of new CTAS sequencing and scheduling algorithms that take into account airline arrival preferences.[8]

II. Priority-Scheduling

For most airlines, the schedule that is determined internally by the airline to satisfy its business and economic objectives is an ideal schedule. This schedule is ideal in the sense that the everyday realities of operating an airline and interacting with the various elements of the National Airspace System (NAS) largely preclude this ideal schedule from ever being achieved. Because of the uncertainties throughout both the airline (equipment breakdowns, maintenance problems, personnel shortages) and the NAS (weather, ground delays, ATM restrictions), aircraft often arrive in the terminal airspace in an order that does not match the ideal order of the airline schedule. Current arrival flow management using FCFS sequencing and scheduling algorithms will likely result in aircraft arriving at the runways in an order that does not match the preferred arrival order. The ability to specify the preferred arrival order within the air carrier's own arrival bank is useful for maximizing bank integrity and minimizing bank time, that is, the exchange of passengers/cargo, and aircraft servicing.[9] It is important to distinguish between "scheduling" or "schedule" in the context of airline operations, and "scheduling" or "schedule" in the context of air traffic control automation. The former refers to the daily scheduled times of departure and arrival that an airline determines for all of its flights, whereas the latter refers to the process of automatically choosing 1) the order or sequence in which the aircraft should land or cross a particular fix and 2) the time that each aircraft in the sequence should pass over a specified fix.[7]

Earlier studies have shown that scheduling aircraft according to an FCFS sequence based on ETA at the runway produces a schedule that is considered to be both fair to air carriers and that is efficient in terms of minimizing delays.[5] These studies also have shown that the resulting scheduled arrival sequence at the runway will, for the most part, match the FCFS sequence that is input to the scheduling algorithm. Because the scheduling algorithm attempts to preserve the input sequence, specifying a preferred sequence will result in a schedule that closely approximates the preferred arrival order. The concept of "priority scheduling" is then defined as the scheduling of a bank of arrival aircraft according to a preferred order of arrival.

III. Scope

The purpose of the present study is to examine a CTAS-derived priority-scheduling technique in terms of the algorithm's ability to produce a specified order of arrival and in terms of its ability to minimize delays. The study is not presented as an operational concept and is not intended to determine the operational viability of the scheduling technique. The output of the simulation is used only to compare the effectiveness and efficiency of the priority and the FCFS scheduling techniques. This is a crucial step in the determination of operational viability for any CTAS-derived scheduling technique; a significant loss in scheduling efficiency would render the technique operationally infeasible.

IV. Fast-Time Simulation

A fast-time simulation originally developed for statistical evaluation of CTAS sequencing and scheduling algorithms has been modified for use in this investigation.[8,10] In contrast to real-time simulation or field tests, which would require on the order of 90 min to examine a single traffic rush period, the fast-time simulation allows examination of large numbers of statistically similar rush periods in a matter of minutes. For each simulated traffic situation, the deviation of a designated bank's scheduled arrival order from the preferred arrival order can be determined. The impact of priority scheduling on delays is also determined by comparing delays for priority scheduling and FCFS scheduling. The fast-time simulation comprises three major components: an airport model, a statistical model of the arrival traffic flow, and the scheduler.

A. Airport Model

The arrival airspace at DFW is divided into Center and TRACON regions, with the TRACON encompassing the airspace within approximately 40 n mile of the airport. Arrival traffic is merged at four waypoints on the Center-TRACON boundary, which correspond to the four primary arrival directions. These waypoints are referred to as feeder gates because during heavy traffic periods aircraft are funneled through these gates as a means of controlling or metering the flow rate into the terminal area.[7] Traffic flowing to each gate is separated into two independent streams that are vertically separated by 2000 ft at the feeder gate. This allows jet and turboprop aircraft, which have significantly different airspeed ranges, to cross the feeder gates independently and avoid conflicts due to overtakes near the gates.

The airport is modeled according to the landing practices at DFW with four feeder gates and three runways available for landing. The runways are considered to be independent so that no stagger requirements are necessary for scheduling. The airport model comprises the minimum flight times from each feeder gate to all landing runways for each independent stream. These TRACON transition times were obtained from an analysis using the minimum flight times measured for several traffic samples.[11] The TRACON transition times vary with feeder gate, aircraft type, runway assignment, and airport configuration. The airport model contains transition times for both airport configurations at DFW: north flow, with arrival traffic arriving/departing in a northerly direction, and south flow, with traffic arriving/departing in a southerly direction. Note that since the data used in this simulation were collected, a fourth arrival runway has been added at DFW. However, the three landing runway configuration is still used during instrument meteorological conditions. Thus, the three-runway model and traffic data are sufficient for purposes of this investigation.

B. Traffic Model

The traffic model is based on actual traffic data recorded during six rush periods at DFW. Although the traffic data were recorded over a span of several months, the mix of aircraft type remained nearly constant for each traffic sample. The data were recorded during the "noon balloon", a daily arrival rush lasting approximately 90 min. The noon balloon was chosen as the basis for the traffic model because during this arrival rush demand exceeds airport capacity and air traffic managers impose time-based metering restrictions through CTAS sequencing and scheduling algorithms. Data recorded during the six rush periods include the aircraft type, aircraft identification, arrival stream, and the ETA at the feeder gate (ETA_{FG}). The average of these ETA for the six rushes is taken as the nominal ETA_{FG}. Errors in aircraft time of arrival in Center airspace are modeled by adding an approximately Gaussian distribution to the nominal ETA at the feeder gate. The maximum range of the variation in the ETA_{FG} is specified as an input to the simulation and is referred to as the Center arrival error.

C. Bank Definition

Although an actual arrival bank of aircraft for an airline may consist of between 30 and 50 aircraft, in this study it is assumed that a bank comprises a single group of up to 20 aircraft belonging to one airline and its subsidiary carrier. With a majority of the flights in the traffic model belonging to American Airlines (AAL) and American Eagle (EGF), these flights are used to form arrival banks. The bank is not a contiguous set of aircraft because aircraft belonging to other airlines are interspersed among the bank aircraft, as would be the case in a real traffic situation. The bank of aircraft is defined by specifying the first member of the bank and the number of aircraft belonging to the bank. For the purposes of this simulation,

Table 1 Bank definition and preferred arrival order

Sequence number	Aircraft identification	ETA_{RWY}, s	Priority
54	DAL910	3000	—
55	DAL428	3034	—
56	UAL359	3036	—
57	**AAL1150**	**3060**	**1**
58	**EGF628**	**3116**	**2**
59	DAL1086	3120	—
60	ASE924	3123	—
61	DAL2062	3180	—
62	**AAL1934**	**3240**	**3**
63	**AAL1428**	**3285**	**4**
64	DAL1670	3300	—
65	**AAL1554**	**3345**	**5**
66	AAL410	3376	—
67	DAL756	3531	—
68	DAL431	3546	—

we assume that the preferred order of arrival at the runway equals the order of arrival based on the minimum ETA at the runway with no Center arrival error. Each of the bank aircraft is assigned a priority ranking that is simply equal to the preferred order of arrival for the aircraft within the bank. The minimum ETA at the runway (ETA_{RWY}) is calculated by adding the TRACON transition times for each of the three runways to the nominal ETA_{FG} and selecting the minimum of the three resulting values. This ETA_{RWY} represents the earliest possible time of arrival for an aircraft provided that the aircraft could fly to the runway with no delay.

For example, consider the list of aircraft shown in Table 1, which represents a portion of a single arrival rush where AAL1150 has been designated as the lead aircraft in the bank, and the number of aircraft in the bank has been specified as five. The number in the first column represents the sequence number or position of the aircraft within the arrival rush when the aircraft are time ordered according to increasing ETA_{RWY}. Each arrival rush or traffic sample consists of 108 aircraft. In the example in Table 1, the aircraft belonging to the defined bank range from the 57th aircraft to the 65th aircraft in the arrival rush (AAL1554). The resulting bank aircraft are denoted by bold text for purposes of illustration. This example shows that aircraft belonging to other airlines are interspersed among the arrival aircraft that comprise the bank. The second column is the aircraft identifier and the third column is each aircraft's corresponding minimum ETA_{RWY}. The fourth column shows the priority ranking assigned to each of the aircraft belonging to the bank based on this preferred order of arrival.

The actual order of arrival for aircraft in a traffic rush period is modeled by adding the Center arrival error to the nominal ETA_{FG}. The Center arrival error represents the uncertainties in the NAS that cause the same flight to arrive in Center airspace at different times on different days. Because the minimum ETA_{RWY} is calculated by

Table 2 Actual arrival order (Center arrival error = ±5 min)

Sequence number	Aircraft identification	ETA_{RWY}, s	Priority
54	DAL428	2972	—
55	**AAL1150**	**3007**	**1**
56	DAL1670	3086	—
57	EGF006	3089	—
58	DAL834	3146	—
59	DAL2062	3206	—
60	UAL359	3206	—
61	ASE924	3212	—
62	**AAL1428**	**3212**	**4**
63	**AAL1934**	**3266**	**3**
64	**AAL1554**	**3272**	**5**
65	**EGF628**	**3300**	**2**
66	AAL410	3326	—
67	DAL431	3441	—
68	DAL756	3624	—

adding a TRACON transition time to the ETA_{FG}, the minimum ETA_{RWY} will also vary. As a result, when the aircraft are ordered according to increasing ETA_{RWY}, the actual order for the bank aircraft will differ from the preferred arrival order. In addition, the number of aircraft interspersed among the arrival bank may vary because the variation in arrival time is modeled for all aircraft in the traffic rush, not only those belonging to the specified bank. Table 2 shows the resulting estimated arrival order for the specified bank when a Center arrival error having a range of up to ±5 min is added to the traffic sample.

D. FCFS Scheduling

The FCFS scheduler is intended to approximate the sequencing and scheduling algorithms presently used in CTAS at the Fort Worth Center. A detailed description of the actual scheduling algorithm can be found in Ref. 7. Aircraft are sequenced and scheduled to be FCFS at both the feeder gates and runways while meeting feeder gate and runway threshold separation constraints. Because scheduling is done in time rather than distance, the prescribed minimum separation criteria are translated into minimum time separations at both the feeder gates and the runway threshold. For aircraft crossing the feeder gate, the minimum in-trail separation requirement for aircraft is 5 n mile, which is translated to a 60-s time separation for purposes of this simulation. The separation criteria at the runway threshold are a function of both aircraft weight class and landing order as determined by the FAA's wake vortex safety rules. Airport acceptance rate (AAR) is taken into consideration by limiting the number of aircraft that are allowed to enter the TRACON in sliding 10-min intervals, and the scheduler balances flights between runways to minimize overall delay.

The FCFS sequence is established by time ordering arrival aircraft according to increasing ETA_{RWY}. Beginning with the first aircraft in the sequence,

**Table 3 FCFS sequence and resulting schedule
(Center arrival error = ±5 min)**

	Priority sequence		Resulting schedule	
Sequence number	Aircraft identification	ETA_{RWY}, s	Aircraft identification	STA_{RWY}, s
54	DAL428	2972	DAL428	3483
55	**AAL1150 (1)**	**3007**	**AAL1150 (1)**	**3507**
56	DAL1670	3086	DAL1670	3540
57	EGF006	3089	EGF006	3593
58	DAL834	3146	DAL834	3601
59	DAL2062	3206	DAL2062	3627
60	UAL359	3206	UAL359	3695
61	ASE924	3212	ASE924	3737
62	**AAL1428 (4)**	**3212**	**AAL1934 (3)**	**3768**
63	**AAL1934 (3)**	**3266**	**AAL1428 (4)**	**3789**
64	**AAL1554 (5)**	**3272**	**AAL1554 (5)**	**3843**
65	**EGF638 (2)**	**3300**	**EGF628 (2)**	**3895**
66	AAL410	3326	DAL431	3952
67	DAL431	3441	AAL410	3953
68	DAL756	3624	DAL756	4005

each aircraft is tentatively scheduled to each of the three runways, while it is ensured that the prescribed minimum time separation between aircraft at the runway thresholds is met for each subsequent aircraft. The runway that results in the earliest scheduled time of arrival for the aircraft at the runway (STA_{RWY}) is then chosen as the landing runway. Scheduling to the runway automatically provides the correct amount of traffic to load the runways equally when traffic is heavy (runway balancing), and directs aircraft to the closest available runway. The STA at the feeder gate (STA_{FG}) is determined by subtracting the sum of the TRACON transition time and any TRACON delay from the previously calculated STA_{RWY}. Finally, if STA_{FG} for two flights are less than the required 60 s apart, the scheduled times will be altered to meet the required separation at the feeder gate.

Table 3 shows the resulting order of arrival when the aircraft are scheduled according to an FCFS sequence. The priority ranking of each bank aircraft is shown in parentheses following the aircraft identifier. The second and third columns in Table 3 show the FCFS sequence that is input to the scheduler, with the aircraft time ordered according to increasing ETA_{RWY}. The fourth and fifth columns are the resulting schedule, with aircraft time ordered according to increasing STA_{RWY}. Note that the resulting scheduled order of arrival at the runway does not precisely match the FCFS sequence based on ETA_{RWY} that is input to the scheduler. Because the schedule must meet in-trail separation criteria at both the feeder gate and the runway threshold, and the separation criteria at the runway threshold are a function of aircraft weight class and landing order, the FCFS sequence may not be preserved at the runway. Among the aircraft belonging to the designated bank, flights AAL1934 and AAL1428 have shifted positions from the sequence that is input to the scheduler (as have aircraft DAL431 and AAL410, which do not belong to the designated bank). In this case, the position shift has resulted in a scheduled sequence that does more closely match the ideal or desired order of arrival than does the input FCFS sequence based on ETA_{RWY}. However, it is purely fortuitous that the resulting schedule more closely matches the preferred order, and depending on the magnitude of the Center arrival error, the scheduled order may actually deviate further from the preferred order.

E. Priority Scheduling

The priority-scheduling algorithm is identical to the FCFS algorithm with one exception: instead of time ordering the aircraft according to increasing ETA_{RWY} prior to scheduling, the arrival aircraft belonging to the designated bank are ordered according to their priority ranking, which establishes the bank aircraft in the preferred arrival order. Note that only the aircraft belonging to the bank are reordered according to their priority ranking and that other aircraft in the traffic sample are still sequenced in an FCFS order based on ETA_{RWY}. By reordering only the bank aircraft and scheduling the remaining aircraft according to an FCFS sequence, the impact of the reordering on scheduling efficiency is minimized. Table 4 shows the resulting order of arrival when the bank aircraft are scheduled according to the preferred sequence of arrival. The second and third columns show the priority sequence that is input to the scheduler, with the bank aircraft ordered according to their priority ranking and the remaining aircraft time ordered according to increasing ETA_{RWY}. The fourth and fifth columns show the resulting schedule time ordered according to STA_{RWY}. As was the case with FCFS scheduling, the resulting order of arrival does not match the sequence that was input to the scheduler because the schedule must meet separation criteria at the runway threshold that

Table 4 Priority sequence and resulting schedule
(Center arrival error = ± 5 min)

Sequence number	Priority sequence		Resulting schedule	
	Aircraft identification	ETA_{RWY}, s	Aircraft identification	STA_{RWY}, s
54	DAL428	2972	DAL428	3483
55	**AAL1150 (1)**	**3007**	**AAL1150 (1)**	**3507**
56	DAL1670	3086	DAL1670	3540
57	EGF006	3089	EGF006	3593
58	DAL834	3146	DAL834	3601
59	DAL2062	3206	DAL2062	3627
60	UAL359	3206	UAL359	3695
61	ASE924	3212	ASE924	3737
62	**EGF628 (2)**	**3300**	**AAL1934 (3)**	**3782**
63	**AAL1934 (3)**	**3266**	**EGF628 (2)**	**3789**
64	**AAL1428 (4)**	**3212**	**AAL1428 (4)**	**3843**
65	**AAL1554 (5)**	**3272**	**AAL1554 (5)**	**3895**
66	AAL410	3326	DAL431	3952
67	DAL431	3441	AAL410	3953
68	DAL756	3624	DAL756	4005

are a function of aircraft weight class and landing order. Although the resulting scheduled bank order does not precisely match the preferred order, it does indeed match more closely the preferred bank order than does the FCFS schedule shown in Table 3.

V. Order Deviation

The purpose of this study is to compare the performance of the FCFS and priority-scheduling algorithms in terms of their ability to produce a preferred order of arrival while minimizing delays. The performance of the priority-scheduling algorithm is not measured in absolute terms, but is measured relative to the performance of the FCFS algorithm, which is considered to be a baseline. For this study, the primary measure of the effectiveness of the algorithm is its ability to produce a preferred order of arrival. To quantify the effectiveness of the priority-scheduling method relative to the FCFS method, we need a measure of how closely the scheduled order of arrival for a designated bank matches the preferred arrival order. We first define a position shift (PS) for an aircraft as the difference between the aircraft position in the preferred bank order and the sequence number in the scheduled bank order as

$$PS = N_{Preferred} - N_{Scheduled}$$

where N is the sequence number of the aircraft within the bank.

Table 5 illustrates the calculation of the PS for each of the aircraft in the bank defined in Table 1. The position shift of each aircraft is calculated for both FCFS scheduling (Table 3) and priority scheduling (Table 4). Note that a positive PS indicates that an aircraft is scheduled ahead of its preferred position in the bank, and a negative position shift indicates that an aircraft is scheduled behind its preferred position in the bank. For example, the sequence number of flight EGF628 in the preferred order of arrival is 2, whereas its sequence number in the FCFS schedule is 5 and its sequence number in the priority schedule is 3. This results in a PS of -3 for the FCFS schedule and -1 for the priority schedule and reflects that EGF628 is scheduled three slots behind its preferred position in the bank using FCFS scheduling, and one slot behind the preferred position using priority scheduling.

Because we are interested in how closely the overall bank order matches the preferred order, we want a single measure that will indicate the deviation from the

Table 5 Calculation of position shift for a bank of aircraft

Sequence number in bank	Preferred order	FCFS schedule	Position shift for FCFS schedule	Priority schedule	Position shift for priority schedule
1	AAL1150	AAL1150	0	AAL1150	0
2	EGF628	AAL1934	-3	AAL1934	-1
3	AAL1934	AAL1428	1	EGF628	1
4	AAL1428	AAL1554	1	AAL1428	0
5	AAL1554	EGF628	1	AAL1554	0

preferred order for a bank of any length. We then define the order deviation (OD) for a bank as the algebraic sum of the absolute value of the PS for each aircraft in the bank divided by the number of aircraft in the bank:

$$OD = \frac{\sum_{\text{no. of bank aircraft}} |PS|}{\text{no. of bank aircraft}}$$

It can be seen from this definition that if the OD for a bank of aircraft equals zero, then the scheduled bank order is the same as the preferred bank order. More importantly, the larger the value of the OD, the further the scheduled bank order deviates from the preferred order. This will allow us to easily compare the relative effectiveness of the FCFS and priority-scheduling methods in producing the preferred order of arrival. Note that the OD measures are used only to indicate the performance of the priority-scheduling algorithm relative to the FCFS algorithm. To determine the operational significance of OD, further studies would have to be performed that, for example, would investigate and quantify the relationship between OD and Center arrival error and determine what magnitude of OD is considered to be operationally significant to an air carrier. However, given the importance of arrival order to airline economic efficiency, a measure such as OD can provide a basis for determining the operational viability of the priority-scheduling algorithm and is a critical topic for further work.

The order deviations for each scheduling method using the example in Table 5 are calculated here. Because the priority-scheduling scheme results in the designated bank arriving in an order that more closely matches the preferred arrival order, the OD for the priority scheduled bank is smaller than that for the FCFS scheduled bank:

$$OD_{\text{FCFS}} = (|0| + |-3| + |1| + |1| + |1|)/5 = 1.2$$

$$OD_{\text{Priority}} = (|0| + |-1| + |1| + |0| + |0|)/5 = 0.4$$

To investigate the statistical performance of the two scheduling methods, a large number of traffic samples are generated for a specified bank. To compare the effectiveness of FCFS scheduling and priority scheduling for a large number of traffic samples, we define the average OD as the sum of the ODs for each traffic sample divided by the number of traffic samples:

$$OD_{\text{Average}} = \frac{\sum_{\text{no. of traffic samples}} OD}{\text{no. of traffic samples}}$$

VI. Simulation Inputs/Outputs

Inputs to the fast-time simulation include the aircraft identifier of the lead aircraft in the bank, the size of the bank, the number of traffic samples, the range in Center arrival error, the airport configuration, and the airport acceptance rate. To determine the statistical performance of the FCFS algorithm and the priority algorithm, 500 traffic samples are generated for each designated bank. Each traffic sample comprises 108 jet and turboprop aircraft, 72 of which are AAL or EGF flights. In this simulation the modeled airport configuration is south flow for DFW. Because the traffic model is limited to a single arrival rush period, and because of the manner in which a bank is defined, banks cannot be formed at or near the end of the arrival rush period. For example, if the bank length is specified as 20 and the designated lead aircraft is the 100th aircraft in the arrival rush, no bank will be formed because there are not enough aircraft following the lead aircraft to form a bank. Although we attempt to form banks across the entire range of the traffic rush period, this cannot be done for the reasons just outlined. The output of the fast-time simulation includes the average OD as well as histograms of the PSs for each bank of aircraft. Total delays and histograms of individual delays for all aircraft in the traffic rush are generated as well. Results can then be compared for the FCFS scheduling algorithm and the priority-scheduling algorithm.

VII. Results and Discussion

The primary measure of the effectiveness of the priority-scheduling algorithm is the closeness of the match between the scheduled order of arrival and the preferred order of arrival. Figure 1 is a plot of the average order deviation for a bank size of 20, a range in Center arrival errors of ± 5 min, and an AAR of 96 aircraft/h. For a designated bank whose lead aircraft has a nominal ETA_{FG} given on the x axis, a corresponding pair of ordinates shows the average OD for the bank using FCFS scheduling and priority scheduling. Figure 1 confirms that the priority-scheduling algorithm substantially reduces the average OD from that of the FCFS-scheduling algorithm. Note, however, that although the OD for each bank is less using the priority-scheduling algorithm, the OD is still nonzero for each bank. In other words, although the resulting bank order using priority scheduling matches much more closely the preferred order than does the FCFS order, the scheduled bank order does not precisely match the preferred order. Because the schedule must meet in-trail separation criteria at the runway threshold and the separation criteria are a function of both weight class and landing order, the preferred order of arrival is not always preserved at the runway.

Figure 1 shows the resulting OD for banks of aircraft beginning at different points in the arrival rush. The average order deviation for the FCFS algorithm first increases and then decreases as the ETA_{FG} of the lead aircraft in the bank

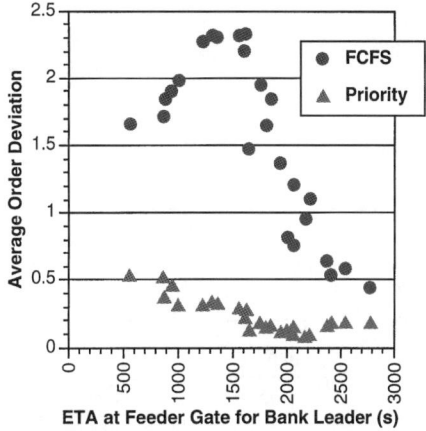

Fig. 1 Average order deviation (Center arrival error = ± 5 min).

increases. The change in average OD for the FCFS schedule is due to changing traffic density and mixture in the arrival rush. As the traffic density increases (ETA are more closely spaced), a given arrival error will cause larger position shifts within a bank and thus larger ODs. By the same token, the traffic mix impacts the OD because if non-AAL/EGF flights are interspersed among the bank aircraft, the aircraft comprising the bank will be spaced farther apart. Then, for a given arrival error, the OD for the bank will be smaller because the aircraft are not as closely spaced. The average OD for the priority-scheduling algorithm also varies with traffic density and mixture and is most effective in a region where some non-AAL/EGF aircraft are interspersed among the bank aircraft.

The effects of AAR, bank size, and Center arrival error on the success of the priority-scheduling algorithm are also examined. For the sake of brevity, no plots are shown, but important results are summarized here. Results show that for a given Center arrival error and bank size, the priority OD tends to decrease with decreasing AAR, meaning that the priority-scheduling algorithm is more effective for a more restrictive AAR. This is actually a characteristic of both the priority scheduler and the FCFS scheduler, and it can be shown that, for a lower AAR, either scheduler is better able to preserve the order in which the aircraft are scheduled. Lowering the AAR effectively reduces the airport capacity (because demand remains constant), requiring that the STAs be spaced farther apart. Because the STAs must be spaced farther apart, differences in crossing times or separation criteria are less likely to cause the resulting order to deviate from the order in which the aircraft are scheduled. Therefore, the resulting schedule for either algorithm will more closely match the sequence in which the aircraft are scheduled. Results also show that increasing the size of the bank of aircraft does not significantly impact the effectiveness of the scheduling algorithm. However, increasing the magnitude of the Center arrival error for a given bank size and AAR does lead to a decrease in the effectiveness of the priority-scheduling algorithm.

For purposes of illustration, Fig. 2 is a histogram of the position shifts for the bank whose lead aircraft has the earliest ETA_{FG} shown in Fig. 1. This histogram, along with the pair of corresponding OD values in Fig. 1, demonstrate the relationship between average OD and the closeness of the match between the scheduled

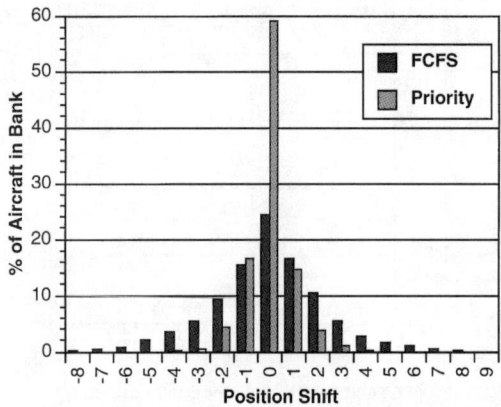

Fig. 2   Histogram of position shifts (Center arrival error = ±5 min).

bank order and the preferred arrival order. Priority scheduling reduces the spread of the position shifts for the designated bank of aircraft. In this case, aircraft belonging to the designated bank are scheduled in the preferred position (position shift = 0) approximately 60% of the time using priority scheduling. Using FCFS scheduling, bank aircraft are scheduled in the preferred position only about 25% of the time. The increase in the number of aircraft scheduled in the preferred position leads to a decrease in average OD for the bank.

The efficiency of the priority-scheduling algorithm is not measured in absolute terms, but is measured relative to the efficiency of the FCFS algorithm, which is considered to be a baseline. The change in average delay per aircraft when priority scheduling is used instead of FCFS scheduling can be measured as

$$\Delta_{\text{Delay}} = \left(\frac{\text{Delay}_{\text{Priority}} - \text{Delay}_{\text{FCFS}}}{\text{Delay}_{\text{FCFS}}} \right) \times 100\%$$

Figure 3 is a plot of the change in the average delay per aircraft for all aircraft in an arrival rush. For each designated arrival bank whose order deviation is shown in Fig. 1, a corresponding pair of points in Fig. 3 shows the change in average delay for the AAL/EGF aircraft in the arrival rush and for the non-AAL/EGF (others) aircraft in the arrival rush. Figure 3 shows that the change in delays due to priority scheduling varies with the position of the bank in the arrival rush and that the greatest delay increase occurs for a bank that starts near the beginning of the arrival rush. This is attributable to the changing traffic density and traffic mixture in the arrival rush and to all aircraft following the bank lead aircraft potentially being impacted by the reordering of the bank aircraft before scheduling. Because a larger number of aircraft may be impacted by the reordering, the aggregate increase in delays will be greater for a bank that begins earlier in the arrival rush. The average delay increase then diminishes as the ETA$_{\text{FG}}$ of the lead bank aircraft increases, and priority scheduling in some instances actually results in a slight decrease in average delay per aircraft. The priority-scheduling algorithm has the smallest impact on scheduling efficiency in regions where arrivals are not closely spaced and banks have non-AAL/EGF flights interspersed among the bank aircraft.

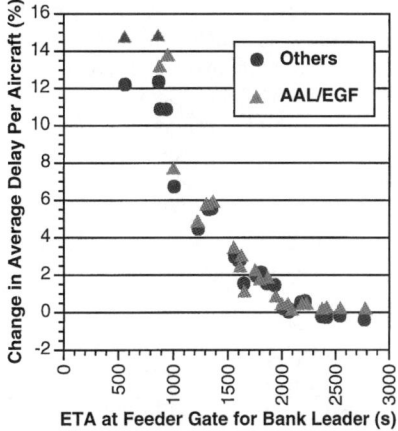

Fig. 3 Change in average delays per aircraft (Center arrival error = ±5 min).

A scheduling method that takes into account user preferences would ideally have no impact on scheduling efficiency when compared with FCFS scheduling. Figure 3 shows that for certain traffic conditions, the priority-scheduling method results in little or no decrease in scheduling efficiency; 18 of 25 banks scheduled had less than a 5% increase in average per aircraft delay. Although this simulation does not provide any information about the controller workload, it can be reasonably assumed that an increase in scheduled delays exceeding a certain threshold (when compared with FCFS scheduling) would be unacceptable to air traffic controllers because of the likely adverse effect on controller workload. Similarly, airlines would likely find an increase in scheduled delays exceeding a certain threshold (when compared with FCFS scheduling) to be unacceptable from the standpoint of increased cost. The amount of delay increase acceptable to controllers and airlines would have to be determined before a priority-scheduling method could be considered practicable. However, because the simulation results show that the priority-scheduling technique can in some instances equal the efficiency of the FCFS scheduler, the present simulation provides initial insight into the efficiency of priority scheduling and indicates that the technique warrants further investigation.

Any type of scheme that allows the introduction of user preferences into the arrival flow management process must ultimately be fair to all air carriers. In light of this, we are particularly interested in determining whether the priority scheduling of flights belonging to the airline whose flights are reordered disproportionately impacts the scheduled delays of aircraft belonging to other airlines. Examination of the delay increases for AAL/EGF flights in Fig. 3 shows that, for most of the banks, the delay increase for AAL/EGF flights in the arrival rush is greater than the delay increase for the non-AAL/EGF aircraft. In practice, an airline may be willing to accept increased delays on some flights if the early arrival of other flights can be achieved through reordering. Reordering only the aircraft belonging to the designated bank and scheduling all other aircraft according to an FCFS sequence minimizes the impact of reordering on both aircraft belonging to other airlines and on overall scheduling efficiency.

The effects of AAR, bank size, and Center arrival error on the change in scheduled delays are also examined. For a given bank size and Center arrival error, when priority scheduling is used instead of FCFS scheduling, the change in average delay per aircraft tends to increase as AAR is increased. Results are similar to those seen in Fig. 3 with the greatest change in delay occurring for banks that begin early in the arrival rush and the change in delays decreasing for banks that are positioned later in the arrival rush. Increasing the magnitude of the Center arrival error for a given bank size and AAR substantially increases the change in delays for banks of aircraft arriving early in the rush period, while not significantly impacting the change in delay for banks arriving later in the traffic period. Finally, results show that the change in delays due to priority scheduling is largely unaffected by an increase or decrease in the size of the arrival bank.

VIII. Conclusions

This chapter investigates a method of scheduling a bank of arrival aircraft according to a preferred order of arrival instead of according to an FCFS sequence based on ETA at the runway. Fast-time simulation is used to evaluate the performance of the priority-scheduling method in terms of the algorithm's ability to produce a preferred order of arrival and in terms of its ability to minimize delays (scheduling efficiency). The output of the simulation is used to compare the performance of the priority-scheduling technique and the current CTAS FCFS scheduling technique. Although the determination of the operational feasibility of the priority-scheduling technique is beyond the scope of this chapter, examination of scheduling efficiency is a crucial step in the determination of the operational viability of any CTAS-derived scheduling technique. Results show that compared with the FCFS algorithm, the priority-scheduling algorithm, for certain traffic conditions, produces a schedule that more closely matches a preferred order of arrival while nearly equaling the efficiency of the FCFS schedule, indicating that the method certainly warrants further investigation.

References

[1] Bond, L., "Global Positioning Sense II: An Update," *Journal of Air Traffic Control*, Vol. 39, No. 4, 1997, pp. 51–55.

[2] Wall, R., Cook, W., DeArmon, J., and Beaton, E., "Self-Managed Arrival Resequencing Tool (SMART): An Experiment in Collaborative Air Traffic Management," *Journal of Air Traffic Control*, Vol. 39, No. 2, 1997, pp. 30–34.

[3] Wambsganns, M., "Collaborative Decision Making Through Dynamic Information Transfer," *Air Traffic Control Quarterly*, Vol. 4, No. 2, 1997, pp. 109–127.

[4] Erzberger, H., Davis, T. J., and Green, S. M., "Design of Center-TRACON Automation System," *Proceedings of the AGARD Guidance and Control Panel 56th Symposium on Machine Intelligence in Air Traffic Management*, AGARD, 1993, pp. 52-1–52-14.

[5] Neuman, F., and Erzberger, H., "Analysis of Delay Reducing and Fuel Saving Sequencing and Spacing Algorithms for Arrival Traffic," NASA TM-103880, 1991.

[6] Swenson, H. N., Hoang, T., Engelland, S., Vincent, D., Sanders, T., Sanford, B., Heere, K., "Design and Operational Evaluation of the Traffic Management Advisor at the Fort Worth Air Route Traffic Control Center," *Proceedings of the 1st USA/Europe Air Traffic Management R&D Seminar*, 1997.

[7] Erzberger, H., "Design Principles and Algorithms for Automated Air Traffic Management," *Knowledge-Based Functions in Aerospace Systems*, AGARD Lecture Series 200, AGARD, Nov. 1995.

[8] Carr, G. C., Erzberger, H., and Neuman, F., "Delay Exchanges in Arrival Sequencing and Scheduling," AIAA Paper 98-4478, Aug. 1998.

[9] Green, S. M., Goka, T., Williams, D. H., "Enabling User Preferences Through Data Exchange," AIAA Paper 97-3682, Aug. 1997.

[10] Slattery, R. A., Cheng, V. H. L., "Sensitivity of En-Route Scheduling to Variable Separation in the Terminal Area," AIAA Paper 97-3736, Aug. 1997.

[11] Ballin, M., and Erzberger, H., "An Analysis of Landing Rates and Separations at the Dallas/Fort Worth International Airport," NASA TM-110397, July 1996.

Chapter 18

Capacity-Related Benefits of Proposed Communication, Navigation, Surveillance, and Air Traffic Management Technologies

Tara J. Weidner*
Seagull Technology, Inc., Los Gatos, California

I. Introduction

RESEARCH and technology development efforts by the U.S. government are under way that will improve existing communication, navigation, and surveillance (CNS) systems as well as implement advanced air traffic management (ATM) decision support tools (DSTs) in the National Airspace System (NAS).[1,2] A key requirement in the implementation of such technologies is a careful analysis of expected benefits, weighed against system costs. This chapter assesses the airport capacity-related benefits of a subset of the CNS/ATM technologies currently proposed for the NAS. The focus is on ATM systems and user-ATM data exchange.

Initially, in Section II, three scenarios of future CNS/ATM technology deployment in the NAS are proposed. These evolutionary phases reflect future implementation of current or under-development programs. Section III discusses the specific capacity-related benefit mechanisms under study in the context of other potential benefits. The analysis methodology is summarized briefly in Section IV. The model inputs and resulting benefit estimates appear in Section V. Section VI offers conclusions and remarks.

II. Assumed Technology Scenarios

A three-phase evolution of CNS/ATM technology enhancements are proposed in Table 1. These phases represent steps toward implementation of proposed technologies.[3–6] In this study benefits will be calculated by comparing these scenarios to a baseline of current system operations.

The three phases include deployment of proposed ATM DSTs, widespread use of flight management system (FMS) flight control and required time of arrival

---

Copyright © 2001 by the American Institute of Aeronautics and Astronautics, Inc. All rights reserved.
*Currently with Parsons Brickerhoff Quade & Douglas.

Table 1 Evolutionary CNS/ATM scenarios

Scenario	CNS/ATM technologies				
	ATM DST	Communication	Navigation	Surveillance	CDM
Current system	None	voice	Non-FMS	radar	None
Phase 1	TMA pFAST	voice	Non-FMS	radar	None
Phase 2	EDA AFAST	ACARS	FMS (LNAV/ VNAV)	radar	Basic DX
Phase 3		ATN	RTA	ADS-B	UPT DX

(RTA) capability in the terminal area, and use of current data link to exchange non-time-critical data and future data link to negotiate user preferred trajectories (UPT) to meet ATM DST-calculated RTAs.

Specifically, the technologies assumed in the three proposed CNS/ATM phases are discussed in the following:

Phase 1 focuses on near-term deployment of the Center-Terminal Radar Approach Control (TRACON) Automation System (CTAS) Build 2. This includes the Passive Final Approach Spacing Tool (pFAST) in the TRACON and the Traffic Management Advisor (TMA) in the air route traffic control center (ARTCC) transition airspace. These tools assist controllers in the sequencing and scheduling of arrival traffic into congested airports, both at arrival fixes and landing runways. Prototypes of both tools are operational at Dallas/Fort Worth airport (DFW).

Phase 2 enhances the Phase 1 ATM DSTs with Active Final Approach Spacing Tool (AFAST) in the TRACON and En Route Descent Advisor (EDA) in the ARTCC transition airspace. Beyond Phase 1 DSTs, these tools provide controllers with maneuver advisories to meet the CTAS sequences and schedules. These tools are currently under development at NASA Ames Research Center. Phase 2 also assumes the widespread use of aircraft FMS flight control, including lateral and vertical navigation (LNAV/VNAV), in the extended terminal airspace, reducing flight variability. Basic user-CTAS data exchange is also assumed. This includes the passive exchange of calibration data, including aircraft weight, planned threshold crossing speed, and wind/temperature measurements/forecasts. The exchange of these data will improve the airborne and ground trajectory prediction models.

Phase 3 is a far-term vision of a four-dimensional user-ATM arrival trajectory negotiation in the extended terminal area. This requires a far-term data link such as the proposed Aeronautical Telecommunications Network (ATN) to host the exchange of time-critical ATM clearance and route negotiation. Additionally, the aircraft must be equipped with RTA capabilities and traffic avoidance equipment such as automatic dependent surveillance broadcast (ADS-B) equipment.

The study was limited to the ATM/CNS improvements shown in Table 1, as these tools will have a significant impact on capacity-related benefits of terminal operations. It is recognized, however, that there are numerous additional ATM

Table 2 Possible ATM/CNS benefits

Benefit type	Stakeholder	Domain/airspace
Delay	Users/airlines	Terminal
Flight efficiency	ATM service provider (FAA)	Transition
Productivity	Flying public	En route
Safety		Oceanic
ATM interruptions		Airport surface

DSTs (e.g., surface management advisor and initial conflict probe), collaborative decision making (CDM) tools (e.g., flight system monitor), and aircraft avionics (e.g., global positioning systems) under study that could also be included in such an analysis. In fact, the included ATM/CNS improvements may have additional benefits, not herewith addressed. As such, this study is a first-cut estimate at a much larger realm of possible ATM/CNS benefits. The larger world of benefits is discussed in Section III.

III. Capacity-Related Benefits Defined

Implementation of future CNS/ATM technologies in the NAS may allow a number of benefits to accrue to various NAS stakeholders in multiple domains, as identified in Table 2.

As discussed in Section II, this study is limited to capacity-related benefits of the terminal/transition airspace, accrued to users in the form of reduced delay and flight efficiencies, measured in direct operating costs of time and fuel. In the remainder of this section, these two benefit mechanisms are discussed in more detail.

A. Increased Airport Throughput

Airport throughput is a function of the spacing achieved between successive aircraft at the runway threshold. Actual spacings, as implemented by air traffic controllers, are generally larger than the minimum separation requirements.[7] Observations indicate that the extra spaces can be assumed to be intentional spacing buffers that serve in part to assure that separation minima are not violated because of trajectory uncertainties. Excess spacing is also generated by uncertainty in the delivery of arrival aircraft at the inbound metering fixes. Deviations from the ATM metering fix crossing schedule, due to timing delivery inaccuracies, require subsequent trajectory adjustments by downstream ATM to prevent violations of separation minima and eliminate extraneous gaps at downstream merge points and the runway threshold. With reductions in predicted trajectory uncertainty under proposed future technologies, the size of the excess spacing buffer that is needed to compensate for trajectory variances can be reduced. A smaller buffer would reduce the spacing applied between successive aircraft, thereby increasing the throughput of the runway system. The increased throughput would reduce aircraft delays and associated user fuel and time direct operating costs.

B. ARTCC/TRACON Delay Distribution Fuel Savings

In the terminal area, a delay distribution function, formalized in future ATM automation, is employed to allocate aircraft delay between ARTCC and TRACON airspace during busy traffic periods. The allocation process works to achieve an optimum balance between fuel burn savings and runway system throughput. The delay distribution function performs a tradeoff between the advantage of absorbing delay at the higher en route altitudes, where fuel efficiency is greater, vs the advantage of packing more aircraft in the terminal airspace to ensure that aircraft are continually available to use the runway system. As trajectory prediction and control accuracy is improved with improved trajectory prediction under proposed technologies, less delay is needed to be absorbed in the TRACON airspace to maintain high runway system throughput. This provides savings as a result of more fuel-efficient arrival trajectories. Whereas excess spacing buffers determine the runway system throughput and the associated total amount of delay, delay distribution determines where and how efficiently delay is to be absorbed.

IV. Analysis Methodology Overview

One way to significantly increase airport throughput is through the optimal scheduling and sequencing of aircraft trajectories in congested terminal airspace.[8] Such algorithms are dependent on accurate predictions of when aircraft will cross downstream waypoints. Thus, a key requirement of advanced terminal area ATM DSTs is a highly accurate prediction of trajectory waypoint crossing times. Other analyses that focus on the benefits of a conflict detection and resolution tool would find trajectory prediction *position* accuracy of paramount interest, providing reliable future aircraft locations to a conflict probe tool.

Trajectory prediction accuracy can be improved through high-fidelity trajectory modeling of flight trajectories in the terminal airspace, as found in the proposed ATM DSTs and aircraft FMSs. Additionally, both the air and ground trajectory models can be improved with the use of more accurate model inputs, such as aircraft weight estimates and wind/temperature forecasts obtained through data exchange. Future avionics may also improve aircraft surveillance and intent information through ADS-B reporting. In the long term, arrival trajectory negotiation using FMS RTA capabilities should allow for highly accurate trajectory prediction and control accuracies. This study analyzes trajectory prediction timing accuracy and its various improvements with future CNS/ATM technologies, and its effects on capacity-related benefit metrics.

Figure 1 shows the overall analysis process used in this study. This analysis process has been used in numerous research efforts[9−12] and is only briefly detailed in the remainder of this section. The analysis process is initiated by identifying the various candidate technologies and their capabilities, as was done in Section II. Each technology is then described in terms of parameters that relate to operational factors that affect aircraft trajectories. Figure 1 lists key parameters impacted by CNS/ATM technologies. These parameters are defined by a stochastic distribution that quantitatively describes the accuracy of each parameter for baseline and proposed CNS/ATM scenarios.

The modeling process encompasses three subcomponents. Initially, the accuracy with which trajectories can be predicted and controlled is modeled using

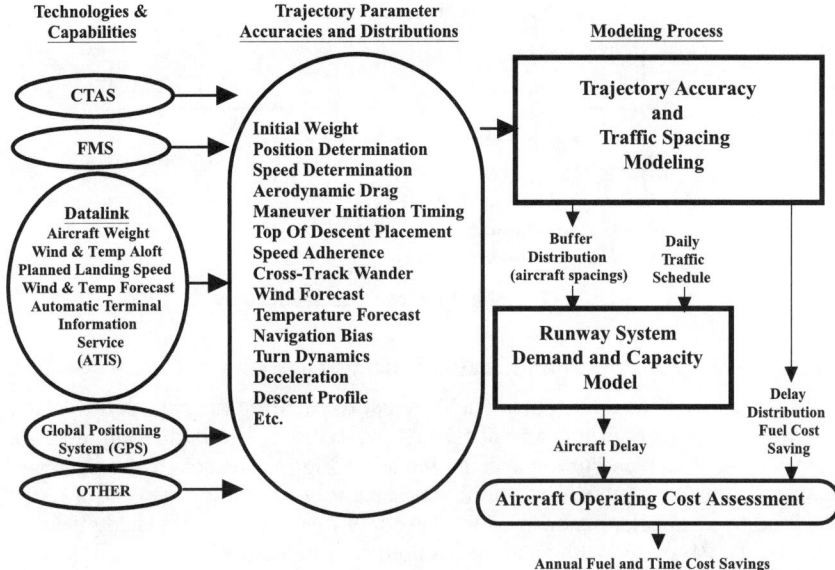

Fig. 1 Analysis process.[10–12]

computer simulation, close-form analytical solutions, and a combination of the two, as appropriate for each phase of flight. The parameter accuracy distributions are inputs to this trajectory modeling process. The outputs include excess spacing buffers applicable to runway system operations and incremental fuel cost due to ARTCC/TRACON delay distribution. Recent CTAS prototype field test observations were used to calibrate the various parameter accuracy distributions in this model.[13,14]

A second computer simulation is used to evaluate airport throughput and determine traffic delay using the threshold excess spacing buffer data and minimum separation requirements as input. The model incorporates data describing daily flight schedules of commercial, general aviation, and military aircraft and detailed configurations of the major domestic airports for instrument flight rules (IFR) and visual flight rules (VFR). Modeling parameters describing separation procedures for the IFR and VFR runway configurations at each airport are adjusted to reflect the operating environment of the various scenarios under study. The model produces average daily traffic delay data by arrival and departure operations and instrument and visual meteorological conditions for each airport under study.

The daily traffic delay data are then extrapolated to annual cost savings by airport using detailed aircraft direct operating cost rates, annual airport traffic forecasts, and meteorological factors. The aircraft direct operating costs used represent fuel, crew, and maintenance costs.

The three models are described further in the remainder of this section. For more information, the reader is directed to Ref. 9–12. The chosen model parameter values and resulting benefits are identified in Section V.

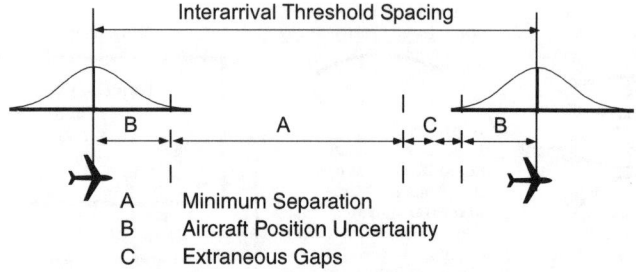

Fig. 2 Threshold interarrival separation.

A. Trajectory Accuracy and Traffic Spacing Model

The interarrival aircraft spacing achieved at the landing runways is the critical metric in analyzing airport arrival throughput. This metric is the output of the trajectory accuracy and traffic spacing model of Fig. 1. The achieved interarrival spacings consists of the ATM minimum separation requirement[15] and a buffer imposed by controllers to ensure that these minimums are not violated given the uncertainty associated with an aircraft's predicted trajectory.

Additionally, delivery inaccuracies in aircraft crossings at the metering fixes or TRACON entry point can contribute to reduced throughput. This results because aircraft may not be in place to fill landing slots, causing gaps in the arrival stream, which reduces overall arrival landing throughput. Although these gaps would likely have a duration of several seconds and occur infrequently, their impact can be approximated by averaging the gaps over all interarrival separations. Likewise, suboptimal runway balancing and sequencing would lead to extraneous gaps in runway utilization that can also be approximated in this way. These threshold spacing components, the required minimum separation, controllers' position uncertainty buffer, and buffer due to extraneous gaps, are shown in Fig. 2. The trajectory accuracy and traffic spacing model calculation of position uncertainty uses fast-time Monte Carlo simulations of nominal arrival trajectories. This is shown in Fig. 3.

The model simulates the accumulation of an arrival aircraft's position uncertainty from cruise to the runway threshold. On final approach, the model ensures that aircraft pairs do not violate minimum separation criteria as they decelerate to their final landing speeds and touchdown.

The calculation of threshold buffers due to extraneous gaps utilizes an analytical model of center/ARTCC contributions to threshold buffer[16] as well as estimates of runway balancing and sequencing impacts from NASA Ames Research Center simulations,[8] confirmed by field test results.[13,14] The center model derives the optimal relationship between a TRACON delay setting and extraneous gaps at the runway threshold. It is based on the premise that scheduling some delay in the TRACON allows TRACON controllers more flexibility to absorb metering fix crossing variability, allowing them to maximize airport throughput. However, delay absorbed in the TRACON requires more fuel than if absorbed in the higher elevation ARTCC. Thus fuel efficiency is improved with less TRACON delay, i.e., at lower TRACON delay settings. As a result of this tradeoff, an optimum TRACON delay setting can be derived, one that minimizes the combined cost of lost throughput and fuel burn.

CAPACITY-RELATED BENEFITS

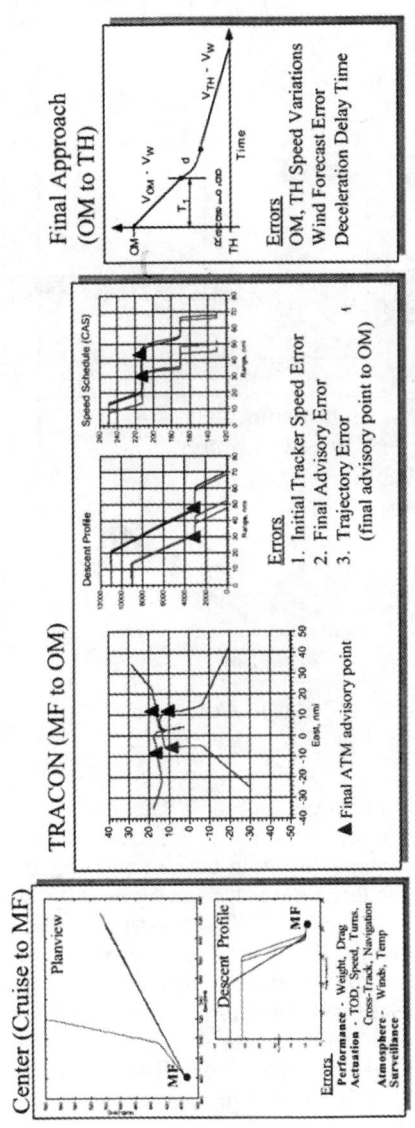

Fig. 3 Position uncertainty trajectory simulation.

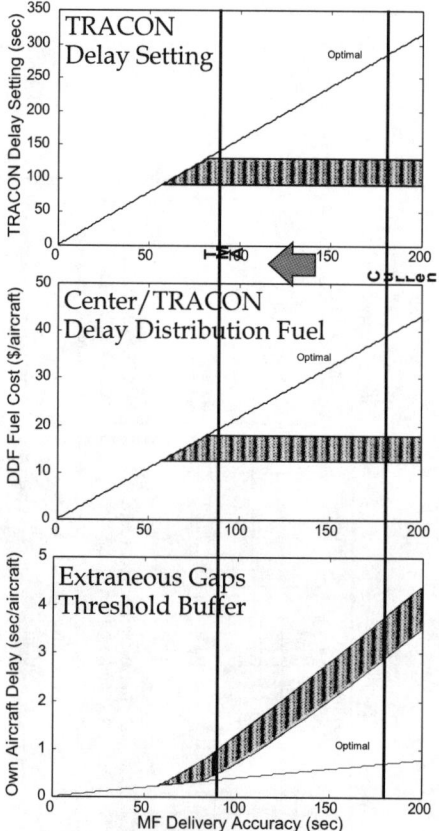

Fig. 4 Center model.

These relationships are plotted in Fig. 4. The figure parametrically relates metering fix delivery accuracy to the optimal TRACON delay setting and the corresponding extraneous gaps threshold buffer and delay distribution incremental fuel costs. Although the threshold extraneous gap buffer is on the order of a few seconds, the delay actually incurred by an aircraft at the end of the rush could accumulate to 20 to 50 times this value depending on the rush size. As the optimal lines in the figures show, improved metering fix accuracy (moving in the direction of the arrow) would result in a reduction in optimal TRACON delay setting with corresponding delay and fuel efficiency savings.

However, based on previous studies of TRACON flight track data,[17,18] TRACON airspace can typically absorb only 100–200 s of delay on average beyond the fastest feasible path to the runway. This is reflective of the airspace geometry and complexity of TRACON air traffic control operations. In this study, we restrict the TRACON delay setting to 100 s. As a result, the buffer and fuel costs change considerably, as shown by the bottom edge of the shaded region in Fig. 4. These relationships are used to determine the best TRACON delay setting for each modeled CNS/ATM scenario, as well as the corresponding extraneous gap threshold buffer and delay distribution incremental fuel cost.

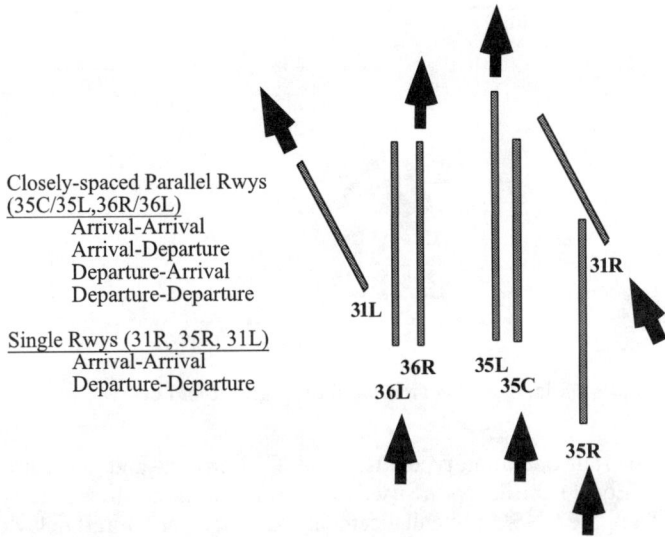

Fig. 5 IFR runway configuration, DFW.

B. Runway System Demand and Capacity Model

Once the interarrival separations (rule and buffer) are determined for each scenario from the trajectory accuracy and traffic spacing model, they are used in simulations of a full day's traffic at candidate airports. The model generates scheduled takeoff/landing times and any associated delays, by simulating the interactions of traffic demand, the runway use configuration assumed at the airport, and the appropriate aircraft separation procedures for instrument and visual flight rules (IFR and VFR). Figure 5 shows the IFR configuration modeled at DFW and lists appropriate separations required to cover the various DFW operations.

Each simulation run depicts a typical day at the subject airport, imposing visual meteorological condition (VMC) for the bulk of the day, with inclement meteorological conditions (IMC) for a 5-h a.m. period. The results of each run are used to calculate an average VMC and IMC delay per operation. These average delays take into account the impact of historical IMC persistence on delay severity.[19] Persistence in this case refers to the time duration of continuous IMC. Thus, locations with only intermittent IMC would expect lower delays, because periods of VMC would be able to clear delays accumulated from prior IMC periods. Figure 6 shows sample scheduled operations, realized throughput, and delay results from DFW airport simulation. As noted, significant delays are accrued during the shaded a.m. IMC period, but they dissipate soon after VMC conditions resume.

C. Aircraft Operating Cost Assessment

The resulting IMC and VMC average delays for each CNS/ATM scenario at each airport are recorded and compared. Equation (1) gives the general formula used to convert these delays into airport cost savings. This equation is

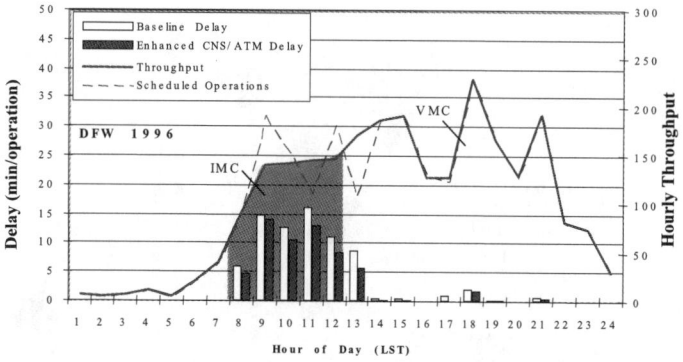

Fig. 6 DFW simulated throughput and delay.

evaluated for four operation types: IFR and VFR arrivals and departures. In the first term of Eq. (1), the delays are used to determine average aircraft delay savings expected from the CNS/ATM enhancements. Average per aircraft delay cost savings are then identified for each scenario by application of aircraft direct operating cost rates [second term in Eq. (1)] to the estimated minutes of aircraft delay savings. These savings estimates account for operating cost differences among aircraft classes, based on the traffic mix at each airport. Finally, these data are extrapolated, using airport-specific annual traffic forecasts and historic meteorological data [third term in Eq. (1)], to estimate the corresponding annual cost savings:

$$\begin{array}{c} \text{Annual Delay} \\ \text{Cost Savings} \\ (\$/\text{year}) \end{array} = \begin{array}{c} (\text{Delay}_{BL} - \text{Delay}_{CNS/ATM}) \\ (\text{min/operation}) \end{array} \times \begin{array}{c} \text{Aircraft Cost} \\ \text{Rate} \\ (\$/\text{min}) \end{array} \times \begin{array}{c} \text{Airport Annual} \\ \text{Traffic Forecast} \\ (\text{operations/year}) \end{array} \quad (1)$$

V. Model Assumptions and Results

Using the models discussed in Section IV, we are able to evaluate the capacity-related benefits of the three evolutionary CNS/ATM phases, described in Section II at 11 candidate airports. This section details both the specific parameter values assumed in modeling the various scenarios as well as the resulting delay and fuel efficiency benefits.

The trajectory accuracy and traffic spacing model uses as input, estimates of parameter accuracies that affect arrival aircraft trajectories. These parameter errors are chosen to reflect the operational environment of the current system and CNS/ATM phases described in Section II. The choice of these parameter values requires the reconciliation of manufacturer and/or research-based accuracies with actual field performance. Assumptions for future technologies must also assume important operational issues will be addressed to not limit implementation locations. Fortunately, recent 1996 prototype field test results of the CTAS Build 2 baseline ATM DSTs at DFW[13,14] provide realistic understanding of the current system and Phase 1 operations. Future phases must rely more heavily on manufacturer's claims and simulation studies.

The parameter error estimates used to represent the three CNS/ATM technology scenarios are detailed in Table 3, along with the resulting metering fix crossing accuracies, threshold spacing buffers, and center/TRACON delay distribution fuel costs, expected under each scenario.

Table 3 parameters reflect Phase 1 improvements due to high-fidelity modeling of TMA and pFAST over cognitive sequencing and scheduling algorithms of the current system. Phase 2 reduces actuation errors with CTAS-calculated maneuver advisories of EDA and AFAST, increases maneuver adherence with FMS flight control, and improves aircraft weight and meteorological forecasts through user-CTAS data exchange. Phase 3 reduces mismatches in CTAS-FMS trajectory prediction by allowing an aircraft to fly its preferred trajectory to meet the CTAS-calculated RTA at a very high accuracy.

The results at the bottom of Table 3, show significant reduction in metering fix delivery error, expected threshold buffer, and center/TRACON delay distribution incremental fuel costs. Interestingly, Phases 1 and 2 reap almost all of the metering fix delivery improvement and corresponding fuel savings, while 45% of the threshold buffer savings comes in Phase 3.

The expected buffer results of Table 3 were then used in the runway system demand and capacity model to determine the average delays for arrivals and departures under IMC and VMC conditions. The resulting delays for the three CNS/ATM enhancement phases are shown in Table 4. Typically, larger savings are expected for arrival savings, as the ATM DSTs under study alter primarily arrival sequencing and scheduling. However, as controllers try to keep arrival and departure delays in balance, all operations benefit. Because the demand at New York LaGuardia (LGA) is much larger than its capacity, it incurs the largest delay savings. Because current conditions do not allow the airport to dissipate the growing delays throughout the day, any improvement in capacity will reduce delays on almost every operation at LGA. Other airports show similar trends.

These delay savings are then used to assess annual aircraft operating cost savings, incorporating the cost rates and annualization factors shown in Table 5. Table 6 details the resulting delay cost savings.

Center/TRACON delay distribution fuel savings use the incremental fuel cost values at the bottom of Table 3 and apply these savings to all peak period rush arrivals. Peak period arrivals account for 30% of all operations at DFW. This percentage was assumed to represent rush arrivals at the other large and medium hub airports under study. As a result, the largest delay distribution fuel savings, shown in Table 7, occur at airports with the highest number of annual operations.

The combined delay and center/TRACON delay distribution fuel cost savings are shown in Table 8 and graphically depicted in Fig. 7. Because these cost components represent two different, but operationally distinct, impacts of CNS/ATM improvements, they were summed. However, the two components were evaluated using different methods (i.e., simulations vs analytical models), and ideally a common methodology would be employed to assure that benefits evaluations are compatible with each other.

VI. Conclusions

This chapter assesses the capacity-related benefits of proposed CNS/ATM enhancements at 11 major U.S. airports. These enhancements were broken into three evolutionary phases of NAS improvements to current system operations:

Table 3 Assumed trajectory parameter accuracies

Trajectory error parameter	Units	Mean value	Current system baseline	Standard deviation CNS/ATM evolution		
				Phase 1	Phase 2	Phase 3
Center flight segment						
Initial weight	%	0	—	10	2.5	2.5
Aerodynamic drag	%	0	—	7	7	7
Top of descent location	n mile	0	—	16	0.5	NA
Speed adherence	kn	0	—	20	4	4
Cross-track wander	n mile	0	—	5	0.25	0.25
Aircraft navigation bias	deg	0	—	2	2	2
Turn dynamics	s	0	—	15	1	1
Wind forecast	kn	0	—	20	4	4
Temperature forecast	°C	0	—	10	1	1
Surveillance	kn	0	—	15	15	NA
RTA open loop	%	NA	NA	NA	NA	33
RTA setting	s	NA	NA	NA	NA	6
TRACON flight segment						
Final advisory delay	s	0	12.12	10.80/9.75	9.4/8.49	NA
Turn variation	s	35	7.0	7.0	1.0	1.0
Deceleration variation	%	0.52	0.120	0.120	0.108	0.108
Descent rate variation	ft/min	1440	160	160	144	144
Speed adherence	kn	0	4.0	4.0	1.2	1.2
Wind forecast accuracy	kn	0	4.7	4.7	4.0	4.0
Tracker speed accuracy	kn	0	3.5	3.5	3.5	NA

CAPACITY-RELATED BENEFITS

				Final approach flight segment			
RTA open loop	%	NA	NA	NA	NA	16.45
RTA setting	s	NA	NA	NA	NA	6
Runway balancing/sequencing[a]	s	2.3	No	No/Yes	No/Yes	No/Yes
Outer marker speed	kn	170	5.0	5.0	1.8	1.2
Threshold speed[b]	kn	120–135	9	9	4	1.2
Deceleration delay time	s	10	12	12	12	NA[c]
Wind forecast error	kn	10	10	4.7	4.0	4.0
MF crossing accuracy[d]	s	0	180	90	14	5
Expected buffer[e–f]	s	0	31.44	26.62/23.54	18.91/15.78	7.23/4.93
Incremental fuel cost	$/arrival[g]	0	13.67	13.6	3.02	1.08

[a] Parameter varies by number of arrival runways, 1–2/3+ runway values shown. Applied per runway usage at each airport.
[b] Speed varies by aircraft size classification, 120, 125, and 135 for small, large, and heavy aircraft, respectively.
[c] Not applicable under RTA operations.
[d] Current system and Phase 1 metering fix crossing accuracy observed in TMA prototype field tests.[14]
[e] Average excess spacing buffer value, weighted by airport's aircraft class distribution.
[f] Current system and Phase 1 buffers calibrated from p-FAST prototype field test results.[13]
[g] Delay distribution fuel cost savings applies only to rush period arrivals.

Note: Shading indicates improvements from prior scenario.

Table 4 Average delay savings, min

Airport	Phase 1					Phase 2					Phase 3				
	IFR		VFR		IFR		VFR		IFR		VFR	
	Departure	Arrival	Departure	Arrival	Departure	Arrival	Departure	Arrival	Departure	Arrival	Departure	Arrival
ATL	1.07	0.94	0.00	1.74	2.78	2.44	0.00	4.51	2.18	2.10	0.15	7.83
BOS	1.89	1.54	1.96	1.83	4.87	3.96	5.06	4.72	8.68	6.95	13.01	12.34
DFW	0.96	1.23	0.00	0.28	1.79	2.32	0.00	0.47	2.67	3.54	0.00	0.53
DEN	0.23	0.50	0.05	0.12	0.50	1.02	0.08	0.21	0.97	1.82	0.11	0.31
DTW	0.26	0.71	0.11	0.71	0.68	1.82	0.28	1.82	1.18	3.41	0.44	3.24
EWR	1.28	1.46	0.12	0.18	3.32	3.78	0.24	0.35	6.79	7.39	0.46	0.56
JFK	0.01	0.08	0.22	0.27	0.01	0.22	0.56	0.69	0.01	0.31	0.97	1.13
LAX	1.55	1.48	0.00	5.80	4.00	3.82	0.00	10.19	7.27	7.57	0.00	12.25
LGA	0.87	1.00	7.95	7.96	2.27	2.59	20.64	20.66	4.64	5.73	30.79	30.68
ORD	0.00	0.00	2.08	3.11	0.00	0.00	3.73	5.59	2.64	3.22	5.05	7.66
SFO	0.78	0.65	1.30	1.21	2.01	1.66	3.35	3.11	4.81	4.57	8.28	8.52

Table 5 Cost rates and annualization factors[20–22,*]

Airport	Annual IMC occurrence 7 a.m.–10 p.m.	1996 Annual operations, 000s	Average operating cost, $/min	
			Departures	Arrivals
ATL	14.2%	855	15.58	22.41
BOS	15.6%	480	11.34	15.92
DFW	8.4%	892	13.30	18.93
DEN	6.0%	499	13.50	19.25
DTW	16.6%	511	13.92	19.82
EWR	16.6%	446	14.12	20.18
JFK	15.0%	348	19.84	28.96
LAX	22.2%	747	15.17	21.82
LGA	16.4%	350	13.05	18.59
ORD	16.1%	914	15.24	21.92
SFO	12.5%	451	15.58	22.47

Note: Analysis also assumes 50% of airport traffic is arrivals.

1) Phase 1: CTAS TMA and pFAST ATM DSTs are deployed to improve arrival trajectory scheduling and sequencing.
2) Phase 2: CTAS EDA and AFAST ATM DSTs are deployed, providing controllers CTAS-calculated maneuver advisories to meet Phase 1 schedules. Additionally, FMS flight control is employed in the terminal area, and user-CTAS exchange of basic calibration data is exercised for use in air and ground trajectory prediction calculations.

Table 6 Annual delay cost savings

Airport	Annual operating cost savings (1998, $ million)		
	Phase 1	Phase 2	Phase 3
ATL	16.66	43.25	70.45
BOS	12.21	31.51	78.08
DFW	3.54	6.19	7.99
DEN	0.86	1.63	2.48
DTW	4.08	10.53	18.68
EWR	2.77	6.59	12.55
JFK	1.88	4.82	8.00
LAX	41.35	76.50	100.55
LGA	38.22	99.14	149.02
ORD	38.78	69.66	103.28
SFO	10.23	26.40	69.14

*Data available online at http://www.apo.data.faa.gov/faatafall.htm [cited 23 March 1998].

Table 7 Annual delay distribution fuel savings

Airport	Annual operating cost savings (1998, $ million)		
	Phase 1	Phase 2	Phase 3
ATL	0	2.40	2.84
BOS	0	1.63	1.93
DFW	0	2.86	3.38
DEN	0	1.64	1.93
DTW	0	1.64	1.94
EWR	0	1.46	1.72
JFK	0	1.15	1.36
LAX	0	2.32	2.74
LGA	0	1.12	1.32
ORD	0	2.86	3.38
SFO	0	1.47	1.73

3) Phase 3: User-CTAS negotiation of four-dimensional arrival trajectories to meet CTAS-calculated RTAs becomes operational. This requires ATN data link and ADS-B surveillance technologies.

The study assessed capacity-related benefits for these scenarios, including 1) reduced delay due to increases in runway system throughput, and 2) increased fuel efficiency with improved distribution of delay between center and TRACON flight segments. These are two of many potential benefits achievable with these proposed CNS/ATM technologies. The discussed methodology should be extended to cover other potential benefits mechanisms for a more complete benefits picture of these technologies.

The resulting capacity-related benefits of the proposed CNS/ATM enhancements are significant and vary considerably by airport. The estimated 1996 operating cost savings associated with runway throughput and delay benefits were found to range

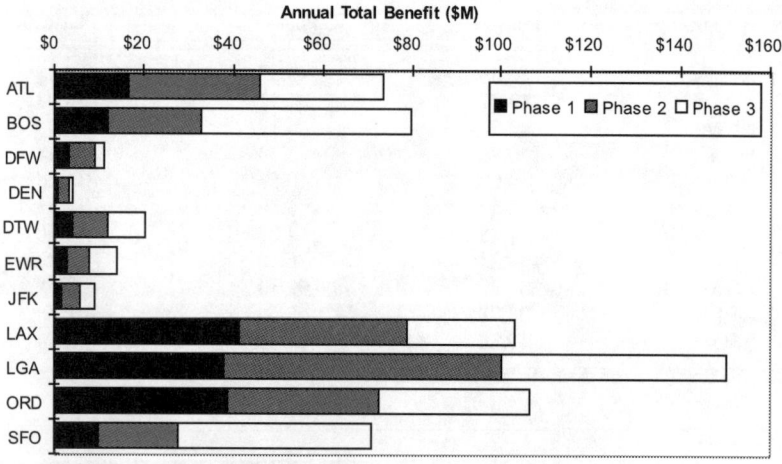

Fig. 7 Total annual cost savings.

Table 8 Total cost savings

Airport	Annual operating cost savings (1998, $ million)		
	Phase 1	Phase 2	Phase 3
ATL	16.66	45.98	73.68
BOS	12.21	33.05	79.90
DFW	3.54	9.04	11.36
DEN	0.86	3.22	4.37
DTW	4.08	12.16	20.61
EWR	2.77	8.01	14.23
JFK	1.88	5.93	9.31
LAX	41.35	78.88	103.37
LGA	38.22	100.26	150.34
ORD	38.78	72.58	106.73
SFO	10.23	27.84	70.85

from $1–38, $2–99, and $3–149 million annually at the 11 airports in Phases 1, 2, and 3, respectively. Better distribution of delay between center and TRACON airspace resulted in no fuel savings in Phase 1, due to the limitation that each airport can absorb only a maximum of 100 s of delay in the TRACON. In Phases 2 and 3, when this restriction is nonbinding, the savings range from $1–3 million annually per airport.

When the two capacity-related benefits are combined, the benefits range from $1–41, $3–100, and $4–150 million annually in Phases 1, 2, and 3, respectively. Total benefits at all 11 airports are estimated at $170, $397, and $575 million annually in each phase, with over 90% resulting from delay savings. The benefits were relatively equal between phases, although Phase 2 had the highest incremental benefit.

The largest overall savings occurred at severely capacity-constrained airports including New York LaGuardia (LGA), Chicago O'Hare (ORD), and Los Angeles International (LAX) airports. These are closely followed by Boston Logan (BOS), Atlanta Hartsfield (ATL), and San Francisco (SFO). As the current capacity restrictions at the remaining airports are lower, they are predicted to have fewer capacity-related benefits.

This chapter has demonstrated a methodology to quantify specific benefits of proposed CNS/ATM technologies. Combined with costs, such analyses are a necessary step in evaluating the overall economic value of these proposed technologies. Such studies can be useful at several stages of NAS development, such as to optimally direct scarce research moneys, to assist in deployment decisions, and to evaluate in-field tool performance.

Acknowledgment

The author would like to acknowledge the efforts of numerous people who helped develop the proposed methodology in previous research efforts. This includes George Couluris, George Hunter, and John Sorensen of Seagull Technology, Inc., as well as contract work done for the FAA AUA-500, FAA FANG, and NASA Terminal Area Productivity programs. The author is also grateful to David Schleicher of Seagull Technology, Inc., for discussions on related topics.

References

[1]"Integrated Plan for Air Traffic Management Research and Technology Development," Federal Aviation Administration and NASA, Sept. 1996.

[2]"Government/Industry Operational Concept of the Evolution of Free Flight, Addendum 1: Free Flight Phase 1 Core Capabilities Limited Deployment," Federal Aviation Administration Modernization Task Force, 1998.

[3]*A Joint Government/Industry Operational Concept for the Evolution of Free Flight*, RTCA Select Committee on Free Flight Implementation, RTCA, Inc., Washington DC, Aug. 1997.

[4]*Air Transportation System (ATS) Concept of Operations for the National Airspace System (NAS) in 2005, Narrative*, Federal Aviation Administration, Sept. 1997.

[5]*ATM Concept Definition, Volume 1, Current and Future Operational Concepts for the National Airspace System*, NASA Advanced Air Transportation Technologies Program Office, Aviation System Capacity Program, Oct. 1997.

[6]*FANG Operational Concept*, FMS/ATM Next Generation (FANG) program, DOT/FAA/ AND-97/7, July 1997.

[7]Ballin, M., and Erzberger, H., "An Analysis of Landing Rates and Separations at the Dallas/Fort Worth International Airport," NASA Ames Research Center, NASA TM110397, July 1996.

[8]Erzberger, H., "Design Principles and Algorithms for Automated ATM," *AGARD Lecture Series 200 on Knowledge Based Functions in Aerospace*, Nov. 1995.

[9]Hunter, C. G., Weidner, T., Couluris, G. J., Sorensen, J., and Bortins, R., "CTAS Error Sensitivity, Fuel Efficiency, and Throughput Benefits Analysis," Seagull Technology, TR96150-02, Los Gatos, CA, July 1996.

[10]Weidner, T., Couluris, G., and Hunter, C. G., "Center-TRACON Automation System (CTAS) Benefits Extrapolation Preliminary Analysis," Seagull Technology, TR98156-01, Feb. 1998.

[11]Weidner, T., Couluris, G., and Sorensen, J., "Initial Data Link Enhancement to CTAS Build 2, Potential Benefits Analysis," Seagull Technology, TR98151-01, Los Gatos, CA, June 1998.

[12]Couluris, G., Weidner, T., and Sorensen, J., "Final Approach Enhancement and Descent Trajectory Negotiation Potential Benefits Analysis," Seagull Technology, TR97142-02, July 1997.

[13]"CTAS Passive Final Approach Spacing Tool (P-FAST) Assessment—Final Report," Crown Communications, Inc., CTASDS-BAPRPT-002, FAA, Dec. 1996.

[14]Seagull Technology, preliminary assessment of CTAS Traffic Management Advisor (TMA) July 1996 Fort Worth ARTCC field tests, June 1997.

[15]*Air Traffic Control*, FAA 7110.65L, Feb. 1998.

[16]Hunter, C. G., and Weidner, T., "Performance and Benefits Modeling of Center Airspace ATM Improvements," *AIAA, Guidance, Navigation, and Control Conference*, AIAA Reston, VA, Aug. 1997.

[17]Dorsky, S., and Hunter, C. G., "Time to Fly in the Boston TRACON," Seagull Technology, TR91120-02, Los Gatos, CA, May 1991.

[18]Shen, M., and Hunter, C. G., "Time to Fly in the Dallas/Fort Worth TRACON," Seagull Technology, TR92120-03, Los Gatos, CA, Nov. 1992.

[19]Clark, D., and Evans, J., "Analysis of Hourly Surface Weather Observations, 1988-1992," [computer data file], MIT-Lincoln Laboratory, Cambridge, MA, 1995.

[20] *FAA Aviation Forecasts, Fiscal Years 1995-2006*, FAA-APO-95-1, Federal Aviation Administration Office of Aviation Policy and Plans, March 1995.

[21] *Economic Values for Evaluation of FAA Investment and Regulatory Programs*, FAA-APO-89-10, Federal Aviation Administration Office of Aviation Policy and Plans, Oct. 1989.

[22] *Ceiling and Climatological Study and System Enhancement Factors*, Federal Aviation Administration Office of Aviation System Plans, June 1975.

Chapter 19

Collaborative Optimization of Arrival and Departure Traffic Flow Management Strategies at Airports

Eugene P. Gilbo*
John A. Volpe National Transportation Systems Center,
Cambridge, Massachusetts
and
Kenneth W. Howard[†]
Arcon Corporation, Waltham, Massachusetts

Nomenclature

a_i = airport total arrival demand for the ith time interval, $i \in I$
$a_{i,r}$ = first priority arrival demand of the rth user for the ith time interval, $i \in I, r \in \Pi$
$a_i^{(s)}$ = total first priority arrival demand for the ith time interval, $i \in I$; $a_i^{(s)} \leq a_i$
d_i = airport total departure demand for the ith time interval, $i \in I$
$d_{i,r}$ = first priority departure demand of the rth user for the ith time interval, $i \in I, r \in \Pi$
$d_i^{(s)}$ = total first priority departure demand for the ith time interval, $i \in I$; $d_i^{(s)} \leq d_i$
I = $\{1, 2, \ldots, N\}$, a set of time intervals
T = time period of interest, consisting of N discrete-time intervals of length Δ (e.g., $\Delta = 15$ min); $T = N\Delta$
u_i = airport arrival capacity at the ith time interval, $i \in I$
v_i = $\varphi_i(u_i)$, airport departure capacity at the ith time interval, $i \in I$, $\varphi_i(u_i) \in \Phi$
X_i = airport arrival queue at the beginning of ith time interval, $i = 1, 2, \ldots, N + 1$
Y_i = airport departure queue at the beginning of ith time interval, $i = 1, 2, \ldots, N + 1$

Copyright © 2001. This material is declared a work of the U.S. Government and is not subject to copyright protection in the United States.
*Senior Operations Research Analyst, Automation Applications Division.
[†]Technical Director.

$\Phi = \{\varphi^{(1)}(u), \varphi^{(2)}(u), \ldots, \varphi^{(M)}(u)\}$, a set of M airport arrival-departure capacity curves that represent the operational limits for each available runway configuration under various operational (e.g., weather) conditions

$\Pi = \{1, 2, \ldots, P\}$, a set of airport users (e.g., airlines)

I. Introduction

THE Federal Aviation Administration's (FAA) traffic flow management (TFM) system manages and controls air traffic and provides the operational resources, or capacity, to serve the traffic. In this sense, the FAA is considered a service provider. The airlines and other users of the National Airspace System (NAS) resources create the traffic demand that can be satisfied (served) without any delay if it does not exceed available capacity. If demand exceeds capacity, some flights will be delayed. TFM strategies attempt to resolve conflicts between demand and capacity, providing smooth traffic flow that is compatible with capacity. Strategic TFM decisions are made for some hours in the future (up to 15 h) on the basis of comparison of predicted traffic demand and capacity.

Severe congestion problems can occur at airports when capacity substantially decreases due to weather deterioration. If the local TFM specialists cannot resolve the problem within their constraints, the FAA's air traffic control system command center (ATCSCC) takes actions of a wider scale, generally involving many airports and air route traffic control centers (ARTCC). For solving airport arrival problems, the ATCSCC often uses a ground delay program (GDP). A GDP is a strategy that releases flights from the ground in a manner that will not greatly exceed the airport arrival capacity, and therefore cause significant airborne holding.

In 1995 the collaborative decision making (CDM) program started as a cooperative effort between NAS users and FAA to make the GDP system work better for all parties involved. The fundamentals of the CDM approach are:

1) Create a common view of the problem that is shared between FAA and users.

2) Create the opportunity and incentive for users to mitigate problems through their own actions and notify FAA of their intentions.

3) Give the users flexibility to satisfy their own priorities within the context of FAA-initiated TFM constraints.

4) Allow the users to participate in the determination of TFM policies and procedures.

A major accomplishment of the CDM program is the development and implementation of enhanced collaborative procedures and tools for GDPs.

A CDM GDP works as follows. Users provide FAA with a continuous data stream updating their operating plans on a flight-by-flight basis. FAA combines the user data with other TFM data to provide an aggregate arrival demand list (ADL) for the airport. The ADL is shared with users and FAA TFM specialists. The ATCSCC establishes the hourly airport acceptance rate (AAR) (also shared with the users), which along with the ADL is input to the flight schedule monitor (FSM) algorithms. Based on the data, FSM computes a GDP strategy that determines which flights should be delayed and for how long. The FAA then issues the flight-specific delays to the users as controlled times of departure (CTDs). There are two main control procedures used in the FSM to determine CTDs: the ration-by-schedule procedure and compression (see Ref. 1).

FSM currently solves arrival problems at airports without taking into account departure operations at the same airports. It is well known, however, that a significant number of airports, such as Boston Logan (BOS), Washington National (DCA), New York John F. Kennedy (JFK), Los Angeles International (LAX), San Francisco (SFO), Lambert-St. Louis (STL), practice a tradeoff between arrival and departure capacity. At these airports, arrivals and departures are interdependent processes and generally cannot be considered independently. The degree of the interaction varies from airport to airport, and even within an airport it may vary for various runway configurations. In the case of substantial interaction between arrivals and departures, the airport arrival and departure capacities vary when the arrival/departure mix varies. Under given weather conditions (or operational category), a runway configuration may have several different values of arrival and departure capacities depending on arrival/departure mix. For each arrival/departure mix, there is a pair of values: arrival capacity and corresponding departure capacity.

Graphically, the airport capacity can be represented on an "arrival capacity—departure capacity" plane by the airport capacity curve shown in Fig. 1 (see Ref. 2). The arrival/departure capacity curve reflects a functional relationship $v = \varphi(u)$ between these two capacities in the entire range of arrival/departure ratios.

In Fig. 1, vertex 1 determines the airport's operational limits with maximum departure capacity V and minimum arrival capacity. Vertex 2 corresponds to maximum arrival capacity U and minimum departure capacity.

The segment between points 1 and 2 shows the tradeoff area between arrival and departure capacities. In this area, arrival capacity can be increased at the expense of decreasing departure capacity and vice versa. The question then arises: if a range of arrival/departure capacities is available for each time interval, what is the best set of choices to accommodate the traffic demand over a given time period?

An answer to this question has been given by Gilbo,[2,3] where optimization problems were formulated to determine the best allocation of airport arrival/departure capacities during periods of congestion at airports. These solutions were based on the goals of minimizing overall delays and maximizing overall throughput, and they assume that all flights are of equal priority. However, optimal strategies that maximize efficiency of NAS utilization (the major objective for the service provider) are not necessarily optimal or even acceptable to users.

Each user creates its own schedule in accordance with its specific multicriteria objectives and optimization procedures. The cost of delaying one flight may vary greatly from the cost of giving another flight the same delay. At the hub airports, for example, an airline may operate in banks, and may assign a higher cost to delays for flights in the banks. Another airline may assign higher costs to delays for flights making international connections. Whatever the criteria, it is clear that

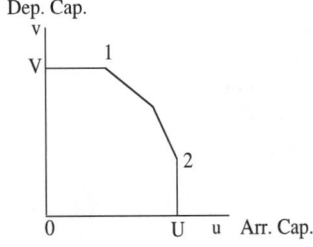

Fig. 1   Airport arrival/departure capacity curve.

the TFM strategies generated by the FAA cannot be considered optimal for the users if they do not include the users' objectives explicitly.

This chapter expands the arrival/departure capacity optimization model that was first presented by Gilbo[2] and applies it to the CDM environment. It first reintroduces joint consideration of arrivals and departures at an airport during a GDP. This makes it possible for more strategic and more efficient assignment of AARs during a GDP. Next, this chapter proposes an arrival/departure optimization model that is capable of taking into account users' priorities in their traffic demand when allocating airport capacities.

To reach the full benefit of this model, additional ADL input data from users are needed. The new data include the number of first priority (i.e., high cost of delay) arrival and departure slots in the total user demand at each 15-min interval. First priority demand is treated in the model as exempt. The model aggregates the first priority demands from all of the users in a subset of the overall demand, both for arrivals and departures. This subset is called the priority demand. The priority demand is used as additional constraints when allocating the arrival and departure capacity at each 15-min interval. These additional constraints may affect the allocation of arrival and departure capacity and, hence, may allow the airlines an opportunity to reduce or eliminate the delays from their first priority flights, at the possible expense of more delays to their other flights.

The objective of this model is not to alter the manner in which delays are assigned to individual flights. Airlines might object if FAA is trying to second-guess their intentions by altering the order of the flights in the airline schedule. Rather, the user priority constraints are used solely in the allocation of the available capacity between arrival and departure demand. This approach can perhaps provide each user with the number of arrival and departure slots necessary to best satisfy its own goals during a GDP. Once the slots have been allocated, the airlines will fill the slots with specific flights using the existing CDM GDP procedures.

The chapter has been organized as follows. Mathematical formulation of the optimization problem is presented in Section II. Illustrative, numerical examples are given in Section III. Section IV contains conclusive remarks.

II. Mathematical Formulation

A. Uncontrollable Flights

There is a fraction of traffic demand, both arrival and departure, that cannot be delayed and must be served at the scheduled slots. These flights are called uncontrollable and comprise an uncontrollable fraction of traffic demand. Those include airborne flights with long en route time and flights with special status, such as international, military, etc. The flights in the first priority demand identified by the users are also considered to be uncontrollable flights.

Let $a_{i,r}$ and $d_{i,r}$ be the first priority arrival and departure demand of the rth user for the ith time interval, respectively. Then by summing first priority demands over all P users, we can determine total priority fraction of arrival and departure demand for the ith interval, respectively:

$$a_i^{(s)} = \sum_{r=1}^{P} a_{i,r}; \; d_i^{(s)} = \sum_{r=1}^{P} d_{i,r}, i \in I \qquad (1)$$

If the priority demand does not exceed airport capacity constraints at all N time intervals, i.e., the points $(a_i^{(s)}, d_i^{(s)})$ are under or on the capacity curve (see Fig. 1), then the priority demand is called a feasible demand and can be completely satisfied without any delay.

If $a_i^{(s)}$ or/and $d_i^{(s)}$ exceed airport capacity constraints at least in one of N intervals, then the priority demand cannot be completely satisfied. This demand is nonfeasible. In this case, a feasible priority demand will be created by delaying some of the priority flights from the original slots. After that the optimization model will provide the best solution that satisfies the modified users' priority demand.

B. Optimization Model

Now we can present the optimization model that provides the best allocation of arrival and departure capacities (u_i, v_i) over N intervals of period $T (i = 1, 2, \ldots, N)$. First, determine arrival and departure queues at the beginning of $(i + 1)$th interval by the following recurrent equations:

$$X_{i+1} = (X_i + a_i - u_i)^+, X_1 = X^0, i \in I \quad (2)$$

$$Y_{i+1} = (Y_i + d_i - v_i)^+, Y_1 = Y^0, i \in I \quad (3)$$

where X^0 and Y^0 are initial conditions that represent residual queues left from the previous time period; symbol $(A)^+$ denotes:

$$(A)^+ = \begin{cases} A, & A \geq 0 \\ 0, & A < 0 \end{cases}$$

Equations (2) and (3) reflect the evolution of arrival and departure queues at each time interval, respectively. In particular, the queue at the beginning of the $(i + 1)$th interval is equal to the sum of the queue at the beginning of the ith interval left from the preceding intervals and traffic demand for the ith interval minus airport capacity.

The regions for arrival and departure capacities are determined by the following inequalities:

$$a_i^{(s)} \leq u_i \leq U_i \quad (4)$$

$$d_i^{(s)} \leq v_i = \varphi_i(u_i) \quad (5)$$

where $a_i^{(s)}$ and $d_i^{(s)}$ are uncontrollable fractions of arrival and departure demands for the ith interval, respectively, that are determined by Eq. (1); U_i is maximum arrival capacity at the ith interval (see Fig. 1).

As an optimization criterion, consider a minimum weighted sum of cumulative arrival and departure queues

$$\sum_{i=1}^{N} X_{i+1} \quad \text{and} \quad \sum_{i=1}^{N} Y_{i+1},$$

respectively:

$$\underset{u,v}{\text{minimize}} \sum_{i=1}^{N}[\alpha X_{i+1} + (1-\alpha)Y_{i+1}], 0 \leq \alpha \leq 1 \qquad (6)$$

where α is a weight coefficient; $u = (u_1, u_2, \ldots, u_N)$ and $v = (v_1, v_2, \ldots, v_N)$ are vectors comprising consecutive values of arrival and departure capacities, respectively.

It is not difficult to show that if the traffic demand has been completely satisfied within the time period T, then the cumulative queue over a period T is equal to a number of time intervals Δ (e.g., 15-min intervals) in the total aircraft flight delay. Therefore, criterion (6) corresponds to minimum weighted total arrival and departure delay. In Ref. 3, it is also shown that criterion (6) is equivalent to maximizing the weighted sum of total arrival and departure traffic flow (i.e., throughput) at the airport. In the case of $\alpha = 0.5$, the optimization criterion (6) corresponds to minimum of total arrival and departure delay or maximum of total airport arrival and departure throughput.

Note that criterion (6) is a special case of more general criterion

$$\underset{u,v}{\text{minimize}} \sum_{i=1}^{N}[\alpha_i X_{i+1} + (1-\alpha_i)Y_{i+1}], 0 \leq \alpha_i \leq 1 \qquad (7)$$

with variable weights α_i at each time interval i, which was introduced in Ref. 2.

In this chapter, we consider criterion (6). Then the optimal sequence of arrival and departure capacity pairs (u_i, v_i) at each interval $i = 1, 2, \ldots, N$ can be found as a solution of the following optimization problem:

$$\underset{u,v}{\text{minimize}} \sum_{i=1}^{N}[\alpha X_{i+1} + (1-\alpha)Y_{i+1}], 0 \leq \alpha \leq 1 \qquad (8)$$

subject to

$$X_{i+1} = (X_i + a_i - u_i)^+, X_1 = X^0, i \in I \qquad (9)$$

$$Y_{i+1} = (Y_i + d_i - v_i)^+, Y_1 = Y^0, i \in I \qquad (10)$$

$$a_i^{(s)} \leq u_i \leq U_i \qquad (11)$$

$$d_i^{(s)} \leq v_i = \varphi_i(u_i) \qquad (12)$$

where u_i and v_i are integers.

After optimal airport capacities have been determined, the optimal number of arrivals w_i and departures z_i that are accommodated at each time interval (we call it traffic flow) can be calculated as follows:

$$w_i = X_i + a_i - X_{i+1}, i \in I \qquad (13)$$

$$z_i = Y_i + d_i - Y_{i+1}, i \in I \qquad (14)$$

Constraints (11) and (12) guarantee that at least the first priority flights will be served at their requested intervals.

The optimization model contains parameter α that determines relative impact of arrivals and departures in the optimization criterion. The parameter may be interpreted as a relative priority rate for arrivals in total airport operations. Allocation of weights between the two components in the objective function affects the optimal solution, making it more favorable to arrivals or departures. Traffic management specialists can use this parameter as a control parameter for generating alternative strategies of allocation of arrival and departure operations at each 15-min interval. By varying the parameter, a TFM specialist and other participants in the collaborative optimization process can perform "what if" experiments to find mutually acceptable TFM strategies.

III. Numerical Examples

Numerical examples in this section are based on 3 March, 2000, historical data for Newark International Airport (EWR). That day, there was a GDP for EWR that was initiated at 1716 and lasted from 2000 to 2300. (Note: time in this section is Zulu time.) Hourly AAR's during the GDP were 40 flights per hour. Examples in this section will be calculated for a time period from 1945 to 0245, which includes three hours of GDP and four hours post-GDP.

Table 1 shows arrival and departure demand predicted at 1700 for each 15-min interval from 1945 to 0245. For both arrivals and departures, there are separate columns for total demand and priority demand for each 15-min interval.

During the time period, there was low visibility and ceiling that determined the instrument flight rules (IFR) conditions at the airport. Single runway 4L/4R was used for both arrivals and departures. The 15-min arrival/departure capacity curve for this runway configuration under IFR conditions is shown in Fig. 2.

The coordinates of the vertices in Fig. 2 clarify the scale of the capacity curve. According to the curve, the tradeoff area comprises six arrival/departure capacity pairs: (7, 14), (8, 13), (10, 12), (11, 11), (13, 10), and (14, 8). Thus, under maximum arrival capacity of 14 flights per 15 min, there is minimum departure capacity equal to 8 flights per 15-min interval. Under maximum departure capacity of 14 flights per 15 min, there is minimum arrival capacity equal to 7 flights per 15-min interval. There are also four arrival/departure capacity pairs of intermediate values available that can be realized at the runway configuration.

The optimization problem (8)–(12) was solved for the predicted traffic demand from Table 1 and the airport capacity curve from Fig. 2 using various values of arrival priority rate α. Some of the results are shown in Tables 2–5. The tables contain the optimal values of arrival and departure capacities, the number of arrivals

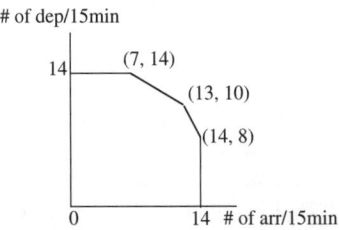

Fig. 2 EWR airport capacity curve.

Table 1 Predicted demand

	Total demand		Priority demand	
Time	Arrival	Departure	Arrival	Departure
1945–2000	17	9	14	8
2000–2015	9	10	9	10
2015–2030	9	3	9	3
2030–2045	14	14	11	11
2045–2100	15	11	14	8
2100–2115	10	15	8	13
2115–2130	9	10	9	10
2130–2145	20	12	10	12
2145–2200	10	9	10	9
2200–2215	12	8	12	8
2215–2230	6	14	6	14
2230–2245	10	14	10	12
2245–2300	9	8	9	7
2300–2315	11	13	11	11
2315–2330	9	16	9	12
2330–2345	15	8	13	8
2345–0000	6	9	6	9
0000–0015	15	9	14	8
0015–0030	8	12	8	8
0030–0045	4	10	4	10
0045–0100	13	5	13	5
0100–0115	7	11	7	11
0115–0130	10	13	10	12
0130–0145	16	12	13	10
0145–0200	5	12	5	12
0200–0215	6	9	6	9
0215–0230	5	11	5	11
0230–0245	11	6	11	6
Totals	291	293	266	267

and departures accommodated in each 15-min interval, and the size of queue at the end of each interval. The tables also show total traffic and cumulative arrival and departure queues over the period from 1945 to 0245.

Tables 2 and 3 represent the optimization results obtained for $\alpha = 0.5$ and 0.4, respectively, without taking into account the users' priorities in the predicted demand, i.e., for $a_i^{(s)} = 0$ and $d_i^{(s)} = 0, i \in I$.

In the case of equal priority for arrivals and departures ($\alpha = 0.5$), the optimal solution provided cumulative queues of 40 arrival and 106 departure flights with total arrival and departure cumulative queue of 146 flights. It can be interpreted that 40 arrival flights and 106 departure flights were moved (delayed) from one 15-min interval to another. (Note: the optimization model deals with 15-min counts and does not consider flight delays within a 15-min interval). It is interesting to notice that in spite of almost equal total demands for arrival and departures (291 and 293,

Table 2 Optimal solutions without users' priorities, $\alpha = 0.5$

Time	Arrival capacity	Departure capacity	Arrival flow	Departure flow	Arrival queue	Departure queue
1945–2000	13	10	13	9	4	0
2000–2015	13	10	13	10	0	0
2015–2030	14	8	9	3	0	0
2030–2045	13	10	13	10	1	4
2045–2100	13	10	13	10	3	5
2100–2115	13	10	13	10	0	10
2115–2130	7	14	7	14	2	6
2130–2145	13	10	13	10	9	8
2145–2200	13	10	13	10	6	7
2200–2215	13	10	13	10	5	5
2215–2230	10	12	10	12	1	7
2230–2245	10	12	10	12	1	9
2245–2300	10	12	10	12	0	5
2300–2315	10	12	10	12	1	6
2315–2330	10	12	10	12	0	10
2330–2345	13	10	13	10	2	8
2345–0000	8	13	8	13	0	4
0000–0015	13	10	13	10	2	3
0015–0030	10	12	10	12	0	3
0030–0045	8	13	4	13	0	0
0045–0100	14	8	13	5	0	0
0100–0115	11	11	7	11	0	0
0115–0130	10	12	10	12	0	1
0130–0145	13	10	13	10	3	3
0145–0200	8	13	8	13	0	2
0200–0215	11	11	6	11	0	0
0215–0230	11	11	5	11	0	0
0230–0245	14	8	11	6	0	0
Totals			291	293	40	106

respectively) and equal priority ($\alpha = 0.5$) for arrivals and departures, the optimization model provided the strategy much more favorable to arrivals. This happens due to the specific arrival and departure demand profiles over the period of interest.

Table 3 shows that reducing arrival priority rate from 0.5 to 0.4 changed the allocation of arrival and departure capacities so that total arrival queue increased from 40 to 53 flights and total departure queue decreased from 106 to 96 flights. This made the strategy more balanced in terms of number of arrival and departure flights delayed, though the sum of total arrival and departure queues increased slightly from 146 to 149.

Different allocation of arrival and departure capacities for $\alpha = 0.5$ and $\alpha = 0.4$ resulted also in different times of complete recovery from congestion. The complete recovery occurs at the 15-min interval after which there is no queue. For $\alpha = 0.5$, the complete recovery for arrivals occurred earlier than for $\alpha = 0.4$ (at 0145 as opposed to 0200). The departure problem in both cases was resolved at the same time interval, 0200.

Table 3 Optimal solutions without users' priorities, $\alpha = 0.4$

Time	Arrival capacity	Departure capacity	Arrival flow	Departure flow	Arrival queue	Departure queue
1945–2000	13	10	13	9	4	0
2000–2015	13	10	13	10	0	0
2015–2030	14	8	9	3	0	0
2030–2045	13	10	13	10	1	4
2045–2100	13	10	13	10	3	5
2100–2115	7	14	7	14	6	6
2115–2130	13	10	13	10	2	6
2130–2145	13	10	13	10	9	8
2145–2200	13	10	13	10	6	7
2200–2215	13	10	13	10	5	5
2215–2230	10	12	10	12	1	7
2230–2245	10	12	10	12	1	9
2245–2300	10	12	10	12	0	5
2300–2315	10	12	10	12	1	6
2315–2330	10	12	10	12	0	10
2330–2345	10	12	10	12	5	6
2345–0000	10	12	10	12	1	3
0000–0015	13	10	13	10	3	2
0015–0030	10	12	10	12	1	2
0030–0045	10	12	5	12	0	0
0045–0100	14	8	13	5	0	0
0100–0115	11	11	7	11	0	0
0115–0130	10	12	10	12	0	1
0130–0145	13	10	13	10	3	3
0145–0200	7	14	7	14	1	1
0200–0215	13	10	7	10	0	0
0215–0230	11	11	5	11	0	0
0230–0245	14	8	11	6	0	0
Totals			291	293	53	96

Now one can determine whether the optimal solutions satisfy the hypothetical first priority flights in the total traffic demand. This can be done by comparing the number of arrivals and departures accommodated by the optimal strategies in each 15-min interval (from the "Arrival flow" and "Departure flow" columns in Tables 2 and 3) with corresponding values of priority demands from Table 1.

It appeared that neither of the optimal solutions completely satisfies the priority demand and some of them would be delayed. For $\alpha = 0.5$, priority arrival demand was not satisfied at five intervals and priority departure demand was not satisfied at four intervals. For $\alpha = 0.4$, priority arrival demand was not satisfied at six intervals and priority departure demand was not satisfied at three intervals.

The optimization problem (8)–(12) was then solved using the values of first priority demand in constraints (11) and (12) taken from Table 1. Optimal solutions for $\alpha = 0.5$ are presented in Table 4. As one can see, the priority demand is completely

Table 4 Optimal solutions with users' priorities, $\alpha = 0.5$

Time	Arrival capacity	Departure capacity	Arrival flow	Departure flow	Arrival queue	Departure queue
1945–2000	14	8	14	8	3	1
2000–2015	11	11	11	11	1	0
2015–2030	14	8	10	3	0	0
2030–2045	11	11	11	11	3	3
2045–2100	14	8	14	8	4	6
2100–2115	8	13	8	13	6	8
2115–2130	13	10	13	10	2	8
2130–2145	10	12	10	12	12	8
2145–2200	13	10	13	10	9	7
2200–2215	13	10	13	10	8	5
2215–2230	7	14	7	14	7	5
2230–2245	10	12	10	12	7	7
2245–2300	13	10	13	10	3	5
2300–2315	11	11	11	11	3	7
2315–2330	10	12	10	12	2	11
2330–2345	13	10	13	10	4	9
2345–0000	10	12	10	12	0	6
0000–0015	14	8	14	8	1	7
0015–0030	8	13	8	13	1	6
0030–0045	7	14	5	14	0	2
0045–0100	14	8	13	7	0	0
0100–0115	11	11	7	11	0	0
0115–0130	10	12	10	12	0	1
0130–0145	13	10	13	10	3	3
0145–0200	8	13	8	13	0	2
0200–0215	11	11	6	11	0	0
0215–0230	11	11	5	11	0	0
0230–0245	14	8	11	6	0	0
Totals			291	293	79	117

satisfied. The price, however, for making it happen is a substantial deviation from the optimal solution of Table 2: both arrival and departure cumulative queues are increased (from 40 to 79 flights for arrivals and from 106 to 117 flights for departures). The sum of cumulative arrival and departure queues increased from 146 to 196, i.e., 34% increase. Note that no other strategy can provide smaller sum of cumulative arrival and departure queues (and therefore total delay) than the one obtained for $\alpha = 0.5$ without additional constraints for priority demand. In spite of increases in queues and delays, the users might be happy because their first priority flights in the traffic demand were satisfied.

The optimal results with users' priorities for $\alpha = 0.4$ are shown in Table 5.

Any of the optimization results were much better than results obtained when the tradeoff between arrival and departure capacity was not utilized. For a constant arrival rate of 13 flights and departure rate of 10 flights per 15 min (this corresponds to a maximum total capacity of 23 flights per 15 min), the cumulative

Table 5 Optimal solutions with users' priorities, $\alpha = 0.4$

Time	Arrival capacity	Departure capacity	Arrival flow	Departure flow	Arrival queue	Departure queue
1945–2000	14	8	14	8	3	1
2000–2015	11	11	11	11	1	0
2015–2030	14	8	10	3	0	0
2030–2045	11	11	11	11	3	3
2045–2100	14	8	14	8	4	6
2100–2115	8	13	8	13	6	8
2115–2130	13	10	13	10	2	8
2130–2145	10	12	10	12	12	8
2145–2200	13	10	13	10	9	7
2200–2215	13	10	13	10	8	5
2215–2230	7	14	7	14	7	5
2230–2245	10	12	10	12	7	7
2245–2300	13	10	13	10	3	5
2300–2315	11	11	11	11	3	7
2315–2330	10	12	10	12	2	11
2330–2345	13	10	13	10	4	9
2345–0000	10	12	10	12	0	6
0000–0015	14	8	14	8	1	7
0015–0030	8	13	8	13	1	6
0030–0045	7	14	5	14	0	2
0045–0100	14	8	13	7	0	0
0100–0115	11	11	7	11	0	0
0115–0130	10	12	10	12	0	1
0130–0145	13	10	13	10	3	3
0145–0200	7	14	7	14	1	1
0200–0215	11	11	7	10	0	0
0215–0230	11	11	5	11	0	0
0230–0245	14	8	11	6	0	0
Totals			291	293	80	116

arrival queue is 29 flights and cumulative departure queue is 429 flights (see Table 6). Moreover, 21 departure flights were not served within the time period considered and were left in outstanding queue for service after 0245. In this case, without using the tradeoff, arrival capacity was underutilized in many time intervals because of insufficient demand. At the same time, departure demand was high enough so that departure capacity of 10 flights per 15 min was consistently insufficient.

In the case of constant arrival and departure capacity of 11 flights per 15 min each during the whole time period, cumulative arrival and departure queues are equal to 155 and 115 flights, respectively, with the sum of these queues equal to 270 flights (see Table 7). This is much worse than any of the optimal tradeoff solutions.

To estimate the potential benefits of optimal arrival/departure tradeoff for the CDM GDPs, the optimal arrival rates for each solution were incorporated into the FSM algorithms and applied to the 3 March 2000, data using the reply feature

Table 6 Solutions without tradeoff for constant airport capacity: 13/10

Time	Arrival capacity	Departure capacity	Arrival flow	Departure flow	Arrival queue	Departure queue
1945–2000	13	10	13	9	4	0
2000–2015	13	10	13	10	0	0
2015–2030	13	10	9	3	0	0
2030–2045	13	10	13	10	1	4
2045–2100	13	10	13	10	3	5
2100–2115	13	10	13	10	0	10
2115–2130	13	10	9	10	0	10
2130–2145	13	10	13	10	7	12
2145–2200	13	10	13	10	4	11
2200–2215	13	10	13	10	3	9
2215–2230	13	10	9	10	0	13
2230–2245	13	10	10	10	0	17
2245–2300	13	10	9	10	0	15
2300–2315	13	10	11	10	0	18
2315–2330	13	10	9	10	0	24
2330–2345	13	10	13	10	2	22
2345–0000	13	10	8	10	0	21
0000–0015	13	10	13	10	2	20
0015–0030	13	10	10	10	0	22
0030–0045	13	10	4	10	0	22
0045–0100	13	10	13	10	0	17
0100–0115	13	10	7	10	0	18
0115–0130	13	10	10	10	0	21
0130–0145	13	10	13	10	3	23
0145–0200	13	10	8	10	0	25
0200–0215	13	10	6	10	0	24
0215–0230	13	10	5	10	0	25
0230–0245	13	10	11	10	0	21
Totals			291	272	29	429

of FSM. (Note: FSM does not currently control departures.) In the case without taking into account the users' priority and for $\alpha = 0.5$, the total delay is 1129 min. The optimal arrival rates that reflect the users' priority provided the total arrival delay of 1737 min, i.e., 54% increase. This is the price for satisfying the users' priority in their traffic demand.

Without a tradeoff, for constant airport arrival rate of 11 flights per 15 min (or hourly AAR of 44 flights), the FSM algorithms provided a total arrival delay of 4054 min. This is 133% higher than for optimal AARs with users' priorities and 259% higher than for optimal AARs without users' priorities in the case of $\alpha = 0.5$.

The examples clearly illustrate the potential benefits from collaborative optimization and show how it could upgrade the level of collaboration between the FAA and NAS users within the CDM program. Collaborative optimization offers more efficient arrival/departure strategies and better utilization of existing airport capacity in combination with users' participation in TFM decision making.

Table 7 Solutions without tradeoff for constant airport capacity: 11/11

Time	Arrival capacity	Departure capacity	Arrival flow	Departure flow	Arrival queue	Departure queue
1945–2000	11	11	11	9	6	0
2000–2015	11	11	11	10	4	0
2015–2030	11	11	11	3	2	0
2030–2045	11	11	11	11	5	3
2045–2100	11	11	11	11	9	3
2100–2115	11	11	11	11	8	7
2115–2130	11	11	11	11	6	6
2130–2145	11	11	11	11	15	7
2145–2200	11	11	11	11	14	5
2200–2215	11	11	11	11	15	2
2215–2230	11	11	11	11	10	5
2230–2245	11	11	11	11	9	8
2245–2300	11	11	11	11	7	5
2300–2315	11	11	11	11	7	7
2315–2330	11	11	11	11	5	12
2330–2345	11	11	11	11	9	9
2345–0000	11	11	11	11	4	7
0000–0015	11	11	11	11	8	5
0015–0030	11	11	11	11	5	6
0030–0045	11	11	9	11	0	5
0045–0100	11	11	11	10	2	0
0100–0115	11	11	9	11	0	0
0115–0130	11	11	10	11	0	2
0130–0145	11	11	11	11	5	3
0145–0200	11	11	10	11	0	4
0200–0215	11	11	6	11	0	2
0215–0230	11	11	5	11	0	2
0230–0245	11	11	11	8	0	0
Totals			291	293	155	115

IV. Conclusions

This chapter introduces a new concept, collaborative optimization, that could advance the CDM GDP procedures and algorithms, improve efficiency of GDP, and offer a constructive approach for optimizing utilization of airport capacity while simultaneously solving arrival and departure congestion problems at airports in the CDM environment.

The core element of the concept is joint consideration of arrivals and departures and optimization of tradeoff between arrival and departure capacity at the airport during GDP. It provides more strategic and more efficient assignment of AARs during GDP. Additionally, it makes it possible for more efficient departure planning at airports that are subject to GDP.

The chapter also proposed an optimization model that is capable of taking into account users' priorities in their arrival and departure traffic demand. Optimal solution provides more efficient TFM strategies at airports with minimum delays and maximum airport throughput that could be achieved while satisfying users' priorities.

The collaborative optimization approach and the underlying optimization model can be used as a basic addition to CDM decision-making processes and decision support tools to improve their efficiency and expand their functionality and applications. Numerical examples presented in the chapter illustrate the benefits that collaborative optimization would provide to advance the CDM procedures and enhance their efficiency and level of collaboration between the FAA and NAS users.

References

[1]Wambsganss, M., "Collaborative Decision Making Through Dynamic Information Transfer," *Air Traffic Control Quarterly*, Vol. 4, No. 2, 1996, pp. 107–123.

[2]Gilbo, E. P., "Airport Capacity: Representation, Estimation, Optimization," *IEEE Transactions on Control Systems Technology*, Vol. 1, No. 3, 1993, pp. 144–154.

[3]Gilbo, E. P., "Optimizing Airport Capacity Utilization in Air Traffic Flow Management Subject to Constraints at Arrival and Departure Fixes," *IEEE Transactions on Control Systems Technology*, Vol. 5, No. 5, 1997, pp. 490–503.

Chapter 20

Analysis, Modeling, and Control of Ground Operations at Hub Airports

Kari Andersson* and Francis Carr[†]
Massachusetts Institute of Technology, Cambridge, Massachusetts

William D. Hall[‡]
Charles Stark Draper Laboratory, Cambridge, Massachusetts
and
Nicolas Pujet[§] and Eric Feron[¶]
Massachusetts Institute of Technology, Cambridge, Massachusetts

I. Introduction

AS THE demand for air travel increases, congestion and delays in the air traffic system become more commonplace. Inherent delay uncertainty makes it difficult for airlines and air traffic service providers to manage passengers, fleets, and crews. In addition, increased congestion at busy airports results in significant financial and environmental inefficiencies. Many of these delays can be directly or indirectly attributed to airports. Thus, several efforts are under way to improve airport congestion management, throughput, and predictability. To achieve the goal of increased predictability and airport efficiency, much research has been undertaken to study both the departure and the arrival processes at busy airports.

All quantitative approaches to predicting and improving airport operations must eventually rely upon mathematical models. Most highly detailed models of airport operations such as SIMMOD, TAAM, or the Airport Machine are based on a detailed, physical modeling of the airport operations.[1] These models can be useful

---

Copyright © 2001 by the Massachusetts Institute of Technology. Published by the American Institute of Aeronautics and Astronautics, Inc., with permission.
* Research Assistant.
[†] Research Assistant.
[‡] Research Engineer.
[§] International Center for Air Transportation, Department of Aeronautics and Astronautics.
[¶] Associate Professor, Laboratory for Information and Decision Systems, Department of Aeronautics and Astronautics. Member AIAA.

to evaluate qualitatively the relative effects of various airport improvements on airport efficiency. However, calibrating and validating them in a formal sense is a very challenging, if not impossible, task. As a consequence, these models require very significant efforts and extensive working knowledge of the particular airport under study to provide quantitative information about the effect of improved airport processes.

In the work of Idris et al.,[2] pilot reports, on-site investigations, and statistical analyses of automatically recorded data indicate that runway capacity is the primary limiting constraint in the departure process at busy airports like Boston Logan International Airport (BOS). For example, substantial congestion was observed at BOS under certain airport configurations, leading to significant environmental and financial inefficiencies. This observation led to the construction of aggregate airport departure models,[3] which were used to predict taxi-out times. It was shown by Pujet et al.[4] that these models could be thoroughly calibrated and validated, and that they could be used to quantify the effects of holding departing aircraft at their gates during periods of taxiway system congestion.

The airport arrival process has been studied intensively for airborne traffic, especially through the development of airport arrival management tools such as the Center Terminal Radar Approach Control (TRACON) Automation System (CTAS), a suite of decision support tools to help the TRACON manage the flow of aircraft arriving at a busy airport. Arrival management tools such as CTAS provide two benefits for congested airports. First, they contribute to increasing airport throughput by achieving efficient runway balancing and regularizing aircraft arrival flows. Second, the powerful model-based trajectory prediction of CTAS enables the accurate prediction of aircraft landing times up to 40 min in advance.[5,6,7] These accurate landing time estimates have the potential to benefit airlines substantially by offering advance information about incoming flights.

In contrast, most studies of the air transportation system do not consider ground operations. In fact, in many models the ground time, which includes all processes and activities from wheels-on to wheels-off, is assumed to be of constant length. This assumption ignores queuing effects arising at airports, and it implies that the airlines cannot influence delays and delay propagation while aircraft are on the ground. However, in practice airlines frequently attempt to shorten the ground times of delayed aircraft to control the downstream impacts of delays.

This chapter is organized as follows. We first extend the observational base of arrival and departure ground operations from a moderate-size, nonhub airport such as BOS to large hub airports such as Dallas/Fort Worth International Airport (DFW), George Bush Intercontinental Airport in Houston, Texas (IAH), and Atlanta Hartsfield International Airport (ATL). Then three models are presented that have been designed to capture the dynamics of ground operations at busy hub airports, including an arrival (taxi-in) model, a ground (aircraft-turn) model, and a departure (taxi-out) model. Finally, the ongoing development of several applications is discussed, including 1) a predictive capability for air transportation system monitoring purposes, 2) a means to evaluate policies aimed at managing airport congestion by queue delay management, and 3) a means to evaluate the potential economic impacts of airline intervention in the aircraft arrival scheduling process.

II. Available Data

A. Airport Layouts

As shown in Fig. 1, DFW is oriented in a north/south configuration with east and west sides running almost independent operations.* On the west are two parallel runways and one diagonal runway, and on the east side are three parallel runways and one diagonal runway. The parallel runways are spaced such that simultaneous operations can occur. A south configuration includes the use of any runways in the set of 18R/L, 13R/L, and 17R/C/L runways. A north configuration includes the use of any runways in the set of 31R/L, 36R/L, and 35R/L runways. At any time, several runways are simultaneously available for departure and arrival operations.

At the time of data collection, ATL had four runways oriented in an east/west configuration,* as shown in Fig. 2. The four runways consist of two sets of parallel runways: two to the north and two to the south. The runways are spaced such that simultaneous operations can occur. An east configuration includes the use of any runways in the set of 8R/L and 9R/L runways. A west configuration includes the use of any runways in the set of 26R/L and 27R/L runways. At any time, several runways are simultaneously available for departure and arrival operations.

IAH has six runways oriented in an east/west overall orientation,* as shown in Fig. 3. The six runways are partitioned as three pairs of parallel runways: two to the north, two to the south, and two diagonal runways. The parallel runways are spaced such that simultaneous operations can occur. An east configuration includes the use of any runways in the set of 8R/L, 9R/L, and 14R/L runways. A west configuration includes the use of any runways in the set of 26R/L, 27R/L, and 32R/L runways. At any time, several runways are simultaneously available for departure and arrival operations.

B. Flight Operations Data

The analyses discussed herein rely on the Airline Service Quality Performance (ASQP) database, which provides information about the jet operations of 10 major passenger airlines: Alaska, American, America West, Continental, Delta, Northwest, Southwest, TWA, United, and US Airways. For most of these airlines' flights, ASQP provides both scheduled and actual pushback, takeoff, landing, and gate arrival times. Note that because the ASQP database includes jet operations only, it does not provide a complete picture of the activity at each airport. For example, it captures approximately 66% of the operations at DFW, with similar percentages at the other airports analyzed.

The accuracy of the ASQP data has been confirmed via independent observations. Visual observations at BOS confirmed ASQP recorded pushback times.[8] A formal validation of takeoff and landing times recorded in the ASQP data was done by crosschecking them against high-resolution, timed radar tracks available at DFW. A threshold location was chosen on the departure path or on the final approach path roughly 5 n miles from the runway threshold, and the time difference between the recorded wheels-off time (available from ASQP) and the time

---

*Data available online at, http://www.faa.gov/ats/asc/ [cited Sept. 2000].

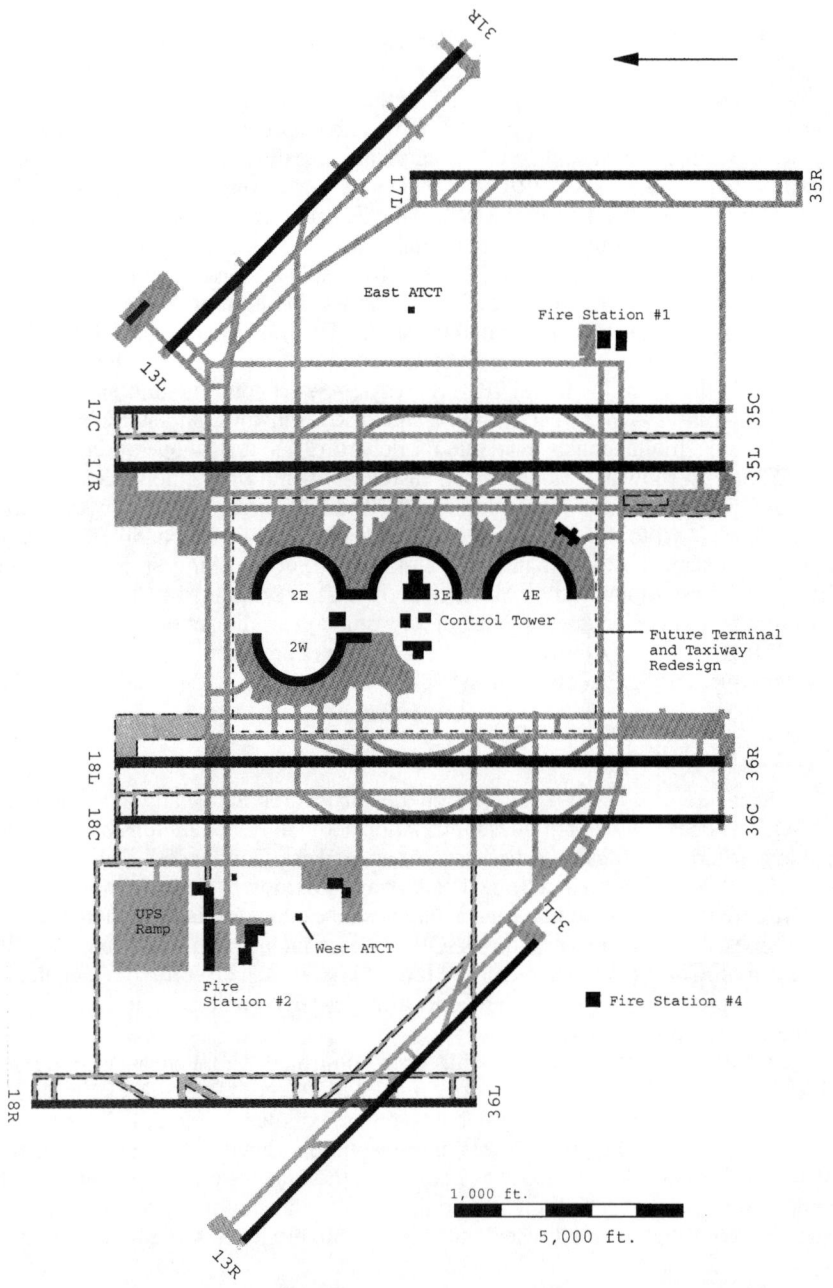

Fig. 1 Map of DFW.

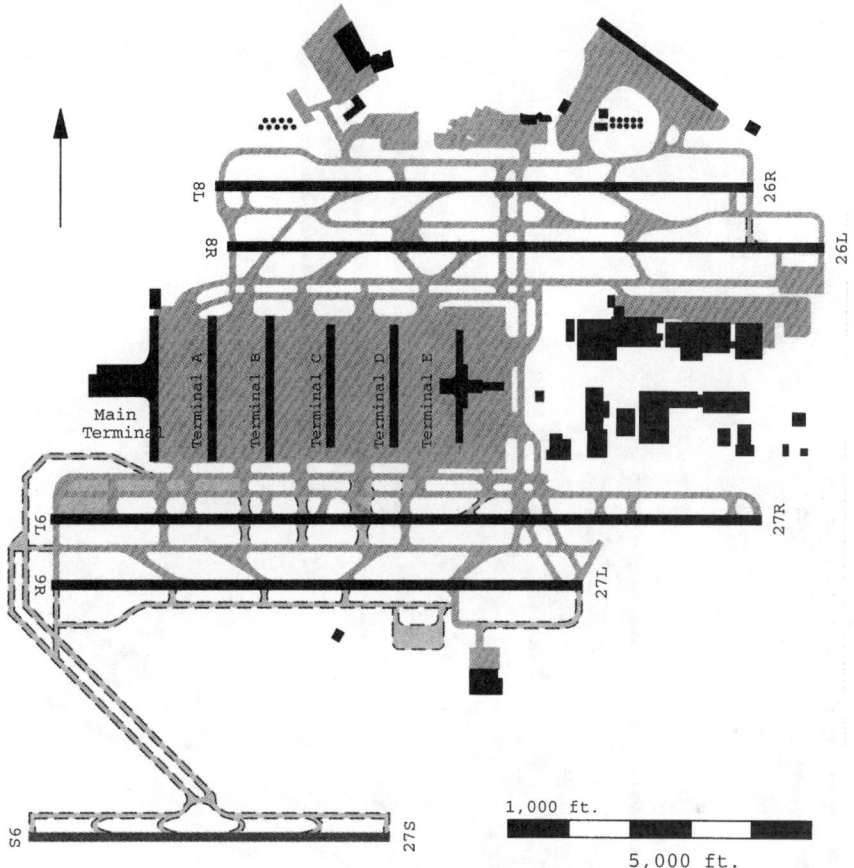

Fig. 2 Map of ATL.

of threshold crossing (obtained by radar track interpolation) was computed and detrended for all jet aircraft that used that particular runway. As can be seen from Fig. 4, the ASQP records closely match estimated takeoff and landing times generated from high-resolution, timed radar tracks provided by CTAS at DFW; the ASQP data are accurate to within a 1-min roundoff error. Finally, it is worth noting that the estimated landing and takeoff times from radar tracks do not rely upon enhanced traffic management system (ETMS) estimates or data in any way.

C. Weather and Airport Configurations

Based on field observations and data analysis at BOS,[9] airport runway configuration is a major determinant of ground operations dynamics. In particular, runway configuration is a major factor to determine airport arrival and departure acceptance rates. Unfortunately, historical runway configuration data are not readily available for the hub airports analyzed here. However, detailed historical wind and

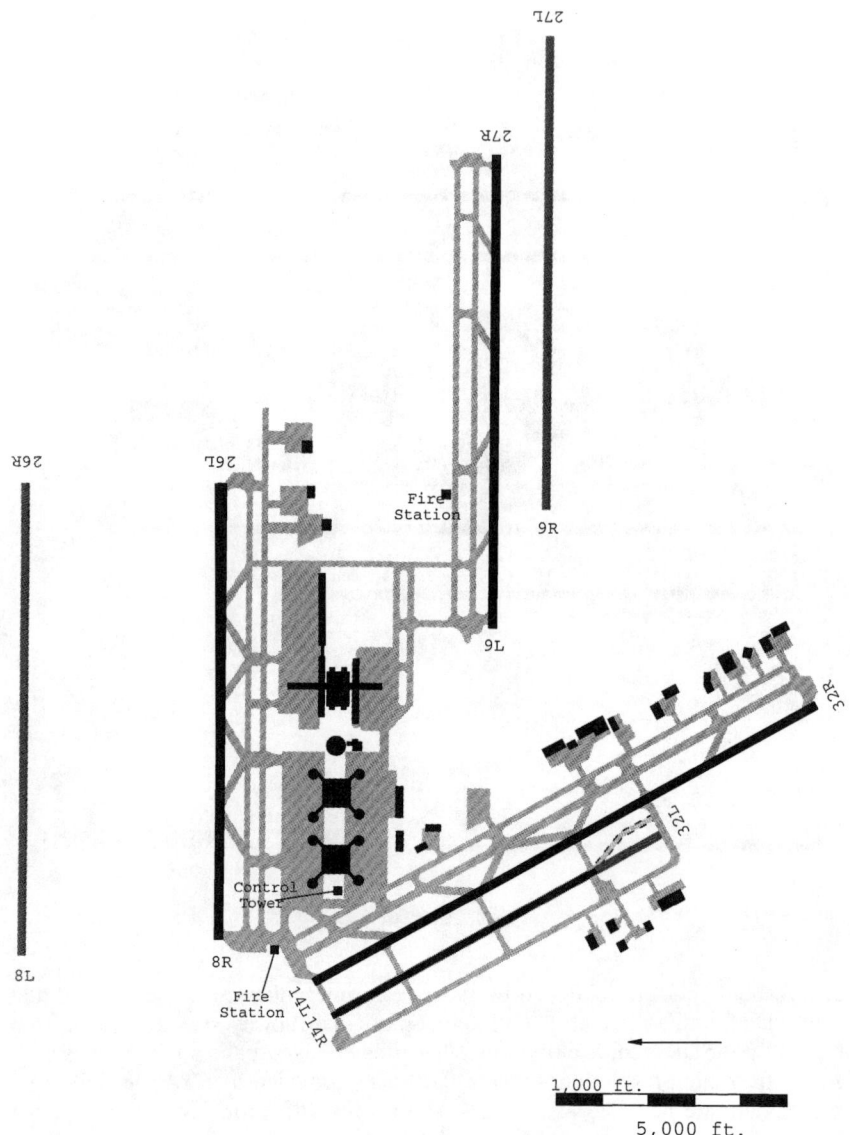

Fig. 3 MAP of IAH.

weather data are available from the Consolidated Operations and Delay Analysis System (CODAS) database, which provides airport-specific weather information over 15-min intervals. The CODAS weather data include wind speed, wind direction, wind gust, temperature, precipitation, ceiling, and visibility. The data set is remarkably complete. For example, at DFW in 1997, only eight 15-min intervals are missing from the records, and only 7% of the temperature data are missing, while all of the other data fields are complete.

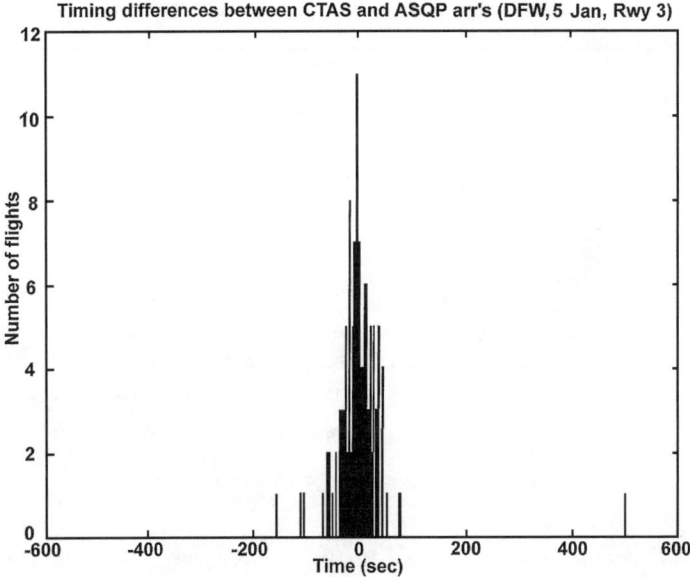

Fig. 4 Landing-time differences between CTAS radar tracks and ASQP sensors.

Although weather conditions alone do not fully determine runway configuration (e.g., in the case of BOS, environmental concerns are also a significant influencing factor), the CODAS data may still be used in conjunction with airport layout information to partition the available operations data into distinct segments; for our purposes, a segment is defined as a particular combination of runway operability and weather conditions, independent of actual runway operations or standard operating procedures. For this analysis, we considered CODAS data from 1997 for DFW and from 1998 for IAH and ATL. A summary of the segmentation methodology and results of this segmentation is provided here for the three airports studied. The same segmentation methodology was used for each airport, with changes to accommodate the different runway layouts.

With the help of an experienced jet pilot employed by a major U.S. airline, a set of standards was developed for runway operability and airport capacity under various weather conditions. A runway was considered operable if the crosswind was less than 20 kn and the headwind was positive. Otherwise, the runway was considered inoperable. If the wind data over an interval were incomplete, a conservative approach was used to estimate the runway configuration: if at either end of the interval a runway was considered inoperable, the runway was considered inoperable during the entire interval. Further segmentation was then conducted to include weather factors such as ceiling, visibility, and precipitation (including thunderstorm activity), all of which are known to influence airport capacity significantly. Each of these weather factors was assigned a threshold at which it was considered to affect the airport operations. Precipitation was considered to affect operations when there was thunderstorm activity or when precipitation was indicated. Ceiling was considered to affect operations when it dropped below

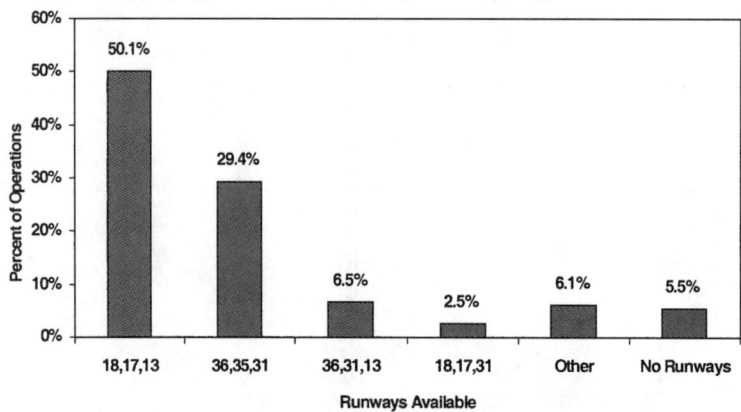

Fig. 5 Configuration breakdown for DEW.

1000 ft. Visibility was considered to affect operations when it dropped below 3 miles.

The first step in the segmentation process was to estimate the runway configuration for each 15-min interval using the CODAS winds data. After determining the runway configurations, the number of operations occurring under each of the configurations was tallied. The percentage of operations occurring under each of the determined runway configurations for each of the airports is shown in Figs. 5–7. Note that the "no runways available" bin includes both times of severely high winds and times of incomplete wind data for which the runway operability was conservatively estimated using the method described previously. The second step in the segmentation process was to consider additional weather factors

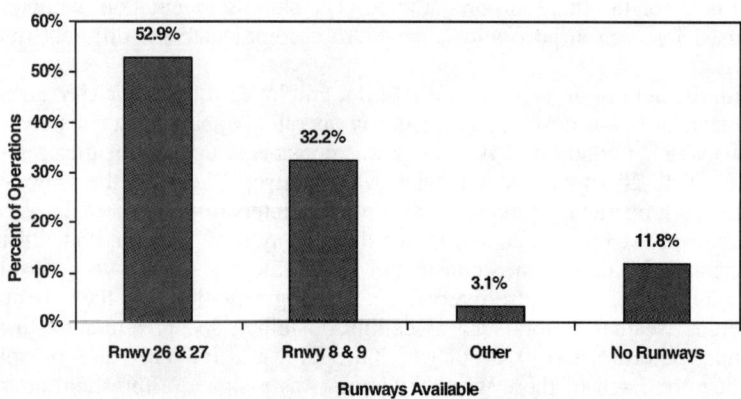

Fig. 6 Configuration breakdown for ATL.

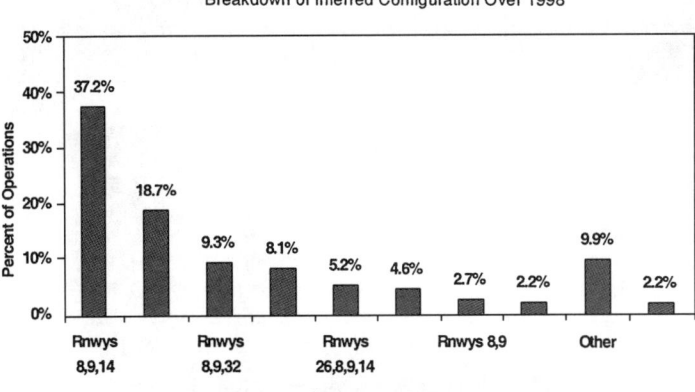

Fig. 7 Configuration breakdown for IAH.

such as ceiling, visibility, temperature, and precipitation. Using the thresholds just described, each of the four weather factors was assessed as to whether it affected airport operations for each 15-min interval. The number of operations occurring under each of the weather conditions was tallied. The percentage of operations occurring under each of the determined weather conditions for each of the airports is shown in Table 1.

The final step in the segmentation process was to link the runway configuration for each 15-min interval to the corresponding weather data to create distinct segments. Even though the number of possible segments is large, operations occur only during a small subset of the possible segments. As might be expected, for the three airports studied the majority of the operations occurred under the segments corresponding to the primary runway orientation of the airport, as defined in Section II.A. At DFW, the segments corresponding to the north/south configuration represent 80% of the operations; similarly at ATL and IAH, the segments corresponding to the east/west or north/south configurations represented 85 and 56% of the operations, respectively. Given this result, the possible segments were summarized into six segment groups: the two primary configurations under good weather conditions; the two primary configurations under inclement weather conditions [where inclement weather was assumed whenever at least one of ceiling, visibility, or precipitation indicated instrument flight rules (IFR) conditions]; indeterminable configuration; and other (in which wind information indicated that

Table 1 Percentage of operations affected by weather

Weather factor	DFW, %	ATL, %	IAH, %
Ceiling	5	9	6
Visibility	3	8	6
Precipitation	3	4	4

Table 2 Final results of segmentation analysis

	DFW		ATL		IAH	
Segment	%	No. operations	%	No. operations	%	No. operations
N/E Good	26.6%	#1	24.2%	#1	31.4%	#2
S/W Good	46.4%	#2	48.2%	#2	16.5%	#1
N/E Bad	2.8%	#3	8.0%	#3	5.9%	#4
S/W Bad	3.7%	#4	4.7%	#4	2.3%	#3
Other	15.0%	#5	3.1%	#5	41.7%	#5
Excluded	5.5%	#6	11.8%	#6	2.2%	#6

no runways were usable). The number of operations occurring in each of these six segments for the three airports is shown in Table 2.

These results have not been explicitly validated for the three airports because historical runway configuration data are not readily available. However, our results support anecdotal reports that the south configuration is the primary configuration for DFW. Similarly at ATL, the segmentation results are consistent with the account that the west operation is considered the most efficient, and hence is the preferred configuration. Further, we were able to validate the segmentation results at DFW in an implicit sense using radar track data. Through an analysis of radar-track data from CTAS, the runways used for takeoffs and landings were determined and then compared against the weather-inferred runway configurations. In general, the weather-inferred configurations were the same as the radar-inferred configurations. The orientation of each configuration (north/south) was identical, but the weather-inferred set of runways often varied slightly from the radar-inferred set of runways, probably because of the sensitivity of the weather-inferred configuration to short-term changes in wind speed and direction and weather.

III. Models

A. Departure Process (Gate Pushback to Takeoff)

1. Modeling Approach

The general approach taken to modeling the taxi-out process is to treat the system as an input/output system with very simple dynamics. The model is intended to capture the observed statistical behavior of the departure process, rather than replicate the exact physically and procedurally constrained dynamics of aircraft motion on the airport surface. This approach has the advantage that particular models can be easily calibrated and validated to describe a wide variety of airports under various traffic and weather conditions, and statistically significant conclusions can be drawn from these models.

2. Observed Behaviors of the Departure Process

Based on extensive field observations at BOS,[9] and analysis of historical data from BOS, ATL, and DFW, there are several major behaviors of the departure process that the model must capture. The taxi-out time of a particular aircraft (from pushback to takeoff) is primarily determined by the departure congestion

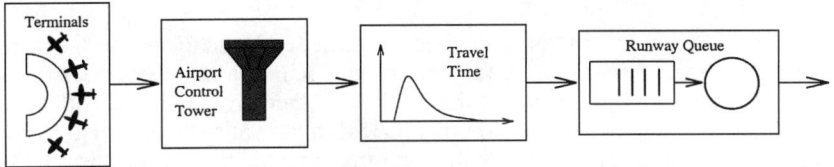

Fig. 8 Proposed queuing model for the departure process.

at pushback, i.e., the number of departing aircraft that are already on the airport's surface but have not yet taken off. When departure congestion is low, a nominal (or unimpeded) distribution of taxi-out times can be observed for aircraft pushing back from a particular gate. Aircraft are often observed to reach the active runways in a different order from their pushback sequence, indicating that the departure traffic flow up to the runway is relatively unconstrained.

In contrast, once an aircraft reaches the runway, it usually enters a runway queue, and its position in the queue becomes fixed. The airport throughput is primarily limited by this bottleneck effect at the runways.[2] Runway configuration and weather are observed to be the primary factors that determine the behavior of the runway queue, including the maximum runway throughput, and the approach to throughput saturation as a function of rising departure congestion.

3. Proposed Model Structure

Based on these behaviors, simple queuing structures are proposed to represent the input/output system dynamics (Fig. 8). Aircraft enter the system after they have called ready for pushback and have been given pushback clearance by the tower; they leave the system at the time they take off.

The initial unconstrained phase of departure traffic flow is modeled as a random delay, in which each aircraft that pushes back is assigned a stochastic taxi-out time to reach the active runways. The probability distribution of these taxi-out times is taken to be the nominal (unimpeded) taxi-out time observed at low congestion levels. Ideally, to capture the differences in travel time due to different gate locations, each gate would be assigned an individual probability distribution. Unfortunately, historical gate-assignment information is not readily available. However, it has been found that the airline for each flight is a reasonable proxy variable because the gates for a particular airline are often clustered at a particular terminal.[10]

Once aircraft complete their nominal taxi-out time, they are assumed to enter the runway queue. This queue is first-come-first-served (FCFS), which captures the bottleneck and sequencing effects observed near the active runways. During each interval of time, a stochastic number of takeoff opportunities is available, and aircraft at the head of the runway queue can exit the system if sufficient opportunities are available. This stochastic behavior is observed under conditions of high departure congestion, when the runway system is almost certainly nonempty. Pujet[10] proposed a similar queuing model for the airport departure process, and the model was extensively calibrated and validated for BOS using several years of historical runway configuration and traffic data. Our current model uses the same queuing structure but proposes several changes to the runway queue model and to the calibration and validation techniques.

4. Calibration Methods

Based on results from Idris,[9] a method has been developed to observe the nominal (unimpeded) distribution of taxi-out times. Each departing flight is assigned an index (denoted N_H) that counts the number of other aircraft that take off while that flight is taxiing out on the airport surface. If a particular flight is held on the airport surface after pushback due to downstream restrictions, mechanical problems, bureaucratic delays, or other effects that are unrelated to departure surface congestion, it will tend to be passed on the taxiway by other departing aircraft, and its N_H index will be large. If a particular flight pushes back and encounters substantial queuing delays near the runway, then its N_H index will be large due to the large number of other departing aircraft that take off while it waits in the queue. Therefore, flights with a low N_H index are assumed to have experienced little delay while taxiing out to the runway, and the nominal (unimpeded) distribution of taxi-out times is estimated from their taxi-out times. It is worth noting that the N_H index cannot be calculated at the time an aircraft pushes back from the gate, and hence it cannot be used directly in real time to predict taxi-out time.

The effect of N_H on the observed distribution of taxi-out times is shown in Fig. 9. The plot shows how the observed distribution increases in both mean and variance as a function of increasing values of the N_H index. Similar results are found for ATL and DFW. Log-normal distributions are fitted to the observed data to approximate the underlying distribution on unimpeded taxi-out time.

The stochastic model for the runway queue behavior is based on the observation that, at a fixed level of departure congestion, the distribution of takeoffs over

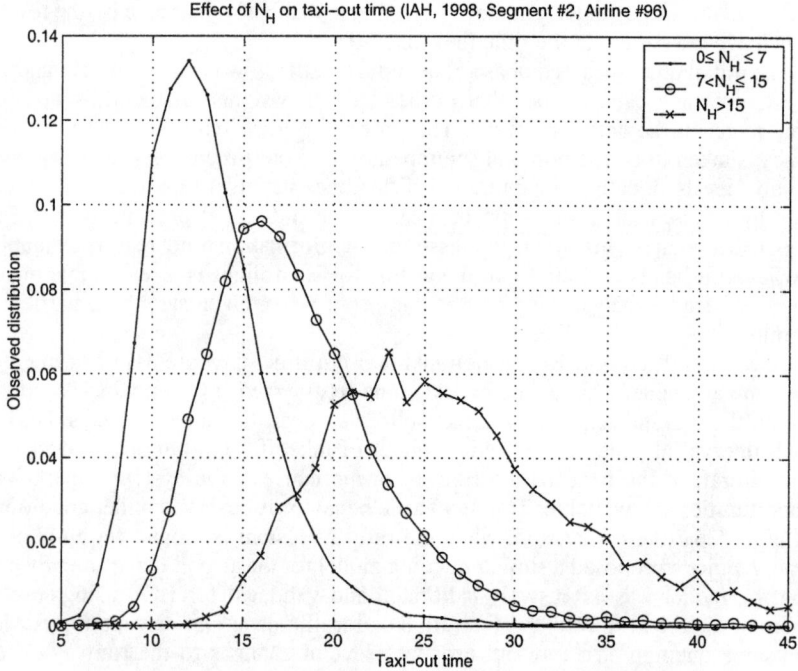

Fig. 9 Effected of N_H on taxi-out time distribution at IAH.

each 1-min interval is well fitted as a Poisson distribution. Further, as the level of departure congestion increases, the rate of the fitted Poisson distribution increases, until a threshold is reached where further increases in departure congestion levels do not result in increased rates. Based on these observations, the runway queue is modeled as providing a stochastic number of takeoff opportunities during each interval of time, in which the distribution of the number of opportunities is Poisson with the maximum observed rate.

A type of runway throughput plot was developed to aid in calibrating this model. At each level of departure congestion, a Poisson distribution (with 95% confidence intervals) is fitted to the observed distribution of takeoffs. Then these fitted rates are plotted as a function of the departure congestion level to yield a throughput plot. Additionally, the number of time intervals at each level of departure congestion is plotted to ensure that sufficient data points are being used in the fitting process. Several of these plots are shown for the various airports we have analyzed.

The first pair of plots (Figs. 10 and 11) was made using data from ATL during those intervals in 1998 when the airport was operating in its secondary runway orientation. The first plot corresponds to good-weather conditions, and the second plot corresponds to inclement-weather conditions. Note that the distribution of takeoffs is fitted very well as a Poisson distribution over a wide range of departure congestion levels. It is apparent that the throughput in good-weather conditions saturates at a higher level of congestion than the throughput in inclement-weather conditions. Overlaid in Fig. 11 are similar statistics collected from the calibrated

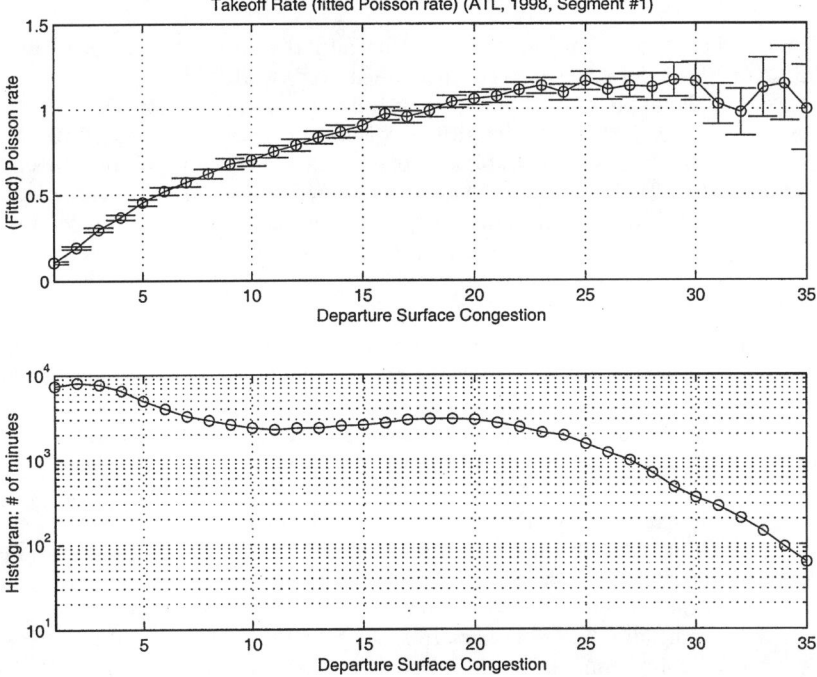

Fig. 10 Throughput at ATL during good weather.

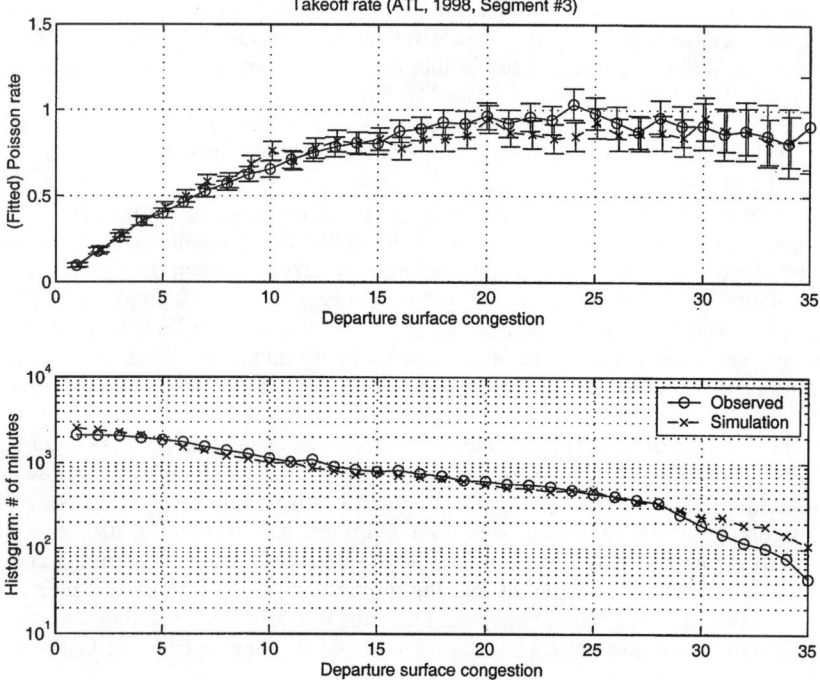

Fig. 11 Throughput at ATL during inclement weather.

queuing model of departure operations for that airport segment. According to those statistics, this model matches experimental data very well.

A second pair of plots (Figs. 12 and 13) was derived using data from DFW during those intervals in 1997 when the airport was operating in its secondary runway orientation. We observe effects similar to those seen at ATL. However, note that the throughput at DFW during good-weather conditions appears to steadily increase as departure congestion increases; there is no observed saturation effect. In contrast, the throughput during inclement-weather conditions shows a clear saturation effect.

5. Work in Progress

To date, there are several important observations that have not yet been successfully incorporated into the departure process model. First and foremost is the competition between arriving and departing aircraft for airport resources. It is intuitively obvious that on a very short time scale there must be some tradeoff between landings and takeoffs on the same runway. This tradeoff is currently treated as an additional source of stochastic noise in the runway behavior, but work is currently in progress to explicitly model this effect in the behavior of the runway queue. Observations indicate that at some airports, departure taxi-out times tend to increase as arrival congestion increases, where arrival congestion (denoted N_A) is measured as the number of arriving aircraft that are taxiing in from the runways when a departing aircraft pushes back from the gate. Figure 14 shows that increasing levels of arrival congestion are clearly related to increasing taxi-out times at ATL. However, the same phenomenon is not readily apparent at DFW (Fig. 15).

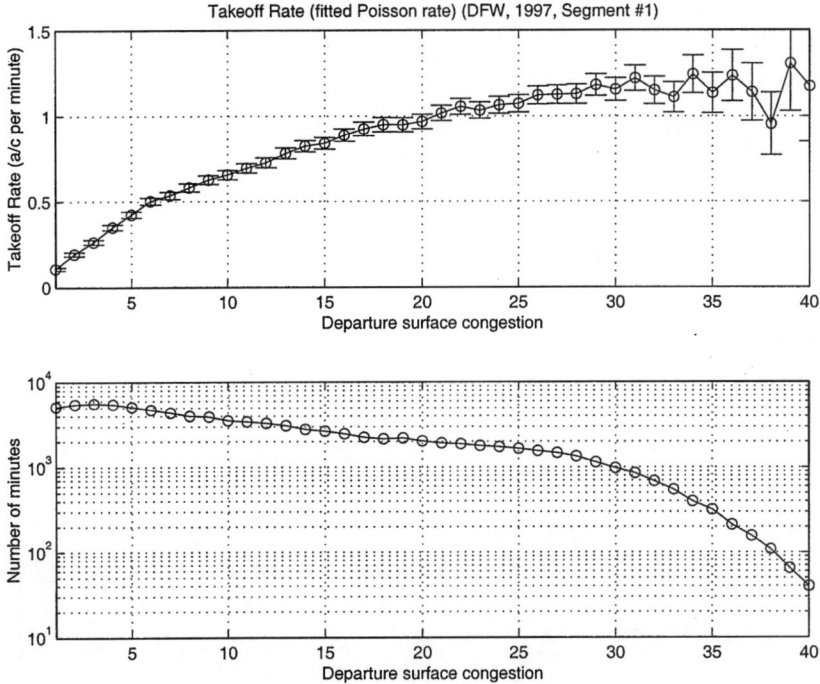

Fig. 12 Throughput at DFW during good weather.

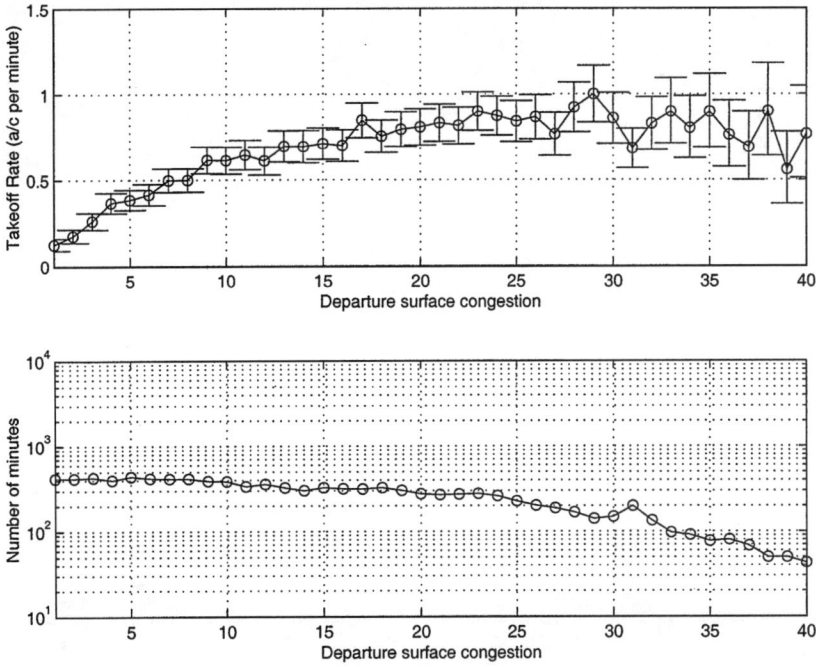

Fig. 13 Throughput at DFW during inclement weather.

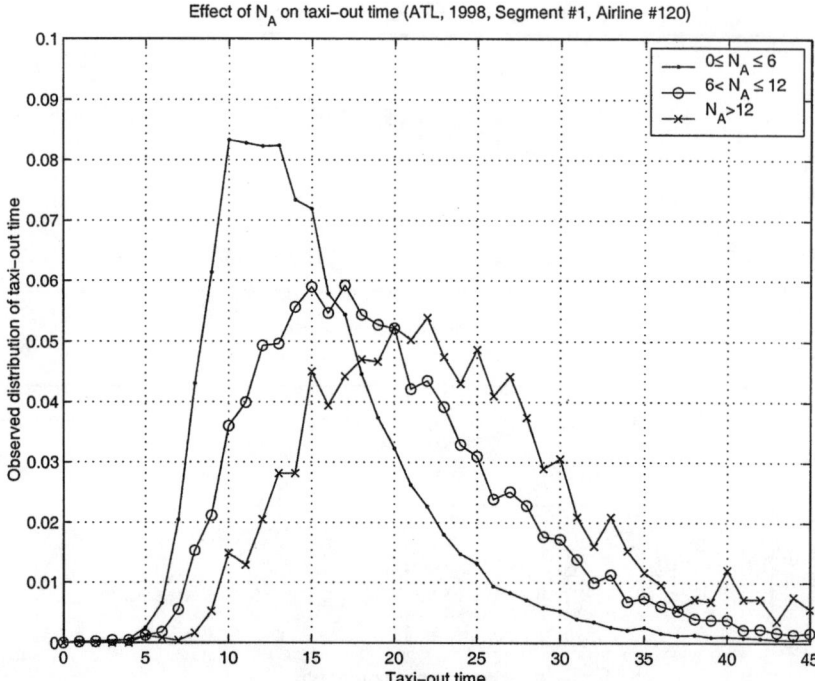

Fig. 14 Arrival congestion influences taxi-out times at ATL.

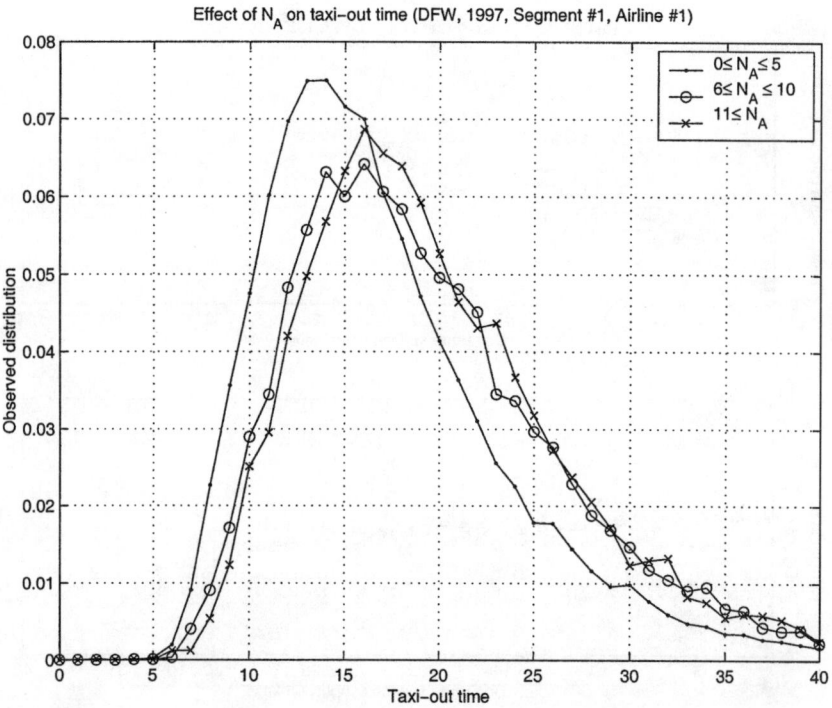

Fig. 15 Arrival congestion has limited influences in taxi-out times at DFW.

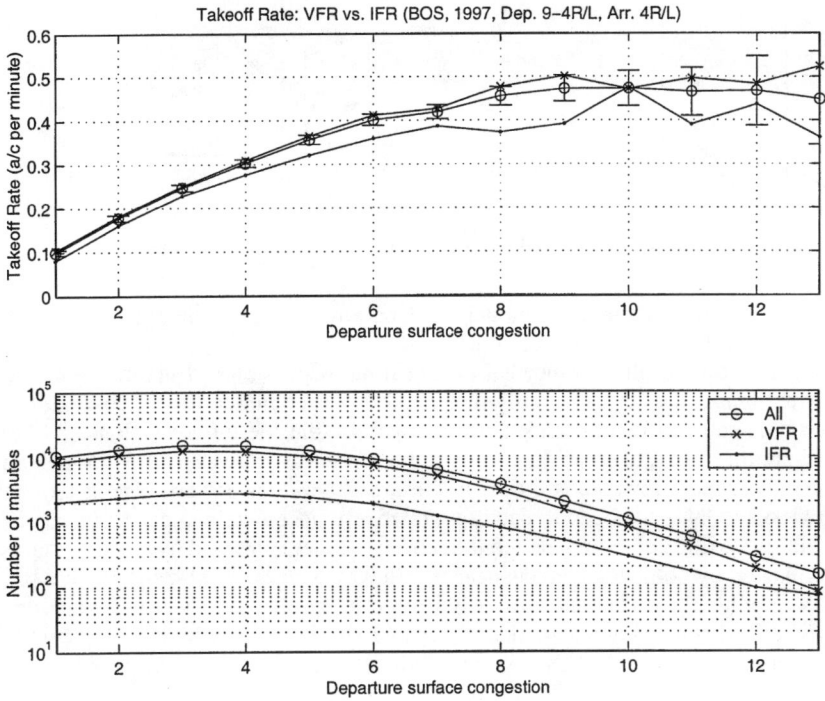

Fig. 16 Evolution of throughput by weather at BOS.

It is not clear if this dependence is the effect of a causal relationship, or if arrival congestion and departure congestion simply have some positive correlation due to the airlines' schedule bunching and block-scheduling at certain hub airports. A joint arrival/departure model is needed to elucidate these issues; it is difficult to propose and test such models, however, without very detailed operations data covering aircraft motion and air traffic control (ATC) communications/control across the entire airport surface.

Several other effects have been observed but not incorporated into the models. Data analysis using weather conditions is not perfectly equivalent to analysis using runway configurations; following observations made at BOS, airport capacity may be affected somewhat by weather within a single configuration, as shown in Fig. 16. The effect of propeller traffic (which is notably absent from the ASQP database) is currently treated as an additional source of stochastic noise in the system, although in principle the current queuing model can be trivially extended to include propeller traffic. Occasionally aircraft experience significantly longer taxi-out times due to downstream restrictions, and work is under way to accommodate these outliers.

B. Arrival Process (Landing to Gate Arrival)

Data analysis at BOS, DFW, and ATL indicates a somewhat surprising result: the statistical behavior of arrival operations can be captured using the same general input/output queuing structures and calibration/validation techniques that are

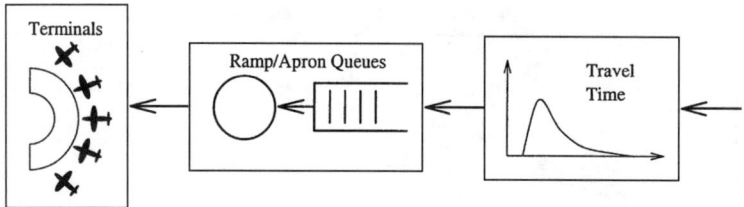

Fig. 17 Proposed queuing model for the arrival process.

currently used to statistically model the departure process. A diagram of the proposed arrival model is shown in Fig. 17.

At first this result is somewhat unappealing, because the structure of the departure model has been explicitly motivated by a specific set of field observations and data-analysis results, and it is not apparent that these observations and behaviors immediately generalize to the arrival process. However, it is possible to view the departure process model in a more general framework. The departure process model is intended to capture a relatively unconstrained period when aircraft are taxiing out unimpeded to the runway queues, followed by a period that is dominated by bottleneck effects near the runway queues. The arrival process follows roughly the same pattern, in which aircraft initially taxi toward the gates and then slow down and queue up near the gates. This effect is especially apparent in airports with physical bottlenecks near certain terminals, such as the corridor-type terminals at ATL and the "horseshoe" at BOS.

Nominal distributions of taxi-in times were obtained using the same method used to obtain nominal distributions of taxi-out times. For the arrival process, the N_H index is defined as the number of arriving aircraft that reach the gates while a particular flight is taxiing in from the runways. Representative distributions of taxi-in time at IAH and DFW are shown in Figs. 18 and 19. Gate throughput curves were similarly obtained, using arrival rather than departure congestion. The gate throughput curves for all three airports are quite similar in character. As might be expected, gate throughput appears unaffected by inclement weather conditions at all of the airports studied (see Figs. 20–23). One interesting observation is that the gate throughput can saturate, similar to the saturation effect in the departure process. These observations indicate that gate throughput saturation may be an effect of very high traffic loads, rather than a degradation of the system performance. Figure 22 shows statistics obtained for a calibrated arrival model of ATL. Again this model performs quite well with respect to the experimental data.

C. Ground Operations (Gate Arrival to Gate Pushback)

1. Modeling Approach

The ground operations model is an optimization model designed to simulate airline operational decisions about aircraft pushback times under resource constraints. The ground operations model considers the departure schedule, aircraft-gate compatibility, gate availability, and ground crew resource availability in determining pushback times that minimize passenger delay given arrival at gate times. As a result, the model can measure how an airline can reduce delays and delay propagation on the ground. This section includes a description of the ground operations

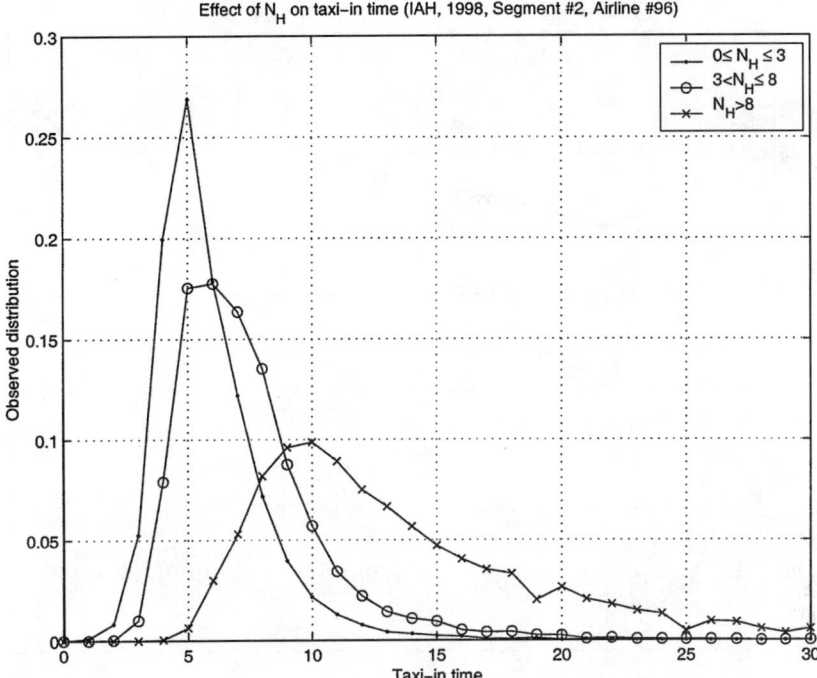

Fig. 18 Effect of N_H on taxi-in time at IAH.

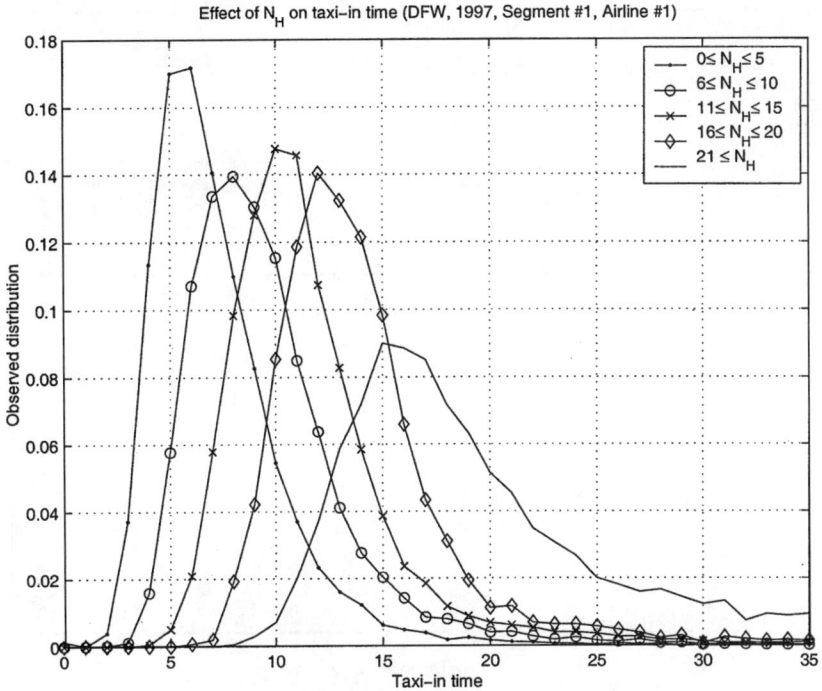

Fig. 19 Effect of N_H on taxi-in time at DFW.

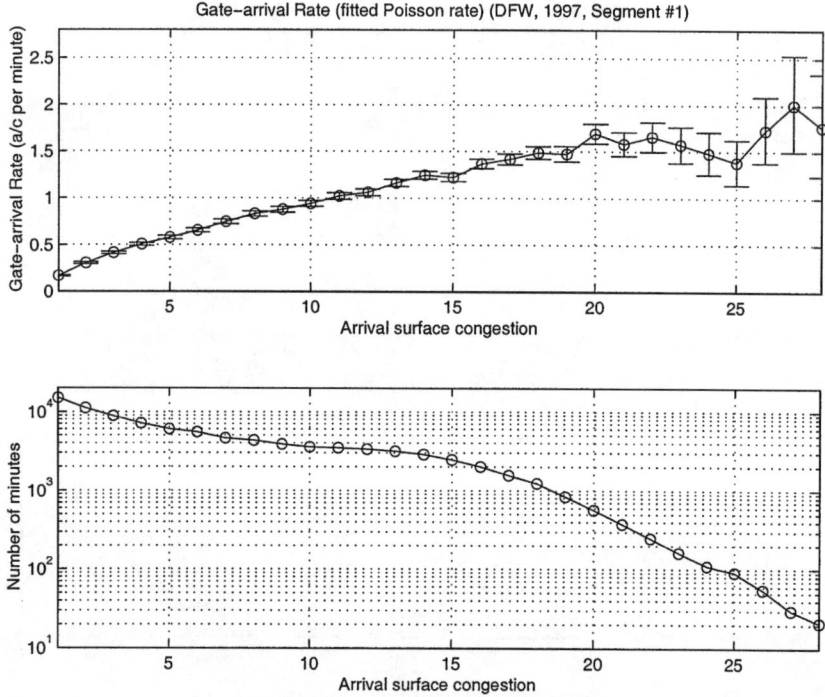

Fig. 20 Gate throughput at DFW (good weather).

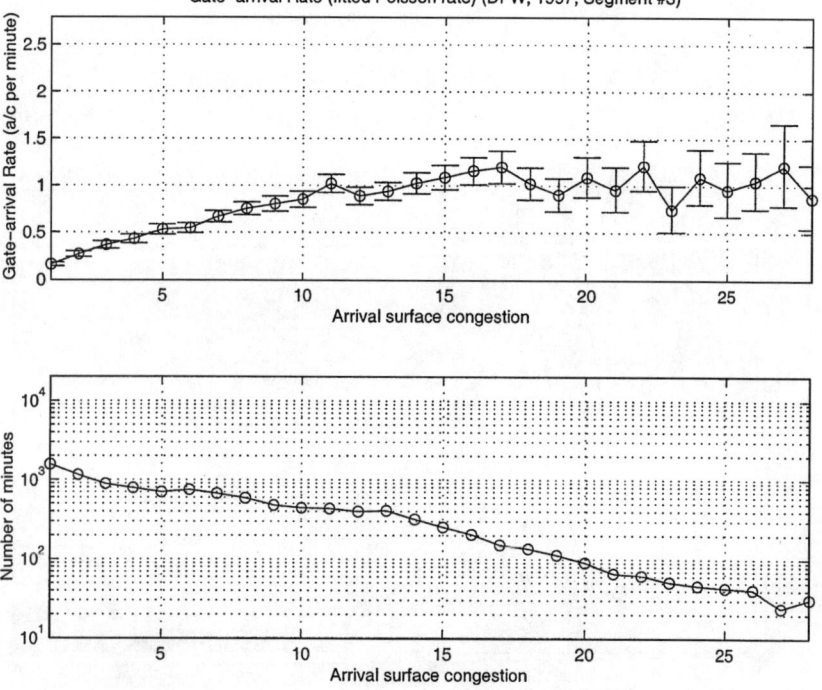

Fig. 21 Gate throughput at DFW (inclement weather).

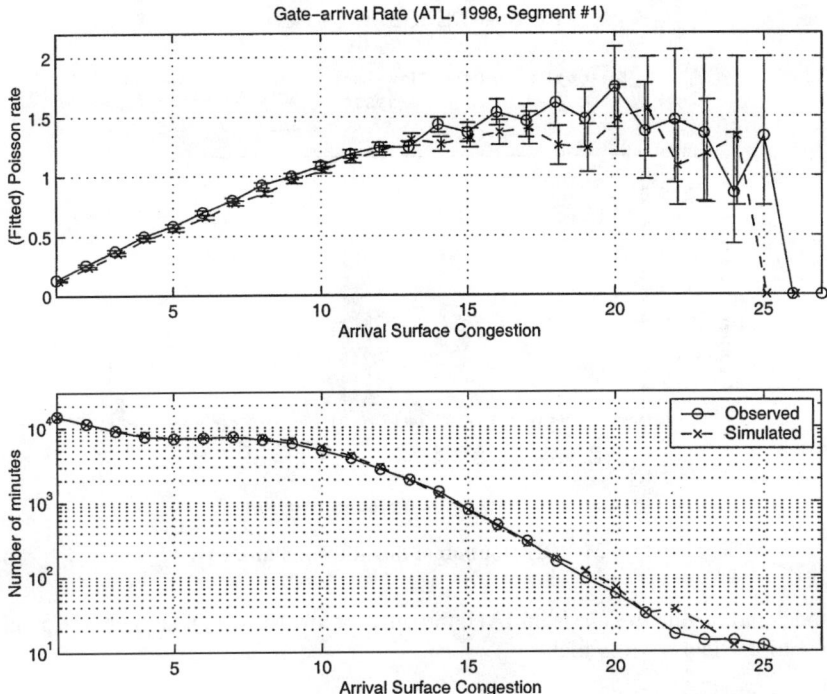

Fig. 22 Gate throughput at ATL and calibrated model (secondary runway orientation).

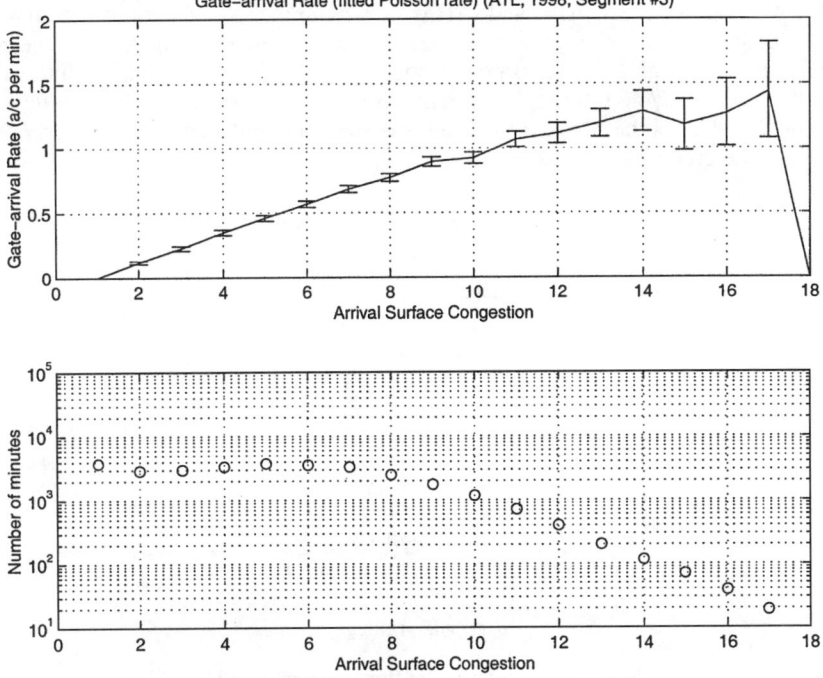

Fig. 23 Gate throughput at ATL (secondary runway orientation, bad weather).

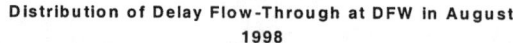
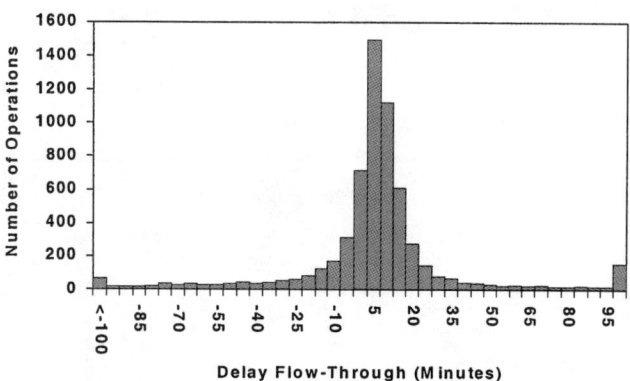

Fig. 24 Distribution of flow-through at DFW.

model and results to date from IAH, the only airport for which sufficient ground operations data were available.

2. Observed Behavior of the Ground Operations

Many factors contribute to departure delays, including arrival delay. To illustrate this, the difference between arrival delay and departure delay was computed; this difference will be referred to as delay flow-through. The distribution of delay flow-through for DFW, ATL, and IAH is shown in Figs. 24–26. Notice that these distributions appear Gaussian, with means greater than zero. In fact, the mean delay flow-through for each airport is positive, and the 95% confidence interval of the mean delay flow-through is strictly positive. This positive mean delay flow-through indicates that the arrival delay was somehow reduced while the aircraft was on the ground.

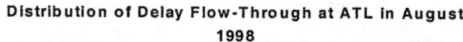
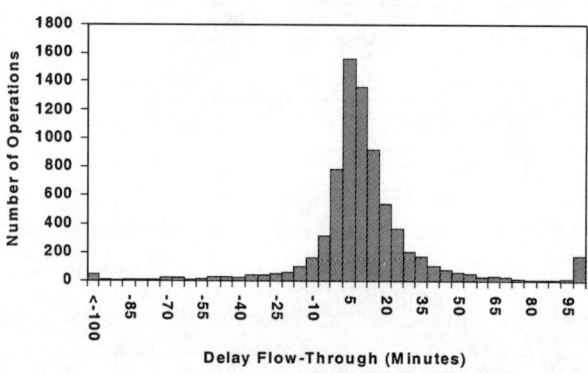

Fig. 25 Distribution of flow-through at ATL.

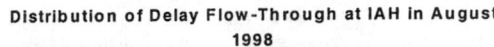

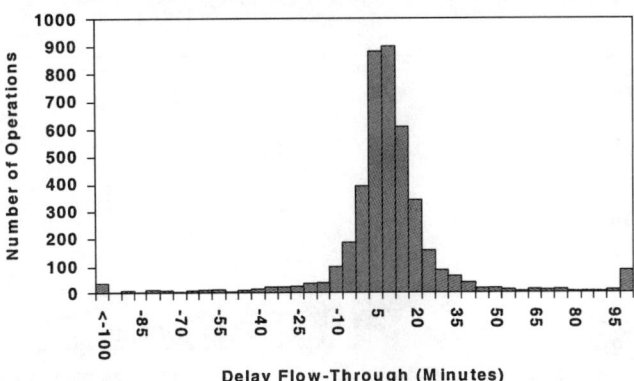

Fig. 26 Distribution of flow-through at IAH.

There are two potential explanations for this observation. First, the "slack" in the arrival and departure schedule may have absorbed the arrival delay. For example, assume an aircraft is scheduled to arrive at 1000 hrs and depart at 1040 hrs and that the scheduled minimum turnaround time for the aircraft is 30 min. In this case, there are 10 min of slack built into the schedule. Therefore, the aircraft can arrive up to 10 min late without affecting the departure time. Second, the airline may have prepared the aircraft for departure ahead of schedule. Continuing with the same example, if the aircraft arrived 20 min late, but departed on time, the airline turned the aircraft around in 20 min, 10 min under the scheduled minimum turnaround time. In this case, the airline reduced the turnaround time of the aircraft.

To understand the extent to which the airline is able to reduce the turnaround time of the aircraft in order to reduce delays, we identify aircraft with departure delay greater than 10 min and with arrival delay greater than departure delay. For these aircraft, the histogram of the actual turnaround time minus the minimum scheduled turnaround time was plotted. The scheduled minimum turnaround time is the turnaround time assumed by the airline in the scheduling process. This plot is shown in Fig. 27 for one airline at one of the hub airport studies (data for other airlines were not accessible). The confidence interval for the mean of the distribution is negative, indicating that the airlines tend to prioritize late arrivals on the ground to reduce the corresponding departure delay.

3. Model Structure

As discussed, an airline can reduce departure delay by reducing turnaround time. However, the results do not indicate exactly *how* an airline achieves the turnaround time reduction. One of the biggest challenges in modeling ground operations is determining which resources and activities to include in the model. The turnaround process involves numerous distinct sets of crews conducting distinct activities, including baggage unloading and loading, catering, cleaning, maintenance, passenger deplaning and boarding, and so forth. During visits to airline ground operations centers, key airline personnel indicated that the baggage

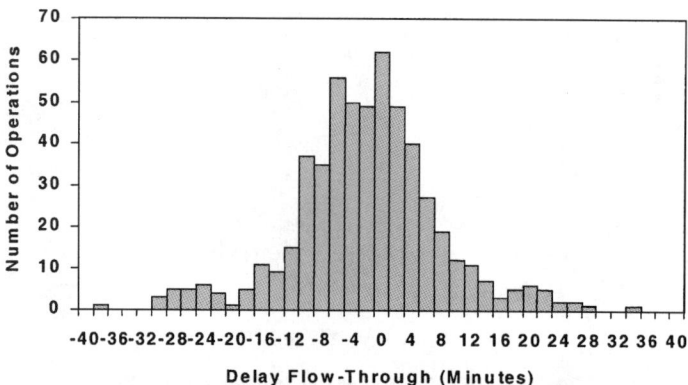

Fig. 27 Distribution of the difference between actual and scheduled minimum turnaround time.

handling process could be one bottleneck in the turnaround process. Therefore, we decided to include baggage handler constraints in the ground operations model. The extent to which these constraints explain the variability in the actual turnaround process is discussed later.

During visits made to airline ground operations, airlines made last-minute decisions to hold departing aircraft to accommodate connecting passengers from a delayed arrival flight. To incorporate this decision process into the model, the ground operations model explicitly considers passenger flows. If a passenger connection is missed, the total delay to that passenger, which is the time until the next departure to the same destination, is included in the objective function. This means that the ground operations model determines the tradeoff of delaying an aircraft to allow for passenger connections and rerouting passengers who miss connecting flights.

Even after carefully deciding which factors to consider, some simplifying assumptions need to be made to maintain the tractability of the problem. First, baggage handlers are assigned to aircraft irrespective of their previous aircraft assignments. This assumption means a baggage handler can be assigned to a different aircraft in every time unit. At most hub airports, however, baggage handlers are assigned in teams to a particular aircraft for unloading and loading. Therefore, the resulting assignment of baggage handlers may not map to a feasible assignment of baggage handler teams.

Second, the ground operations model is a deterministic model, meaning there is no stochasticity incorporated in its design. In particular, the taxi-in times of the aircraft are assumed constant. As discussed previously and seen in Figs. 18 and 19, this is not true in practice. The extent to which this assumption affects the model remains to be addressed.

Finally, the objective function is measured in passenger-minutes, which is not a metric directly linked to the airline's cost structure. The translation of this metric to dollars is difficult. However, the metric does link both operational efficiency and the passenger experience, both of which have an effect on the profitability of the airline. Some sensitivity analyses with respect to the objective function are discussed by Andersson.[11]

In practice, the run time of the ground operations model has shown to be acceptable. Problems including about 80 aircraft and covering a 3-h time horizon solve in about 1 min. The specifics behind the formulation of the integer programming model, as well as a detailed description of problem size and run time, are not within the scope of this chapter; interested readers are referred to Ref. 11 for further details.

4. Model Calibration and Validation

To determine whether the ground operations model is effective in predicting departure time, its pushback time estimates were compared to those from a simpler, "naive" model. The naive model is designed with a constant turnaround time, based on the minimum scheduled turnaround time. The basic difference between the models, therefore, is that the ground operations model provides more flexibility and considers ground crew resources and passenger flows in determining departure time.

Because the departure times of the aircraft in a particular scenario are interdependent in the ground operations model (the aircraft share finite resources), multiple independent scenarios were considered to compare the two models. The metric considered is departure error, defined as the model's prediction of departure time minus the actual departure time. For each scenario and for each model, the average departure error and the mean-squared departure error were calculated, with the results given in Table 3. The data included in the analysis are for 12 days in January 1998 from 1600 to 1915 hrs.

Table 3 Error comparison of ground operations model vs naive model

Scenario	Ground operations model departure error		Naive model departure error	
	MSE	Average	MSE	Average
1	168.66	−0.33	882.21	6.28
2	46.95	−2.29	113.31	0.98
3	46.00	−0.08	107.54	3.74
4	33.04	−1.71	157.02	3.95
5	697.38	−5.27	1347.09	−3.42
6	563.72	−9.08	1160.26	−5.86
7	288.54	−3.73	646.04	1.88
8	12.67	−0.67	53.27	1.78
9	16.78	1.07	396.37	8.22
10	122.09	−5.00	382.53	2.98
11	572.92	−7.81	2409.85	0.69
12	113.02	−4.47	1757.24	6.07
Average	223.48	−3.28	784.39	2.28
Std. error	248.32	3.17	752.20	3.98
Lower bound	82.98	−5.08	358.81	0.02
Upper bound	363.98	−1.48	1,209.98	4.53

Notice that the mean-squared departure errors for the two models are significantly different; the ground operations model errors are generally significantly smaller than the naive model's errors. In fact, a Wilcoxon signed rank test confirms that the MSE values for the model are less than those for the naive model with a significance level of 0.2%. This implies that the additional factors considered in the ground operations model are influencing the turnaround process and are improving the departure time predictions. However, the confidence interval of the average departure delay for the ground operations model does not cover zero. In fact, the confidence interval contains only negative numbers. This means the ground operation model's departure time estimates tend to be earlier than the actual departure time, implying there is some bias in the predictions.

To better understand this bias, the distributions of the actual departure delay and the delay predicted by the ground operations model were plotted. This aggregated analysis is necessary because the passenger connection data used in the ground operations model are simulated rather than observed data, meaning the departure time decisions made for a particular aircraft are likely to deviate from actual. However, we would expect the delay decisions to be similar over the entire *set* of aircraft. The distributions for the actual delay and the ground operations model delay are shown in Fig. 28.

It is important to note that the data set used to generate Fig. 28 excludes aircraft for which the difference between actual departure delay and actual arrival delay exceeds 40 min. These data points were excluded because delays of that magnitude (greater than 40 min) are unlikely to be caused by ground crew resource issues, gate availability, or passenger connections. Therefore, some factor(s) external to the ground operations model influenced the departure time. Despite the omission of these identified data points, 5% of departures experienced delays exceeding 40 min, while the ground operations model predicted only 2% of departures would incur such delays. However, these excessive actual delays are still likely attributable

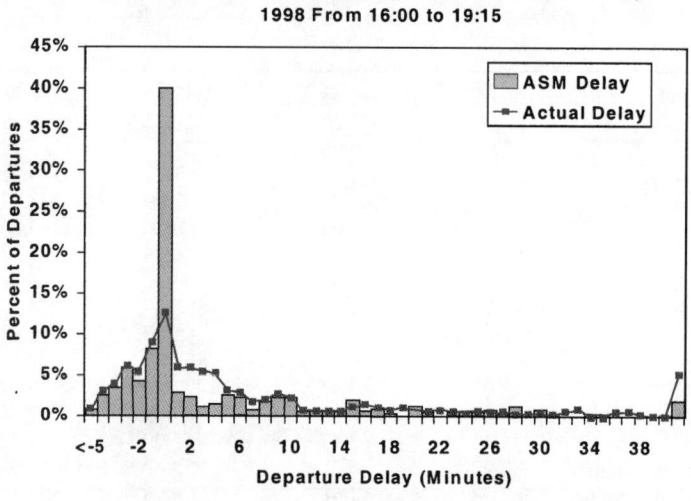

Fig. 28 **Comparing predicted and actual on-time departures.**

to factors external to the ground operations model. For example, mechanical problems, ground delay programs, and delayed cockpit crews can all lead to delays of 20 min or more. If an aircraft is already delayed 20–30 min, the departure delay exceeds 40 min but is still included in the model.

The major difference in the distributions exists at departure delays of 1–4 min. The ground operations model assigns on-time departures to aircraft that were actually delayed 1–4 min. The departure process is an extremely complex process involving the synchronization of many resources and subprocesses. Before an aircraft is ready for departure, the passengers must deplane the arrival and board the departure, baggage handlers must unload and load the baggage, caterers and cleaning crews must remove rubbish and replenish food and beverage supplies, the aircraft must be checked for flight safety and refueled, the cabin and cockpit crews must arrive and prepare for departure, and so forth. Variability exists in each of these subprocesses. A delay of a few minutes could be caused by numerous factors external to the ground operations model. Recent improvements in the computational performance of the model formulated so far will enable inclusion of some of these factors.

It is possible to tune the parameters of the ground operations model to reduce the differences in the results. However, optimizing the parameters, meaning setting the parameters to yield departure time predictions close to actual departure times, presents substantial challenges. The model is sufficiently complicated so that it is impossible to determine a priori how the parameter changes will affect the solution. Further, and more importantly, it is difficult to identify an optimal parameter setting. It is unknown how much of the deviation from actual departure times is attributable to the use of suboptimal parameters and how much is attributable to including insufficient information about the ground operations process in the model. This is a fundamental and new research problem encountered in many other types of operations (e.g., military operations) and is currently being addressed by other work.

IV. Applications

The three models discussed have or are being applied to current issues in the air transportation system. This section describes three such applications. First, the models can be integrated to improve predictions of aircraft movement times on the ground. Second, the departure model can be used to evaluate the impact of congestion control on the airport surface. Finally, a semi-integrated arrival and ground operations model is being used to quantify the benefits of procedural changes and decision support tool enhancements.

A. Prediction

One immediate potential application for the three models developed for airport operations is to create or extend existing predictive capabilities to factor in delays due to airport operations. The purpose of such predictive capabilities is to predict anticipated congestion periods better so that appropriate measures may be taken. The necessary elements for building such a system include the ability to incorporate new information as it becomes available (e.g., knowledge of a pushback request or knowledge of a takeoff), and the ability to propagate the evolution of the airport system into the future. Conceptually, building such a predictive delay capability is

Table 4 Direct operating cost estimates at the gate and in runway queues

Jet aircraft type	$/min. at gate			$/min. in queue		
	Medium	Large	Heavy	Medium	Large	Heavy
Fuel	0	0	0	2	4	9
Flight crew	2.5	4.5	6	6	12	20
Maintenance	0	0	0	5	9	25
Total	2.5	4.5	6	13	25	54

not new; for example, Shumsky presents departure delay prediction algorithms.[3] The value of such a tool depends upon the quality of the models used. It also is fundamentally limited by the amount of stochastic noise present in the system, which is important in the framework of our present analysis. As a consequence, planned delay predictive capabilities relying upon the presented models as well as other empirical models* will necessarily be probabilistic and will include mean values as well as standard deviations. Building such a tool for most major U.S. airports is the object of current research and development efforts.

B. Departure Congestion Control via Gate-Holding Queues

As seen in the departure throughput plots presented here, the departure surface congestion is often substantially higher than the level necessary to saturate the runway system at maximum throughput. A simple control scheme based on this observation has been proposed and investigated at BOS.[4,10,12] Under this scheme, departing flights are held at the gate if the departure surface congestion exceeds some control threshold (denoted N_C); the held flights are immediately given pushback clearance when the departure surface congestion drops to an acceptable level. This approach is formally identical to the window flow control mechanism used in packet-switched data networks such as TCP/IP networks.[13] The underlying motivation is to minimize departure congestion and trade runway queueing delay for delay at the gates; this tradeoff reduces both environmental impacts due to fuel burn while taxiing, and direct operating costs (DOC) to the airline. We present a summary of previous results and then investigate several variations of the original control law.

1. Summary of Previous Results on Window Control

There are substantial financial and environmental incentives to reduce runway queueing delays. U.S. airlines are required to report DOC data to the Department of Transportation.† Estimated DOC values (based on 1992 and 1995 data, and averaged over all major U.S. airlines) are shown in Table 4 for three different aircraft types.‡ The mix of jet aircraft departure traffic at BOS was estimated from

---

*Private communication with J.-P. Clarke, Massachusetts Inst. of Technology, Cambridge, MA, March 2000.

†Form 41, web site of the Office of Airline Information, Bureau of Transportation Statistics, http://www.bts.gov/oia/source [cited 1999].

‡Lecture notes from Air Transportation Economics course Massachusetts Inst. of Technology, Cambridge, MA, 1996.

Table 5 Jet aircraft engine emissions

Aircraft/engine	Emissions, g/min			Boston jet operations, %
	HC	CO	NO$_x$	
B-757 / PW2037	38.24	390.85	74.45	19.4
MD-80 / JT8D-209	63.01	220.47	54.73	16.2
B-727 / JT8D-9A	74.30	336.73	69.06	13.3
B-737 / CFM56-3-B1	31.19	470.59	53.35	11.4
DC-9 / JT8D-9A	49.53	224.49	46.04	11.2
B-737 / JT8D-9A	49.53	224.49	46.04	8.3
A320 / V2500-A1	3.27	115.47	87.94	6.9
B-767 / CF6-80C2A2	237.69	1043.51	89.59	3.7
A300 / PW4060	42.43	519.38	125.24	3.0
B-747 / CF6-80C2A2	988.85	2803.25	163.30	2.0
DC-10 / CF6-50C	843.66	2391.66	139.32	1.6
L-1011 / RB211-R22B	2671.02	3806.93	110.32	1.5
Weighted average for Boston	85.58	416.49	67.20	

ETMS data collected in June 1998.[12] Medium-sized jets represented about 65% of the jet traffic, whereas large jets represented 35% and heavy jets the remaining 10%. Combining the data in Table 4 with this jet traffic mix yields an average cost of $3.3 per min at the gate and $18.7 per min in the runway queue, for an average DOC savings of $15.4 for each minute of runway queueing time transferred to the gates. Similarly, Table 5 shows the estimated rate of pollutant emissions for the jet aircraft mix at BOS, including unburnt hydrocarbons (HC), carbon monoxide (CO), and nitrogen oxides (NO$_x$).[12] Note that aircraft engines typically contribute 45% of the combustion pollutants emissions at an airport,[14] and thus there is a very strong incentive to reduce aircraft engine emissions on the surface of airports.

Using a departure (taxi-out) model (Fig. 29) similar to the one proposed previously, Pujet and Feron have estimated that jet aircraft runway queueing at BOS in 1996 cost the airlines $6.1 million in DOC and produced the pollution equivalent of 9440–22,330 cars visiting the airport every day of the year.[12] To evaluate the benefits of the window control scheme, they modified the basic model to incorporate the control law, and then ran Monte Carlo simulations of jet aircraft traffic using

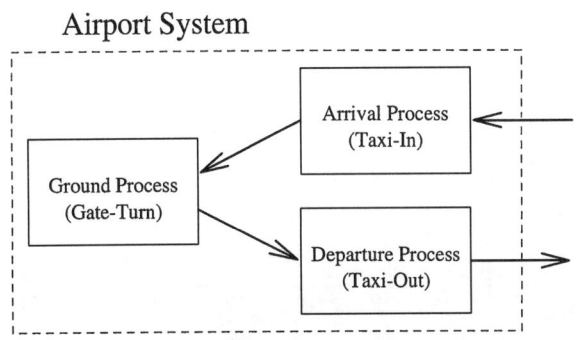

Fig. 29 Developing an integrated ground operations model.

demand data from 1996 at BOS. These simulations indicated a 7.8% reduction in runway queueing with only a 3.2% increase in total (gate plus runway) queueing, leading to a DOC savings of approximately $361,000 and the elimination of the equivalent pollution of 740–1750 car visits per day throughout the year.

Despite its potential benefits, the window control scheme has several potential drawbacks. Because runway queueing delay is transferred to the gates, airlines may find themselves faced with gate shortages. Airlines must also be concerned that delayed pushbacks are not measured or perceived as undesirable delays by passengers, who typically measure on-time performance according to actual pushback and gate-arrival epochs relative to the airline schedule. Pujet and Feron's simulations showed that at BOS in 1996, the window control scheme would require an additional gate for only 192 min in 1996, and only 2.9% of flights would be held more than 5 min, indicating that these drawbacks are not insurmountable.

2. Variations and New Analysis of Window Control

As noted, all of the airports we have investigated show a departure throughput saturation effect similar to that observed at BOS, suggesting that the window control scheme may be applicable. We first tested the basic window control law at ATL for weather segment 3 (secondary runway orientation during inclement weather). The tradeoffs between runway queuing, gate-hold queueing, and total queuing delay are shown in Fig. 30. Note that the simulation results suggest that, at least in the case of ATL under the specified conditions, it may be possible to directly reduce runway queuing by 40% without increasing total queuing

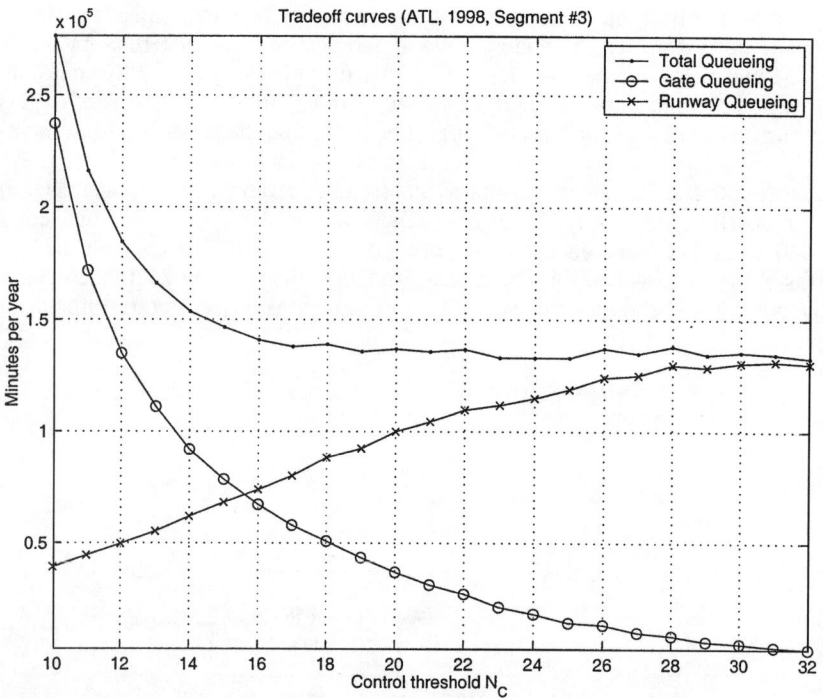

Fig. 30 Effect of window control scheme on delay distribution.

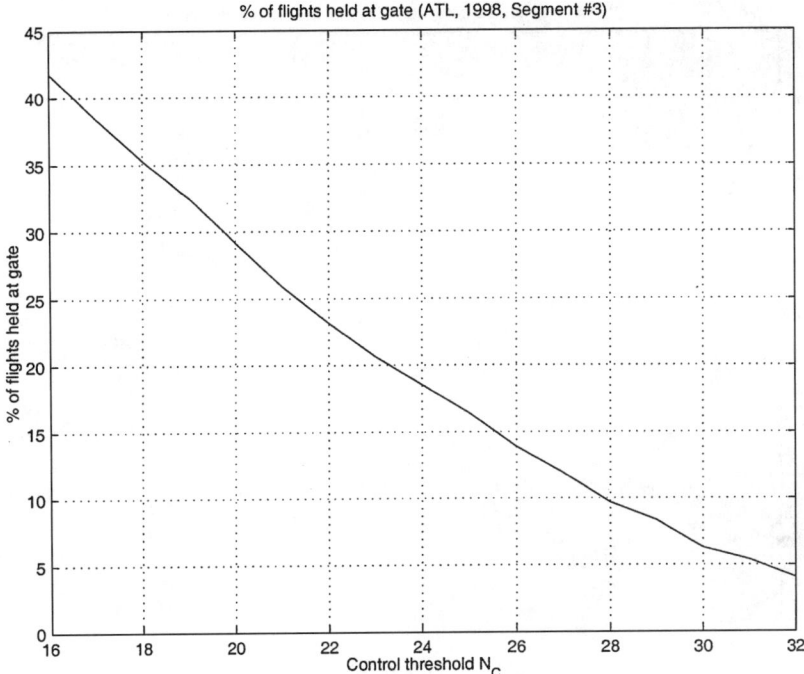

Fig. 31 Percentage of flights held at the gate.

delay, and further reductions in runway queuing are possible at the expense of increased total delay. This is a very substantial reduction, especially in contrast to the comparatively limited success of the window control scheme at BOS, and it confirms earlier estimates for potential delay reductions at ATL.* The percentage of flights that are held at the gate for any length of time is indicated in Fig. 31.

The original form of the window control scheme was intended for implementation by air traffic controllers in the airport tower. We have also investigated the possibility of a decentralized implementation, in which a single airline would window-control its pushbacks according to the current aggregate departure congestion. This scheme has been tested via Monte Carlo simulation for Delta Airlines (DL) traffic at ATL in 1998 (DL controlled 84% of ATL jet operations), and for Continental Airlines (CO) traffic at EWR in 1998 (CO controlled 61% of EWR jet operations). The results are shown in Figs. 32 and 33 for departure operations from the secondary runway orientation during inclement weather. Although the results at ATL look quite promising, the tradeoff between runway queueing and gate queueing delay at EWR is penalized by an unacceptable increase in total delays. Two possible explanations for this negative result are the CO does not control enough traffic at EWR to influence congestion, and that EWR traffic does not have a bank structure (in contrast to the strong bank structure at ATL) to absorb any gate queueing delays that may accumulate.

---

*Private notes of J.-P Clarke, International Center for Air Transportation, Massachusetts Inst. of Technology, Cambridge, MA, Sept. 1999.

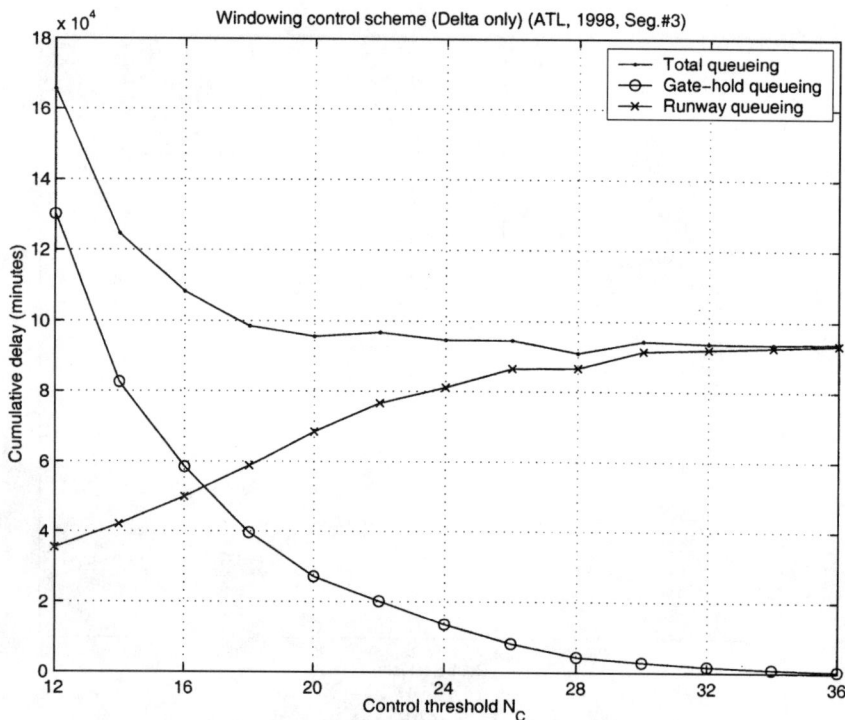

Fig. 32 Queueing tradeoffs for decentralized window control (ATL).

In response to the negative result at EWR, a further refinement of the window control scheme has been proposed and tested. The Co-only scheme was modified as follows: whenever the control law had held an aircraft at the gate for 15 min, that aircraft was immediately pushed back regardless of the current departure congestion. The 15-min cutoff threshold both prevents gate queueing delays from accumulating during a long period of congestion and ensures that the pushback time of any given flight is not excessive (airlines and passengers typically consider a flight to be on time if it pushes back and arrives with a delay of no more than 15 min). In addition to Monte Carlo simulation to test the queueing tradeoffs of the control scheme, a careful accounting of the gate usage by arriving and departing CO flights was conducted to determine when extra gates (in excess of the observed gate usage) would have been required. The results for a representative time period are shown in Figs. 34 and 35. Note that the gate usage under the control scheme never exceeds the daily peak usage, indicating that in principle there is sufficient airport capacity to accommodate the control-induced gate queueing. It is also important to note, however, that Fig. 35 indicates an aggregate CO gate usage; although sufficient gate capacity is always available, some incoming flights might need to be reassigned to unoccupied capacity, causing some distress for passengers and airline personnel. More accurate operations data (e.g., the gate assignments of incoming and departing flights) are required to fully evaluate the potential level of gate reassignment.

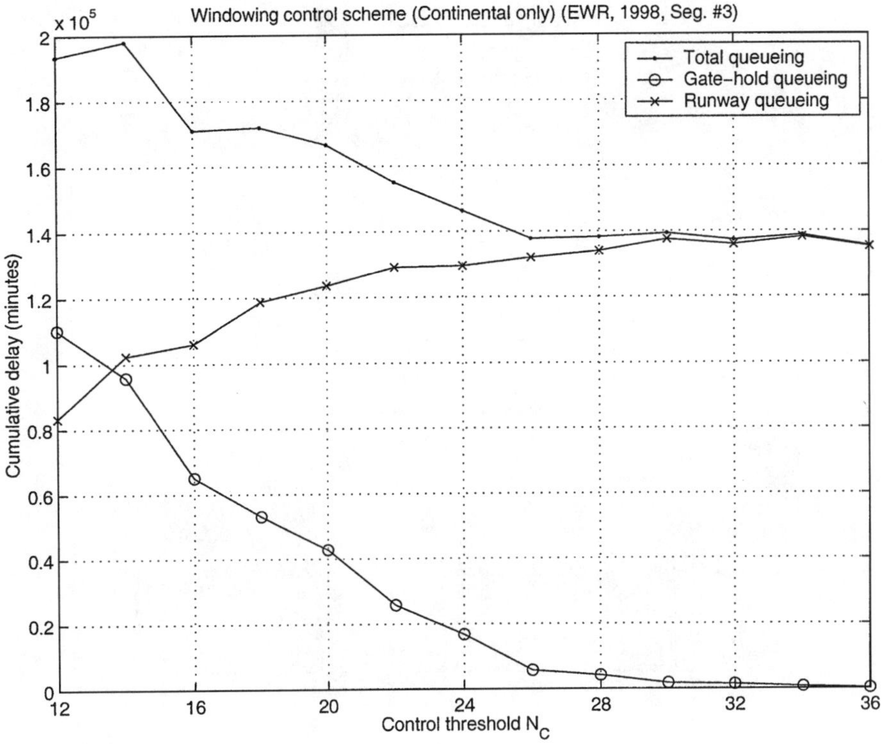

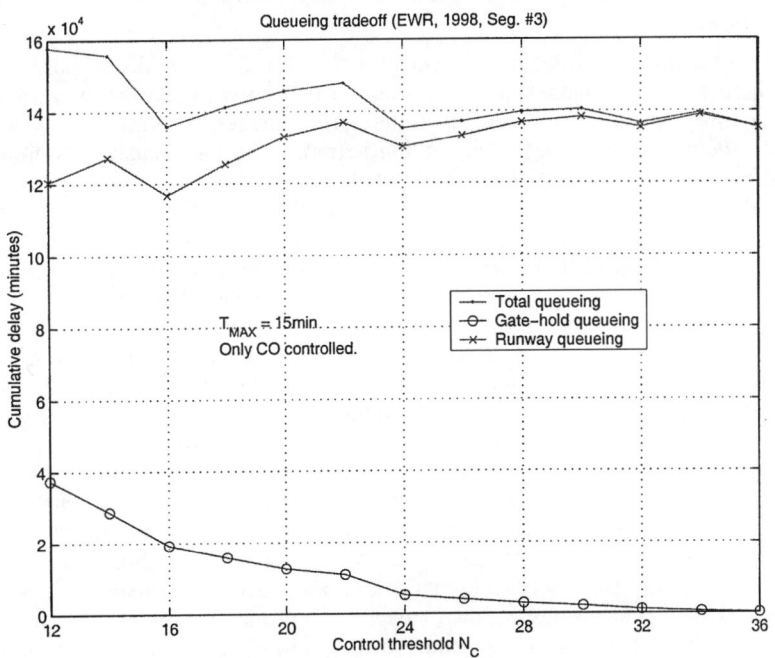

Fig. 33 Queueing tradeoffs for decentralized window control (EWR).

Fig. 34 Queueing tradeoffs by CO at EWR under modified window control.

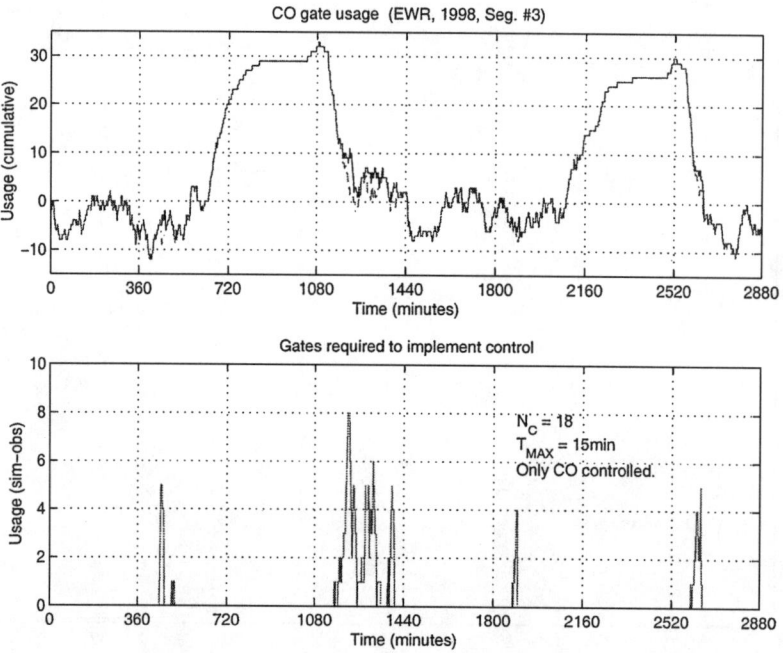

Fig. 35 Excess gate usage by CO at EWR under modified window control.

C. Benefits of Alternate Procedures and Improved Decision Support Tools

As noted in the Introduction and shown in Fig. 36, CTAS produces more accurate arrival time estimates than the airlines currently use to manage their ground operations.[3] The improved accuracy of the arrival time estimates could translate to more efficient use of ground resources. Furthermore, procedural changes combined with new or modified decision support tools could take airline sequence preferences into account when merging arriving aircraft. Therefore, an airline could potentially influence the order in which its arriving traffic landed.

To measure the potential benefits of sharing improved arrival time estimates with the airlines and of incorporating airline preferences in the sequence, an integrated arrival and ground operations model is necessary. This integrated model would determine the times of landing, arrival at gate, and pushback from gate to minimize delays under resource constraints. Ideally, the model would integrate the model structures of the arrival model and ground operations model already discussed. In effect, a queuing model would determine the taxi-in and gate arrival times, while the optimization model would determine each aircraft's movement times including landing, arrival at gate, and pushback from gate. However, the implementation of this concatenated model is not presently feasible. The queuing model requires as input the congestion levels of arriving aircraft, meaning the landing times would have to be given. The optimization model, on the other hand, solves for landing, arrival at gate, and pushback times, given taxi-in time as an input. It is impossible to solve these problems simultaneously. A heuristic approach wherein the models are solved iteratively until they converge on an optimal solution is under development.

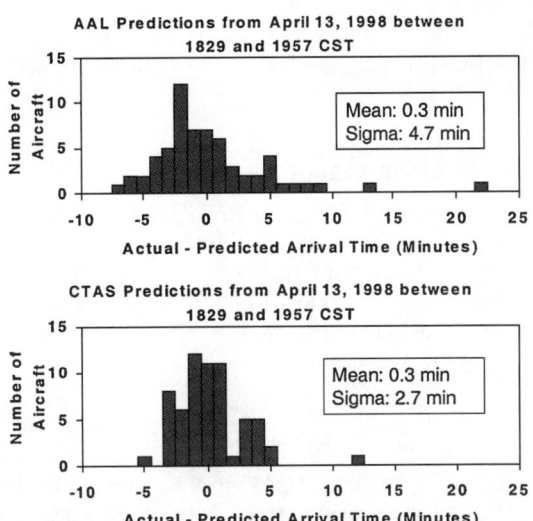

Fig. 36   Prediction error of airline and CTAS arrival time estimates.

For now, we have designed the airline sequencing model (ASM) (an optimization model based on the aircraft turnaround model presented earlier), which considers departure schedule, physical gate resource, and ground crew resource constraints in determining an arrival sequence that minimizes passenger delay. In this model, the taxi-in time is assumed to be constant. All of the constraints considered in the ground operations model discussed in the previous section are included in ASM. A number of additional constraints are included in ASM to restrict the landing and arrival at gate times. First, the model prevents an aircraft from arriving at the gate until it has landed and taxied to the gate. Another important consideration in the model is airline fairness, meaning ASM guarantees that an airline does not improve its operational performance at the expense of another airline's. Fixing the airline's landing times in the model enforces airline fairness; an airline is allowed to shuffle aircraft landing times only within its set of input landing times. Finally, ASM considers gate compatibility and availability. An arriving aircraft can come to the gates only if a gate compatible with its aircraft type is available. Further details of this model and the results from the analysis are included in Ref. 11.

It is also important to note that ASM could eventually be used by an airline to manage its arriving aircraft. Assuming that new procedures incorporated the capability for preferential arrival sequencing, an airline could use ASM to determine its optimal sequence. Because ASM solutions are generated quickly,[12] its solutions can be incorporated into the models currently used by the airlines to help manage gate and ground crew resources.

V.   Conclusions

We have considered modeling operations at busy hub airports. Models of aircraft arrival, gate-turnaround, and departure operations have been proposed that account for the dominant airport dynamics at each stage. These models have been extensively calibrated and validated using historical operations data. It was shown how these models can be concatenated to build an airport congestion prediction capability and how these models can be used to evaluate some improvements in airport operations.

Acknowledgments

This research was supported in part by Honeywell, by a Massachusetts Institute of Technology teaching fellowship, and by NASA under grant NAG 2-1128 and through the National Center of Excellence for Aviation Operations Research (NEXTOR), Stephen Atkins, Technical Monitor. The authors would also like to thank Shawn Engelland and Doug Isaacson from NASA for their help with some of the data sets used for this research.

References

[1] Odoni, A. R., Bowman, J., Delahaye, D., Deyst, J. J., Feron, E., Hansman, R. J., Khan, K., Kuchar, J. K., Pujet, N., and Simpson, R. W. "Existing and Required Modeling Capabilities for Evaluating ATM Systems and Concepts," Final Rept., International Center for Air Transportation, Massachusetts Inst. of Technology, Cambridge, MA, 1997.

[2] Idris, H., Delcaire, B., Anagnostakis, I., Hall, W., Pujet, N., Feron, E., Hansman, R. J., Clarke, J. P., and Odoni, A. R., "Identification of Flow Constraint and Control Points in Departure Operations at Airport Systems," *AIAA Guidance, Navigation, and Control Conference*, AIAA, Reston, VA, Aug. 1998; also *Air Traffic Control Quarterly*, Vol. 7, No. 4, 2000.

[3] Shumsky, R. A., "Dynamic Statistical Models for the Prediction of Aircraft Take-Off Times," Ph.D. Dissertation, Massachusetts Inst. of Technology, Cambridge, MA, 1995.

[4] Pujet, N., Delcaire, B., and Feron, E., "Input-Output Modeling and Control of the Departure Process of Congested Airports," *Air Traffic Control Quarterly*, Vol. 8, No. 1, 2000, pp. 1–32.

[5] Zelenka, R., Beatty, R., and Engelland, S., "Preliminary Results of the Impact of CTAS Information on Airline Operational Control," *AIAA*, 1998.

[6] Ballin, M., and Erzberger, H., "Benefits Analysis of Terminal-Area Air Traffic Automation at the Dallas/Fort Worth International Airport," *Proceedings of the AIAA Guidance, Navigation, and Control Conference*, AIAA, Reston, VA, 1996.

[7] Green, S., and Vivona, R., "Field Evaluation of Descent Advisor Trajectory Prediction Accuracy," *Proceedings of the AIAA Guidance, Navigation, and Control Conference*, AIAA, Reston, VA, 1996.

[8] Delcaire, B., and Feron, E., "Dealing with Airport Congestion: Development of Tactical Tools for the Departure Flows from a Large Airport," Tech. Rept., International Center for Air Transportation, Massachusetts Inst. of Technology, Cambridge, MA, 1998.

[9] Idris, H., "Diagnosis and Analysis of Departure Operations at a Major Airport," Ph.D. Dissertation, Massachusetts Inst. of Technology, Cambridge, MA, 2000.

[10] Pujet, N., "Modeling and Control of the Departure Process of Congested Airports," Ph.D. Dissertation, Massachusetts Inst. of Technology, Cambridge, MA, 1999.

[11] Andersson, K., "Potential Benefits of Information Sharing During the Arrival Process at Hub Airports," S.M. Thesis, Massachusetts Inst. of Technology, Cambridge, MA, 2000.

[12] Pujet, N., and Feron, E., "Input-output Modeling and Control of the Departure Process of Busy Airports," *AIAA Conference on Guidance, Navigation, and Control*, AIAA, Reston, VA, Aug. 1999.

[13]Bertsekas, D. P., and Gallager, R., *Data Networks*, Prentice-Hall, Englewood Cliffs, NJ, 1987.

[14]"Control of Air Pollution from Aircraft and Aircraft Engines; Emissions Standards and Test Procedures; Final and Proposed Rule," Environmental Protection Agency, Direct Final Rule, 40 CFR Part 87 [AMS-FRL-5821-3], Federal Register 62(89), 1997, pp. 25355–25367.

Chapter 21

Conceptual Design of a Departure Planner Decision Aid

Ioannis Anagnostakis,* Husni R. Idris,* John-Paul Clarke,† Eric Feron,†
R. John Hansman,‡ Amedeo R. Odoni,§ and William D. Hall¶
Massachusetts Institute of Technology, Cambridge, Massachusetts

I. Introduction

AVIATION researchers in the United States [NASA, Massachusetts Institute of Technology (MIT)] and Europe (e.g., DLR, German Aerospace Research Center) have recognized the need for enhanced airport surface traffic control systems. The departure planner (DP) project at MIT and the taxi and ramp management and control (TARMAC) system developed by DLR are both research efforts in this direction with the objectives of optimizing departing traffic, closing unnecessary gaps between arrivals and departures, and establishing an integrated ground movement planning system.[1–3**]

The development of (possibly automated) decision support tools for air traffic controllers calls for a thorough understanding of links and interactions in air traffic management (ATM) operations and requires constant evaluation and assessment. Furthermore, the design of a high-level architecture for airport departure management should be based on a thorough analysis of the airport system and understanding of the needs and constraints in current airport operational procedures. To this end, a significant set of field observations has been done at Boston Logan International Airport (BOS),[4–6††] the results of which are summarized in Section II.

The observations and analyses discussed in previous work introduced significant issues that should be accounted for in the design phase of the DP.[6] Section III presents the suggested general architecture, which is expected to be sufficient for

---

Copyright © 2001 by the Massachusetts Institute of Technology. Published by the American Institute of Aeronautics and Astronautics, Inc., with permission.
*Research Assistant, Dept. of Aeronautics and Astronautics.
†Faculty Member, Dept. of Aeronautics and Astronautics. Senior Member AIAA.
‡Faculty Member, Dept. of Aeronautics and Astronautics. Associate Fellow AIAA.
§Faculty Member, Dept. of Aeronautics and Astronautics.
¶Research Scientist, Charles Stark Draper Laboratory.
**Data available online at http://snipe.mit.edu/dp/documents.html [cited 17 April 1998].
††Data available online at http://atm-seminar-98.eurocontrol.fr/finalpapers/track2/finalt2.htm [cited 1999].

all aspects of departure management. The DP is designed to include several components. Each of these components addresses a certain aspect of the departure process and its interaction with the arrival flow. The architecture proposed here describes the control function of each component of the planning system, its inputs and outputs, as well as identifies the point along the departure flow where this function could potentially be introduced. In particular Section III.B highlights the virtual queue manager (VQM) as the architectural component that is actually assigned the task of runway operations planning and coordination between arrivals and departures that use the same runway resources. Implementation issues are also addressed.

II. Departure Process—Results from Field Observations

Analysis of departure and arrival operations at BOS and other major U.S. airports, such as Chicago O'Hare (ORD), Atlanta Hartsfield (ATL), and Dallas/Fort Worth (DFW), revealed significant operational delays and environmental impacts associated with the departure process.[4] Analysis also revealed that, although there are many similarities between the departure and arrival processes, there are also significant differences that affect the way in which improvements may be effected. For example, beyond a certain entry fix point in the terminal airspace, the arrival stream is quite determined, and there is not much opportunity for sequence adjustments. On the ground, however, there is little observability and high volatility associated with departure operations,[5,6] and controllers are presented with more opportunity to affect the final runway operations sequence. Therefore, because controllers prefer, for safety reasons, to keep aircraft on the ground rather than in the air for flow management purposes, it will be potentially beneficial for ATM operations to design a decision-aiding system to assist controllers in handling and optimizing departure operations.

Different components of the airport were identified as flow constraints, which introduce delays and inefficiencies and contribute to the low prediction capability associated with departures.[4-6] The flow constraints identified were associated with the main airport system elements: 1) the gates complex, 2) the ramp area, 3) the taxiway system, and 4) the runway system. The flow constraints manifest physically through the aircraft queues that form at these elements. In that sense, an airport system can be modeled as a complex interactive queueing system in which departures and arrivals are highly coupled. Figures 1 and 2 illustrate all of the different types of aircraft queues that form on the BOS airport surface in the configuration 22/27.

As shown in Fig. 1, in this configuration runway 27 is usually dedicated to arrivals, runway 22R is used only for departures, and runway 22L is used primarily for arrivals. Often, pilots who specifically request a longer runway for takeoff, use the latter for departure, in which case they line up and wait on the south taxiway segment between 22L and 22R. When a large number of departures are expected, the airport switches to accelerated departure procedures (ADP), in which case runway 22L is used only for departures and all arrivals are routed to runway 27.

The flow of arriving and departing aircraft through the airport system and the various queues forming on the airport surface in configuration 22/27 can be abstracted as in Fig. 2. The physical elements of the airport system (gates, taxiways, runways) are depicted in the middle part of the figure, and their interactions with the airport queues are shown as dashed arrow lines. Each bar between different

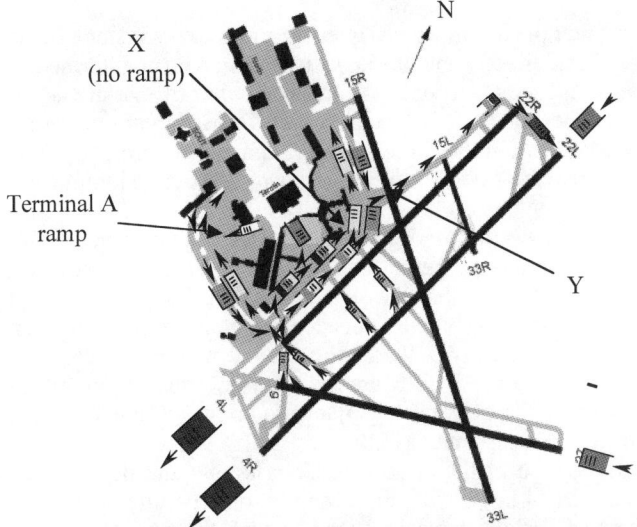

Fig. 1 Taxiway and runway queues at BOS under configuration 22/27.

queues represents a transition from one queue to another. Solid arrows represent the aircraft flow, and dashed arrows associated with a specific airport resource represent use of that resource for the queue transitions. Following the aircraft flow in Fig. 2, an arriving aircraft queues on final approach and, after landing on runways 27 or 22L, it joins a runway crossing queue waiting to cross runway 22R. After crossing, it joins other aircraft in taxiing queues, which include arriving and/or departing aircraft. Upon arrival at its assigned gate, it may have to wait for the gate to be released from the previous aircraft. When the gate becomes available, the aircraft joins a pushback queue according to its scheduled departure time. Figure 2 depicts two different types of gates that were observed. In one case (far right side of Fig. 2), aircraft that push back from these gates enter the ramp area (e.g., BOS terminal A ramp in Fig. 1) and wait in a ramp queue for air traffic control (ATC)

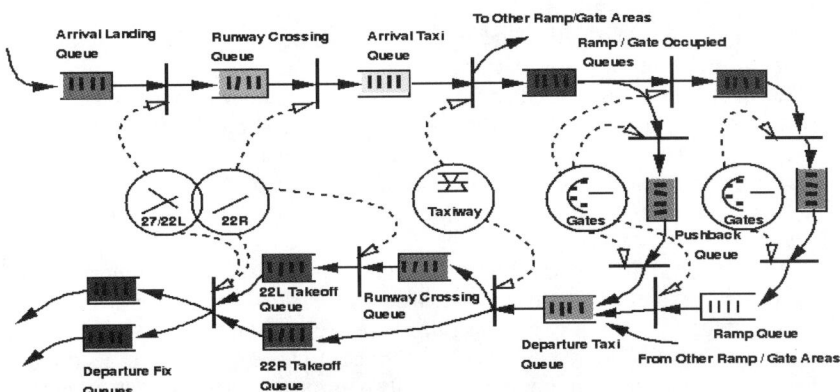

Fig. 2 Queuing model for BOS under configuration 22/27.

clearance to enter the departure taxi queue in the taxiway system. In the other case, aircraft push back directly onto the taxiway system with no intermediate ramp area (e.g., point X in Fig. 1). At point Y, departing aircraft that are assigned runway 22R for takeoff enter the 22R takeoff queue. Departing aircraft assigned to take off on runway 22L join the same taxiway segment but are considered to enter a runway crossing queue in order to cross runway 22R before joining the 22L takeoff queue.

Figure 1 also shows the possible location of these queues on the surface of BOS in the case of configuration 22/27. The two arrival queues on runways 27 and 22L are easily identified, as well as the departure queue that is formed on the taxiway segment adjacent to runway 22R. This departure queue includes aircraft that line up to take off from runway 22R and aircraft that will cross 22R to take off from 22L. Operations on runway 22R are impeded not only by aircraft departing on 22L but also by arriving aircraft that queue in taxiway segments between the two parallel runways to cross runway 22R.

This identification of the airport flow constraints and the associated queuing network was critical in studying the dynamics of departure operations. It enabled the definition of various control points in which the departure operations could be affected by control actions, and it also helped in determining the DP control options.

The observed operations described are tactically under the control of the air traffic controllers in the control tower. The air traffic controller has to clear each aircraft to use an airport resource such as a ramp, a taxiway, or a runway, by delivering a clearance instruction. By observing the aircraft queuing process and the control process, it was possible to identify control functions that occur at specific control points. The transitions in Fig. 2 are examples of such control actions and control points. Most aircraft operations are under the control of the tower, and, therefore, control actions and points are associated with the controllers' clearances. However, some aircraft operations, such as engine-start, are not currently under the tower's control at BOS.

The following control functions were identified based on the observations at BOS airport:

1) Pushback clearance (for jets) or taxi clearance (for propeller-driven aircraft).

2) Clearance to enter the taxiway system of the airport from the ramp area gate where the aircraft is waiting.

3) Runway and taxi-path allocation, i.e., the process of routing aircraft to a specific runway through a predetermined taxiway path.

4) Sequencing of aircraft takeoff, i.e., merging of aircraft into the same takeoff queue or mixing between aircraft from multiple queues (e.g., one jet aircraft and one turboprop aircraft queue) at the same takeoff runway.

5) Takeoff release of each aircraft. If a runway is used by departures, landings, and runway crossings, takeoff release also involves mixing of operations on that runway.

It was observed that a control function could be exercised at different times and locations. For example, aircraft sequencing can be performed at the gate (pushback control), at the taxiway entry points as aircraft are released into the taxiway system, and up to the physical point beyond which the aircraft have to commit to a particular takeoff queue. Once the aircraft are physically present at the runway end, the takeoff sequence is hard to modify. Therefore, notionally, a control point is defined as the last opportunity that the controllers have to apply a particular control function to the departure queues. A control point can be a physical point on the airport surface, or it can be a point in time during the departure process, when the aircraft transitions from one state to another. For example, a control point exists at the gates when aircraft are cleared to push back into the ramp area. A possible control point is also the instant when aircraft, while taxiing, are handed off to a specific controller who handles a particular set of runways. At that point in time these aircraft are committed to enter a runway queue, with much less room for further adjustments than in the taxiing phase. The main control points associated with the observed operations outlined are 1) the gate, 2) the point of entry from the gate or ramp into the airport taxiway system, 3) the point of commitment to a specific queue (temporal or spatial), and 4) the point of entry to an active runway (exit from a takeoff queue).

The control points and the control functions are generic, but there are airport specific and runway configuration specific differences. ATL and BOS were also compared.[4] BOS has one runway system and minimal ramp area, whereas ATL has two runway systems and a well-defined ramp area. Therefore, BOS is usually operating with one departure queue. The structure of this queue can be primarily determined at the gate (pushback clearance control function), which in most cases is the point of entry into the taxiway system, because the intermediate ramp area in BOS is almost nonexistent. On the other hand, ATL has at least two departure queues in most cases, as well as a larger controlled ramp area than BOS does. This means that the structure of the departure queues can be affected at the gate control point, but also to a larger extent at the taxiway entry points and through the mixing of aircraft from different queues.

One of the main objectives of the DP is to mitigate the existing inefficiencies and reduce the observed delays resulting from the flow constraints that have been described. However, an airport system is a multiobjective environment with several stakeholders, such as airport users (airlines, passengers) and ATM service providers. Each of them attaches different weights to competing system objectives, which makes it hard to define a single objective function for the DP system. Some system objectives that have been identified are 1) comply with safety and separation requirements, 2) maximize system throughput, 3) minimize taxi time (aircraft engine emissions), 4) consider noise regulations and constraints, 5) balance the load on all runways, 6) maintain the controllers' workload at acceptable levels, and 7) provide fair treatment for all airport users. In an effort to satisfy these objectives, the DP is designed to include several components, based on the observed airport operations. These components are examined in detail in the following sections.

III. Overview of the Proposed Departure Planner Architecture and Operational Context

Field observations, as a method of airport system identification, helped in identifying the control points and functions mentioned previously. The results from this analysis were combined with documented airport and ATC operations to generate the system architecture presented here.

Figure 3 illustrates the two principal parts of the architecture: the strategic planner and the tactical planner. The strategic planner is essentially a configuration

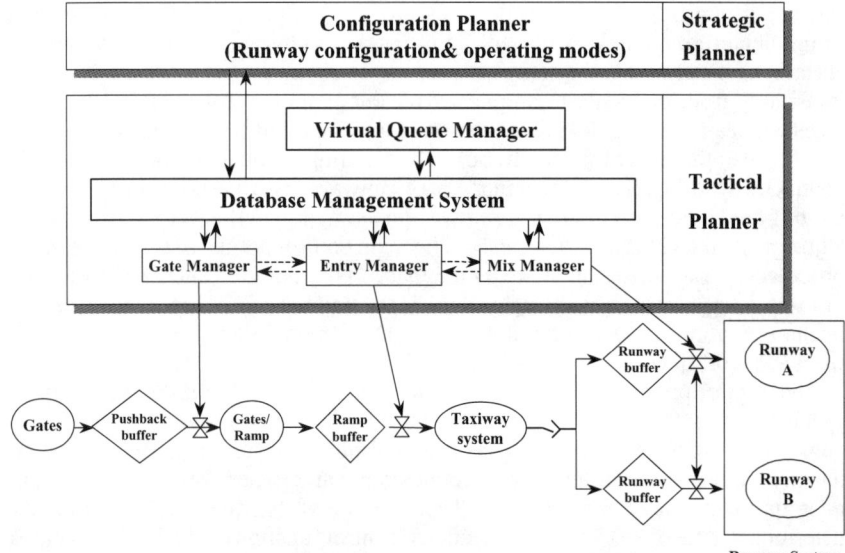

Fig. 3 Overview of the departure planner system hierarchy.

planner that would typically have an approximately 3–4-h time horizon and would perform configuration management tasks. The tactical planner has an approximately 15–30-min time horizon, performs tactical planning of runway operations under a specific runway configuration, and exercises appropriate control to implement the generated plans.

The most critical tactical component introduced in the system is the VQM. The remaining three tactical system components (starting from the gates and following the departure flow to the runway takeoff queues) are 1) the gate manager, which is introduced in order to support the controllers in managing the pushback schedule given the unpredictability (uncertainty) inherent in airline gate operations and schedules; 2) the taxiway entry manager, which modulates the release of aircraft for entry into the taxiway system; and 3) the mix manager, which is introduced in order to manage the arrival/departure mix onto active runways. Detailed descriptions of the various DP components are provided in the following sections.

In a generic framework that can be applied to any airport, each of the four tactical components is designed to exercise control and address inefficiencies at specific control points along the departure process. Each strategic and tactical component can be linked (mapped) to a group of objectives drawn from the family of the general functional objectives of the decision-aiding system that were presented earlier. At the same time, all components are envisioned to communicate and exchange data with each other directly or via a common database management system (Fig. 3). The latter is designed to ensure that all components have access to the same consistent information. It should interact with several specialized databases containing necessary airport specific data, such as 1) airport topology; 2) ATC procedures, regulations, and restrictions, e.g., airport-specific arrival and departure routes, such as standard terminal arrival routes (STAR) and standard instrument departures (SID); 3) aircraft performance data, such as EUROCONTROL's base of aircraft data (BADA), as well as dynamically generated data, such as flight plans and aircraft identification information; 4) ATC constraints that are dynamically introduced, e.g., flight priorities, ground delay programs.

Generalizing, the main tasks of all components should be to:

1) Implement the designed virtual queue through control actions, taking into account specific information and constraints for the area of operations that each component is associated with, i.e., gates, ramp, taxiways, runways, and specific settings input in the system by the controller responsible for that area.

2) Distribute information to and between airport operators, such as the controllers, and airport users, such as the airlines (external interface). An example of such external communication is the distribution of information to controllers that cannot be directly accessible by them, but is contained in other, in some cases remote, technical systems, such as central airline operations' management systems or local airline station control networks. Information should also be distributed between system components (internal interface). Examples of such internal communication is the distribution of information about all performed control actions of a specific component, which are relevant to other components of the architecture, or the communication of new events that make certain system solutions infeasible.

In any busy airport, even with a fairly simple runway geometry, managing ground operations and planning the allocation of runway time to the various types of runway operations can be a very challenging task. Any effort to design a decision-aiding tool deployed to assist air traffic controllers with this task must take into consideration several operational requirements, which arise primarily due to the presence of human operators in the process. For example, in managing airport ground operations, there are two main tasks to be performed, planning and control, and both of them are inevitably performed in a distributed fashion because of the many different parties involved in most planning and control functions. Therefore, a successful design of an automated decision-aiding system that will be under airport-wide use must take into account integration issues between the varying objectives, constraints, information requirements, and control inputs from all stakeholders involved in airport ground operations, such as the airlines, airport authorities, airport service operators, passengers, and the Federal Aviation Administration (FAA) personnel in the air traffic control tower and the Terminal Radar Approach Control (TRACON) room. As an example, consider the information needs of the gate personnel and station managers of a specific airline at an airport, as opposed to the information needs of the air traffic controllers in the tower cab. The latter have to deal with all incoming and outgoing airport ground traffic, not only the aircraft of a particular airline, which means that their information needs are apparently larger than those of the specific airline operators. In addition, the controllers' objectives are naturally more system oriented than the understandably profit-oriented objectives of an airline operations control center.

It is uncertain how feasible it is for an automated system to take into account the objectives of all involved parties. Therefore, it is likely that the solutions suggested by such a system may not always be optimal. In addition, runway operations plans generated by the decision-aiding tool may even be unimplementable due to its inability to incorporate very dynamic last minute changes of the system state (e.g., traffic situation) that it was not aware of when those plans were calculated.

Real-world applicability of the generated plans requires that the controllers have ample planning and control flexibility to modify the plans for factors and events not visible to the decision-aiding tool and that the solutions generated are presented to controllers as operational guidelines and not as inviolable laws that they have to abide by.

On a final note, careful consideration must be given to usability and the interaction between the controller and the decision-aiding tool in the context of the operational environment and standard procedures within the ATC tower. The underlying structure of the decision aid and optimization process must be clear to the controller, and the results must be consistent with the controllers' heuristic estimates. In addition, the interface limits "head-down" time for controllers. It should also be noted that the tool could be used as a communication device between controllers. Controller interface devices, which support these functions in the tower cab environment, are under development.

A. Configuration Planner

The configuration planner attempts to match the airport's operational capacity as closely as possible to the scheduled demand, always taking into consideration limitations imposed by the forecasted terminal weather and by environmental constraints (aircraft engine noise and emissions), such as noise restrictions.

1. Functionality

The main task performed by the configuration planner is the development of the runway configuration plan for the airport so that all arrivals and departures expected to utilize the airport runway resources can be handled. It must be designed to take into account the stochasticity associated with weather. Accurate terminal weather and wind forecasts in conjunction with the pertinent noise abatement rules are used to define the set of feasible configurations for the airport. Based on these, the configuration planner determines the number of hourly operations that the airport can handle, and the expected arrival and departure demand over successive intervals during the planning horizon is then matched to each of these configurations (level 1 in Fig. 4), in order to design the best configuration strategy throughout the day.

To accommodate short-term demand fluctuations, the configuration planner should be able to suggest discrete operating modes within the time horizon of each of the planned configurations (level 2 in Fig. 4), especially because the response time of the airport to a change from one operating mode to another within the same runway configuration is quicker than the airport response time in the case of a full change of runway configuration.

2. Supporting Analysis

Figure 5 presents examples of four capacity envelopes that were generated based on airport throughput data over the course of 15 days. The data were collected from the air traffic control tower records as well as enhanced traffic management system (ETMS) records from the FAA's Consolidated Operations and Delay Analysis System (CODAS) database.[7] Evidently, different runway configurations at BOS yield different capacities. In their effort to match the levels of demand expected, the controllers try to use the configurations with the highest arrival and departure capacity. However, the final selection is dependent on the weather and wind

DESIGN OF A DEP. PLANNER DECISION AID

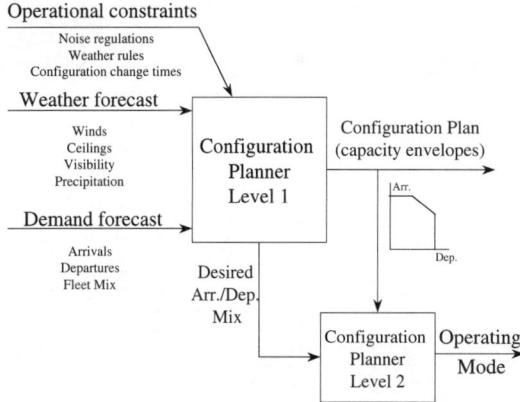

Fig. 4 The configuration planner.

conditions at the airport. In addition, there are several noise abatement rules that limit the use of certain runways at certain times of the day. These weather and noise constraints are taken into account by the controllers in their decision-making process for selecting the airport runway configuration. In fact, at BOS, there are certain runway configurations that are recommended at night hours.

Matching different possible configurations to the schedule takes into account the time required for transitioning between configurations, which can take values up to 20 min for a busy period at major airports like BOS. When the arrival flow is very high, it takes longer to implement a configuration change because it is harder

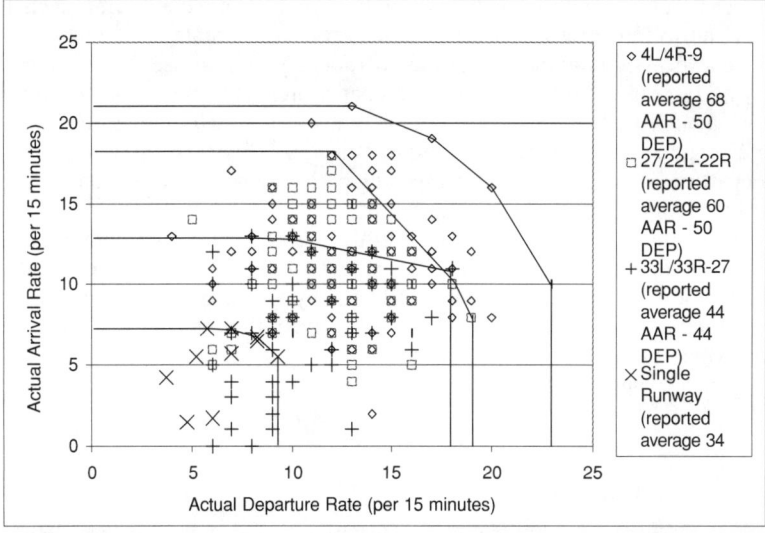

Fig. 5 Runway configuration capacity envelopes (source: ETMS/tower records, 0700–0900 hrs, 1600–2000 hrs, 1–15 July 1998, except Saturdays, BOS).

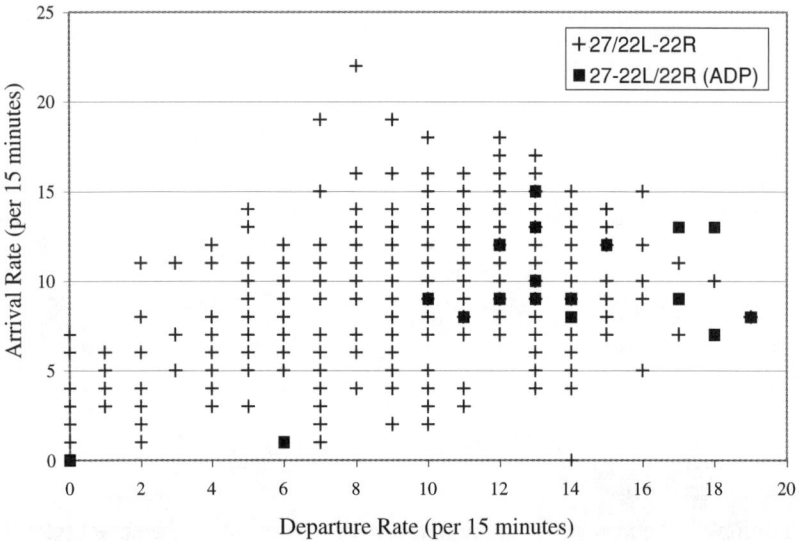

Fig. 6 Accelerated departure procedure (ADP) (source: ETMS, 1-15, July 1998, BOS).

to interrupt the arrival stream on final approach. For example, at BOS, switching to a high-capacity configuration is usually attempted before periods of expected high traffic.

Short-term fluctuations in the arrival/departure mix drive the airport in "departure push" or "arrival pull" mode. In these cases, the air traffic controllers perform short-term configuration changes by adjusting the operations that are assigned to utilize each runway within the current configuration. These configuration changes correspond to transitions between different operating points on the airport's capacity curve.[6] For example, in normal operations within the 22/27 configuration, runway 22L is used both for arrivals and departures. However, when BOS is in a departure push mode, runway 22L is sometimes used only for departures (together with 22R) and all arrivals are assigned to runway 27. Figure 6 shows the effect on the departure/arrival mix by superimposing periods when ADP was used over the capacity envelopes presented in Fig. 5.[7]

In matching the scheduled demand to the set of possible configurations, the configuration planner should take into account the uncertainty inherent in departure operations. Departure demand is affected by airline decisions on delays and cancellations, which are not always known sufficiently in advance. Collaborative decision making (CDM) is a step toward addressing this problem. It has been shown that advance cancellation notices have improved noticeably after the introduction of CDM.[8]

B. Virtual Queue Manager

The VQM proactively manages the airport's virtual queue so that DP objectives are met and therefore airport resources (runways, taxiways, and gates) are efficiently utilized.

1. Virtual Queue Definition

A virtual queue can be defined as a notional waiting line of departing aircraft arranged, at any instant of time, according to the order in which they are expected to take off. It consists of:

1) A physical part, which involves aircraft that are or will shortly be physically present at a certain location on the airport surface, with no further chance for resequencing; therefore, these aircraft have a fixed ("frozen") position in the virtual queue.

2) A virtual part, which involves aircraft that are scheduled to occupy a particular position in the sequence of aircraft that will take off, but are not physically present in the takeoff queue yet. Position assignments in this virtual part of the virtual queue are very much subject to revision.

In other words, the virtual queue can be seen as an extension of the notion of a physical queue that depicts the final takeoff sequence of all scheduled departures as the DP has planned it up to the current point in time.

If two or more departure runways are currently in use, or are expected to be shortly, then multiple virtual queues (one for each departure runway) will be in use. As an alternative, in such cases there might be a single virtual queue with each aircraft in the queue being "tagged" to indicate which departure runway it will use.

2. Functionality

In Fig. 3, the VQM is hypothesized to reside in the system hierarchy at one level above the tactical DP elements. It interacts separately with the strategic configuration planner, and also, acting as a central processing function, it coordinates the three tactical DP components. It incorporates all the requests from various physical queues in the system and relays back to them information about generated runway operation plans, the virtual queue, and the required control actions to implement it. The major challenge is to design the optimum size of the virtual queue (minimum buffer size) in such a way that the aircraft queues in the system (especially the runway takeoff queue) are not consistently "starved" or saturated.

Because the runway was observed to be the main flow constraint, a possible design of the virtual queue is generated assuming that its physical part resides at the runway threshold. In this case, the two parts of the virtual queue would be:

1) The physical part, including flights whose position in the queue may be frozen a few (10 or 15) minutes before their assigned takeoff time, and

2) The virtual part, in which the scheduled departure time and the sequencing of some aircraft may be subject to change due to the fact that there is still considerable time to go, e.g., more than 15 min until the actual departure event.

The mode of interaction between the VQM and the other DP tools is still a research issue. One possibility is a "master-slave" relationship, in which the optimization logic is entirely included in the VQM. Each of the DP components simply relays information and communicates its specific requests to the VQM with the hope that the system status will allow its requests to be satisfied. Another possible design philosophy is for each of the DP components to carry its own optimization logic and perform a local optimization dealing with a specific subproblem of the overall problem. Subsequently, the VQM takes all the individual optimization results from the DP components and attempts to combine all the "local" solutions into a "global" one. This process may involve iterations and re-optimization until a feasible solution is achieved.

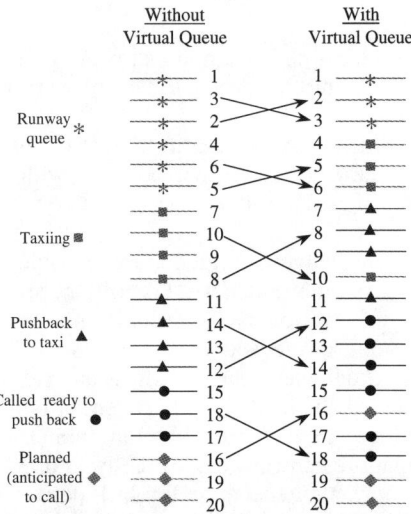

Fig. 7 Managing the departure sequence of the same 20 scheduled flight, with and without the implementation of a virtual queue (right-hand side indicates the current optimal sequence).

Figure 7 provides a hypothetical means to visualize the virtual queue and understand its potential benefits. Each side of the figure represents a snapshot of the takeoff sequence as it is currently projected in the future. The left-hand side represents the observed behavior in which aircraft are transitioned from one state to the next mainly in a first-come-first-served (FCFS) order and where the queue buffer sizes are not controlled, resulting in unnecessarily overloaded takeoff queues and taxiway congestion. The right-hand side in Fig. 7 represents the behavior with the implementation of the virtual queue, which provides a tool for effectively controlling the number of aircraft in each state at each point in time and regulating the timing of aircraft transitions from one state to the next. Aircraft move from the gate to the ramp onto the taxiway system and into one of the takeoff queues, following the timing and sequence schedule commanded by the VQM. This optimal (or near-optimal) sequence is determined based on the system-wide objectives and constraints that were discussed earlier.

Each line corresponds to a departing flight scheduled to take off within the time span that the virtual queue covers. In each case, the state of the departing aircraft is represented by a specific symbol as follows: 1) physically present at the runway threshold, 2) taxiing, 3) pushed back from the gate but not yet released into the taxiway system (ramp area), 4) waiting for pushback clearance from the tower, after having called "ready for pushback," and 5) expected to call "ready for pushback" within a predetermined time horizon.

Because of high workload, it is very hard in most cases for air traffic controllers to mentally determine the appropriate timing and sequence of departures, while at the same time keeping in mind all constraints and satisfying all system objectives. The existence of the virtual queue may assist controllers to determine possible "aircraft takeoff swaps" within the same state or even between different aircraft states (arrows in Fig. 7) in order to optimize departure operations. The virtual queue may point out some of the optimal sequences that the controllers may not

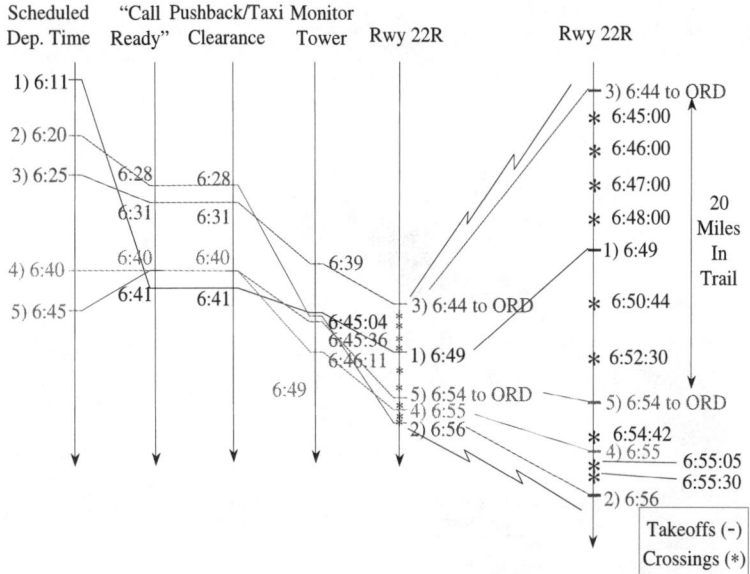

Fig. 8   Miles-in-trail (MIT) restrictions.

realize under heavy workload (see examples in Figs. 8 and 9). In addition, the virtual queue may be used to convert taxi delays to gate delays, which are less costly both for the airlines and the environment. In addition, operational flexibility for the airlines can be increased without sacrificing fairness.

The VQM can be designed to exercise two levels of control on the allocation of runway time to different operations: 1) simple control of the size and/or sequence of takeoff queues, and 2) time-based control. In the second case it is expected that the solution quality will be enhanced, but at the same time the computational complexity of the problem may be increased.

3. Supporting Analysis

The following analytical examples are extracted from observations of real-world operations at BOS. They are used to describe actual inefficiencies in runway utilization due to ATM operational constraints, and they also demonstrate the potential benefits that the airport system can have from efficient utilization and sequencing of the runway resources. Figures 8 and 9 present actual operations at BOS on 2 February 1999, under configuration 22/27. Figure 8 displays the progress of five departing flights through time. From left to right, the following time events are given: 1) scheduled departure time, 2) time the pilot called ready to leave the gate, 3) time the clearance to pushback (for jets) or taxi (for props) was granted by the tower, and 4) the monitor tower time, which (in this configuration) is the time that each departing aircraft is cleared to join the takeoff queue. This is the point where controllers usually exercise sequencing control among departures.

The two remaining timelines in each of Figs. 8 and 9 describe the takeoff and crossing operations on runway 22R (presented once in scale and once magnified).

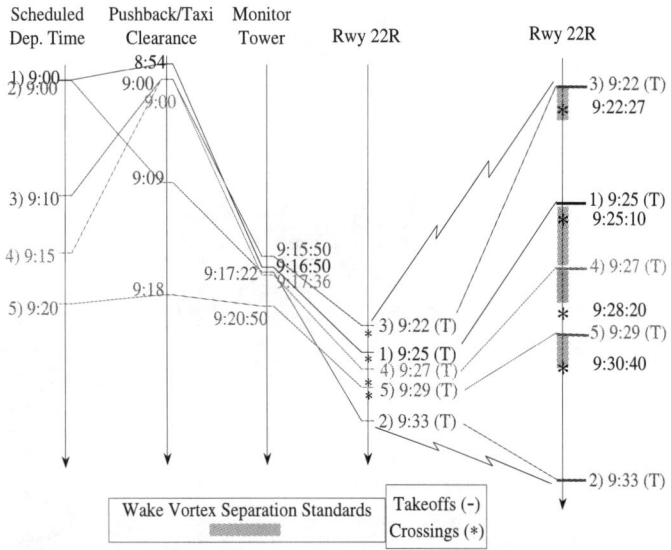

Fig. 9   Takeoff sequence adjustments.

Each departure and crossing event has a record of the time the aircraft was cleared to take off or cross runway 22R.

In Fig. 8 Chicago-bound flights (ORD) were under a 20 miles-in-trail (MIT) restriction. Hypothetically, departure 1, which called ready for pushback at 0641 hrs after departure 5 (0640 hrs) was sequenced to take off before the latter, so that ORD-bound departure 5 maintains a 20 MIT separation from prior ORD-bound departure 3. Departures 2 and 4, which called ready for pushback before or together with departure 5, ended up behind the latter in the takeoff sequence. Flights 2 and 4 could have been sequenced to take off before flight 5, which was already delayed due to the MIT restriction, resulting in more efficient runway throughput. Note that in a 5-min period between flights 1 and 5, the runway was used only for two aircraft crossings (asterisk symbols on the runway 22R timeline). The data collected showed that there were:

1) A runway idle time gap of 104 s between departure 1 and the first crossing event that occurred immediately after it (0649 to 0650:44 hrs).

2) A gap of 106 s (0650:44 to 0652:30 hrs) between the two crossings.

3) A gap of 90 s (0652:30 to 0654 hrs) between the second crossing and the next, departure 5. The utilization of runway 22R suggests that there is room for assisting air traffic controllers in designing takeoff sequences under several operational constraints, with an objective to avoid such unnecessarily large utilization gaps and achieve higher takeoff throughput.

Similarly, in Fig. 9 we can observe large time gaps between successive takeoffs with only one crossing operated within each gap. For simplicity, the call ready timeline has been omitted. The five departures examined here call ready and pushback in the order 1, 3 and 4, 2, and 5. Departure 5 called ready for pushback (0918 hrs) after departure 2 (0909 hrs). However, it was sequenced to take off before the latter, probably due to the fact that in this configuration the estimated taxi out time for flight 5 is lower than the one for flight 2, based on the distance from each flight's originating terminal to the runway 22R takeoff point. From this example, it is obvious that controllers actually implement certain sequencing rules, based primarily on their vast operational experience. Nevertheless, there seems to be room for improved sequences to avoid excessive runway idle time. Focusing on the takeoff/crossing timeline in Fig. 9, we can see that in each of the four time intervals between the five departures, there was only one crossing operated. Each crossing occurred very soon after the preceding takeoff, and for the remaining time until the next takeoff, the runway remained idle. Based on the wake vortex separation requirements for the aircraft types present, only the gap between departures 1 and 4 needed to be as long as it actually was.

Note that, even if a takeoff sequence is very well designed based on the latest information available to the controllers, there are always unexpected events that can cause serious deviations from the planned takeoff order. For example, departure 2 may have been carefully sequenced after departure 5, as mentioned. However, it was cleared to take off 140 s (0930:40 to 0933:00 hrs) after the previous crossing aircraft cleared the runway and 4 min (0929 to 0933 hrs) after the previous takeoff occurred on the same runway. This could have possibly happened because the takeoff weight and balance calculations were not available for departure 2, even though it was first in the takeoff queue. In other cases, scheduled or nonscheduled priority flights (e.g., lifeguard flights) or aircraft in the departure queue unable to take off can perturb the take off sequence.

C. Gate Manager

The gate manager is the DP component that assists the controllers in determining the pushback schedule, subject to the uncertainty associated with airline gate operations. Initial runway assignments for departing flights may also be an important part of the gate manager's task.

1. Functionality

Being the first DP component that can have an effect along the departure flow, the gate manager incorporates and processes data generated from the rest of the DP system components, as depicted in the free body diagram shown in Fig. 10. Note that arrows pointing inward toward the gate manager carry information (flight status data, system constraints) coming from other elements, which are adjacent to the gate manager in the system architecture, or from other National Airspace System (NAS) databases that exchange data with the departure planner [e.g., surface movement advisor (SMA), Center-TRACON Automation System (CTAS)]. On the other hand, arrows pointing outward from the DP component convey to the rest of the system commands and requests generated by the gate manager function. A similar convention is used to read the free body diagrams presented for the remaining system components that are described in the following sections.

Initially, based on traffic information from the gates, the ramp area, and the taxiway system (Fig. 10, top left and right data blocks), the gate manager assesses the current airport situation and suggests a feasible pushback schedule within a predetermined planning horizon. At this point, data necessary for an accurate estimate of the current and projected demand (possibly obtained from the SMA database) are airline specific data, such as hangar status, current towing operations,

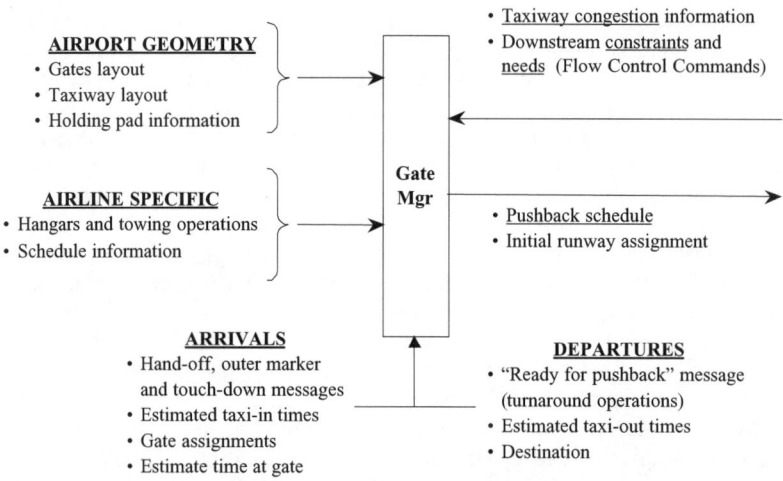

Fig. 10 The gate manager.

and flight-specific data local to each gate, such as destination, turnaround readiness messages, and taxi-out time estimates (Fig. 10, middle left and bottom data blocks). In addition, downstream constraints such as gate holds and ground delay programs (Fig. 10, top right data block), which usually involve many cancellations, delays, and gate rescheduling, must be communicated to the gate manager as soon as the related information is available from the FAA central flow control (system command center).

Many of the airline operations, especially the ones performed before aircraft are actually ready to push back from their gates, are not observable by the controllers. For example, often aircraft will call ready for pushback before their gate operations are actually complete, anticipating a delay between the call for pushback and the actual time that a clearance is granted. Sometimes, delays and cancellations due to inclement weather or mechanical problems result in aircraft being held at their gates and cause unexpected gate blockages. With DP, such situations can be observed by controllers through the exchange of pertinent information between the gate manager and the database management system.

2. Supporting Analysis

One control strategy is to limit the number of aircraft released in the system by holding some of them at their gates. This method is fairly easy to implement; however, it may raise gate capacity issues because it transfers runway queue delays to gate delays.[6] Controlling the gate release times with the help of the gate manager provides the controllers with a unique opportunity to implement the control strategy previously discussed to control the size of takeoff queues and the sequencing of aircraft within the queues.

In view of this control option, airline service quality performance (ASQP) data were studied,[9] and a simple dynamic queuing model for departure operations was developed and used to analyze airport surface operations. Based on the analysis results, simple departure control strategies were suggested for the purpose of

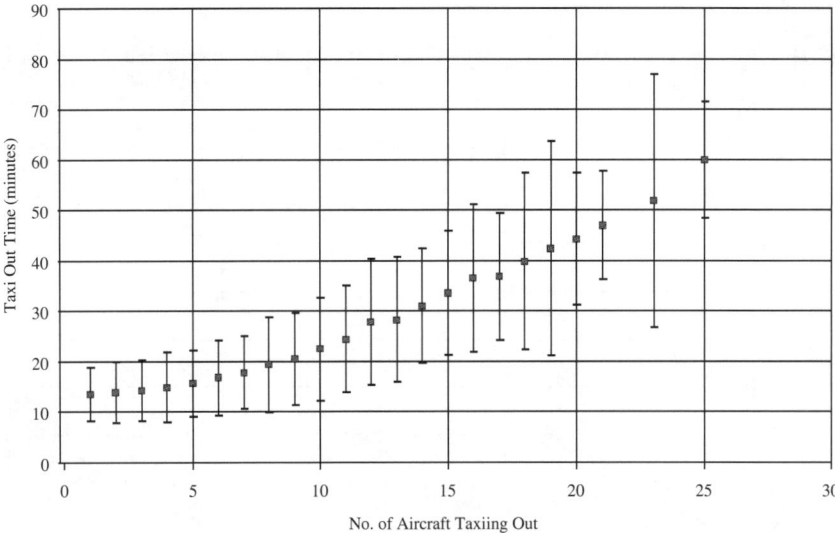

Fig. 11 Taxi out time as a function of airport congestion (BOS: 1997/09–1998/08).

alleviating surface congestion. An important parameter in the problem of departure control appears to be the number N of aircraft that are present in the system[9] while an aircraft is taxiing out from its gate to the assigned departure runway. It was realized that, as the number N increases, the expected value and variance of the taxi-out time also increases, as shown in Fig. 11 for one year of data from BOS.

In addition, for each airport and under any configuration, there is a runway saturation point beyond which there is no significant gain in takeoff rate even if controllers keep releasing more aircraft in the system, as shown in Fig. 12. It is observed that, as controllers release more aircraft in the system, fewer than the released number of aircraft actually take off, and the remaining form congestion queues on the airport surface.

Based on downstream requests, the schedule can be adjusted through gate release control to feed the takeoff buffers with the requested number of aircraft. The system-wide objective of maximizing airport throughput is addressed, and preallocated departure slots can be met. Engine-running times are also minimized, and compliance with environmental emissions regulations is achieved, while gate-blocking delays are significantly reduced. Furthermore, the airlines benefit from fuel savings and late passenger/baggage accommodation by remaining at the gate until they can actually be accepted in the taxiway system, as opposed to pushing back on time and being delayed in holding pad areas or in taxiway queues.

D. Taxiway Entry Manager

The taxiway entry manager determines the sequence and timing of release from the ramp into the taxiway system for aircraft that have pushed back from their gates and entered the ramp buffer in Fig. 3. It considers system objectives related

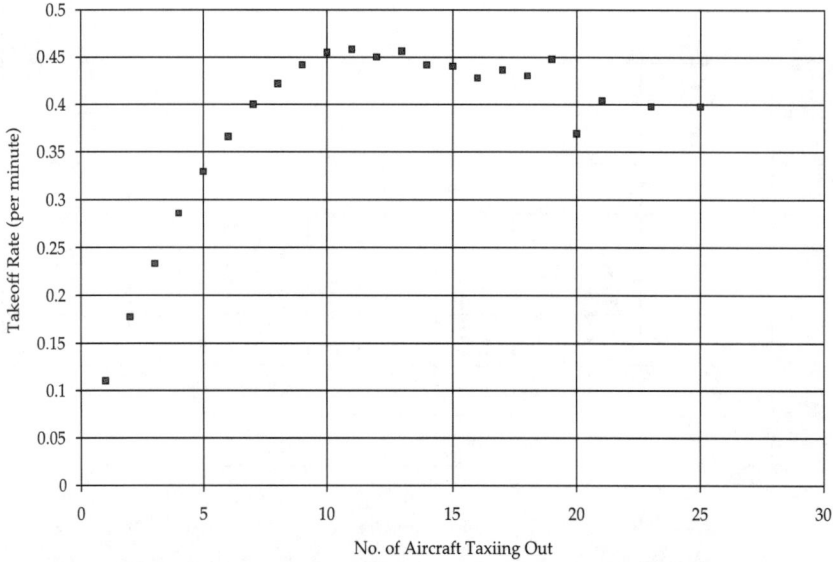

Fig. 12 Takeoff rate saturation point (BOS: 1997/09–1998/08).

to the total time that each aircraft spends on the ramp or taxiing with its engines running.

1. Functionality

The taxiway entry manager can affect departure operations by regulating the flow of aircraft through the interface between the gates and the taxiway system, which was identified as another possible control point in the departure flow.[5,6] The taxiway entry manager provides a means of indirectly controlling the runway takeoff queues, by controlling the total number of departing aircraft that will be distributed to them. Note that, depending on the specific airport geometry and complexity, this interface can take various forms. At BOS, there is a set of entry points to the taxiway system with little or no ramp area around the various terminals, whereas other airports, such as ATL or ORD, have a ramp area of considerable size adjacent to the terminals.

The current and projected taxiway situation (congestion levels) feeding back from simple observations or from sophisticated airport surveillance systems (in the future) and the takeoff queue (buffer) size feeding back from the mix manager are the most critical pieces of information for the taxiway entry manager (Fig. 13, top right data block). Accurate short-term estimation of pushback operations and prediction of the demand to enter the taxiway system must also be performed and the results fed into the entry manager to avoid overloading the entry points (Fig. 13, bottom left data block). All the above information is processed under the constraints of environmental regulations on aircraft engine emissions (Fig. 13, top input). The outcome of this system element (Fig. 13, bottom right data block) could be a feasible schedule of release times for aircraft to enter the taxiway system, which also meets the system objective of minimizing aircraft taxi times and therefore engine-running times, emissions, and airline direct operating costs.

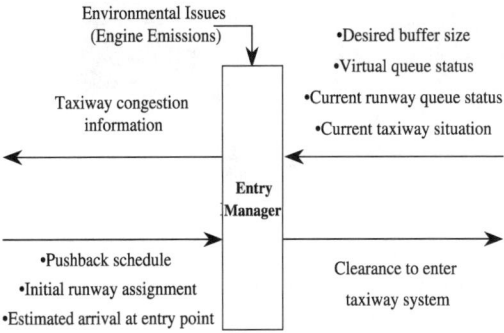

Fig. 13   The taxiway entry manager.

Engine-start time control is an additional issue pertaining to the environmental impact from aircraft engine noise and emissions, which deserves further examination. In current operations, only pushback and taxiway entry clearances are commanded by terminal ATC, and the exact time that aircraft engines are started is left entirely to the pilot's discretion. The gate and entry managers could possibly schedule the movement of aircraft under the additional objective of postponing engine start times until as close to the taxiway entry clearances as possible. This is an example of the kind of coordination between two tactical subcomponents, the gate manager and the taxiway entry manager. At BOS, where ramp space is limited, the two subcomponents can be merged into a single tool.

2. *Taxi Path Modifications—Runway Reassignments*

It should be noted that departing flights push back from their gate with an initial runway assignment, which they usually maintain until takeoff. Nevertheless, initial runway assignments and taxi paths are not always constant. When, for example, a taxiway segment is unexpectedly blocked, or there is a short-term or scheduled runway configuration change, or the load in a certain takeoff queue is high, a runway reassignment may then be needed. Had the controllers had prior knowledge of the particular circumstances that led to the need for a runway reassignment, their pushback and taxiway entry clearance deliveries could have been different, and also the additional workload imposed on them and additional cost and delays imposed on the airlines could have been prevented. In the context of the DP, reassignment and rerouting may also be necessary to implement the virtual queue planned sequence, which is very dynamic.

Potentially, an additional subcomponent could be inserted at this point in the system architecture to assist controllers with their runway reassignment and taxi routing decisions and to help them implement the virtual queue. Such a component is not currently included in the proposed DP architecture because it is believed that the taxi ground operations are not amenable to automation and are better handled by the experienced air traffic controllers. In the current operational environment, operations at the gates/ramp and at the runway are supported by automation systems, such as CTAS and SMA, which are already in place and can be interfaced to the proposed DP components. On the contrary, automation during taxiing is hindered by the lack of surveillance information. In fact, in managing taxi operations, air traffic controllers obtain (visually) downstream information regarding the

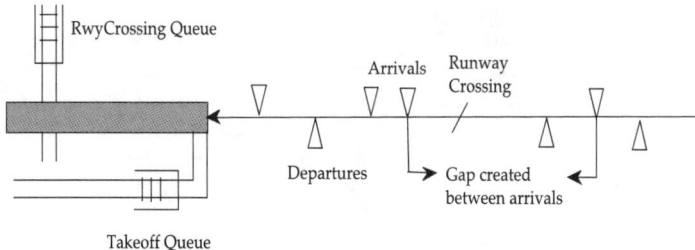

Fig. 14 Runway operations schedule: mixing of arrivals, departure, and runway crossings.

size and sequence of each takeoff queue, as well as the current status of each of the available runway queues. Such information can possibly come from downstream DP components such as the mix manager (see Section III. E).

E. Mix Manager

The mix manager regulates the release of departing aircraft from each runway buffer onto the corresponding runway (runway A or B in Fig. 3), as well as controls the release of aircraft from the runway crossing queues building up on the taxiway segments. The coordination of operations on dependent runways and the mixing of arrivals, departures, and crossing operations on a single runway are its main tasks.

1. Functionality

Air traffic controllers usually prefer to assign arrivals and departures to different runways. However, this is not always feasible, especially in tightly constrained airports such as BOS. For many configurations, the runway resource utilized by departing aircraft is shared with arriving aircraft, which in most cases have priority over departures. In addition, the runway system is frequently shared with taxiing aircraft that have to cross active runways. As illustrated in Fig. 14, the controllers often have to introduce gaps in the arrival stream in an effort to accommodate departures between arrivals and to allow taxiing aircraft to cross active runways. Sharing of the runway resources introduces a strong coupling between the arrival and the departure streams.[4-6] This suggests that we must consider and manage airport runway resources as sets of dependent runways, as opposed to individual runways.

Figure 15 describes the interaction of the mix manager with the rest of the aircraft flow at an airport system. As suggested, it is the connective component between terminal airspace traffic (departures ascending within the terminal airspace and arrival flow approaching the airport) and airport surface traffic (the set of departing or arriving aircraft that are physically present on the taxiway system).

As shown in Fig. 15, working under a given runway configuration and a specific mode of operations (Fig. 15, top data block), the mix manager processes the following inputs:

1) Projected takeoff demand information, based on inputs from the actual and projected pushback schedule (Fig. 15, bottom left data block).

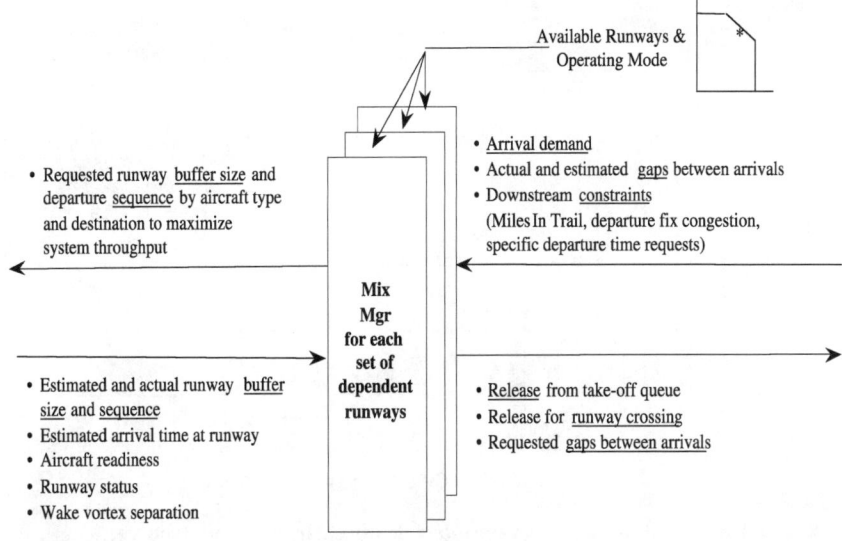

Fig. 15   The mix manager.

2) Projected landing demand information from the final approach arrival queues that are forming in the terminal area (Fig. 15, top right data block).

3) Data on downstream constraints, such as MIT and departure fix capacity (Fig. 15, top right data block).

CDM can play a vital role at this point, in providing accurately updated demand information (cancellations and delays) to the air traffic controllers and to the mixing function of the mix manager.

The main output generated from the mix manager is the suggested schedule of aircraft release from the takeoff and runway crossing queues (Fig. 15). Requests for gaps in the arrival flow could be given to the TRACON controllers to implement the suggested takeoff releases. In addition, specific tactical suggestions on the sequence and size of takeoff queues can be communicated to the tower controllers as a basis for carrying out efficiently the gate pushback and taxiway entry processes.

2. Supporting Analysis

In BOS configuration 22R-22L-27, which was presented in Fig. 1, arrivals using runways 27 and 22L have to cross runway 22R to reach the terminal area. Crossing aircraft queue in the taxiway segments between runways 22R and 22L, but when there is no more space for queuing aircraft, the departure stream on 22R has to be interrupted for crossings to occur and for making runways 22R and 27 available for further landings.

Figure 16 clearly demonstrates how takeoff operations on runway 22R are affected by the presence of crossing aircraft. When one or more crossings were operated between successive takeoffs on 22R, the mean time between the two takeoff clearances for the data sample presented is 1 min and 59 s. The mean value drops to only 45 s when successive operations occur with no crossings interjected between them. The overall mean time for the whole data set is 1 min and 10 s.

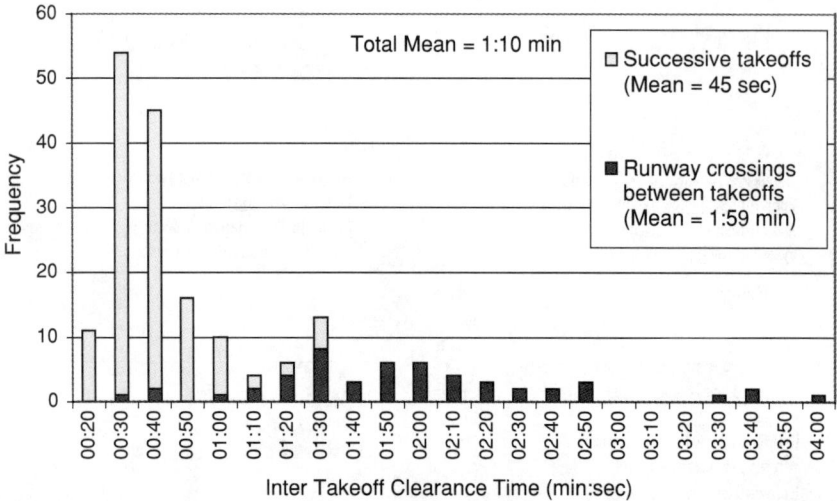

Fig. 16 Runway crossing effect on intertakeoff clearance time (source: local controller clearances, runway 22R, BOS, 2 Dec. 1998, 1600–2100 hrs).

The DP cannot be developed independently from CTAS or other arrival automation tools, which carry information critical to DP for successful configuration planning and arrival/departure mixing. In fact, the arrival-departure interaction introduces a new complex challenge for existing tools such as CTAS, which will now have to be enhanced to take departures into account. DP can have important inputs to CTAS and especially Active Final Approach Spacing Tool (AFAST), such as the runway crossing and takeoff queue information. These inputs can then be used to determine the most appropriate sequencing and tactical spacing of arrivals (introducing the necessary gaps in the arrival stream, Fig. 14).

IV. Conclusions

The DP, as presented in this chapter, is intended to assist short-term planning operations at major commercial airports. Its emphasis will be on supporting ATM in the next 30–45 min from the current time, but it also has a component that does planning with a time horizon of a few hours. It consists of a set of functional components, some used for strategic ATM operations such as configuration planning and some for tactical departure planning. DP is not necessarily viewed as a fully automated system. These components could potentially become automation tools used by the controllers to manage the various physical queues existing in the flow of departing aircraft, without increasing workload levels.

Acknowledgments

This research was supported in part by NASA under grant NAG 2-1128. The authors would also like to acknowledge the cooperation and courtesy of the BOS tower and TRACON in providing access for field observations.

References

[1]Völckers, U., Brokof, U., Dippe, D., and Schubert, M., "Contributions of DLR to Air Traffic Capacity Enhancement within a Terminal Area," AGARD meeting on 'Machine Intelligence in Air Traffic Management,' AGARD-CP-538, May 1993.

[2]Delcaire, B., and Feron, E., "Dealing with Airport Congestion: Development of Tactical Tools for the Departure Flows from a Large Airport," S.M. Thesis, Technology and Policy Program, Massachusetts Inst. of Technology, Cambridge, MA, June 1998.

[3]Feron, E., Hansman, R. J., Odoni, A. R., Cots, R. B., Delcaire, B., Feng, X., Hall, W. D., Idris, H., Muharremoglu, A., and Pujet, N., "The Departure Planner: A Conceptual Discussion," Massachusetts Inst. of Technology, Dec. 1997.

[4]Idris, H., Delcaire, B., Anagnostakis, I., Hall, W. D., Pujet, N., Feron, E., Hansman, R. J., Clarke, J.-P., and Odoni, A. R., "Identification of Flow Constraint and Control Points in Departure Operations at Airport Systems," *AIAA Guidance, Navigation, and Control Conference,* 98-4291, Vol. 2, AIAA, Reston, VA, Aug. 1998, pp. 947–956.

[5]Idris, H., Delcaire, B., Anagnostakis, I., Hall, W. D., Hansman, R. J., Feron, E., and Odoni, A. R., "Observations of Departure Processes at Logan Airport to Support the Development of Departure Planning Tools," 2nd USA/Europe Air Traffic Management R&D Seminar, Orlando, FL, 1–4 Dec. 1998.

[6]Idris, H., Anagnostakis, I., Delcaire, B., Hansman, R. J., Clarke, J.-P., Feron, E., and Odoni, A. R., "Observations of Departure Processes at Logan Airport to Support the Development of Departure Planning Tools," *Air Traffic Control Quarterly* Vol. 7, No. 4, 1999, pp. 229–257.

[7]Idris, H., "Diagnosis and Empirical Analysis of Departure Operations at a Major Airport," Ph.D. Thesis, Dept. of Aeronautics and Astronautics, Massachusetts Inst. of Technology, Cambridge, MA, Feb. 2001.

[8]Ball, M., and Hoffman, R., "Collaborative Decision-Making in Air Traffic Management: A Preliminary Assessment," National Center of Excellence in Aviation Operations Research, University of California at Berkeley, Berkeley, CA 94720-1720, Aug. 1998.

[9]Pujet, N., "Modeling and Control of the Departure Process of Congested Airports," Ph.D. Thesis, Dept. of Aeronautics and Astronautics, Massachusetts Inst. of Technology, Cambridge, MA, Sept. 1999.

Chapter 22

Modeling Air Traffic Management Automation Metering Conformance Benefits

Tara J. Weidner*
Seagull Technology, Inc., Los Gatos, California
and
Steve Green[†]
NASA Ames Research Center, Moffett Field, California

I. Introduction

AIR traffic controllers must occasionally interrupt flights (deviations from the user's preferred trajectory) to avert traffic conflicts and manage downstream airspace congestion. The large number of interruptions associated with current air traffic operations have led airspace users to strongly advocate for industry initiatives such as free flight.[1,2] Strong international efforts are under way to develop and deploy new air traffic management (ATM) decision support tools (DSTs) to assist controllers in reducing the frequency and impact of ATM-based interruptions.

It is critical to consider flow-rate or metering restrictions as well as conflicts in aircraft separation. The benefits associated with a focus on separation alone are unrealistic in that many ATM interruptions are due to dynamic capacity overloads that result in flight delays independent of conflict occurrences. The benefit of reductions in conflict deviations and route restrictions for any one flight will be negated if downstream congestion forces the flight to be delayed anyway. A hybrid approach is needed to model the impact of, and interactions between, ATM interruptions for conflicts and flow-rate restrictions due to congestion.

Such a methodology has been developed to quantify the benefits of reduced and more efficient ATM flight interruptions. Prior work documented the application of this methodology to the modeling of conflict deviations for separation assurance.[3]

Copyright © 2001 by the American Institute of Aeronautics and Astronautics, Inc. No copyright is asserted in the United States under Title 17, U.S. Code. The U.S. Government has a royalty-free license to exercise all rights under the copyright claimed herein for Governmental purposes. All other rights are reserved by the copyright owner.
*Currently with Parsons Brickerhoff Quade & Douglas.
[†]Manager, En Route Systems and Operations. Member AIAA.

The focus of this chapter is to introduce the application of this methodology to the modeling of controller conformance to metering restrictions due to downstream congestion. Important linkages between the integration of metering conformance and separation assurance functions are also discussed.

An example analysis is presented to illustrate the benefit potential of advanced en route DST capabilities for managing en route traffic in a high-density extended terminal area. Results are presented comparing metering conformance ATM interruption costs for a baseline and advanced DST case. The baseline loosely represents the operations associated with Free Flight Phase 1 (FFP1) conflict probe and arrival metering consisting of the User Request Evaluation Tool Core Capabilities Limited Deployment (URET CCLD),[4] and the Center-Terminal Radar Approach Control (TRACON) Automation System (CTAS) Traffic Management Advisor (TMA).[5] An advanced DST case represents future operations employing the CTAS En Route Descent Advisor (EDA) integrated with TMA.[6]

The remainder of this chapter discusses the metering conformance ATM interruptions modeling approach and illustrative example. Section II describes the overall ATM interruptions model methodology, including key model components. Section III presents the example application of the model to realistic ATM cases. Assumed input parameters, estimated ATM interruptions savings, and extrapolation of single-day, single-airspace simulation results to annual benefits at candidate deployment regions are discussed. Section IV offers closing remarks.

II. ATM Interruptions Model

The ATM interruptions model is shown conceptually in Fig. 1. The focus of this chapter is on the metering conformance model components illustrated in the figure. Initially a set of air traffic demand trajectories (including arrival, departure, and overflight traffic) are simulated for a typical day within a block of en route airspace. This airspace simulation generates a set of four-dimensional undelayed trajectories, representing what each flight would do if left alone to fly the user's preferred trajectory. These trajectories define the conflict and arrival congestion traffic scenario to be evaluated.

The metering conformance ATM interruption model component begins by analyzing traffic to determine the natural sequence and level of congestion at the

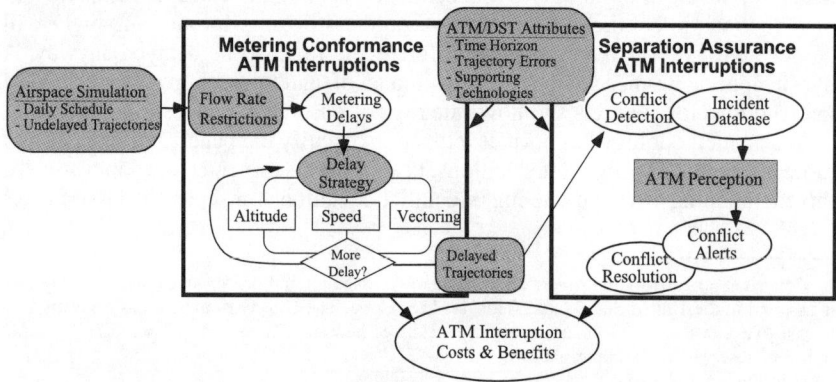

Fig. 1 ATM interruptions model approach.

airport. Arrival-metering operations are modeled and scheduled times of arrival (STA) are assigned for each arrival flight. These scheduled crossing times resolve downstream airport congestion. A second set of arrival flight trajectories is then generated, incorporating delay maneuvers necessary to meet the STA. The particular strategies used to absorb the delay depend on the ATM technology and procedures employed. The aircraft-specific strategies used and their associated interruption costs are tabulated.

The separation assurance ATM interruption modeling components initially identify and record conflicts and near-conflicts from the metered (delayed) traffic scenario (output from the metering conformance model) in a conflict-incident database. Near-conflicts are included to allow the analysis of false alerts. These incidents are then filtered through an ATM perception model to identify whether ATM would perceive the incident as a conflict requiring interruption. This perception model reflects the level of conflict probe technology in terms of trajectory prediction accuracy, time horizon, and separation criteria. As such, it can account for various combinations of DST capabilities, supporting technologies [e.g., data exchange, flight management system (FMS)], and controller procedures. A resolution is identified for each separation assurance ATM interruption including tabulation of daily resolution fuel costs. Daily fuel savings of both metering conformance and separation assurance ATM interruptions are then extrapolated to annual and National Airspace System (NAS)-wide benefits.

The functions of key model components are discussed in the following sections.

A. Airspace Simulation

In this study the Fort Worth Air Route Traffic Control Center (ZFW) airspace was analyzed, including arrival, departure, and overflight traffic operations between 40 and 250 n miles at or above 10,000 ft from Dallas/Fort Worth International Airport (DFW). Enhanced traffic management system (ETMS)-based flight trajectories for a typical day (Friday, 14 June 1996) were used to generate nominal trajectories for approximately 2500 DFW arrivals and departures.[7] Sample-day arrival, departure, and overflight operations are illustrated in profile view in Fig. 2.

Standard departure and arrival routes, commonly known as standard instrument departure (SID) and standard terminal arrival routes (STAR), are published procedures to aid in the coordination and routing of air traffic between center and TRACON airspace. Aircraft typically follow SIDs and STARs to/from major airports. These routes are characterized by specific waypoints, headings, speeds, and other parameters. The modeled undelayed ZFW trajectories followed DFW SID and STAR routings, as shown in Fig. 3.

B. Arrival Delays

During peak periods controllers meter DFW arrival flights to meet airport capacity restrictions. A simplified model of TMA metering was developed to estimate metering delays for each ZFW arrival. Meter-fix STA at the TRACON boundary, and associated delays, were based on maximum TRACON entry rates and minimum interarrival fix separations, as shown in Table 1.

Figure 4 shows a distribution of the arrival delays required to meet the Table 1 constraints over the course of the sample day. For reference, Fig. 4 also shows overall DFW arrival throughput.

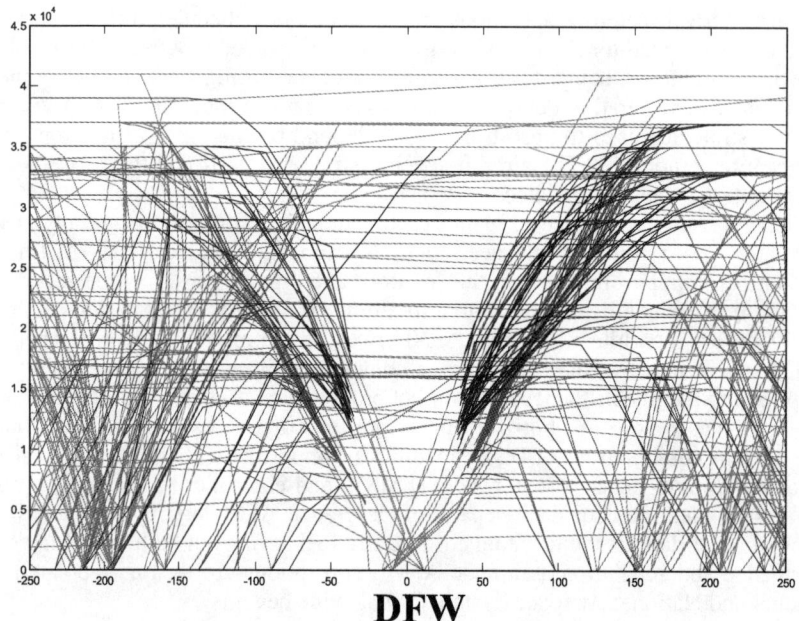

Fig. 2 Profile view of DFW study-day operations.

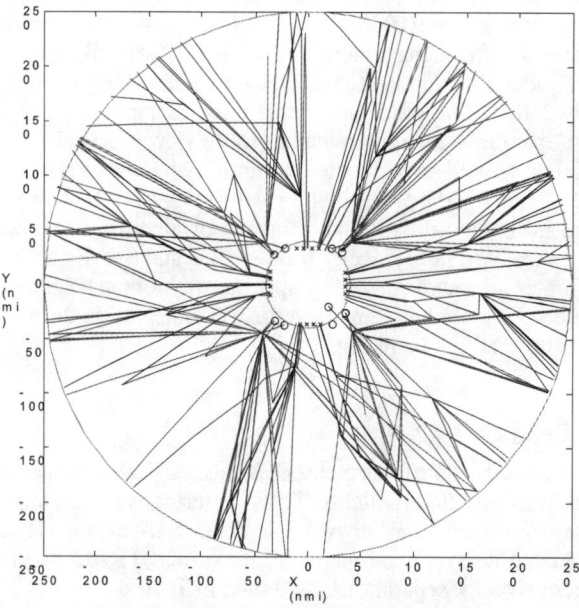

Fig. 3 DFW study-day STAR arrival and SID departure operations.

Table 1 DFW scheduling criteria

Scheduling criteria	Assumed value
Arrival metering-fix separation	5.50 n miles
TRACON arrival rate (four runways)	150 aircraft/h

C. Metering Delay Absorption

The absorption of arrival metering delay is essentially the resolution of intrastream or metering conformance conflicts produced by traffic intending to converge on the same meter fix. Arrival metering delay was absorbed by altering the initial arrival trajectory with a combination of changes to the speed profile (cruise and descent), cruise altitude, and routing (vector/path stretching).

Figure 5 illustrates the general methodology employed to clear an aircraft to meet an arrival-fix STA. The figure employs a strategy ordering of speed, altitude, and vectoring, where the maximum amount of delay is absorbed by each method before moving onto the following method. Thus, the STA of Fig. 5 is met by delaying the flight with a change in cruise speed (CAS_1 to CAS_2), a reduction in cruise altitude (h_1 at CAS_2) to h_2 at CAS_3), and the remaining delay absorbed with vectoring.

The effectiveness of the delay absorption model will depend on the amount of delay to be absorbed by any one flight, the time available to absorb the delay (i.e., effective time horizon), and the delay absorption strategy. For this analysis, the traffic scenario defines the flight delay (i.e., each arriving flight is subject to the same delay in the baseline and EDA cases). Differences in the delay-absorption performance of the baseline and EDA are modeled through differences in the effective time horizon and employed delay strategy.

The affect of time horizon is illustrated in Fig. 5. Note that at larger time horizons (right figure), speed and altitude changes can absorb more delay. As the effective time horizon decreases (left figure), the need for more expensive vectors (path stretching) increases because the speed and altitude changes cannot absorb as much delay.

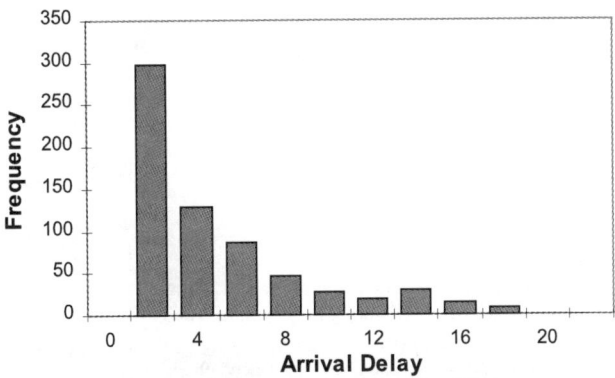

Fig. 4 DFW study-day arrival delays.

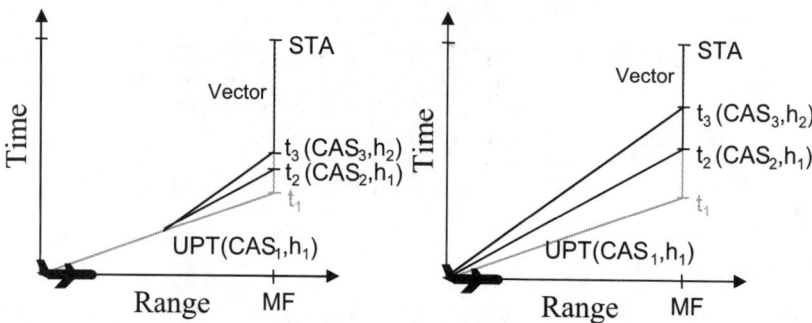

t_1 = User Preferred Trajectory (UPT) = Undelayed Metering Fix (MF) crossing time (CAS_1, h_1)
t_2 = Crossing time if cleared to minimum speed (CAS_2)
t_3 = Crossing time if cleared to minimum speed at a lower altitude (h_2)
t_4 = STA = crossing time if cleared to minimum speed, lower altitude, and vectored

Fig. 5 En route arrival delay absorption for different time horizons.

D. Metering Conformance Methods

Four possible metering conformance methods are used to alter the trajectory of particular flights so that the proper amount of delay is absorbed. The four methods are summarized in the following and graphically illustrated in Fig. 6 (profile view) and Fig. 7 (plan view):

1) Speed control. Reduce aircraft cruise and descent *CAS* speed along the initial routing and altitude profile. Chosen speeds are limited by aircraft performance-based minimum speeds and subject to ATM controller rounding/increment limitations. In this study, the descent speed is set to essentially "balance" cruise and descent *CAS* speeds. The higher of cruise/descent *CAS* is initially decremented until both speeds are equal. Then each speed is alternately decremented. Although actual controller techniques may not be so precise, this approach conservatively represents controller actions. Reduction in speed profile results in an earlier TOD location.

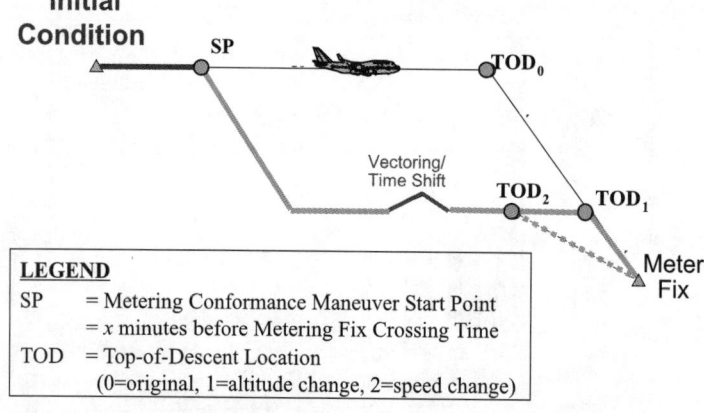

Fig. 6 Metering conformance ATM interruption (vertical profile).

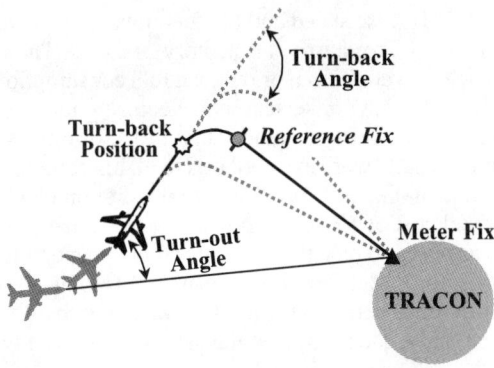

Fig. 7 Modeled vectoring method.

2) Altitude change. Descend and maintain a new cruise altitude (until final top of descent) down to floor of the high-altitude sector airspace [flight level (FL) 230/240]. With future technology cases, speed may also be allowed to change at the new altitude, providing an optimal combined speed/altitude approach.

3) Vectoring. Increase path length, using simple out-and-back vectors, at constant altitude and speed, up to a maximum heading change. [Although actual controller vectoring is more complex (e.g, S-turns), this approach captures the primary impact of vectoring for the purposes of this chapter.] An error is imposed on the timing of the final turnback vector to reflect ATM clearance limitations that may lead to arrival-fix STA deviations, as shown in Fig. 7. A turnback error is modeled as a random sample from a distribution, with bounds reflecting ATM/DST accuracy.

4) Time shift. A last resort method assumes delay is absorbed in additional vectoring at cruise altitude/speed, essentially shifting metering-fix crossing times to absorb any remaining delay.

A specific delay strategy is defined by the ordering of these methods in addition to time horizon and clearance accuracy parameters.

E. Metering Conformance Cost Model

Metering conformance ATM interruptions, which delay arrival aircraft to meet airport capacity constraints, result in both time and fuel penalties. Time costs were calculated directly from the TMA-required delay combined with FAA-based time cost rates.[8] Time costs include both crew and maintenance components, which vary by aircraft type.

Fuel costs were calculated using Eq. 1:

$$Arrival\ FuelCost = Fuelburn\ Rate \times Distance_{Cruise}/Speed_{Cruise} \qquad (1)$$

Equation (1) arrival fuelburn rates were based on high-fidelity simulations of a B737 aircraft under various conditions normalized to determine the fuelburn rate (lb/min) at various altitudes and airspeeds.[9] Thus delay strategies that reduced speed or altitude, would employ different fuelburn rates. Vectoring or timeshift methods increased fuel costs by increasing the time (distance) spent at constant speed/altitude with its associated fuelburn rate. Additionally, the change in TOD location under modified cruise speed and the fuel impact of the vectoring turnback error were added to the delayed arrival trajectory fuel cost. The fuel impact of the new TOD location leads to additional or reduced fuel consumption on the extended (faster) or shortened (slower) cruise segment. Vectoring turnback error impacted fuel costs as increased vectoring distance (late turn), or on descent (early turn). The B737 simulation results were extrapolated to all aircraft classes by applying a scale factor derived from Federal Aviation Administration (FAA)-based airborne cost rate data.[8] A fuel cost of $0.10 per pound was assumed.

This approach was used to calculate the total fuel expended for each simulated delayed arrival trajectory. Metering conformance ATM interruption benefits are calculated as the difference between the total arrival delay costs of the various technology cases under study. This reflected the savings at the study airport/airspace over a single day.

F. Separation Assurance ATM Interruptions

Because the focus of the chapter is on metering conformance ATM interruptions, the model's separation assurance ATM interruptions methodology is only briefly summarized here. However, it is important to note the need to consider the cumulative affect of decisions on a flight to produce accurate results from the modeling effort. Integrating functions (e.g., metering and conflict probe) captures important coupling interactions. One coupling involves the inefficiency of solutions that do not consider the entire problem domain. For example, accommodating a faster route/UPT just to reach a metering situation does not necessarily improve flight fuel efficiency. Second, not knowing the outcome of one DST function may limit the effectiveness of other functions. For example, lack of aircraft intent from not knowing metering conformance flight changes degrades trajectory prediction used by the conflict probe. As shown in Fig. 8, this may lead to missed and false alerts.

The separation assurance model component detects and resolves conflicts among the metered arrival, departure, and overflight trajectories. Conflict detection models are employed to develop an incident database of potential conflicts. ATM is assumed to intervene and interrupt conflicting trajectories that are perceived to violate acceptable controller spacing. Integration with metered flight changes provides improved perception, leading to fewer false and missed conflict alerts. For each perceived conflict recorded in the incident database, the model uses conflict

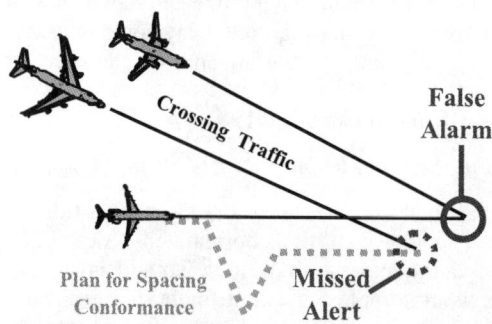

Fig. 8 Improved trajectory intent with integrated metering and conflict probe.

separation assurance resolution algorithms to tally the cost of resolution. The separation assurance ATM interruptions model application and the resulting metrics of missed/false alert rates and conflict probe interruptions costs are discussed in detail in Refs. 3 and 10.

G. NAS Extrapolation

Using the interruption rates and resolution costs found in the single airport daily simulation, the model employs a simple extrapolation, shown in Eq. (2–3), to estimate annual ATM interruptions benefits at candidate deployment regions. A similar extrapolation method would be applied to separation assurance ATM interruptions, as discussed in Refs. 3 and 10.

$$Annual\ Costs = (Annual\ Ops) \times (Interrupt\ Rate) \times (Cost\ Per\ Interrupt) \quad (2)$$

$$Annual\ Savings = Annual\ Cost_{BL} - Annual\ Cost_{EDA} \quad (3)$$

where
- $Annual\ Ops$ = annual airport operations, 100s
- $Interrupt\ Rate$ = number of interruptions per 100 operations, $Interrupt\ Rate \times Airport\ Factor$
- $Airport\ Factor$ = factor accounting for local airport rush arrival frequency relative to DFW, based on FAA delay data
- $Cost\ Per\ Interrupt$ = average cost per interruption, $/operation

The metering conformance ATM interruption rates are also adjusted by airport to account for variations in congestion, based on historical delay statistics.[11] It is assumed that airports with fewer overall delays would require disproportionately fewer metered arrivals. Airports are broken into four delay classes with 60–130% of DFW's portion of metered arrivals.

III. Illustrative Application

Researchers at the NASA Ames Research Center are developing en route tools within the CTAS to include the EDA. The following illustrative example is used to show the potential benefits of CTAS EDA over a Free Flight Phase 1 (FFP1) baseline. Using the model, potential metering conformance ATM interruption costs were calculated for both technology cases using a detailed ZFW single-day traffic scenario. The resulting benefits were then extrapolated to annual NAS-wide candidate deployment regions.

A. Case Definitions

The following cases describe a system baseline, reflecting FFP1 capabilities, and an advanced system based on CTAS EDA capabilities. The EDA case is shown to lead to more efficient ATM interruptions by improving the metering conformance delay absorption clearances, per assistance from the EDA-calculated advisories and longer time horizon (facilitated by DST advisories).

Both cases are assumed to employ the CTAS TMA to schedule arriving aircraft. TMA creates an optimum time-based arrival schedule for an airport complex and establishes scheduled times of arrival STA at TRACON-boundary meter fixes to control the flow into the TRACON airspace. The TMA schedule is continually updated from radar returns flight data from the air route traffic control center (ARTCC) host computer system in response to changing events, until an aircraft's metering-fix estimated time of arrival (ETA) is within 19 min (the "freeze horizon"), at which point the aircraft's STA is frozen. TMA STA are distributed to each en route sector managing arrival traffic. The STA and TMA estimates of delay to be absorbed are displayed directly on the controller's display system replacement (DSR) in an alphanumeric meter list.

Other attributes of the two cases are discussed in the following.

1. Case 1: Free Flight Phase 1 Baseline

The modeled baseline reflects en route operations aided by FAA FFP1 arrival metering and conflict probe. CTAS TMA schedules and meters arrival flights, separately from a URET CCLD conflict probe and trial-planning tool. TMA sets an arrival aircraft metering-fix crossing schedule at the center/TRACON boundary and displays flight-specific delay advisories to the controller. The controller cognitively creates a strategy to absorb the specified delay to meet the TMA schedule. As each arrival progresses toward the terminal area, and is delayed by the controller, TMA updates the displayed delay estimate to provide feedback to the controller as to the effectiveness of the employed delay strategy.

The FFP1 baseline reflects current ZFW metering conformance methods, based on discussions with NASA ATM experts familiar with ZFW en route airspace.* A time horizon of 16 min (before the undelayed metering-fix crossing time) is assumed, allowing a 3-min lag after the TMA STA and delay advisories are displayed to the controller.

The baseline delay strategy assumes controllers first employ altitude control by descending aircraft to the floor of the high-altitude airspace. Additional delay is absorbed using speed reductions, based on controller experience, down to a minimum speed applicable to most aircraft types. A speed error is added to the optimal case to represent cognitive limitations in developing the metering conformance clearance without automation assistance. Finally, vectoring is used to absorb any residual delay. (Controllers typically employ holding patterns for vectoring delays in excess of 8 min. Although not modeled geometrically, the time-shift method adequately models the economic affects of such vectoring.) The vectoring turn-back error reflects controller cognitive limitations in identifying the optimal vector turnback location/time.

2. Case 2: CTAS EDA

Case 2 future en route operations are defined by an ATM system with the integrated capabilities of both the CTAS TMA and EDA tools. This includes TMA arrival scheduling, as in the baseline case, and EDA high-fidelity trajectory modeling to predict future aircraft positions for conflict probe and metering conformance maneuver advisories. EDA's integration of automated metering-conformance advisories with conflict probe reduced conflict-probe false-alarm and missed-alert

---

*Personal communication with Harry Swenson, NASA Ames Research Center, Moffett Field, CA, April 2000.

rates. The EDA maneuver advisories assist controllers in formulating and executing a traffic delay strategy to meet the TMA schedule, allowing the controller to quickly and accurately assess the impact of various delay strategies. As such, a longer 18-min time horizon is assumed.

The EDA delay strategies[6] are modeled in the following way. With a longer time horizon, speed control can be used more effectively, and because of its fuel efficiency, it is attempted first. If speed control alone is not sufficient, a combination of altitude/speed adjustments are used instead. Here, EDA advises an optimal speed/altitude combination, difficult to calculate without EDA data and computational assistance. Vectoring, the least precise and least efficient strategy, is reserved for large delays. EDA vectoring advisories are designed to bring the flight within speed-control range using precise "turn-back" advisories to reduce uncertainty.[12]

Table 2 presents the parameters used to model the delay absorption strategies for both cases. Key attributes are the ordering of the strategy, the assumed time horizon, and various accuracy parameters assumed within each delay absorption method.

Table 2   Assumed delay strategy parameters

	FFP1 Case	EDA Case
	General Criteria	
Strategy order	Altitude	Speed
	Speed	Altitude/speed
	Vectoring	Vectoring
	Time shift	Time shift
Time horizon	16 min	18 min
	Speed Strategy	
Speed increments	10 kn	5 kn
Speed error	+10 kn	None
Minimum cruise speed[a]	BADA	BADA, 10 kn
Minimum descent speed[a]	BADA	BADA, 20 kn
	Altitude Strategy (jets only)	
Permitted altitudes[b]	Minimum altitude	FAR altitudes
Minimum altitude	FL230/FL240	FL230/FL240
	Vectoring Strategy	
Heading increment	1 deg	1 deg
Maximum vector angle	60 deg	60 deg
Turnback error	± 60 s	±30 s

[a]The minimum cruise/descent speeds for FFP1 used Eurocontrol Base of Aircraft Data (BADA) model[13] "low" cruise speeds, reflecting a lack of automation to help controllers identify efficient speeds. EDA minimum speeds were modeled as 10 kn (20 kn in descent) lower than BADA, closer to best endurance speed.
[b]FAR = Federal Aviation Regulations permitted altitudes: 2000 ft increments at/above FL290, 1000 ft increments below.

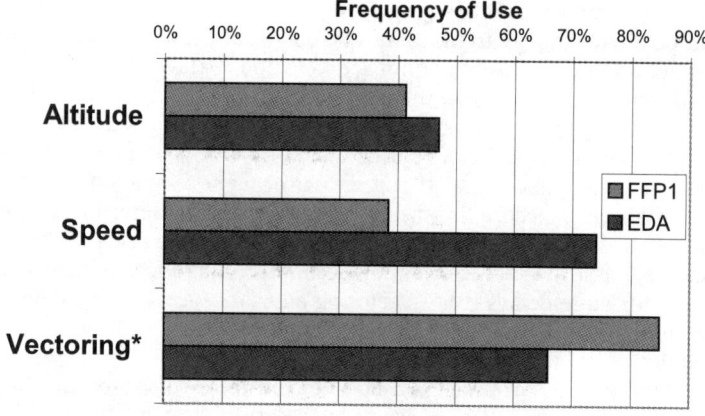

*Vectoring includes Time Shift method.

Fig. 9 **Comparison of employed metered arrival delay strategy.**

B. ATM Metering Conformance Benefits

The resulting metering conformance ATM interruptions and frequency of each delay method employed by case are shown in Fig. 9. This figure shows the increased use of fuel efficient speed control methods and reduced reliance on the more expensive vectoring methods in meeting the TMA metered schedule.

On average, the metered arrivals were delayed by 3–5 min. The range of delay absorbed with each method is shown in Table 3. The table compares the various delay absorption methods employed in both the baseline and EDA cases. The results show that EDA appears to replace the altitude method with the less intrusive and more cost-effective speed adjustment method.

Table 4 compares the resulting fuel consumption tied to the FFP1 baseline and EDA cases. The table points out the fuel efficiency of speed delays, where delays absorbed with speed control can actually reduce the overall flight cost.

As both cases used the same traffic scenario, each flight was subject to the same time delays in the baseline FFP1 and EDA cases. As a result, EDA savings reflect improved fuel efficiency in absorbing the common TMA metering delay. Figure 10 graphically shows the distribution of per operation EDA metering conformance fuel savings.

C. Annual/National Benefits Extrapolation

The simulated 1996 daily ATM cost savings at ZFW, due to more efficient metering conformance ATM interruptions, was extrapolated to an annual level and to other candidate regions using Eqs. (2–3). The DFW metering conformance ATM interruption rates and costs per operation observed in the simulation are shown in Table 5.

NAS-wide benefits were calculated assuming en route/transition airspace deployment of the scenario technologies at 37 candidate airport sites. An airport's assumed share of delayed arrivals, relative to DFW, based on the 1996 Aviation Capacity Enhancement Plan[11] delay data follows:

Table 3 Metered arrival delay comparison

	Delay, min					
	FFP1 Baseline			CTAS EDA[a]		
	Range	Average	Total	Range	Average	Total
Altitude	0–2.5	0.2	160	0–10.2	1.3	845
Speed	0–4.6	0.4	275			
Vectoring[b]	0–18.1	3.4	2247	2.8–17.9	2.7	1808

[a]CTAS EDA combines speed and altitude methods.
[b]Vectoring includes time shift method.

Table 4 Metered arrival fuelburn comparison

	Fuelburn, lb					
	FFP1 Baseline			CTAS EDA		
	Range	Average	Total	Range	Average	Total
Altitude/speed	(168)–579	14	9502	(533)–590	13	8645
Vectoring[a]	0–2659	244	161,422	0–2440	183	120,866
Total	(133)–2793	258	170,924	(533)–2773	196	12,951

[a]Vectoring includes time shift method.

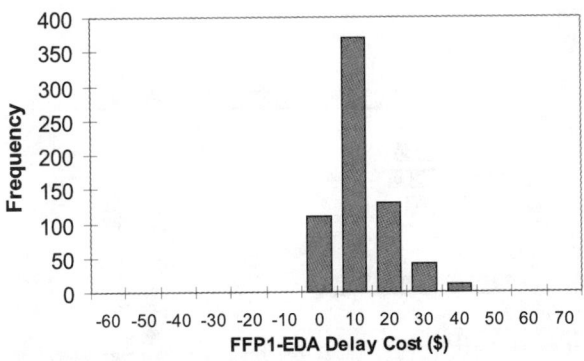

Fig. 10 CTAS EDA savings per operation.

Table 5 DFW metering conformance simulated interruption rates

Case	Rate	Average cost, $
FFP1	30.4/100 ops	$104.66/op
CTAS EDA	36.5/100 ops	$97.50/op

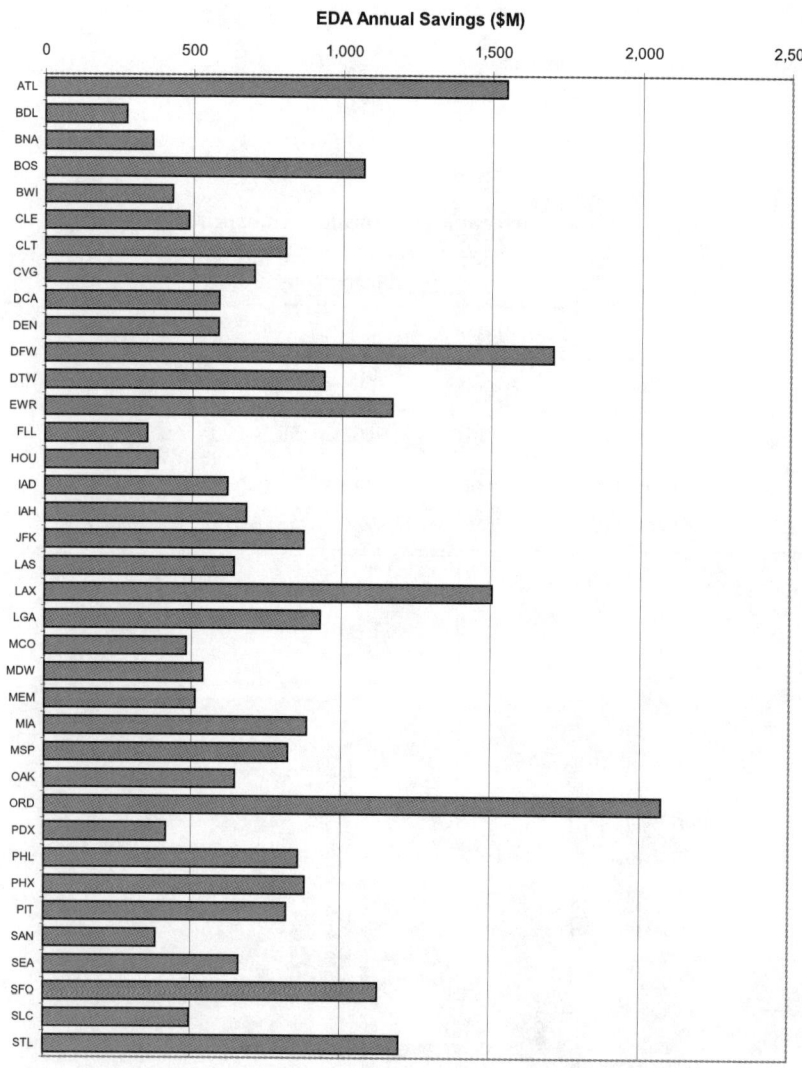

Fig. 11 Annual NAS-wide EDA metering conformance benefits.

130%	EWR, SFO, LGA
115%	ORD, STL, JFK, BOS
100%	LAX, ATL, DFW, PHL
80%	IAH, CVG, MSP, DTW, PHX, IAD, MIA, MDW, PIT, CLT, DCA, SEA
60%	CLE, MCO, LAS, BWI, SLC, SAN, HOU, PDX, DEN, FLL, BDL, BNA, MEM, OAK

Using 1996 annual arrival operations,* annual savings by airport were calculated using Eqs. (2–3). Annual savings reflect the difference between the annual ATM interruptions cost of the FFP1 baseline and CTAS EDA case. The total benefit at all 37 airports is $25.1 million annually. The annual savings for various deployment airports are plotted graphically in Fig. 11. The benefits vary significantly by airport due to different activity levels and extent of existing delays. The largest savings occur at the capacity-restricted hub airports of Chicago O'Hare (ORD), Dallas/Fort Worth (DFW), Atlanta Hartsfield (ATL), and Los Angeles (LAX).

IV. Conclusions

This chapter presented a new methodology for assessing the performance of en route DST technologies for reducing the frequency and impact of ATM-based deviations to the user's preferred trajectories. The methodology provides an approach to evaluating the trajectory costs of en route ATM interruptions by modeling specific controller metering and conflict resolution actions, aided by automated DST technology. The model is sensitive to the complex interactions of time horizon, controller delay strategy, ordering, and accuracy used to absorb arrival metering conformance delays. Additionally, the importance of integrating metering conformance and separation assurance (e.g., conflict probe) functions was discussed. Furthermore, a simple method was identified to extrapolate results from a single day's traffic simulation at ZFW to annual NAS-wide technology deployment benefits. This methodology was illustrated by a brief example of the CTAS EDA tool relative to FFP1 operations. The example revealed the methodology's value for use in the concept development and validation of ATM automation concepts in support of NAS free flight initiatives. Additionally, it estimated EDA metering conformance benefits of $1.6 million annually at DFW and $25.1 million if employed at 37 likely airport locations.

Acknowledgment

The authors would like to acknowledge those who helped develop the ATM interruptions model in previous research efforts. The model development would not have occurred without the collaborative efforts of T. Golpar Davidson and Susan Dorsky of Seagull Technology, Inc., as well as integral contract support from NASA Ames En Route Systems and Operations Group.

1997 Terminal Area Forecast (TAF) System, Federal Aviation Administration, Office of Aviation Policy and Plans. Data available online at http://www.apo.data.faa.gov/faatafall.htm [cited Oct. 1998].

References

[1]*Final Report of RTCA Task Force 3: Free Flight Implementation*, RTCA, Inc., Washington, DC, Oct. 1995.

[2]Erzberger, H., "Design Principles and Algorithms for Automated ATM," *AGARD Lecture Series No. 200 on Knowledge-Based Functions in Aerospace Systems*, AGARD, Germany, Nov. 1995.

[3]Weidner, T., Davidson, G., Coppenbarger, R., and Green, S., "Modeling ATM Interruption Benefits," *AIAA Guidance, Navigation and Control Conference*, AIAA Paper 99-4296, June 1999.

[4]Celio, J., Bowen, K., Winokur, D., Lindsay, K., Newberger, E., and Sicenavage, D., "Free Flight Phase 1 Conflict Probe Operational Description," Technical Rept. MTR 0W00000100, MITRE Corp., Washington, DC, March 2000.

[5]Swensen, H., Hoang, T., Engelland, S., Vincent, D., Sanders, T., Sanford, B., and Heere, K., "Design and Operational Evaluation of the Trafffic Management Advisor (TMA) at the Fort Worth ARTCC," *1st USA/Europe ATM R&D Seminar*, France, June 1997.

[6]Green, S. M., and Vivona, R., *En route Descent Advisor (EDA) Concept*, Advanced Air Transportation Technologies (AATT) Project Milestone 5.10 Rept., NASA Ames Research Center, Sept. 1999.

[7]"Traffic Demand Scenarios," computer data files, CSSI, Inc., Washington, DC, Sept. 1998.

[8]*Economic Values for Evaluation of FAA Investment and Regulatory Programs*, FAA-APO-98-8, Office of Aviation Policy and Plans, Federal Aviation Administration, June 1998.

[9]Mukai, C., "Design and Analysis of Aircraft Dynamics Models for the ATC Simulation at NASA Ames Research Center," TR92119-02, Seagull Technology, Los Gatos, CA, March 1992.

[10]Weidner, T., Davidson, T. G., and Dorsky, S., "En Route Descent Advisor (EDA) and En Route Data Exchange (EDX) ATM Interruption Benefits," TM98188.26-01f, Seagull Technology, NASA AATT TO26, Dec. 2000.

[11]*1997 Aviation Capacity Enhancement Plan*, Office of System Capacity, Federal Aviation Administration, Dec. 1997.

[12]Green, S., Vivona, R., and Grace, M., "Field Evaluation of Descent Advisor Trajectory Prediction Accuracy for En Route Clearance Advisories," *AIAA Guidance, Navigation, and Control Conference*, AIAA paper 98–4479, Aug. 1998.

[13]*User Manual for the Base of Aircraft Data (BADA) Revision 3.1*, EED Note No. 23/97, Eurocontrol Experimental Centre, Paris, Oct. 1998.

Section V: Airspace Operations and Constraints

Six chapters are presented in this section emphasizing the development of air traffic management (ATM) metrics with a primary focus on airspace constraints and performance. The first two chapters present metrics and performance assessment methods for evaluating ATM system changes. Chapter 23 describes the theoretical benefits of more efficient en route flight trajectories. Chpater 24 presents specific metrics that can be monitored to assess the benefits of ATM performance enhancement measures.

The next four chapters are performance analysis of specific decision support systems that have been deployed and have undergone preliminary evaluation. The chapters describe measures of performance enhancement to both the U.S. and European ATM systems. In general, the measured performance to date has been less than anticipated based upon prior simulation studies.

Chapter 23. The Effect of Direct Routing on ATC Capacity, S. A. N. Magill, DERA, 1998.

Chapter 24. Performance Measures for Future Architecture, Steve Bradford, Diana Liang, and David Knorr, FAA, 2000.

Chapter 25. Analytical Identification of Airport and Airspace Capacity Constraints, William R. Voss, FAA; and Jonathan Hoffman, MITRE Corporation, 2000.

Chapter 26. Operational Assessment of Free Flight Phase 1 Air Traffic Management Capabilities, Dave Knorr, FAA; Joseph Post and Jeff Biros, CNA Corporation; and Michelle Blucher, MITRE Corporation, 2000.

Chapter 27. CENA-PHARE Experiment: Requirements for Evaluation of Novel Concepts in Air Traffic Control, Didier Pavet, CENA, 2000.

Chapter 28. Restriction Relaxation Experiments Enabled by User Request Evaluation Tool, Michael J. Burski and Joseph Celio, FAA, 2000.

Chapter 23

Effect of Direct Routing on Air Traffic Control Capacity

S. A. N. Magill*
Defence Evaluation and Research Agency, Malvern, Worcestershire, United Kingdom

I. Introduction

ON BUSY days in the summer of 1998 the air traffic control (ATC) systems in many parts of Europe operated at or near capacity, and both passengers and airlines complained about excessive delays. Yet traffic demand is forecast to grow with a mean rate in the region of 3–4% per annum,[1] leading to an increase of 23–32% by 2005 and 65–95% by 2015. Thus the question naturally arises: where is the additional capacity to come from?

The traditional response of ATC service providers to increasing demand has been to resectorize the airspace to apply more controllers in parallel to the total control problem. However, this process cannot be continued indefinitely. When sectors become too small, the gains from applying more controllers in parallel are outweighed by the increase in coordination needed between neighboring sectors. The airspace in the busier parts of Europe is now close to this state.[2] It is widely accepted that human air traffic controllers (as opposed to automated systems) will continue to have the central role in the control process for the foreseeable future. It is also widely accepted that the traffic-handling capacity of most types of airspace is limited by the maximum amount of work that can safely be assigned to human controllers. These considerations have motivated the development of a number of experimental systems that aimed to assist controllers by providing them with computer-based prediction and monitoring tools. Although such systems have demonstrated some potential capacity gains, these have generally been small compared with what is needed.[3,4]

In current practice most air transport flights operate along a fixed network of airways rather than flying directly from departure airport to arrival airport. Until relatively recently, point-to-point navigation was the only method available, but with the development of area navigation and satellite navigation systems, this is

---

Copyright © 2001 by DERA. Published by the American Institute of Aeronautics and Astronautics, Inc., with permission.
*Principal Scientist.

no longer the case. Direct routing has obvious economic attractions for aircraft operating companies. It leads to shorter flight times and reduced fuel costs, but it might also offer capacity advantages for ATC systems.

There are two mechanisms by which direct routing might be expected to increase capacity:

1) By reducing flight times, direct routing will reduce traffic density for a given frequency of departures, or produce the same traffic density for a higher frequency of departures.

2) Airways operation forces all traffic into a restricted volume of airspace while leaving the remainder unoccupied, whereas direct routing is likely to spread traffic more evenly throughout the airspace. The latter situation might be expected to lead to fewer separation problems for the same traffic demand, or to support a higher traffic demand for the same frequency of separation problems.

This chapter reports the results of a fast-time simulation study that aimed to quantify these effects in the context of the whole of European airspace with the traffic demand forecast for the year 2005.

However, the limits of what is currently possible with fast-time simulation must be recognized. Computer models of human mental activity (and of air traffic control in particular) have not yet developed to the point where they can predict how human controllers will respond to the change from airways to direct routes. Ultimately human workload and traffic-handling capacity can be measured only from real-time simulations or operational systems. However, real-time simulations are very costly exercises. They require many air traffic controllers and a considerable amount of time to train them in the new methods to be tested. They also require pseudopilots and supporting infrastructure as well as a high-fidelity simulation of the operating environment. A fast-time simulation study can estimate an upper bound for the capacity gain that might be obtained from direct routing, and can do this over a much larger geographical region (with a greater variety of traffic conditions) than would be practicable for real-time simulation. This information can then be used to inform a decision about whether or not to embark on a costly real-time simulation.

II. Workload and Capacity

We would like to use fast-time simulation to compare the traffic handling capacity of the following two ATC systems: 1) a system that uses airways for all flights; and 2) a system that uses direct routing for all flights, but is the same as system 1 in all other respects. Unfortunately, there is no way of measuring capacity directly in a simulation, and so we must approach the problem indirectly. It was pointed out in Section I that, in most types of airspace, capacity is determined by the amount of work that can safely be assigned to human air traffic controllers. Therefore, the problem of making capacity comparisons can be transformed into one of making comparisons of required control workload.

A. Characterizing Workload

We have chosen to characterize control workload in terms of the frequency of separation problems that require attention from air traffic controllers. Any measure of workload has its pros and its cons, and this one is no exception. Against, it might be argued that there are control tasks that do not relate directly to separation problems, and that the separation problems that would occur in a system based on airways are likely to be much more stereotyped (and hence easier to deal with) than those that would occur in a system based on direct routes. For, it might be countered that, given that the main purpose of ATC is to keep aircraft safely separated, frequency of separation problems is a measure of *what* must be done to achieve this purpose without specifying *how* it will be done; the measure is therefore independent of the fine detail of exactly how ATC will operate in future. This author is persuaded by the latter argument.

B. Interaction Frequencies

Having decided to count separation problems, the next question is: what constitutes a separation problem that contributes to controller workload? At the extreme there are infringements of allowed separation minima (typically, less than 5 n miles horizontal separation simultaneously with less than 1000 ft vertical separation), which always demand avoiding action. However, there are many less severe separation problems that nevertheless absorb controller attention and thereby contribute to workload. These include situations in which controllers take avoiding action because they allow margins for uncertainty in predictions of minimum separation, situations in which controllers begin to plan avoiding action even though it probably will not be needed, and situations in which they simply decide to monitor more intensively. It seems likely that the less serious separation problems contribute as much or more to the total workload because there are many more of them. The separation thresholds that delimit these various possibilities are not constant; they vary greatly from one situation to another. Consequently, results are presented for a range of separation values.

We use the term *trajectory interaction* (or simply *interaction*) rather than *conflict* or *proximity* to include all potential separation problems ranging from the most severe to no problem. An interaction is a situation in which two aircraft would violate a given separation threshold if no avoiding action were taken. The vertical component of the separation threshold is assumed to be 1000 ft throughout this study, and so results are presented in terms of interaction frequency as a function of horizontal separation threshold.

III. Simulation

A. Software

The study made use of a package of simulation software known as flexible airspace modeling environment (FLAME) which was developed by Defence Evaluation and Research Agency (DERA) for use in air traffic management (ATM) research applications. It has previously been used for several such projects.[5,6] FLAME simulates the movement of individual flights through the airspace, collects statistics on quantities of interest, and provides traffic displays for scenario validation purposes.

FLAME models traffic demand and aircraft profiles in some detail, but it does not attempt to model conflict resolution at its present stage of development. Aircraft horizontal speeds are obtained by converting calibrated air speed and Mach values to true air speeds, and combining these with wind vectors. Modeling of climb and descent rates is based on an analysis of radar for a large sample of real traffic.[7] Altitudes of aircraft in level flight are exact multiples of 1000 ft; cruise levels are allocated according to reduced vertical separation minimum (RVSM) and the semicircular rule. Aircraft on airways fly along the airway centerlines.

B. Airspace

The geographical region in which traffic was simulated was that bounded by the meridians at 10°W and 30°E, and the parallels at 36°N and 60°N. This large region includes practically all of Europe and contains a wide variety of conditions and traffic densities. Using such a large area also minimizes edge effects. For flights entering and leaving the region, only the portions inside the region were simulated. To avoid the cost of modeling the fine detail of departure and arrival procedures at several hundred airports, greatly simplified terminal area structures were used, and the collection of results was restricted to airspace at or above 10,000 ft.

Details of the positions of airports, waypoints, and airway segments were obtained in electronic form from the Jeppesen Flight Planning Database. For each airport, two entry/exit fixes were identified where traffic to/from the airport would leave/join the airways system in the airways simulations (these were usually points where airways merged or crossed), and simplified paths were constructed between runways and these fixes. The same set of entry/exit fixes was used in both the airways simulations and the direct-routes simulations.

For each flight the simulation first determined the departure airport's exit fix nearest to the arrival airport (A) and the arrival airport's entry fix nearest to the departure airport (B). For a flight in a direct-routes simulation, the route from A to B was simply the great circle through the two points, but for a flight in an airways simulation, determining the route was more complicated. The simulation found a sequence of airway segments that joined A with B. It made use of a backtracking algorithm that attempted to find the shortest route by searching the directed graph formed by all airways and their crossing/merging points. The depth of backtracking was limited (so that simulation runs would complete in reasonable periods of time), and so the algorithm did not necessarily find the shortest route, but it did find good approximations to the shortest routes. It is perhaps worth noting that airways flights in the real world do not necessarily fly the shortest routes.

C. Traffic Scenario

The results presented in Section IV were derived from five pseudorandom traffic samples, each 12 h in length. For each sample, results were discarded from the traffic buildup period (first 2 h) and the tail-off period (after 12 h) so that the system could be considered to be in a steady state for a period of 10 h.

The pseudorandom traffic samples referred to in the preceding paragraph were generated from a statistical summary of European traffic. This summary was obtained by analyzing flight-plan data recordings for the month of April 1996, which were obtained from the European Central Flow Management Unit (CFMU). Some small simplifications were made during the summarizing process. For example, it was found that about 4% of traffic operated into or out of airports with less than 15 airways movements per month; by reassigning these flights to nearby larger airports, the total number of airports in the simulation could be reduced by a factor of almost 3.

The statistical summary of European traffic thus obtained was scaled to match traffic demand forecasts for 2005. It was assumed that traffic will on average grow by 3.8% per annum (the *High* scenario from Ref. 1), which gives a growth factor of 1.4 between 1996 and 2005; this was increased to 1.5 to allow for the fact that April is not the busiest time of year. For the region simulated this gave a rate of traffic generation of 1730 flights per hour.

D. Analysis

Any parameter estimate obtained from a simulation with random inputs is itself a random number. One of the traditional ways of dealing with this problem[8] is to take the mean of the results from several independent simulation runs and estimate a confidence interval for the mean. That procedure has been applied in the study reported here, and that is why five independent traffic samples were used. Although confidence intervals are not reported in Section IV, they were calculated and were seen to be small compared with the parameter differences reported.

The FLAME trajectory generator was run twice for each of the five traffic samples referred to, once with airways and once with direct routes; this gave 10 files of trajectories. Each file of trajectories was then analyzed by a program that identified all pairs of flights that simultaneously came within 20 miles and 1000 ft of one another; this gave 10 files with more than 10,000 trajectory pairs in each. A number of simple programs were written to analyze this data in the ways shown in Section IV, to estimate traffic densities, to count interaction frequencies, to find the distribution of relative track angles, etc.

IV. Results

A. Traffic Density

Flight times for traffic on direct routes will generally be shorter than those for traffic on airways, and this will lead to a lower traffic density for a given frequency of departures. To find the relative traffic densities for the two routing scenarios, the number of flights at or above 10,000 ft in the whole of the region simulated was counted every quarter hour throughout the 50 h of simulated data available for each scenario. The result is shown in Table 1. The ratio of mean squared traffic densities will be needed in Section IV.D, and so it too is shown. Thus, direct routing reduces the mean traffic density at and above 10,000 ft by almost 15%.

Table 1 Relative traffic densities

Ratio of mean traffic density for direct routes to that for airways	0.853
Ratio of mean squared traffic density for direct routes to that for airways	0.727

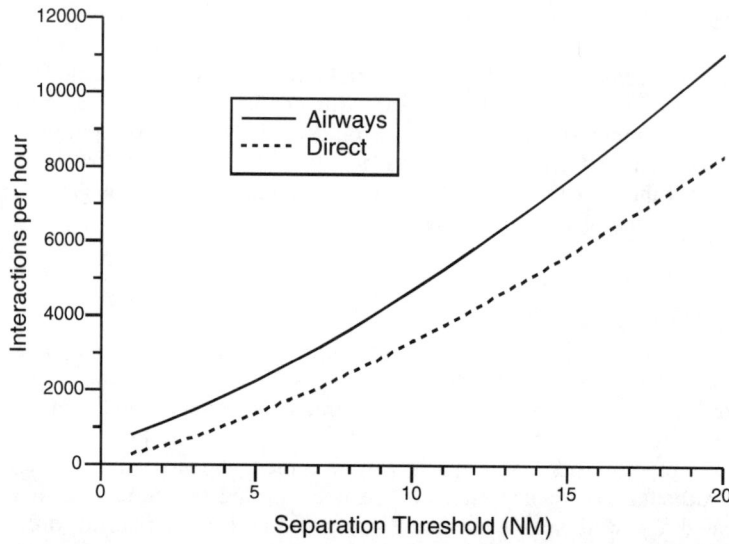

Fig. 1 Interaction frequency and separation threshold.

B. Horizontal Separation Threshold

The frequency of occurrence of pairs of aircraft that were simultaneously separated by less than s n miles and less than 1000 ft was counted for both routing scenarios. This was done for values of s between 1 and 20 n miles at 1-mile intervals. The result is shown in Fig. 1. This is the main result of the study, and we will return to it several times.

The graph shows that use of direct routing does indeed give rise to lower interaction frequencies as expected, but that the extent to which it does so varies considerably with separation threshold value. It might be helpful to think in terms of the percentage by which direct routing reduces interaction frequency compared with airways. The percentage reduction is plotted against separation threshold in Fig. 2.

This curve shows that the reduction of interaction frequency from direct routing is about 67% at a separation threshold of 1 n mile. It falls rapidly at first as separation threshold is increased, but flattens out to about 25% at 20 n miles. Assuming that the allowed minimum separation is 5 n miles simultaneously with 1000 ft, then in the absence of avoiding action by controllers or pilots, direct routing will produce 39% less conflicts than airways.

C. Shapes of Graphs

The shapes of the lines in Fig. 1 are of some interest. Although interaction frequencies were not estimated for separation thresholds below 1 n mile, if we extrapolate toward 0 miles we can see that the direct-routes curve will pass through the origin, whereas the airways curve will definitely not. The reason for this difference is presumed to be as follows: in our airways simulations aircraft fly along the airway centerlines, and so they can pass through one another (so that there will be some interactions with zero horizontal separation). Although this will be

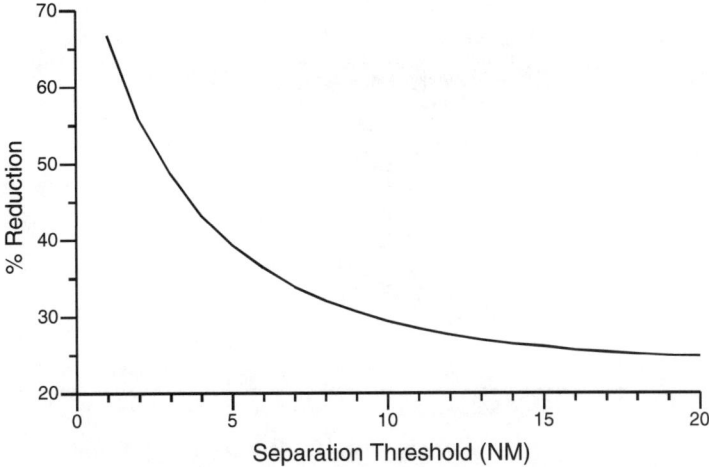

Fig. 2 Reduction in interaction frequency from direct routing.

a relatively infrequent occurrence in level flight because of the operation of the semicircular rule, it will be much more common in climb and descent.

Both lines in Fig. 1 exhibit a small but definite curvature. The direct-routes curve fits closely the shape $y = ax^{4/3}$, which has been reported by other authors,[9] and the fit is especially close for separation values greater than 5 n miles. When the airways curve has been displaced parallel to the interaction frequency axis so that its extrapolation passes through the origin, it too fits closely the same function, but with a different value for the a coefficient. These results are useful for calculating the effects of postulated changes to separation criteria for controller action.

D. Traffic Density and Spatial Distribution

We might reasonably expect the difference between the two curves in Fig. 1 to be caused by two distinct mechanisms:

1) Use of direct routes leads to lower traffic densities, which in turn leads to lower interaction frequencies.

2) Use of direct routes tends to spread the traffic out over more airspace, whereas use of airways tends to concentrate the traffic into some parts of the airspace and leave other parts unused. The former might be expected to lead to lower interaction frequencies.

Thus, the question naturally arises: how much contribution does each mechanism make to the difference between the two curves in Fig. 1? To answer this question, each data point on the airways curve in Fig. 1 was "corrected" to remove the effect of the density difference. Interaction frequency is approximately proportional to the square of traffic density,[10]* and so the correction was applied by multiplying each airways interaction frequency by the mean squared ratio of traffic densities from Table 1. The result in Fig. 3 shows the effect of spatial distribution alone.

---

*Data available online at http://atm-seminar-97.eurocontrol.fr/andrews.htm [cited 1997].

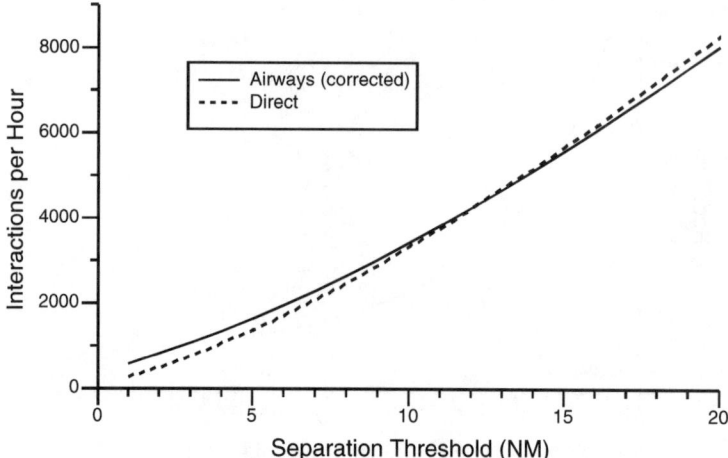

Fig. 3 Same as Fig. 1, but corrected for difference in traffic density.

When the effect of traffic density has been removed (Fig. 3), there is much less difference between the two curves. This indicates that most of the difference between the curves in Fig. 1 arises from traffic density. The effect of spatial distribution alone (Fig. 3) is rather unexpected: at the lower separation threshold values, interaction frequencies for direct routing are lower than those for airways as expected, but at higher separation threshold values, the converse is true. The separation threshold value where the two curves in Fig. 3 cross one another is about 12.4 n miles.

E. Relative Track Angles

The effect of routing scenario on the relative track angles of interacting pairs of trajectories was investigated. This was done by recording the track difference at the point of closest approach for all pairs of aircraft separated by less than distance d horizontally and 1000 ft vertically at this point. The value of d was chosen to be 12.4 n miles for reasons that will shortly become clear. The result is shown in Fig. 4.

Figure 4 shows that airways operation produces higher interaction frequencies for most relative track angles, but that the difference is much more marked for angles close to 0 and 180 deg (overtaking and head-on) than for other angles. It could be argued that the interactions with these relative track angles are the ones that contribute most to controller workload: overtaking situations because each tends to persist for a long time; head-on because they happen so quickly. As in the previous subsection, we might ask how much of the effect shown in Fig. 4 arises from difference in traffic density and how much arises from difference in spatial distribution. To answer this question, the interaction frequencies for airways operation were corrected in the same way as in the previous subsection, by multiplying each by the ratio of mean squared traffic densities from Table 1. The result is shown in Fig. 5.

The value of d used in Figs. 4 and 5 was chosen to be 12.4 n miles (the value where the curves in Fig. 3 cross), so that the total interaction frequencies in Fig. 5

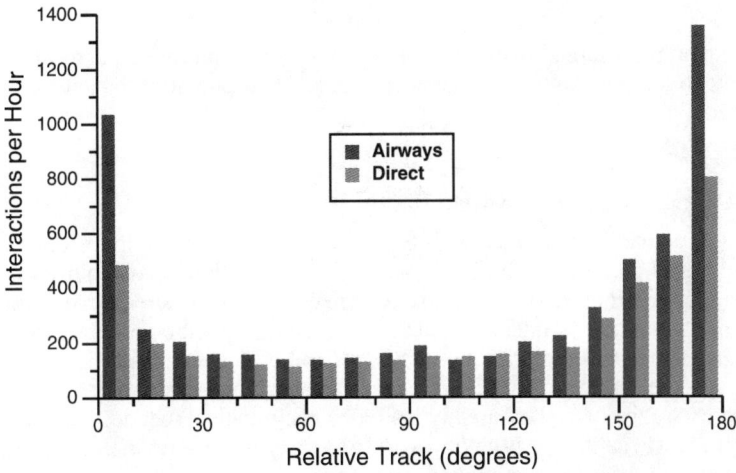

Fig. 4 Interaction frequency and relative track angle.

(summed over all angles) for the two routing scenarios would be nearly equal; this highlights the effect of spatial distribution alone on relative track angle. From Fig. 5, the effect of spatial distribution alone is as follows: for relative track angles close to 0 and 180 deg, airways operation produces higher interaction frequencies, but for all other angles direct routing produces higher interaction frequencies. Thus, for those angles between 10 and 170 deg where Fig. 4 shows higher interaction frequencies for airways operation, this is an effect of traffic density rather than spatial distribution.

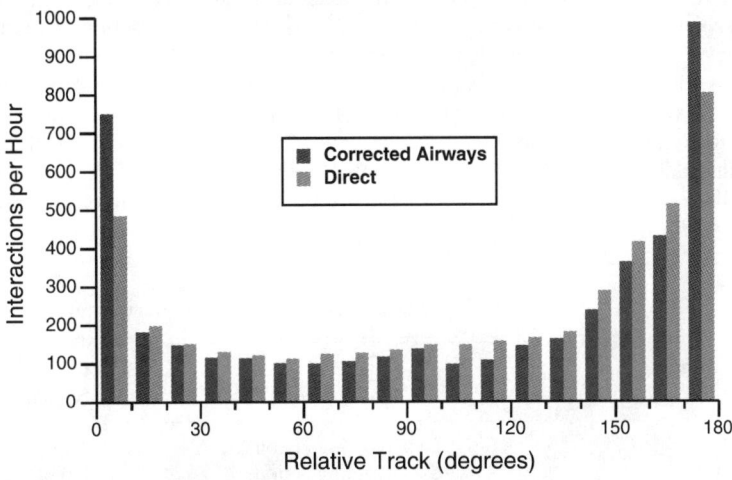

Fig. 5 Same as Fig. 4, but corrected for difference in traffic density.

V. Discussion

The results presented in the previous section can be interpreted in terms of both the potential capacity gain from direct routing and the potential capacity gain from reduced horizontal separation thresholds.

A. Capacity Gain from Direct Routing

It was pointed out in Section II.B that the range of separation problems that contribute to controller workload extends from situations that require avoiding action at one extreme to situations that require no more than increased monitoring at the other. It was also pointed out that the less serious separation problems are likely to contribute as much or more to total workload because there are more of them. Although the threshold values that delimit the various categories of separation problems are likely to vary greatly from one geographical location and operating environment to another, it is helpful to keep some typical values in mind to interpret the results discussed in the previous section.

In many regions of airspace aircraft are not permitted to be separated by less than 5 n miles and 1000 ft simultaneously. However, to allow for uncertainties in their knowledge of predicted positions, controllers are likely to take avoiding action when they or their supporting automation predict a horizontal separation somewhat greater than these values, perhaps at 8 n miles horizontally. They are likely to begin planning avoiding action when the separation is predicted to be larger still, say 10 miles, and to institute more intensive monitoring at an even greater predicted minimum separation, say 12 miles. Although these figures are no more than informed guesses of average values, the graph in Fig. 2 is relatively flat in this region, and so the precise values are not critical.

If the frequency of interactions that require controller attention is reduced by direct routing, then the potential capacity gain is the increase in traffic density that would restore this frequency to its original value. Assuming that the frequency of interactions for any given separation criterion is approximately proportional to the square of traffic density,[10] it follows that the factor by which capacity is increased is the reciprocal of the square root of the factor by which interaction frequency is reduced. These assumptions lead to the values shown in Table 2.

Thus, the potential capacity gain from direct routing is in the region of 17%. However, if the separation problems in a system using all direct routes are more difficult to deal with than those in a system using airways, the actual gain will be less than this, and might even be negative. On the other hand, if most of the 17% is available, this is well worth having. Such a figure for potential capacity gain would be

Table 2 Illustration of potential capacity gain from direct routing

Predicted minimum separation	Controller action	Potential capacity gain, %
8	take avoiding action	21.3
10	plan avoiding action	19.0
12	increase monitoring	17.5

Table 3 Effect of changing separation thresholds

Change separation threshold, n miles	Int. freq. reduction, %	Potential capacity gain, %
12–10	21.3	12.8
12–8	41.4	30.7

ample justification for large-scale real-time simulations to quantify how much of the potential gain can be realized.

B. Capacity Gains from Reduced Separation Thresholds

The potential 17% capacity gain from direct routing, while useful, is still small compared with the 65–95% needed by 2015, and so where is the rest to come from? Further resectorization of the airspace and use of computer assistance tools will provide some of the deficit, but these are unlikely to provide all of it.

Another possibility is to reduce the separation thresholds at which trajectory interactions contribute to controller workload. This might be done, for example, by use of computer-based prediction and monitoring tools to reduce uncertainty about predicted positions. Table 3 illustrates the effects of two relatively modest changes to the 12-mile threshold used in Table 2. Again, these potential capacity gains are well worth having and should encourage efforts to reduce the various thresholds.

VI. Concluding Remarks

A fast-time simulation study has been described that quantified the potential increase in traffic-handling capacity that might result from use of direct-routing instead of traditional airways operation. Data were collected from the simulation for the volume of airspace at and above 10,000 ft, which covers practically all of Europe. The main results and conclusions are:

1) Curves relating trajectory interaction frequencies to horizontal separation thresholds, for both direct-routes and airways operation, are shown in Figs. 1 and 2. These show that direct-routes operation produces substantially lower interaction frequencies for any given separation threshold, but that the reduction is much more marked for smaller threshold values.

2) A decomposition of the reduction in interaction frequency into the component that arises from reduced traffic density and the component that arises from spatial distribution is discussed in Section IV.D. This shows that most of the frequency reduction from direct routing results from reduced traffic density. The spatial distribution component contributes positively to the frequency reduction at low separation thresholds, but contributes negatively to it at high thresholds (see Fig. 3).

3) The distribution of relative track angles at points of closest approach for direct-routes and airways operation is shown in Figs. 4 and 5. Direct routing significantly reduces the overtaking and head-on interactions.

4) Interpretation of the results in Fig. 1 indicates a potential capacity benefit of about 17%. However, it must be emphasized that this is a *potential* capacity

benefit. How much of it can be realized in practice can only be determined by real-time experiments involving human air traffic controllers.

5) A possible source of further capacity gains is indicated. The potential capacity gain that would result from reducing the threshold for what constitutes a problem requiring controller attention can be estimated from Fig. 1 (see Section V.B).

The potential capacity benefit from direct routing was estimated to be about 17%. This could make a useful contribution to the need for additional ATC capacity and would seem to be ample to justify much more costly but more precise real-time simulation studies. However, an additional 17% is not going to solve Europe's capacity problem. In the author's opinion, the solution will ultimately require a reduction in the threshold values of what constitutes a separation problem needing controller attention. The potential capacity benefit from any such reduction can be determined from the results presented.

Acknowledgments

The author wishes to thank the EUROCONTROL Central Flow Management Unit, and Monsieur de Muelenaere in particular, for providing the traffic data which this study was based. The author also wishes to thank his DERA colleagues for reviewing drafts of the chapter and for making helpful suggestions.

References

[1] "EATMS User Requirements Document, Volume 1," EUROCONTROL EATCHIP Doc. FCO.ET1.ST04.DEL01, EUROCONTROL, Brussels, 1995.

[2] Mandal S., and Overend, V., "Assessment of Capacity Shortfall in the ECAC Airspace Through 2000-2015," R&D Rept. 9614, U.K. National Air Traffic Services London, 1996.

[3] "PD/1 Final Report," Programme for Harmonised ATM Research in Eurocontrol (PHARE), EUROCONTROL Doc. 96-70-24, EUROCONTROL, Brussels, Jan. 1997.

[4] Garot, J.-M., "Message from the Director," Annual Rept. for 1996, EUROCONTROL Experimental Centre, Bretigny, France, 1996.

[5] Magill, S. A. N., "Trajectory Predictability and Frequency of Conflict-Avoiding Action," *Confederation of European Aerospace Societies 10th European Aerospace Conference*, Amsterdam, 20-21 Oct. 1997.

[6] "Estimation of the Effect of the Quality of Meteo Information on the Traffic-Handling Capacity of ATM Systems," *4MIDaBLE Project Work Package 5, Final Rept.*, European Commission DG VII, Brussels, 1998, pp. 332–354.

[7] Magill, S. A. N., "On the Vertical Speeds of Airways Traffic," *Journal of Navigation*, Vol. 49, No. 1, 1996, pp. 58–71.

[8] Law, A. M., and Kelton, W. D., *Simulation Modeling and Analysis*, 2nd ed., McGraw-Hill, 1991.

[9] Warren, A. W., and Schwab, R. W., "A Methodology and Initial Results Specifying Requirements for Free Flight Transitions," *Air Traffic Control Quarterly*, Vol. 5, No. 3, 1997, pp. 133–156.

[10] Andrews, J. W., and Welch, J. D., "Workload Implications of Free Flight Concepts," *FAA/EUROCONTROL ATM R&D Seminar*, Saclay, France, June 1997.

Chapter 24

Performance Measures for Future Architecture

Steve Bradford,* Dave Knorr,[†] and Diana Liang[‡]
Federal Aviation Administration, Washington, D.C.

I. Introduction

THE goal of the Federal Aviation Administration (FAA) is to provide safe, equitable, and efficient air traffic services to the users community. To gauge the effectiveness of the delivered services, a set of metrics has been used to measure the performance of the system. Data are collected and results are published periodically.

The historically predominant metrics have been delay and safety. The safety of the National Airspace System (NAS) is based on the number of accidents and incidents in the air and on the airport surface. When considering delay, system delay increases and decreases are assumed to measure the impact on the users in terms of profit and loss. These are very high-level metrics that may indicate trends but may not actually answer any questions regarding the quality of service and the health of the NAS and its users. For instance, a reporting of system delay does not provide insights into where and why the delay occurred. Delays may occur due to weather, airline activities, or air traffic control (ATC) services; therefore, more detailed information is needed to aid the service provider in identifying and addressing a direct cause and not an easily observed symptom.

As the FAA moves forward with modernizing the NAS, several programs are in place to incrementally implement enhanced capabilities through fielding of new systems. One such program is Free Flight Phase 1 (FFP1), in which selected sites in the continental United States (CONUS) are scheduled to receive new air traffic management (ATM) systems. How efficient are these systems? Do they deliver the services as forecast? How does it impact the NAS overall? How is it best measured or measured at all?

This material is declared a work of the U.S. Government and is not subject to copyright protection in the United States.
*Chief Scientist, NAS Concept Development Branch.
[†]Metrics Team Manager, Free Flight, Free Flight Phase I Office.
[‡]Ops Research Analyst, NAS Concept Development Branch.

II. Architecture

To support NAS modernization, the FAA, working with the aviation community, has developed a strategic plan (operational concepts) for the future of the NAS. The plan outlines the services and capabilities that the agency will provide and the role of the service provider in the future NAS. Along with this strategic plan, a roadmap (architecture) has also been developed to outline the systems and support activities (people, procedures, training, etc.) that are needed to deliver these capabilities. A schedule and resources are included in the roadmap to explain when the capabilities will be available.

The architecture lays out the capabilities listed in the operational concepts. Each capability is then mapped to a set of systems and the nonsystems. Each system and nonsystem component has a schedule and funding stream associated with it. This traceability allows the decision maker to conduct trades as the FAA is faced with a new budget crisis or to comprehend the impact to the capability when an individual system slips in schedule. The architecture shows the building blocks needed to provide a capability.

Each capability contributes value to the user. Performance measures are used both to establish the relative value of new capabilities and to measure the actual system impacts. Collecting performance data as new capabilities are implemented is key to future decision making, as implementation priorities in the architecture are refined.

III. Metrics

As the FAA proceeds forward with NAS modernization and implementing the future architecture, the agency recognizes the need to modernize and update the metrics that it uses to measure and report its performance. As a result, new metrics such as predictability, flexibility, access, and efficiency have joined the old standbys of delay and safety. How are these metrics being used? What sort of data need to be collected? Can we measure real changes in actual performance, and are we doing so? From past experience we know that the NAS is a complex and adaptive system, quickly moving to respond to any changes. With this high degree of adaptation and variability, the agency needs to be vigilant to assure it is measuring itself against the right performance metrics.

For metrics to capture the real impact of NAS modernization, a more sophisticated approach is required that considers a broader spectrum of measures and a focus on individual phases of flight. The FFP1 program made progress towards an expanded set of metric categories and definitions. These categories relate directly to objectives developed in collaboration with NAS system users and will be used as a reference point for this chapter:

1) Access—ability of users to enter airspace/airports on demand. Runway throughput during peak periods is an example of a specific metric under the access category.

2) Delay/Efficiency—grouped together to capture all changes in mean flight times as well as redistribution of flight times resulting in reduced fuel burn (i.e., spending more time at preferred altitudes). A specific example for Center Terminal-Radar Approach Control (TRACON) Automation System (CTAS) transition is "time from 200 n miles to the TRACON meter fix."

3) Predictability—focuses on the variation in the ATM system as experienced by the user. Predictability includes both variability in flight times and arrival rates.

4) Flexibility—focuses on the user's ability to meet specific flight goals when the preceding metrics do not capture intent. Included are events in which a reduced flying time or direct route is not desired, or when a flight is of such high priority that an air carrier might prefer greater total delays (and increased variability) to allow one critical flight to make connections.

Safety is a difficult value to measure directly because the typical measurements of fatalities or damage are rare events. Safety can be measured by looking at adherence to norms. It can also be measured qualitatively with respect to perception of safety. Because it is both the fact and the perception of safety that we seek to improve in reducing the accident rate, looking for measurement of perception should be an objective of the metrics development work.

Each of these categories contains specific measures for flight segments. Each of these measures are influenced by conditions surrounding the observation. To properly measure or estimate the impact of system enhancements on the NAS, we must have not only context information on weather and demand but also a better understanding of user intent and traffic flow restrictions. Without this information, the measurements discussed in the preceding paragraphs become polluted with data points with sets having completely different objectives and constraints.

Collecting metrics data in concert with weather and demand, user intent, and system restrictions will allow educated tradeoffs between individual flight objectives and overall NAS objectives. This approach will also support extrapolation of the benefits of new ATC enhancements to other sites or regions.

How do we establish such a set of data, which addresses user intent, restrictions, and phase of flight?

IV. Architecture and Performance

Similar to the way in which the architecture is built, performance measures may also be built upon building blocks we will refer to as traffic management phases. The traffic management phases are the typical transitional states that individual flights experience in moving through the NAS. It is the management of the individual states' objectives that is inherently a part of each controller's job. Controllers do not focus on the movement of each flight and try to optimize each flight; rather they focus on the process associated with the phase—can they meter aircraft across a fix hitting the spacing required aircraft after aircraft; can they move aircraft out a departure runway with minimal interdeparture time, etc. Although the individual flight statistics end-to-end, from request-for-taxi to engine shutdown, is a measure of the NAS-level performance, they currently have not been focused on the individual state objectives and their interactions. A clear understanding of this model of operations and its derivation are essential to attempt measuring the future NAS depicts a set of phases applicable for gate-to-gate or cruise-to-cruise concept.

What are the parameters, time and space, associated with the traffic management phases? In the United States, a typical flight distance is 400 miles, and the aircraft crosses two to three centers. The flight includes flight planning, ramp, surface, departure, dispersion, cruise, collection, arrival, surface, and ramp again (Fig. 1). The aircraft departs from the gate, taxis to the runway, takes off, follows a standard instrument departure (SID), enters the jetway, exits the jetway, queues for the standard terminal arrival route (STAR), approaches, lands, and taxis off the runway to the gate.

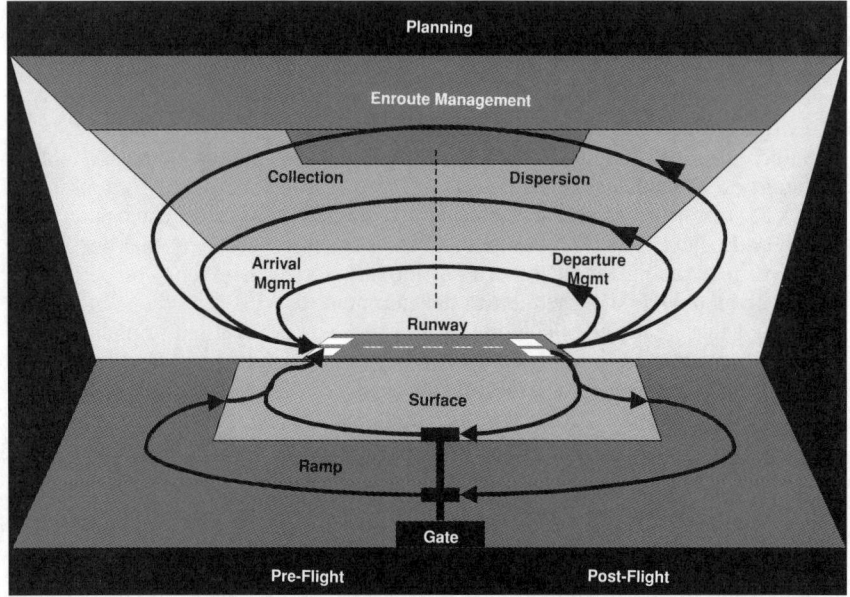

Fig. 1 NAS operation—traffic management phases.

There are also flights with distances flown less than 400 miles in the high-density area. For example, flights between Washington, DC (DCA); and Newark, New Jersey (EWR); may never reach cruising altitude. Before the flight can leave DCA for EWR, an arrival slot assignment may be required. The aircraft departs from the gate, taxis to the runway, takes off, follows a SID, queues for a STAR, approaches, lands, and taxis off the runway to the gate. For shorter flights, the aircraft will go directly from departure to arrival.

Each flight goes through these phases as it services a certain city pair. Each individual phase impacts its neighbors. An inefficient flow from the collection phase at an airport will impact the arrival phase for that airport. Similarly, it would also impact the previous phases that feed it, such as cruise, dispersion, or departure phase. These phases will have to modify their operations because the aircraft they supply cannot be accepted by the collection phase. Understanding the relationship of these phases is the first step in measuring the future NAS.

The generic NAS operation is now partitioned into smaller phases. The individual phases represented have a clear objective and therefore can be measured. For example, what is the right metric for measuring the dispersion phase? Possible metrics are:

1) Throughput, does the throughput per time unit of the dispersion phase match the potential throughput of the departure phase?

2) Efficiency, can aircraft file their preferred trajectory for this phase?

3) Efficiency, can aircraft achieve their preferred trajectory in this phase?

4) Flexibility, is the phase tuned to demand, is the time in phase/extent of the phase tuned to the flow?

5) Predictability, is the performance across different days and configurations stable?

The comparison of performance also includes a review of services: Is the airspace tuned to the needs of this phase? Are the letters of agreement with respect to this airspace aligned with the flow and operations? Are the procedures adequate or is the cognitive support required available to dynamically manage the flow most efficiently? Do the characteristics of the flow increase complexity for the controller and impact the separation service? Each of these can also be reviewed with respect to the preceding metrics.

V. Analysis
A. Case Study 1: FFP1, Measuring Direct Routes in En Route Airspace, Cruise vs Collection Process

1. Introduction

FFP1 represents a joint FAA/industry collaboration to accelerate the deployment of new air traffic management capabilities to limited locations in the NAS. The intent of FFP1 is to provide near-term benefits to airspace users while collecting data to determine the cost and benefits of a more robust or national deployment of each capability.

One of the key components of the FFP1 initiative is the User Request Evaluation Tool (URET), which supports air traffic controllers in granting user-preferred trajectories and/or direct routings. A prototype version of URET is currently operating in the Indianapolis, Indiana (ZID), and Memphis, Tennessee (ZME), air route traffic control centers (ARTCCs). The FFP1 performance metrics team has been studying the impacts of URET at both of these sites.

URET provides controllers with automatic conflict detection and trial planning capabilities in which controllers are able to view alternative aircraft routes and implement those that are most efficient and free of potential conflicts. As of July 1999, the trial planning capability was enhanced to include automated amendments to the host computer through the URET interface (a two-way interface to the host). The expectation was that automated amendment entries through URET would increase the number of direct routings given by controllers, because URET would support much less time-consuming amendment inputs.

From discussions with controllers at ZID, we have learned that all routes through ZID airspace do not equally lend themselves to improvements in direct routings. Some traffic flows are restricted by congestion at destination airports or subsequent en route centers, which may mask overall improvements. To effectively evaluate URET's operational impact, it is imperative to consider the flight phase.

2. Cruise vs Collection State

Measurement of en route performance without consideration of the traffic management phase can lead to misinterpretation of new technology impacts and fail to identify priorities in capabilities supporting the future architecture. To properly assess URET's impact on the air traffic control operation, the data set was stratified to identify those aircraft in the collection phase from those aircraft in a true en route phase. Both states occur within ZID en route airspace, but they are significantly different in their ability to support direct routes.

Direct routing opportunities in the collection phase become much more limited than in the en route phase. By separating aircraft into these two phases and applying metrics appropriate for each, we can refine our ability to predict and measure the impact of new capabilities.

The Indianapolis (ZID) center has some of the most complex airspace in the NAS because of both volume and the amount of crossing traffic. ZID also has significant constraints on flows heading north and east to busy airports like Chicago O'Hare (ORD), Detroit Metropolitan (DTW), Cleveland Hopkins International (CLE), Newark International (EWR), New York-John F. Kennedy International (JFK), and Philadelphia International (PHL). For these airports traffic flows can be set up with miles-in-trail (MIT) restrictions going back as far as 400 miles from the final destination airport.

Cleveland (ZOB) and Washington (ZDC) centers border ZID on the east. They have airspace equally as complex as ZID and have established restrictions on traffic crossing from ZID into their airspace. Centers to the south and west of ZID tend to have less complex airspace with fewer capacity constraints.

Indianapolis controllers have reported an increase in the number of direct routing amendments for aircraft with destinations south and west of the Indianapolis center. We have attempted to validate this anecdotal information by comparing flights within city pairs traveling north from ORD with the same city pairs leaving ORD and heading to airports in the south or west (e.g., STC, DFW, or LAX). Sample city pairs chosen for this analysis are presented in Table 1.

Table 1 Sample city pairs

South-West	North-East
SIL	PHL
BNA	CLE
BNA	ORD
BNA	PIT
BNA	BWI
STL	BWI
CLT	ORD
STL	DCA
DFW	PHL
BNA	DTW
PIT	DFW
STL	PIT
IAH	PIT
IAH	DTW
IAH	EWR
STL	EWR
LAX	IAD
SFO	IAD
STL	IAD
LAX	JFK
MCO	ORD
MIA	ORD
RDU	ORD
STL	DTW

3. Methodology

Several approaches and data sets were considered for measuring URET's impact on direct routings in ZID. The availability of data associated with the pre- and postimplementation period, however, is a constraining factor. The approach discussed in this chapter uses enhanced traffic management system (ETMS) data for two months (October 1998 and 1999) before and after implementation of URET's two-way capability. We have focused this study on the Indianapolis center, where controllers have had more exposure to URET and are using the capabilities of the tool more frequently.

To calculate the delta (excess) distance metric, we use x,y coordinates from ETMS to calculate the actual distance flown through ZID (in n miles). We then use the entry and exit points of each flight going through the center to calculate the great circle route (assuming it is the optimum distance). We use the difference between the actual distance and the great circle distance to calculate excess distance. We are interested in measuring changes in this excess distance associated with the implementation of the two-way host in URET.

We recognize that using distance as a measure of efficiency fails to capture impacts of wind or speed restrictions put on the aircraft. We feel, however, that the impact of wind over the distances flown in ZID airspace would be minimal. Speed restrictions placed on aircraft to achieve spacing and meter feeds to adjacent centers are not captured with the distance metric.

In its favor, wind generates less variability in a distance metric than it would in a metric like flight time. For this reason, we believe distance is a suitable metric for assessing the impact of URET on direct routes in the Indianapolis center. Furthermore, we believe the great circle route through the ZID entry and exit points is a reasonable measure of optimal distance.

Our metric—the difference between the actual distance flown through ZID and the great circle distance (referred to as excess distance)—is intended to capture the difference between the optimal route that airlines *prefer* to fly and the actual route they *do* fly. With *the hypothesis that URET is contributing to more direct routings through Indianapolis airspace*, a favorable change would identify itself as a reduction in excess flight distance through ZID in 1999 from the 1998 data. The analysis of the entire data set, however, indicates no statistically significant change (1999 from 1998 excess distance).

When sample city pairs are segregated into flows moving north and east (NE) of ZID from those moving south and west (SW), it is clear that the *collection* vs *en route* state phenomena exists (as observed by ZID controllers) shows the additional distance required to traverse the Indianapolis center when heading to NE destinations vs the same city pairs with flows to SW destinations. The statistics presented in Table 2 identify a difference in excess distance of approximately 0.917 n miles. This difference is shown to be significant at the $\alpha = .05$ level. The data used for this test include the data sets for both 1998 and 1999 and are segregated by NE and SW only.

This additional distance for flows NE is an indication that flights heading to capacity-constrained airports are, in fact, in the collection phase. Flights heading through ZID airspace to destinations SW exhibit less excess flight distance and are more representative of aircraft in a true en route environment. Similar trends exist for traffic heading to Atlanta, which is heavily congested (but is actually 14 south of ZID).

Table 2 Excess flight distance NE vs SW: *t*-test assuming unequal variances

	NE	SW
Mean	3.761365	2.844028
Variance	33.91213	16.14122
Observations	12156	11996
Hypothesized mean difference	0	
df	21613	
t stat	14,26513	
$P(T< = t)$ one-tail	2.92E-46	
t critical one-tail	1.644923	
$P(T< = t)$ one-tail	5.84E-46	
t critical one-tail	1.960075	

4. URET's Impact: Collection vs En Route Phases

We are testing the hypothesis that the controller's ability to use URET's direct routing capability is measurable when flights are in a true en route phase. Conversely, flights already in the collection phase have constraints, which may override URET's direct routing capability. Initially we analyze the data set going through ZID to the NE. Table 3 presents the various statistics calculated for NE bound flights in October 1998 vs October 1999. As expected, these statistics show that the average excess distance for NE flights has increased. This difference of nearly a half mile is statistically significant at the $\alpha = .05$ level. Additionally, the observed variance has also increased, indicating less predictability in flight distance and time.

Notably the number of operations in ZID increased by nearly 8% in October 1999 from the previous year. Additionally, the number of operations at facilities NE of ZID also increased. It is reasonable that these increases in demand where capacity is already limited would override any improvements from URET.

In contrast, Table 4 presents similar results for SW bound flights for the same two periods. The statistics show that the average delta distance for SW flights has

Table 3 Excess flight distance north or east (NE) bound: *t*-test assuming unequal variances

	1998	1999
Mean	3.533928	3.97389
Variance	29.67187	37.78622
Observations	5872	6284
Hypothesized mean difference	0	
df	12120	
t stat	−4.182301	
$P(T< = t)$ one-tail	1.45E-05	
t critical one-tail	1.64498	
$P(T< = t)$ one-tail	2.91E-05	
t critical one-tail	1.960161	

Table 4 Excess flight distance south or west (SW) bound: *t*-test assuming unequal variances

	1998	1999
Mean	3.00891	2.692897
Variance	18.75015	13.70472
Observations	5737	6259
Hypothesized mean difference	0	
df	11334	
t stat	4.277526	
$P(T< = t)$ one-tail	9.53E-06	
t critical one-tail	1.644989	
$P(T< = t)$ one-tail	1.91E-05	
t critical one-tail	1.960175	

fallen by approximately 0.316 n miles in 1999 vs 1998. Again, this difference is statistically significant at the $\alpha = .05$ level. Variance has also fallen slightly between the two periods.

With the data segregated to focus only on those flights in an en route state, we can see the expected impacts of the URET capability (measured in reduced excess distance). Our statistical test alone is not conclusive that URET is responsible for this improvement. Discussions with ZID controllers and data collected on use of the URET tool indicate URET is contributing to an increase in direct routes. As with any statistical analysis, expert observation should support the data analysis. Furthermore, we believe the amount of improvement measured may be understated given the increase in ZID traffic from 1998 to 1999.

Analysis of additional ZID data will continue, recognizing that improvements will likely manifest themselves in flows to the SW where a purer en route environment exists. In the collection phase, URET benefits may be related to removal of altitude restrictions as opposed to direct routes. Similar approaches that consider traffic management phases will be used as URET is deployed to subsequent en route centers. As important, traffic management phases must be considered when analyzing potential candidates for future URET deployments and setting expectations for results.

B. Case Study 2: Dry Heat Departure, Dispersion Phase

Case study 2 looks at the performance of the dispersion phase. This example is instructive because the "bottleneck" is not in the much-maligned runway or departure phase, but rather in the ability to get aircraft out of the terminal airspace in a manner that meets the inherent capacity of the airports. This example is taken from airspace planning activities conducted by the FAA's System Capacity Office (ASC) with the western region (AWP).

Dry heat is an example in which the dispersion phase is a hindrance to overall performance. The different management phases have not been balanced to utilize the maximum peak throughput of the airport runways.

The Phoenix International Airport (PHX) has been experiencing growth in airport traffic. With the growth, congestion and inefficient operations were beginning

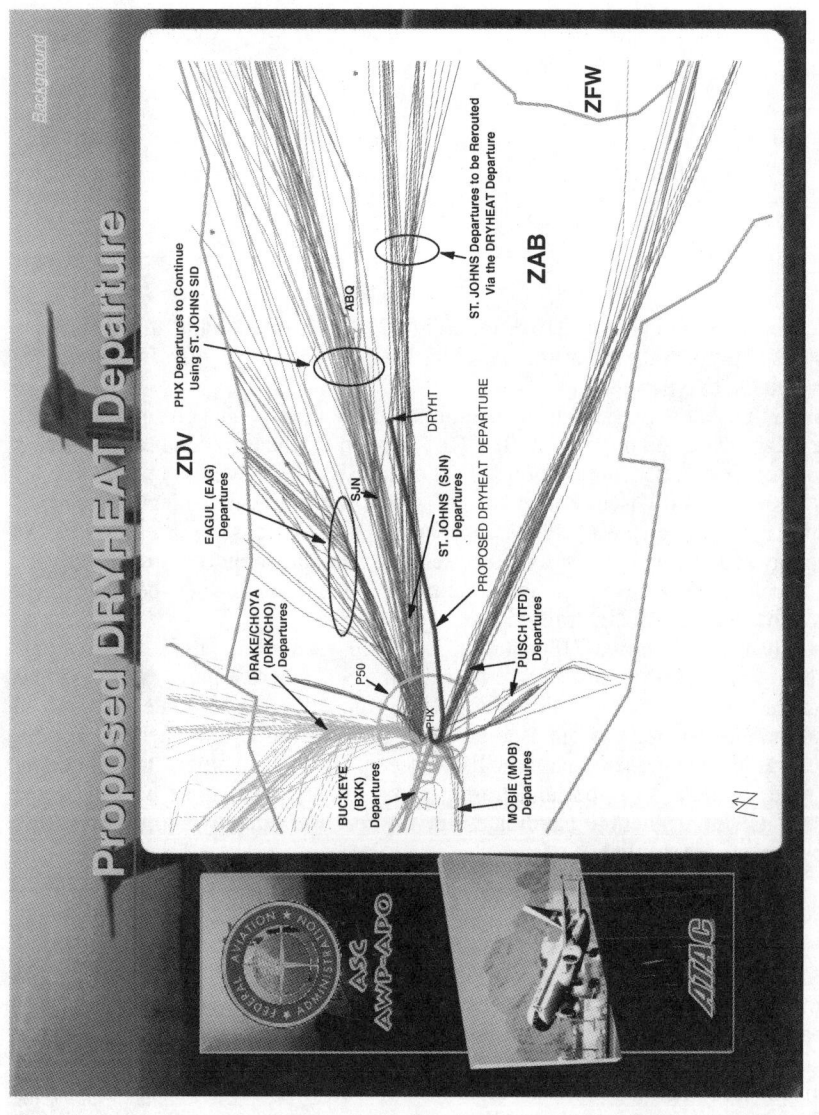

Fig. 2 Proposed Dry Heat departure.

to have an impact on the airlines' daily operations. As the radar data show, there is significant flow of traffic to the east and northeast from Phoenix.

In the original airspace configuration for the 26R/L runway configuration, the flow to the east was assigned to 26R and departed over the St. John's fix. A fix is a geographical position identified for the purpose of defining routes and procedures. Fixes are named for ease of use and reference. In this airspace structure the departure fix count for St. John's was centered around 160 aircraft per day. For all departures on the north runway, the peak day counts were centered around 375, with the south departure runway counts around 250. The imbalance is clear; and so, would an airspace redesign for the flow to the east better match the dispersion phase of the potential departure throughput?

As part of the redesign effort, the Dry Heat departure was proposed and assessed by the FAA and its support contractor. The Dry Heat departure is a procedure defined by headings, altitude, and distance for aircraft departing Phoenix Airport. The results reported here are not based on model projection but on a data collection of the area before and after the implementation of Dry Heat departure fix. With the development of a south departure for the east flow, the distribution of departures across the runways is more balanced with the north side peaks near 340 and the south peaks at 305.

The redesign resulted in improvements in both efficiency and predictability. With the Dry Heat departure added, the complexes with significant east flow pushes had reduced total travel times to facility boundaries of 1 min (20.9 to 19.9) and a reduction in the standard deviation of 1 min. The improved flow is also evident in the taxi-out time numbers. The taxi-out numbers include surface movement and the queue time for departure runway. The decreased values reflect that the departure phase's inherent performance characteristic, i.e., potential airport runway throughput, is now matched by the airspace structure of the dispersion phase. Average taxi-out times were reduced by 1.5 min (13.5 to 12.0) and have lower standard deviations of 1.9 min (7.7 to 5.8).

VI. Conclusions

The NAS is a large, complex system that will continue to change and adapt to new infrastructure enhancement, technologies, or procedures. The FAA is working with the aviation community to develop a set of metrics for reporting the health of the NAS and projecting the benefits of improvements. Of equal importance to the improved measures is the identification of not only the operator's objective for a flight but also the system's flow objectives at play at various times during the flight.

By looking at the system as a series of interacting traffic management phases, we can partition the end-to-end performance into smaller and more manageable states to assess the problems and proposed solutions, and measure the success of the new implementation (procedures or technologies). For instance, the FFP1 program is fielding decision support tools to various centers. In some cases the tools will coexist geographically. However, each of the tools focuses on enhancing the operation or objective of a particular management phase in the NAS. By partitioning the flights into the phases and understanding the objectives of the phases, the tool developers can be more effective in identifying the operational needs and evaluating their solutions.

Chapter 25

Analytical Identification of Airport and Airspace Capacity Constraints

William R. Voss*
Federal Aviation Administration, Washington, D.C.
and
Jonathan Hoffman[†]
MITRE Corporation, McLean, Virginia

I. Introduction

IDENTIFICATION of airport congestion problems is well understood. As aircraft line up to use runways, standard queuing theory applies, to a good approximation. The idea of capacity is well defined, albeit dependent upon details of the arriving and departing traffic. Away from airports, the airspace also can become congested. In this case, though, the great volume of space (in the dynamical sense, meaning both position and velocity) through which aircraft can move, and the sensitive dependence on the controller's workload, make a simple capacity calculation less applicable. When capacity calculation is problematic, identification of a congestion problem lacks a sound analytical basis. This chapter demonstrates a theoretical framework for extending the idea of identifying congestion problems from airports to airspace.

Traffic patterns change, and so over time airspace designs gradually become less efficient. Inefficient airspace design leads to congestion, but airspace congestion is not permitted to exist for long because it overworks controllers and threatens safety. To alleviate this sense of congestion, air traffic flow management (ATFM) was introduced. ATFM is generally successful at managing airspace congestion in its most literal sense. The result is the kind of airspace congestion problems that we see in today's air traffic control systems. An inefficient design is not cured by ATFM—the congestion is turned into ground delays.

In addition, it has recently been recognized[1] that inefficiencies other than delay cause penalties to the users of the airspace. These are the secondary effects of congestion. Reducing ground delays requires correcting the airport or airspace

---

Copyright © 2001 by the MITRE Corporation. Published by the American Institute of Aeronautics and Astronautics, Inc., with permission.
*Director, Office of Air Traffic System Development.
[†]Principal Scientist, Center for Advanced Aviation System Development.

problem that led to them. This provides the motive force for resolving airspace problems. Unfortunately, institutional memory of the cause of a particular sector regulation (Europe), restriction (United States), or routing (both) quickly fades and is frequently not well communicated outside the original facility. Therefore, an analytical method for identifying airspace problems from their secondary effects is essential to guide any attempt to correct the problems with new procedures, sectorization, or technology. Thus, the technical question that led to this research: Given an airspace in which inefficiencies are suspected, how can the true problems be identified analytically?

II. Background

The way that airspace studies are typically initiated and conducted shows the importance of an analytical process to identify airspace problems. When contemplating an airspace study, one is confronted with at least two difficult questions:
1) How large a study is needed to find a solution to the problem at hand?
2) How can the problem be solved without creating another one?

In practice, the scope of an airspace study is more likely to be defined by the sphere of influence of the organization running the study rather than by the nature of the actual problem at hand. For example, if a local group is chartered to conduct a study of arrival delays at a specific airport, then the scope of inquiry will normally be limited to arrival routes in the terminal airspace and adjacent en route centers. This will lead to a local solution even though a subtle adjustment in regional traffic flow may have produced a far more satisfactory result. The same problem also occurs in reverse. If a national authority addresses system-wide inefficiencies, then its inquiry will likely focus on large-scale models and identify strategic solutions when a number of local changes may in fact be far more effective. Decision makers need analytical tools that ensure that the scope and granularity of a study is appropriate to the problem at hand.

Ensuring that a local solution does not cause problems elsewhere is even more difficult. In practice, each entity involved in a study tends to solve problems in a way that is "good" for its own people. For controllers, getting aircraft out of their airspace (and into somebody else's) in the most efficient way is desirable, even though this may cause a worse problem somewhere else. Solutions obtained in this way inevitably lead to disagreements among air traffic management (ATM) facilities. Therefore, implementation of airspace changes is often dominated by negotiation rather than analysis.

To solve this problem, decision makers need a quantitative, objective definition of what they are trying to achieve with the study. The Federal Aviation Administration's (FAA) *Airspace Management Handbook*[2] describes important steps to follow to accomplish a practical reconfiguration of ATM resources. The first step in the process is analytical problem identification. The work described here is an example of that first step.

III. How to Find Airspace Problems

There are two ways to identify airspace problems: the predictive and the deductive. Each has its advantages and disadvantages. Emphasis has traditionally been given to the predictive method.

A. Predictive Method

The predictive method uses simulation to remove ATFM ground delay programs and fly the aircraft as pilots would prefer. The simulated volume of airspace must be large because the objective is to find problems, and problems tend to exist at the interfaces between ATM facilities. For airport problems, one should look for large taxiway delays on departure or excessive airborne holding of arrivals (45 min, the amount of holding fuel required in the United States, is a convenient standard for "excessive"). For airspace problems, using some workload metric, one should identify "red sectors," where the density of traffic is unacceptably high. This is unbeatable in principle, but developing the simulation means balancing the need for a wide-scale survey of the airspace with the need for detailed modeling to accurately reproduce the operating environment of each single sector.

Efforts to develop such a large-scale simulation are under way in various places around the world. NavCanada, at their headquarters in Ottawa, have simulated the entire Canadian airspace using the Total Airport and Airspace Modeller (TAAM). Simulating this enormous area is possible because of the relatively low total number of flights in that airspace, about 9700. At the Center for Advanced Aviation Systems Development in McLean, Virginia, six air route traffic control centers are being simulated, a volume of 800 by 700 n miles, containing some 45,000 flights per day. At the EUROCONTROL Experimental Centre in Bretigny-sur-Orge, France, the airspace from Britain to Poland, north to Scandinavia and south to Italy, about 1000 n miles^2, is being simulated using the Reorganized ATC Mathematical Simulator (RAMS).

The biggest advantages to this approach are that it gives the analyst nearly complete control over variables in the system, it can predict future congestion, and it can identify limits to growth. It can be a tool for active management of airspace, not just reactive adaptation once problems are identified. The biggest disadvantage is that it is slow and expensive—the steady increase we have seen in computer power is not accompanied by a corresponding increase in the ease of modeling the unique details of each ATM facility.

B. Deductive Method

The deductive method of identifying airport and airspace problems is data intensive, but less work than validating a large simulation. The precise definition of metrics is determined partially by available data, as long as the categories defined in the preceding sections are covered. The ATFM system is monitored on an appropriate scale: airlines change their schedules in the United States about once a month, and so long-term trends are based on monthly averages. The time series are inspected for patterns in the evolution of the metrics that are symptomatic of airport or airspace congestion.

After identifying patterns, operational personnel in both the ATM facilities and the primary users' dispatch offices should be consulted. This step is essential. Because the ATM system serves many needs, it must be ascertained whether an identified phenomenon is a problem, or a solution put in place to cure something worse (see Section VI). This method is relatively fast and inexpensive; our work covered flows to 14 airports in 6 months.

In practice, of course, a combination of the two methods offers best power and speed. Deductive problem identification can identify parts of the ATM system that do not cause problems for users (e.g., because excess demand is decreasing). Simulation offers a confidence in assigning causes to observed effects that lends credence to multifacility airspace problems, which are not completely verifiable by a single domain expert.

IV. Definition of an Airspace Problem

Our operating principle is that runways should be the limiting factor in an efficient air traffic control (ATC) system. Runways are expensive to build and difficult to alter once built. They are the least flexible part of the ATM system. Airspace design, by contrast, is a matter of agreement between ATM facilities. Apart from navigation aids, whose location is becoming less constraining as area navigation becomes more common, there is nothing physically fixed about airspace. Therefore, airspace should adapt to the runways. If it does not, that is, if runways are not being used to capacity, you have the archetypal airspace problem, measured in terms of throughput. Problems of this sort are so obvious that ATM personnel are constantly making adjustments to the airspace to prevent them. The U.S. Air Traffic Control System Command Center has recently begun efforts[3] to compare the agreed-upon flow rates at a number of U.S. airports to the throughput actually achieved. These comparisons, which will be available to managers on a next-day basis, will institutionalize the use of throughput as a diagnostic performance metric.

Consulting with users of the airspace, however, shows that there are concerns other than these, which sometimes do not get addressed so efficiently. Metrics other than throughput[4] can be used to reveal other kinds of inefficiencies in the airspace. Combinations of trends in these metrics may point to airspace design problems, to flow management procedures mismatched to the traffic situation, or to other problems in the ATFM system.

A. Signatures of Airspace Problems

1. Delay Signatures

A certain amount of delay is necessary to the proper functioning of any transportation system.[5] If an airspace route is desirable to the flying public, more aircraft will fly on it until it becomes congested and delays result. However, if delay appears without a trend of increasing traffic, it is a sign that something is wrong. Delay in this context may be either a departure delay imposed on the ground by ATFM, an increase over several years in the time to complete an operation, or a significant change in the number of an airline's flights that arrive on time with respect to schedule. The notion of excess demand (see the following section) provides a measure of whether runways are the limiting factor, or something else.

2. Predictability Signatures

Predictability of schedule is just as important to a commercial carrier as the reduction of delays (though less important to other classes of users). Predictability is usually measured as the variation in some kind of movement time. In this work, the predictability of en route time (time from wheels up to wheels down) is of interest. At times of low traffic, this time will be short; at times of high traffic, speed controls, offloading of traffic, and holding will add to this time. To minimize the effect of winds, a single day is analyzed at a time.

Because airspace designs are created with a particular flow in mind, the en route times from each connecting airport to a common destination are collected for statistical analysis. The interquartile range of en route times from each origin is calculated. For the purpose of summarizing flows to the airport, a histogram has been most useful.[6] It should be remembered, though, that the variability associated with an airport might not be due to the airport, but to a congestion point upstream. For this reason, variabilities are evaluated separately by direction of flow. In no case evaluated in our initial efforts, however, was there significant variation with direction.

3. Flexibility Signatures

Just as ATFM turns a very bad thing (excess workload) into a less-bad thing (ground delays), stratification of traffic turns congestion into inflexibility. This represents a loss of flexibility in choosing altitudes for the users, which leads to inefficient fuel usage, increased wear on aircraft, and increased pollution. Although these can be serious, these are of secondary importance to users (compared to arriving on schedule). However, if a certain kind of flow is continually given preference, it can lead to inequitable treatment of different classes of users. Measuring altitude times from top of descent to touchdown, and times from wheels up to top of climb, can give an idea of which users are benefiting from a particular airspace organization and which are bearing the penalties.

B. Things That Are Not Airspace Problems

1. Excess Demand

If demand increases beyond the capacity of runways, delays will result that can be treated only by capital improvements (add runways) or reduced separation standards (e.g., via active wake vortex detection). The excess demand metric compares the scheduled traffic at an airport to a theoretical measure of capacity, and adds up the flights that are scheduled above the capacity of the runways to handle them. It is recognized that the "true" capacity is difficult to capture; however, because the objective is to identify changes in excess demand over several years, the exact value of the theoretical capacity is not very important. In this work, simulation results were used where available, engineered performance standards (EPS) were used where not.

The U.S. Air Traffic Control System Command Center's next-day throughput assessment, described in the preceding section, will yield as a by-product a record of the flow rates airports agree to accept, including airport configuration and meteorological conditions. As the numbers from this evaluation become available, they may replace the EPS, providing a substantial improvement of this metric.

2. Non-ATFM Factors

Low throughput may be mandated by statutory limits on airport traffic, as at Washington National Airport; LaGuardia Airport and John F. Kennedy International Airport in New York; and Chicago O'Hare International Airport. If these limits are based on considerations other than delay reduction, these metrics will show signs of a capacity problem, usually in the form of arrival and departure throughput below runway capacity.

V. Data Sources

For long-term tracking, the Airline Service Quality Performance (ASQP) database was used. Times of pushback, wheels up, wheels down, and arrival at gate are available for all domestic flights by the 10 largest airlines in the United States. For the flows of interest in the eastern United States, more than half of the traffic is represented. ASQP data were used when averages over traffic were sufficient. Data are available from January 1994 to the present.

When trajectory data are needed, the enhanced traffic management system (ETMS) provides radar data with a four-minute update rate for all flights under U.S. or Canadian control. The version used here was the Aircraft Situation Display to Industry, version 4.2.

This analysis was primarily looking for performance shortfalls under the best possible operating conditions, and so days of good weather were chosen. Thursday, 1 October 1998, was a heavy traffic day of generally good weather (only 1 h of fog at Boston). Friday, 23 October 1998, was a day of clear weather across the entire continental United States. No major equipment problems were reported on either day. When total traffic numbers are necessary, as in the case of the actual throughputs, these two days are shown.

VI. Results

A. Overview

Generally across the country, increases in delay were seen from 1994 to 1997. The delay performance of the system improved significantly in 1998. A survey of Official Airline Guide (OAG) schedules showed that a large part of this improvement was due to changes in aircraft schedules, reducing the excess demand. At several of the biggest airports, which were examined here, there were several cases in which the delays did not improve, even given reduced excess demand.

Low arrival throughputs are seldom seen. As the most obvious airspace problem, low arrival rates tend to be fixed as soon as they appear.

Taxi out time is increasing, often in conjunction with low departure throughputs. This is a conspicuous sign that traffic flow management programs are working.

Examination of the throughputs at various points in the airspace is facilitated by a table that includes a positive/negative indicator for each stage of aircraft flight (Table 1.) The table properly has 64 rows, corresponding to all possibilities. Rows that do not correspond to observed behavior in the eastern United States were omitted for brevity in this chapter. A number of interesting phenomena observed at individual airports are described in detail in the following sections.

B. ORD

Chicago O'Hare International Airport (ORD) is a singularly busy airport, serving as a hub for two of the largest air carriers in the United States. Demand for space at ORD is so great that neither the airport nor the airspace can handle it. The airport is

Table 1 Throughput and delay summary

Airports	Delay increase without excess scheduling?	Increasing taxi-out time?	Departure throughput below EPS?	Large en-route time variability	Center-to-TRACON throughput below EPS?	Possible airport or airspace problem?	Comments
Atlanta Hartsfield International (ATL)				x		A/S	Look for inefficient routing
Boston Logan International (BOS)	x	x	x			maybe	Early stages of airspace problem
Charlotte/Douglas International (CLT), Orlando International (MCO)			x				Not much traffic, watch out if demand increases
Greater Cincinnati International (CVG)	x	x	x		x	A/S	TFM over-resolving airspace congestion
Washington Reagan National (DCA), New York-John F. Kennedy International (JFK)		x	x		x	A/S	Airspace congestion over-solved (both slot controlled)
Newark International (EWR)		x	x	x		maybe	Could be airport arrival capacity or TRACON problem
New York-La Guardia (LGA)	x	x		x		A/S	Insufficient en-route capacity
Miami International (MIA)		x				A/P	a) departure capacity limits, or b) TFM decongesting airspace
Chicago-O'Hare International (ORD)	x	x	x	x	x	A/P, A/S	Airport and airspace problems combined
Philadelphia International (PHL)	x	x	x	x		A/S	TRACON departure capacity problem or en route congestion
Greater Pittsburgh International (PIT), Detroit Metropolitan (DTW)							No sign of problems from these metrics

415

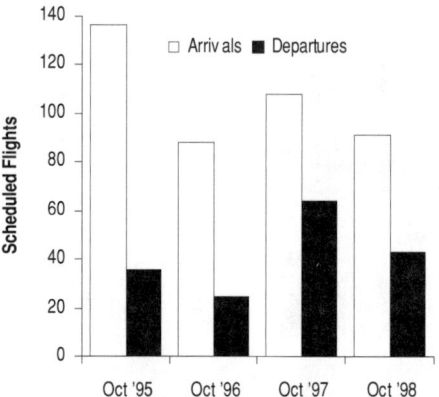

Fig. 1 Excess scheduling at ORD.

slot controlled, but this did not prevent all of the metrics in this study from showing performance shortfalls. Arrival and departure delays have been rising since 1997, even though the excess scheduling has decreased (Fig. 1). Seasonal fluctuations in average delay are large. The most pronounced performance shortfall is in average taxi-out time (Fig. 2). Since 1995 there has been a steady trend toward longer taxi times, from 16 to 19 min per flight on average. Given that ORD has about 1100 departures per day, this excess taxi-out time represents about 500 h of extra time on the taxiways per day.

ORD has less departure throughput than its runways can handle, even on a day when there is no inclement weather at its connecting airports. This is a sign of aggressive traffic flow management to avoid capacity problems.

C. EWR

The Newark International Airport (EWR) in New Jersey was one of the 27 airports that experienced more than 20,000 h of annual delay in fiscal year 1997. Almost 6% of the EWR operations experienced delays of 15 min or more, making EWR the worst airport in the nation for percentage of operations experiencing these delays. A number of factors combine to produce the high delays at EWR. The most obvious is demand. The New York metropolitan area is served primarily by EWR,

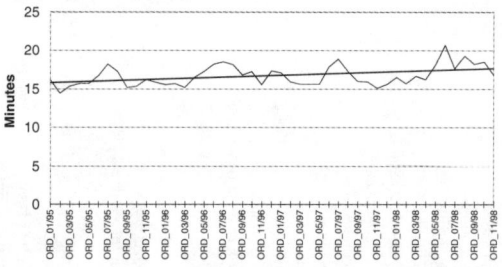

Fig. 2 Taxi-out time at ORD.

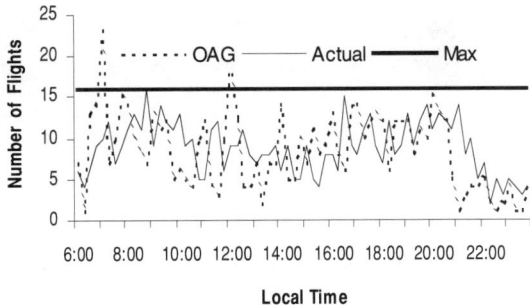

Fig. 3   Departures from EWR.

LaGuardia (LGA), and John F. Kennedy (JFK) airports. Of these, LGA and JFK are slot controlled, so that traffic cannot increase there. Any airline wishing to add flights to New York must either fly to EWR or one of the smaller airports in the vicinity.

A second, and perhaps more important, factor is the highly congested airspace around New York City. The presence of two other major hubs deprives EWR of the airspace it would ordinarily use to organize and separate arrival and departure flows. The result, as can be seen in Fig. 3, is restrictions on departures. There are three lines on this chart. The heavy horizontal line is the departure runway capacity in its preferred departure configuration, according to the Engineered Performance Standards. The dotted line (OAG) is the scheduled departure demand, according to the Official Airline Guide for October 1998. The solid line (Actual) is the departures recorded by the enhanced traffic management system for Friday 23 October 1998, a day of generally good weather.

The runway capacity is not fully utilized, even on this day of good weather. Demand is there: the scheduled arrival line shows several times when demand exceeds capacity (e.g., near 0700 and 1200 hrs). Even worse, between 2100 and 2200 hrs local time, there is little scheduled demand but significant departure throughput. These are flights with an hour or more of departure delay. Some part of the ATM system is limiting throughput, and it is not the runway. This is the clearest sign of airspace capacity limits in the current system. Arrivals at EWR do not show unused capacity, primarily because holding arrivals would aggravate airspace congestion.

D.   ATL

In Atlanta Hartsfield (ATL), the various throughput and delay metrics show no deviation from theoretically expected behavior. However, the predictability of en route times is poor compared to the national average (Fig. 4). Variabilities in en route time of 6–12 min are common. The variability is due to extensive airborne holding, as can be seen by inspection of flight trajectories. Taken by itself, this would be a sign of airspace congestion. However, the lack of delay means that further investigation is needed.

In this case, interviews with ATC personnel revealed that Delta Air Lines, the primary carrier at ATL, preferred airborne holding to ground delays. Because

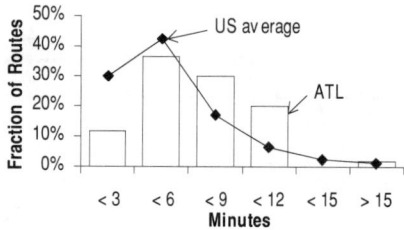

Fig. 4   En route time variabilities for ATL.

Atlanta is far from any other large airport, there were no major flows conflicting with the airspace reserved for holding, and so it was possible for the en route center and approach control to oblige. Therefore, this is not an airspace problem—it is an example of how the system was designed to work. The benefits to users are seen in Fig. 5. This chart, similar to Fig. 3, shows theoretical, planned, and actual arrival throughputs at ATL. The theoretical capacity was frequently exceeded by the actual arrivals, possibly as a result of the fact that approach controllers have a reservoir of holding flights that they can insert into arrival streams with great efficiency.

E.   CVG

Greater Cincinnati International Airport (CVG) is a relatively new hub for Delta Air Lines and its regional partner ComAir. Since 1995, traffic has grown considerably, and an additional 73% growth is expected in the next 15 years. From mid-1997 through 1998, delays have decreased, mostly because excess demand has decreased. The airlines have been expanding their scheduled flight times for some reason, presumably a capacity limit. Arrival throughput into the terminal matches runway capacity, but departure throughput is below capacity. As in the preceding, this is a sign that en route airspace is the limit in the system.

CVG is particularly interesting because of its altitude flexibility metric (Table 2). Jets in the northeastern departure flow take, on average, almost 10 min longer than the other flows to reach cruising altitude. Upon inspection of climb profiles, it was found that regional jet departures were filing lower altitudes in their flight plans, presumably to avoid the famously congested airspace in the area above 23,000 ft. In this case, the airline has traded one form of penalty for another, accepting

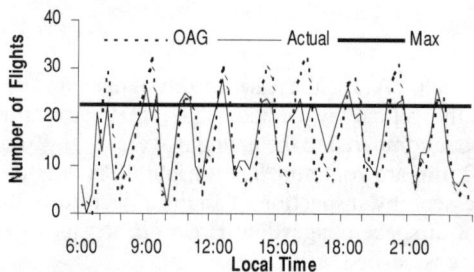

Fig. 5   Arrival throughputs at ATL.

Table 2 Median descent and climb times, in min

Arrivals	Descent	Departures	Climb
Northeast	32	North	32
Southeast	34	NE	41
Southwest	36	SE	32
Northwest	40	South	26
		SW	31
		West	32
Mean for eastern United States	34.9	Mean for eastern United States	21.4

increased fuel consumption to reduce delay. The multidimensional metric approach to analyzing system performance has exposed the fact that the throughput score at CVG is actually misleadingly high.

VII. Conclusions

The balance between predictive and deductive analysis of ATFM problems has long been tilted in favor of predictive simulations. This chapter has outlined a method by which deductive analysis can use widely available data to identify airport and airspace problems to support airspace redesign and guide simulations. We have shown examples of how individual impacts can be extracted from the complex network of the ATM system.

This work has been presented to many of the facility liaison teams involved in the national airspace redesign. It provides a synoptic view of the system, reaching across facility boundaries to ensure that individual facilities' efforts to ameliorate their own problems can be done with due consideration of the impacts on neighboring airspace. Ultimately, it is hoped that this work can guide simulation studies of proposed redesigns, making it possible to estimate the relative benefits of alternative airspace organizations that include the second-order benefits and penalties that occur as users adapt to the new system.

Acknowledgments

We would like to acknowledge the invaluable contributions of Thor Abrahamsen, Samuel Bowden, Debra Moch-Mooney, and Tess Brothersen, who performed much of the research on which this work is based. The contents of this material reflect the views of the authors. Neither the Federal Aviation Administration nor the Department of Transportation makes any warranty or guarantee, or promise, expressed or implied, concerning the content or accuracy of the views expressed herein.

References

[1] *ATS Performance Plan*, Federal Aviation Administration, Washington, DC, 1999.

[2] *Airspace Management Handbook*, Federal Aviation Administration, Washington, DC, 1999.

[3] Abrahamsen, T., and Cherniavsky, E., "National Airspace System Report Card," MITRE Corp., MP99W106, McLean, VA, June 1999.

[4] Hoffman, J., and Voss, W., "Overview of the ASC System Performance Management Project," Proceedings of the 1st U.S./EUROCONTROL ATM Seminar, sponsored by the EUROCONTROL Experimental Centre, Bretigny-Sur-Orge, France, June 1997.

[5] Zimmerli, W. C., "Future Traffic—A Challenge to Human Intelligence and Social Values," *Advanced Technologies for Air Traffic Management*, Springer-Verlag, Bonn, 1994, pp. 1-1—1-20.

[6] Hoffman, J., Abrahamsen, T. R., Bowden, S. J., Brothersen, T., and Moch-Mooney, D., "Initial Eastern United States Airspace Problem Identification," MITRE Corp., MTR99W32, McLean, VA, April 1999.

Chapter 26

Operational Assessment of Free Flight Phase 1 Air Traffic Management Capabilities

Dave Knorr*
Federal Aviation Administration, Washington, D.C.

Joseph Post[†] and Jeff Biros[‡]
CNA Corporation, Alexandria, Virginia
and
Michelle Blucher[§]
MITRE Corporation, McLean, Virginia

I. Introduction

THE Federal Aviation Administration's (FAA) Free Flight Phase 1 (FFP1) program will deploy air traffic management (ATM) capabilities that can provide early benefits to National Airspace System (NAS) users and service providers, leveraging proven technologies with needed procedural enhancements and appropriate standards. FFP1 capabilities have been developed by the FAA, in concert with the user community, and are intended to assist controllers and airlines with decision making, thereby increasing the efficiency of operations. FFP1 tools will be deployed at a limited number of sites between 1998 and 2002 and evaluated to determine their operational effectiveness, allowing for informed decisions regarding future system development and acquisitions.

An evaluation of the operational effectiveness of FFP1 capabilities will be accomplished collaboratively by the FAA and industry stakeholders. This chapter describes the collaborative approach used to develop FFP1 performance metrics, outlines the proposed measurement process, and presents some sample evaluation results for one of the capabilities. We believe a joint FAA/industry approach for gauging the success of FFP1 capabilities is an important step in maintaining the original FFP1 consensus and preparing for future deployment decisions.

Copyright © 2001 by the American Institute of Aeronautics and Astronautics, Inc. All rights reserved.
*Manager, Performance Metrics, Free Flight Phase 1 Program Office.
[†]Project Director, Free Flight Operations Analysis.
[‡]Research Analyst.
[§]Project Team Manager, Systems Analysis, Center for Advanced Aviation System Development.

II. System Description

The FFP1 program will field five different systems between 1998 and 2002 at a limited number of sites. The operational evaluation of this core capability limited deployment (CCLD) will provide the basis for decisions regarding national deployment and further development of these systems. The flight domains in which the tools operate are depicted in Fig. 1.

A. Collaborative Decision Making

Collaborative decision making (CDM) is a set of tools and procedures that allows the airlines and FAA to improve operations through information sharing. Ground delay program enhancements (GDP-E), the initial focus of CDM, started prototype operations at San Francisco and Newark airports in January 1998. Under GDP-E, participating airlines send operational schedules and changes to schedules to the air traffic control systems command center (ATCSCC) on a continuous basis. This schedule information includes, but is not limited to, flight delay information, cancellations, and newly created flights. The ATCSCC uses this information to better implement and manage ground delay programs (GDPs).

In addition to improving the execution of GDPs, CDM has been found to have application to other air traffic management problems, such as airspace congestion due to heavy traffic or en route weather. CDM's collaborative routing (CR) function is intended to provide better information to airspace users about potential flow problems that are likely to require rerouting or other flow management actions. This may allow users to prepare for possible effects on their operations in advance. The NAS status information (NASSI) function will provide a mechanism to share critical safety and efficiency data with NAS users. A recently formed group has been tasked to determine what these data are and how to set priorities for getting the data distributed.

B. User Request Evaluation Tool

The User Request Evaluation Tool (URET) is a decision support system developed by the MITRE Center for Advanced Aviation System Development (CAASD) for use by en route center controllers. URET provides aircraft-to-aircraft and aircraft-to-airspace conflict detection and trial planning of proposed air traffic control (ATC) solutions to ensure that they are conflict free. These capabilities will be used primarily by data (or D-side) controllers for strategic problem detection. The basis for URET strategic planning capabilities is aircraft flight plan information, track data, forecasted winds and temperatures, aircraft performance characteristics, and facility information. Using this information, the progress of an aircraft is continuously monitored, problems are detected, and controllers are notified of possible conflicts between the current flight and other aircraft and/or airspace. In addition, when a pilot requests a new clearance, the controller can use URET to identify any possible conflicts.

C. Center-TRACON Automation System

The Center Terminal Radar Approach Control (TRACON) Automation System (CTAS) is a set of decision support tools that assist air traffic managers and

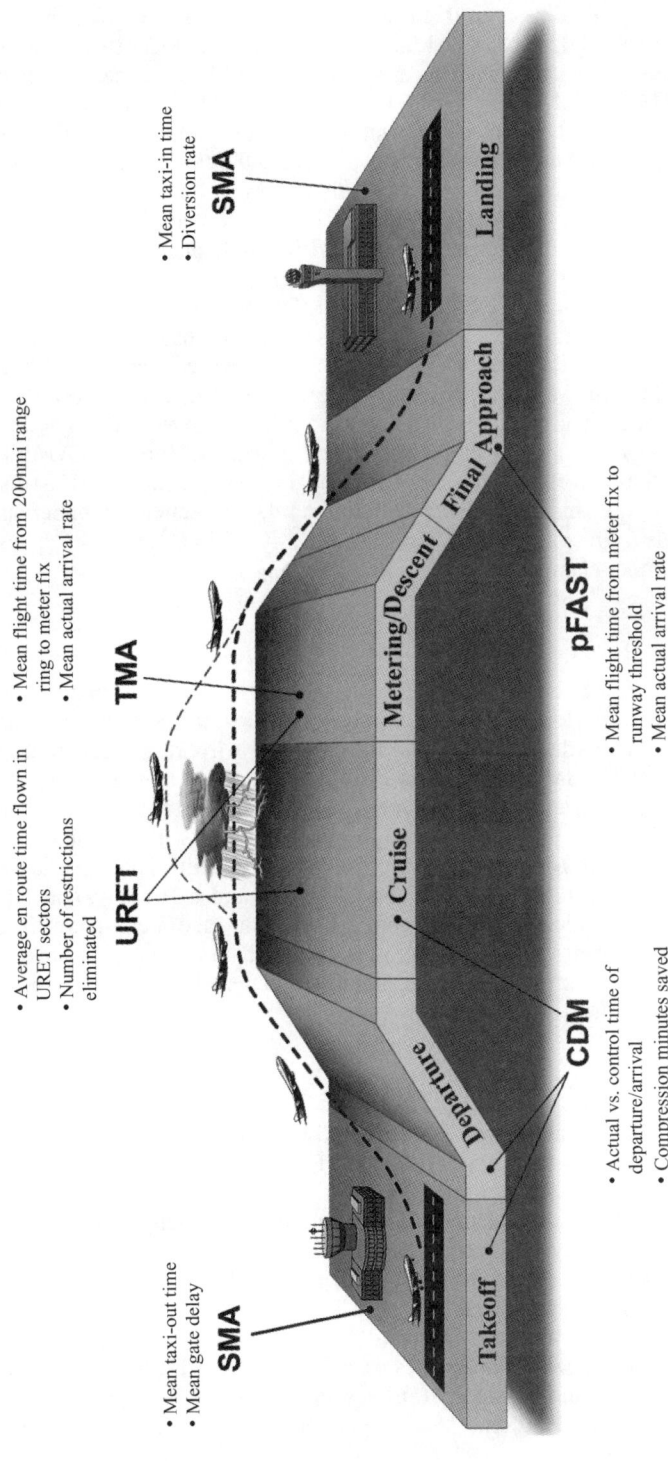

Fig. 1 FFP1 tools, associated flight domains, and principal metrics.

controllers by generating aircraft schedules and advisories to regulate aircraft arrivals to a runway complex. Traffic Management Advisor (TMA) assists controllers in the en route cruise and transition airspace managed by air route traffic control centers (ARTCC). TMA provides ARTCC personnel with a means of optimizing the arrival throughput of capacity constrained airports. Inputs to the system include real-time radar track data, flight plan data, and a three-dimensional grid of wind vectors. TMA's trajectory models use this information, updated every 12 s, to compute routes and optimal schedules to the TRACON meter fixes for all arriving aircraft that have filed instrument flight rules (IFR) flight plans, with consideration given to separation, airspace, and airport constraints.

The Passive Final Approach Spacing Tool (pFAST), the other CTAS tool being fielded, is used by controllers and air traffic managers to manage the flow of arrivals in terminal airspace. Passive FAST computes a relative sequence for each arrival aircraft for each runway at a particular airport. The system calculates a runway assignment for each aircraft in such a way as to minimize overall flight delay, with consideration given to aircraft type, speed, and trajectory. Runway advisories are displayed to the controller on the automated radar terminal system (ARTS) display. The controller may manually override both the relative sequence number and the runway advisory displayed by pFAST, and the system automatically adjusts to sequence number changes.

D. Surface Movement Advisor

The concept for a surface movement advisor (SMA) sprang from development work by NASA on a tool to aid ramp controllers with gate management. This tool was prototyped at Atlanta's Hartsfield International Airport, and continues to be used there. For the SMA CCLD, the FAA will distribute filtered ARTS data to participating airlines and service providers via the Collaborative Decision Making Network (CDMNet). FAA is also providing software that will present a planform display of arriving and departing aircraft (similar to a radar display). Several airlines (including Northwest Airlines, Southwest Airlines, and USAirways) have begun using this display in their operations centers to better manage operations at the SMA airports. These airlines also intend to integrate this data into their respective gate management tools, so that ramp controllers will benefit from timely information on arriving aircraft.

III. Collaborative Approach

FFP1 capabilities and sites were originally selected by the Radio Technical Commission for Aeronautics (RTCA) Free Flight Steering Committee (RTCA, Inc., is a private, not-for-profit corporation that advises the FAA by developing consensus-based recommendations on communications, navigation, surveillance, and air traffic management issues).[1] The collaboration on FFP1 between industry stakeholders (as represented by the RTCA) and the FAA will continue through deployment and include an evaluation of FFP1's operational impact. The RTCA and FAA developed a joint workgroup to define a core set of metrics to assess the impact of each capability. This metrics workgroup included expert representatives from the airspace user community (airlines, cargo carriers, and general aviation), as well as service providers and CAASD. Quantitative measures were derived from the desired impacts outlined in the operations concept and aligned with the FAA's operational goals.[2] Mapping the FFP1 metrics to the RTCA's operational concept and the FAA's goals was a key step in reaching an FAA/industry agreement on the metrics.

In addition, these metrics were assessed for their measurability (practicality of data collection) and meaningfulness (interpretability in terms of a positive or negative value). Since approval of the core set in April 1999, the FFP1 Metrics Team has continued to work with the RTCA, developing a detailed operational evaluation plan to assess the metrics.[3] This evaluation plan further refines the set of metrics following more detailed discussions with air traffic controllers and managers at the FFP1 prototype sites, discussions with other FAA analysts, preliminary analyses of prototype data, and continued interface with industry stakeholders. Although the plan defines metrics for each tool, we recognize that the metrics may need to be adjusted as we gain more experience. The plan specifically identifies data sources and methodologies used to calculate each metric.

Stakeholders' data inputs, and their interpretation and validation of impacts, are a vital link to operational impacts. The FFP1 Program Office (PO) will make data available for stakeholder review and will provide stakeholders with consistent information from reliable sources to assess operational performance. This information will also facilitate future decisions about system enhancements, site proliferation, and funding. The FFP1 PO will implement both formal and informal reporting mechanisms to share the results of operational evaluations with stakeholders. Formal mechanisms will include quarterly reporting to the RTCA Free Flight Steering Committee, which will be coordinated with the RTCA Select Committee on Free Flight Implementation.

IV. Metrics Definitions

The FFP1 metrics have been grouped into five categories aligned with the FAA's operational goals: safety, access, delay/efficiency, predictability, and flexibility. All of the revised FFP1 performance metrics are tabulated in Table 1.

A. Safety

Safety may be defined as the ability to maintain the standards specifying safe spacing distances, both between aircraft and between aircraft and terrain or obstructions. FFP1 safety metrics are the changes in the number of operational errors and operational deviations.

B. Access

Access can be defined as the ability of users to enter the ATC system and obtain services on demand. For FFP1, access focuses on maximizing the use of existing runways for arrivals and departures. For CDM, access also includes system throughput related to improved information. Clearly, improved access to airspace and runways will have a direct relationship to delays. However, as demand increases, runway throughput may increase while delays remain constant or even increase. This phenomenon is well known in surface transportation; when a lane is added to a highway, drive times initially decrease, but then increase as traffic

Table 1 FFP1 performance metrics

Outcome category	CDM	URET		TMA	pFAST	SMA
			Safety			
			Change in operational errors			
			Change in operational deviations			
User access	Number of unused slots			Actual arrival rate	Actual arrival rate Actual arrival rate for each runway	Diversion rate
Delay/efficiency	Mean flight time Compression minutes saved	Mean en-route time Mean distance flown Mean air distance flown Mean fuel usage % time spent at or near desired altitude Number of restrictions eliminated Aggregate degrees turned		Mean flight time from 200 n mile range ring to meter fix Mean arrival delay Mean fuel usage from 200 n mile range ring to meter fix Variability of fuel usage	Mean flight time from meter fix to runway threshold Mean fuel usage from meter fix to threshold Variability of fuel usage from meter fix to threshold	Mean taxi-in time Mean taxi-out time Mean gate delay

Category				
Predictability	Integrated predictive error Rate control index EDCT[a] compliance ratio Number of GDPs canceled near start Number of GDP revisions	Planned vs actual en route time Planned vs actual distance flown	Mean error in predicted meter fix arrival time Variability in error Variability of actual arrival rate Mean difference between AAR and actual arrival rate Variability of time from 200 n mile range ring to meter fix	Mean difference between AAR and actual arrival rate Variability of flight time from meter fix to threshold	Variability of taxi-in time Variability of taxi-out time Variability of gate delay Gate reassignment rate
Flexibility	Mean distance flown Control time of arrival	Mean en-route time and distance flown (late departures)			
Productivity		Number of aircraft per sector per unit time Change in monitor alert threshold	Mean actual arrival rate/throughput per sector or position	Distribution and throughput of operations per runway/position	

[a]Estimated departure clearance time.
[b]Airport acceptance rate.

increases. For this reason it is important to have specific measures for access (throughput) and delay.

Specific FFP1 access metrics include mean peak period actual arrival rates per airport and per runway, diversion rate, and the number of unused slots.

C. Delay/Efficiency

We have elected to combine delay and efficiency metrics into one category because some metrics in these areas are closely related or even equivalent. In the past delay has been defined in three different ways:

1) the amount of time beyond expectations that it takes to complete a flight or flight segment,
2) the difference between actual and scheduled arrival times, and
3) the additional transit time above the optimal or unimpeded time.

We intend to consider all of these definitions of delay because each has a unique impact on the NAS user's value function.

Definitions of efficiency have centered around fuel efficiency for a given flight as well as reductions in flight times. Efficiency has often incorporated all reductions in delay. From our perspective, both definitions have unique components that are valued separately as well as a common component. Both definitions will be captured under this metric category.

FFP1 delay/efficiency metrics include mean gate delay, mean flight time, mean taxi-out time, mean flight time in URET sectors, mean flight time from the 200 n mile range ring to the meter fix, mean flight time from the meter fix to the runway threshold, mean taxi-in time, mean arrival delay, GDP compression minutes saved, and the mean and variance of fuel usage per flight segment by aircraft type.

D. Predictability

Predictability measures the variation in the ATM system as experienced by the user. Our definition of predictability focuses on the dispersion (specifically, the variance) associated with flight segment times. Commercial airlines may benefit as much from a reduction in the variance (or an improvement in the consistency) of flight/taxi times as they would from a reduction in average flight times. System predictability allows for improved scheduling and more efficient bank operations.

Specific predictability metrics include variability of flight time from the 200 n mile range ring to the meter fix, variability of flight time from the meter fix to the runway threshold, mean difference between actual arrival rate and airport acceptance rate (AAR), variability of actual arrival rate, variability of taxi-in and taxi-out times, variability of gate delay, number of GDPs canceled, and time spent at or near desired altitude for specific city pairs.

E. Flexibility

Ultimately flexibility measures the ability of the ATC system to meet users changing needs in their efforts to optimize daily operations. For example, commercial air carriers may prefer increased delay in exchange for an on-time arrival on a specific flight. We have focused flexibility metrics on capturing anything an airline would like to accomplish on an individual flight not already captured in the previous metrics, including faster routes (flight times) to make up lost schedule, slower routes to reduce taxi-in delay, and requests for altitude changes for passenger comfort.

In practice, it is extremely difficult to establish airline intent on an individual flight basis. In fact, within an airline the pilot and dispatcher may have different objectives. For this reason, our approach to measuring flexibility is to segregate flights delayed at departure from those departing on time. Our supposition is that those aircraft leaving late will desire to make up time en route. Flexibility will be measured by an airline's ability to make up time (i.e., to keep to schedule). Other measures of flexibility will be developed after obtaining feedback from users on perceived changes or improvements in service.

V. Measurement Process

An extensive data collection effort is planned to fully assess the impact of the FFP1 tools at each of the CCLD sites. Various data sources from the FAA, Department of Transportation, NASA, the National Oceanic and Atmospheric Administration (NOAA), and the airlines will be combined to provide a complete picture of NAS operations and FFP1 tool performance, both prior and subsequent to fielding of the tools at each of the CCLD sites.

Data for each metric will be collected for at least one year prior to initial daily use (IDU) at each site so that a robust baseline can be established (for CDM metrics data are collected nationally). Data for each metric will then be collected for a period of at least one year following planned capability available (PCA). In this way seasonal factors may be fully removed. Between IDU and PCA, operations will be observed and trends in the metrics reported to understand any learning-curve effects, and to provide feedback to system developers as to local adaptation. Local environmental, airport configuration, and airport demand data will also be collected for this length of time so that we may better isolate the effects of the particular FFP1 tool from those of changes in these conditions.

For each capability, the evaluation (and consequently the data collection effort) will focus on the flight segments that are expected to be most affected by the particular tool (see Fig. 1). For example, the pFAST evaluation will focus on flight times in the terminal area (specifically, flight times from the meter fix to the runway threshold) and airport arrival rates. However, because NAS operations are tightly coupled, the evaluation will also consider upstream and downstream effects when appropriate. As an example, one of the primary metrics for SMA will be mean taxi-out time. Additionally, we intend to examine upstream gate delay for SMA (gate delay is defined here as the difference between scheduled and actual gate departure times).

Wide-ranging data relating to the local environment and conditions at each FFP1 site will be collected for the same time periods that performance metric data are collected. This data includes, but will not be limited to, airport configuration; surface weather, including visibility, ceiling, and precipitation rate; winds aloft; arrival demand (i.e., actual arrivals per unit time); and departure demand. These data are essential to ensure that "apples to apples" comparisons of system performance are made before and after the deployment of a capability. For tools that operate in the extended terminal area, arrival demand data are particularly critical, because flight times can be expected to be larger when there is a high level of arriving traffic.

To normalize for differing distributions of these conditions pre- and post-deployment, two different analytical techniques will be employed. The first and most transparent approach will be to group data into "bins" with similar local conditions. This approach is simplest, but it does not take full advantage of all the information content in the data. To remedy this limitation, multivariate regression techniques will also be used. These statistical techniques are well suited to problems in which an output variable (e.g., flight time) may be influenced by many "exogenous" factors.

A number of data sources will be used to compute the metrics and obtain the associated environmental data, including existing FAA databases, airline data, and new data sources. Actual arrival rates and coarse flight time and distance information will be collected from the Enhanced Traffic Management System (ETMS), Airline Service Quality Performance (ASQP), and Consolidated Operations and Delay Analysis System (CODAS). More precise flight times and distances will be obtained from local ARTSs for the terminal domain, and from the host computer for the en route domain. In some cases flight tracks and associated data will be obtained from log files produced by the various FFP1 tools. AARs will be obtained from facility and ATCSCC logs. Fuel usage will be modeled using airline-provided equipment-specific fuel consumption data and observed flight trajectories. Finally, safety data will be obtained from the National Airspace Incident Monitoring System (NAIMS) databases.

VI. Preliminary pFAST Results

In February 1999 air traffic controllers at the Dallas/Fort Worth International Airport (DFW) TRACON began using pFAST to help sequence and assign runways to arriving aircraft. Initially pFAST was used only by a cadre of controllers, but over the past year pFAST usage has gradually increased, to the point now that all controllers in the TRACON are using the tool.

We have conducted a preliminary analysis of the impact of pFAST on operations at DFW, examining airport acceptance rates, actual arrival rates and operations (i.e., arrivals plus departures) rates, TRACON flight times, taxi times, and runway balancing. Our results suggest that pFAST usage has indeed led to an increase in airport acceptance rates, which has thereby led to an increase in peak arrival rates. An observed improvement in runway balancing has also resulted in an increase in total operations rates. While taxi-in times have slightly increased, there has been an offsetting and larger decrease in taxi-out times. Finally, these improvements have occurred without any measurable change in TRACON flying times. We present here summary results for acceptance rates, actual arrival rates, runway balance, and operation rates.

To determine if pFAST usage has led to an increase in airport acceptance rates at DFW, we performed a regression analysis of the airport acceptance rate and various environmental variables that, in the judgment of experienced air traffic controllers, should affect the AAR. Specifically, we regressed the number of arrival runways in use, the type of approaches being used (visual or instrument), the natural logarithm of ceiling, the square of crosswind component, and a pFAST dummy variable on AAR. We also included a dummy variable that accounts for a procedural change implemented in July 1999 whereby the TRACON accepts an unlimited number of aircraft for specified times on a fifth arrival route (a so-called unlimited dual route). We included data in 10-min increments from 20 February through 31 December, 1999, for a total of approximately 220,000 observations.

The results of this regression analysis are presented in Table 2. All of the variables included in this model were found to be significant at the 5% level, and the signs of the coefficients were all as we would expect. For example, when DFW uses three arrival runways (rather than four), the acceptance rate is reduced ceteris paribus by approximately 22 aircraft per hour. Similarly, when the ceiling increases from 100 to 1000 ft, the acceptance rate increases by $0.97 \cdot ln\,(1000 - 100) \approx 6.6$ aircraft per hour. After controlling for all of these factors, we found that pFAST usage resulted in an increase in acceptance rates of approximately 2.5 aircraft per hour.

Next we examined actual airport arrival rates to determine if controllers were able to land aircraft at the higher rates that the TRACON is now requesting from the en route center. We collected aircraft arrival counts in 10-min intervals for

Table 2   DFW acceptance rate regression analysis

Independent variables	Unstandardized coefficients		Standardized coefficients		
	B	Std. error	Beta	t	Sig.
(Constant)	127.936	.173		739.002	.000
3_Runways	−21.871	.060	−.444	−366.191	.000
IFR	−13.978	.038	.540	−368.845	.000
UnltdDuals	1.382	.030	.058	45.793	.000
Ln_Ceiling	.970	.017	.085	58.499	.000
NorthFlow	−.936	.031	−.036	−29.775	.000
CrosswindCompSq	−.01196	.000	−.049	−40.816	.000
pFAST	2.486	.030	.098	82.543	.000

Dependent variable:	
AAR	Airport acceptance rate (arrivals/h)
Independent variables:	
3_Runways	0-four arrival runways
	1-three arrival runways
IFR	0-visual approaches
	1-instrument approaches
UnltdDuals	0-Feb. 20–June 30, 1999
	1-July 1–Dec. 31, 1999
Ln_Ceiling	natural logarithm of ceiling in feet
NorthFlow	0-south flow
	1-north flow
CrosswindCompSq	square of crosswind component in knots
pFAST	0-pFAST off
	1-pFAST on

Model summary

R	R square	Adjusted R square	Std. error of the estimate
.859	.793	.793	6.0243

DFW from 20 February 1999, through 29 February 2000. To diminish the impact of any possible sampling error in these rates, we used a sliding 30-min window to calculate the arrival rates (thus we calculated the number of aircraft arriving in a 30-min period every 10 min). Once we had "filtered" the arrival rate data in this manner, we used an algorithm to identify the eight highest arrival peaks per day (the choice of eight peaks per day is somewhat arbitrary). Figure 2 presents a box plot of these peak arrival rates, segregated into instrument and visual arrival conditions, with and without pFAST in use. The dark central line within each box indicates the median flight time for that particular set of conditions. The bottom and top of the central box represent the 25th and 75th percentiles of the data, respectively. The lower whisker (the line distending downward from the box) extends from the 25th percentile to the value equal to or greater than the 25th percentile minus 1.5 times the interquartile range. Similarly, the upper whisker extends upward from the box to the value equal to or less than the 75th percentile plus 1.5 times the interquartile range. Values more than 1.5 but less than 3 times the interquartile range from the box are represented by circles, and those more than 3 times the interquartile range from the box are depicted by asterisks.[4] The sample size for each box plot is indicated at the bottom of the plots.

Figure 2 suggests that there has been a measurable increase in median peak arrival rates at DFW associated with pFAST usage. The median peak arrival rate has increased from 58 arrivals per 30-min period to 59.5 when running instrument approaches. When using visual approaches, the median arrival rate has increased from 61 to 62 arrivals per 30-min period. A chi-square type test of these medians indicates that the differences are statistically significant at the 5% level.[5]

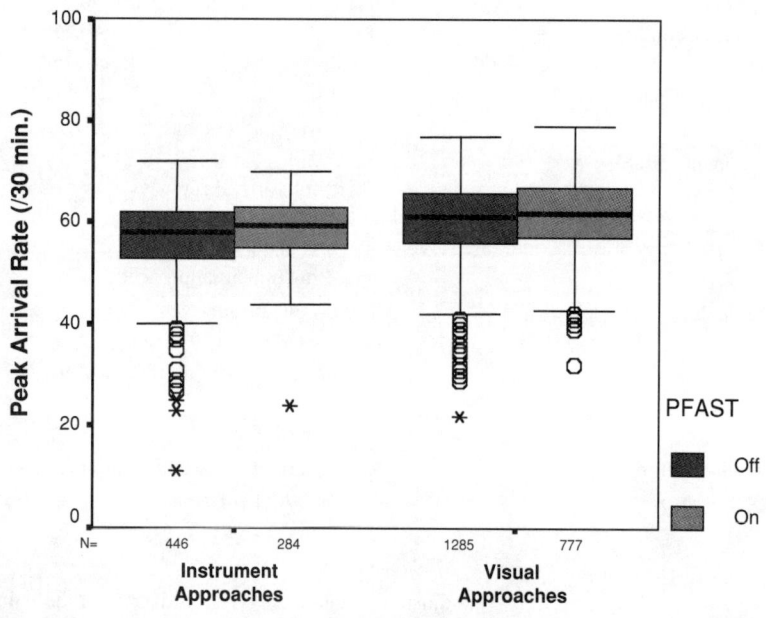

Fig. 2 DFW peak arrival rates.

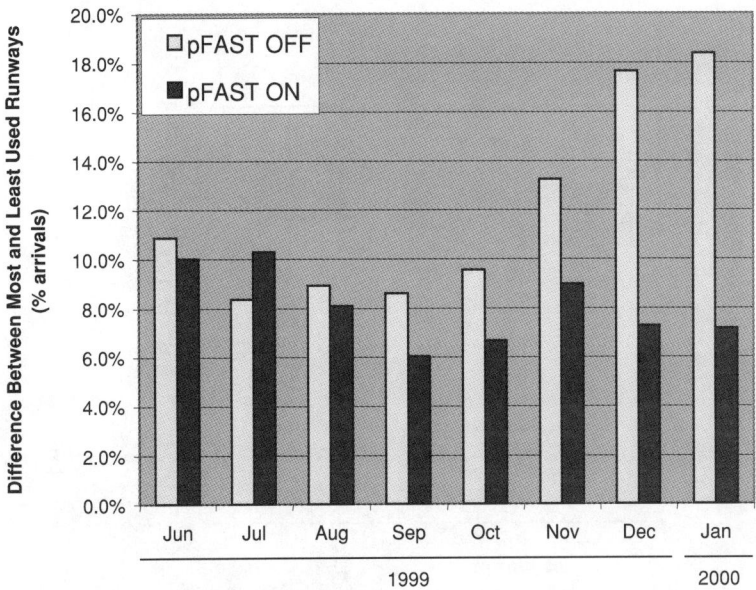

Fig. 3 DFW arrival runway balance.

We also examined the balancing of the runways at DFW. Runways are considered to be balanced if the arrival rates on the individual runways are approximately equal. By balancing the runways, the overall arrival rates should be able to be increased, and surface congestion reduced. Our relatively simple measure of the degree to which the arrival runways are balanced is the difference in the percent of arrivals handled by the most used and least used runways. The data used for this calculation are the same data used in the preceding, namely the 10-min arrival counts. We limit the data sample to periods when the airport was in a south flow configuration, because for most of the period under examination pFAST was used only in this configuration. Additionally, we included only 10-min time periods in which there were at least four total arrivals. All of the arrivals are then summed by month, and the difference between the most used and least used runways is expressed as a percentage of total arrivals. The result of this calculation, displayed in Fig. 3, indicates that the difference between the most and least used arrival runways is reduced when pFAST is in use.

The final metric examined here is the peak rate of operations. The improvement in runway balancing made possible with pFAST has resulted in less surface congestion on DFW taxiways at peak times. This decrease in congestion has led to a corresponding increase in departure rates. Figure 4 illustrates the distributions of peak operation rates (arrivals plus departures) from 22 April 1999, through 29 February 2000. We filtered the operation rates using a 30-min window, selected the eight highest peak operation rates for each day, and segmented these peak rates by type of approaches being flown and pFAST status (the identical procedure as that used for arrival rates). The median peak rate of operations has increased

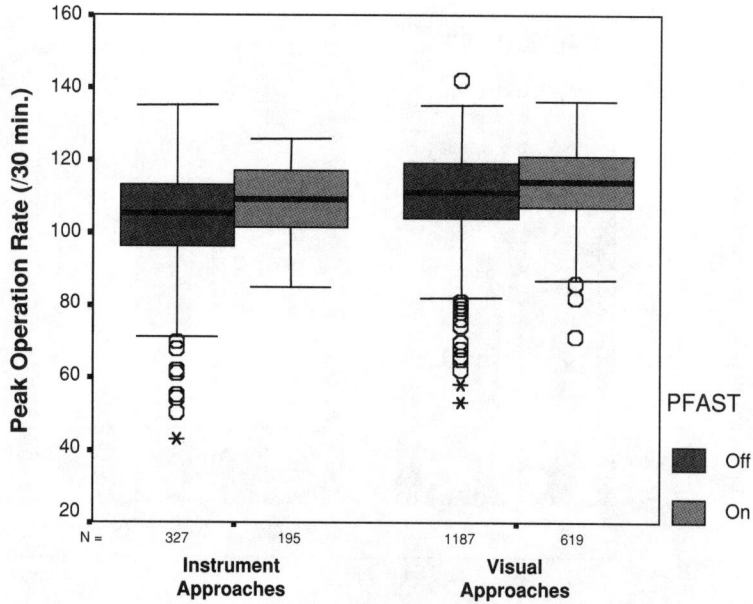

Fig. 4 DFW peak operation rates.

from 105 to 109 operations per 30-min period under instrument approaches, and from 111 to 114 operations per 30-min period under visual approaches. These differences are statistically significant at the 5% level.

VII. Conclusions

The FFP1 operational evaluation represents a departure from the FAA's usual approach to system acquisition, in which impact studies are limited to a specified operational evaluation period. We recognize the manifold challenges inherent in this approach. To mitigate these risks, we have assembled an expert team of engineers, analysts, and air traffic controllers with considerable experience in conducting operational evaluations. In addition, we will draw on the experience of our industry partners to help guide our efforts and interpret our results. By studying data and experiences from the FFP1 prototype sites, we will be well prepared to accommodate the operational data. We are confident that we will be able to conduct an objective assessment of FFP1 capabilities that will consider the interests of NAS users and service providers alike, and will aid developers in refining their products and facilitate informed national deployment decisions.

Acknowledgment

We would like to thank Ed Meyer of the Free Flight Phase 1 Program Office for his invaluable assistance in conducting this work.

References

[1] "Government/Industry Operational Concept for the Evolution of Free Flight, Addendum 1: Free Flight Phase 1 Core Capabilities Limited Deployment," RTCA Select Committee for Free Flight Implementation, RTCA, Inc., Washington, DC, Aug. 1998.

[2] "Air Traffic Services Performance Plan for Fiscal Years 1998–2000," Office of Air Traffic Services, Federal Aviation Administration, Washington, DC, 1997.

[3] "FFP1 Performance Metrics: An Operational Impact Evaluation Plan," Free Flight Phase one Program Office, Federal Aviation Administration, Washington, DC, 12 Aug. 1999.

[4] Tukey, J.W., *Exploratory Data Analysis*, Addison Wesley Longman, Reading, MA, 1977.

[5] Conover, W. J., *Practical Nonparametric Statistics*, Wiley, New York, 1980, Chap. 4.

Chapter 27

CENA-PHARE Experiment: Requirements for Evaluation of Novel Concepts in Air Traffic Control

Didier Pavet*
CENA, Athis-Mons, France

I. Introduction

THE Program for Harmonized Air Traffic Management Research in EUROCONTROL (PHARE) was launched in 1989 to investigate ways to increase capacity in the most busy areas of European airspace. Indeed, EUROCONTROL traffic growth scenarios predicted at that time, and confirmed by update-to-date figures, showed that in 2005, traffic demand would increase by 56% compared to that observed in 1995 and by around 121% in 2020. Whereas air traffic services should adopt drastic strategies to cope with the demand (airspace and airways restructuration, negotiation with military authorities, etc.), PHARE focused efforts on the most limiting factor of the air traffic management (ATM) system, i.e., ability of air traffic controllers to handle the traffic with the required level of safety within a given cell of airspace. The effort has been in designing new operational concepts, new tools, and procedures that could be applied within the targeted time frame 2005–2015.

To measure the performance of the new concepts taking advantage of enhanced technologies, a series of real-time simulations, or PHARE demonstrations (PD), was planned. These demonstrations addressed the evaluation of computer assistance tools and the introduction of data link in en route (PD/1), in terminal maneuvering area (TMA) (PD/2), and combined en route/TMA environment (PD/3), each demonstration being led by one or more PHARE partners. These large-scale validation activities, comprising integrated ground system, air system, and air/ground data-link facilities were the last step in the validation process consisting of functional testing, basic evaluation of individual tools, and partial validation of subsystems of increasing complexity.

The first PHARE demonstration (PD/1) was hosted by the National Air Traffic Services, Ltd. (NATS). The trial was conducted over a period of eight weeks

---

Copyright © 2001 by DGAC/CENA. Published by the American Institute of Aeronautics and Astronautics, Inc., with permission.
*Head, Controller System Interface.

toward the end of 1995 (see Ref. 1). The primary aim of PD/1 was to investigate the introduction of computer assistance tools, data-link equipped aircraft, and a four-dimensional flight management system (FMS) within en route airspace.

The second PHARE demonstration (PD/2) was hosted by the Deutsches Zentrum für Luft-und Raumfahrt (DLR) in Germany (see Ref. 2). The trial was conducted over a period of eight weeks in late 1996, early 1997. PD/2 pursued the assessment of the introduction of computer assistance tools and aircraft equipped with four-dimensional-data-link flight management system in a TMA environment, using the Frankfurt TMA as simulated airspace. The main computer assistance tool to be demonstrated during the trial was the arrival manager (AM), which sequenced arrival traffic into a conflict-free stream at a designated point before the runway (the metering fix).

PD/3 concluded the series of PHARE demonstrations; CENA, NLR, and EEC were the main participating research organizations that hosted the PD/3 demonstrations, assisted by NATS (for evaluation aspects notably). Three real-time simulations (one for each site) were settled to evaluate PHARE concepts on a quite complete volume of airspace stretching from the en route environment to the extended TMA of several airports. The different site specific operational environments, gathered as a whole, were approximately the airspace from Amsterdam/Schiphol to Paris/Charles de Gaulle plus the overall en route airspace between the two airports. Key input parameters were common to the three simulations: 1) focusing on an airspace representative of the Europe core area airspace, 2) using the year 1996 as a reference for operational environment and traffic level, and 3) using traffic samples matching the predicted traffic level according to the expected time frame 2005–2015. Note that this chapter concentrates on the CENA/PD3 experiment. This experiment took place in May and June 1998, and the subsequent analysis resulted in a detailed report extracting the major findings.[3] The experiments hosted by NLR and ECC have been documented in other respects (see Refs. 4, 5).

CENA PD/3 provided evaluation of a future ATM concept for the time period 2005–2015, which supports the transitional introduction of four-dimensional and data-link equipped aircraft by combining 1) air/ground integration, 2) introduction of new tools to support the controllers, and 3) keeping the man in the loop following a human-centred approach. CENA PD/3 was applied in an operational environment stretching from a TMA/Departure sector, an Extended Terminal Maneuvering Area (ETMA)/Departure sector, to an en route sector, thereby covering areas left unexplored by previous PHARE demonstrations.

For the en route environment, CENA PD/3 intended to demonstrate the capacity increase and productivity benefits of the core PD/3 operational philosophy, i.e., the traffic organization planning philosophy, including the following progressive air traffic control (ATC) enhancements: 1) introduction of advanced assistance tools among which are cooperative tools aimed at organizing the traffic in a human-in-the-loop philosophy, and 2) introduction of four-dimensional trajectory negotiation and four-dimensional planning.

In a similar way for the TMA environment, PD/3 intended to cover the experimental domains related to the traffic organization planning philosophy with the following ATC enhancements: 1) introduction of advanced assistance tools, 2) introduction of planning functions, such as the departure manager tool, and 3) introduction and extension of the concept of four-dimensional trajectory negotiation and planning.

The CENA PD/3 main phase demonstration took place over seven weeks in May and June 1998. One week was devoted to a successful airborne demonstration in collaboration with the U.K. Defence Evaluation and Research Agency (DERA), using the NATS BAC1-11 live aircraft equipped with an experimental flight management system (EFMS) and using an experimental Aeronautical Telecommunications Network[6] (ATN) based on the International Maritime Satellite Organization (INMARSAT) satellite.

II. Evaluation Methodology

A. Experimental Design

The process of evaluation was conducted in a comparative manner: an advanced scenario supported by an advanced system and a baseline scenario supported by a baseline system were defined (see Refs. 7–9). The baseline system was settled to lead simulations in the same conditions as the ones led with the advanced system, giving reliable reference measurements. This system did not mimic trustfully a real system because no paper strips were provided to the controllers. Instead, the information was presented through interactive track data blocks or simple computer aided assistance.

As shown in Table 1, added to the baseline scenario, the advanced scenario was utilized according to three subscenarios; these so-called A0, A30, A70 scenarios have allowed measurement of the effects of the introduction of PHARE advanced tools and a new ground-human machine interface on the one hand, and the progressive introduction of aircraft equipped with four-dimensional FMS and data link on the other hand. The traffic samples elaborated for the experiments were derived from the 21 June 1996 traffic data [summer peak for French Area Control Centers (ACCs)] this level of traffic ($\times 1.0$) was set as the reference level (low level). The medium traffic samples corresponded to this traffic $\times 1.5$, and the high traffic samples corresponded to this traffic $\times 2.25$.

Each controller participated in two weeks of measured runs and performed alternate roles (either as planning or tactical controller) on the same sector for each run. A sophisticated rotation between controllers has been designed so that statistical tests could be run reducing the effect of variability between controllers while avoiding a learning effect as far as possible.

B. Measurements

As prepared by the CENA experimental team (assisted by the PHARE validation group), an experimental protocol has defined a quite complete set of data to be recorded for post-run analysis. This protocol (see Refs. 10, 11) has been set to enable evaluating the concept toward major external criteria, such as workload, safety of control, quality of service (capacity being a parameter), and to evaluate more thoroughly the operational working of the couple human machine.

These measurements were divided as:

1) Subjective measurements. Subjective measurements consisted of some kind of self-assessment performed by the controllers. Measuring the effect of the PD/3 operational concepts on the controller's workload was one of the main aims of PD/3. Two major subjective measures of workload were used to record and analyze the controller's workload: the instantaneous self-assessment (ISA) and the NASA

Table 1 Summary of scenarios

Scenarios	Traffic sample	PHARE advanced stools	Procedures
Baseline	Three-dimensional equipped aircraft	None	Controller plans ahead; limited computer assistance for detection of conflicts.
	Three traffic volumes	Limited planning aid tools	Procedures to suit strip-less system.
A0	As above	Cooperative tools (CT), departure manager (DM), trajectory editor and problem solver (TEPS)	Advanced planning; computer assistance looks up to 20 min ahead to conflict-free trajectories.
A30	Mixed population: 30% four-dimensional equipped aircraft	As above	As above plus procedures for negotiating with four-dimensional FMS and data-link equipped aircraft.
	Two traffic volumes (medium and high)		
A70	Mixed population: 70% four-dimensional equipped aircraft Two traffic volumes (medium and high)	As above	As above

developed Task Load Index (TLX). Questionnaires were administered at the end of each trial to collect the controllers' opinions and comments. The questionnaires were designed to make possible the quantitative analysis and significance testing of the controllers' answers.

2) Objective measurements. Objective measurements were automatically recorded by the ground system. During the PD/3 trial a large number of these objective data were collected that contributed to assess the PD/3 main topics (workload, safety, sector capacity and quality of service) and also the use of the tools.

To allow the autoconfrontation with the controllers just after a run, video and audio recording provided the human factors specialists with an efficient means to more deeply investigate the overall strategy for using one specific human

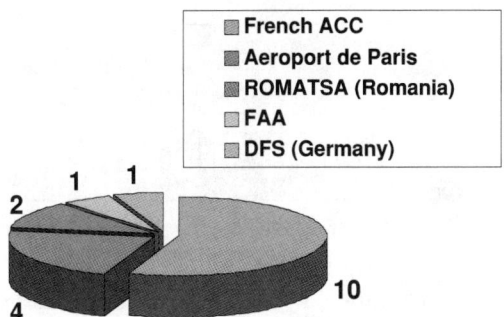

Fig. 1   Controllers' professional origin.

machine interface (HMI) object or to better interpret the reason why a specific event occurred.

Finally, direct observations (supported by analysis chart) by one human factor expert dedicated to each control suite were an invaluable source of information.

C. Controllers

The two main phase trials for PD/3 were carried out over a six week period with nine controllers participating per session of three weeks. The controllers selected for the PD/3 trial covered a range of ages, nationalities, and backgrounds as shown in Fig. 1.

III.   Lessons Learned

A.   Training Aspects

The first PHARE demonstration has underlined the requirement to reinforce training especially on the working method. As an anwer, the controllers were taught a six-day course. A full week was devoted to theoretical lessons supported by slides shows (450 slides were designed, see Ref. 12) and practical training either on a very convenient stand-alone system or on the full system. These sessions were led in a very guided manner. The day before the measured runs, all the controllers performed two refresher runs.

The excellence of the preparation and the very directive policy to interlace theory and practice counterbalanced the still too short period of time allocated for training. This method, inspired by what has been applied for several years on the En Route Air Traffic Organizer (ERATO) project at CENA/Toulouse (see Ref. 13) has enabled the controllers to catch HMI, concepts, sharing of tasks, and airspace. The controllers' answers to questionnaires illustrate this point (see Fig. 2).

B.   External Evaluation

Here are the main trends emerging from the evaluation of the external criteria that will be useful to illustrate the following discussion.

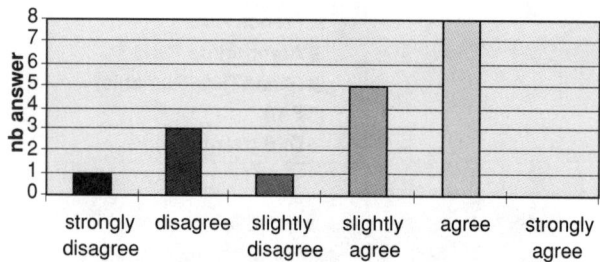

Fig. 2 Controllers' answers to the question: "there was enough training to get familiar with the HMI tools and functions."

1. Perceived Workload

The combination of ISA, TLX indicators and controllers feedback showed that the advanced scenarios promoted a transfer of workload from the tactical controller to the planning controller. The traffic load had a major influence on the PC's workload whatever the percentage of four-dimensional Flight Management System and Data-Link (4D/DL) equipped aircraft. On the other hand, we can consider that the TC's workload remained extremely high. Although a light drop has been observed for a high percentage of 4D/DL (70%), the mixed situation (30% DL equipped aircraft) emerged as the worst situation to manage.

On a very subjective point of view, frustration and at a lesser extent time pressure were the two major TLX indicators badly scored by controllers for advanced scenarios.

The reduction of very high frequency (VHF) activity (and thus the drop of time pressure) explains the gain TC has felt while operating for advanced scenario with 70% of 4D/DL aircraft.

The increase of workload for PC was expected. The introduction of planning activities—manifested in concrete form by quite a long time spent to edit trajectory (20–40 s knowing that 20 flight plans per hour were concerned) and management of air/ground communication—has taken up a lot of PC's time.

The fact, as a finding, that TC's workload remained higher in all advanced scenarios than in the baseline scenario, has stemmed from the difficulty to build and maintain a structured working method. Especially, the feeling of being powerless in front of silent equipped aircraft (lack of an immediate awareness of what has been decided by the PC) and the difficulties of managing the prepared clearances for unequipped aircraft have strengthened their frustration. Moreover, these elements have lessened the ability of the TC to intervene tactically when needed. As a consequence of the increase of the PC's workload, the TC was not always assisted by the PC when requested.

Some "parasite workload" due to long response times appeared during the experiment. Though some improvement could be obtained on long response times requested to edit a trajectory (as a magnificent HMI should promise it), times requested to compute a trajectory (as a trajectory predictor empowered by huge computing capacity could offer), times requested to exchange data through the air (as a "lightning speed" data-link network would allow), the delay required for the pilot to acknowledge, analyze, and validate will still remain in the loop. As a

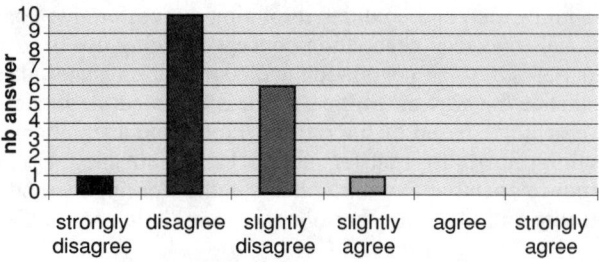

Fig. 3 Controllers' answers to the question: "it was always possible to keep aircraft well separated."

consequence, even if answers to uplinked requests from ATC would become a high priority to the crew members in the future, which is not the case in today's practice, basing any new operational scenario on the fact that air-ground negotiation would become immediate is utopian.

2. Safety of Control

Based on the occurrence of safe separation infringements, an in-depth analysis showed that safety was less ensured in advanced scenarios than in baseline scenarios. While this topic remains hard to study (the occurrence of event remains low in every case), these results added to the controllers' feedback (see Fig. 3) demonstrate that the situation awareness of the tactical controller was modified.

3. Quality of Service

PD/3 did not allow assessment of interesting objective results on this topic. The use of anticipated planning could enhance the systematic optimization of trajectories and this at a multisectors level which could yield some benefit for the airlines. On the other hand, the anticipated planning forces the controller to take larger margins; quoting the controllers' wording, "We feel less sharp today on radar separation with PD/3 advanced system." This topic deserves to be specifically studied in the future.

4. Capacity

Capacity level is an input factor that experimental experts tune as the experimental protocol has defined it. Capacity has been assessed in PD/3 measuring whether the desired level of operators' workload, safety of control, and quality of services could be maintained or improved compared with the ones obtained with the baseline scenarios. At this stage, this evaluation of the PHARE concept does not predict an increase in capacity.

C. PD/3 Concepts: Consequences of the role of Man

1. New Methods to Accomplish the Work

The capability to plan and to implement in advance a new trajectory (concept of advanced planning—trajectory negotiation) showed a great potential for trajectories optimization and conflicts resolution in comfortable conditions. The four-dimensional trajectory negotiation concept allowing use of airborne calculated trajectory as a basis of negotiation and coordination is an indisputable plus that could enhance the way air traffic control is operated. A greater accuracy in the forecast trajectory (based on aircraft trajectory) and a significant decrease in VHF occupation (mostly for transfers action) can be obtained. However, this attempt to introduce operationally this concept in the working method of a team of tactical controller and planning controller puts into light some major matters of concern:

1) The use of an asynchronous and silent medium obliges establishing a strict sharing of tasks between controllers inside the same sector and defining strictly the rules to manage the planning authority (right to use DL for planning the trajectory). Letting controllers manage resources like a distributed database does not meet time pressure constraints in ATC. Probably, a strict procedural system allowing operators to intervene one at a time in a preordered manner is a clue to set a scenario from which ambiguity is withdrawn.

2) The introduction of DL management into the planning controller framework is not a minor issue: the experiment shows that under heavy traffic load, the planning controller is more and more absorbed by DL activities and ahead in the time frame, becoming less available to follow behavior of the traffic in real time inside the sector. An operational gap, both on temporal and spatial dimensions, appears between the two actors; this fact contrasts with the current strategy of planning controllers to become a tactical assistant during peak traffic;

3) The performances of the system, especially when air-ground data link applications intervene, remain crucial if one wants to introduce this medium in the framework of tactical operators for whom time is critical. It means that no overhead due to pure calculations should participate at the overall response delay.

2. New Mental Awareness

The PHARE time scope (2005–2015) addressed clearly a transitional period in which new equipped aircraft will share the sky with traditional aircraft, and although new modes of operation could appear, "old-fashioned radar control" will still be applied. Therefore, PHARE and PD/3 settled the human-centred approach as a major principle, meaning that the controllers should keep in mind the current and future situations as they do today. Experiment showed that clearly situation awareness of the tactical controller was changed, for many reasons: as previously mentioned, an unclear sharing of tasks spoiled a good transfer of knowledge between the planning and tactical operators; basic unit tasks that contribute to the memorization process were withdrawn in too carelessly a way regarding mental awareness.

One of the PD/1 experiment recommendations was "to either improve how the system helps TCs to maintain situation awareness or fully persuade them that it is no longer necessary." Unfortunately, efforts at the HMI side did not allow to make some progress to help TCs to maintain situation awareness. No persuasive elements have been brought up until now guessing that controllers could get rid of a clear awareness of what happens in the cell of airspace they have the responsibility to manage.

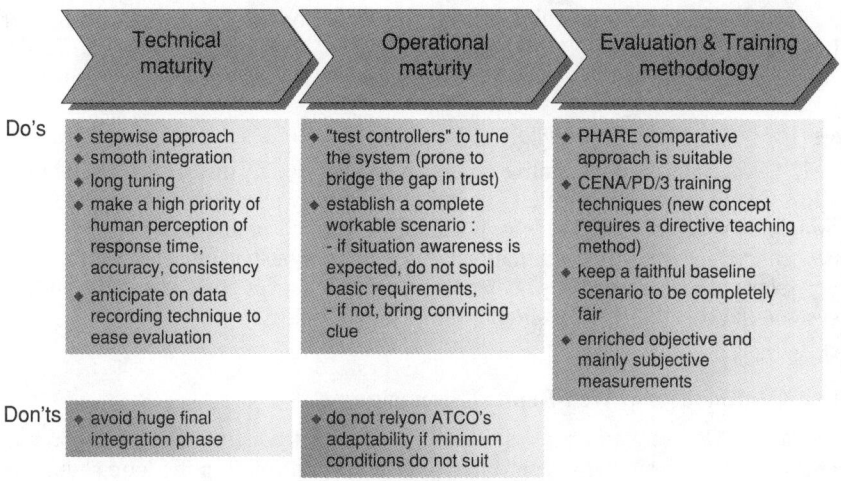

Fig. 4 Summary of requirements.

IV. Discussion: Requirements for Future Developments of Novel Concepts

Figure 4 highlights the requirements explained in the following sections.

A. Technical Maturity

This is a "one-thousands-times-written requirement," but any attempt to evaluate such new concepts must gather the minimum conditions to allow an operational evaluation. In this area, PD/3 taught us that performances (response times, quality of information) at an operator level point of view (human machine interface) as an utmost priority, puts strong requirements on the technical management of the project.

It is now obvious that a stepwise approach (in which technical components are integrated smoothly) is more prone to the building of a technically mature experimental system (as opposed to a "one shot big bang" venture).

B. Trial Controllers To Reach Operational Maturity

In the path through the real assessment of new concepts, a first step should be to enlist one team of trial controllers capable of being involved in the design, of tuning the system, and proving that the system is workable, efficient, and safe. This first target, which probably should take several months to achieve once technical maturity is obtained, is a compelled step.

A second step would be to perform evaluation/validation of the system by current controllers in order to measure external criteria with the suitable experimental procedure, validation tools. This second step would surely benefit from the PD/3 experience to organize both an evaluation methodology and a training phase. Having a set of trial controllers would allow to bridge the gap in trust that new controllers may face when discovering any new philosophy to perform his/her job.

C. Evaluation Methodology

The experimental comparative approach as proposed by the PHARE/Validation Tools Group and set up for PD/3 turned out perfectly adequate with regard to the principles used. This methodology should be applied in any attempt to validate new concepts in real-time simulations. To be more robust, using a baseline system very close to the current ones is mandatory: if paper strips are a part of the current system, the baseline system should mimic this working.

D. Complete New, Well-Defined Scenario

The PHARE scenario kept the idea that while introducing a new method and new tools for the air traffic controller, letting the controller in the loop should be pursued because he or she remains a key element to ensure security in case of a non-nominal event. Although it could sound too definitive to conclude this, we can, however, question the possibility of building a system capable of both offering air-ground integration principles and helping to maintain the same level of situation awareness as controllers should maintain currently. Reference 14 concluded on situation awareness: *"··· If the policy is that the maintenance of the controllers' picture is not required, however, then all the consequences of this must be defined and allowed in advance: they are likely to include the inability of the human controller to intervene quickly and successfully in an emergency ···"* (p. 69).

Despite this, PHARE did not investigate if the requirement of maintaining the man in the loop could be escaped, still addressing the targeted time frame. This idea could lead to study in more detail what could be potential strategies capable to cope with non-nominal situations. The goal would be to provide operators with a ready-to-use temporary but immediately applicable set of solutions, letting them later analyze the problem with the required delay.

This chapter suggests that this track was not deeply followed yet and could bring some interesting spin-offs.

V. Conclusions

Two research and development directions are followed to enhance ATC to cope with the traffic demand:

1) The first category attempts to improve the same skills that currently are required to work as an air traffic controller: anticipation and memorization mainly. These projects promote a reinforcement of controllers' cognitive capabilities. In a certain way, these projects push to the edge the capability of the same engine: the brain of the controller. Their indisputable advantage is to ensure continuity with today's practices. The standard mode of operation helps to feed controller's awareness. Thus, it is respectful of cognitive resources which will be required from the controllers during emergency situations.

2) The second category explores ways to provide an automated system in which the controllers' role—in the scenario in which there are still some controllers—becomes routine interlaced with exception management. One identified flaw lies in

the risk that nominal mode of operation does not respect the building of cognitive resource that will be required in case of emergency. Even if one could assess that in the long run ATC will be performed like that, this project will require additional effort to assure the transition between today's practices and the target scenario.

The concepts proposed by PHARE on the one hand promote skills of the controller that are different from those required today: the required skills are no longer to monitor and construct mentally a situation, but to plan and to solve ahead in advance ATC problems, and a procedural mode is more suitable to these tasks. On the other hand, as they have been applied hitherto, the PHARE concepts are likely to diminish the ability of controllers to maintain the current picture of the traffic.

To conclude, we definitely count PHARE concepts among the projects of the second category; it is not an appalling finding. It just means that any project applying the essence of PHARE concepts should prove, through a complete scenario taking into account non-nominal solution and transition aspects, that the overall system is workable. This task is not impossible, but additional efforts should be made to make this second category result in operationally usable systems.

Beyond all doubt, PHARE and PD/3 have delivered brand new operational concepts, strengthened research and development on data-link technology and computer-aided ATC tools, and procured innovative HMIs, evaluation methodology, and teaching methods. PHARE also seized, the extent of what remains to be performed to build a new workable mode of operation in future ATC.

Acknowledgments

PD/3 was by nature a collaborative venture, gathering efforts from many participants among the ATM research and development community. CENA PD/3 could not have taken place without the cooperation of the PHARE partners: DERA, DLR, EEC (the Eurocontrol Experimental Centre), NATS, and NLR.

References

[1] "PD/1 Final Report," DOC 96-70-24 - PHARE/NATS/PD1 - 10.2 /SSR, 1.1.

[2] "PD/2 Final Report," DOC 97-70-13—PHARE/DLR/PD/2-10.2/SSR, 1.2.

[3] "Centre D'Etudes Navigation Aérienne—PHARE Demonstration #3—Final Report," DOC 99-70-01, Vol. 2, PHARE/CENA/PD/3-2.4/FR/2.0, May 1999.

[4] "EEC PHARE Demonstration #3 Final Report," DOC 99-70-01, Vol. 3, PHARE/EEC/PD/3-3.20/FR/1.0, Feb. 2000.

[5] "NLR—PHARE Demonstration #3 Final Report," DOC 99-70-01, Vol. 4, PHARE/NLR/PD/3-4.4.2/FR/1.0, Feb. 2000.

[6] "PATN Final Report," DOC 99-70-03, PHARE/CENA/PATN-5.2/FR; 2.3, Aug. 1999.

[7] "PHARE Demonstration #3 Operational Scenario Document," Vol. I, DOC 97-70-04—PHARE/PD3-1.1.3.2.2/OSD1;2.2, Jan. 1997.

[8] "PHARE Demonstration #3 Operational Scenario Document," Vol. II, DOC 97-70-08—PHARE/PD3-1.1.3.2.2/OSD1;2.0, March 1997.

[9] "PHARE Demonstration #3 Ground Human Machine Interface Specifications v 1.2," PHARE/GHMI 4.6, May 1997.

[10] Garron, J., and Chabrol, C., "CENA Experimental Protocol for PHARE Demonstration #3," CENA/NR98-156, May 1998.

[11] "PHARE Demonstration #3 Measurement and Analysis Specification," DOC 98-70-08—PHARE/NATS/VAL-4.4.2/WP005;1.0, Feb. 1998.

[12] Chabrol, C., "PHARE Demonstration #3 Training Course Support," CENA/NT98164, April 1998.

[13] Abdesslem, S., and Leroux, M., "Training Evaluation Approach of a New ATC Tool," 3rd USA/Europe Air Traffic Management R&D Seminar, 13–16 June 2000.

[14] "Role of the Man Within PHARE," DOC 93-70-35, June 1993.

[15] "PD/1 Final Report," DOC 96-70-24—PHARE/NATS/PD1—10.2 /SSR;1.1

Chapter 28

Restriction Relaxation Experiments Enabled by User Request Evaluation Tool

Michael J. Burski[*]
Federal Aviation Administration, Washington, DC
and
Joseph Celio[†]
MITRE Corporation, McLean, Virginia

I. Introduction

THE User Request Evaluation Tool (URET) was developed by the MITRE Corporation's Center of Advanced Aviation System Development (CAASD) as a tool to assist the controllers in managing operations in a free flight environment. The tool was derived from the advanced en route automation (AERA) research and is designed to support the sector team strategic planning function. The tool uses flight plan, track, and wind data as the basis upon which to build trajectories of the projected flight of controlled aircraft and to indicate possible conflicts up to 20 min into the future. The tool functionality and computer-human interface evolved from its initial installation at the Indianapolis air route traffic control center (ARTCC) until it was deemed useful and necessary to do large-scale evaluations. The large-scale evaluations required the availability of the tool at all positions in both the Indianapolis ARTCC and neighboring Memphis ARTCC on a daily basis. The primary goal of the daily use URET was to exercise the functionality of the tool to its fullest and to identify enhancement/fixes required to ensure its usability on a national basis. A secondary goal of the daily use URET was to train a proficient controller workforce on the use of the new technology so that the strategic planning capabilities of the tool and the associated benefits to the flying public could be assessed. This secondary goal involves an evolution toward separating aircraft from aircraft rather than separating aircraft from airspace.

The functionality embodied by the URET operation in Indianapolis and Memphis is being refined and will be installed in five additional sites as part of the Free Flight Phase 1 (FFP1) program. Figure 1 graphically depicts the sites that will

---

This material is declared a work of the U.S. Government and is not subject to copyright protection in the United States.
[*]Project Lead, Aeronautical Data Link System (ADLS) Build II Program.
[†]Project Team Manager.

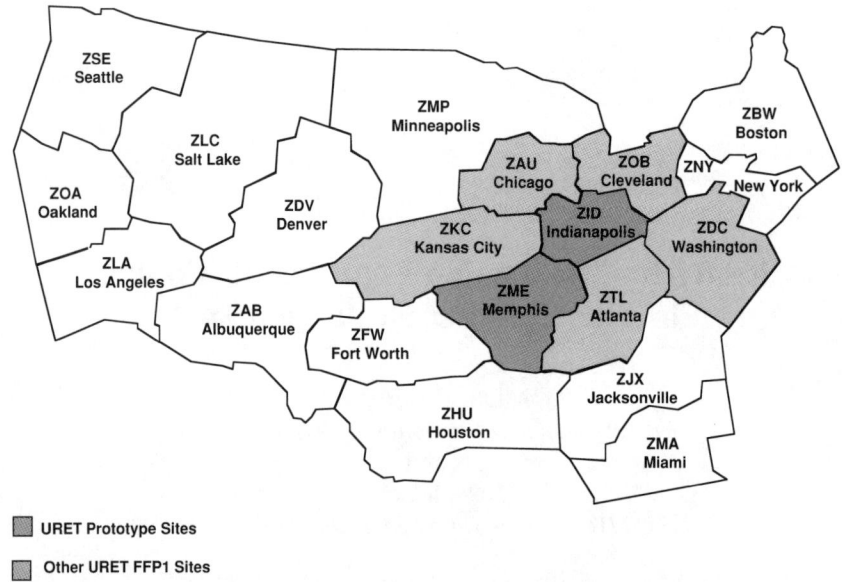

Fig. 1 URET FFP1 implementation sites.

have URET capability by the end of 2002. The methodology developed and tested using the URET prototype to assess benefits will be employed at all the URET FFP1 sites. By the time the five new sites approach operational usage, a plan for achieving user benefits at these facilities will already be in place.

This chapter describes efforts to measure the benefit from relaxing certain restrictions due to the availability of a strategic planning conflict detection tool such as URET. The chapter will briefly discuss the full range of benefits a conflict detection tool could provide to the aviation community. It will concentrate on the user benefits from relaxing restrictions and specifically on the efforts at the Indianapolis ARTCC and Memphis ARTCC to relax altitude restrictions.

For more detailed information concerning the functions and operational use of URET, refer to Ref. 1. Within this chapter, the Indianapolis ARTCC and Memphis ARTCC will be referred to by their three-letter abbreviations, ZID and ZME, respectively.

II. URET Utilization

To determine what benefits URET is providing, it is important to examine how URET is being used. URET has been in daily use at ZID and ZME since the fall of 1997. Metrics on various URET capabilities are collected and updated on a monthly basis, based on the daily files generated by URET. Over a two-year period, URET has grown from a single workstation to full center operations at both sites. Controller teams recently requested URET be available full time (24 h a day/7 days per week). Starting in February 2000, URET was made available 22 h a day/7 days a week at both ZID and ZME. URET is now

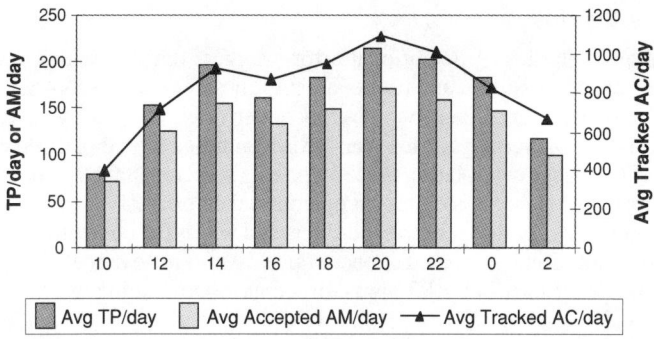

Fig. 2 Daily average counts of trial plans, amendments, and tracked aircraft for ZID in February 2000.

considered a key component of the data (D-side) controller's toolset. The D-side controller routinely uses URET functionality (e.g., trial planning) to support strategic planning to improve the air traffic flow through the National Airspace System (NAS).

In July 1999 tools became available to examine URET usage of the trial planning function and flight plan amendment submission in correlation with traffic counts. It was an often-spoken belief that URET usage went down as traffic count went up, the rationale being that, when traffic density increases, the D-side controller does not have time for strategic planning. Figure 2 shows that usage is almost in direct correlation with traffic volume, at least since this measurement began (after two-way communications for amendments had started).

The metric is calculated by summing all trial plans, amendments, and tracked aircraft during a 2-h interval every day of the month if URET is running. That sum is divided by the number of days URET was available during that time range. For the month of February at ZID, between 1400 and 1600 (time in Zulu hours), on average 195 trial plans were generated, 155 of which were entered and accepted by the host computer as flight plan amendments. During that same 2-h interval, an average of 927 flights were tracked at ZID.

The URET prototype has become an integral part of daily-use operations at ZID and ZME. Operational personnel have started to use the reduction in workload provided by the general use of the URET prototype at the centers to provide benefits to the NAS user community.

III. URET Benefits

The objectives of Free Flight are to provide greater flexibility and cost savings to the users without compromising safety. The evolution to Free Flight and the increased reliance on URET and other capabilities are intended to provide tangible benefits early to users during the system's life cycle. The use of URET will enhance safety and Federal Aviation Administration (FAA) productivity, and enable user cost saving benefits[2-4]. Although the safety and productivity benefits will be identified, the focus of this chapter is in user cost saving benefits enabled by restriction relaxation.

A. Safety Benefits

A number of studies of operational errors suggest that the widespread use of URET can produce significant improvements in safety of the NAS.[5] En route operational errors occur when two aircraft under radar control by en route controllers violate specified separation standards (5 n miles lateral or 1000 or 2000 ft vertical). If the cause of violating these standards was a mistake on the part of the radar controller, this event is called an operational error.

URET predicts aircraft-to-aircraft conflicts and warns the controller in the sector in which the violation is expected to occur up to 20 min in advance of the predicted loss of separation. Study of URET alerts for scenarios containing recent operational errors from ZID and ZME[6] shows that URET dependably provides alerts with substantial warning times. This after-the-fact analysis indicated that in 16 out of 19 cases, URET provided the D-side controller a warning time of between 2.5 and 13 min with an average of 7 min.

B. FAA Productivity Gains

The productivity gains and reduction in sector workload that URET provides sector control teams are critical for the effective use of the tool. Relief from routine tasks and more efficient management of sector workload are essential aspects of URET that create the opportunity to carry out the strategic planning tasks that will achieve user benefits. Reduced workload includes: 1) less physical movement to manipulate and write on strips, 2) reduced mental projection of flight paths to determine possible conflicts, and 3) quicker entry of route amendments into the host computer.

URET is the primary source of flight data for the sector. The flight trajectory is a more accurate model of an aircraft's predicted flight path than what is presented on a paper strip. The trajectory is continually adjusted using host computer track information, wind and temperature. These changes are automatically made to the displayed information.

The conflict probe and trial plan results generated by URET provide new, accurate, continuously updated future situation awareness data. When notification of a predicted conflict is received, the controller may elect to trial plan potential solutions. The controller is able to see the impact of potential solutions in terms of creating other conflicts. Although the sector planning decision process with URET remains a mental assessment and judgment by controllers, URET relieves controllers from performing routine, recurring and often time-consuming manual calculations to predict and compare future positions of aircraft.[7,8]

URET is used to enter flight data amendments and flight plans into the host computer. The graphic display of an aircraft's trajectory and the point and click capabilities of the URET computer human interface (CHI) provide a significant time savings for controllers using URET to generate a route amendment. URET automatically formats a flight amendment for the host computer and provides the option to submit it to the host computer with a single action. The ease with which route amendments can be generated and submitted to the host computer is expected to result in flight plans that better reflect the intent of the aircraft.

The results of these productivity improvements are being examined for their effects on throughput. Preliminary data on traffic loads indicate that with URET use, sector air traffic counts have increased while sector transit times have either held steady or have decreased. This aspect of the results is still under investigation and may be reported on at a later date.

C. Restriction Relaxation

URET has been postulated as a tool that will enable aircraft to fly more user-preferred flight profiles. Because of its extended look ahead, the tool will enable aircraft to fly longer direct routes and enable proposed flight amendments to be evaluated on a real-time basis. At sector and facility boundaries, there may be the potential to relax certain restrictions during all or portions of the day due to the better data inherent in URET and the capability of the tool to project aircraft trajectories into and out of the sector/center airspace. This enables controllers to separate aircraft from aircraft rather than aircraft from airspace. Altitude restrictions force aircraft to fly at a less than optimum altitude, thus increasing fuel burn, time in the air, and operating costs.

Restrictions are in place to better manage flows between sectors or centers. There are about 190 static altitude restrictions in ZME airspace, and about 370 in ZID airspace. They primarily help with the separation of aircraft. They also support an orderly transition of traffic through sectors and help controllers at each of the sectors handle the altitudes they are responsible for without making decisions for the sectors below them. Restrictions are a form of coordination that everyone understands; they keep aircraft climbing and descending in a way that everyone expects.

URET capabilities provide better and more accurate information to the D-side controller, which may enable the controller to dynamically relax restrictions in accordance with existing and planned traffic flows. The URET-provided conflict information will alert controllers to potential conflict situations in a free-flow environment that could thereby reduce the need for static restrictions imposed on the flows.

D. Operational Evaluations of Restriction Relaxation

The purpose of the operational evaluation of restriction relaxation is to determine if the strategic planning capabilities of URET will allow certain static restrictions to be relaxed during a portion of the day at ZID and ZME without adverse impact to controller operations or to the overall air traffic flow. The goal of the first evaluation, held in May 1999, was to determine if URET enables controllers managing the arrival stream to Nashville Metropolitan (BNA) and Louisville Standiford Field (SDF) to relax crossing restrictions and if there is measurable benefit in relaxing the restrictions.

1. First Evaluation: 27 May 1999

On 27 May 1999, an evaluation of the lifting of altitude restrictions between ZID and ZME took place. Arrival restrictions into BNA in ZME airspace and into SDF in ZID airspace were lifted for a 2-h period, from approximately 1000 to 1200 hrs local time. For a complete analysis of this evaluation, refer to the CAASD report.[9] The following summarizes the main conclusions of the evaluation.

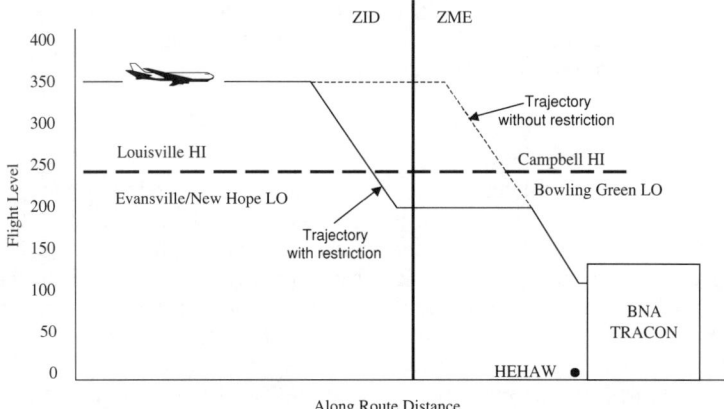

Fig. 3 Approximate altitude profile for trajectories with and without ZID/ZME crossing restriction for Nashville arrivals.

Jet aircraft going to BNA are normally restricted to Flight level (FL) 200 or below crossing from ZID to ZME airspace (see Fig. 3). Typically under normal operations with the restriction in place, the aircraft would traverse either the Pocket City or Louisville sector and be handed off to the low-altitude ZID sector Evansville/New Hope. The controller working Evansville/New Hope would clear the aircraft to cross the facility boundary at 16,000 ft. The interfacility handoff would be between Evansville in ZID and Bowling Green in ZME. Bowling Green would then clear the aircraft down to cross the fix designated as HEHAW at 10,000 ft.

During the conduct of the evaluations (URET operational and restriction relaxed) for BNA aircraft, the D-side of the controlling high-altitude sector would do a "show-all," in which all predicted conflicts would be depicted on his URET screen. (Normally only conflicts predicted to occur within his own sector are highlighted on his URET display.) If the show-all revealed no conflicts, the controller issued a clearance to the pilot of the BNA inbound flight for a discretionary descent to cross HEHAW at 10,000 ft. The aircraft would be handed off (within ZID) to the Evansville/New Hope low sector (even if the aircraft remained at FL310). As the aircraft approached the ZID/ZME facility boundary, the aircraft would be handed off from the Evansville/New Hope sector to the Bowling Green ZME sector. If the aircraft remained at altitude and would enter the Campbell high sector in ZME, the Louisville high sector controller would make the request for point-out to the ZME Campbell high sector.

The air traffic recorded data from 27 May 1999 were analyzed in terms of URET trajectory comparisons and track comparisons. These comparisons were made between URET trajectories with and without the subject restrictions, and recorded tracks with and without these restrictions.

There were 12 flights eligible for the URET trajectory analysis. The top of descent along route distance shows that aircraft stayed higher for an average of 38.7 n miles when the restriction was turned off than with it on. The along-route time shows aircraft stayed higher for an average of 5 min farther down route when the restriction was turned off.

Track analysis compared seven common flights on 20, 26, and 27 May 1999. It took the same flights on three different days. Two of the days, 20 and 26 May, the aircraft were kept at the restricted altitude. On 27 May, during the evaluation, the aircraft descended at their own discretion. Actual comparison of the tracks shows that flights on 27 May crossed the restriction location an average of 5000 ft above the same flights on 20 and 26 May 1999. In addition, they stayed at altitude 3–3.5 min longer before beginning their arrival descent.

The results of the analysis that followed the temporary relaxation of restrictions were positive and encouraging, both from the traffic analysis and from the operational perspectives. Operational personnel were willing to review other altitude restrictions as candidates for testing and possible removal. Controller feedback from the evaluation provided excellent insights into the conduct of future evaluations. Some significant results were:

1) Controllers were uncomfortable clearing aircraft to an altitude they did not control. Giving discretionary descents below the floor of their airspace was an uncomfortable concept. Through post-evaluation discussion, it was determined that for the next evaluation, the clearance should be limited to the floor of the high-altitude airspace, and then an early hand-off should be effected to the low-altitude sector.

2) The impact of the high-altitude sector receiving the point-out needs to be monitored. During the low traffic conditions under which the evaluation was undertaken, the point-outs were handled. When traffic gets busy and the controller "unables" a point-out, the aircraft will have to descend quickly in order to not enter the high-altitude sectors' airspace.

3) Future selection of restriction relaxation candidates should consider the rationale behind the restriction.

Subsequent to this evaluation it was decided to implement facility teams comprised of operational personnel to evaluate how the use of URET will enable them to change their way of doing business and possibly relax restrictions during portions of the day on a routine basis. Procedures and benefits teams were established at the two sites. These teams were comprised of sector controllers, supervisors, traffic managers, union representatives, and plans and procedures specialists. The teams were chartered to review the static altitude restrictions, identify candidates for temporary relaxation of the restriction that would be enabled by use of URET, plan structured evaluations of the restriction relaxation, evaluate the results, and determine the impact of lifting the restriction on controller workload and on user benefits. If the results were positive, the teams would recommend that a restriction be relaxed during all or a portion of the day when URET was operational.

Both teams assessed relaxing static restrictions with the following characteristics:

1) Internal restrictions. The relaxation of internal restrictions does not require coordination with other centers. (In the future, ZID and ZME are planning to cooperate in lifting restrictions that are in place between the two facilities.)

2) Restrictions in high and super-high sectors. Restrictions in low-altitude sectors, especially those adjoining approach control areas, are not candidates because aircraft have to be descended for landing and sequencing for orderly arrival into the airport.

Table 1 Sample set of ZID altitude restrictions from 18 November 1999, 10-h period

Restriction ID	Restriction description	Number of flights	Average passing length, n miles	Gallons saved per day
A.01	BNA_A_80/81_290	8	66	90
A.17.1	CVG_A_87/23_VIA_BOWRR_240	50	23.8	440
A.23	CVG_A_84/83_VIA_DRESR_240	57	31.1	510
A.25	CVG_A_80/35_VIA_JEANE_240	44	24.7	390
A.26	CVG_A_81/82_240	67	16	547
A.36	IND_A_84/82_310	9	39.2	84
A.37	IND_A_87/88_310	5	20.3	15
A.73	CMH_A_86/85_290	6	50	86
C25	SDF_A_ZME/19_SWEWO_110	7	16.4	114

3) Arrival restrictions. There are better candidates than departure restrictions. Internal departure restrictions frequently have little impact on departing aircraft. Regional jets, especially in warm weather, do not reach the departure altitude restriction at the time the aircraft crosses the restriction boundary.

The procedures and benefits teams utilized the output of the CAASD Analysis of Restrictions Tool (ART) to help identify candidates for restriction relaxation. The ART software was run for various days on which URET was active. The output lists all of the restrictions in the facility, all of the traffic subject to each restriction for the time that URET was functional, and the average nautical miles that the aircraft were constrained to in the restricted altitude.

Table 1 contains the fields in the analysis that are significant in identifying candidates for restriction removal. The table is a subset of the complete list of restrictions for ZID on the specified day. Each of these restrictions meets the criteria set down by the teams as possible relaxation candidates. The restriction describes the location where aircraft must be at or below the crossing altitude. The fields in Table 1 are:

1) Restriction ID: This field gives the internal tracking number used to identify, sort, and refer to each restriction.

2) Restriction description: This field provides the arrival airport, boundary crossing definition, and the altitude to cross at or below.

3) Number of flights: This field specifies the number of flights that were subject to the restriction during the analysis period. Other aircraft may have met the criteria but were below the altitude.

4) Average passing length: This field represents the average distance in n miles that the flights were in level flight at the restriction altitude.

CAASD analyzed the traffic subject to the restrictions in Table 1 and identified aircraft type and cost penalty per restriction. The aircraft type data came from the data collection capability of the URET prototype. The cost penalty came from figures provided by airlines on fuel burn for aircraft type at various altitudes. Estimates of the fuel burn penalty were obtained for B737, B727, B757, B767, MD80, and Canadair regional jet. These are the most flown aircraft types within ZID airspace. From the information from the ART analysis and the fuel burn penalty data, the amount of gallons saved per day was estimated. Table 1 also contains the penalty of each restriction by the per-day cost in gallons.

The teams at each site began meeting to plan restriction relaxation evaluations in October 1999. The evaluations undertaken as a result of this effort are described in the following sections.

2. Second Evaluation: 29–30 December 1999

On 29 December, the restriction A.36, which requires Indianapolis terminal arrivals to cross the sector 84 to sector 82 boundary at FL310 or below, was lifted from 0800 to 1130 hrs local time. The period from 0800 to 1000 hrs was extremely busy in all participating sectors. Controllers managed traffic tactically; they had little opportunity for strategic planning in the first two hours of the restriction removal evaluation. The workload for the receiving sector controller in the super-high sector with the restriction removed was dramatically increased during the busy period.

Ten Indianapolis arrivals entered sector 84 during the evaluation. Six aircraft met the criteria for evaluation of the altitude restriction. Of these, two aircraft were descended for traffic considerations to the restricted altitude of FL310 during the busy period. Four aircraft stayed at altitude, above FL310, crossing the 84/82 sector boundary. Table 2 lists the flights that stayed at altitude and the fuel savings estimate.

The initial evaluation by the team was that, until the center controllers move to a more strategic operation, there was no real benefit in lifting this particular restriction during busy periods. However, when there was not heavy traffic, restriction removal was extremely beneficial. Traffic moved through both sectors at altitude for substantial distances with no problems encountered.

On 30 December the restriction A.01, which requires Nashville terminal arrivals to cross the sector 80 to 81 boundary at FL290 or below, was lifted. Four Nashville arrivals entered sector 80 during the time that the test was conducted. One was descended to FL290 at the request of the receiving controller in sector 81. Two were at a lower altitude than the restriction. One aircraft remained at altitude longer, not descending to FL290 (see Table 2).

The procedures and benefits team determined that strategic planning in this instance was properly carried out and that most of the aircraft, had they been at

Table 2 Example: Impact of ZID restriction removal test

Restriction number	Number of flights	Aircraft type	Crossing altitude	Remain at altitude, n miles	Estimate of fuel savings, gal
A.01	1	B727	FL350	73	20.4
A.36	4	MD80	FL350	70	13
		C525	FL390	72	13.5
		B737	FL350	70	13
		A310	FL350	25	4.5

altitude, would have remained at altitude for an appreciable period. It was agreed that this particular crossing restriction is a perfect candidate for extended testing and possible elimination. This restriction is affected by the Memphis operation; the ZID team suggested that the removal of the restriction could, possibly, extend into ZME airspace.

3. Third Evaluation: 24–25 February 2000 at ZID

A third relaxation evaluation took place at ZID on 24 and 25 February. On 24 and 25 February, the restriction A.37, which requires Indianapolis terminal arrivals to cross the sector 87 to 88 boundary at FL310 or below, was relaxed for several hours on each day. Eighteen aircraft were subject to the restriction. Of these 18 flights, only 11 saved fuel by staying at a higher altitude. The estimate of fuel savings for all 18 flights is 144 gal.

On 24 February, the restriction A.73, which requires Columbus terminal arrivals to cross the sector 85 to 86 boundary at FL290 or below, was relaxed for several hours. Ten aircraft were subject to the restriction. All of these flights were kept at a higher altitude. The estimate of fuel savings for all 10 flights is 70 gal.

A summary of each restriction evaluation is shown in Table 3. The fuel savings are the average savings per flight. The savings for restriction A.37 was averaged over all 18 flights even though 7 of the flights saved no fuel.

Further estimates were made concerning the possible savings from removing restrictions A.37 and A.73 permanently. Based on the traffic levels estimated from the CAASD ART analysis and site estimates, these flights represent about 40% of the daily traffic over the two restrictions. The average fuel saved per restriction per day during the evaluation was 80 gal. This extrapolates to 400 gal saved per day and 144,000 gal saved for the entire year if just these two restrictions can be removed.

4. Future Evaluations

The Memphis procedures and benefits team has examined all of ZME's internal arrival restrictions. Five Nashville arrival restrictions are candidates for relaxation based on the criteria mentioned in the preceding sections. ZME intends to lift all five of these restrictions for a 30-day period starting in mid-March 2000 and then evaluate the results. ZME will also relax three Louisville arrival restrictions starting 1 April 2000. ZID intends to evaluate two additional restrictions starting in April for 14 days: one for Indianapolis arrivals and one for Louisville arrivals.

The procedures and benefits teams at both sites will continue to relax altitude restrictions for the foreseeable future. ZID has about 70 restrictions that meet the criteria for relaxation, and they will continue to review and evaluate. ZME and ZID will also start to discuss relaxing restrictions between facilities. CAASD will continue to support both facilities in their efforts through analysis of traffic and continued reporting of the potential user savings.

The airlines are a key player in this restriction relaxation effort. A clear understanding from the airline perspective as to the impact of particular restrictions is needed. An effort was initiated in January 2000 to communicate with the airlines to determine the high-impact restrictions affecting their aircraft traversing ZID and ZME. These data in conjunction with the site procedure and benefit teams will drive the sequence of restriction relaxation evaluations.

Table 3 Summary of restriction relaxation evaluations

Date	Facility		Restrictions		Number of aircraft	Impact	
	ZID	ZME	Restriction description	Local time		Average longer distance at altitude per aircraft, miles	Average fuel saved per aircraft, gal
5/27/99	X	X	FL200 or below for Nashville arrivals	1000–1200 hrs	9	38.7	15
12/29/99	X		FL310 or below for Indianapolis arrivals	0800–1130 hrs	4	59	11
12/30/99	X		FL290 or below for Nashville arrivals	0800–1000 hrs	1	73	20.4
2/24–2/25/00	X		FL310 or below for Indianapolis arrivals	0800–1200 hrs	18	57	8
2/25/00	X		FL290 or below for Columbus arrivals	0800–1200 hrs	10	54	7

IV. Conclusions

The relaxation of selected restrictions during some parts of the day is enabled by the use of the more accurate URET information to support strategic planning (as compared with non-URET operations) and well-defined procedures. Widespread use of the tool in support of strategic planning will enable restriction relaxation and be a step toward free flight. The resulting benefits of staying at altitude longer will be direct fuel savings to the aircraft operator. More evaluations in using this tool to relax restrictions are required before applying the tool for this purpose on a standard use basis. Continued evaluations will define the problem of how much relaxation may be available through use of the conflict probe tool. The future of relaxing restrictions is bright, but it must be carefully managed.

References

[1] Celio, J. C., Bowen, K. A., Winokur, D. J., Lindsay, K. S., Newberger, E. G., and Sicenavage, D., "Free Flight Phase 1 Conflict Probe Operational Description," MTR 00W000001000, The MITRE Corp., McLean, VA, March 2000.

[2] Walker, M. G., "Initial Altitude Restriction Analysis for URET Benefits During Free Flight Phase 1," WN 99W0000063, The MITRE Corp., McLean, VA, Sept. 1999.

[3] Brudnicki, D. J., and McFarland, A. L., "User Request Evaluation Tool (URET) Conflict Probe Performance and Benefits Assessment," MP-97W0000112, The MITRE Corp., McLean, VA, June 1997.

[4] Brudnicki, D. J., McFarland, A. L., and Schultheis, S., "Conflict Probe Benefits to Controllers and Users: Indications from Field Evaluations," MP-96W0000194, The MITRE Corp., McLean, VA, Aug. 1996.

[5] Rodgers, M. D., Mogford, R. H., and Mogford, L. S., "The Relationship of Sector Characteristics to Operational Errors," DOT/FAA/AM-98/14, Office of Aviation Medicine, U.S. Department of Transportation, Federal Aviation Administration, Washington, DC, May 1998.

[6] Sharma, R., "Operational Error Analysis," WN99W0000032, The MITRE Corp., McLean, VA, June 1999.

[7] "Air Traffic Conflict Probe Team, Report #1, Results of Team Efforts August 1998–January 1999," Appendix C, URET Core Capability Limited Deployment (CCLD) Operational Concept/Air Traffic Requirements, Federal Aviation Administration, Washington, DC, Nov. 1998.

[8] "Air Traffic Conflict Probe Team, Report #1, Results of Team Efforts August 1998–January 1999," Federal Aviation Administration, Washington, DC, Feb. 1999.

[9] Ricker, M. L., and Walker, M. G., "Evaluation of Altitude Restriction Relaxation for Standiford and Nashville Arrivals Using URET," WN 99W0000055, The MITRE Corp., McLean, VA, May 1999.

Section VI: Safety and Free Flight

Safety has always been the overarching objective of the entire aviation system, but until recently, there has been very little progress in a systematic methodology for ensuring that new air traffic management (ATM) system developments are indeed safe. A great deal has been learned about how to build and operate safe systems from the nuclear industry and even from the aircraft industry, and we are now beginning to see the transfer of this to ATM system development.

However, as Blom points out in the first chapter of this section, the tools that have been successful in these arenas are not well suited to the assessment of safety in a complex, dynamic, interactive environment like ATM. Thus Blom and his coworkers developed a safety risk assessment model, the Traffic Organization and Perturbation Analyzer (TOPAZ). The methodology includes heuristic and stochastic system analysis and also takes into consideration the "human element" in the safety formula. They apply their model to airborne separation assurance, comparing this to traditional air traffic control (ATC) separation assurance. In the second chapter, Blom and his coworkers extend the TOPAZ methodology with a mathematical model of the cognitive performance of the air traffic controller. The model is an excellent start at factoring human performance into the equation, but it does not provide a mechanism for understanding human performance risk when other system elements exhibit "failures." These are important interactions that must be analyzed because the science of safety of highly reliable systems has clearly established that accidents generally are the result of multiple failures of different elements of the total system. In the third chapter of the section, Blom and his coworkers present the results of an extension of the TOPAZ models to probabilistic risk assessment of wake vortex induced accidents. The TOPAZ methodology used in these three chapters represents the state-of-the-art in rigorous communication, navigation, surveillance, and air traffic management (CNS/ATM) safety analysis.

The fourth chapter in this section, by Hoekstra et al. develops a case for the need of a distributed system to permit free flight in high-density airspace. The results of analysis and simulation are presented to make the case that it is precisely in high-density airspace that the power of a distributed system (as envisioned by free flight airborne separation) is required. The authors argue that a distributed system for separation is inherently safer (more reliable and more robust) than a centralized, ground-based ATC system.

The fifth chapter presents a comprehensive safety assessment methodology, developed under International Civil Aviation Organization (ICAO) auspices, that will help lead to certification of Airborne Separation Assurance System (ASAS) applications. The work is based on a Radio Technical Commission for Aeronautics (RTCA)/European Organization for Civil Aviation Equipment (EUROCAE) methodology for operational safety assessment of CNS/ATM systems. The assessment is built around a comprehensive operational hazard analysis that takes into account the severity and likelihood of hazards.

The final chapter proposes a methodology for evaluating changes to separation standards. To ensure safety in a free flight environment it will be necessary to develop the appropriate separation standards, and a methodology such as the one proposed in this chapter might be used for this.

The chapters in this section represent the state-of-the-art in addressing the safety considerations that must be taken into account as we move toward a free flight paradigm, but much more work is required in this area. The challenge becomes even greater than in the past as we see an increase in the complex interaction between humans and machines in the cockpit with the ground-based elements (human and machine) of the CNS/ATM system.

Safety experts agree that systems with the complexity and component interaction of the CNS/ATM system elements will fail. The challenge is to design a system that minimizes interactions and builds in "defenses" to guard against and to mitigate the effects of failures. A broad-based pragmatic approach to safety that considers all aspects of the system, including the human operator, is required. There must be a focus on safety from early conceptual development through system design, development, test, and ultimately operation. Techniques such as the ones presented in this section must be augmented with analysis and fast- and real-time simulation to ensure a thorough understanding of system behavior under a range of normal and abnormal conditions.

Chapter 29. Accident Risk Assessment for Advanced Air Traffic Management, H. A. P. Blom, G. J. Bakker, P. J. G. Blanker, J. Daams, and M. H. C. Everdij, NLR, 1998.

Chapter 30. Human Cognition Modelling in ATM Safety Assessment, Henk A. P. Blom, Jasper Daams, and Herman Nijhuis, NLR, 2000.

Chapter 31. Probabilistic Wake Vortex Induced Accident Risk Assessment, J. Kos, H. A. P. Blom, L. J. P. Speijker, M. B. Klompstra, and G. J. Bakker, NLR, 2000.

Chapter 32. Free Flight in a Crowed Airspace?, J. M. Hoekstra, R. C. J. Ruigrok, and R. N. H. W. van Gent, NLR, 2000.

Chapter 33. Managing Criticality of Airborne Separation Assurance Systems Applications, Andrew D. Zeitlin, MITRE Corporation; and Beatrice Bonnemaison, CENA, 2000.

Chapter 34. Analysis of Aircraft Seperation Minima Using a Surveillance State Vector Approach, Tom G. Reynolds and R. John Hansman, MIT, 2000.

Chapter 29

Accident Risk Assessment for Advanced Air Traffic Management

H. A. P. Blom,* G. J. Bakker,† P. J. G. Blanker,‡ J. Daams,§ M. H. C. Everdij,¶ and M. B. Klompstra¶
National Aerospace Laboratory NLR, Amsterdam, The Netherlands

I. Introduction

AIR traffic management (ATM) is the result of complex interactions between human operators, procedures, and technical systems (hardware and software), all of which are highly distributed. Providing safety is more than making sure that each of these elements function properly and safely. The complex interactions between the various elements of ATM significantly determine safety. Therefore, it is imperative to understand the safety impact of these interactions, particularly in relation to non-nominal situations. Traditional ATM design approaches tend first to design advanced ATM that provides sufficient capacity, and next to extend the design with safety features. The advantage of this approach is that ATM developments can be organized around the clusters of individual elements, i.e., the communication cluster, the navigation cluster, the surveillance cluster, the automation tools cluster, the human machine interfaces (HMIs), the advanced procedures, etc. The key problem is that safety effects stay unclear. A far more effective approach is to try to design an ATM system that is inherently safe at the capacity-level required. From this perspective, safety assessment should be one of the primary filters in ATM concept development. An early filtering of ATM design concepts on safety grounds can potentially avoid a costly development program that turns out to be ineffective, or an even more costly implementation program that fails. Although understanding this idea is principally not very difficult, it can be brought into practice only when an ATM safety assessment approach is available that

---

Copyright © 2001 by NLR. Published by the American Institute of Aeronautics and Astronautics, Inc., with permission.
*Group Leader, Air Traffic Management Modeling, Air Transport Division.
†ATM Modeling Researcher, Air Transport Division.
‡ATM Modeling Researcher, Air Transport Division; currently with Marine Safety Research.
§ATM Modeling Researcher, Air Transport Division; currently with LVNL.
¶ATM Modeling Researcher, Air Transport Division.

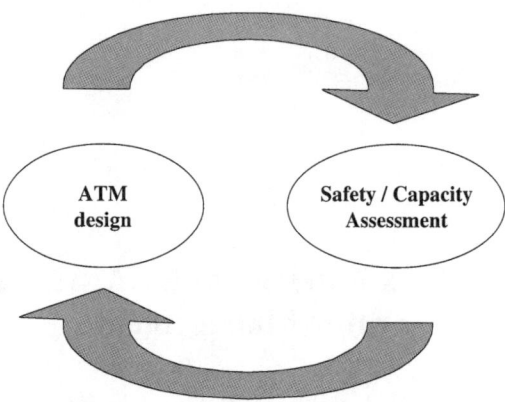

Fig. 1 Safety feedback based ATM design.

provides appropriate feedback to the ATM designers at an early stage of the concept development (Fig. 1). Such an approach has been presented by Ref. 1 and this chapter is based on this.

This feedback should not only provide information on whether the design is safe enough, but it should also identify the safety-capacity bottlenecks. By now, consensus is building that appropriate ATM safety modeling approaches are needed to understand the mechanisms behind designing advanced ATM. It is also recognized that, once such an ATM safety modeling approach is available, a safety feedback based design approach of future ATM will become feasible.[2-4]

Safety is a general notion that is typically studied from one of three different perspectives:

1) Safety perception (e.g., by pilot, controller, passenger, human society, etc.). An ATM design that is perceived as being unsafe will not easily be accepted by the humans involved. A positive perception about the safety of an ATM design is an implementation-critical requirement. By its very nature, however, safety perception is a subjective notion, and therefore insufficient to really approve safety-critical changes in ATM.

2) Dependability of a technical system (e.g., of a computer program, an aircraft navigation system, a satellite-based communication system, etc.). Dependability metrics are definitively objective. They are widely studied in the literature (e.g., Refs. 5 and 6). However, they have been developed to cover technical systems only (e.g., Refs. 7-9), and not the human operators and procedures of ATM (Ref. 10).

3) Accident risk (e.g., for the first, second, and third parties in air transport). Accident risk metrics definitively are objective and are commonly in use for other human-controlled safety-critical operations such as in the chemical and nuclear industries (Ref. 11). Two well-known International Civil Aviation Organisation (ICAO) adopted accident risk metrics are for collision of an aircraft with another aircraft during the en route phase, or with fixed obstacles during landing. A recent review of various accident risk metric possibilities in air transport is given in (Ref. 12).

In view of the ATM safety assessment needs, the accident risk perspective has the best joint characteristics: 1) it implies the use of objective risk metrics; 2) it

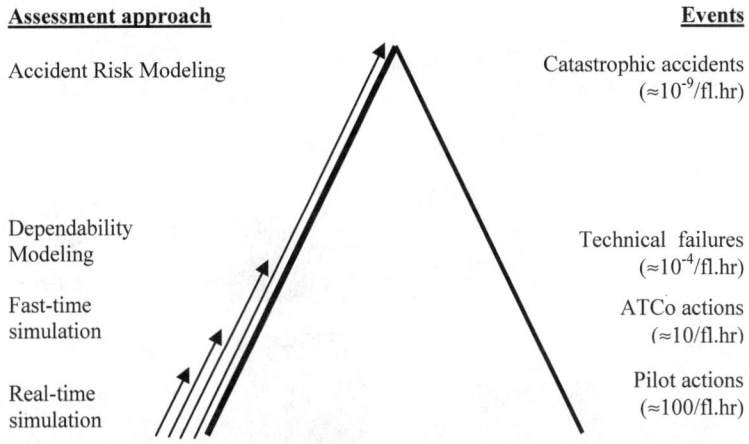

Fig. 2   ATM safety iceberg.

has proven its usability to human-controlled safety-critical operations; and 3) it is supported by ICAO. As such, in this chapter ATM safety will be considered from an accident risk perspective, with emphasis on risk of collision between two aircraft.

For air traffic the fatal accident risks should be on the order of $10^{-7}-10^{-10}$ per aircraft flight hour. To develop some feeling of the difficulty to assess such rare events, it is quite helpful to understand why the well-known fast-time simulators like National Airspace Systems Performance Analysis Capability (NASPAC), Reorganized ATC Mathematical Simulator (RAMS) or Total Airport and Airspace Modeller (TAAM) fall short for that purpose. One major shortcoming of these tools is that they are not really capable of modeling the aviation safety-critical combinations of non-nominal events; they often do not even model the single non-nominal events. Another major shortcoming is that an accident rate of, say, 10^{-9} per aircraft flight hour cannot in a practically reasonable way be reached through a straightforward simulation, because this would require a simulation of 10^{10} aircraft flight hours. This problem is well illustrated by the ATM safety iceberg (Fig. 2). To assess a catastrophic accident rate, one really needs to decompose the risk assessment problem into an effective hierarchy of simpler conditional assessment problems, in which simplicity means an appropriate combination of scope (e.g., volume of airspace) and depth (i.e., level of model detail) at each conditional assessment level. Indeed, tools like TAAM apply to assessments that address a broad scope in combination with a low level of non-nominal detail.

In general, the accident risk assessment problem has been widely studied for other safety-critical operations, such as for the nuclear and chemical industries, and for these applications, numerous techniques and tools have been developed. To take maximal advantage of this existing body of knowledge, we made a thorough study of the applicability of these techniques to accident risk assessment in air traffic.[13] A large variety of techniques has been identified, varying from qualitative hazard identification methods such as preliminary hazard analysis (PHA), common cause analysis (CCA), and failure mode and effect analysis (FMEA), through static assessment techniques such as fault tree analysis (FTA) and event tree analysis (ETA), to dynamic assessment techniques such as Petri net and Markov chain

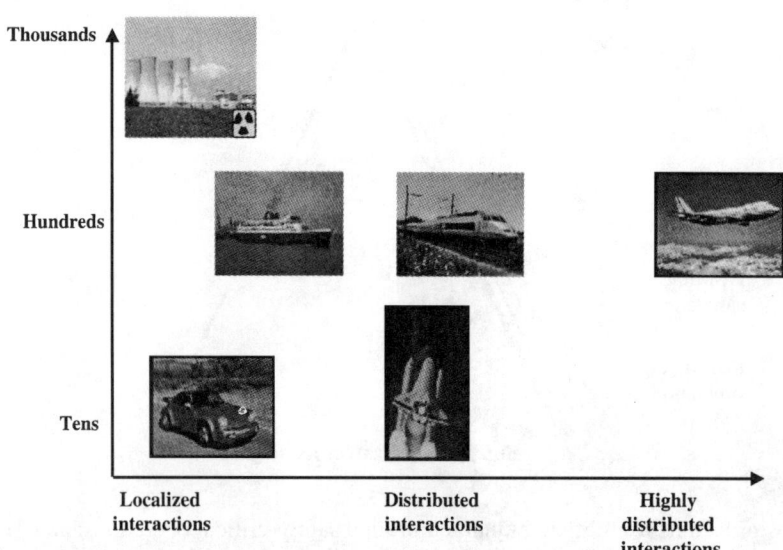

Fig. 3 Potential fatalities and distribution level of ATM and other safety critical activities.

modeling, dynamic event trees, etc.[14] Each of these techniques has advantages and disadvantages, but these appear to be minor in comparison to what is required for modeling ATM- related risk. The key finding is that the established techniques fail to support a systematic approach toward modeling stochastic dynamical behavior over time for complex interactions of highly distributed ATM (see Fig. 3). The established techniques would therefore force one to adopt a rather heuristic type of argumentation in trying to capture the complex interactions inherent to ATM.

The basic ATM safety assessment needs have already been identified in Ref. 15. This finding motivated the development of an adequate safety assessment approach within a project named Traffic Organization and Perturbation Analyzer (TOPAZ). The scientific basis for this was the idea to explore a stochastic analysis framework[16,17] that supports stochastic models in which both discrete and continuous variables evolve over continuous time, possibly affected by probabilistic disturbances, and the knowledge that this framework would be sufficiently general to properly model and evaluate ATM safety problems.

In the meantime, from parallel conducted studies on advanced ATM, it became clear that without an appropriate accident risk model it would be difficult to ever manage a cost-effective design of advanced ATM. In these studies three complementary perspectives have been considered: 1) the selection of route structures perspective,[18] 2) a stochastic dynamical game perspective,[19] and 3) an ATM overall validation perspective.[20]

The accident risk assessment results obtained through stochastic analysis studies have initially been exploited toward the assessment of accident risk for staggered landings on converging runways at Schiphol.[21,22] All of this contributed to the development of both the novel accident risk assessment methodology and a growing suite of supporting tools. In this chapter, emphasis is on the former, for the reason that an effective usage of the suite of tools requires firm background in the novel methodology.

Recently, by a joint effort of EUROCONTROL and the Federal Aviation Administration (FAA), in collaboration with some key developers of aviation risk assessment tools, an overview has been produced that outlines the relevant approaches currently in development and/or in use for the safe separation assessment of advanced procedures in air traffic.[23] In addition to TOPAZ, four other collision risk directed approaches, Analytic Blunder Risk Model (ABRM), Airspace Simulation and Analysis for Terminal instrument procedures (ASAT), ICAO's Collision Risk Model (CRM), and Reduced Aircraft Separation Risk Assessment Model (RASRAM),[24] have been identified and reviewed. TOPAZ appeared to be most advanced in going beyond established approaches.

This chapter is organized as follows. Section II gives an overview of the advanced methodology. Section III outlines the principles of the underlying stochastic dynamical framework. Section IV presents some example scenarios for the results of accident risk assessments. Section V gives concluding remarks on the methodology.

II. Accident Risk Assessment Methodology

The accident risk assessment methodology has been developed to provide designers of advanced ATM with safety feedback following a (re)design cycle. An illustrative overview of how such safety feedback is obtained during an assessment cycle is given in Fig. 4.

During such an assessment cycle, two types of assessments are sequentially conducted: first a qualitative safety assessment (illustrated by the upper drawings in Fig. 4), and then a quantitative safety assessment (illustrated by the middle and lower drawings in Fig. 4). The qualitative assessment starts with a systematic gathering of information about nominal and non-nominal behavior of the concept design considered, concerning the human roles, the procedures, the technical systems, etc., and with involvement of all relevant experts. For the gathering of non-nominal information, explicit use is made of structured hazard identification sessions with a variety of experts and hazard databases. The resulting list of identified potential hazards is subsequently analyzed using established qualitative hazard analysis techniques to identify the safety-critical encounter scenarios and associated hazards, to select one or more of those safety-critical encounter scenarios for quantitative safety assessment, and to develop a modular system engineering type of representation of the ATM design (see upper right corner of Fig. 4). Such modular representation is easily recognizable and understandable for ATM designers, thus supporting an effective communication between ATM designers and safety analysts.

From this point on, the accident risk assessment cycle continues with the quantitative phase, which is based on stochastic modeling, stochastic analysis, and numerical evaluation. First, an appropriate stochastic dynamical model instantiation is developed in an iterative way and with verification against the results of the qualitative safety assessment phase. Next, the accident risk is assessed for this stochastic dynamical model, and the safety criticalities are identified. Finally, these results are given back to the designers (see lower left corner of Fig. 4).

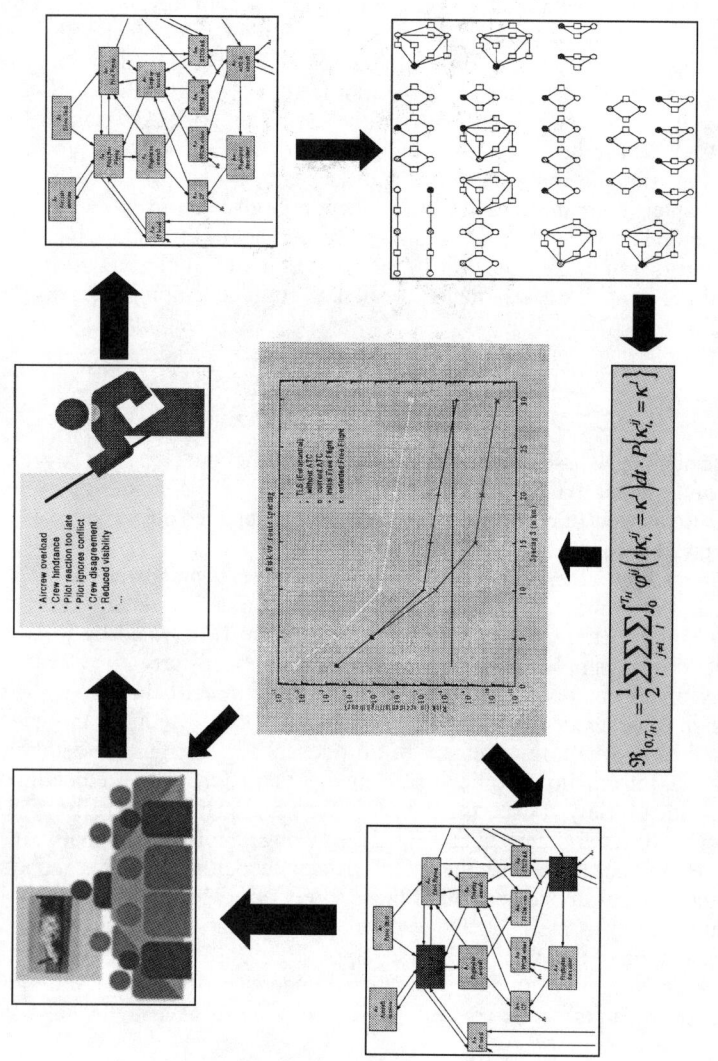

Fig. 4 TOPAZ accident risk assessment cycle.

To form a natural balance between the creative mode of the designers and the critical mode of the safety analysts, we have identified a definitive need for the safety analysts to use a conservative approach when adopting assumptions during the risk analysis. Obviously, the design team need not always agree with these conservative assumptions and should be aware that a negative outcome of a conservative assessment cycle does not mean that the design is unsafe; it just means that sufficient safety has not been proven during that cycle. This natural balance between designers and safety analysts means that both parties should be open to accept each other's views as being of mutual use. Conservatism could be reduced by refining the instantiated stochastic dynamical model on the appropriate issues identified by the designers. For the designers it could even be more effective to relax potential safety criticalities through redesign, rather than awaiting a potential accident risk modeling based improvement.

Underlying an accident risk assessment cycle is a stochastic analysis framework, which allows distinguishing the following five activities: 1) develop a stochastic dynamical model for the situation considered; 2) where necessary develop appropriate cognitive models for human operators involved; 3) perform the stochastic analysis necessary to decompose the risk assessment; 4) execute the various assessment activities (e.g., through Monte Carlo simulation, numerical evaluation, mathematical analysis, or a combination of these); and 5) validate the risk assessment exercise. More details on these five activities are given in the following sections.

A. Develop a Stochastic Dynamical Model

The aim of this development is to represent for the selected encounter scenarios the results from the qualitative safety assessment in the form of a stochastic differential equation (SDE) on a hybrid state space. The reason for aiming for such SDE representation is twofold: 1) it provides a very widely applicable class of causal models for stochastic dynamical situations such as in ATM; and 2) it allows the exploration of powerful mathematical tools from the theory of stochastic analysis (e.g., Refs. 25, 26, 16). Unfortunately, the direct identification of the SDE model would be very complicated for most ATM situations. In addition to a very large state space of the corresponding SDE, there are many interactions between the many state components. This requires a systematic approach to develop an SDE instantiation for such complex situations. Such an approach has been introduced through the development of a specific type of Petri net,[27,28] which we refer to as the dynamically colored Petri net (DCPN). Through a DCPN instantiation, an SDE instantiation can be done systematically while the result is transparent. Once a DCPN instantiation has been completed, the result defines an SDE on a hybrid state space. Obviously, a logical part of the DCPN instantiation is to verify the resulting DCPN against the information that is gathered during the qualitative safety assessment phase.

B. Cognitive Human Modeling

When assessing ATM safety, a key role is played by procedures, human operators, and their responsibilities. At present, the view on human reliability has shifted from a context-free error centered approach in which unreliability is modeled through failures of human information processing, toward a contextual perspective in which human actions are the product of human internal states, strategies, and the environment. By now, it is a widely accepted belief[29–31] that for modeling of the human the established human reliability analysis (HRA) techniques fall short for complex situations, and that one should rather aim for contextual performance models that are based on generally applicable human cognition and responsibility principles. It should also be noted that in the HRA widely used skill-, rule-, and knowledge-based errors[32] essentially fall short in paying proper attention to situations that fall beyond procedures. For example, situations in which the operator chooses to let an even more urgent problem receive attention when the subjectively available time is short or when high workload causes one to make quick decisions, without bothering excessively about the quality of those decisions. It should be noted that these effects are inextricably bound with human flexibility and the ability of humans to deal with unforeseen situations. When assessing ATM safety, it is necessary to take these aspects of human performance into account.

The main benefits expected from contextual models is that they provide better feedback to designers and that they remove the need to use overly conservative individual submodels for relevant operator actions that may blur understanding of how safety is achieved in ATM. To develop appropriate models for this, mathematicians and psychologists are jointly developing high-level models of cognitive human performance, through a sequence of studies (e.g., Refs. 33 and 34). At this moment this collaboration has led to a novel contextual human task-network model, which is formulated in terms of a DCPN and which effectively combines the cognitive modes of Hollnagel[30] with the multiple resources theory of Wickens,[35] the classical slips/lapses model[32], and the human capability to recover from errors.[29] In addition, we have developed a model for the evolution of situational awareness errors. Compared with those considered in a recent study,[36] our approach appears to be an innovative one.

C. Perform Stochastic Analysis

Although it definitively is possible to realize a straightforward Monte Carlo simulation of the SDE model, it is clear from the earlier discussion that this will not be really effective for the assessment of catastrophic risks in aviation. To develop an effective approach to the numerical evaluation of an SDE model, the SDE should be analyzed first by mathematicians with the appropriate background in the theory of stochastic analysis. At this moment, this is done on a case-by-case basis. For each case, the aim is to analyze the SDE model so that its numerical evaluation can be done by decomposition into a logical sequence of fast-time simulations, Monte Carlo simulations, and/or analytical evaluations. The aim always is to first decompose the risk assessment problem into several conditional assessment problems for which appropriate assessment techniques are available or feasible. The main principle we are using for identifying an appropriate decomposition is the following: under quite general conditions, the solution of an SDE is a strong Markov process. This means that the Markov property also holds true for stopping times (sometimes called Markov times). These stopping times serve as the mathematical powertool to decompose the risk assessment for an SDE model. So far, this approach appears to work satisfactorily for all situations evaluated.

D. Execute the Various Assessment Activities

Typically, the resulting sequence of conditional assessments is as follows:

1) Run a conventional fast time simulation (e.g., with TAAM) to identify traffic densities and encounter type frequencies.

2) Input these traffic densities and encounter type frequencies to a safety-directed human simulator to identify appropriate pilot and/or controller characteristics.

3) Input these conditional human characteristics to a Monte Carlo simulation that identifies and statistically analyzes critical conditional events, such as incidents.

4) Input these critical conditional event characteristics to a Monte Carlo simulation that identifies potential accident characteristics.

5) Input these potential accident characteristics to a conditional collision risk analyzer.

6) Transform all results from the preceding conditional assessments into appropriate safety metrics.

7) Identify the safety-separation and/or safety-modeling bottlenecks of the specifically modeled ATM concept/scenario.

For each of these activities, except activity 1, dedicated computer tools have been and are being further developed within the TOPAZ project. The splitting of activities 3, 4, and 5 from each other usually appears to be the most challenging one, for the very reason that often there are many dependencies between various elements of a hazardous air traffic situation. To handle this in a valid way, we make use of a mathematical framework, the basis of which is explained in Section III.

E. Validation of the Risk Assessment Exercise

A crucial issue concerns the validation that a risk assessment exercise is performed to an acceptable degree, without the need to first employ very expensive large-scale real-time simulations of new concepts. Because of our underlying stochastic analysis framework, such a validation can be done through executing the following activities:

1) Judge the level of conservatism of the assumptions adopted for the development of the DCPN instantiation for the situation considered. This should be done through active involvement of operational and design experts.

2) Verify the correctness of the instantiated DCPN vs the results of the qualitative assessment and the assumptions adopted. This should be done by stochastic analysis experts, with at least one who has not been involved with the DCPN instantiation.

3) Verify the correctness of the mathematical transformations applied to the instantiated stochastic dynamical model. This should be done by applying mathematical tools from stochastic analysis theory.

4) Verify that the various assessment activities have been executed according to the unambiguous mathematical model developed, including the decomposition. This should be done by stochastic analysis experts.

III. Mathematical Framework

Each DCPN instantiation can be represented by an SDE on a hybrid state space,[28] which has a strong Markov process $\{\xi_t\}$ on a hybrid state space as its unique solution. The hybrid state process $\{\xi_t\}$ has two components, i.e., $\xi_t = (x_t, \theta_t)$, with x_t the component assuming values in a Euclidean space and with θ_t the component assuming values in a discrete space. From the theory of Markov processes it then follows that it is possible to characterize the evolution of the density distribution $p_{\xi_t}(\xi)$ of the joint process through a well-defined differential equation in function space:

$$\frac{d}{dt} p_{\xi_t}(\xi) = \mathcal{A} p_{\xi_t}(\xi)$$

with $\mathcal{A}$ an operator defined by the Markov process $\{\xi_t\}$. Because of the strong Markov property, this differential equation also applies under the condition of an $\{\xi_t\}$-adapted stopping time τ (also referred to as Markov time):

$$\frac{d}{dt} p_{\xi_t|\tau}(\xi) = \mathcal{A} p_{\xi_t|\tau}(\xi), \quad \text{for } t > \tau$$

It is particularly relevant to notice that these equations are well known for Markov chains, i.e., Markov processes with discrete state space, which have shown to be very useful in the development of advanced dependability and performability assessment methodology (e.g., Refs. 37 and 38). For hybrid state Markov processes, this equation is well known in Bayesian estimation theory (e.g., Ref. 16) and this for example has led to advanced multitarget multisensor tracking applications (e.g., Ref. 39).

The preceding equations imply that once the scenario to be assessed on collision risk has been represented through a DCPN instantiation, all probabilistic properties are well defined, including the collision risk. Let y_t^i and v_t^i be the components of x_t that represent the three-dimensional location and three-dimensional velocity of aircraft i, $i \in \{1, \ldots, n\}$. Let $y_t^{ij} = y_t^i - y_t^j$, let $v_t^{ij} = v_t^i - v_t^j$, and let D^{ij} be the area such that $y_t^{ij} \in D^{ij}$ means that at moment t the physical volumes of aircraft i and j are not separated anymore (i.e., they have collided). Each time the process y_t^{ij} enters the area D^{ij}, we note that an incrossing occurs, and each time the process y_t^{ij} leaves the area D^{ij}, we note that an outcrossing occurs. The first incrossing for the pair (i,j) is a collision for that pair. If we assume that the relative speed v_t^{ij} is very rapidly going to zero as long as y_t^{ij} resides in D^{ij}, the chances are zero that there is more than one incrossing per aircraft pair, and thus the expected number of incrossings equals the expected number of collisions. Following Ref. 40, the expected number $\Re_{[0,T]}$ of incrossings, or collisions, between aircraft pairs in the time interval $[0,T]$ satisfies:

$$\Re_{[0,T]} = \sum_{i=1}^{n} \sum_{j>i}^{n} \int_0^T \varphi^{ij}(t)\, dt$$

with $\varphi^{ij}(t)$ the incrossing rate, which is defined by:

$$\varphi^{ij}(t) = \lim_{\Delta \downarrow 0} Pr\{y_t^{ij} \notin D^{ij}, y_{t+\Delta}^{ij} \in D^{ij}\}/\Delta$$

In Ref. 40 it is also shown that $\varphi^{ij}(t)$ is well defined and can be evaluated under nonrestrictive assumptions as a function of the probability density of the joint relative state (y_t^{ij}, v_t^{ij}). In general, a characterization of this probability density is complex, especially since there are combinatorially many types of non-nominal events. A plausible way out of this is by conditioning on classes of non-nominal events, where those non-nominal events are placed in the same class if they have a similar impact on the subsequent evolution of the relative state process $\{y_t^{ij}, v_t^{ij}\}$.

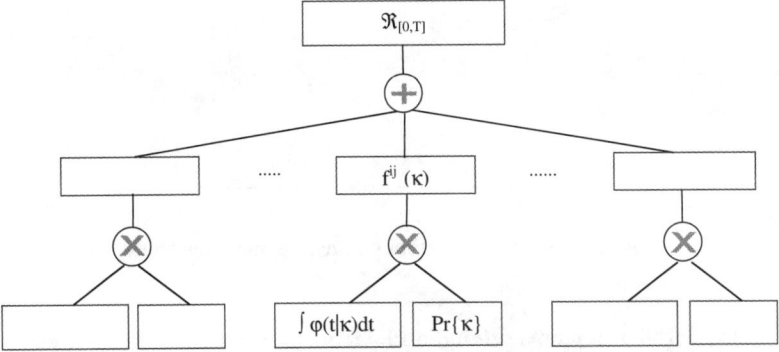

Fig. 5 Collision risk tree.

This is done through 1) defining an appropriate event sequence classification process $\{\kappa_t\}$, such that the joint process $\{\xi_t, \kappa_t\}$ is a strong Markov process as well, and 2) subsequently identifying an appropriate $\{\xi_t, \kappa_t\}$-adapted stopping time τ^{ij} such that there is a zero probability that the pair (i, j) collides before τ^{ij}. With this, the preceding equations can be transformed into:

$$\Re_{[0,T]} = \sum_{i=1}^{n} \sum_{j>i}^{n} \sum_{\kappa} \int_{\tau^{ij}}^{T} \varphi^{ij}(t \mid \kappa_{\tau^{ij}} = \kappa) \, dt \cdot Pr\{\kappa_{\tau^{ij}} = \kappa\}$$

with $\varphi^{ij}(t\mid\kappa_{\tau^{ij}} = \kappa)$ the conditional incrossing rate, being defined for $t \geq \tau^{ij}$ by:

$$\varphi^{ij}(t \mid \kappa_{\tau^{ij}} = \kappa) = \lim_{\Delta \downarrow 0} Pr\{y_t^{ij} \notin D^{ij}, y_{t+\Delta}^{ij} \in D^{ij} \mid \kappa_{\tau^{ij}} = \kappa\}/\Delta$$

In Fig. 5, the equation for $\Re_{[0,T]}$ is presented in the form of a tree, in which $f^{ij}(\kappa)$ is short for

$$\int_{\tau^{ij}}^{T} \varphi^{ij}(t \mid \kappa_{\tau^{ij}} = \kappa) \, dt \cdot Pr\{\kappa_{\tau^{ij}} = \kappa\}$$

This tree has some resemblance with the well-known fault tree. However, because of the underlying stochastic and physical relations, our new tree differs significantly and is called a collision risk tree.

For the quantification of the boxes in the collision risk tree, use is made of three types of evaluations:

1) Monte Carlo simulations of the DCPN to quantify $Pr\{\kappa_{\tau^{ij}} = \kappa\}$ and the statistical properties of the relevant DCPN components at the stopping time τ^{ij}.

2) Evaluations of the evolution of the relative aircraft states from stopping time τ^{ij} on, and for each $\kappa_{\tau^{ij}} = \kappa$. If complexity requires, this process can even be done for a sequence of increasing stopping times.

3) Numerical evaluation of

$$\int_{\tau^{ij}}^{T} \varphi^{ij}(t \mid \kappa_{\tau^{ij}} = \kappa) \, dt$$

using the generalized Reich equation of Ref. 40; see also Ref. 41.

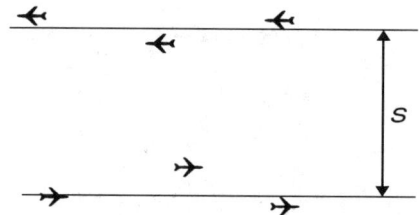

Fig. 6 Opposite direction traffic in a dual lane route.

IV. RNP1 in Conventional and Airborne Separation Assurance Scenario Examples

In this section, the accident risk assessment approach is used to evaluate a simple scenario of two en route traffic streams, flying in opposite directions, all at one single flight level. This rather hypothetical scenario has been developed by EUROCONTROL with the aim to learn how ATC influences accident risk, and how far the nominal separation S between opposite RNP1 traffic streams can safely be reduced. The specific details of this scenario are[42]:

1) Straight route, with two traffic lanes (Fig. 6),
2) Flight plans contain no lane changes,
3) Parameter S denotes distance between the two lanes,
4) Opposite traffic flows along each lane,
5) Aircraft fly at one flight level only,
6) Traffic flow per lane is 3.6 aircraft/hour,
7) All aircraft nominally perform RNP1 with 95% of the time within 1 n mile,
8) None of the aircraft is equipped with the Traffic Alert and Collision Avoidance System (TCAS),
9) Target level of safety is 5×10^{-9} accidents/flight hour.[43]

This simple scenario is considered for the following four ATM concepts:

A) Procedural separation only. In this case, there is no air traffic control (ATC) surveillance system. This is the type of situation encountered with traffic over the North Atlantic.

B) ATC based only on short-term conflict alert (STCA). In this case there is radio telephony (R/T) communication, but it is assumed that ATC is doing nothing unless its STCA system issues an alert, thus assuming no monitoring by the air traffic controller (ATCo). It should be noted that this differs significantly from conventional ATC, in which an executive controller autonomously monitors and issues corrective actions, while STCA is a safety net only.

C) Basic airborne separation assurance. In this case, there is automatic dependent surveillance broadcast (ADS-B) and R/T between aircraft, but there is no ATC. For this concept it is assumed that aircraft behave cooperatively, in the sense that when an aircraft's conflict detection and resolution (CDR) system detects a conflict with another aircraft, then its pilot will try to make an avoidance maneuver. Thus, in most cases both pilots will try to make an avoidance maneuver.

D) Negotiated airborne separation assurance, a design that is explicitly due to the feedback received from accident risk assessments conducted for concepts A, B, and C. For this concept, it is assumed that aircraft also behave cooperatively during conflict-free trajectory planning. Thus, in addition to ADS-B surveillance

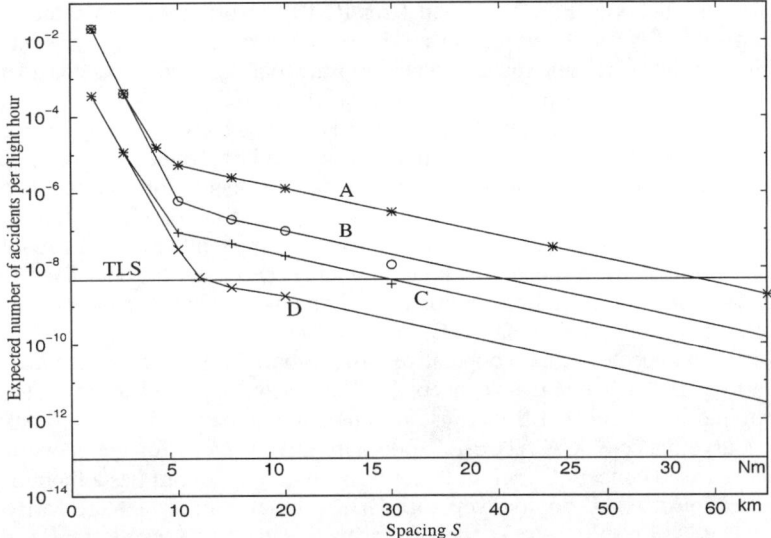

Fig. 7 Accident risk for the opposite traffic scenario, as a function of spacing parameter S, for the four ATM concepts considered: A) procedural separation, B) STCA-based ATC, C) basic airborne separation assurance, and D) negotiated airborne separation assurance. The accident risk unit used is from ICAO, where one collision between two aircraft counts for two "accidents."

and R/T, there is also a data link between aircraft to exchange and negotiate conflict-free trajectory plans that are assumed to extend 5 min or more into the future.

Obviously, for each of these four ATM concepts there are various traffic navigation and encounter scenarios that deserve an accident risk evaluation. We believe, however, that it is most effective to understand the safe separation issues for a simple traffic navigation and encounter scenario first, before considering other more complicated scenarios.

For each of the four ATM concepts, the accident risk assessment methodology and supporting tool set have been used to conservatively assess accident risk for the preceding scenario, as a function of the spacing parameter S. The resulting accident risk curves are presented in Fig. 7. Because all four curves are based on conservative modeling assumptions for the ATM situations considered, they provide an upper bound for the true accident risk.

These results are obtained over a period of two years during three subsequent studies. The first en route study[42] was conducted for EUROCONTROL and covered ATM concepts A and B. The assessment of concept A was rather straightforward and could also have been done with ICAO's CRM. For the assessment of the other three concepts, however, full use has been made of the accident risk assessment methodology. Concept B has been assessed during an initial study for EUROCONTROL.[42] Concept C has been developed[44] and assessed[45] during studies within NASA's Free Flight research program. The safety assessment results from concepts A, B, and C have subsequently been fed back[46] to enable the development of the design concept D and subsequently to assess it on accident risk.[47]

The risk curves in Fig. 7 show that for RNP1 performing aircraft, the ATM concept may have quite an impact on the selection of the spacing parameter S within a straight dual lane route structure. For the four ATM concepts considered, it has been shown that the spacing S can safely be reduced to 31, 22, 16, and 7 n miles for ATM concepts A, B, C, and D, respectively. The large value of 31 n miles for concept A is not a real surprise; such large values are well known for procedural traffic situations over the ocean. The results for concept B show that STCA really is a safety net that provides at least a factor 15 in safety when compared with concept A for sufficiently large S. Apparently, this STCA safety net alone falls short to support the kind of spacings necessary for busy fixed route traffic situations. This finding confirms the prior expectation that concept B is not representative for conventional ATC.

Rather unexpectedly, the cooperative basic airborne separation concept C appears to perform better than concept B. The reason appeared to be that with the ground-based concept B there is one single monitoring and decision-making loop (surveillance-STCA-ATCo-R/T-pilot-aircraft), whereas for the cooperative airborne-based concept C each of the two encountering aircraft has a monitoring and decision-making loop (surveillance-CDR-pilot-aircraft) which are partly independent. As a result, the safety net of concept C leads to a factor 5 lower risk than concept B for the same spacing, or allows to safely reduce S from 22 n miles to 16 n miles. Obviously, such improved safety net still falls short to support the kind of spacings necessary for busy fixed route traffic situations. Thus in view of their safe spacing values of 22 and 16 n miles, concepts B and C do not support spacings that are required for busy fixed route situations over the continent.

Finally, the cooperative negotiated airborne separation assurance concept D allows such low spacing values. This is not a coincidence, but it is the result of effectively making use of accident risk assessment feedback from A, B, and C. It appeared that for all three of these concepts, the safe spacing was determined by the effects of the exponential tails of large deviations due to non-nominal situations. Thus, the design objective for concept D was to reduce those non-nominal effects to a level below the target level of safety (TLS). To accomplish this, the two monitoring and decision-making loops of concept C have been extended with a largely independent and cooperative conflict-free-planning loop. The curve for concept D shows that this worked out successfully, by which the safe spacing value for concept D is governed by the RNP1-Gaussian navigation error characteristics, rather than by the exponential tails due to non-nominal situations.

V. Concluding Remarks

This chapter has given an outline of an accident risk assessment methodology to assess advanced ATM on midair collision risk and has illustrated that this approach may provide effective feedback to designers of advanced ATM. From this outline it has become clear that this methodology exhibits several remarkable features, such as:

1) It applies established techniques during a qualitative assessment phase only.
2) Quantification is based on stochastic dynamical modeling.
3) It uses powerful tools from the theory of stochastic analysis.
4) It handles complex interactions between different ATM elements.
5) It incorporates advanced human cognitive modeling.
6) It incorporates the generalized Reich collision risk model.

7) It provides effective feedback to ATM concept designers.
8) Validation of a risk assessment exercise forms part of the methodology.

It has also become clear that currently a high level of expertise in stochastic analysis is required for an effective application of the methodology. One should, however, be aware that the need for sophisticated mathematical expertise is well accepted in other complex design areas of civil aviation, such as the area of aerodynamic optimization of aircraft structures.

Obviously, within an overall ATM concept, a large variety of relevant aircraft encounter scenarios can be identified. As such, it is important to notice that our DCPN instantiation for a particular ATM concept mainly depends on the ATM concept and only marginally on the encounter scenario. Thus, the DCPN instantiations for the four RNP1 based ATM concepts in Section IV can be extended relatively simply to other encounter scenarios. This also means that it should be possible to identify classes of encounter scenarios so that it is sufficient to perform an accident risk assessment for one scenario from each class only.

In this chapter, the methodology of accident risk assessment has been concentrated on the risk of midair collision. Because of the generality of the methodology, however, it is also applicable to other accident risks in air traffic, such as risk induced by runway incursion, controlled flight into terrain, etc. We have, for example, already made good progress in the extension of the accident risk assessment methodology with a probabilistic model for wake vortex induced accident risk.[48]

References

[1]Blom, H. A. P., Bakker, G. J., Blanker, P. J. G., Daams, J., Everdij, M. H. C., and Klompstra, M. B., "Accident Risk Assessment for Advanced ATM," *2nd USA/Europe Air Traffic Management R&D Seminar*, FAA/EUROCONTROL, 1998; also NLR Rept. TP-99-015, Amsterdam, 1999.

[2]Haraldsdottir, A., Alcabin, M. S., Burgemeister, A. H., Lindsey, C. G., Makins, N. J., Schwab, R. W., Shakarian, A., Shontz, W. D., Singleton, M. K., van Tulder, P. A., and Warren, A. W., "Air Traffic Management Concept Baseline Definition," NEXTOR Rept. RR-97-3, Boeing, Seattle, WA, 1997.

[3]Odoni, A. R., Bowman, J., Delahaye, D., Deyst, J. J., Feron, E., Hansman, R. J., Khan, K., Kuchar, J. K., Pujet, N., and Simpson, R. W., "Existing and Required Modeling Capabilities for Evaluating ATM Systems and Concepts," sponsored by NASA under Grant no. NAG2-997, Massachusetts Inst. of Technology, Cambridge, MA, March 1997.

[4]EVAS, "EATMS Validation Strategy Document," Ed. 1.1, Eurocontrol, Brussels, June 1998.

[5]Randell, B. (ed.), *Predictably Dependable Computing Systems*, Springer-Verlag, Berlin, 1995.

[6]"DAAS Dependability Approach to ATM Systems" Final Report, European Commission, Brussels, 1995.

[7]"ARP 4761, Guidelines and Methods for Conducting the Safety Assessment Process on Civil Airborne Systems and Equipment," S-18 Committee, Society of Automotive Engineers, Inc., March 1994.

[8]"ARP 4754, Certification Considerations for Highly-Integrated or Complex Aircraft Systems," Systems Integration Requirements Task Group AS-1C, Avionics Systems Division, Society of Automotive Engineers, Inc., Sept. 1995.

[9] "Air Navigation System Safety Methodology," Ed. 0.4, Working Draft, EATCHIP, Eurocontrol, Brussels, 1996.

[10] Klompstra, M. B., and Everdij, M. H. C., "Evaluation of JAR and EATCHIP Safety Assessment Methodologies," National Aerospace Laboratory NLR, NLR Rept. CR 97678 L, Amsterdam, 1997.

[11] "Risk Assessment, Report of a Royal Society Study Group," Royal Society, London, 1983.

[12] Moek, G., Blom, H. A. P., Klompstra, M. B., Beaujard, J. P., Kelly, C., Mann, J., Clarke, L., and Marsh, D., "Methods and Techniques," National Aerospace Laboratory NLR, GENOVA WP3, NLR Rept. TR-98595-PT-1, Amsterdam, 1997.

[13] Everdij, M. H. C., Klompstra, M. B., Blom, H. A. P., and Fota, O. N., "Evaluation of Hazard Analysis Techniques for Application to En-Route ATM," National Aerospace Laboratory NLR, MUFTIS Final Rept. on Safety Model, Part I, NLR Rept. TR 96196 L, Amsterdam, 1996.

[14] Aldemir, T., Siu, N. O., Mosleh, A., Cacciabue, P. C., and Göktepe, B. G. (eds.), *Reliability and Safety Assessment of Dynamic Process Systems*, Springer-Verlag, Berlin, 1994.

[15] Blom, H. A. P., "The Layered Safety Concept, an Integrated Approach to the Design and Validation of Air Traffic Management Enhancements," National Aerospace Laboratory NLR, NLR Rept. TP 92046 L, Amsterdam, 1992.

[16] Blom, H. A. P., "Bayesian Estimation for Decision-Directed Stochastic Control," Ph.D. Dissertation, Delft University of Technology, Delft, Netherlands, 1990.

[17] Everdij, M. H. C., Klompstra, M. B., and Blom, H. A. P., "Development of Mathematical Techniques for ATM Safety Analysis, MUFTIS Final Report on Safety Model, Part II," National Aerospace Laboratory NLR, NLR Rept. TR 96197 L, Amsterdam, 1996.

[18] Blom, H. A. P., and Bakker, G. J., "A Macroscopic Assessment of the Target Safety Gain for Different En Route Airspace Structures Within SUATMS," National Aerospace Laboratory NLR, NLR Rept. CR 93364 L, Amsterdam, 1993.

[19] Blom, H. A. P., Klompstra, M. B., and Bakker, G. J., "Air Traffic Management as a Multi-Agent Stochastic Dynamic Game Under Partial State Observation," *Proceedings of the 7th IFAC/IFORS Symposium on Transportation Systems*, International Federation of Automatic Control, 1994, pp. 249–254.

[20] Blom, H. A. P., Hendriks, C. F. W., and Nijhuis, H. B., "Assess Necessary Validation Developments," National Aerospace Laboratory NLR, VAPORETO WP3 Final Rept: NLR Rept. CR 95524 L, Amsterdam, 1995.

[21] Bakker, G. J., Blom, H. A. P., and Everdij, M. H. C., "Collision Risk Evaluation of the Dependent Converging Instrument Approach (DCIA) Procedure Under Gaussian Deviations from Expected Missed Approach Paths," National Aerospace Laboratory NLR, NLR Rept. CR 95322 L, Amsterdam, 1995.

[22] Everdij, M. H. C., Bakker, G. J., and Blom, H. A. P., "Application of Collision Risk Tree Analysis to DCIA/CRDA Through Support of TOPAZ," National Aerospace Laboratory NLR, NLR Rept. CR 96784 L, Amsterdam, 1996.

[23] Cohen, S. and Hockaday, S., (eds.), "A Concept Paper for Separation Safety Modeling, an FAA/EUROCONTROL Cooperative Effort on Air Traffic Modeling for Separation Standards," FAA and EUROCONTROL, Brussels, May 1998.

[24] Sheperd, R., Cassell, R., Thava, R., and Lee, D., "A Reduced Aircraft Separation Risk Assessment Model," *Proceedings of the AIAA Guidance, Navigation, and Control Conference*, AIAA, Reston, VA, Aug. 1997.

[25] Elliott, R. J., *Stochastic Calculus and Applications*, Springer-Verlag, New York, 1982.

[26]Davis, M. H. A., "Piecewise Deterministic Markov Processes: A General Class of Non-Diffusion Stochastic Models," *Journal Royal Statistic Society (B)*, Vol. 46, 1984, pp. 353–388.

[27]Everdij, M. H. C., Blom, H. A. P., and Klompstra, M. B., "Dynamically Coloured Petri Nets for Air Traffic Management Safety Purposes," *Preprints 8th IFAC Symposium on Transportation Systems*, edited by M. Papageorgiou and A. Pouliezos, Intelligent Technological Systems Lab., Technical University of Crete, 1997, pp. 184–189; also National Aerospace Laboratory NLR, NLR Rept. TP-97-493, Amsterdam, 1997.

[28]Everdij, M. H. C., and Blom, H. A. P., "Piecewise Deterministic Markov Processes Represented by Dynamically Coloured Petri Nets," National Aerospace Laboratory NLR, NLR Rept. TP-2000-428, Amsterdam, 2000.

[29]Amalberti, R., and Wioland, L., "Human Error in Aviation," *International Aviation Safety Conference*, edited by H. Soekkha, VSP, Utrecht, The Netherlands, 1997, pp. 91–108.

[30]Hollnagel, E., *Human Reliability Analysis, Context and Control*, Academic Press, London, 1993.

[31]Bainbridge, L., "The Change of Concepts Needed to Account for Human Behaviour in Complex Dynamic Tasks," *1993 International Conference on Systems, Man and Cybernetics*, 1993, pp. 126–131.

[32]Reason, J., *Human Error*, Cambridge Univ. Press, Cambridge, MA, 1990.

[33]Biemans, M. C. M., and Daams, J., "Human Operator Modeling to Evaluate Reliability, Organisation and Safety," National Aerospace Laboratory NLR, Rept. TR 98073, Amsterdam, 1997.

[34]Daams, J., Nijhuis, H. B., and Blom, H. A. P., "Human Operators Controllability of ATM Safety, ARIBA," National Aerospace Laboratory NLR, Rept. TR-99575, Amsterdam, 1999.

[35]Wickens, C. D., *Engineering, Psychology and Human Performance*, Merrill, Columbus, OH, 1992.

[36]"A Designers Guide to Human Performance Modeling," Advisory Group for Aerospace Research and Development, AGARD, Advisory Rept. 356, Neuilly-Sur-Seine, France, Dec. 1998.

[37]Pattipati, K. R., Li, Y., and Blom, H. A. P., "A Unified Framework for the Performability Evaluation of Fault-Tolerant Computer Systems," *IEEE Transactions on Computers*, Vol. 42, No. 3, 1993, pp. 312–326.

[38]Fota, O. N., Kaaniche, M., and Kanoun, K., "A Modular and Incremental Approach for Building Complex Stochastic Petri Net Models," *First International Conference on Mathematical Methods in Reliability*, 1997, pp. 151–158.

[39]Blom, H. A. P., Hogendoorn, R. A., and Van Doorn, B. A., "Design of a Multisensor Tracking System for Advanced Air Traffic Control," *Multitarget-Multisensor Tracking*, edited by Y. Bar-Shalom, Vol. II, Artech House, Norwood, MA, 1992, pp. 31–63.

[40]Bakker, G. J., and Blom, H. A. P., "Air Traffic Collision Risk Modeling," *Proceedings of the 32nd IEEE Conference on Decision and Control*, Vol. 2, Institute of Electrical and Electronics Engineers, New York, 1993, pp. 1464–1469.

[41]Bakker, G. J., Kremer, H. J., and Blom, H. A. P., "Geometric and Probabilistic Approach Towards Conflict Prediction," *3rd USA/Europe Air Traffic Management R&D Seminar*, FAA/ EUROCONTROL, 2000.

[42]Everdij, M. H. C., Bakker, G. J., Blom, H. A. P., and Blanker, P. J. G., "Demonstration Report in Preparation to Designing EATMS Inherently Safe," National Aerospace Laboratory NLR, TOSCA II WP4 Phase I Rept., Amsterdam, 1997.

[43]"Annex 11—Air Traffic Services," 12th ed., incorporating amendments 1–38, green pages, attachment B, paragraph 3.2.1., International Civil Aviation Organisation, Montreal, July 1998.

[44]Hoekstra, J. M., Ruigrok, R. C. J., and Van Gent, R. N. H. W., "Conceptual Design of Free Flight Cruise with Airborne Separation Assurance," National Aerospace Laboratory NLR, Rept. TP 98252, Amsterdam, 1997.

[45]Daams, J., Bakker, G. J., and Blom, H. A. P., "Safety Evaluation of an Initial Free Flight Scenario with TOPAZ," National Aerospace Laboratory NLR, NLR Rept. TR 98098, Amsterdam, 1998.

[46]Van Gent, R. N. H. W., Hoekstra, J. M., and Ruigrok, R. C. J., "Free Flight with Airborne Separation Assurance," *Proceedings CEAS Symposium*, 1997; also National Aerospace Laboratory NLR, Rept. TP-98286, Amsterdam, 1998.

[47]Daams, J., Bakker, G. J., and Blom, H. A. P., "Safety Evaluation of Encounters Between Free-Flight Equipped Aircraft in a Dual Route Structure," National Aerospace Laboratory NLR, Rept. TR-99577, Amsterdam, 1998.

[48]Kos, J., Blom, H. A. P., Speijker, L. J. P., Klompstra, M. B., and Bakker, G. J., "Probabilistic Wake Vortex Induced Accident Risk Assessment," *3rd USA/Europe Air Traffic Management R&D Seminar*, FAA/EUROCONTROL, 2000.

Chapter 30

Human Cognition Modelling in Air Traffic Management Safety Assessment

Henk A. P. Blom,* Jasper Daams,† and Herman B. Nijhuis‡
National Aerospace Laboratory NLR, Amsterdam, The Netherlands

I. Introduction

OVER decades, the aviation industry has been able to compensate the increase in air traffic with a decrease in accident risk per flight hour. In view of the rapid growth of air traffic and the technological and organizational complexity of it, this has been a major accomplishment. Unfortunately, the point has been reached where it is unclear how to continue such compensation. The reason is that in the past the decrease of risk per flight hour has come in large part from technology-driven improvements of safety. The effect of this technology-driven approach is shown through the accident statistics; they reveal that the relative share of human-related causes is approximately 80%. This means that the historical air traffic safety compensation process can be continued if one learns to understand how the human and procedure-related accidents could be reduced. This should be accomplished by learning the principles behind human-related accident causes in aviation.

If we try to understand these principles on the basis of an evaluation of incidents and accidents alone, then several difficulties arise. The number of incidents and accidents is limited, whereas the situations that caused them are quite complex and reports are not free from discussion. Because of the limited availability of data and the questionable validity of data, statistical analysis alone is not sufficient to model safety in complex situations with multiple human involvement. By now there is a broad consensus that appropriate safety models are needed to understand the mechanisms behind the remaining accident risk in relation to separation criteria and near-misses.[1] It is also recognized that such a safety modeling approach should be useful in optimizing advanced air traffic management (ATM) operations.[2-4]

Copyright © 2001 by NLR. Published by the American Institute of Aeronautics and Astronautics, Inc., with permission.
*Group Leader, Air Traffic Management Modeling, Air Transport Division.
†ATM Modeling Researcher, Air Transport Division; currently with LVNL.
‡Human Factors Researcher, Flight Division.

Most existing studies on ATM safety either focus on hazard analysis techniques or on collision risk analysis. Studies with thorough hazard analysis results generally use simplified collision risk models, advanced studies on collision risk between aircraft usually do not take into account hazards or non-nominal events (except in adapted tails of probability density functions). It appears that most established techniques fall short in integrating hazard analysis techniques with advanced collision risk analysis techniques. In a series of studies at the National Aerospace Laboratory NLR, this problem has been addressed with the development of an accident risk assessment methodology and supporting evaluation tool set [both named Traffic Organization and Perturbation Analyzer (TOPAZ)] that takes an integral approach toward ATM safety assessment.[5] Recently it has also been shown how this approach effectively supports safety management and the building of modern safety cases for advanced operations in ATM.[6]

At the basis of this accident risk assessment approach lies the use of a very general class of mathematical models for describing the ATM process. The models used are hybrid state Markov processes, which describe stochastic evolution of both continuous and discrete variables over time. This means that both aircraft three-dimensional position and velocity and operator states can be described as a function of time, including their interactions and the effects of probabilistic disturbances. To accomplish this, existing and newly developed models, such as the generalized Reich collision risk model[7] and the high-level Petri net model,[8,9] were combined to model and evaluate ATM operations on safety.

In parallel with its development, the methodology is applied to a variety of accident risk assessment studies, e.g., converging runways,[10] free flight equipped aircraft,[11,12] and wake vortex induced accident risk.[13] Another type of application considered is conventional en route traffic in a scenario of two parallel opposite direction lanes. In Ref. 14, for this scenario, risk has been evaluated under two operational concepts. In the first concept, named "No ATC," there is no air traffic control (ATC) surveillance of the traffic at all; in the second concept, named "STCA-only based ATC," the tactical air traffic controller sends deviating aircraft back to their lane if and only if there has been a short-term conflict alert (STCA). Although the demonstrated possibility to obtain accident risk results for such complex operations as ATM is quite promising in itself, several operational experts pointed out that an STCA-only based ATC concept is overly conservative as a representation of conventional ATM concepts, because routine monitoring and anticipation are not incorporated. Therefore, the follow-up was to develop an appropriate human performance model for risk assessments of such routine monitoring situations.

A crucial issue in ATM safety assessment is how the human factor is incorporated into the risk model. Hence, there is a clear need for a modeling approach to assess and understand accident risk in relation to the performance of the human operators involved. This means that appropriate human performance models are required that describe human cognitive and responsibility principles up to the level of accident risk. This chapter aims to present the developments of such a human cognition/performance model for a tactical controller within the context of conventional en route ATC, and it is based on a series of studies.[15–19]

At present, the view on human reliability has shifted from a context-free error centered approach, in which unreliability is modeled as failures of human information processing, toward a contextual perspective in which human actions are the product of human internal states, strategies, and the environment.[20-22] From this viewpoint, safety critical human actions should be modeled in their relation to the other activities of the operator and the environment. Thus for a proper description of human reliability, it is necessary to include the cognitive processes that underlie the operator actions. As a result, one obtains a comprehensive model of the operator performing his job.

The main benefits expected from contextual models for safety assessment are that they provide better feedback to designers and that they remove the need to use overly conservative individual submodels of relevant operator actions that may complicate understanding of how safety is achieved in aviation.

The chapter is organized as follows. Section II provides the background of three complementary psychological models on which human cognitive performance modeling in this chapter is based. In Section III we explain how these three psychological models are jointly used in a mathematical human cognition/performance model for a tactical en route controller. This is largely done on the basis of human factors ATC expertise. Next, in Section IV this mathematical model is reduced to a simpler model on the basis of clearly defined model aggregation steps. In Section V, the reduced human cognition/performance model is used to evaluate a conventional en route ATC situation with respect to accident risk and air traffic controller actions. Finally, in Section VI we discuss the results obtained.

II. Human Modeling Approaches

The mathematical human performance/reliability model development in this chapter is based on the following three complementary psychological models: 1) multiple resources model, 2) contextual control mode model, and 3) human error modeling.

In this section, we outline these three psychological models. One should be aware that several other psychological human error type of models exist that have potential application in ATM (see, e.g., Ref. 23).

A. Multiple Resources Model

The main reference used here is Ref. 24. The multiple resource model reflects the idea that humans have several different mental capacities with resource properties. In this view, task interference depends on the extent to which tasks use the same resources: two difficult tasks may be time-shared easily if they use different types of resources. The multiple resources approach has been well developed both for military applications[25] and for ATM (e.g., Refs. 26, 27). The principal idea behind the model is that human cognitive effort can be divided over several activities. This is called the resources metaphor.[28] Because human cognitive effort is limited, the resources metaphor may readily account for failures in time-sharing between competing activities. The underlying assumption of the resources metaphor is that the human is an information processing system with limited processing capacity. The model focuses on how this limited processing capacity can be used to time-share several processing tasks.

When two or more tasks are to be successfully time-shared, the first important aspects are the efficient scheduling and switching between activities. If sufficient time is available, the operator can occupy himself with one task at a time, although this does not necessarily mean that the tasks are performed sequentially. However, if the available time is not sufficient to apply this strategy, concurrent task performance becomes necessary. With respect to concurrent task performance, Wickens mentions three performance influencing task characteristics:

1) Confusion. When an element of one task is similar to an element of a concurrently performed other task, the elements may become confused, leading to a decrease in performance.

2) Cooperation. In some cases, the similarity between performance routines for elements of two tasks leads to cooperation between the routines. It is even possible that the two task elements can be merged into one new task.

3) Difficulty. The task difficulty highly influences whether a second task can be performed concurrently.

The confusion and cooperation aspects are closely related: both emerge from the similarity between tasks. However, cooperation is associated with similar processing *routines*, whereas confusion emerges from similar input *material*. It will appear that by taking into account the confusion and cooperation aspects of concurrent task performance leads to multiple resources dimensions.

On the basis of a large number of dual-task studies, Wickens proposes a three-dimensional resource quantity, with dichotomous dimensions. The dimensions are:

1) Information processing stages. Dimension with early and central processing on the one extreme (sensory processing, encoding and perception of stimuli) and late processing on the other (deciding on the best response and its execution). For example, the requirement for an air traffic controller to give a response to each change in aircraft state (late processing) is predicted not to disrupt the ability to maintain an accurate mental model of the radar display (early processing).

2) Modalities. Input modalities differentiate between the encoding of auditory and visually presented stimuli. It is easier to divide attention between the eye and ear than between two auditory or two visual stimuli. Response modes refer to the choice between a vocal and a manual response. The reason that manual and vocal outputs can be efficiently time-shared is probably due to the separation of spatial and verbal information processing resources (manual responses are spatial in nature, whereas vocal ones are verbal).

3) Processing codes. Human controllers can rely on two working memory codes, namely a spatial and a verbal one. Each is used to process or retain qualitatively different kinds of information (spatial and visual vs temporal, verbal and phonetic) and each can be disrupted by different concurrent activities. Resources underlying spatial processing and left-hand control reside predominantly in the right hemisphere of the brain. Resources underlying verbal processing, speech-responses, and right-hand control reside more in the left hemisphere.

A note should be made about the modality dimension. In Ref. 29 it is pointed out that the resources metaphor does not readily apply to input modality. Instead, pre-emption and attention-switching seem to dominate cross-modal time-sharing. However, these effects are relatively small in comparison to the effort required for the extra scanning activity that is generally involved with intramodal time-sharing.

Figure 1 is a representation of the multiple resources theory. Although the theory does not pretend to account for all influences on multiple-task performance and time-sharing, research showed that the identified dimensions account for a

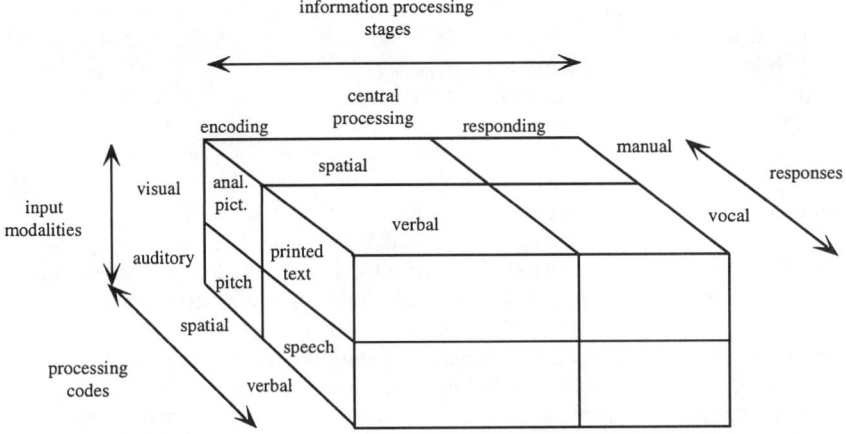

Fig. 1 Proposed structure of rsource dimensions (free after Wickens[24]).

reasonably large proportion of these influences and can be used in predicting task interference. Ideally, the loads on the dimensions must be established a priori. Input modalities are easy to define, as are vocal and manual output modes. Information processing stages and processing codes will cause more problems. Sometimes task analysis can reveal memory requirements.

We can now model the operator's ability of time-sharing by evaluating which resources are used during the simultaneous execution of the tasks. Heavy concurrent demands upon the same resource then reduce time-sharing, whereas tasks using different resources can be time-shared easily. However, the Wickens model does not further describe this.

B. Contextual Control Mode Model

A major trend in human performance modeling is the cognitive viewpoint.[21] Within this approach, human behavior is looked upon as a cyclic process, where human action is determined as much by the context as by inherent traits and mechanisms of human cognition. In this view humans do not passively react to events, they actively look for information and act based on intentions as well as external developments.

This approach in human performance modeling is in accordance with concepts from ecological psychology. Ecological psychology studies the information transaction between living systems and their environments, especially as they pertain to the perceived significance of environmental situations for the planning and execution of purposive behavior.[30]

Based on the cognitive viewpoint, as stated above, Hollnagel[21] describes a new approach that focuses on different control modes of the human operator's cognition, which reflect different control strategies in operator behavior.

1. Control Modes

The specific four control modes that are described by Hollnagel[21] characterize in more detail regions of the continuum of control and can be specified as follows:

1) Scrambled. Scrambled control denotes the case in which the choice of the next action is completely unpredictable or random. The scrambled control mode constitutes the extreme situation of zero control.

2) Opportunistic. Opportunistic control corresponds to the case in which the next action is chosen from the current context alone, and mainly based on salient features rather than on more durable intentions or goals. It is opportunistic in the sense that the operator takes a chance, not because he is deliberately exploring an alternative, but because there is no time or possibility to do anything better.

3) Tactical. Tactical control is characteristic for situations in which the operator's performance is based on some kind of planning. Hence, the operator more or less follows a known procedure or rule. The planning is limited in scope and/or limited in range, and the needs taken into account may sometimes be ad hoc.

4) Strategic. Strategic control means that the operator is considering the global context, i.e., using a wider event horizon and looking ahead at higher level goals: either those that have been suspended and have to be resumed or those that, according to experience and expectations, may appear in the near future. This mode should provide a more efficient and robust performance.

An obvious question that arises is what determines the degree of control an operator has of a situation in a particular control mode and how the control mode changes. These topics are discussed next.

2. Control Mode Characteristics

In the Hollnagel approach, the control mode model consists of a high-level description of human behavior, rather than a description of human tasks like planning, monitoring, and decision making. To stress the high-level character of these activities, they will be called meta-activities. Following Ref. 21, the following four meta-activities are briefly elaborated on: 1) number of simultaneous goals, 2) availability of plans, 3) event horizon, and 4) mode of execution.

1) Number of simultaneous goals. This variable describes whether the operator considers only a single goal at a time or whether possible actions from multiple goals are considered. It is not the same as considering multiple choices or actions that may lead to the same goal (i.e., evaluating the effects of several possible lines of action, such as different ways of avoiding a conflict).

2) Availability of plans. This variable describes whether the operator can refer to predefined or preexisting plans (action templates) as a basis for choosing the next action. A plan can either be made on the spot, have been learned by experience, or have been defined explicitly in advance, e.g., as a written procedure. In either case, the availability of plans requires that the situation is familiar. A plan can either be followed rigidly or serve as a guideline for actions to be taken.

3) Event horizon. This variable refers to how much of the past and how much of the future are taken into consideration when a choice of action is being made. The event horizon is described in terms of the number of steps, moves, or items that are considered, rather than in subjective or objective time. The extent of the past is referred to as the history size, while the extent of the future is referred to as the prediction length.

4) Mode of execution. The mode of execution can vary from feedforward to feedback driven. In the feedforward driven execution the operator carries out the steps of a chosen plan until either a predefined checkpoint has been reached or until external conditions force an interrupt. The better a feedforward is, the longer

the uninterrupted period may continue. In the feedback driven execution, each step (or group of steps) is followed by the evaluation of the feedback before the plan is continued. Even with a feedback mode of execution there may be sets of actions that are carried out in a feedforward way. Thus feedforward and feedback modes of execution define two ends of a continuum of possibilities.

3. Control Mode Changes

It is reasonable to assume that several factors will determine how and when an operator's control mode changes from one to another. Two of the obvious candidates that are described by Hollnagel are the amount of subjectively available time and the outcome of the previous action in terms of success and failure. These two main control parameters are briefly elaborated on:

1) Subjectively available time. This involves a consideration of the number of activities that remains to be carried out, e.g., suspended actions, the number of simultaneous goals, the predicted changes and developments in the process and the environment (hence "objective" time), the level of arousal, the level of familiarity of the situation, etc. The estimation ranges from being quite detailed and precise to guessing and "gut feeling".

2) Determination of outcome (of previous actions). This is not just a matter of ascertaining whether the previous action succeeded or failed. On the contrary, the determination of the outcome is different for each control mode and may vary between a rudimentary detection of noticeable changes in the scrambled mode to a detailed evaluation of the feedback in the strategic mode. Complicating factors are, for example, the possible delays in outcome (for systems with large time constants), ambiguous or incomplete state indications, and equivocal rules for interpretation.

C. Human Error Modeling

The main reference used here is Ref. 31. For safety-critical operations like nuclear power plants, an attempt has been made to take into account the human factor in probabilistic risk assessment (PRA) approaches. In general a PRA starts with an identification of hazards that might compromise the plant's safety. Next, the propagation of the possible consequences of the identified hazards through the plant to the level of accidents is described. Frequently, this is done by means of fault and event trees. After quantification of the frequency of occurrence of the identified hazards, plant accident risks are evaluated using the fault and/or event trees. Human error modeling approaches consist of two main elements: human error identification (HEI) and human error probability (HEP) assessment. The results of these then fit within the fault/event tree analysis framework.

1. Human Error Identification

At the basis of human reliability modeling lies the (structured) identification of human operator related hazards. Generally, this involves a task analysis and a human error analysis. During the task analysis, the required operator actions during the various (sub-) processes of the plant are identified. In this stage, the equipment, interfaces, procedures, and (trained) skills that are related to these actions are also identified.

After the task analysis, the HEI considers systematically what can go wrong. Commonly, the following types of errors are considered:

1) Error of omission. Failing to carry out a required action.

2) Error of commission. Failing to carry out a required act adequately: insufficient accuracy, wrong timing, actions performed in a wrong sequence.

3) Extraneous action. Unrequired act performed instead of, or in addition to, required act.

4) Error-recovery opportunities. Actions that can recover previous errors.

Underlying the HEI is a taxonomy of human error. As an example of a taxonomy of human error, we describe the framework of skill-, rule-, and knowledge-based behavior[32] in the next subsection.

2. Skill-, Rule-, Knowledge-Based Taxonomy

In Ref. 33 human error is related to cognitive processes that underlie human performance. Here, the human operator is looked upon as an information processor. The human information processor receives stimuli from the outside world, then processes these stimuli and finally responds. In this framework, human error emerges when during this scheme a deviation from normal processing routine occurs. One of the more influential models of human (erroneous) performance is the step ladder model of Rasmussen.[32] This model distinguishes three levels of human information processing: skill-based level, rule-based level, and knowledge-based level. These levels induce the following taxonomy of human errors:

1) Slips and lapses. Slips and lapses are unintended deviations from planned actions due to execution or memory failures.

2) Rule-based errors. These are errors resulting from erroneous intentions due to the application of bad rules or due to the misapplication of good rules.

3) Knowledge-based errors. These are errors due to wrong reasoning about the to-be-controlled process. These mistakes may emerge from wrong or incomplete knowledge of the process or the bounded rationality of the operator.

3. Human Error Probability Assessment

The second step in human error modeling is to quantify the probability of occurrence of the identified errors, for which many methods exist. The classical example is the human as a technical system approach of the technique for human error rate prediction (THERP) (see Ref. 34). Other examples are the success likelihood index methodology (SLIM),[35] which relies mainly on expert judgment or human error assessment and reduction technique (HEART),[36] which focuses on the effects of identified error-producing conditions.

III. Modeling for En Route ATC

The three psychological models of Section II are now used to develop a single mathematical model of a tactical en route air traffic controller (ATCo) performing his job at a high (cognitive) level. Detail is given only when necessary. The model focuses on the following aspects of the interaction between the controller and the ATM process: 1) maintaining situational awareness, 2) timely taking of safety critical actions, 3) effectiveness of safety critical actions, and 4) occurrence of hazardous situations that involve the controller.

This section is organized as follows. First, the tactical controller task is described in terms of a suitable set of subtasks. Subsequently, the performance of the identified subtasks is related to the context in which the tasks are performed. Next, the scheduling of subtask performance is discussed, and it is explained how clearance errors that are initiated by the controller are incorporated. Finally, the resulting mathematical model is described.

A. Description of Controller Task

The idea is to decompose the controller's task into several subtasks. This decomposition has been carried out along two dimensions. First is a generic dimension, where the task is decomposed into cognitive activities at a general level that is independent from the scenario and operational concept. Second the task is decomposed according to a scenario/concept specific dimension, in which the controller task is described at the level of operational functions in the scenario. This twofold decomposition of the controller task allows flexibility in incorporating detail into the model: in this setup we can restrict detail in the task description along the scenario/concept specific dimension to subtasks relevant for the problem under consideration, while the overall interaction between controller and ATM process may still be properly modeled using the task description at the generic dimension.

First, a task decomposition along the generic dimension has been identified from Ref. 15. The resulting subtasks originate in Ref. 37, however, in Ref. 15 it was merged with several existing task-analyses.[38–41] The following subtasks resulted:

1) Sensing, to gather all information that is needed to get an overview over the air traffic situation.

2) Integration, to connect the gathered information, thus forming a more global air traffic picture.

3) Prediction, to use the more global picture to anticipate future situations and events.

4) Complementary communication, to pass the information to the aircraft to improve the pilot's understanding of the situation.

5) ATC problem solving planning, to use the understanding gained from the more global perspective to plan and prioritize aircraft actions.

6) Executive action, to communicate information and priorities as instructions to the aircraft in the system.

7) Rule monitoring, to ensure that the active components of the system behave in accordance with the rules; monitoring and taking corrective actions for exceptions.

8) Coordination, to coordinate laterally with other parts of the ATC organization.

9) Overall performance, to ensure that the objectives of the operation are achieved, and that the infrastructure functions correctly.

10) Maintenance and monitoring of nonhuman part, to ensure that all systems supporting the controller work correctly.

Second, subtasks are also defined along the en route ATC specific dimensions, where attention is focused on safety-critical actions in the definition of the subtasks. This leads to the identification of three en route context specific tasks: A) anticipate aircraft deviating from intentions, B) react to automation alerts, and C) perform other control activities.

We are now in the following position: the ATCo's task has been decomposed into subtasks along two dimensions, one relating the task to generic cognitive activities and the other relating the task to specific situations in the scenario and operational concept considered. We next identified the task overlap *across* the dimensions in Table 1. This leads to 19 combinations across the dimensions, and thus a decomposition into 19 combined ATCo subtasks.

Table 1 Task overlap across the generic cognitive activities
and the en-route ATC specific tasks

	A, Anticipate	B, Alerts	C, Others
1. Sensing	X		X
2. Integration	X		X
3. Prediction	X		X
4. Complementary communication			X
5. Problem solving /planning	X	X	X
6. Executive action	X	X	X
7. Rule monitoring	X	X	X
8. Coordination			X
9. Overall performance			X
10. Maintenance			X

B. Task Performance and Control Modes

In modeling the influence of the context on performance, we adopt a mathematical model that incorporates two control modes: tactical control and opportunistic control. We identify the characteristic influence of these control modes on the performance of the A and B subtasks.

Because we may look upon subtask C as representing a range of subtasks other than A and B along the en route ATC specific dimension, it suffices to describe differences in tactical and opportunistic control mode at a general level only (see Ref. 17). First, we characterize *subtasks related to anticipation*:

1) Sensing:

Tactical: Whenever possible, the controller scans his display to detect possible deviations from ATC intentions. The controller divides the display into regions of interest and assesses these regions in a particular order. If scanning is interrupted at some time instant, the controller will resume scanning starting at the region that he was scanning when the interruption took place. Further information may also be obtained through radio telephony (R/T) communication.

Opportunistic: Whenever possible, the controller scans his display to detect possible deviations. The controller scans in a random fashion.

2) Integration:

Tactical: The ATCo systematically integrates the information derived from scanning to improve his mental picture of the traffic situation. When some relevant information is not available, the ATCo may return to sensing to actively seek information to improve his assessment of the situation.

Opportunistic: The ATCo integrates the randomly obtained information. An incomplete or even distorted mental picture may develop.

3) Prediction:

Tactical: The ATCo extrapolates his mental picture to the future traffic situation. On the basis of the assessment of the situation, the ATCo decides whether a problem may occur in the midterm future.

Opportunistic: The assessment of the future situation is restricted to a short time horizon and is based on incomplete information. It is assessed whether or not a problem may be expected in the short-term future.

4) Problem solving/planning:
Tactical: On the basis of the assessment of the (future) situation, the ATCo decides on a resolution to the expected problem. In principle, the resolution involves replanning the aircraft trajectories in an optimal fashion with respect to safety and efficiency.
Opportunistic: The resolution is aimed at solving the imminent problem only.

5) Executive action:
Tactical: The controller gives a series of R/T instructions to the aircraft involved. He verifies whether or not the pilot(s) read back these instructions correctly.
Opportunistic: The verification of correct read back may be omitted.

6) Rule monitoring:
Tactical: After the R/T communication, the controller verifies whether or not the aircraft comply with his clearances.
Opportunistic: This may be omitted or be performed less thoroughly.

Next, we characterize subtasks related to alerts:
1) Problem solving/planning:
Tactical: On the basis of the assessment of the situation, the ATCo decides on a resolution for the conflict. The resolution may range from vectoring both aircraft to doing nothing.
Opportunistic: Same as in tactical control mode.

2) Executive action:
Tactical: The controller gives the necessary R/T instructions to the aircraft involved. He verifies whether or not the pilots read back these instructions correctly.
Opportunistic: The verification of correct read back may be omitted.

3) Rule monitoring:
Tactical: After the R/T communication, the controller verifies whether or not the aircraft comply with his clearance.
Opportunistic: Monitoring may be done less thoroughly or even be omitted.

C. Scheduling of Subtasks

In this subsection, the scheduling strategy applied will be defined for the subtasks. The scheduling strategy is expressed in the following (input) task parameters:

1) Preemption. For each subtask an assumption is made about whether or not it may preempt another subtask.

2) Concurrency. For each subtask it is known whether or not it may be performed concurrently with another subtask.

3) Initiation. For each subtask the circumstances under which the subtask should be performed are known.

The assumptions concerning preemption and concurrency are implemented according to Tables 2 and 3. These tables have been identified on the basis of ATC human factors expert knowledge.

Tables 2 and 3 should be read as follows. Consider subtasks C4 (general communication) and A3 (prediction with respect to deviations). It follows from Table 2 that these two subtasks cannot be performed concurrently. Next, inspect the row C4 in Table 3 at the column corresponding to A3, we see that C4 preempts A3. Thus if A3 is carried out and C4 is initiated, execution of A3 will stop and C4

Table 2a Concurrent performance of subtasks (A1-C10) × (A1-B7)

	A1	A2	A3	A5	A6	A7	B5	B6	B7
A1	—	y	y	y	n	y	n	n	y
A2	y	—	y	y	n	y	n	n	y
A3	y	y	—	y	n	y	n	n	y
A5	y	y	y	—	n	y	y	n	y
A6	n	n	n	n	—	n	n	n	n
A7	y	y	y	y	n	—	n	n	y
B5	n	n	n	y	n	n	—	n	n
B6	n	n	n	n	n	n	n	—	n
B7	y	y	y	y	n	y	n	n	—
C1	y	y	y	y	n	y	n	n	y
C2	y	y	y	y	n	y	n	n	y
C3	y	y	y	y	n	y	n	n	y
C4	y	y	n	n	n	y	n	n	y
C5	y	y	y	y	n	y	y	n	y
C6	y	y	n	n	n	y	n	n	n
C7	y	y	y	y	n	y	n	n	y
C8	y	y	n	n	n	y	n	n	y
C9	y	y	y	y	n	y	n	n	y
C10	y	y	y	y	n	y	n	n	y

Table 2b Concurrent performance of subtasks (A1-C10) × (C1-C10)

	C1	C2	C3	C4	C5	C6	C7	C8	C9	C10
A1	y	y	y	y	y	y	y	y	y	y
A2	y	y	y	y	y	y	y	y	y	y
A3	y	y	y	n	y	n	y	n	y	y
A5	y	y	y	n	y	n	y	n	y	y
A6	n	n	n	n	n	n	n	n	n	n
A7	y	y	y	y	y	y	y	y	y	y
B5	n	n	n	n	y	n	n	n	n	n
B6	n	n	n	n	n	n	n	n	n	n
B7	y	y	y	y	y	n	y	y	y	y
C1	—	y	y	y	y	y	y	y	y	y
C2	y	—	y	y	y	y	y	y	y	y
C3	y	y	—	n	y	n	y	n	y	y
C4	y	y	n	—	n	n	y	n	y	y
C5	y	y	y	n	—	n	y	n	y	y
C6	y	y	n	n	n	—	y	n	y	y
C7	y	y	y	y	y	y	—	y	y	y
C8	y	y	n	n	n	n	y	—	y	y
C9	y	y	y	y	y	y	y	y	—	y
C10	y	y	y	y	y	y	y	y	y	—

Table 3a   Preemption between subtasks (A1-C10) × (A1-B7)

	A1	A2	A3	A5	A6	A7	B5	B6	B7
A1	—	n	n	n	n	n	n	n	n
A2	n	—	n	n	n	n	n	n	n
A3	n	n	—	n	n	n	n	n	n
A5	A5	A5	A5	—	n	A5	n	n	A5
A6	A6	A6	A6	A6	—	A6	n	n	A6
A7	n	n	n	n	n	—	n	n	n
B5	B5	B5	B5	B5	n	B5	—	n	B5
B6	B6	B6	B6	B6	B6	B6	B6	—	B6
B7	n	n	n	n	n	n	n	n	—
C1	n	n	n	n	n	n	n	n	n
C2	n	n	n	n	n	n	n	n	n
C3	n	n	n	n	n	n	n	n	n
C4	C4	C4	C4	n	n	n	n	n	n
C5	n	n	n	n	n	n	n	n	n
C6	C6	C6	C6	n	n	C6	n	n	C6
C7	n	n	n	n	n	n	n	n	n
C8	C8	C8	C8	n	n	n	n	n	n
C9	n	n	n	n	n	n	n	n	n
C10	n	n	n	n	n	n	n	n	n

Table 3b   Preemption between subtasks (A1-C10) × (C1-C10)

	C1	C2	C3	C4	C5	C6	C7	C8	C9	C10
A1	n	n	n	n	n	n	n	n	n	n
A2	n	n	n	n	n	n	n	n	n	n
A3	n	n	n	n	n	n	n	n	n	n
A5	A5	A5	A5	A5	A5	A5	A5	A5	A5	A5
A6	A6	A6	A6	A6	A6	A6	A6	A6	A6	A6
A7	n	n	n	n	n	n	n	n	n	n
B5	B5	B5	B5	B5	B5	B5	B5	B5	B5	B5
B6	B6	B6	B6	B6	B6	B6	B6	B6	B6	B6
B7	n	n	n	n	n	n	n	n	n	n
C1	—	n	n	n	n	n	n	n	n	n
C2	n	—	n	n	n	n	n	n	n	n
C3	n	n	—	n	n	n	n	n	n	n
C4	C4	C4	C4	—	C4	n	C4	n	C4	C4
C5	C5	C5	C5	n	—	n	C5	n	C5	C5
C6	C6	C6	C6	C6	C6	—	C6	C6	C6	C6
C7	n	n	n	n	n	n	—	n	n	n
C8	C8	C8	C8	C8	C8	n	C8	—	C8	C8
C9	n	n	n	n	n	n	n	n	—	n
C10	n	n	n	n	n	n	n	n	n	—

will be performed first. If concurrent performance were possible (i.e., there would be a "y" in Table 2), then preemption would mean that C4 and A3 are performed concurrently, with C4 as the primary and A3 as the secondary task. In terms of a stack of to-be-performed subtasks, this scheduling principle can be formulated generically as the following two rules:

Rule 1: An initiated subtask will be placed in the stack before the subtasks that it may preempt.

Rule 2: If the first two subtasks of the stack can be processed concurrently, this will be done (subtask duration will be slightly longer, however).

D. Errors in Flight Plans and Intents

An important safety issue is that for one single aircraft there may be all kinds of differences between the flight intents on the ground and in the air, and the ATCo and pilot awareness of those intents, i.e.: 1) tactical ATCo's awareness of the flight intent, 2) flightplan in the ATC system, 3) pilot's awareness of the flight intent, and 4) Flight plan used by the flight management system (FMS). To allow for these differences, the following mathematical modeling approach is adopted:

1) ATCo. The tactical ATCo's awareness of the flight intent is assumed to be ATC's true reference. The quality of ATC's true reference is in one of the following two discrete modes: a) the true reference provides separation, b) the true reference does not provide separation. In general, the latter mode value may be reached if an ATCo has made a knowledge-based error.

2) ATC. The quality of the flightplan in the ATC system may be in one of the following two discrete modes: a) agrees with ATC's true reference, b) differs from ATC's true reference. The latter is due to an ATCo input error, or an ATC database error.

3) Pilot. The quality of the pilot's awareness of ATC's true reference is in one of the following two discrete modes: a) agrees with ATC's true reference, b) differs from ATC's true reference. The latter may happen due to a clearance error. There are two types of clearance errors: 1) intended clearance given to wrong aircraft or 2) wrong clearance given to intended aircraft. The causing factor may be with the ATCo, or the pilot, or both, and may be knowledge based, rule based, or skill based.

4) FMS. The quality of the flightplan used in the FMS is in one of the following two discrete modes: a) agrees with ATC's true reference, b) differs from ATC's true reference. The latter happens if pilot awareness differs from ATC's true reference or is due to a pilot input error or an FMS database error.

In elaborating the preceding approach, it is assumed that all the ATCo related errors may occur at random during performance of subtasks A6, B6, or C6 (executive action), where the frequency of occurrence depends on the control mode the controller is in. Furthermore, such errors may be detected and corrected during rule monitoring subtasks A7, B7, or C7, also depending on the control mode (e.g., Ref. 20).

E. Mathematical Model of Tactical ATCo

To establish the connections with the other ATM processes, in this subsection we describe the mathematical model of the ATCo from an input-output point of view. First we describe how initiation of cognitive activity is modeled, and then the implementation of the task description and controller performance is described. The Petri net of the tactical ATCo model is shown in Fig. 2.

HUMAN COGNITION MODELLING IN AIR TRAFFIC MANAGEMENT 495

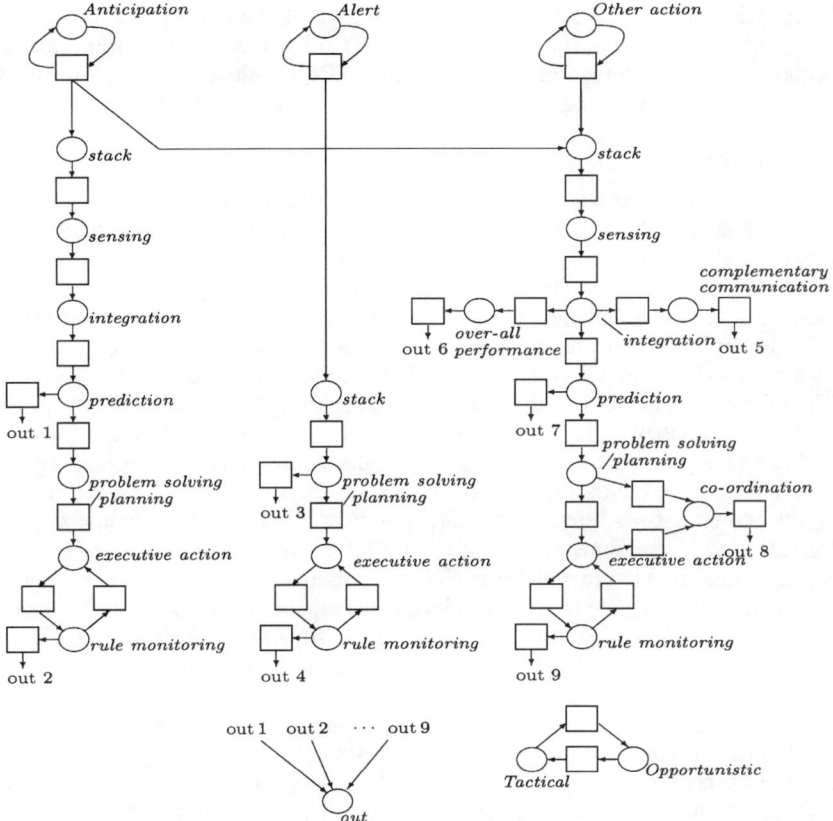

Fig. 2 Petri net of tactical ATCo model.

1. Initiation

Three stimuli for ATCo cognitive activity are identified: ATCo's anticipation, automation alerts, and other actions. Activity triggering situations that first have to be detected by the operator (like an aircraft severely deviating from its route) are not considered as an initiation stimulus, because general sensing is modeled as a part of the operator's task, and therefore the sensing activity has to be initiated first. For the occurrence of certain stimuli, various other ATM modules may need to function properly, such as the ATCo human machine interface (HMI) and surveillance for an automation alert.

Within the Petri net each stimulus is modeled as a place, connected with one transition that fires if initiation of the corresponding cognitive activity takes place. These transitions produce two tokens: one token returning to the stimulus place for future generation of cognitive activity and one token in a *stack* place. The *stack* places represent the situation in which the respective initiated cognitive activity has to wait until the operator has completed other (more important) tasks. The places *Anticipation, Alert,* and *Other action* represent initiation of cognitive activity by own initiative, automation alerts, and other action (e.g., a pilot request), respectively. Preconditions on occurrence of these stimuli are modeled within the respective transitions: if the preconditions are not met, the transition does not fire. For example, the proper functioning of the ATCo HMI as a precondition for the occurrence of an automation alert triggering ATCo cognitive activity is modeled as a precondition for firing of the transition connected to the *Alert* place.

2. ATCo Subtasks

The ATCo task has been divided into several subtasks that are each defined as the combination of a scenario-specific purpose and a generically described cognitive activity. Three context-specific purposes are modeled: ATCo to detect and correct deviations of aircraft from ATCo intentions, ATCo to react to automation alerts (initiated by automation tools), and ATCo to perform other control activities (initiated by own initiative or through other actions). Each subtask is represented by a place in the Petri net, which is named after the cognitive activity it represents. The tokens then model cognitive activity on the subtask that corresponds to the place in which they reside. Some cognitive activities may be performed for several purposes, leading to several places with the same name. In the following we describe the places with respect to the cognitive activities that they represent.

The places *sensing* represent the situation in which the ATCo is gathering information to improve his picture of the traffic situation. The places *integration* represent the situation in which the ATCo incorporates the newly obtained information into this mental picture. The place *communication* represents the situation in which the ATCo makes his knowledge of the situation available to the pilots. The place named *over-all performance* describes the evaluation of sector performance as a whole. In the *prediction* place, the ATCo extrapolates his picture of the traffic to the future, whereas in the *problem solving/planning* place he synthesizes solutions to possible (future) problems. In the *executive action* place, the operator gives clearances to aircraft, followed by the *monitoring* places where it is verified whether the aircraft adhere to these clearances. In the *out* place, the tokens are collected after performance.

Whenever one subtask is logically performed after performing another (e.g., *prediction* is performed after *integration*), and they have the same scenario-specific purpose, a transition is drawn between those two subtasks.

3. Subtask Scheduling

We next incorporate the scheduling rules. Scheduling depends on the relative priority of a subtask and the possible concurrent performance of two subtasks. The relative priority is modeled as a color type that is associated with the tokens that represent cognitive activity on subtasks. This color type is a number 1,2,... where low numbers correspond to high priority. The priority colors are updated whenever a new token is initiated and when a token is collected in the *out* place, according to a suitable set of assumptions. According to the scheduling rules, either the token that has priority 1 is performed exclusively or the tokens with priority 1 and 2 are performed concurrently, with the token with priority 2 being the secondary task.

We assume that for each subtask the time needed to complete it has a certain probability density, given the current control mode of the ATCo and possible concurrent performance of another subtask. In the Petri net, the duration of performing a subtask is modeled as a delay in the firing of the transition that has the subtask as input place. Transitions with a token in the input place that does not have priority 1 or 2 have "infinite" delays. Transitions with a token in the input place that has priority 1 has a delay corresponding to the normal duration of the subtask, given the control mode. Delays of transitions with a token in the input place that has priority 2 either have an infinite delay or a delay that may be longer than when the corresponding subtask is performed exclusively. This depends on the extent to which the subtasks with priority 1 and 2 may be performed concurrently. Hence in the Petri net, each transition has a delay that is a function of the priority of the token in the input place, the current control mode, and the place in which the token with priority 1 resides.

The ATCo's executive actions (i.e., the clearances given) are also modeled as a color type associated with the tokens in the subtasks. This color type is a set of paired numbers describing the type of clearance given and the aircraft to which the clearance is given. The decision to give no clearance at all is also modeled as an executive action and has a separate color value. In the present model, it is assumed that the type of clearance given is determined during the *executive action* subtask only and that it depends on the control mode only. Thus the firing of the transitions after the *executive action* places also affects the Petri nets of other ATM modules: completion of executive action means that a decision to give a clearance to an aircraft has been carried out and therefore the firing of these transitions describes the ATCo control actions.

4. ATCo Control Modes

In the model, ATCo performance depends on the control mode, scheduling rules, and results in a clearance. In the DCPN model of the ATCo, two control modes are identified, which are each represented by a place in the Petri net: the place *Tactical* models the situation in which the controller has a relatively high degree of control and the place *Opportunistic* models a relatively low degree of control. The control mode may influence ATCo performance in all aspects. The switching between control modes is modeled by transitions between the *Tactical* and *Opportunistic* places. The resulting subnet contains one token, the place of which defines the current degree of control. The firing of the transitions between the control modes depends on the number of tokens in the *stack* places (indicating the subjectively available time) and the number of times that *monitoring* was followed by another *executive action* during the last few minutes (indicating the outcome of previous actions measured as the number of clearances that the controller considers to be insufficiently effective). Details for this type of modeling appeared to be available through human factors ATC expert knowledge.

IV. Reduction of the ATCo Model

In this section, we explain how the ATCo model that was developed in Section III is reduced by applying appropriate model aggregations. The motivation for this reduction is that the complexity of the original model results from a detailed modeling that is judged unnecessary for the application at hand. This makes the resulting reduced model interesting in its own right.

First, we explain how the subtasks are clustered into a new set of subtasks, and how scheduling simplifies accordingly. Second, the Petri net for the ATCo reduced model is given. Third, within an en route context we compare the relevant model characteristics to verify that the model based on the reduced task description is indeed an appropriate approximation.

A. Aggregation of Subtasks

In the previous section, Tables 2 and 3 show that, because of the possibility of concurrent performance of subtasks, the number of required assumptions concerning concurrent subtask performance equals $\frac{1}{2}n(n-1)$ and the number of required assumptions concerning preemption equals $n(n-1)$, with n the number of identified subtasks. For the present 19 subtasks, this means a total of 342 rules concerning task scheduling in the model. This large number of rules may severely complicate the stochastic analysis that is required for risk evaluation. Therefore, it is desirable to reduce the complexity of the model without compromising conservativeness or psychological validity.

This reduction of the full model is achieved by decreasing the level of detail at which the air traffic control task is described and the way performance of these tasks is scheduled according to single-task performance.

The approach taken is to group the 19 subtasks into a smaller number of clusters of subtasks. The clusters are identified in Table 4.

Next, we need to identify how task scheduling at the level of clusters of subtasks takes place. First, concurrent performance of clusters of subtasks is investigated using Table 2. This is done conservatively using the principle that if one combination of the clustered subtasks cannot be performed concurrently, then the whole cluster of subtasks cannot be performed concurrently. Application of this principle yields Table 5.

In a similar fashion, we identify a new table (Table 6) for the preemption between clusters of subtasks. The following rule is applied: if any subtask in some cluster A preempts all subtasks in some other cluster B, then cluster A preempts cluster B. Otherwise, cluster A does not preempt cluster B.

Table 6 implies that the cluster $Misc_C$ does not preempt any other cluster. Moreover, $Misc_C$ is preempted by all other clusters, except $Monitoring_A$. Furthermore, it follows from Table 2 that $Monitoring_A$ and $Misc_C$ can be performed concurrently. From this, we conclude that performance of the subtasks in the cluster $Misc_C$ does not conflict with other subtasks at cluster level. Because the cluster $Misc_C$ itself does not contain subtasks that are directly relevant for safe separation, we can therefore discard this cluster in the model without compromising conservativeness. Therefore, we do not take into account this cluster in the sequel.

Now inspect Tables 5 and 6 again. Perhaps surprisingly, we see that concurrent performance of the remaining clusters of subtasks is not possible. Moreover, the

Table 4   Clustering of the subtasks

Cluster	Initial subtasks
$Monitoring_A$	A1-A3
$Communication_A$	A5-A7
$Communication_B$	B5-B7
Complementary $Communication_C$	C4
$Communication_C$	C6
Co-ordination$_C$	C8
Miscellaneous$_C$	C1-C3, C5, C7, C9, C10

Table 5 Concurrent performance of clusters of subtasks, derived from Tables 2 and 4

	Mon_A	Com_A	Com_B	CpC_C	Com_C	$Coor_C$	$Misc_C$
Mon_A	——	n	n	n	n	n	y
Com_A	n	——	n	n	n	n	n
Com_B	n	n	——	n	n	n	n
CpC_C	n	n	n	——	n	n	n
Com_C	n	n	n	n	——	n	n
$Coor_C$	n	n	n	n	n	——	n
$Misc_C$	y	n	n	n	n	n	——

remaining preemption rules boil down to a fixed priority list where Monitoring$_A$ has lowest and Communication$_B$ has highest priority. Apparently, similar principles underlie Tables 2 and 3, although the construction of these tables was done before and independently from the subtask clustering analysis.

We conclude that at the level of clustered tasks, the complexity of the scheduling principle is reduced significantly, without compromising conservativeness. In summary, the main model simplifications are 1) 19 subtasks are reduced to 6 clusters of subtasks, 2) concurrent task performance is simplified into single task performance, and 3) preemption rules for each combination of subtasks are simplified into a fixed priority list.

B. Reduced ATCo Model

On the basis of the aggregation, a reduced model of the ATCo can now be developed. Six main ATCo cognitive tasks are identified, which describe the operator performance at a cognitive level. For each task, we assumed a relative priority ranking, an average duration, and the percentage of his time that the operator would spend on the task if uninterrupted (Table 7).

The ATCo performs these tasks one at a time, according to the given priorities. Task scheduling is kept straightforward: high priority tasks are performed first, possibly preempting a low priority task. Two important aspects of performance are incorporated as well, the influence of the control mode and the possibility of

Table 6 Preemption between clusters of subtasks, derived from Tables 3 and 4

	Mon_A	Com_A	Com_B	CpC_C	Com_C	$Coor_C$	$Misc_C$
Mon_A	——	n	n	n	n	n	n
Com_A	Com_A	——	n	Com_A	Com_A	Com_A	Com_A
Com_B	Com_B	Com_B	——	Com_B	Com_B	Com_B	Com_B
CpC_C	CpC_C	n	n	——	n	n	CpC_C
Com_C	Com_C	n	n	Com_C	——	Com_C	Com_C
$Coor_C$	$Coor_C$	n	n	$Coor_C$	n	——	$Coor_C$
$Misc_C$	n	n	n	n	n	n	——

Table 7 Six main cognitive tasks

Task	Priority	Description
$Monitoring_A$	6	Visual anticipation and detection of deviations from the ATCo intention
$Communication_A$	2	Communicate clearance with an aircraft that deviated severely visually from ATCo intention
$Communication_B$	1	Communicate clearance with aircraft for which an automation alert was issued
$Complementary\ communication_C$	5	General complementary communication with pilots
$Communication_C$	3	General communication of executive action (i.e., clearances)
$Coordination_C$	4	General coordination with planner controller, controllers of other sectors.

erroneous clearances. The Petri net describing the discrete modes for the ATCo model is given in Fig. 3.

Two control modes are considered, *Tactical* and *Opportunistic*, which reflect the degree of control. In the *Tactical* mode, the ATCo takes his time and makes little errors. In the *Opportunistic* mode, the general tasks (marked subscript C) are performed faster, but the chances of errors are also larger. The switching between the control modes depends on the subjectively available time (measured as the number of tasks waiting to be performed) and the outcome of previous actions (measured as the number of corrective actions, i.e., $Communication_A$ and $Communication_B$, taken by the ATCo during the last 2 min). If the subjectively available time is short or if the outcome of previous actions is bad, then the ATCo switches to *Opportunistic* control mode.

ATCo erroneous clearances are taken into account as follows: the ATCo may give a different clearance than he intended (e.g., switching heading and speed), or he may give the clearance to a different aircraft than he intended (call-signs

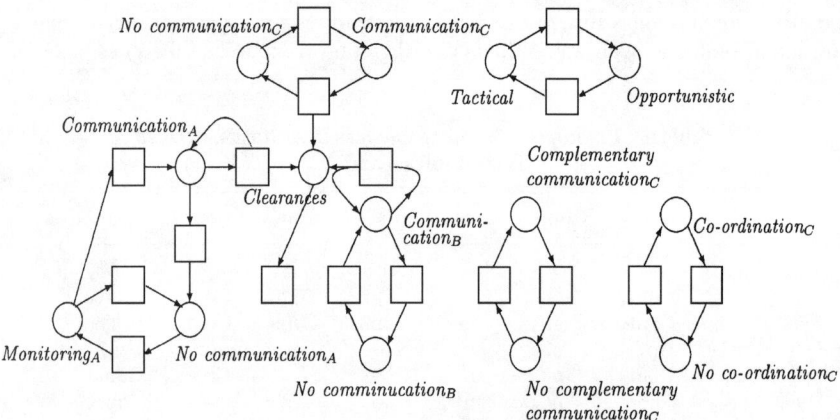

Fig. 3 Petri net of reduced ATCo model.

mixed up). These errors are incorporated as random variations in the ATCo actions. The error types are represented as a color value of the tokens in the place *Clearances*.

The switching between modes is affected by several other modules, such as aircraft evolution, surveillance, ATC system, R/T local, R/T global, and performance of pilot. Surveillance output (i.e., the estimated aircraft state) is input for the visual detection of severe deviations by the ATCo. The ATC system must be *Working* for the ATCo to be able to do his job. The R/T modules and pilot module together form the decision making loop or DM-loop. If all modules in the DM-loop are *Working*, *Relaxed*, *Delaying*, or *Busy* for a given aircraft, then the ATCo is able to give a clearance to that aircraft.

C. Comparison Against Statistical Data

Next we evaluated for the ATCo routine monitoring concept the period to detect severe deviations so that a comparison with available statistical data is possible.[42]

A full and reduced ATCo performance model was developed on the basis of the cognitive principles identified in Section II and integrated with appropriate Petri net models for the other relevant components in conventional ATC (see Fig. 4).

Comparison with the model-based results (Fig. 5) shows that the detection time results of both the original and the reduced ATCo model agree quite well with the measured data. It should be noted that in Ref. 42 only very few detection

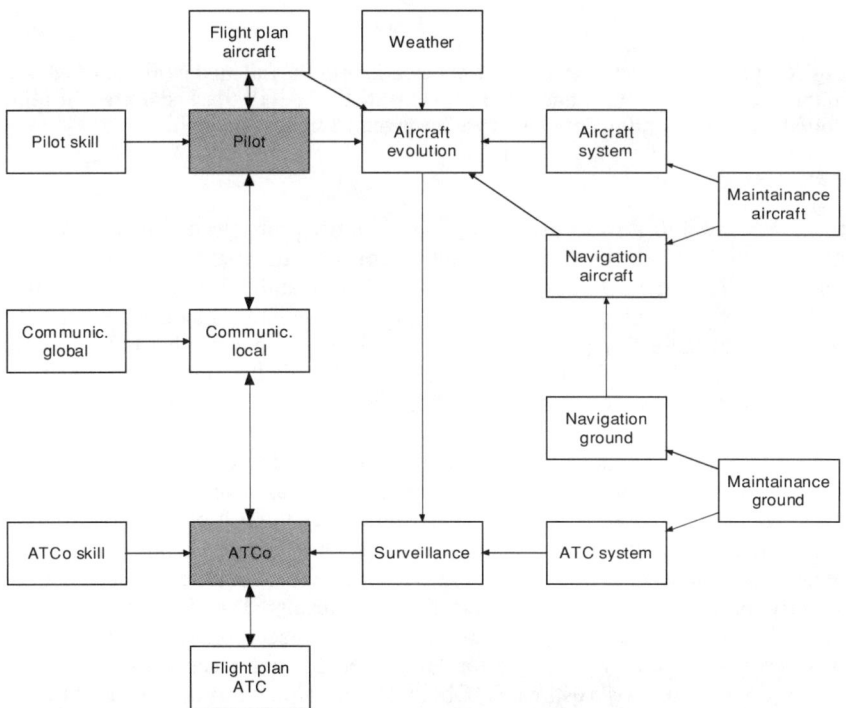

Fig. 4   Functional representation of conventional ATC.

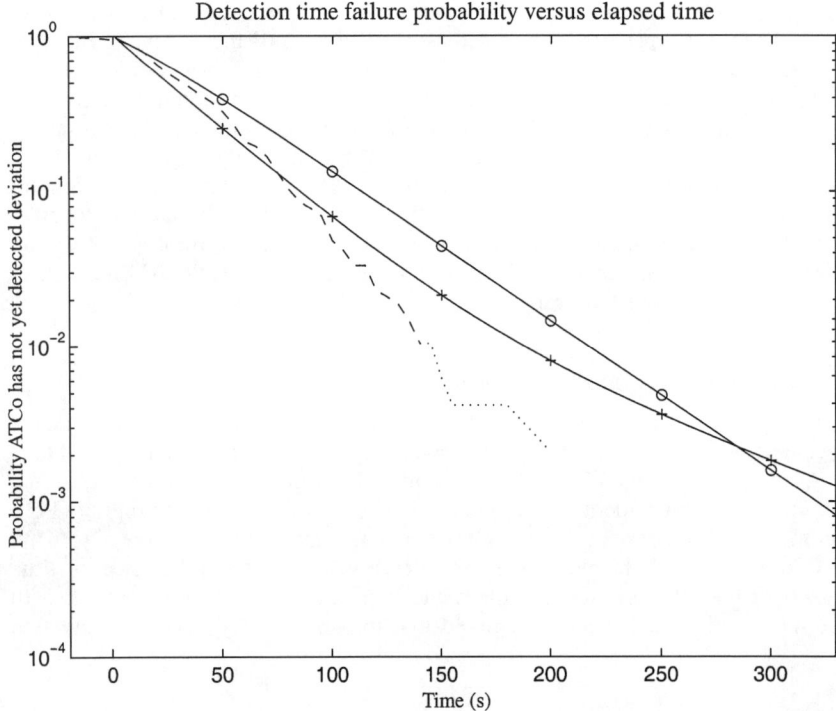

Fig. 5 ATCo detection time of severe of deviations of the full model (line marked +), of the reduced model (line marked o), and of statistical data[42] (dashed/dotted line, the dotted part representing data based on fewer than 5 measurements).

times beyond 150 s were measured. This is most probably due to the limited number of measurements made in combination with the low probability of such long detection times. Although they have low probability, the longer detection times add significantly to the risk, and Fig. 5 shows that model-based results do extend to these low probability values. We may conclude that both the full and the reduced model curves agree quite well with the statistical data. This clearly contributes to gaining confidence in the model-based approach taken.

In this section, we have shown how to derive a reduced model of the ATCo performance from a more detailed ATCo model that was developed in Section III. This reduction is based on using a less detailed decomposition of the air traffic control task and simplifying concurrent task performance into single task performance (i.e., one task at a time). From Fig. 5 it appears that this reduced model yields slightly more conservative ATCo detection time results. Therefore, we conclude that for the particular application considered here, incorporation of concurrent task processing into the ATCo performance model is not necessary for avoiding overly conservative risk estimates. Obviously, incorporation of concurrent processing into human performance models may be essential for other applications such as detailed workload assessment.

V. Example Application

In this section we show TOPAZ-based assessment results for accident risk and ATCo actions for a hypothetical ATM scenario that consists of two en route traffic streams, flying in opposite directions, all at one single flight level.

A. Hypothetical ATC Example

The rather hypothetical example has been developed by EUROCONTROL with the aim to learn how ATC influences accident risk, and how far the nominal separation S between opposite RNP1 traffic streams can safely be reduced. The specific details of this scenario are:
1) There is a straight route, with two traffic lanes.
2) ATCo expects all aircraft to stay on these lanes.
3) Parameter S denotes distance between the two lanes (see Fig. 6).
4) Opposite traffic flows along each lane.
5) Aircraft fly at one flight level only.
6) Traffic flow per lane is 3.6 aircraft/h.
7) All aircraft nominally satisfy required navigation performance with 95% of time less than 1 n mile deviation (RNP1).
8) None of the aircraft is equipped with a traffic collision avoidance system (TCAS).
9) Target level of safety (TLS) is 5×10^{-9} accidents/flight h.[43]
10) 15 aircraft per sector/ATCo.
11) There are no military aircraft.

This exemplar scenario is considered for the following three ATM concepts:

A) Procedural separation only. In this case, there is no ATC surveillance system. This is the type of situation encountered with traffic over the North Atlantic.

B) STCA-only based ATC. In this case there is a radar-based surveillance and R/T communication, but it is assumed that ATC is doing nothing with this information unless its STCA system issues an alert, thus assuming no monitoring by ATCo.

C) Routine monitoring based ATC. The same as in B, but now without the STCA system. Thus, aircraft deviations are identified only through routine monitoring.

B. Accident Risk

For each of the three ATM concepts, the TOPAZ accident risk assessment methodology and tool set have been used to assess accident risk for the preceding

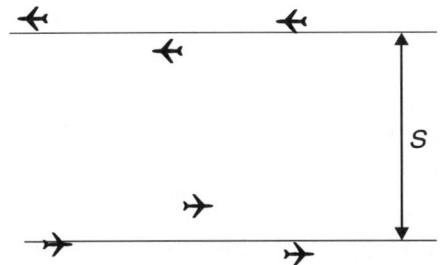

Fig. 6 Opposite direction traffic in a dual lane structure.

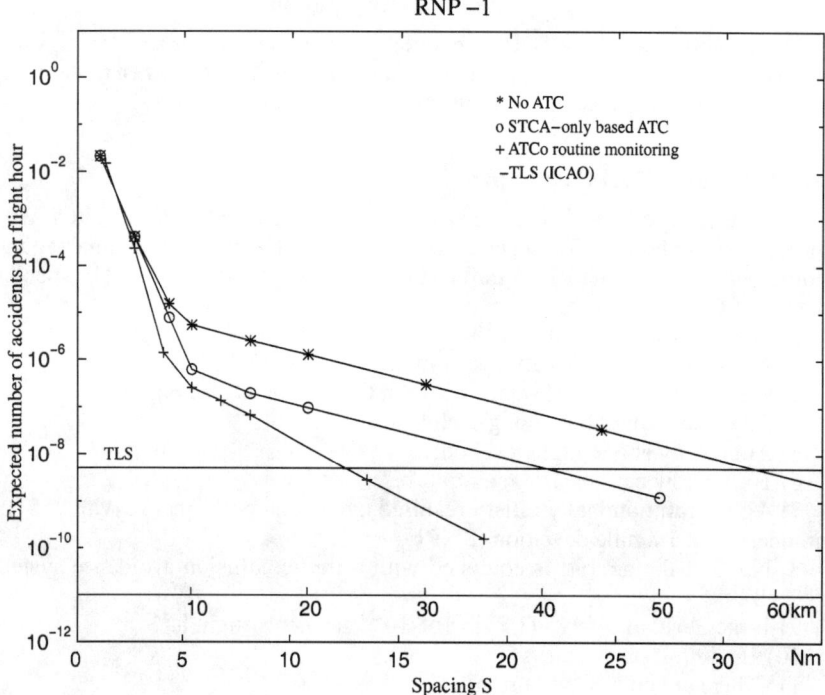

Fig. 7 Accident risk vs route spacing, the graphs marked No ATC and STCA-only based ATC have been taken from Ref. 8. The graph ATCo routine monitoring is from Ref. 17. The TLS value used is defined in Ref. 43.

scenario, as a function of the spacing parameter S. The accident risk for the newly developed model result is presented as the graph marked 'ATCo routine monitoring' in Fig. 7. In Fig. 7, there also is a horizontal line that represents the target level of safety (TLS).

1. Qualitative Uncertainty Analysis

Absolute usage of the risk curves without taking into consideration existing bias and/or uncertainty can inspire undue conclusions. Because of a model-based quantitative risk assessment approach, it is possible to bring the model assumptions made to the foreground and subsequently perform an uncertainty analysis of the model vs reality.

With the TOPAZ methodology, the starting point for such uncertainty analysis consists of the following: 1) description of nominal operation and procedures, 2) list of hazards identified for the operation considered, 3) list of assumptions made when building the Petri net, 4) Petri net specification (local and interactions), and 5) list of parameters and values used during the numerical evaluation, and their sources.

For the routine monitoring concept there are more than 200 hazards (about 50% are human related), about 25 model assumptions, and about 100 model parameters (about 20% for the reduced ATCo model).

The qualitative uncertainty analysis that can be performed works as follows. First, for each hazard it is specified how it is incorporated in the Petri net or not (due to a model assumption listed). The result is that for each parameter and for each assumption, the related hazards are identified. The subsequent steps are: 1) per assumption, perform a qualitative assessment of its uncertainty impact on the risk; 2) per parameter value, perform a qualitative assessment of the uncertainty in relation to the applicable hazards; and 3) per parameter value, assess the impact of this uncertainty impact on the risk.

At this moment, this qualitative uncertainty analysis has not yet been applied to the preceding evaluated en route examples. However, it has successfully been applied in a Wake vortex risk assessment study (see Ref. 13). On the basis of this experience, we expect that the main contribution to uncertainty will come from unmodeled hazards (either due to model assumptions or due to missing hazards), rather than from parameter value uncertainty. For the curves in Fig. 7 this means that, for the time being, they should be interpreted in a relative way only.

2. Analysis of Risk Curves

Inspection of Fig. 7 yields that the TLS is reached for a route spacing of about 24 km (13 n miles), which is a significant improvement of the values of the No ATC curve [TLS reached at about 58 km (~32 n miles)] and the STCA-only based ATC curve [TLS reached at about 40 km (~22 n miles)]. Obviously, for busy fixed-route situations over the continent, procedural separation is not very helpful with STCA-only based ATC. The improvement provided by the routine monitoring shows that it is much more effective in safely managing deviations from centerline than reacting to STCA alerts only. Apparently, STCA really is a safety net only.

We also observe that the risk reduction provided by monitoring based ATC increases as route spacing increases. This is in contrast to the STCA-only based control strategy, in which the ATCo prevents a fixed ratio of the deviating aircraft that reach the other route from collision. The reason for this increasing risk reduction is that the number of severe deviations that are detected before the aircraft reaches the other route increases faster with route spacing than the decrease in the number of deviating aircraft that reach the other route. Hence, the slope of the risk figure depends on the slope of the ATCo detection time instead of the slope of the non-nominal lateral deviation probability density function. Consequently, accident risk may be further reduced by changing the ATM design and in particular the role of the controller such that the ATCo detection time is improved.

3. Safety Criticality Analysis

Further evaluation showed that safety criticality lies with the *Sharp turn* type of deviations. This is caused by the fact that during the *Sharp turns* the aircraft deviates from the route much faster than in the case of a general *Non-nominal* deviation. For $S = 24$ km (13 n miles) our evaluations showed that the risk involved with the *Sharp turns* to be a factor 15 higher than for the *Non-nominal* deviations.

In the present model, the *Sharp turns* are caused by erroneous ATCo clearances and aircraft flightplan errors, whereas the *Non-nominal* deviations are caused by degraded navigation systems, degraded aircraft systems, etc. Hence, from the safety criticality result we conclude that the most risky situations originate in the human factor rather than in degraded performance of technical systems.

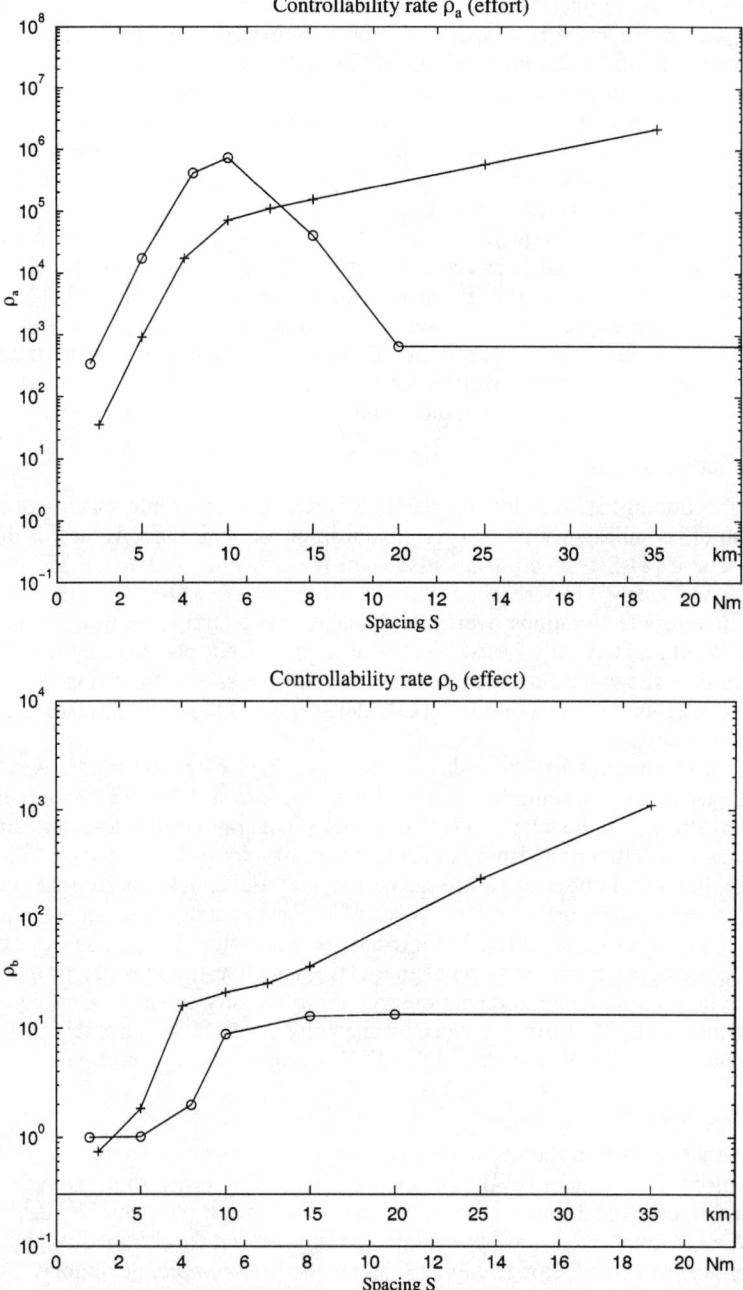

Fig. 8　Effort ρ_a and effect ρ_b for routine monitoring (+ graph) and STCA-only based ATC (o graph).

C. ATC Effort and Effect

The ATC effort is related to the number of ATC actions normalized by the theoretical minimum of ATC actions required for averting all accidents (i.e., one action per accident that would occur if there were no ATC). This is approximately equal to the number of ATC actions required to avert one accident (as almost all potential accidents should be averted):

$$\text{Effort: } \rho_a = \frac{\text{ATC actions}}{\text{Accidents, without ATC}} \left(\approx \frac{\text{ATC actions}}{\text{Averted accidents}} \right)$$

Next, we express the ATC effect as the factor of accident risk reduction achieved by ATC:

$$\text{Effect: } \rho_b = \frac{\text{Accidents, without ATC}}{\text{Accidents, with ATC}}$$

Graphs for the metrics ρ_a and ρ_b for STCA-only based ATC and routine monitoring are given in Fig. 8.

From the ρ_b curves in Fig. 8, we conclude that the monitoring strategy yields a better risk reduction for *all* spacings. Second, inspection of the ρ_a graphs yields that for small spacings, monitoring requires even less effort than STCA-only based ATC. For larger spacings monitoring requires more effort than STCA-only based ATC. We conclude that for small spacings, monitoring is preferred (more effect, less effort), whereas for larger spacings, the situation is less clear (more effect, but also more effort).

A remark should be made concerning spacings below 2 n miles (where the ρ_b has negative values for monitoring). Notice that these very low spacings are not realistic for the monitoring concept, since for these spacings aircraft may collide while remaining within the safe boundary around the lanes (whence the ATCo does not take action to prevent these collisions). We therefore disregarded these very small spacings.

VI. Concluding Remarks

This chapter applied state-of-the-art psychology in human cognition/performance modeling for application to accident risk modeling. This led to the development of mathematical human cognition/performance models for a tactical ATCo in a conventional en route ATC situation. This model is shown to be of great use in the evaluation of accident risks for an ATM scenario with the tactical ATCo performing routine monitoring to detect and correct for severely deviating aircraft.

In this work, we took a model-based approach toward the assessment of concepts such as accident risk and controllability in ATM situations. This makes the approach a formal one: for the model, accident risk and ATCo effort and effect indicators are unambiguously defined. If numerical evaluations of the model are carried out in a verifiably correct way, then the validity of the results depends on the verifiability of the model only.

The main problem thus is how to verify that the model "matches" reality sufficiently well with respect to the intended use of the model. It should be stressed that an absolute match is not feasible; however, this is also not necessary. Instead, a case that the model is sufficiently realistic for its purposes should be built, by testing both the assumptions made during model development and relevant characteristics of the eventual model. The confidence in the model should then be based on the quality of the arguments for its validity (i.e., the test results). This model validation approach is currently under development. On the basis of the human cognition modeling and the controllability results in this chapter, we recognize a contribution to this approach, which consists of comparing relevant model characteristics with human-in-the-loop measurements in the case of human controllability evaluation. Such comparison should always be treated with care, as the results may be sensitive to the context.

For the present model, three tests of its validity have been carried out. First, in Ref. 17 the human performance modeling approaches that underlie the ATCo model used have been shown to be sufficiently powerful to explain ATCo-related hazards in en route ATM. Second, in Fig. 5 ATCo detection time, which is a relevant model characteristic for accident risk, was compared against controller-in-the-loop-based data from the literature. Third, during the whole development of the ATCo model, a human factors specialist has been actively involved, and the results have been reviewed by an operational expert. Obviously, further confidence building can and should be done, e.g., on the basis of detailed reviews with a number of experts and comparison of a range of model characteristics with additional empirical data.

When designing advanced ATM, it is important to understand the safety issues already at a conceptual level. Because of the extremely low probability of accidents in existing ATM practice, statistical data from practical situations are limited and analyzing accident reports alone is not sufficient to understand safety at the level of the interactions between the various ATM components. For advanced ATM designs, data concerning unsafe events may even be lacking. Therefore, some kind of modeling approach is required to optimize for capacity and separation criteria without compromising safety.

Because in about 80% of the reported accidents humans were part of the cause, it is imperative to properly incorporate the human factor into the models used for risk assessment. In this chapter, we therefore investigated three complementary psychological models, and we combined them into a single mathematical model of a tactical ATCo in a conventional en route context.

Because monitoring activity is typically performed as an integrated part of the tactical ATCo job, it is necessary to also take into account other ATCo activities that may interfere with monitoring. This was accomplished through our contextual model of ATCo performance that takes into account the interfering tasks at a cognitive level, thus minimizing the level of modeling detail required to take into account the interfering tasks. We also showed that this advanced ATCo performance model can be included in an accident risk model for the conventional en route ATC situation considered, and that the time needed for the ATCo to detect a severe deviation as predicted by the model agrees rather well with statistical data. We also demonstrated that we could use the model to evaluate accident risk for the ATM scenario, and that the results provide valuable insight and feedback to ATM designers.

We conclude that the use of advanced psychological models in accident risk modeling is feasible, thus extending the applicability of the accident risk modeling approach to situations in which isolated models of individual human actions do not suffice.

Acknowledgments

This research has been performed at the National Aerospace Laboratory NLR within a series of projects with support from the European Commission and the Netherlands Civil Aviation Authority (RLD). This chapter has been presented as a paper at the FAA/Eurocontrol ATM2000 Conference.

References

[1] Cohen, S. and Hockaday, S. (eds.), "A Concept Paper for Separation Safety Modelling, An FAA/EUROCONTROL Cooperative Effort on Air Traffic Modeling for Separation Standards," FAA/EUROCONTROL, May 1998.

[2] Haraldsdottir, A., Alcabin, M. S., Burgemeister, A. H., Lindsey, C. G., Makins, N. J., Schwab, R. W., Shakarian, A., Shontz, W. D., Singleton, M. K., van Tulder, P. A., and Warren, A. W., "Air Traffic Management Concept Baseline Definition," NEXTOR Rept. RR-97-3, Boeing, Seattle, WA, 1997.

[3] Odoni, A. R., Bowman, J., Delahaye, D., Deyst, J. J., Feron, E., Hansman, R. J., Khan, K., Kuchar, J. K., Pujet, N., and Simpson, R. W., "Existing and Required Modelling Capabilities for Evaluating ATM Systems and Concepts," NASA NAG2-997, Massachusetts Inst. of Technology, Cambridge, MA, March 1997.

[4] Wickens, C. D., Mavor, A. S., Parasuraman, R., and McGee, J. P. (eds.), The *Future of Air Traffic Control, Human Operators and Automation*, National Academy Press, Washington, DC, 1998.

[5] Blom, H. A. P., Bakker, G. J., Blanker, P. J. G., Daams, J., Everdij, M. H. C., and Klompstra, M. B., "Accident Risk Assessment for Advanced ATM," *2nd USA/Europe Air Traffic Management R&D Seminar*, FAA/EUROCONTROL, 1998.

[6] Blom, H. A. P., Everdij, M. H. C., and Daams, J., "Consolidation of Results for Safety Certification in ATM; Part II: Safety Cases for a New ATM Operation," National Aerospace Laboratory NLR, Rept. TR-99587, Amsterdam, 1999.

[7] Bakker, G. J., and Blom, H. A. P., "Air Traffic Collision Risk Modelling," *Proceedings of the 32nd IEEE Conference on Decision and Control*, Vol. 2, Institute of Electrical and Electronics Engineers, New York, 1993, pp. 1464–1469.

[8] Everdij, M. H. C., Blom, H. A. P., and Klompstra, M. B., "Dynamically Coloured Petri Nets for Air Traffic Management Safety Purposes," *Preprints 8th IFAC Symposium on Transportation Systems*, edited by M. Papageorgiou and A. Pouliezos, Intelligent Technological Systems Lab., Technical University of Crete, 1997, pp. 184–189; also National Aerospace Laboratory NLR, Rept. TP-97-493, Amsterdam, 1997.

[9] Everdij, M. H. C., and Blom, H. A. P., "Piecewise Deterministic Markov Processes Represented by Dynamically Coloured Petri Nets," National Aerospace Laboratory NLR, Rept. TP-2000-428, Amsterdam, 2000.

[10] Everdij, M. H. C., Bakker, G. J., and Blom, H. A. P., "Application of Collision Risk Tree Analysis to DCIA/CRDA with Support of TOPAZ," National Aerospace Laboratory NLR, Rept. CR 96784, Amsterdam, 1996.

[11] Daams, J., Bakker, G. J., and Blom, H. A. P., "Safety Evaluation of an Initial Free Flight Scenario with TOPAZ (Traffic Organization and Perturbation AnalyZer)," National Aerospace Laboratory NLR, Rept. TR-98098, Amsterdam, 1998.

[12] Daams, J., Bakker, G. J., and Blom, H. A. P., "Safety Evaluation of Encounters Between Free Flight Equipped Aircraft in a Dual Route Structure," National Aerospace Laboratory NLR, Rept. TR-99577, Amsterdam, 1999.

[13] Kos, J., Blom, H. A. P., Speijker, L. J. P., Klompstra, M. B., and Bakker, G. J., "Probabilistic Wake Vortex Induced Accident Risk Assessment," *Proceedings of the 3rd USA/Europe Air Traffic Management R&D Seminar*, FAA/EUROCONTROL, 2000.

[14] Everdij, M. H. C., Bakker, G. J., Blom, H. A. P., and Blanker, P. J. G., "Demonstration Report in Preparation to Designing EATMS Inherently Safe," National Aerospace Laboratory NLR, Rept. TOSCA II WPR/4/01 Part I, Amsterdam, 1997.

[15] Buck, S., Biemans, M. C. M., Hilburn, B. G., and Van Woerkom, P. T. L. M., "Synthesis of Functions," National Aerospace Laboratory NLR, Rept. TR 97054 L, Amsterdam, 1996.

[16] Biemans, M. C. M., and Daams, J., "HOMEROS: Human Operator Modelling to Evaluate Reliability, Organisation and Safety," National Aerospace Laboratory NLR, RHEA WP6 Subtask Rept. NLR Rept. TR-98073, Amsterdam, 1997.

[17] Daams, J., Nijhuis, H. B., and Blom, H. A. P., "Accident Risk Assessment with a Human Cognition Model Using TOPAZ (Traffic Organization and Perturbation AnalyZer)," National Aerospace Laboratory NLR, Rept. LL-99-030, Amsterdam, 1998.

[18] Daams, J., Nijhuis, H. B., and Blom, H. A. P., "Human Operators Controllability of ATM Safety," National Aerospace Laboratory NLR, Rept. ARIBA WP4, NLR Rept. TR-99575, Amsterdam, 1999.

[19] Daams, J., Blom, H. A. P., and Nijhuis, H. B., "Modelling Human Reliability in Air Traffic Management," *PSAM5—Probabilistic Safety Assessment and Management*, edited by S. Kondo and K. Furata, Vol. 2/4, Universal Academy Press, Inc., Tokyo, Japan, 2000, pp. 1193–1200.

[20] Amalberti, R., and Wioland, L., "Human Error in Aviation," *International Aviation Safety Conference*, edited by H. Soekkha, VSP, Utrecht, The Netherlands, 1997, pp. 91–108.

[21] Hollnagel, E., *Human Reliability Analysis, Context and Control*, Academic Press, London, 1993.

[22] Bainbridge, L., "The Change of Concepts Needed to Account for Human Behaviour in Complex Dynamic Tasks," *Proceedings of the 1993 International Conference on Systems, Man and Cybernetics*, Vol. 1, 1993, pp. 126–131.

[23] Isaac, A., and Ruitenberg, B., *Air Traffic Control: Human Performance Factors*, Ashgate, Aldershot, 1999.

[24] Wickens, C. D., *Engineering, Psychology and Human Performance*, Merrill, Columbus, OH, 1992.

[25] "A Designers Guide to Human Performance Modelling," AGARD, Advisory Group for Aerospace Research and Development, Advisory Rept. 356, Neuilly-Sur-Seine, France, Dec. 1998.

[26] Corker, K. M., Pisanich, G., and Bunzo, M., "Human Factors in Advanced ATM System Simulation Studies," *Proceedings of the 1st USA/Europe Air Traffic Management R&D Seminar*, FAA/EUROCONTROL, 1997.

[27] Kilner, A., Hook, M., and Duck, R., "Workload Assessment," NATS, Rept. TOSCA II WPR/8/01 Part I, London, 1997.

[28] Norman, D., and Bobrow, D., "On Data-Limited and Resource Limited Processing," *Journal of Cognitive Psychology*, Vol. 7, 1975, pp. 44–60.

[29] Wickens, C. D., "Attention and Skilled Performance," *Human Skills*, edited by D. Holding, Wiley, New York, 1989, Chap. 4.

[30] Gibson, J. J., *The Ecological Approach to Visual Perception*, Lawrence Erlbaum Associates, Mahway, NJ, 1986.

[31] Kirwan, B., *A Guide to Practical Human Reliability Assessment*, Taylor and Francis, London, 1994.

[32]Rasmussen, J., "Skills, Rules and Knowledge; Signals, Signs and Symbols, and Other Distinctions in Human Performance Models," *IEEE Transactions on System, Man and Cybernetics*, Vol. 13, No. 3, 1983, pp. 257–266.

[33]Reason, J., *Human Error*, Cambridge Univ. Press, Cambridge, MA, 1990.

[34]Swain, A. D., and Guttman, H. E., "A Handbook of Human Reliability Analysis with Emphasis on Nuclear Power Plant Applications," NUREG/CR-1278, U.S. Nuclear Regulatory Commission, 1983.

[35]Embrey, D., Hyphreys, P., Rosa, E., Kirwan, B., and Rea, K., "SLIM-MAUD: An Approach to Assessing Human Error Probabilities Using Structured Expert Judgement," NUREG/CR-3518, U.S. Nuclear Regulatory Commission, 1984.

[36]Williams, J. C., "A Data-Based Method for Assessing and Reducing Human Error to Improve Operational Performance," *Proceedings of the IEEE 4th Conference on Human Factors and Power plants*, Institute of Electrical and Electronics Engineers, Piscataway, NJ, 1988, pp. 436–450.

[37]Jackson, A., "The Role of the Controller in Future ATC Systems with Enhanced Information Processing Capabilities," EEC Rept. No. 224, Bretigny, France, 1989.

[38]Ammerman, H. L., Fairhurst, W. S., Hostler, C. M., and Jones, G. W., "FAA Air Traffic Control Concepts Volume VI: ARTCC/HOST En Route Controllers," Report DOT/FAA/AP/87-01, FAA, Washington, DC, 1987.

[39]Cox, M., "Task Analysis of Selected Operating Positions Within UK Air Traffic Control", Royal Air Force Institute of Aviation Medicine, Rept. No. 749, Farnborough, 1994.

[40]"Model for Task and Job Descriptions of Air Traffic Controllers," EATCHIP, Rept. HUM.ET.ST01.1000-REP-01, EUROCONTROL, Brussels, 1996.

[41]Endsley, M. R., and Rodgers, M. D., "Situation Awareness Information Requirements for En Route Air Traffic Control," Department of Industrial Engineering, Texas University, Lubeck, TX, 1994.

[42]George, P. H., Johnson, A. E., and Hopkin, V. D., "Radar Monitoring of Parallel Tracks, Automatic Warning to Controllers of Track Deviations in a Parallel Track System," EEC, Rept. 67, Bretigny, France, 1973.

[43]"ICAO Annex 11—Air Traffic Services," 12th ed. incorporating amendments 1-38, green pages, attachment B, paragraph 3.2.1., International Civil Aviation Organization, Montreal, July 1998.

Chapter 31

Probabilistic Wake Vortex Induced Accident Risk Assessment

J. Kos,* H. A. P. Blom,† L. J. P. Speijker,‡ M. B. Klompstra,§ and G. J. Bakker §

National Aerospace Laboratory NLR, Amsterdam, The Netherlands

I. Introduction

NEW air traffic management (ATM) concepts for departure and landing on busy airports with multiple runways might have a major impact on capacity if the wake vortex induced risks are better understood. In particular, there is a need for identifying the conditions under which the present wake vortex separation standards would limit capacity too much. In Ref.1, an initial probabilistic methodology has been developed to assess such safe separations in case of a single runway. There are several reasons why there is a need to extend this initial methodology, e.g.:

1) To guide and incorporate ongoing developments in wake vortex induced risk modeling;

2) To generalize its application from a single runway to closely spaced runways;

3) To allow the evaluation of advanced ground and/or airborne procedures that make use of wake vortex detection and decision support systems;

4) To allow the evaluation of safe separation standards for new aerodynamic aircraft designs;

5) To integrate it with a methodology that assesses risk of collision with other aircraft in the air or on the ground.

These reasons provided a clear motivation to extend the accident risk assessment methodology to enable the determination of safe separations for advanced ATM.[2] This extension resulted in a complementary and novel probabilistic methodology for the assessment of wake vortex induced accident and incident risks.

The aim of this chapter is to show how the accident risk assessment methodology can be used to support the assessment of wake vortex induced accident and

---

Copyright © 2001 by NLR. Published by the American Institute of Aeronautics and Astronautics, Inc., with permission.
*Mathematical Modeling Researcher, Information and Communication Technology Division.
†Group Leader, Air Traffic Management Modeling, Air Transport Division.
‡Mathematical Modeling Researcher, Information and Communication Technology Division.
§ATM Modeling Researcher, Air Transport Division.

incident risks, and how advantage is taken from recent progress in wake vortex research, e.g. Refs. 1, 3, and 4. The methodology is also illustrated for an example of a medium-weighted aircraft landing behind a heavy-weighted aircraft on a (single) runway.

The chapter is organized as follows. Section II outlines the risk assessment methodology and the supporting tool sets. In Section III the wake vortex specific models are explained in more detail. Section IV illustrates the application to the specific single runway example. Section V draws conclusions.

II. Risk Assessment Methodology

As the basis for the development of a wake vortex risk induced assessment approach, use is made of a stochastic analysis and modeling approach based accident risk assessment methodology for advanced ATM operations.[2] This methodology supports the spiral development cycle that is part of modern safety case building for new ATM operational concepts.[5,6] Such a cycle is typically of the form:

A) Design of an ATM operational concept.

B) Assessment of the ATM concept, resulting in a cost-benefit overview.

C) Detailed analysis of the assessment results, which results in recommendations for improvements of the ATM concept.

D) Review of ATM concept development strategy and plan.

E) Back to A: adapted and/or more detailed ATM concept design using the results from C resulting in a new or optimized ATM concept.

This stochastic analysis and modeling based accident risk assessment methodology has been developed to provide designers of advanced ATM with safety feedback following a (re)design cycle and is referred to as the Traffic Organization and Perturbation Analyzer (TOPAZ), see Fig. 1.

During the assessment cycle four stages are sequentially conducted:

1) Stage 1: Identification of operation and hazards (upper box in Fig. 1). Information about nominal and non-nominal behavior of the ATM concept or procedure is gathered through hazard identification sessions with a variety of experts.

2) Stage 2: Mathematical modeling (lower right box in Fig. 1). A stochastic dynamical model of the operation is developed that incorporates all of the nominal and all of the non-nominal events of the operation. During this stage, all model assumptions made are systematically specified.

3) Stage 3: Accident risk assessment (middle box in Fig. 1). The mathematical model of stage 2 supports an effective procedure, consisting of a number of steps to be followed, to quantify the accident risk. In addition to such a numerical approach, a qualitative analysis of the model assumptions is performed.

4) Stage 4: Feedback to operational experts (lower left box in Fig. 1). The results of the quantitative safety assessment are given back to and discussed with the designers and operational experts, who can use the results to redesign or optimize their proposed ATM design if necessary.

For the second and third stages use can be made of the following TOPAZ tools:

1) SIMULATOR is a tool set used to specify and implement the mathematical model and to subsequently run Monte Carlo simulations with that implementation. SIMULATOR can simulate all aspects of operations, including the stochastic non-nominal aspects.

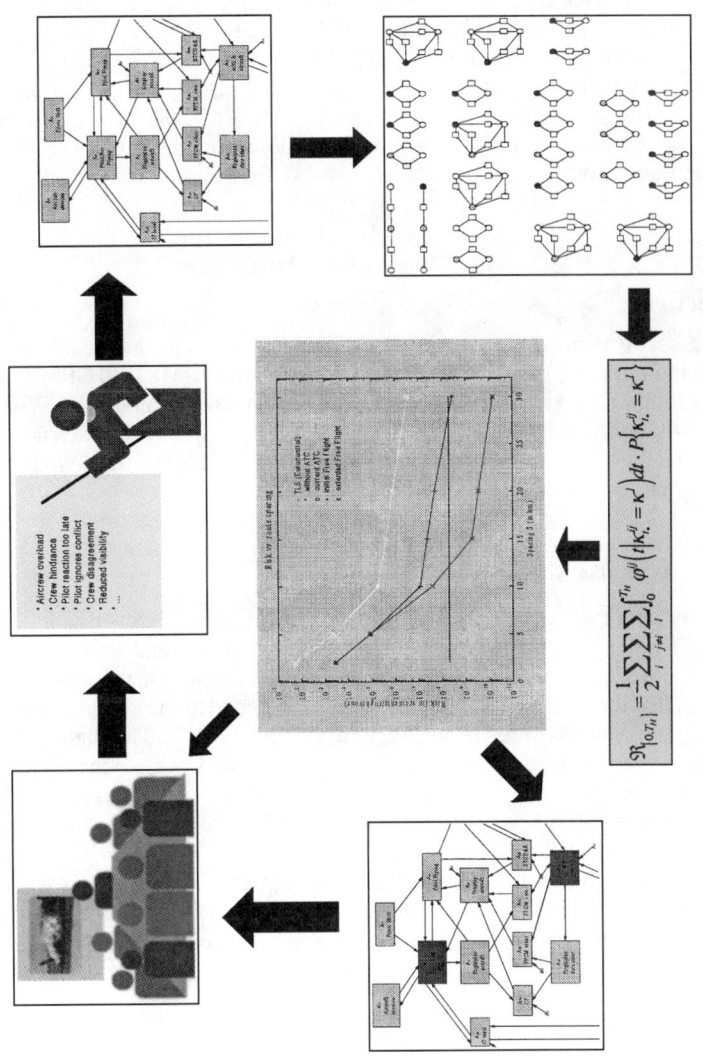

Fig. 1 TOPAZ accident risk assesment cycle.

2) COLLIR is a methodology and tool set that supports the evaluation of collision risks in the terminal manoeuvring area (TMA) and en route.

3) WAVIR is a methodology and tool set that supports the evaluation of wake vortex induced accident risk.

4) TAXIR is a methodology and tool set that supports the evaluation of accident risks at the airport.

5) CRITER is a risk criteria framework that is needed to judge the acceptability of the risks that are assessed by COLLIR, WAVIR, and TAXIR.

The methodological parts of COLLIR, WAVIR, and TAXIR incorporate the evaluation of statistical data that are obtained either through empirical data collections or through Monte Carlo simulations (e.g., SIMULATOR). One should be aware that for each of the tools further extensions are ongoing at the National Aerospace Laboratory NLR.

III. Wake Vortex Risk Assessment

A. Overview

For the assessment of the wake vortex induced accident and incident risks, the following tools are used: 1) flight path evolution (in SIMULATOR), 2) wake vortex evolution and decay (in WAVIR), 3) wake encounter model (in WAVIR), 4) integration and risk evaluation model (in WAVIR), and 5) risk criteria framework (in CRITER). These tools are described in Sections III.B–III.F together with references that give more details about the models used.

B. Flight Path Evolution

The flight path evolution model yields the following stochastic variables: 1) the lateral and vertical coordinates of the leader if its longitudinal co-ordinate x is given, 2) the period of time elapsed between the generation of the wake and the time instant that the trailer has longitudinal position x, and 3) the lateral and vertical coordinates of the trailer when it has longitudinal coordinate x. The flight path evolution model is a stochastic dynamical model, which incorporates the established International Civil Aviation Organization-Collision Risk Model (ICAO-CRM)[7] as baseline, and which has been further developed to handle the dependent usage of closely spaced runways.[8]

The flight path evolution model is represented in a form[9] that allows a straightforward extension of the SIMULATOR tool set for new air and/or ground procedures and advanced vortex detection and decision-support systems.

C. Wake Vortex Evolution and Decay

The wake vortex evolution model yields the position and strength stochastic variables of the wake vortex at any time instant after the generation of the wake vortex. The wake vortex evolution model is mainly based on Refs. 3 and 4. The models in the latter references have been probabilitized, and they have been completed for application to the wake vortex induced risk assessment.

This wake vortex evolution model is able to take into account probabilistic models for stratification, atmospheric turbulence, ground effects (rebound, divergence),

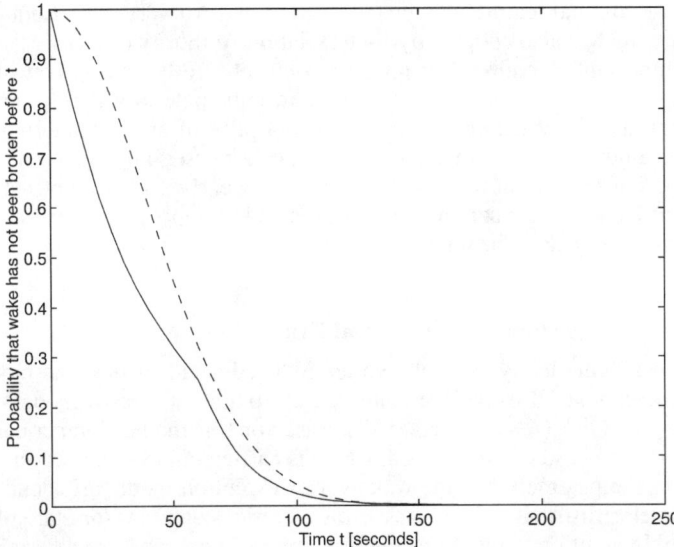

Fig. 2 Observed vortex residence time distribution for B-747 vortices with initial height 30 m (solid line), initial strength of the wake of 600 m^2/s and wingspan 60 m (from Fig. 2 in Ref. 10); Rayleigh density adopted in WAVIR for the vortex bursting or linking period (dashed line).

and crosswind (advection, shear). It can also handle probabilistic models for the vertical and horizontal wind fields and their impact on wake evolution.

During many landings, the trailer aircraft does not meet any wake vortex. In the wake vortex evolution model, two possible causes are distinguished: 1) the wake vortex has disappeared due to a gradual diminishing of its strength, and 2) the wake vortex has disappeared due to a sudden bursting or linking. In Ref. 3, for the latter an analytical model has been proposed that assumes bursting and linking to happen in time as a function of some meteorological parameters. To better account for observed data, in WAVIR the probabilistic bursting and linking period is modeled independently of the vortex evolution and decay as a stochastic variable with a Rayleigh density, the mean of which is assumed equal to 50 s. This Rayleigh density is depicted in Fig. 2 together with empirical data for vortex residence period. This Rayleigh density modeling also differs significantly from the theoretical probability density model of Kuzmin.[1]

D. Wake Vortex Encounter Model

The wake vortex encounter model yields the probability that the wake vortex induced rolling moment is larger than the maximum control capability—in terms of rolling control moment—of the encountering aircraft.

The wake vortex control capability model is based on Refs. 1 and 11 and accounts for the fact that the encountering aircraft tries to compensate for the wake vortex flow field generated by the leader only. In line with this, the key effect is to reduce the rolling moment calculated with the wake vortex evolution model. The aircraft control capability model is also based on Ref. 1 and assumes a value twice as high as the minimal requirement by the British Civil Airworthiness Requirements (BCAR), which is also adopted by Joint Aviation Authority (JAA).[12]

It is important to realize that this approach inherently involves an important modeling assumption: the pilot is *not* able to anticipate in a timely manner on the first signs of a wake vortex. In practice, a pilot of an encountering aircraft might respond with the immediate initiation of a missed approach when its aircraft experiences a slight roll upset, i.e., in a very early stage of a possible wake encounter. Hence, this Kuzmin model is expected to imply a pessimistic effect on the quantified risk near the threshold.

E. Model Integration and Numerical Evaluation

To assess numerically the wake vortex induced accident risk, the models described in Sections III.B–III.D are integrated through a stochastic model for wake vortex induced risk (see the Section VI). Based on this model a numerical assessment procedure has been developed, which is carried out in seven steps:

Step 1: The parameters in the wake vortex evolution model are identified and the parameter distributions are based on empirical data and/or state-of-the-art literature. In addition, a set of relevant longitudinal positions x is determined.

Step 2: Run Monte Carlo simulations with the wake vortex evolution model for the case in which the wake vortex is generated when the leading aircraft has longitudinal position x. The position, strength, and core radius of the wake vortex are obtained at the time instant that it has the same longitudinal coordinate as the trailer aircraft. The latter time instant follows from Monte Carlo simulations with the SIMULATOR tool.

Step 3: The simulation results from Step 2 are analyzed. Based on this analysis, a dedicated probability density fitting procedure is identified that accounts for dependencies between the position coordinates, the strength, and the core radius of the wake vortex. The probability density fitting procedure is carried out and the joint distribution of the wake vortex position, strength, and core radius is obtained.

Step 4: Monte Carlo simulations are carried out to simulate the wake vortex encounter. In this step the joint distribution from Step 3 is used and distributions of the position of the trailer aircraft obtained with the SIMULATOR tool set are used.

Step 5: Step 5 concerns the numerical evaluation of the wake induced accident risk due to a wake vortex that is generated when the leading aircraft was at position x.

Step 6: The wake induced accident risk is obtained by maximizing over x the risk obtained in Step 5.

Step 7: Perform a qualitative evaluation of the influence of the modeling assumptions on the estimated accident risk.

The results for these seven steps are illustrated for the case of a single runway example in Section IV.

F. Risk Criteria Framework

To judge whether a newly proposed ATM concept is safe or to determine more appropriate safe separation distances, a suitable metric for quantification of wake vortex induced risk is required. Up to now several metrics have been used to

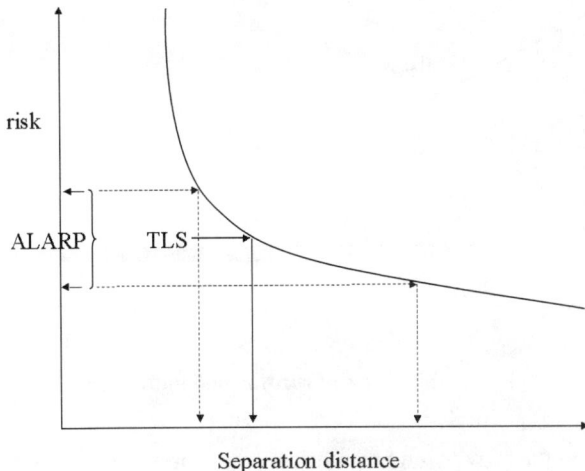

Fig. 3 Wake vortex risk vs separation distance.

quantify the risk imposed by wake vortices, e.g., bank angle, roll angle, roll rate, and roll control ratio. However, because they do not relate to the safety perception of involved interest groups (e.g., crew, passengers, controllers, regulators, people living in the airport vicinity), they are felt to be insufficient. Other possible risk metrics are the risk probability per movement and the risk probability per year.

In Ref. 13 some initial guidelines are developed for the assessment of safety requirements. It discusses two possible safety management approaches: the as-low-as-reasonably-practicable (ALARP) approach and the target level of safety (TLS) approach. Ranges are suggested from which to adopt a TLS for the risk event *probabilities per movement*.

For the adoption of applicable risk criteria, it is clear that policy makers definitively have to be involved, and also the relation with existing wake vortex induced incident and accident frequencies should be clearly identified.

G. From Risk to Safe Separation

By assessing accident risks for various separation distances, one arrives at a curve that shows the risk as a function of the separation distance between successive aircraft. Figure 3 illustrates how such a curve subsequently maps an ALARP region in terms of risk into one in terms of separation distances.

IV. Single Runway Approach
A. Boeing 737-400 Behind Boeing 747-400

To illustrate the novel wake vortex induced accident risk assessment methodology, we consider a (single) runway on which a Boeing 737-400 aircraft, which is in the ICAO medium-weight class, is landing behind a Boeing 747-400 aircraft, which is in the ICAO heavy-weight class, with controller expected separation distance of

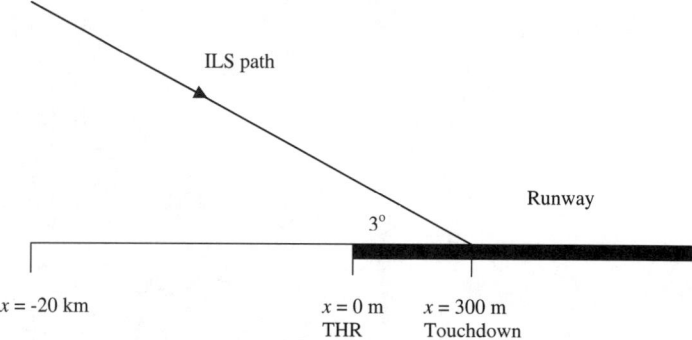

Fig. 4 Side view of runway and glide slope.

5 n miles when the heavy-weight aircraft is at the threshold. For both aircraft, it is assumed that the approach is instrument landing system (ILS) Category I.

The landing phase starts at about 20 km before the threshold, and ends at touchdown, which is 300 m beyond threshold. Figure 4 shows the side view of the runway and glide slope, where the x-axis is along the runway centerline and positive in runway direction.

Because of its stochastic dynamical modeling basis, the novel wake induced risk assessment methodology clearly brings the assumptions made to the foreground. For the example considered, the following main assumptions have been adopted:

1) Long landings (landings far beyond threshold) do not happen.

2) A wake vortex induced accident event is characterized by the wake induced rolling moment being larger than the aircraft control capability.

3) A pilot does not respond with the initiation of a missed approach when its aircraft experiences a slight roll upset.

4) Bursting and linking probabilities are modeled by a Rayleigh density with mean 50 s.

5) There is no head wind, no tail wind, and no vertical wind. The wind speed in lateral direction is normally distributed with expectation 0 and standard deviation 1.5 m/s.

6) There are no wind shear layers.

7) Turbulence of the air is 10% of the wind speed.

In addition to these main assumptions, several other assumptions have been made. It would go beyond the scope of this chapter to list all of these assumptions.

B. Numerical Results

With support of the SIMULATOR and WAVIR tool sets, the wake vortex induced accident risk is evaluated for the single runway scenario. Figures 5 and 6 show data plots of the left vortex for the case in which the wake vortex is generated at 4 km before threshold (cf. output of Step 3 in Section III.E).

Subsequently, Fig. 7 shows the results for the wake induced accident risk resulting from a wake that is generated at $-x$ km before the threshold. The vertical axis has a logarithmic scale. The $+$ signs indicate the values of x (cf. output of Step 5 in Section III.E).

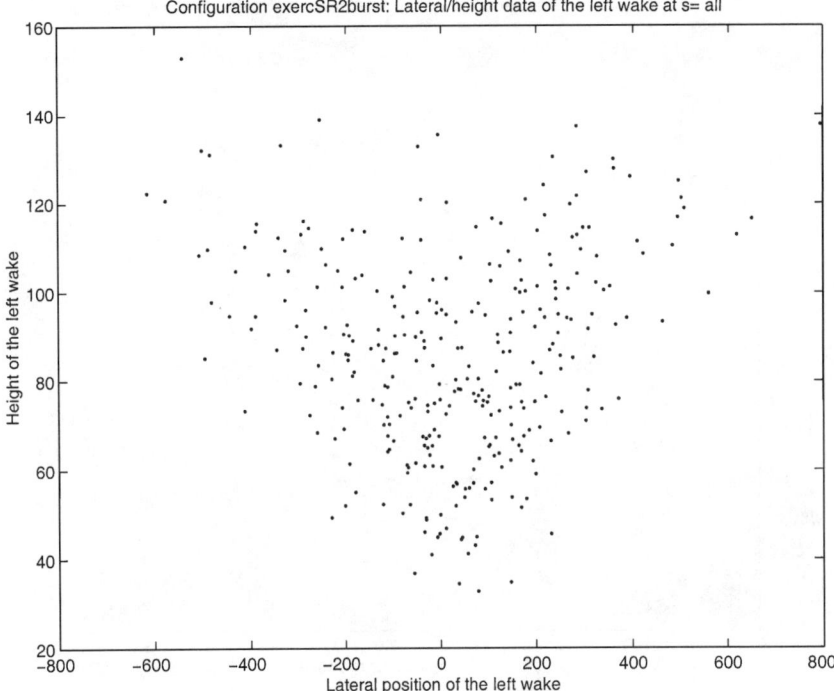

Fig. 5 Monte Carlo simulation results of the lateral and vertical coordinate (m) of the center of the left wake that is generated at 4 km before the threshold.

Figure 7 shows that the estimated values for the accident risk that is instantaneously induced by wake vortices along the glide slope decrease from 20 km until approximately 4 km before the threshold. The decrease is due to the descent of the wake and the higher navigation precision (in height) of the trailer. At shorter distances from the threshold, the instantaneous risk increases due to the rebound of wakes near the ground.

C. Qualitative Uncertainty Analysis

A straightforward maximization over x for the curve in Fig. 7 would lead to an overall maximum risk at the threshold. However, prior to the maximization, one should take into account that the calculated wake vortex induced accident risk curve may bear significant bias and/or uncertainty both in positive and negative directions. Usage of such a curve without taking into consideration existing bias and/or uncertainty can inspire undue conclusions.

To understand the impact of the assumptions on the wake vortex induced risk, assumptions 1–7 have been analyzed in a qualitative way. The results are given in Table 1. The first column refers to the assumptions. The second column gives for each assumption the expected direction of the effect on the wake vortex induced risks (optimistic, pessimistic, or neutral), and the last column gives the expected magnitude (major or significant). A pessimistic expected

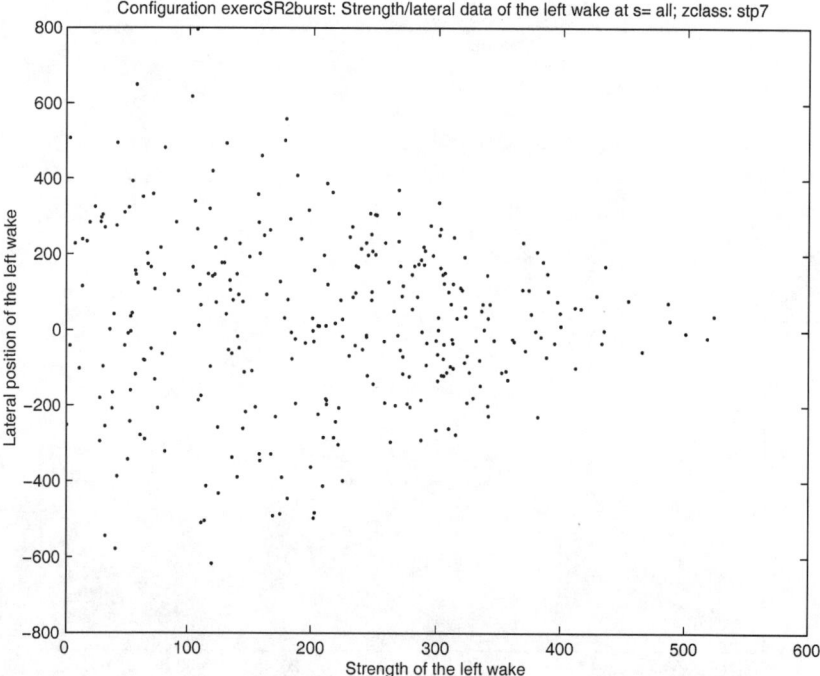

Fig. 6 Monte Carlo simulation results of the lateral coordinate (m) and the strength (m^2/s) of the center of the left wake that is generated at 4 km before the threshold.

direction means that the modeled risk increases due to the assumption. An optimistic expected direction means that the modeled risk reduces due to the assumption. A neutral direction means that there exists uncertainty about the direction.

This qualitative analysis has also been applied to all other assumptions. Because their effect on the wake induced accident risk has been estimated as being either minor or negligible, these assumptions are not listed in this chapter.

Table 1 Expected effects of the main assumptions on assessed risk

Assumption	Expected direction of effect on wake vortex induced accident risk	Expected magnitude
1	Optimistic	Significant
2	Neutral	Significant
3	Pessimistic	Major
4	Neutral	Significant
5	Neutral	Significant
6	Optimistic	Major
7	Pessimistic	Significant

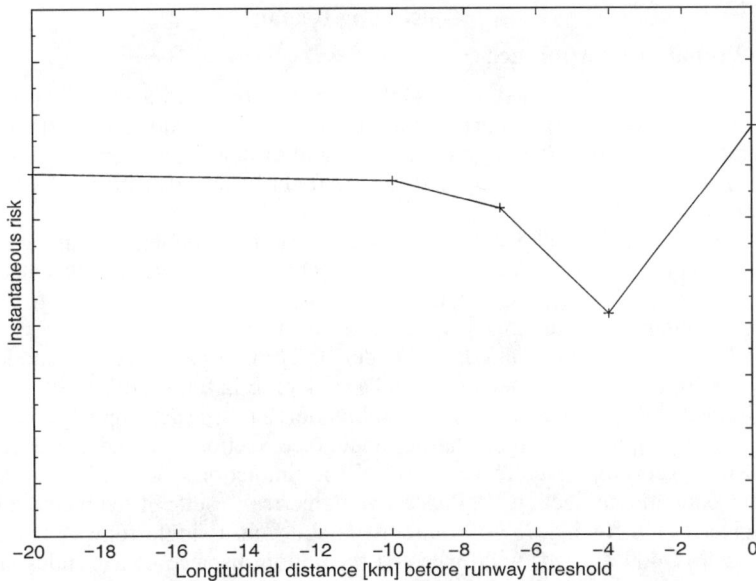

Fig. 7 Estimated values for the severe risk that is instantaneously induced by wake vortex along the glide slope. The vertical axis has a logarithmic scale.

D. Discussion of Results

If one takes into account the impact of assumptions 1–7 then the curve in Fig. 7 shows that there are two distinct areas where the instantaneous wake vortex induced accident risks along the ILS are not negligible: 1) near the threshold, this is due to the ground effect on the wake evolution, and 2) at distances larger than 10 km from the threshold, due to larger ILS navigation errors at further distances from the threshold.

Near the threshold, the effect of assumption 6 is negligible. Assumption 3 is the only assumption that has a major impact (pessimistic). The net effect of all assumptions is that the very right part of the curve in Fig. 7 has a major level of uncertainty with a clear bias in the pessimistic direction.

At distances larger than 10 km from the threshold, the effect of assumptions 1 and 3 is negligible. Assumption 6 is the only assumption that has a major impact (optimistic). The net effect of all assumptions is that the left part of the curve in Fig. 7 has a major level of uncertainty with a clear bias in the optimistic direction.

The example shows that there are a few directions that specifically deserve the development of improved wake vortex induced risk models. These directions can be placed in the following two groups:

1) General modeling is needed to mitigate the need for the neutral and pessimistic assumptions 2, 3, 4, and 7.

2) Airport specific modeling is needed to mitigate the need for the neutral and optimistic assumptions 1, 5, and 6.

V. Concluding Remarks

A. Probabilistic Approach

This chapter has outlined a probabilistic methodology to assess wake vortex induced accident risks. The aim is to understand the safety evaluation of established separation standards for current operations, and of new separation standards for new operational concepts and aircraft designs for busy airports with closely spaced runways.

The probabilistic methodology, the tool sets, and the models initially adopted have been outlined in Sections II and III, respectively. The novelty of this methodology is the probabilistic integration of the models that are available from the complementary domains that play a key role in wake vortex risk assessment: 1) wake vortex evolution and decay model[3,4]; 2) flight path evolution model[7,8]; 3) wake vortex encounter model[1,11]; and 4) risk criteria framework.[13]

The probabilistic integration of these submodels has been accomplished through first developing an integral stochastic model (see Section VI), and subsequently using this to develop a hierarchical Monte Carlo simulation scheme (Section III.E).

Subsequently, in Section IV, illustrative numerical results of the methodology are given for a Boeing 737-400 aircraft landing on a single runway behind a Boeing 747-400 aircraft, with expected separation distance of 5 n miles at the threshold. The numerical and complementary qualitative results obtained for this example clearly show that the methodology is able to identify the key bottlenecks in developing advanced wake vortex procedures. In the current situation, the safety of the established operations is insufficiently understood in a few key areas. Most of these areas ask for general modeling effort. A few areas only ask for airport specific modeling effort.

B. General Modeling Areas

There are four key general modeling areas that deserve significant modeling effort. A relatively simple one is to improve the probabilistic modeling of navigation performance and long landing models under different navigation modes and various wind conditions. The basis for this activity is one of collecting statistical data of aircraft navigation performance.

The second area is to improve the modeling of autopilot reactions to wake induced roll upset, and missed approaches initiated by pilots as a reaction to experiencing a roll upset during the ILS approach. The basis for this activity seems to be one of analyzing pilot incident reports on missed approaches. In addition, flight simulation data should be used to develop models that represent the pilot behavior during wake vortex encounter.

The third area is to improve the modeling of bursting and linking phenomena. Because of the existing uncertainty about bursting and linking, and their dependency on weather condition, it is strongly recommended to model bursting and linking with appropriate probability distributions. The distributions are to be validated with real experiments or state-of-the-art computational fluid dynamic models for wake vortex evolution.

The fourth area is the further development of a risk management framework for wake vortex induced accident and incident risks so that it becomes clearly connected to existing wake vortex incident and accident data. This modeling effort can only be concluded in discussions with regulatory authorities and other relevant interest groups (e.g., pilots and controllers). In the current European constellation, this process still is very much airport specific.

C. Airport Specific Modeling Areas

The key airport specific modeling area that deserves significant modeling effort is weather. It is important to realize the major influence of specific weather conditions, in particular wind fields, turbulence, stable stratification, and wind shear.[14] These weather conditions are so airport specific that the developed models have to be tuned for the airport under consideration.

Another key airport specific modeling issue is that each particular airport runway geometry may lead to all kinds of dependencies between runway usage. The particular geometry of an airport layout often leads to airport specific dependencies that involve combinations of wake vortex induced accident risks and risk of collision with another aircraft.

It should also be taken into account that, due to these airport specific modeling needs, different appropriate and safe separation distances might result for different airports.

D. Ongoing Developments

The results obtained so far form a clear motivation to continue the development of the TOPAZ/WAVIR methodology toward the assessment of wake vortex induced accident and incident risks. The results obtained with the methodology in its current state already provide clear overall insight into the large variety of wake vortex subproblems.

Apart from the single runway example illustrated in this chapter, a closely spaced runway example has also been evaluated with the TOPAZ/WAVIR methodology (this was under a study contract with the DFS). This study also provided valuable overall insight into the wake vortex subproblems in case of parallel flying aircraft.

Since January 2000, the National Aerospace Laboratory NLR is leading a major three-year project [named Assessment of Wake Vortex Safety (S-WAKE)] for the European Commission in which key European wake vortex experts are collaborating to develop solutions for the outstanding modeling areas.

VI. Appendix: Stochastic Wake Vortex Model

A. Introduction

This appendix presents the stochastic model that has been developed to allow the integration of the wake vortex submodels from the various domains. We assume a situation of a sequence of aircraft that fly on parallel tracks toward multiple closely spaced parallel runways. For the position and velocity components of aircraft i, this means there is a well-defined stochastic process $\{x_t^i, y_t^i, z_t^i, \dot{x}_t^i, \dot{y}_t^i, \dot{z}_t^i\}$ for the three-dimensional position and speed components of aircraft i's state. In addition, this means that there is a well-defined stochastic process $\{w_{x,t}^i, w_{y,t}^i, w_{z,t}^i\}$ for the wind speed components acting locally on aircraft i.

For the evaluation of the wake vortex induced risk, it is necessary to present an appropriate stochastic dynamical model characterization of the wake vortex induced incident and accident risks. This is done as follows. In subsection B, we present a causal stochastic model for the wake vortex random field. Subsequently, in subsection C, we extend this with a stochastic model for the wake vortex induced roll moment for the following aircraft, together with a model for the compensation of this roll moment. Next, in subsection D, we use our stochastic dynamical model to present a model for the wake vortex induced risk.

B. Wake Vortex Random Field Model

The left and right centers of the vortex at moment s that are generated by aircraft j at moment t are represented by the two parameter random fields $\delta^{j-}(t,s)$ and $\delta^{j+}(t,s)$, with $s \geq 0$, each of which assumes (y, z) values in $\mathbb{R}^2$. At moment $t + s$, the strengths of the left and right vortices that are generated by aircraft j at moment t are represented by the two parameter random fields $\gamma^{j-}(t,s)$ and $\gamma^{j+}(t,s)$, each of which assumes strength values in $\mathbb{R}$.

To shorten our notation in the sequel, we place these components into a joint $\mathbb{R}^6$-valued random field $v^j(t,s)$:

$$v^j(t,s) \triangleq \text{column}\{\delta^{j-}(t,s), \delta^{j+}(t,s), \gamma^{j-}(t,s), \gamma^{j+}(t,s)\}$$

Research is ongoing to develop differential equations for the motion and decay of the components of the joint random field $v^j(t,s)$. Widely known equations in current literature are the ones given by Ref. 3, which are largely based on those of Refs. 15 and 16. When adding a straightforward extension for non-zero wind velocity in z direction, these equations are of the form:

$$\frac{dv^j(t,s)}{ds} = f[v^j(t,s), \kappa^j(s)]$$

with $\kappa_t^j(s)$ denoting local external influences such as the local crosswind $w_{y,t}^j$ and the local vertical wind $w_{z,t}^j$ at moment t, the drag coefficient C_D^j of j's "wake oval," the local Brünt-Väisälä frequency $N^j(t+s)$, and the RMS velocity of atmospheric turbulence $q^j(t+s)$.

To uniquely define the solution of the latter differential equation for $s \geq 0$, we have to characterize the components of $v^j(t, 0)$. From Refs. 5 and 17, we know

$$\delta_y^{j\pm}(t,0) = y_t^j \pm \tfrac{1}{2} b_0^j$$

$$\delta_z^{j\pm}(t,0) = z_t^j$$

$$\gamma^{j\pm}(t,0) = \pm \Gamma_{0,t}^j \equiv \pm \frac{m^j g}{b_0^j \rho_t^j} (\dot{x}_t^j - w_{x,t}^j)^{-1}$$

with b_0^j the initial spacing between the primary vortex centers, m^j the mass of aircraft j, g the gravitational acceleration, and ρ_t^j the local air density.

Next we have to characterize the moment in time that a wake generated at longitudinal position x by aircraft j will arrive at the longitudinal position of aircraft i. To do so we assume that the velocity in longitudinal direction of the wind acting on the wake is constant. Then that moment in time is a stopping time τ_x^{ij}, which is defined by:

$$\tau_x^{ij} = \tau_x^j + \inf_s \{s > 0 \mid x^i_{\tau_x^j + s} = x + sw^j_{x,\tau_x^j}\}$$

with

$$\tau_x^j \equiv \inf_t \{t \mid x_t^j = x\}$$

By the very nature of the situation we consider, the airspeeds of both aircraft in x direction are bounded and either both strictly positive or both strictly negative. In view of this, in the latter equation this means that $\{\tau_x^{ij}\}$ is a strictly increasing continuous process. Hence, we can define an IR^6-valued stochastic process $\{v_t^{ij}\}$ to represent the actual contribution to aircraft i of the wake vortex generated by aircraft j, as follows:

$$v_{\tau_x^{ij}}^{ij} \equiv v^j\left(\tau_x^j, \tau_x^{ij} - \tau_x^j\right)$$

Our next step is to characterize how the latter stochastic process $\{v_t^{ij}\}$ induces a rolling moment process $\{\mu_t^{ij}\}$ for aircraft i.

C. Induced Roll Moment and Compensation Model

For the characterization of how the vortex stochastic process $\{v_t^{ij}\}$ induces a rolling moment process $\{\mu_t^{ij}\}$ for aircraft i, we adopt the approximate model developed by Ref.1 for a rectangular wing. Then the non-dimensionalized induced rolling moment satisfies:

$$\mu_{\tau_x^{ij}}^{ij} = \frac{C_\alpha^i}{2\pi b^i} \max_\pm \left\{\gamma_{\tau_x^{ij}}^{ij\pm} f\left(\delta_{\tau_x^{ij}}^{ij\pm}, y_{\tau_x^{ij}}^i, z_{\tau_x^{ij}}^i, b^i, r_{core,\tau_x^{ij}}^j\right)\right\} \times \left(\dot{x}_t^j - w_{x,t}^j\right)^{-1}$$

with C_α^i the lift curve slope of aircraft i (assuming values between 1.4 and 1.6), b^i is i's wing span, $r_{core,\tau_x^{ij}}^j$ is the radius of the vortex core generated by j at longitudinal position x and arriving at moment τ_x^{ij} at the longitudinal position of aircraft i, while f satisfies:

$$f(\delta_y, \delta_z, y, z, b, r,) \equiv 1 - f'\left(\sqrt{\frac{(\delta_y - y)^2}{b}}, \sqrt{r^2 + \frac{(\delta_z - z)^2}{b}}\right)$$

$$f'(y, z) \equiv z \arctan\left(\frac{\frac{1}{2} - y}{z}\right) + z \arctan\left(\frac{\frac{1}{2} + y}{z}\right) - \frac{y}{2} \ln\left\{\frac{z^2 + \left(\frac{1}{2} - y\right)^2}{z^2 + \left(\frac{1}{2} + y\right)^2}\right\}$$

Based on the Corjon and Poinsot models, we adopt the following characterization for the vortex core radius:

$$r_{core,\tau_x^{ij}}^j = \max\left\{r_{core,0}^j, \frac{1}{80}\sqrt{\Gamma_{0,\tau_x^j}^j\left(\tau_x^{ij} - \tau_x^j\right)}\right\}$$

with $r_{core,0}^j$ the initial radius of aircraft j's vortex cores.

We get

$$\mu^{ij}_{\tau^{ij}_x} = \frac{C^i_\alpha}{2\pi b^i} \max_{\pm} \left\{ \gamma^{ij\pm}_{\tau^{ij}_x} f\left(\delta^{ij\pm}_{\tau^{ij}_x}, y^i_{\tau^{ij}_x}, z^i_{\tau^{ij}_x}, b^i, r^j_{core,\tau^{ij}_x}\right) \right\} \times \left(\dot{x}^j_{\tau^{ij}_x} - w^j_{x,\tau^{ij}_x}\right)^{-1}$$

Kuzmin [Ref. 1, Eq. (3.3)] also gives a characterization of the maximum non-dimensionalized rolling moment μ_{max,τ^{ij}_x} for which it is possible to compensate by aircraft i

$$\mu_{max,\tau^{ij}_x} = \frac{1}{12} b^i C^i_\alpha \, p_{max} \left(\dot{x}^j_{\tau^{ij}_x} - w^j_{x,\tau^{ij}_x}\right)^{-1}$$

with the maximum steady roll rate p_{max} assumed to be twice as large as the roll rate p_{REQ} minimal required by the British Civil Airworthiness Requirements:

$$p_{max} = 2 p_{REQ}$$

with

$$p_{REQ} = \frac{\phi_{BCAR}}{T_{BCAR} - 1}$$

where ϕ_{BCAR} and T_{BCAR} are specified by the BCAR and by JAA[12] to satisfy $\pi/3$ and 7 s, respectively (i.e., a roll over 60 deg from 30 deg is required to be possible within 7 s). Thus $p_{REQ} \approx 0.175$ s^{-1}.

As long as the vortex induced rolling moment and downwash can be compensated, there will be no reason for a vortex induced accident. Thus it is reasonable to assume that there is no triggering of a rolling induced accident as long as

$$\left|\mu^{ij}_{\tau^{ij}_x}\right| \leq \mu_{max,\tau^{ij}_x}$$

Through substitution of our characterizations for the right and left terms in this inequality, and subsequent evaluation, we get

$$\max_{\pm} \left\{ \left|\gamma^{ij\pm}_{\tau^{ij}_x}\right| f\left(\delta^{ij\pm}_{\tau^{ij}_x}, y^i_{\tau^{ij}_x}, z^i_{\tau^{ij}_x}, b^i, r^j_{core,\tau^{ij}_x}\right) \right\} \leq \frac{\pi}{3} p_{REQ} (b^i)^2$$

Thus our test on the possibility to compensate the wake vortex induced rolling moment at a time-moment simplifies to

$$\xi^{ij}_{\tau^{ij}_x} \leq 1$$

with

$$\xi^{ij}_{\tau^{ij}_x} \equiv \left[\frac{\pi}{3} p_{REQ}(b^i)^2\right]^{-1} \max_{\pm} \left\{ \left|\gamma^{ij\pm}_{\tau^{ij}_x}\right| f\left(\delta^{ij\pm}_{\tau^{ij}_x}, y^i_{\tau^{ij}_x}, z^i_{\tau^{ij}_x}, b^i, r^j_{core,\tau^{ij}_x}\right) \right\}$$

A similar expression for compensation of downwash is not available in this initial model.

D. Induced Risk Model

The risk measures to be characterized are:
1) Probability p^{ij}_I of an incident for i, induced by j's wake.
2) Probability p^{ij}_A of an accident for i, induced by j's wake.
3) Probability p^{ij}_C of a crash of i into terrain, induced by j's wake.

We start the characterization for *accident risk*. As long as ξ_t^{ij} assumes values on the interval $[-1,1]$, then at moment t the vortex induced rolling moment can be compensated, and there will be no reason for a vortex induced accident. Thus

$$p_A^{ij} \equiv \Pr\{\exists t \mid |\xi_t^{ij}| > 1\}$$

In terms of probability at moment t we define

$$p_A^{ij}(t) \equiv \Pr\{|\xi_t^{ij}| > 1\}$$

Evaluation yields

$$p_A^{ij}(t) = 1 - \int_{-1}^{1} p_{\xi_t^{ij}}(\xi) \, d\xi = \int_0^\infty \Pr\{|\xi_t^{ij}| > 1 \mid z_t^i < z'\} p_{z_t^i}(z') \, dz'$$

where $p_{\xi_t^{ij}}(\xi)$ denotes the density of ξ_t^{ij}.

This means we have the following upper bound characterization for p_A^{ij}

$$p_A^{ij} \leq \hat{p}_A^{ij} \equiv \max_t p_A^{ij}(t)$$

Because aircraft i and j are flying in the same direction, both $\{\xi_t^{ij}\}$ and $p_A^{ij}(t)$ have local peaks that are rather flat. This implies that our upper bound characterization also is a rather accurate approximation, thus:

$$p_A^{ij} \approx \hat{p}_A^{ij}$$

Our subsequent characterization for *incidents* follows from a similar approach. As long as ξ_t^{ij} assumes values on the interval $[-\varepsilon, \varepsilon]$, then at moment t the vortex induced rolling moment will not be noticed as being largely uncomfortably for a certain value of ε, and there will be no reason to speak of a vortex induced incident. Thus

$$p_I^{ij} \equiv \Pr\{\exists t \mid |\xi_t^{ij}| > \varepsilon\}$$

In terms of probability at moment t and its upper bound, this yields

$$p_I^{ij}(t) = 1 - \int_{-\varepsilon}^{\varepsilon} p_{\xi_t^{ij}}(\xi) \, d\xi = \int_0^\infty \Pr\{|\xi_t^{ij}| > \varepsilon \mid z_t^i = z'\} p_{z_t^i}(z') \, dz'$$

$$p_I^{ij} \leq \hat{p}_I^{ij} \equiv \max_t p_I^{ij}(t)$$

Again, because aircraft i and j are flying in the same direction, both $\{\xi_t^{ij}\}$ and $p_I^{ij}(t)$ have local peaks that are rather flat. This implies that our upper bound characterization also is a rather accurate approximation, thus

$$p_I^{ij} \approx \hat{p}_I^{ij}$$

Finally, our characterization for *crashes* becomes

$$p_C^{ij} \equiv \Pr\{\exists t \mid |\xi_t^{ij}| > 1, z_t^i < z_{\min}^i\}$$

where $z^i_{\min}$ is a random variable, assuming a Weibull probability density function

$$p_{z^i_{\min}}(z) = \kappa z^{-1} \left(\frac{z}{z^i_{\text{nom}}}\right)^\kappa \exp\left\{-\left(\frac{z}{z^i_{\text{nom}}}\right)^\kappa\right\}$$

with parameters $\kappa \geq 1$ and $z^i_{\text{nom}} > 0$.

Next, similar as before, we define

$$p^{ij}_C(t) \equiv \Pr\{|\xi^{ij}_t| > 1, z^i_t < z^i_{\min}\}$$

by which we get

$$p^{ij}_C \leq \hat{p}^{ij}_C \equiv \max_t p^{ij}_C(t)$$

with $z^i_{\min}$ the minimal height above ground level at which an escape from a crash is possible for an aircraft of type η_i. Straightforward evaluation yields

$$p^{ij}_C(t) = \int_0^\infty \int_0^z \Pr\{|\xi^{ij}_t| > 1, z^i_t < z'\} p_{z^i_t}(z')\,dz'\, p_{z^i_{\min}}(z)\,dz$$

It should be noted that from an incident point of view, the critical moment in time is:

$$\hat{t}^{ij}_I = \arg\max_t p^{ij}_I(t)$$

From an accident point of view, the critical moment in time is:

$$\hat{t}^{ij}_A = \arg\max_t p^{ij}_A(t)$$

From a crash-into-terrain point of view, the critical moment in time is:

$$\hat{t}^{ij}_C = \arg\max_t p^{ij}_C(t)$$

In general, one could expect significant differences between realizations of these three moments in time. In this chapter, simulation results for the accident risk characterization have been given.

References

[1] Kuzmin, V. P., "Estimation of Wake Vortex Separation Distances for Approaching Aircraft," *Trudy TsAGI*, Vol. 2627, 1997, pp. 209–224.

[2] Blom, H. A. P., Bakker, G. J., Blanker, P. J. G., Daams, J., Everdij, M. H. C., and Klompstra, M. B., "Accident Risk Assessment for Advanced ATM," *2nd USA/Europe Air Traffic Management R&D Seminar*, FAA/EUROCONTROL, 1998.

[3] Corjon, A., and Poinsot, T., "Vortex Model to Define Safe Separation Distances," *Journal of Aircraft*, Vol. 33, No. 3, 1996, pp. 547–553.

[4] Corjon, A., and Poinsot, T., "Behavior of Wake Vortices Near Ground," *AIAA Journal*, Vol. 35, No. 5, 1997, pp. 849–855.

[5] Blom, H. A. P., and Nijhuis, H. B., "Safety Certification in ATM, ARIBA Consolidation of Results for Safety Certification, Part I," National Aerospace Laboratory NLR, Rept. TR-99576, Amsterdam, 1999.

[6]Blom, H. A. P., Everdij, M. H. C., and Daams, J., "Safety Cases for a New ATM Operation, ARIBA Consolidation Results for Safety Certification, Part II," National Aerospace Laboratory NLR, Rept. TR-99587, Amsterdam, 1999.

[7]"ICAO Manual on the Use of the Collision Risk Model (CRM) for ILS Operations," International Civil Aviation Organization, ICAO-Doc-9274, Montreal, 1980.

[8]Everdij, M. H. C., Bakker, G. J., and Blom, H. A. P., "Application of Collision Risk Tree Analysis to DCIA/CRDA with Support of TOPAZ," National Aerospace Laboratory NLR, Rept. CR 96784 L, Amsterdam, 1996.

[9]Everdij, M. H. C., Blom, H. A. P., and Klompstra, M. B., "Dynamically Coloured Petri Nets for Air Traffic Management Safety Purposes," *Preprints 8th IFAC Symposium on Transportation Systems*, edited by M. Papageorgiou and A. Pouliezos, Intelligent Technological Systems Lab., Technical University of Crete, 1997, pp. 184–189, also National Aerospace Laboratory NLR, Rept. TP-97-493, Amsterdam, 1997.

[10]"Wake Vortex Reporting Scheme and Meteorological Data Collection System," APAS Rept., European Commission, Brussels, 1996.

[11]Woodfield, A. A., "Roll and Lift Disturbances due to Wake Vortices,"NATS, Rept. CAA-CS-9504, London, 1995.

[12]"Directional and Lateral Control, Joint Aviation Requirements," JAA, JAR 25.147, Oct. 2000.

[13]Speijker, L. J. P., Blom, H. A. P., and Kos, J., "Assessment of Wake Vortex Safety to Evaluate Separation Distances," National Aerospace Laboratory NLR, Rept. TP-99454, Amsterdam, 1999.

[14]Darracq, D., Moet, H., and Corjon, A., "Effects of Cross Wind Shear and Atmospheric Stratification on Aircraft Trailing Vortices," AIAA paper 99-0985, Jan. 1999.

[15]Greene, G., "An Approximate Model of Vortex Decay in the Atmosphere," *Journal of Aircraft*, Vol. 23, No. 7, 1986, pp. 566–573.

[16]Liu, H., "Tow Tank Simulations of Vortex Wake Dynamics," *FAA Proceedings of the Aircraft Wake Vortex Conference*, FAA, Washington, 1991, pp. 32.1–32.26.

[17]Stuever, R. A., and Greene, G. C., "An Analysis of Relative Wake-Vortex Hazards for Typical Transport Aircraft," AIAA 94-0810, 1994, Paper.

Chapter 32

Free Flight in a Crowded Airspace?

J. M. Hoekstra,* R. C. J. Ruigrok,†
and R. N. H. W. van Gent‡
*National Aerospace Laboratory NLR, Amsterdam,
The Netherlands*

I. Introduction

IN THE report of the Radio Technical Commission for Aeronautics (RTCA) Free Flight Task Force, Free Flight is presented as a range of concepts, allowing self-optimization of the routes by the airlines.[1] The document also describes a mechanism for airborne separation as a part of the Free Flight concept. Whether or not airborne separation is allowed depends on, among other things, the so-called dynamic density. This dynamic density is a measure of the density and the complexity of the traffic pattern.

According to the original airborne Free Flight concept as described in the RTCA document, when the dynamic density is too high, less freedom will be allowed and separation will remain on the ground. This is based on the assumption that central coordination is required in these cases.

However, this chapter will try to demonstrate, via analysis and simulation results, why it is precisely high-density situations that require the power of a distributed system. It describes the concept as designed in the National Aerospace Laboratory NLR study as an example of the implementation of the Free Flight concept. The lessons learned apply to most airborne separation concepts. The effect of a distributed air traffic management (ATM) system will be illustrated by looking at some sample scenarios, the robustness, and the conflict rates.

II. Free Flight

The Free Flight concept was designed to allow free routing. Self-optimization by the airlines was thought to be more effective than a global optimization by

---
Copyright © 2001 by NLR. Published by the American Institute of Aeronautics and Astronautics, Inc., with permission.
*Research Scientist.
†Research Scientist.
‡Head of Human Factors Department.

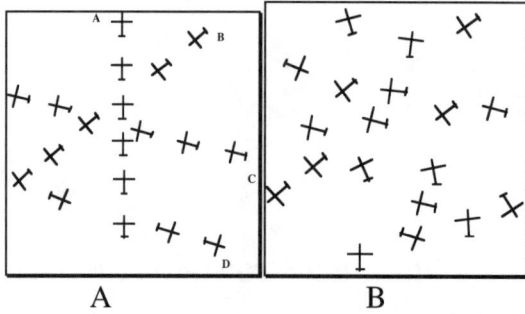

Fig. 1 Schematic representation of how airways aid organization of traffic (A = airways, B = direct routing).

a controller.[1] One airline might give a higher priority to fuel instead of time, depending on the schedule, company strategy, or other factors only known to the airline and crew. Instead of flying along airways, it is clearly beneficial to fly direct routes only to divert from the great circle routes to take advantage of tailwinds. The resulting traffic pattern is chaotic (see Fig. 1).

The lack of structure in the traffic pattern impedes ground-controlled separation when the traffic density is high, especially when the aircraft are also free to fly at any altitude between the normally rounded flight level values. This, together with the drive toward self-optimization, formed the idea of airborne separation. The combined concept of free routing and airborne separation is now commonly referred to as Free Flight. Though a beautiful name for the concept, the freedom for the crew was in fact never the goal. Economic benefits because of more efficient flying were the main goal of Free Flight.

III. Air Traffic Growth

Air traffic growth is predicted to grow exponentially 5–6% per year.[1] Using the average density of upper airspace over western Europe of 13 aircraft per area of 100 × 100 n miles (Refs. 2, 3, and 4), Table 1 indicates the effect of this annual growth (using the conservative 5%). The number also is an indication of the number of aircraft under control of an en route sector of this size, which is not uncommon.

If Free Flight would be applicable only for relatively low densities, it would hardly be worth the effort. After all, what we now call low density airspace will become very scarce in the coming years.

Currently 75% of the delays in Europe are caused by en route ATM problems. The year before it was only 30% of the delays. Although traffic has grown only 5% over Europe, the delays have grown 40%.[5] This indicates that the current en route ATM system is saturated. This requires revolutionary measures. Introducing Free Flight is such a revolutionary measure. A gradual introduction of Free Flight might provide an evolutionary path to the mature Free Flight concept.

IV. NLR Free Flight Study

NLR has investigated the Free Flight concept in collaboration with NASA, the Federal Aviation Administration (FAA) and the RLD (Dutch Civil Aviation

Table 1 Upper airspace traffic densities

Year	Number of aircraft per 100 × 100 n miles	Experiment name
1997	13	Single
2000	15	
2005	19	
2010	25	Double
2015	31	
2020	40	Triple
2025	51	
2030	65	

Authorities). The study started in 1997 and consisted of a number of substudies, such as 1) conceptual design, 2) safety analysis, 3) scenario analysis and generation, 4) man-in-the-loop experiment phase I, 5) avionics requirements and reliability, 6) critical conflict geometries, and 7) man-in-the-loop experiment phase II. Issues that have been addressed in these studies are 1) conflict detection and resolution methods, 2) complex conflict geometries, 3) pilot workload, 4) pilot acceptability, 5) display symbology, 6) safety (both objective and subjective), and 7) mixed equipage procedures.

For the conflict detection and resolution module (Fig. 2), a state-based solution has been implemented. This means no flight plan data are used. The state-based conflict detection algorithm uses a look-ahead time of 5 min. The resolution algorithm uses the geometry of the closest point of approach instead of negotiation to prevent counteracting maneuvers. Both aircraft maneuver cooperatively. The calculated positions at the closest point repel each other, similar to the way charged particles repel each other. This is why the method is often referred to as voltage potential. The repelling force is converted to a displacement of this predicted position in such a way that the minimum distance will be equal to the required separation. This avoidance vector is converted into advised heading and speed changes. The same principle is used in the vertical situation, resulting in

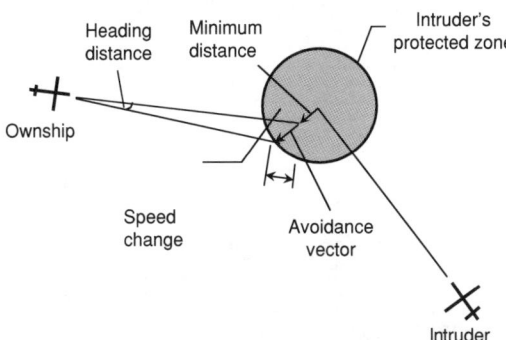

Fig. 2 Resolution advisory is based on geometry of closest point of approach.

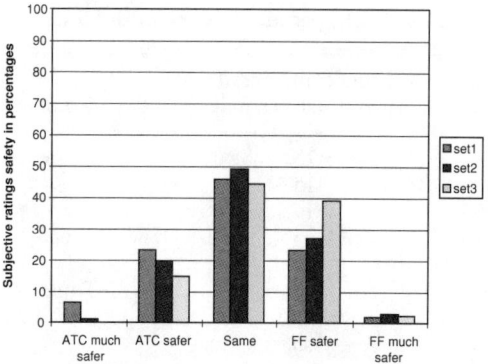

Fig. 3 Rating of safety by subject pilots in comparison with ATC for each set (1, 2, or 3) of six runs in the experiment.

an advised vertical speed. Initially no action of the other aircraft is assumed. This introduces a fail-safe element. Normally the conflict symbology will disappear halfway through the maneuver, indicating that no further avoiding action is required. The resulting maneuvers are on the order of a few degrees heading change or 200 ft/min vertical. Passenger comfort is not affected by these shallow maneuvers. This is very different from the Traffic Alert and Collision Avoidance System (TCAS), which uses a look-ahead time of only 45 s, resulting in drastic evasive maneuvers.

Off-line traffic simulations comprising up to 400 aircraft simultaneously were used to validate several methods for conflict detection and resolution. This simulated traffic densities up to 10 times today's average western European density. The resolution method that proved to be most effective was based on a publication of Eby.[6] Additionally, complex geometries and restrictions were used to test the robustness of the method.

This method has been developed further into an airborne separation assurance system (ASAS). This ASAS includes a human-machine interface that has been tested in several flight simulator trials. Airline pilots have been exposed to scenarios

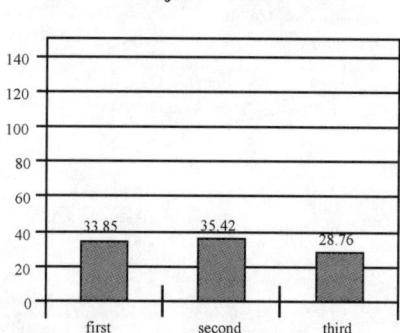

Fig. 4 Rating of workload on scale of 0–130. The third sesion shows a workload rating very close to the 27 found for a comparable ATC situation.

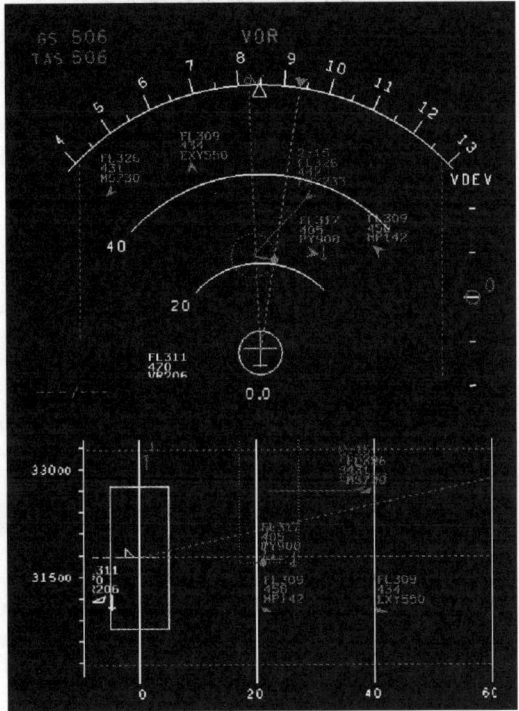

Fig. 5 Coplanar traffic display as used in the study. The symbology indicates a conflict and the resolution advisory.

replicating current densities (single) up to three times the western European (triple) density. It is worth noting that both density and conflicts were tripled, resulting in a nine-fold increase in conflict rate. Training only lasted a few hours. No significant increase in workload has been found during the cruise phase. The acceptability was surprisingly high, and, further, the subjective safety was equal to or better than today's situation[7] (see Figs. 3 and 4).

These results were obtained using a resolution method based on using position and velocity information only. No flight plan information, coordination procedures, priority rules, or ground-based systems were used. An extra system called Predictive ASAS has been developed, alleviating the need for exchanging flight plan information. Because of the simplicity of the architecture and the resolution method, the system was transparent to the crew, allowing a display design as shown in the Fig. 5. The display shows both a horizontal and vertical resolution advisory to the pilot, and he is free to choose one.

In summary, none of the substudies could refute the feasibility of airborne separation, even under extremely dense and constrained traffic situations.

V. Distrust in Distributed System

When people, experts or not, are confronted with the Free Flight concept, the first reaction often is that it sounds like a dangerous idea. This probably is a result of the way human nature reacts to the chaotic nature of the traffic flow.

Chaos is usually associated with danger. Throughout the National Aerospace Laboratory NLR study, the making of conflicts proved much harder than avoiding conflicts. In other words, a random, chaotic scenario, even using existing route structures, was unlikely to have many conflicts. In today's traffic density, applying direct routing (horizontally and vertically) will result in a conflict rate of about one per 50 min per aircraft. A carefully, precisely constructed scenario was required to develop problems on the order of complex geometries like the "wall" or the "super-conflict" (see Section VII). These scenarios are much more orderly but also much more dangerous. The concentration of traffic at airways is also artificially increasing the local traffic density. Even though this increases the collision probability, this orderly pattern is reassuring to the human observer.

What is the reason for this distrust in chaos? This needs to be understood. Acceptance by aviation authorities, pilots, air traffic controllers, and the public is required before the concept can be further developed and gradually introduced. Apart from the conditioned negative association of chaos, there is a reason behind this reaction. The main reason is probably the unpredictability of a distributed system with this high level of interaction.

VI. (Un)Predictability of a Distributed System

A one-on-one encounter can be analyzed with some calculation, and the maneuvers as advised by the resolution algorithm can be derived and understood. To check all one-on-one situations already becomes harder because there are quite a lot of different possibilities with respect to the three-dimensional position and three-dimensional velocity of the aircraft. Still, by sampling several initial conditions, it can be analyzed with some accuracy.

However, the stability of a high-density traffic scenario really is a problem that is of a different order. This touches on the field of mathematics called cellular automata,[8] which deals with the mathematics of interacting units. A famous example of cellular automata is "Conway's Life". This is a simulation in which every state is derived from the previous one with a fairly simple, discrete rule. It uses a two-dimensional matrix field consisting of cells. A cell is either dead or alive. By counting the number of living cells in the eight neighboring cells, the state of the cell in the next state is determined. If the total is 0 or 1, then the cell dies of "starvation". If the number of living cells equals 2, the state of the cell remains the same (stable). If the total is 3, then a new cell originates independent of the previous state (growth) and if the total is higher, 4 to 8, the death of the cell results due to being overcrowded. This rule is much simpler than a geometrical conflict resolution rule. However it yields some surprising higher order effects. Some examples are shown in Fig. 6.

The "windmill" of three cells in Fig. 6 is easy to understand. The "floater" of only five cells moves one cell up and one cell left in five steps. This is something that is already a consequence not easily seen from the simple preceding rule. In fact, most patterns have been discovered in random patterns instead of being designed. The "acorn" illustrates the effect of a structure of only seven cells after 10 and after another 100 iterations.

This life program, which was often used as a screen saver in the old days of computing, is an analogy of how an extremely simple mathematical formula or

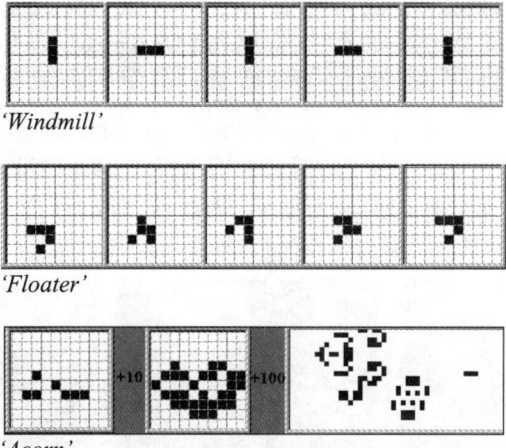

Fig. 6 Examples of life patterns and their evolution: the flipping Windmill, still quite predictable, the Floater that moves up and left in four steps, and the Acorn, showing an explosive growth and complexity from a relative simple starting pattern.

law of nature can result in fantastic unforeseen effects. It has some philosophical aspects, which are not relevant here. However, the behavior of these patterns has puzzled mathematicians for decades, and there is still no theory available that describes the phenomena shown. Even logical AND, OR, and NOT ports have been built (or discovered) in the life program. This means that at a large life-field a complete computer can be built using these cells, while at the low level the simple rule still applies. It is a dramatic illustration of the magnitude of the challenge to analyze the behavior of a distributed system.

A traffic pattern using a Free Flight conflict resolution algorithm is not a discrete but a continuous system, with a geometric interaction as well as scheduling and reaction time effects. It is now, and will be for a long time, impossible to guarantee the stability or risk associated with the behavior of a large number of aircraft in any configuration. The characteristics of an aggregation level below the behavior of the pattern, for example a prescripted one-on-one conflict, are more predictable. The large-scale effects of traffic patterns can only be studied using simulations. The risks of introducing a distributed system can only be analyzed by comparing the effect of the change in structure between a centrally controlled system and a distributed system.

VII. Complex Geometry Examples

To investigate the robustness of the system, several critical scenarios have been developed using the off-line traffic simulator (Traffic Manager). This program is able to simulate up to 400 aircraft simultaneously on a PC or workstation. A demonstration version can be downloaded from the internet at the NLR Free Flight Website*. The aircraft models contain performance models of more than

---

*Data available online at http://www.nlr.nl/public/hosted-sites/freeflight. [cited 2001].

200 types, a flight management system, pilot models, and ASAS. It is possible to view the overall situation on a map, but it is also possible to watch a traffic display of one aircraft with all the navigation data and resolution advisories. With this program, critical conflict geometries have been tested. The algorithms already contained the exception handlers required for singularities like the exact head-on collision, as well as filters preventing conflict alerts from turning aircraft far away. However, the effect of large-scale pattern can only be investigated with simulation. Two examples are shown in Figs. 7 and 8. The aircraft symbols are circles with a

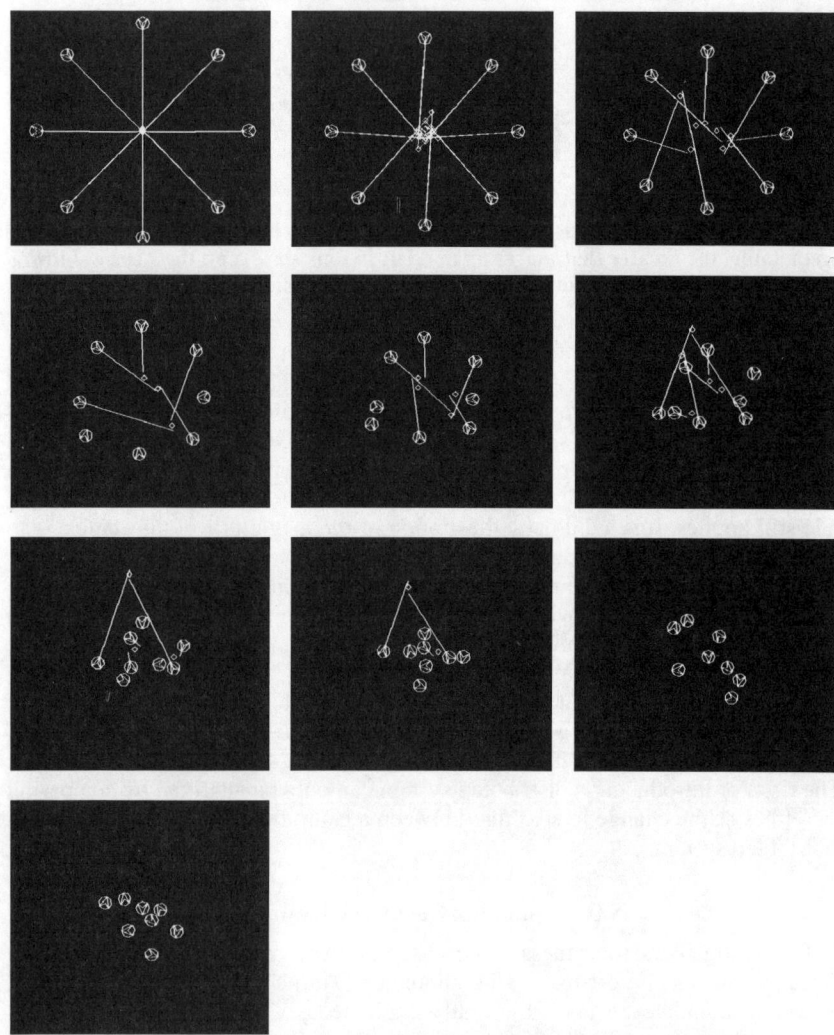

Fig. 7   Superconflict with eight aircraft with the vertical resolution disabled (note the circles have a radius of 2.5 n miles so that touching circles mean the separation is still 5 n miles).

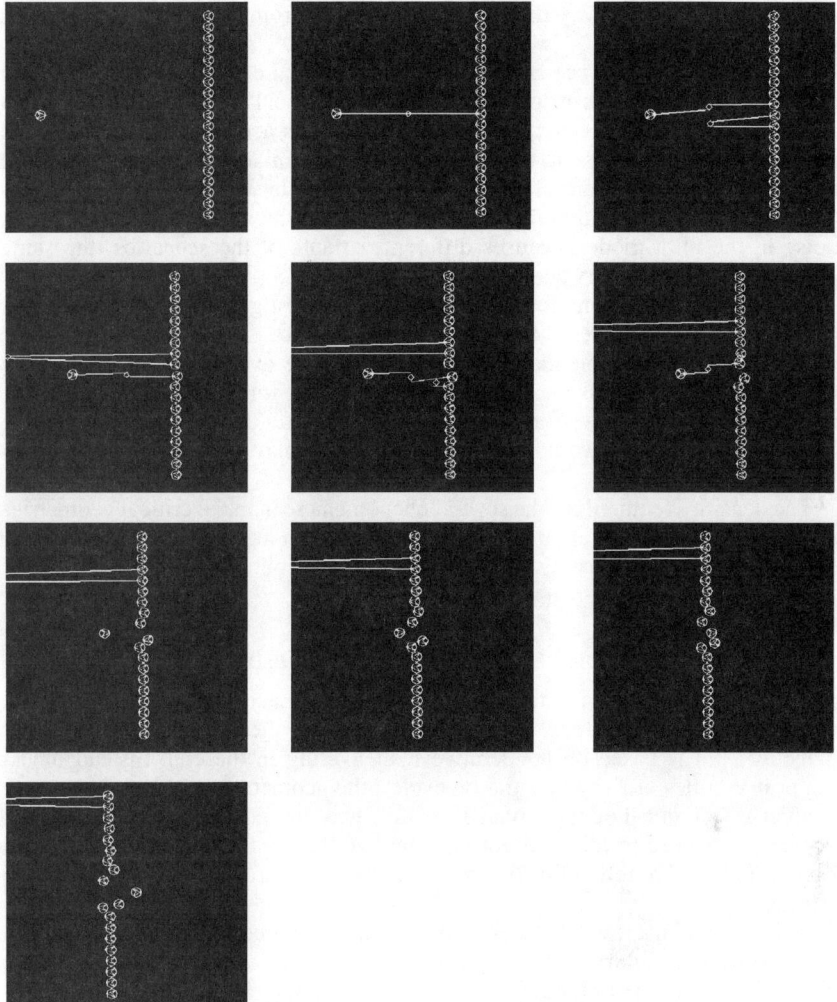

Fig. 8 One aircraft heading at a wall of traffic at separation distance. The initial conflict causes a wave in the wall, creating a hole for the opposing aircraft.

diameter of 2.5 miles, equal to half the required separation. This means the circles are allowed to touch but not overlap.

The super conflict of order eight consists of eight aircraft all flying at the same flight level, with the same speed toward the same point in space and time. Normally this type of conflict is resolved with a vertical maneuver, which is quick and efficient. In this case we have disabled that possibility in the pilot models and forced the pilot models to stay at the same altitude. After some initial conflict alerts, the situation solves itself very quickly. It is a clear illustration of the power of a parallel system: the overall picture contains several conflicts at the same time. Per aircraft, the number of conflicts is quite low. All actions take place at the same time in a way that could never result from one central controlling node.

In the horizontal wall scenario, vertical resolutions are again inhibited. Though every aircraft follows its resolution advisories, a global solution to the problem arises. Some aircraft decelerate slightly while others accelerate. This will yield space to the left and right to maneuver. The wall wrinkles to make a hole in the center for the opposing aircraft to pass through. A similar effect is seen in the three-dimensional version of this scenario. Switching on the noise models (including noise in the pilot models) causes different variants of the scenarios, the same principles appear in every scenario.

A variety of these scenarios has been tested to investigate the robustness. Most can be viewed as analogies of extremely limited airspace. One of the striking results was that the voltage potential algorithm divides the available airspace equally among the aircraft. If this means that there is no space for a 5.0-n mile separation, it will become 4.9 n miles for all aircraft. This graceful degradation is preferable over an algorithm that would just "give up" and display an "Unable to maintain separation" message.

The densities required for the several phenomena seen in the critical geometries study, such as the wall, are unrealistically high. Even in scenarios with 10 times the current densities, excessive numbers of traffic encounters were not observed.

VIII. Robustness and Redundancy of a Distributed System

The conceptual design of the National Aerospace Laboratory NLR Free Flight concept uses an algorithm, which initially assumes the other aircraft does not maneuver but is based on cooperative maneuvering in the end. Instead of using priority rules or risky negotiation cycles, the geometry ensures a cooperative conflict resolution. For this concept to fail, two conflict detection and resolution modules need to fail, and not just any two, but two of the aircraft that will have a conflict. For reliability, this means failure rates are squared. There are two effects:

1) Higher robustness than a central system, compared to a central system in which only one failure of the central node (e.g., the air traffic control (ATC) computer) is a problem for any conflict of any two aircraft in the sector (macro-effect robustness).

2) Higher reliability, in which the failure probability is squared as a result of the cooperative maneuvering. This means that when the airborne equipment is just as accurate and reliable as the ground equipment, and there is no reason why it should not be, the collision probability is squared (it becomes, for example, 10^{-18} instead of 10^{-9}). This also means the ASAS should receive and transmit independently to have completely separated decision loops (micro-effect robustness).

When both the transmitting and receiving function of the data fail, the aircraft becomes blind and invisible at the same time. There are several safety nets that could take over. First, the transmitting and receiving functions should be separated completely, even for the power supply or battery. Further, this system becomes critical and should be made redundant by using two or three systems. If they all fail, there still is the voice radio telephony to broadcast the global position while exiting the free flight airspace. If there is a ground station with a traffic information service broadcast (TIS-B) available, it could transmit the radar data to make sure the aircraft is visible on the traffic displays of the other aircraft. While leaving the free flight airspace, the last received position and velocity vectors of the aircraft could be extrapolated on the display to enhance the situational awareness for this short time. If all this fails, the TCAS collision avoidance system, which is independent of the ASAS, could prevent collision and maintain a certain separation.

IX. Effective Conflict Rate for Air and Ground

The higher capacity of free flight compared to the current en route ATC system can be shown when looking at a direct routing scenario. Suppose the probability of two aircraft having a conflict when flying a direct route in a sector is p_2. This is independent of traffic density and whether the separation task is on the ground or in the air. The global conflict probability as a function of the number of aircraft N in a sector can be calculated assuming p_2 is known. It is the product of the number of combinations of two aircraft times the probability of conflict between two aircraft:

$$p_{c_{\text{ground}}} = \binom{N}{2} p_2 = \frac{N!}{(N-2)!2!} p_2 = \frac{1}{2}N(N-1)p_2 = \frac{1}{2}N^2 p_2 - \frac{1}{2}Np_2$$

When N increases as traffic grows, the probability, and therefore the effective conflict rate as experienced by the controller, increases quadratically with the number of aircraft in the sector.

For the airborne conflict probability, this is different: it is simply the product of the number of aircraft with the probability of meeting that aircraft. The number of other aircraft is $(N-1)$, and so the formula becomes:

$$p_{c_{\text{air}}} = (N-1)p_2 = Np_2 - p_2$$

This probability and the perceived conflict rate increase linearly. The probabilities are equal for $N = 2$, in this case any conflict is also perceived by all (both) aircraft.

For European airspace the conflict rate for single density ($N = 13$, see Table 1) proved to be one per 50 min per aircraft. This yields an example p_2:

$$p_{c_{\text{air}}} = (13-1)p_2 = \frac{1}{50} \; 1/\text{min} \Rightarrow p_2 = \frac{1}{600} \; 1/\text{min} = 0.1 \; 1/\text{hour}$$

The difference between the curves is shown in Fig. 9. One should compare the number on the x axis with the table of traffic growth (Table 1) to see the effect of time.

From this figure, one can observe the effect of increasing traffic on the central and the distributed system. Traffic growth probably will make Free Flight more acceptable over time. Other measures such as improving the ATC user interface or decreasing sector size will change only the slope of the curve but not the quadratic nature.

X. Conclusions

Traffic growth is responsible for a dramatic increase in delays. Recent numbers indicate 75% of these delays are caused by en route ATC. Radical measures are

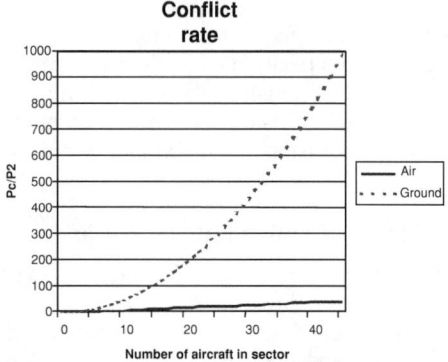

Fig. 9 Relation number of aircraft N in a sector and effective conflict rate for air and ground.

required. Free Flight could provide a solution by changing the structure of the system from a centrally controlled system to a distributed system.

For any change, stronger arguments are required for keeping a system in place. When faced with the concept of Free Flight, often an exactly quantified collision risk is requested. The nature of a distributed system, with the high amount of geometric interaction, may inhibit an absolute risk analysis on a macro-scale that covers all cases. It is possible to do a micro-separation equipment risk analysis based on one-on-one scenarios, similar to the TCAS system. Because the change is more fundamental than an equipment change, this is only a partial answer. Comparing the structure and the nature of the ATM system is equally, if not more, important.

These comparisons indicate that, when the technology in the air is equally as reliable as on the ground, the distributed Free Flight ATM concept features a safety and airspace capacity that is magnitudes higher than the current en route ATM system, in spite of the apparent chaotic traffic pattern. In other words, better a safe chaos than a dangerous order.

References

[1]"Final Report of the RTCA Task Force 3, Free Flight Implementation," RTCA, Inc., Washington, DC, 1995.

[2]"Annual Report 1998," EUROCONTROL Central Office for Delay Analysis (CODA), 1999.

[3]Magill, S. A. N., "Trajectory Predictability and Frequency of Conflict Avoiding Action," DERA (UK), CEAS Free Flight Conference, 1997.

[4]Fitzpatrick, G., and Sagnier, Y., "PHARE AHEAD... New ATM concept demonstration in EUROCONTROL," Centre de l'Etudes de la Navigation Aerienne, Oct. 1996.

[5]EUROCONTROL PRC, *Annual Report of Eurocontrol Performance Review Committee 1999*, EUROCONTROL, 1999.

[6]Eby, M. S., "A Self-Organizational Approach for Resolving Air Traffic Conflicts," *The Lincoln Laboratory Journal*, Vol. 7, No. 2, 1994.

[7]Hoekstra, J. M., Ruigrok, R. C. J., Van Gent, R. N. H. W., Visser, J., Gijsbers, B., Valenti Clari, M. S. V., Heesbeen, W. M. M., Hilburn, B. G., Groeneweg, J., and Bussink, F. J. L., "Overview of NLR Free Flight 1997–1999," National Aerospace Laboratory NLR, Amsterdam, Project/NLR Technical Paper 2000-227, May 2000.

[8]Wolfram, S., "Cellular Automata as Models of Complexity," *Nature*, Vol. 311, Oct. 1984, pp. 419–424.

Chapter 33

Managing Criticality of Airborne Separation Assurance Systems Applications

Andrew D. Zeitlin*
MITRE Corporation, McLean, Virginia
and
Béatrice Bonnemaison[†]
CENA, CS-SI, Toulouse, France

I. Introduction

THE aviation community is paying close attention to the prospect of using airborne traffic display devices for many new purposes. With the concurrent development of automatic dependent surveillance broadcast (ADS-B) air-to-air broadcast technology, in addition to ground-to-air data-link technologies such as traffic information service (TIS) or traffic information service broadcast (TIS-B), the ability is close at hand to receive and display position and additional information for nearby traffic. This has obvious application to enhancing safety, as it improves the aircrew's traffic situational awareness. Moreover, numerous concepts are being proposed and developed that would make use of this data in the cockpit in giving the flight crews some further abilities and responsibilities related to aircraft separation. The International Civil Aviation Organization (ICAO) has named this Airborne Separation Assurance Systems (ASAS).[1]

At present, aircraft separation is a well-defined function performed by air traffic control (ATC) that plays a basic role in assuring the safety of flight. Any time a change in its operation is considered, it is always studied closely to ensure that no degradation of safety results and all essential or critical factors are known and safeguarded. This process is intensive when studying just a single system onboard an aircraft or on the ground, or for considering a procedural change. It could be much more complex for systems that communicate among aircraft and multiple air traffic services.

Copyright © 2001 by the MITRE Corporation. Published by the American Institute of Aeronautics and Astronautics, Inc., with permission.
*Principal Engineer for ATM/Avionics, Center for Advanced Aviation System Development (CAASD).
[†]ACAS and ASAS specialist.

With this concern in mind, a joint committee of Radio Technical Commission for Aeronautics (RTCA) SC-189 and European Organization for Civil Aviation Equipment (EUROCAE) WG-53 has developed a methodology[2] for performing operational safety assessment (OSA) of communications, navigation, and surveillance/air traffic management (CNS/ATM) systems that span multiple institutions. Within the prospect of developing and certifying ASAS applications, this methodology is an excellent candidate for performing a full risk assessment, spanning the pertinent aspects of the aircraft segment and the ground ATC segment. This chapter describes the work performed to date in this area and indicates the direction foreseen for continuing work aimed at completing and introducing ASAS for useful operational service.

The OSA methodology is first described within the context of ASAS safety assessment. Then, the operational safety assessment process is illustrated through some pertinent examples from several ASAS applications. The allocation of safety objectives and requirements for ASAS procedures and systems is discussed, as well as the roles of ATC and the airborne collision avoidance system (ACAS) during ASAS operations. Ongoing ASAS simulations and trials, and their role in the criticality assessment process, are also mentioned.

II. Operational Safety Assessment of ASAS

The ICAO Secondary Surveillance Radar (SSR) Improvements and Collision Avoidance System Panel (SICASP) has determined that ASAS consists of equipment, protocols, airborne surveillance and other state data, and flight crew and ATC procedures. Thus, these elements must be considered as a whole in determining the safety and criticality of each candidate application. The OSA methodology performs the following steps, which are described in the context of evaluating ASAS.

These steps are not required to be performed in serial order. It is more practical to work on several of them in parallel, and it is useful to feed results back to the preceding steps. In many respects, the order is not critical. For example, safety objectives and mitigating factors may be known well before hazards have been enumerated. Depending on the development stage of the application of ASAS, the different steps can be used to refine or validate the operational procedure and the functional and technical characteristics of the ASAS system.

A. Operational Environment Definition

The operational environment definition (OED) describes how and in what context an application of ASAS is expected to operate. It includes the responsibilities of the flight crew and ATC, when and how the application begins and ends, the basic information that supports the conduct of the application, resulting displays and alerts, and communications and decisions that are routinely part of the application.

The OED also describes the environment for this use of ASAS. This identifies the type of airspace for which the application is intended, including the degree to which the aircraft population is expected to be equipped with ASAS and any other pertinent equipment, the nature of ATC service, any requirement for radar surveillance, and any other special characteristics of the airspace (e.g., track system, air routes).

B. Operational Hazard Analysis

The operational hazard analysis (OHA) enumerates operational hazard events that could pertain to the application described in the OED. At this level, the OHA is not concerned with how an operational hazard could occur, but only focuses on its effects. The OHA describes the worst-case effect and assigns a level of severity to this effect. This step also lists mitigating factors that support safety even in the presence of the hazard event.

The hazard classification matrix in Table 1, derived from that already proposed in the OSA methodology, provides a scheme for the severity assignment of each hazard depending on its effects on ASAS operations.

At this stage, it is important to consider all feasible hazards, so that adequate protection may be explicitly given during the development process. In the final operational system, some protection against hazards may be provided by equipment or procedural design.

It is important to consider the severity separately from the event's likelihood of occurrence (the next step). The severity depends only on the effects that the hazard could cause and upon the presence or absence of mitigating factors.

C. Likelihood of Occurrence

A translation is performed between each hazard's severity level and the maximum likelihood of occurrence permitted for that hazard. There is a standard, qualitative relationship illustrated in Fig. 1 that gives the greatest likelihood allowed for each level of severity (naturally, the more severe the hazard, the less frequently it is tolerated).

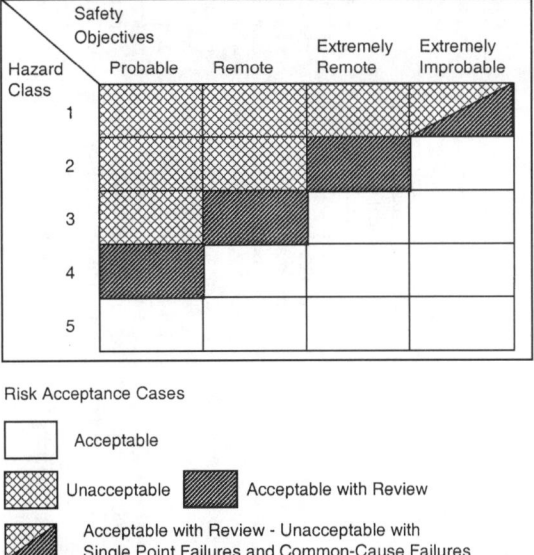

Fig. 1 Relationship between hazard classification and likelihood of occurrences.

Table 1 Hazard classification matrix for ASAS applications

1 Catastrophic	2 Hazardous	3 Major	4 Minor	5 No Effect
Complete loss of safety margin	Large reduction in safety margins	Significant reduction in safety margins	Slight reduction in safety margins	No effect on safety margins
An operational hazard that has the potential for one or more catastrophic accidents. Complete loss of separation from another aircraft, terrain, objects, or obstacles. Operational hazard results in a complete loss of flight control. No independent source of mitigation, such as ATC intervention and/or flight crew procedures could reasonably be expected to prevent a catastrophic accident. Many fatalities and/or hull loss.	An operational hazard that has the potential for one or more aircraft to deviate from their cleared route of flight such that collision or terrain avoidance maneuvers are required to avoid a catastrophic accident. Large reduction in separation as, for example, in a near midair collision. Small number of fatalities, numerous severe injuries, and/or major aircraft or system damage.	An operational hazard that has the potential for one or more aircraft to deviate from their cleared route of flight such that surveillance and communication combined with ATC or flight crew procedures provide the capability to detect and correct the deviation. Significant reduction in separation between aircraft. Minor injuries and/or minor damage to aircraft or systems.	An operational hazard that in itself has no direct impact on the safety of flight operations but has the potential to affect safety either indirectly or in combination with other hazards, for example, by increasing the workload of the controller, flight crew, or by degrading a functional capability needed in the provision of an air traffic service used in the mitigating string for an operational hazard. Slight reduction in separation. Physical discomfort and/or negligible damage to aircraft or systems.	An operational event that can result in no hazardous condition, that is, has no potential for direct or indirect impact to the safety of flight operations.

Analyses of each hazard, taking into account trials data where they exist, must determine whether its likelihood conforms to the allowed maximum level. If it does not, steps must be taken either to mitigate its severity or to reduce its likelihood (or possibly both factors).

D. Allocation of Safety Objectives and Requirements

The next step is the allocation of safety objectives and requirements (ASOR). The level of operational safety that is sought is quantified. From examination of the hazards and their mitigations, a list of functions to be performed by equipment, flight crews, and ATC to achieve safe ASAS operations is developed. This allows designers to determine the elements of the application that assure safety.

E. Ongoing Design and Development

Designers investigate the specific causes of hazardous events, both technical and human, and provide sufficient safeguards in equipment and procedures to satisfy the safety objectives allocated to them. Test and evaluation activities provide assurance that these objectives are met and that the OHA has not overlooked key elements bearing on safety.

F. Certification and Entry into Operational Service

Approval authorities ascertain that safety and performance requirements are met. Equipment and operational approvals should reflect any limits of use that are requisite to safety goals.

G. Monitor Operational Performance

Continued monitoring is desirable to assure that the application is performing as was anticipated during earlier analysis. Lessons learned should be fed back to refine the OHA and, if necessary, the equipment and procedures.

III. Operational Environment of ASAS Applications

ASAS applications range from modest increments that supplement present operations with information, to more ambitious ventures that could change ATC functions in significant ways.[3] At the Air Traffic Management R and D Seminar, ATM-98, in Orlando, Florida the authors presented several papers discussing useful classification schemes for applications and various steps in developing and evaluating them to achieve some benefits in the near-term and progress towards potentially more complex uses.[4,5]

A. Traffic Situational Awareness

Some examples of simple applications fall in the category of traffic situational awareness. These provide the flight crew with information about the surrounding traffic, but no delegation for separation responsibility from the ground to the airborne side is envisaged. Thus, these applications can be defined within the scope of existing ATC practices. An interesting example is aimed at increasing safety for mixed-use visual flight rules/instrument flight rules (VFR/IFR) traffic. Indeed, because of the increasing speed of aircraft, the poor external visibility in modern cockpits, and the pilot's workload in some phases of flight, the limits of the "see and avoid" principle have been reached.

B. Delegation of Separation Assurance

Many applications are proposed in the category of cooperative separation. For these, flight crews will be expected to act upon information provided by ASAS, in the context of specific new procedures. In so doing, the flight crew could accept some aspect of aircraft separation responsibility that has traditionally been provided by ATC alone. The purpose is to alleviate ATC constraints by involving the flight crew in the separation assurance process, where they may be able to exploit better information. Also, it may be possible to reduce the applicable separation minimums. Actually, the determination of the airborne separation minimums is a complex process that needs to take into account various criteria, including the operational procedure and communication, navigation, and surveillance capabilities.

One example application in this category is the ASAS crossing procedure initially presented[6] at the Air Traffic Management R and D Seminar, ATM-97, in Saclay, France. This procedure would enable the pilot flying under IFR within en route controlled airspace to assure his separation from another traffic in accordance with airborne separation minimums. This is analogous to the oceanic in-trail/lead climb or descent, which can allow vertical maneuvers in an oceanic track system at closer longitudinal spacing than previous rules allowed. Although a version based on the Traffic Alert and Collision Avoidance System (TCAS) already is used in some flight information regions, ASAS should alleviate some potential hazards.

A more ambitious cooperative application is for independent closely spaced parallel approaches in instrument meteorological conditions (IMC). Instead of requiring ATC to closely monitor separation, ASAS would detect departures from standard approach paths and alert flight crews when an evasive maneuver was required. This application clearly has a significant set of hazards that involve its own alerting logic, as well as the potential for undesirable interactions between ASAS and ACAS.

C. Airborne Self-Separation Assurance

Another complex application is for conflict detection and resolution at longer look-ahead times than are used by ACAS. This capability is proposed as an enabling capability for user-preferred trajectories and free flight. The concept of ASAS-provided separation in lieu of ATC involves issues that surpass the other applications. The OSA will be crucial to the definition and development of such a capability.

IV. Operational Hazards and Mitigating Factors Associated with ASAS

While the OED descriptions of ASAS applications continue to be developed, progress also is being made on hazard analyses. These hazards should address the functions, interfaces, and procedures associated with ASAS. The hazards related to non-ASAS functions or procedures that might be affected by an ASAS event would also be investigated.

A. Operational Hazard Identification

The operational hazards should be identified by considerations including airborne or ground system failures, human failure to respond appropriately to system failures, human error or omission during normal ASAS operations. These system failures or human-mode errors may concern an aircraft involved in the ASAS procedure or another. Additionally, system failures that are detected or not by the flight crews or the air traffic controllers may result in different hazards. The timing of these operational events (e.g., prior, during, or after the ASAS procedure, different phases of flight) may also have to be considered.

Some aspects common to most ASAS applications include concerns about aircraft lacking proper equipage, or whose equipment has failed. Erroneous information, from identification, position or velocity reports, or for other pertinent data such as intent, clearly must be considered. Also, human factors analysis must address the potential for errors by the flight crew in interpreting ASAS displays or following ASAS procedures.

Other hazards depend on the intended use of ASAS. Interactions between ASAS and ACAS are more pertinent for aircraft that are close enough for ACAS to act.

B. Hazard Severity Assignment

The same operational hazard may be caused by different single operational events or by a combination of events. It is important to classify the operational hazards depending on their effects on ASAS operations to decide on the risk mitigation strategy that can be developed based on the elements of the operational environment.

As illustrated in Table 2 for one hazard, the mitigating factors may either be systems/infrastructure characteristics or procedure/operating practices elements.

The role of the controller will influence the definition and disposition of some hazards. If ASAS is supplementing ATC, the ASAS likely is not critical, whereas if it replaces ATC, criticality falls mainly on the ASAS. More precisely, the criticality required for an ASAS equipment and procedure depends on the level of delegation for separation assurance. In case of full delegation to the flight crew, the highest criticality needs to be considered for the ASAS application.

When aircraft are in close proximity, the effect of an error may be more severe because there is little time to achieve separation in an alternate way. This level of severity could be mitigated if the ground system supports contingency procedures as a backup, for example through ground surveillance. In that case, the criticality of airborne separation assurance is highly dependent on the comparison between the ground and airborne separation minimums. In particular, higher criticality could be expected in airspace where procedural control is applied, if the airborne separation minimums are made much smaller than current procedural separation minimums.

Depending on the operational environment, some risk mitigations may not apply or may be limited. For instance, ACAS may not fully operate in some phases of flight (e.g., below 1000 ft in altitude) or may be inhibited in some applications

Table 2 OHA table for ASAS crossing procedure

Operational hazard	Effects on ASAS operations	Severity	Mitigating factors
Incorrect identification of target aircraft at the procedure initialization.	Potential loss of separation because the pilot (of own aircraft) may accept the ASAS procedure with respect to a wrong aircraft (in close proximity) and not execute the required separation maneuver with respect to the correct target aircraft.	Major	*Infrastructure/systems:* CDTI features enhance pilot's situational awareness about surrounding traffic. Traffic advisory provided by ACAS/TCAS as conflict increases between own aircraft and correct target aircraft. If airborne and ground separation minimums are compatible, short-term conflict alert raised to ATC between aircraft involved in the procedure. *Procedures/operating practices:* Procedural requirement for the pilot to report any change in altitude, direction, or speed during the procedure.

based on intentional close proximity. Another example is the ATS surveillance that does not apply, or is limited to ADS, outside radar coverage. Finally, the use of ADS-B for both the airborne and ground surveillance may constitute a common point of failure, not compatible with the level of safety required for ASAS operations.

C. Hazard Likelihood Analysis

The likelihood of occurrence of each hazard highly depends on the considerations and assumptions made when defining the operational environment of ASAS operations, and in particular, on the required equipment and CNS functions that support the conduct of the ASAS procedures by the flight crew and ATC. The determination of an operational hazard likelihood of occurrence requires the identification of its possible causes (either system failures or malfunctions, human errors or omissions). Indeed, the avoiding factors contributing to reduce the likelihood of each hazard depend on the operational events or combination of events leading to the hazard. Table 3 illustrates the relationship between operational hazards and their causes, and the necessity to distinguish between them when analyzing whether or not their likelihood conforms to the maximum allowed for the hazard's severity level.

Table 3 Hazard likelihood analysis for ASAS applications

Operational hazard
Incorrect identification of target aircraft at the ASAS procedure initialization
Possible causes

Corrupted ADS identification report by other aircraft	Corrupted track correlation by own airborne surveillance	Corrupted aircraft identification displayed in the cockpit of own aircraft	Air traffic controller error when initializing the ASAS procedure	Pilot (from other aircraft) error when entering aircraft identification transmitted through ADS	Pilot error when identifying target aircraft on the CDTI

Likelihood of occurrences
Avoiding factors[a,b]

A) Integrity requirement for ADS report (RCP), and consistency check of ADS identification report by the ground surveillance system (either manually or automatically)	A) Integrity requirement for airborne surveillance (RSP) B) Procedural requirement for the use of target identification in conjunction with a traffic information for consistency check by the pilot	A) Integrity requirement for cockpit display B) Procedural requirement for the use of target identification in conjunction with a traffic information for consistency check by the pilot	B) Procedural requirement for the use of target identification in conjunction with a traffic information for consistency check by the pilot	A) Consistency check of ADS identification report by the ground surveillance system (either manually or automatically) B) Cross-check procedure within the cockpit	A) Ease-of-use of CDTI and highlighting of aircraft identification of selected target B) Cross-check procedure within the cockpit

[a] Infrastructure/systems.
[b] Procedures/operating practices.

V. Allocation of Safety Objectives and Requirements for ASAS Applications

A. Safety Objectives for ASAS Operations

Safety objectives for ASAS operations need to be agreed on at the policy level. They should be compatible with the target level of safety (TLS) normally required for air traffic control, i.e., 1×10^{-9} midair collisions per flight hour. However, depending on the airspace characteristics, this TLS could correspond to an unacceptable number of collisions per year in specific areas, typically in high traffic density areas in continental airspace.

1. Airborne Separation in Collision Risk Management

For ASAS applications in which the flight crew is required to assure some airborne separation, midair collision would result from the combination of an ASAS failure (i.e., loss of airborne separation) and the inability of the backups to avoid the collision. Moreover, in case of a complete loss of airborne separation without any possible ATC or flight crew intervention, the collision risk increases with higher traffic density and lower airborne separation minimums. The establishment of appropriate separation minimums will be an essential product of the overall safety assessment of ASAS operations.

As long as there is no common mode of failure between ASAS and its backups, and no induced risk of collision due to the backups when the ASAS performs correctly, it would be possible to allocate specific safety objectives for the ASAS systems and procedures independently from the backups. These safety objectives, expressed in terms of losses of separation per flight hour, would be more easily monitored during the entry into operational service of ASAS. Furthermore, the reduction in collision risk expected through ATC backup or ACAS II intervention would have to be defined and assessed from the initial ASAS development stages to validate that the overall TLS is achieved.

2. ACAS II Safety Contribution

The worldwide ACAS mandate was not based on an assessment of the absolute level of collision risk with and without ACAS II (the airborne collision avoidance system for transport category aircraft), but only on theoretical analysis and the practical experience that ACAS II reduces the collision risk. In 1989, SSR Improvements and Collision Avoidance System Panel (SICASP)/4 expected[7] a reduction of collision risk of 1×10^{-1} collisions per flight hour with the introduction of ACAS II. More extensive simulations led SICASP/6 to adopt performance-based standards for ACAS that specifically focus on the risk reduction. It still has not been demonstrated that these objectives were met.

Furthermore, the ACAS II safety contribution is mainly based on its independence from the primary means of surveillance. Therefore, ACAS II must remain independent from these primary means, wherever the separation assurance is performed from the ground as it is currently, or in the air as it is envisaged with ASAS. Otherwise, the reduction of collision risk achieved by ACAS II and its supposed contribution to the overall safety objectives would have to be reconsidered.

Finally, it is currently accepted that the ACAS carriage by aircraft shall not be a factor in determining the need for air traffic services (see ICAO Annex 11).[8] This statement would have to be revisited if ACAS II is used in the risk management strategy within ASAS operations.

B. Safety Requirements for Both Airborne and Ground Segments

From the analysis of operational hazards, their possible mitigating and avoidance factors, the elements required to achieve the acceptable level of safety of ASAS operations need to be explicitly identified. These elements include all of the equipment and procedural requirements for both the airborne and ground segments of the CNS/ATM system, and also the required communications, navigation, and surveillance performances (RCP, RNP, and RSP). This allocation of safety requirements is illustrated in Table 4 for representative hazards in several applications.

1. ASAS Equipment Characteristics and Performances

Airborne surveillance is one major component of ASAS equipment, and required surveillance performances will have to be defined for each ASAS application. The ASAS equipment also includes the processing of the airborne surveillance data and navigation data for airborne separation assurance.

During the overall safety analysis, the suitability of ADS surveillance data for ASAS needs to be considered carefully.[9] In particular, the accuracy, availability, and integrity of the navigation and surveillance data, as well as the update rate and surveillance range validation, need to be assessed with respect to the safety requirements.

It is anticipated that airborne surveillance will rely on ADS-B, but not exclusively. Air-air data-link and air-ground data-link such as TIS-B are also envisaged. The minimum operational performances of the correlation and fusion of data from different sources need to be defined.

When active procedural use of information displayed on the cockpit display of traffic information (CDTI) by the flight crew is invoked, the criticality of some hazards related to airborne separation assurance may require the development of an alerting system for airborne separation monitoring. The reliability and adequacy of such alerts would have to be assessed in further safety analysis.

2. Role of ATC in ASAS Operations

For cooperative ASAS applications, one major issue that needs to be addressed during the safety analysis is the sharing of separation responsibilities between ATC and the flight crew. Indeed, misunderstanding or incorrect implementation of these responsibilities could lead to unsafe situations incompatible with the overall safety objectives.

Depending on the relationship between the airborne separation minimums applicable by the flight crew and the ground separation minimums used by ATC, different risk mitigation strategy for ATC intervention may be developed. If applied airborne separation is lower than ground separation minimums, the ability for ATC to maintain the safety margins may be compromised, particularly in case of high traffic density. Otherwise, contingency procedures based on ATC backup could be developed with slight increase in controller workload.

3. Role of ACAS II in ASAS Operations

ACAS II is an airborne system based on secondary surveillance radar (SSR) technology, which acts as a last resort safety function when the primary means

Table 4 ASOR table for ASAS applications

Hazard description	Consequences	Avoidance factors A) Infrastructure/systems B) Procedures/operating practices	Mitigating factors A) Infrastructure/systems B) Procedures/operating practices
ASAS Crossing Procedure			
Incorrect identification of target aircraft at the procedure initialization	Potential loss of separation since the pilot may accept the ASAS procedure with respect to a wrong aircraft	See Table 3	See Table 2
Enhanced Visual Approach			
Pilot misjudges in-trail spacing or closing speed	Loss of separation	A) CDTI features displaying range and closing speed B) Pilot training and procedures	A) ASAS alert to pilot B) ATC monitors spacing and may issue warning
Conflict Detection & Resolution			
Pilot receives incompatible instructions from controller and from ASAS	Pilot unable to comply with both. Potential loss of separation with original threat or with another	B) Procedures must govern the responsibility for resolving conflicts	B) Procedures must resolve this situation
Pilot receives simultaneous ASAS resolution and ACAS Resolution Advisory	Pilot may be unable to comply with both. Potential loss of separation	A) ASAS design should ensure issuing its conflict resolution prior to ACAS RA, and should defer to ACAS when it generates RA	A) ASAS must remove its resolution or change to one compatible with the RA. B) Pilot training and procedures establish reliance on ACAS

of separation assurance has failed. However, current TCAS II equipment is not designated as critical equipment (the minimum equipment list specifies that failed ACAS II equipment shall be fixed within 10 days). Therefore, the use of ACAS II as a mitigating factor for some hazards during ASAS operations needs to be considered carefully and may require further investigation. Any such requirements allocated to ACAS may not be fulfilled by the present generation of equipment.

Another limitation for the reliance on ACAS II mitigation will be the compatibility between the airborne separation minimums and the collision avoidance logic. This issue is particularly crucial for ASAS applications such as the closely spaced parallel approaches, where a significant reduction in aircraft separation is expected to occur.

Despite the fact that ACAS and ASAS are independent by nature, they might share some components of the airborne architecture. In particular, the safe combination of ACAS and ASAS features needs to be investigated when designing a CDTI. Nevertheless, the loss of the ASAS functions must not be detrimental to the ACAS function. This is necessary for ACAS to remain the last resort for collision avoidance in case of navigation failure or separation assurance failure.

Similarly, the use of ACAS data for airborne surveillance purposes could also compromise the ability for ACAS to act as an independent safety net during ASAS operations, and this issue needs to be addressed when designing the ASAS equipment for airborne surveillance.

VI. ASAS Simulations and Trials

Some aspects of the system and applications have begun to be demonstrated through simulations and flight testing. At the MITRE Corporation, laboratory simulations have explored the feasibility of procedures and displays used in enhancing visual acquisition, enhanced visual approaches, and in-trail climb and descent. The National Aerospace Laboratory NLR in Amsterdam has demonstrated a form of conflict detection and resolution. The Eurocontrol Experimental Center in Bretigny, France has conducted real-time simulations of various cooperative separation applications in both en route and terminal airspace.[10] The French Research Center for Civil Aviation (CENA) in Toulouse, France, has investigated the interest in applying the ASAS crossing procedure using simulations on the basis of French radar data.[4] NASA Langley Research Center has simulated independent closely spaced parallel approaches.

Operational trials in Europe have given airlines some experience in situational awareness. Within the North European ADS-B Network (NEAN) Update Program, CENA is involved in experiments for enhanced "see and avoid" between VFR (light aircraft and helicopters) and IFR flights at the second major French airport. An OED has been developed, including procedures mainly dedicated to airspace classes D and E, and should be used for an operational hazard analysis to be performed before the end of the year 2000. The Federal Aviation Administration (FAA) and Cargo Airline Association conducted an experiment in which 24 aircraft conducted enhanced visual approaches and a variety of other applications. Although most of these are quite preliminary, the experience with line flight crews provides data useful for validating performance relevant to some hazards. Further trials are planned for 2000 in the Ohio Valley and in Alaska.

VII. Conclusions and Future Work

Some safety analyses of ASAS applications have been initiated, but studies are still required to validate that the relationship between the airborne separation assurance systems, the associated procedures, and safety nets will ensure that the overall required safety objectives are achieved. This ongoing work developing operational safety assessments is an essential part of the development process of ASAS applications. The objective is to highlight the major criticality issues and also the possible mitigations that need to be taken into account to support safe ASAS operations.

Using the recognized OSA methodology, continuing work is required: 1) to refine the definition of the operational environment of the selected ASAS applications including the air/ground CNS facilities, the ASAS equipage, and the airborne and ground separation minimums; 2) to validate the ASAS procedures through the identification of the hazards and their mitigations, and constructively contribute to the development process; these procedures should define all the flight crew and ATC actions required to satisfy the safety requirements; and 3) to refine the functional and technical characteristics of the ASAS systems used to perform these ASAS procedures while taking into account the safety objectives. As this work proceeds, various test and validation activities will refine technical and procedural details and will develop user community confidence in the new uses of cockpit information. We can foresee the need for computer simulation to evaluate certain hazards, particularly when alerting logic plays a critical role in maintaining separation. This may resemble the work performed to evaluate ACAS logic. The need for standardization provides a sound basis for coordinated efforts among developers, users, and air traffic service providers.

References

[1] "Appendix A to the Report on Agenda Item 6—The ASAS Concept," Secondary Surveillance Radar Improvements and Collision Avoidance System Panel, International Civil Aviation Organization, SICASP/6-WP/44, Montreal, Feb. 1997.

[2] "Guidance for the Approval of the Provision and Use of Air Traffic Services Supported by Data Communications," RTCA, Inc., RTCA SC-189/EUROCAE WG-53, Position Paper P-PUB-22 Rev. I, Washington, DC, Feb. 2000.

[3] Bonnemaison, B., and Casaux, F., "ASAS Concept Considerations," International Civil Aviation Organization, ADSP/5, Montreal, Oct. 1999.

[4] Bonnemaison, B., Casaux, F., and Miquel, T., "Operational Assessment of Co-Operative ASAS Applications," U.S./Europe R&D Seminar, Federal Aviation Administration and EUROCONTROL, Dec. 1998.

[5] Zeitlin, A. D., Hammer, J., Cieplak, J., and Olmos, B. O., "Achieving Early CDTI Capability with ADS-B," U.S./Europe R&D Seminar, Federal Aviation Administration and EUROCONTROL, Dec. 1998.

[6] Casaux, F., and Hasquenoph, B., "Operational Use of ASAS," U.S./Europe R&D Seminar, Federal Aviation Administration and EUROCONTROL, June 1997.

[7]"Report on Agenda Item 3," Secondary Surveillance Radar Improvements and Collision Avoidance System Panel, International Civil Aviation Organization, SICASP/4, Montreal, April 1989.

[8]"Air Traffic Services," Annex II to the Convention on International Civil Aviation, 12th ed., International Civil Aviation Organization, Montreal, July 1998.

[9]"Position Paper on the Use of ADS-B Data for Airborne Separation Assurance Systems (ASAS)," Secondary Surveillance Radar Improvements and Collision Avoidance System Panel, International Civil Aviation Organization, SICASP/WG2/WP774, Montreal, May 1999.

[10]Zeghal, K., Hoffman, E., Nicholaon, J.-P., Cloerec, A., and Grimau, I., "Initial Evaluation of Limited Delegation of Separation Assurance to the Cockpit," World Aviation Conference, Oct. 1998.

Chapter 34

Analysis of Aircraft Separation Minima Using a Surveillance State Vector Approach

Tom G. Reynolds* and R. John Hansman[†]
Massachusetts Institute of Technology, Cambridge, Massachusetts

I. Introduction

CHALLENGES exist to increase both the safety and the capacity of airspace in the next few decades.[1] The safety of the system is designed to be assured by separation minima that govern the minimum distance that controllers must maintain between aircraft under their control during various phases of flight[2] (see Table 1). However, airspace capacity is fundamentally constrained by these same separation minima, and this is becoming a limiting factor to operations in some key points of the system as traffic grows. Many of the proposed strategies to achieve capacity increases either implicitly or explicitly call for reductions in separation minima. Examples include the reduced vertical separation minima (RVSM) procedures for certain oceanic operations,[3,4] Free Flight implementation recommendations,[5] independent parallel approaches to closely spaced runways,[6] and reduced wake vortex spacing requirements between aircraft on final approach.[7] However, the process of reducing separation minima is difficult due to the lack of a rigorous basis for analyzing its impact on safety, which must not be degraded when new procedures are introduced.

Many of the separation minima shown in Table 1 appear to have been based on precedent, consensus, or engineering judgment rather than on well-documented analytical design. For example, there is little published documentation of the rationale behind the early radar separation criteria used in the en route environment, but they appear to have been based upon radar accuracy, display target size, and controller and pilot confidence in the minima at the time.[8,9] Vertical separation standards were apparently dictated by the accuracy of the barometric altimeters and static pressure calibration at cruise altitude. Yet, these en route radar and vertical separation minima have not changed since they were first introduced in

---

Copyright © 2001 by the Massachusetts Institute of Technology. Published by the American Institute of Aeronautics and Astronautics, Inc., with permission.
*Research Assistant, International Center for Air Transportation, Department of Aeronautics and Astronautics.
[†]Professor, International Center for Air Transportation, Department of Aeronautics and Astronautics. Associate Fellow AIAA.

Table 1 Current U.S. separation minima (simplified)[2]

Flight phase	Separation minima	Conditions
Oceanic: North Atlantic	60–120 n miles lateral or	Depending on speed and route
	2000 ft vertical	Above FL290 (non-RVSM)
	1000 ft vertical or	At or below FL290 (or RVSM)
	10–60 min longitudinal at track entry	Depending on speed and route
En route: continental U.S.	5 n miles lateral (radar separation)	Below FL600 and either aircraft more than 40 n miles from radar
	3 n miles lateral (radar separation) or	Below FL180 and both aircraft within 40 n miles on radar
	2000 ft vertical	Above FL290
	1000 ft vertical	At or below FL290
Landing: final approach wake vortex	2.5–6 n miles longitudinal	Depending on leader/follower aircraft types and airport equipage
Landing: independent parallel approach	No parallel runway restrictions exist if conditions are met	Runways 9000+ ft apart[a] Runways 4300–9000 ft apart[b] Runways 3400–4300 ft apart[c]
Landing: runway occupancy	No multiple occupancy of active runways	Exception for land and hold short operations (LAHSO)
Visual flight rules	No mandated separation minima	Responsibility on pilots to maintain separation

[a]Requires operational Airport Surveillance Radar (ASR) radar system.
[b]Requires operational ASR radar system and Final Monitor Aid (high resolution display and alerting logic.
[c]Requires high update rate Mode S radar system and Final Monitor Aid (i.e., precision runway monitor system).

the 1950s despite significant advances in the technologies within the system. The wake vortex minima (used to protect aircraft from wake vortex turbulence produced by aircraft landing in front of them) were not implemented at all before the late 1960s.[9] However, after the introduction of large commercial aircraft such as the Boeing 747, turbulent upset incidents were experienced and conservative wake vortex separation minima were introduced. The absence of a known criteria for the original establishment of these (and other) separation standards has made it difficult to prove that incorporation of new technologies or procedures would allow reduced separation at an equivalent or increased level of safety. Hence, a fundamental understanding of the issues surrounding separation standards and methods of modeling proposed changes to them are essential aspects that this chapter will begin to address.

II. Model of a Separation Assurance Budget

A representation of the factors required for separation assurance is shown schematically in Fig. 1. They make up a separation assurance budget, representing

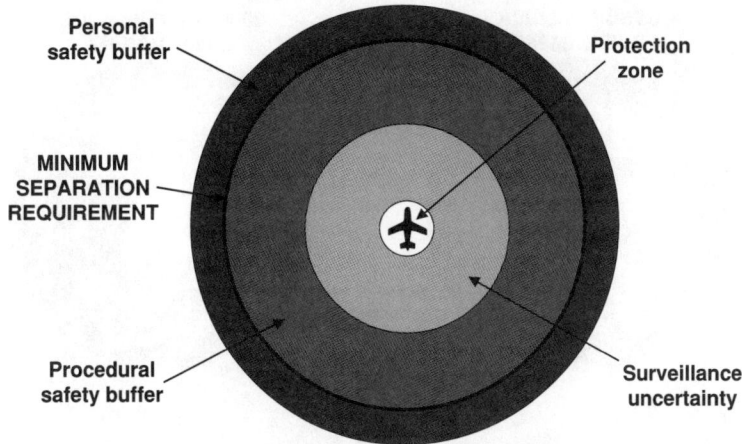

Fig. 1 Separation assurance budget components.

the time or space required around an aircraft to ensure its safety from other aircraft or hazards such as wake vortices. The aircraft is surrounded by a *protection zone,* which should never be penetrated by another object if aircraft safety is to be assured. In the limit, this is represented by the skin of the aircraft. The *surveillance uncertainty* component accounts for the uncertainty in the aircraft state information being output to a controller from the surveillance systems in use (e.g., due to limited accuracy and finite update rate). The third budget element is the *procedural safety buffer,* which is a representation of the formal (proceduralized) buffer required for separation criteria within the operational system. It represents the difference between the existing separation minima and the sum of the protection zone and the surveillance uncertainty components. In some cases it is thought to account for the space and or time required to carry out the recovery actions necessary once a potentially hazardous situation has been detected (e.g., in the closely spaced instrument flight rules (IFR) parallel approach environment).[6] Elements that may be considered within this category include controller detection and comprehension of the developing situation, formulation and communication of an appropriate response to the pilot, pilot detection and comprehension of the communication from the controller, pilot action based on the communication, and aircraft dynamics based on the pilot's action. It should be noted that there is little formal documentation for the basis of the procedural safety buffer, but its role appears to have evolved with time, as discussed later. In addition to these formal separation budget components, there is a *personal safety buffer* that represents the amount of time or space over and above the minimum separation requirement applied by the controller to assure that the separation standard is not violated. For example, in a study by Ballin and Denery at Dallas/Fort Worth International Airport, the personal safety buffer was commonly observed to be greater than a mile for final approach controllers.[10]

In many environments, there is an implied dependence between separation minima and the quality (accuracy and timeliness) with which surveilled state information is made available to the controller. For example, the highly conservative minima in oceanic airspace (where there is no radar coverage and controller knowledge of state information is based primarily on infrequent high frequency radio updates

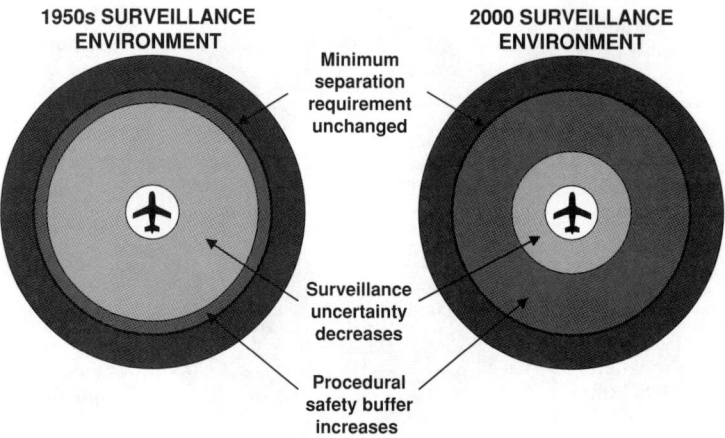

Fig. 2 Implicit increase in procedural safety buffer in some environments (e.g., en route).

from the pilots) are significantly larger than the separations allowed in the domestic en route environment (where there is radar coverage and hence surveillance information is more accurate and timely). This implies that the level of state knowledge (especially positional knowledge) available to the controller plays a key role in determining acceptable separation minima. However, although this dependence implicitly exists, significant advances in surveillance system performance in terms of positional accuracy have not always been reflected in reduced separation minima (e.g., radar improvements in the en route environment). This implies that the procedural safety buffer has increased in these environments (see Fig. 2) and led to greater implicit dependence on the procedural safety buffer without a clear understanding of its basis or role. Hence, further improvements to the accuracy with which traditional aircraft states (such as position) are surveilled may not be sufficient to enable significant reductions in separation minima to be achieved in the en route environment.

III. Need for Surveillance of Intent

In addition to traditionally surveilled states such as radar position, knowledge of likely future behavior, i.e., intent, is an important factor for separation minima. Minima in environments in which intent is highly predictable are often significantly lower than environments in which it is less predictable. For example, the effective separation minima during independent parallel instrument approaches are much lower than the minima in the terminal area prior to the final approach path even though the same surveillance radar is used in both cases. In the approach case, the lateral separation between two aircraft can be as low as the closest runway separation allowed [4300 ft (0.7 n miles)], whereas the terminal area lateral separation minimum (within 40 n miles of the radar) is 3 n miles. The differences between the two situations can be attributed to the differences in the expected intent of aircraft within the two environments. During instrument approaches, the future behaviors of the aircraft are well defined by procedure, and the navigation tolerance is very high. Conversely, in the terminal area away from the final approach path, the trajectories are less well defined. This clearly has implications for safe separation minima.

It is proposed that there is an implied dependence of separation minima on both accuracy of surveillance of current behavior and also on knowledge of intent. Investigating the benefits of formally surveilling intent in future air traffic environments is therefore an important area of research in the context of separation minima reductions.

IV. State Vector Modeling Approach

To formalize the issues discussed in the preceding sections regarding the components necessary for a separation assurance budget and the importance of knowledge of both current and future states, a state vector modeling framework has been developed. It is described in this section.

A. General State Vector Approach

The state vector framework combines states representing the current behavior with intent components representing the future behavior. At any given instant of time t, the aircraft behavior can be defined by the state vector $X(t)$ representing the current and future dynamic states as defined by Eq. (1):

$$X(t) = \begin{Bmatrix} Position, R(t) \\ Velocity, V(t) \\ Acceleration, A(t) \\ Intent, I(t) \end{Bmatrix} \quad (1)$$

$R(t)$ is the position vector, $V(t)$ is the velocity vector, $A(t)$ is the acceleration vector, and $I(t)$ is an abstraction of the higher-order intent states. In general, the states can be considered as being increasingly higher order, interacting in a manner illustrated in Fig. 3 whereby higher states drive those below them.

As discussed in the preceding section, the accuracy and timeliness with which state information is provided to the controller via surveillance systems play a key role in determining safe separation requirements. A surveillance state vector $X_S(t)$ contains measurements of those components of the "true" state vector $X_T(t)$ that are available to the controller through surveillance systems. In many cases, only a subset of the $X_T(t)$ states will be directly measured to populate the surveillance state vector: the controller must either infer those state estimates not directly surveilled or control without consideration of those state components. The surveillance system will also introduce random and systematic

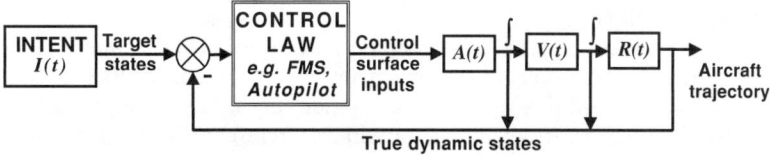

Fig. 3 Interaction of state vector elements.

errors into those state components that are measured. To account for these factors, uncertainties and/or errors in the surveilled state variables of $X_S(t)$ relative to their true values from $X_T(t)$ are represented by $\delta X(t)$, which is defined by Eq. (2):

$$\delta X(t) = X_T(t) - X_S(t) = \begin{Bmatrix} \delta R(t) \\ \delta V(t) \\ \delta A(t) \\ \delta I(t) \end{Bmatrix} \qquad (2)$$

Each of the traditional dynamic state components of position, velocity, and acceleration will be examined in the following subsections in context of how (and how well) they are surveilled. The intent states are also fundamental to the aircraft separation issue and will be discussed in detail in the next section. The implications of the uncertainty and errors in the states contained in the $\delta X(t)$ vector for separation assurance are discussed in the section after that.

B. Surveilled Position Vector $R_S(t)$

The surveilled position state vector of the aircraft, $R_S(t)$, is the three-dimensional position of the aircraft in a given reference frame at time t as it is output from the surveillance system, e.g., $R_S(t) = [R_{\text{latitude}}(t), R_{\text{longitude}}(t), R_{\text{altitude}}(t)]^T$. The elements that make up the surveilled position vector may be provided to the controller from different systems (e.g., radars providing lateral states, aircraft transponder providing vertical state). A future global positioning system/automatic dependent surveillance broadcast (GPS/ADS-B) surveillance environment could provide much higher quality and more timely position state information potentially enabling improved operating practices.[11]

C. Surveilled Velocity Vector $V_S(t)$

The surveilled velocity state vector $V_S(t)$ contains information on horizontal and vertical velocities of the aircraft at a given time. With conventional radar systems, the velocity is generally not measured directly. Rather, the horizontal velocity components are obtained from α-β trackers in the radar system, which estimate aircraft velocity based on previous positional information about the target.[12] The vertical velocity component can be obtained in a similar fashion based on previous transponder altitude returns, but this is far less accurate due to the truncation errors in the reported altitude. In the future, aircraft-derived velocity data could be transmitted to the ground in an ADS-B surveillance environment.

D. Surveilled Acceleration Vector $A_S(t)$

Although not directly surveilled with current radar systems, acceleration data from onboard sensors (e.g., accelerometers) may be available in future ADS-B environments. Under visual separation operations, pilots appear to use aircraft attitude components as surrogates for acceleration states. The enhanced (feed forward) ability to predict the dynamics of proximate aircraft by observing the attitude is thought to be a key factor in pilot acceptance of close parallel approaches under visual conditions.

V. Intent States $I(t)$

The intent components of the state vector $I(t)$ represent the higher-order states that drive the future aircraft trajectory after time t. Example sources of intent are the flight plan (which gives the aircraft's planned route of flight and destination); the currently active air traffic control (ATC) instructions and standard procedures that are followed at a given location and/or time [such as standard instrument departures (SIDs) and standard terminal arrival procedures (STARs)].

Although there is no formal definition of intent in the aviation environment, it is typically understood to be a representation of planned future behavior of the aircraft. This intent is articulated via a command structure between controllers and pilots (e.g., the active flight plan, vectors, etc.) and between the pilot and the flight automation system. These control loops and command structures are observed within the air traffic control environment in use today, as illustrated in Fig. 4. The aircraft intent as it is understood by the ATC system is represented by the currently active flight plan (initial filed flight plan plus amendments), which is presented to the controller in the form of a flight progress strip. It defines the expected route in terms of a series of target states, which together form a planned trajectory from the origin airport to the destination, as represented abstractly in Fig. 5. A definition of intent for the aviation environment that is consistent with this discussion is: the future actions of the aircraft that *can be formally articulated and measured in the current ATC/flight automation system communication structure,* which is represented by Eq. (3):

$$I(t) = \left\{ \begin{array}{l} \text{Current target states, } C(t) \\ \text{Planned trajectory, } T(t) \\ \text{Destination, } D(t) \end{array} \right\} \quad (3)$$

At any point in time, the aircraft trajectory is controlled to a specific set of currently active target states defined by $C(t)$: examples include airspeeds, altitudes,

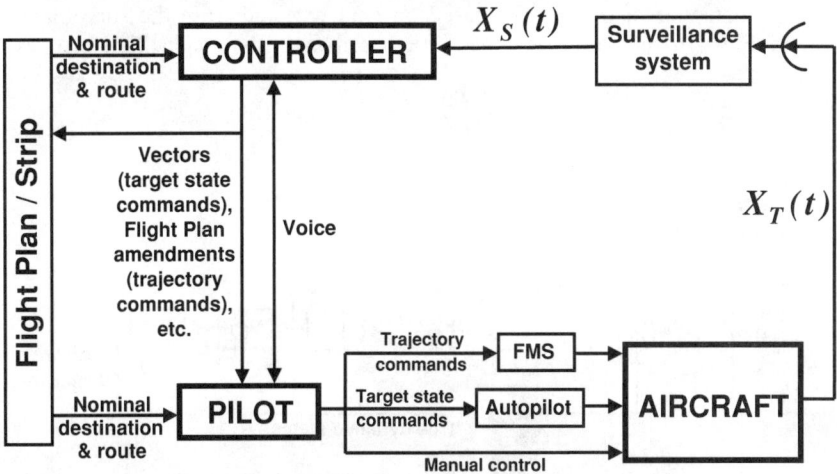

Fig. 4 ATC control loops and command structures.

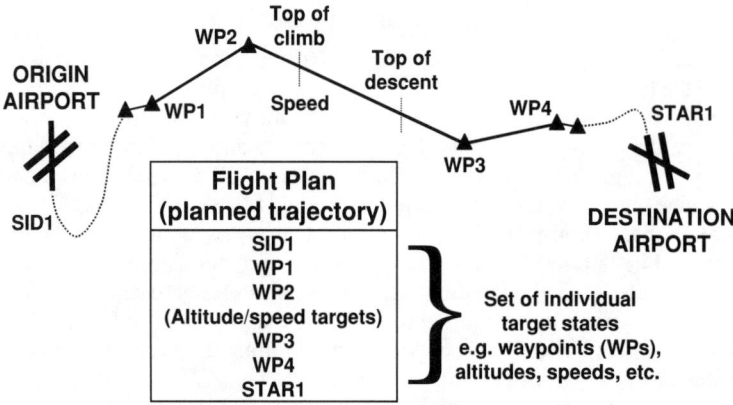

Fig. 5 Abstraction of the flight plan.

headings, or attitudes. Future target states such as later waypoints are contained within the planned trajectory $T(t)$. The last point in the planned trajectory defines the airport that is the desired destination $D(t)$. In light of this definition for intent, the full state vector becomes as shown in Eq. (4) and the components interact as shown in Fig. 6:

$$X(t) = \left\{ \begin{array}{c} \text{Position, } R(t) \\ \text{Velocity, } V(t) \\ \text{Acceleration, } A(t) \\ \text{Current target states, } C(t) \\ \text{Planned trajectory, } T(t) \\ \text{Destination, } D(t) \end{array} \right\} \quad (4)$$

The traditional view of the output of a control system being controlled to a specific target state is maintained by this representation. The controller compares tracking behavior of the aircraft based on data received from the surveillance system (Fig. 4) to the presumed target states based on the flight plan, flight progress

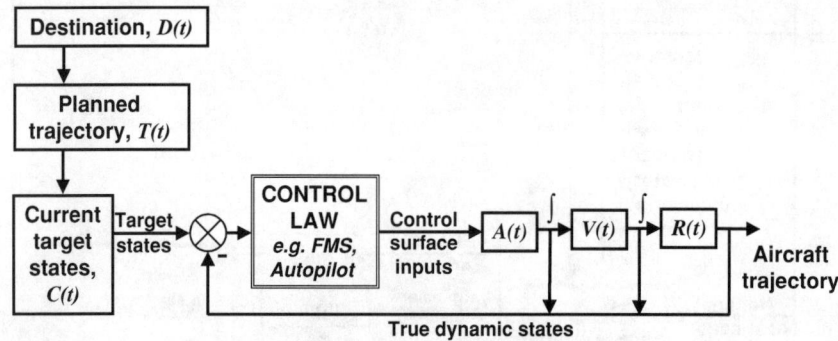

Fig. 6 Interaction of full state vector components.

strip, or clearance amendments to determine if the plan is being followed. This conformance monitoring process is a key separation assurance task of the controller and will be discussed in more detail later.

VI. State Uncertainty

The uncertainty states $\delta X(t)$ contain the error or uncertainty present in the $X_S(t)$ component states. As discussed earlier, not all states may be surveilled to the same degree of accuracy and some states may not be directly measured at all. In these instances, the delta states for those elements may be large with potentially important consequences for separation criteria. This section will discuss the implications of state uncertainty in each of the components of $\delta X(t)$. The last part of the section compares separation examples using the uncertainty states as a method to understand differences in separation minima.

A. Position Uncertainty $\delta R(t)$

Because of inaccuracies inherent in surveillance systems, the measured position contains uncertainty, $\delta R(t)$. Radars have been the main surveillance technology in the ATC system to date. Since most radars have much higher accuracy in the range, ρ measurement than in the azimuth, θ, the horizontal positional uncertainty, $\delta R_h(t)$ of a target at a given range is primarily dependent on the angular accuracy, $\delta \theta$ of the radar as approximated by Eq. (5) below:

$$\delta R_h(t) \approx \rho(\delta \theta) \qquad (5)$$

Although the determination of a radar's angular accuracy can be a complex function of many variables,[13] early ATC primary radar specifications called for an angular accuracy of ±1.0 deg.[14] The technical characteristics of the medium range Airport Surveillance Radar-3 (ASR-3) and the longer range Air Route Surveillance Radar-1 (ARSR-1) introduced in the early 1950s were reported at this value.[15] However, the performance of modern digital radars (e.g., ASR-9, ARSR-4) is significantly improved through beam sharpening, track estimation and other digital processing techniques, such that the angular accuracy of these radars is less than 0.2 deg.[16] Figure 7 shows a plot of the variation of positional uncertainty with range for the angular accuracies quoted above, along with the operating points (in terms of angular accuracy and maximum range) for the ATC radars used in the United States in the 1950s and today. The en route lateral radar separation minima are also indicated for comparative purposes. The correspondence of the variation in the separation minima with range and the variation of position uncertainty of the 1950s radars with range indicates that radar accuracy was likely to have been a primary factor in the establishment of the en route radar separation minima. However, there has been approximately a factor of five improvement in the azimuthal accuracy performance of primary en route radars since then (and an even bigger improvement if secondary radars are considered), without a concomitant reduction in radar separation minima, as illustrated in Fig. 8. This provides support for the observation that the procedural safety buffer has increased in this environment, as discussed earlier.

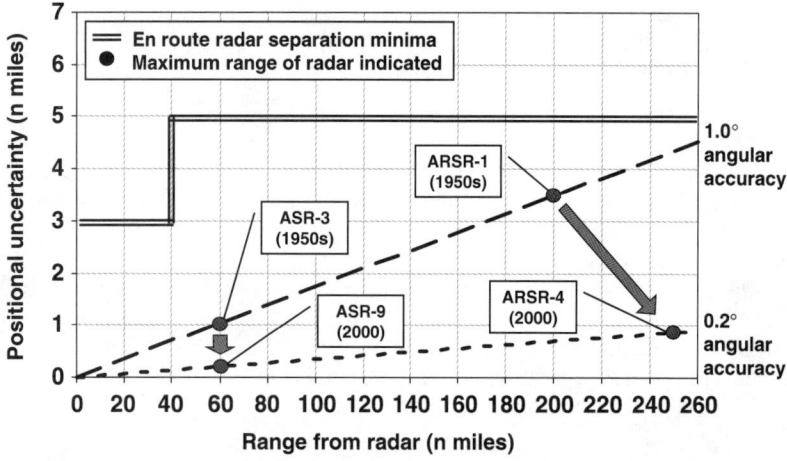

Fig. 7 Variation of position uncertainty with range and angular accuracy plus ATC radar operating points.

The position uncertainty $\delta R(t)$ is also influenced by the update rate of the radar. The scan time for radars used within the United States range from 1 s for the ASDE (used at airports for ground surveillance) to 12 s for the long range ARSR en route radars. Between updates, the controller infers position based on the last position update, heading, speed, and presumed intent. Because all contain their own uncertainties, the inferred position uncertainty grows beyond the basic measurement uncertainty discussed earlier, until position is updated with the next sweep of the radar.

The vertical uncertainty component of the position state vector $\delta R(t)$ can be attributed to several factors that influence the altitude data received from the aircraft's

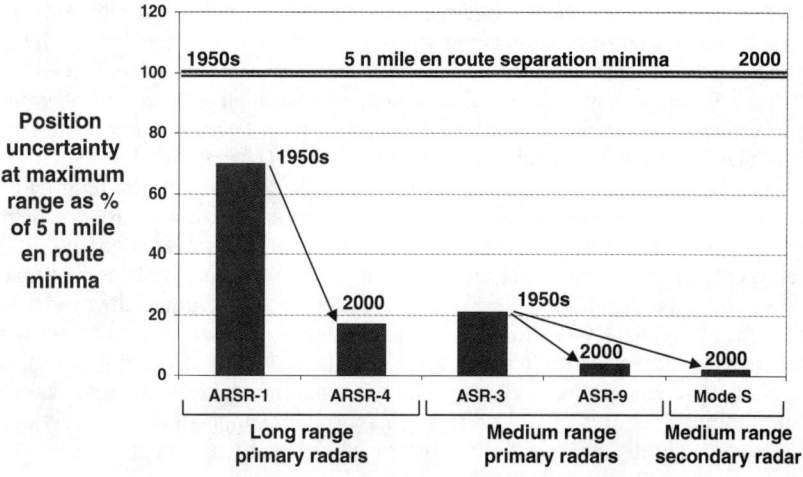

Fig. 8 Improvements in radar system performance.

transponder. These include the 100-ft discretization of the aircraft altitude, calibration errors of the altitude encoders, and static system errors.

In a GPS/ADS-B environment, the positional uncertainty would be significantly reduced due to the higher measurement accuracy (5–10 m) of GPS and the higher update rate of the ADS-B system (1 s).

B. Velocity Uncertainty $\delta V(t)$

In the ATC radar systems, the horizontal velocity is normally inferred from previous positional state measurements, and hence is subject to the same error and uncertainty components. The velocity estimate output from the α-β tracker is also delayed by the filtering process itself, which results in a lag of several radar scan cycles.[12] This significantly adds to the velocity uncertainty. In the vertical domain, the reported altitude is normally discretized to the nearest 100 ft by the transponder. Hence, there is significant noise in the vertical velocity based on these measurements, and typically only a climb/descend indicator is supplied on the surveillance display in the current system.

In ADS-B systems, the expected horizontal velocity uncertainty will be much lower. If velocity derived from onboard systems is broadcast directly, the uncertainty will be at the level of the velocity measurement precision with a maximum of 1 s update delay. If only position is broadcast, then the velocity uncertainty will still be reduced due to the higher precision in the position measurement and the more rapid update rate.

C. Acceleration Uncertainty $\delta A(t)$

In most radar environments, acceleration is not directly measured and can only be inferred by the rate of change of the velocity as measured by the tracking system. The uncertainty is significant, and acceleration is not normally used as an ATC control variable. In the ADS-B environment, direct measurement of acceleration or attitude may be made and transmitted. In the visual surveillance environment, acceleration is estimated based on aircraft attitude. This is effective at indicating direction of acceleration, but significant uncertainty in magnitude remains.

D. Intent Uncertainty $\delta I(t)$

The uncertainty in the traditional dynamic states of $R(t)$, $V(t)$, and $A(t)$ are defined relative to true dynamic states of the aircraft. Because intent is a prediction of future trajectory at a given time, there may always be a certain amount of uncertainty due to unplanned and unexpected events. Intent *errors*, however, result from a totally different understanding of intent between two "agents" in the ATC environment, e.g., pilot, controller, or aircraft automation. The notation $I^N(t)$ is used to define the intent of the aircraft as assumed by agent N at time t. Hence, $I^A(t)$ is the intent as programmed into the aircraft's automation systems (e.g., autopilot, flight management system, etc.) at time t, $I^P(t)$ is the pilot's intent at time t for the future behavior of his or her aircraft, and $I^C(t)$ is the intent of a specific aircraft being controlled to the extent it is known or assumed by the controller at time t. The set of intent states among the various users may be very different in form

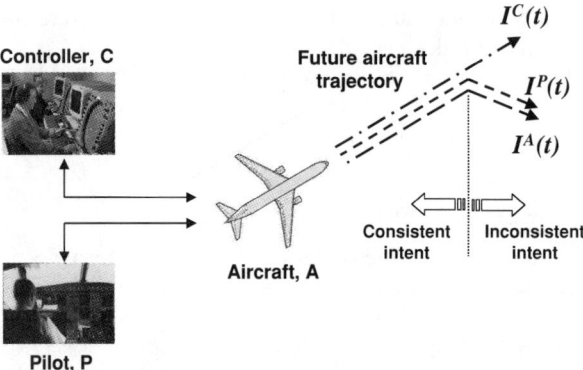

Fig. 9   Compatibility of intents among system agents.

but contain the same composition of intent components as defined in Eq. (3). For example, the aircraft intent may be a pure electronic abstraction of the flight plan as programmed into the flight management system while the pilot and controller intents are mental or textual abstractions of the future trajectory of the aircraft. In the nominal operating environment, all of these intents across the various agents in the system should be compatible with each other, i.e., $I^A(t) = I^P(t) = I^C(t)$: the aircraft flies the trajectory that the pilot and controller have planned for and expect. Errors occur when incompatibilities exist among the various $I^N(t)$ representations (as illustrated in Fig. 9), which can occur for a number of different reasons. For example, the pilot could misunderstand or the controller could misspeak a clearance $[I^A(t) = I^P(t) \neq I^C(t)]$, or the autoflight system could be misprogrammed by the pilot $[I^C(t) = I^P(t) \neq I^A(t)]$. Errors in the controller's intent are fundamentally important from a separation standpoint when it is the controller who is responsible for maintaining separation between aircraft (as in the current environment).

VII.   Relationships Between State Uncertainty and the Current Separation Minima

Separation minima appear to be related to the collective state uncertainties $\delta X(t)$. Hence, it is useful to visualize the relative uncertainty in the different state components in a representation such as a bar chart. The value of each of the uncertainty state components in $\delta X(t)$ are depicted on a notional scale varying from no uncertainty (perfect knowledge) to total uncertainty (no knowledge) of the given state element. Note that the relative uncertainties presented in the following examples are not calibrated on an absolute scale but are estimates used purely to illustrate the impact of different relative levels of state knowledge.

A.   Example 1: En Route Radar Environment (Conforming)

Figure 10 presents an example of the relative state uncertainty in the en route radar control environment for the case of an aircraft conforming to its flight plan. This is used as a baseline example to which other environments are compared. As previously discussed, the radar position has uncertainty due to the limited

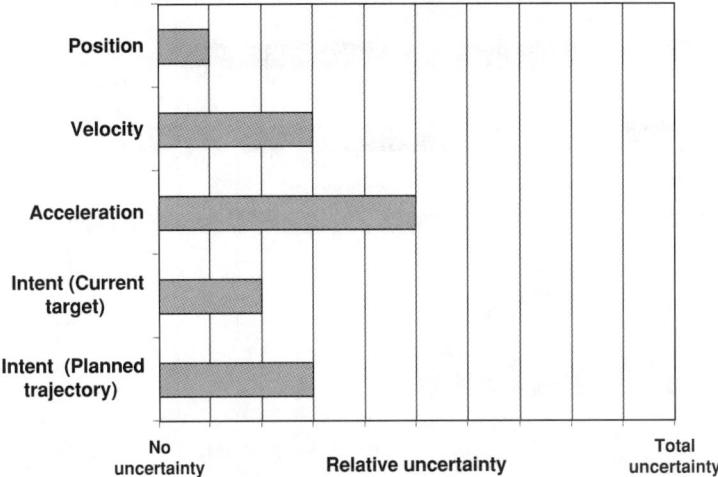

Fig. 10 En route environment state uncertainty (conforming aircraft).

angular accuracy, finite update rate, and discretized altitude data. The velocity state is not directly surveilled, but is estimated from an α-β tracker in the radar. As such, it is known with less certainty than position because it is subject to both the position uncertainty and filter performance/lag errors. The acceleration can only be deduced from the velocity data, and so it is subject to even more uncertainty than the velocity state. Because the aircraft is conforming to the expected trajectory, the current target intent can be inferred quite well from the position and heading states of the aircraft relative to the filed flight plan or current clearance. There is somewhat more uncertainty in the trajectory after the next target state due to unpredictable events, but nominal future behavior is also quite well established from the active flight plan.

B. Example 2: Oceanic Environment

The lateral separation minima in the oceanic environment are much more conservative than in the en route environment (see Table 1). The reasons for this can be viewed in terms of the much larger state uncertainties as illustrated in Fig. 11. Because oceanic surveillance is limited to pilot radio reports of position, which occur at infrequent intervals (typically once every 10° of longitude/latitude or about once an hour), the position uncertainty between updates can be very high. Velocity and acceleration uncertainties are even higher as they can only be inferred from these position reports (although default cruise speeds are filed in the flight plan). Nominally, the intent states (both current target and planned trajectory) are known to the same degree of certainty as in the en route environment using the cleared flight plan.

C. Example 3: ILS Approach Environment with PRM

An example in which the separation minima are less conservative than in the en route environment is provided by an instrument landing system (ILS) approach

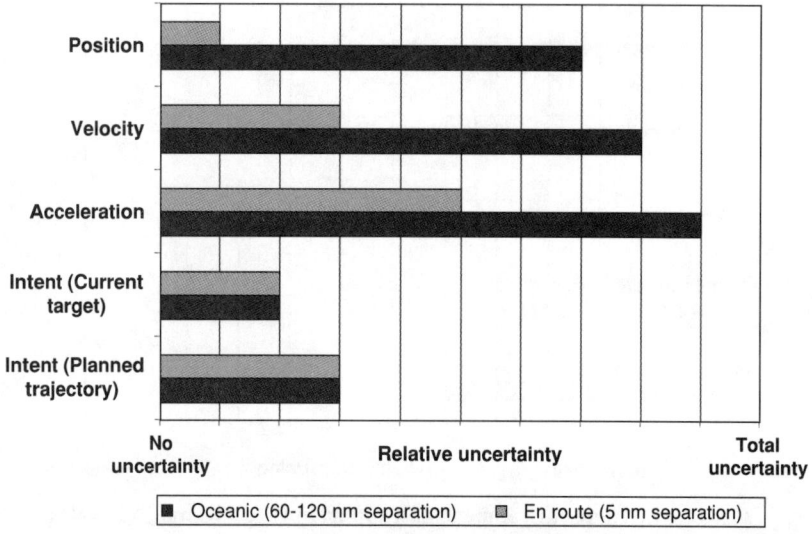

Fig. 11 Oceanic environment state uncertainty.

with the precision runway monitor (PRM) system.[6] Here, the PRM radar has higher azimuthal accuracy and update rate than in the en route ASR/ARSR radars. As a result, the uncertainties in position (and hence estimated velocity and inferred acceleration) are lower in this environment compared to the en route environment, as shown in Fig. 12. Because the target trajectory is well established (i.e.,

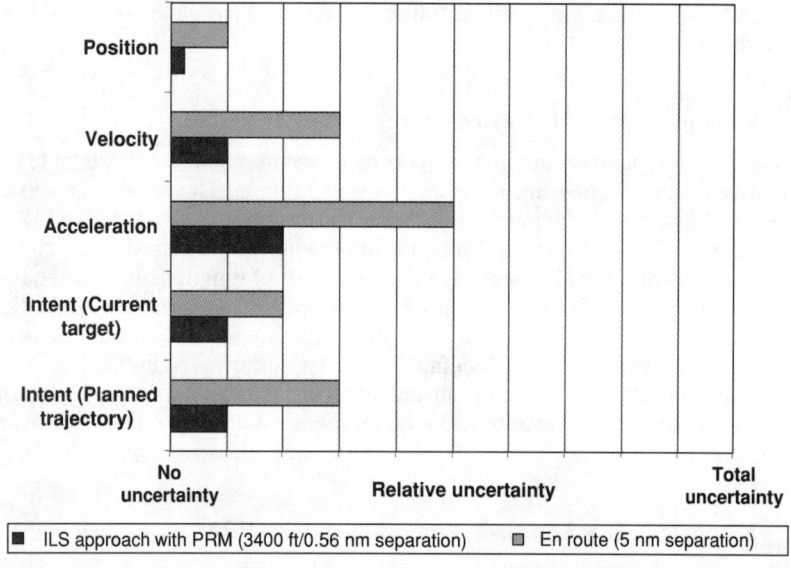

Fig. 12 ILS/PRM approach environment state uncertainty.

maintain trajectory along the defined approach path), and the future trajectory does not exist beyond touchdown, the intent components are also known with lower uncertainty.

D. Example Conclusions

These examples illustrate the relationship between state uncertainty and separation minima within the current operational environment. Exploring the impact on the state uncertainties of new technologies or procedures enables qualitative assessments of their likely implications for safe separation minima.

VIII. Conformance Monitoring

In normal tactical ATC operations, the controller develops a mental picture of the current states and the future dynamics of all aircraft under his/her control. This dynamic projection is based in some part on the understanding of the intent of each aircraft. It is speculated that the controller operates at two levels of focus. At the sector level, the controller monitors the overall pattern of traffic flow within the sector. At this level of focus, the controller is principally concerned with the intent state level as the dynamics are projected forward to determine any potential conflicts or other problems. The uncertainty states are used to establish how much buffer needs to be built into the traffic plan for the sector and where monitoring resources may need to be concentrated. For example, special monitoring attention is reserved for those aircraft for which the controller has a larger intent uncertainty $\delta I(t)$. Examples of such cases include: 1) aircraft in transitioning flight (e.g., vertical or lateral maneuvering); 2) aircraft having emergency status (e.g., with primary system failures); 3) aircraft piloted by inexperienced or foreign crews with language difficulties in which miscommunication may be likely; and 4) aircraft not tracking within acceptable limits.

At the individual aircraft level, the controller monitors the conformance of each aircraft to determine if the surveilled behavior is consistent with the expected or cleared path to maintain the planned separation with other aircraft. However, because of noise in the surveillance systems, atmospheric effects such as winds, and the dynamics of the aircraft navigation systems, small deviations off the expected path could be expected even when the aircraft is following the "correct" intent. These expected deviations can be used to place bounds on observed behavior for conforming or nonconforming aircraft. These bounds act as threshold levels in the n-dimensional state space of $X(t)$, which can define a "hypertube" about the target trajectory, as shown in Fig. 13. An aircraft conforming to the path should stay

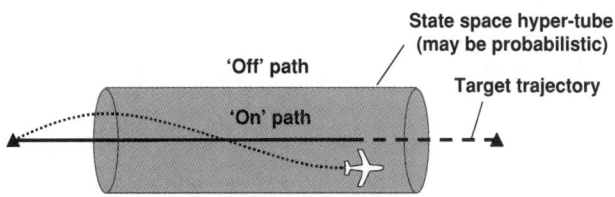

Fig. 13 Hypertube thresholds as a basis for conformance monitoring.

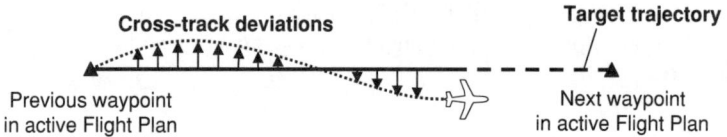

Fig. 14   Cross-track deviation analysis.

within the hypertube $[I^C(t) = I^P(t)]$, while those deviating outside the hypertube can be considered nonconforming $[I^C(t) \neq I^P(t)]$, and this could be a basis for the controller to take corrective action.

An analysis was undertaken to explore factors that might affect these bounds of expected aircraft behavior during a lateral adherence task as illustrated in Fig. 14. The intent was represented by the trajectory segments, which were formed by joining the waypoints within the currently active flight plan. Cross-track deviations from this path were then analyzed. Host computer system data containing active flight plan, radar track, and equipage data were analyzed for 1500 flights within Memphis Air Route Traffic Control Center (ARTCC) over a period of a few hours.* An example analysis for a general aviation (GA) aircraft is shown in Fig. 15. This figure illustrates many of the issues that are formalized within the surveillance state vector approach presented in this chapter. The noise and concomitant position/velocity uncertainty in the radar track is visible. The cross-track deviation is a combination of the ability of the pilot and any onboard navigation systems to maintain the path (i.e., the flight technical error). In the first 5 min, the radar track is diverging from the assumed intent path before it turns back. Although the position states initially indicate divergence from the flight plan, it is unclear

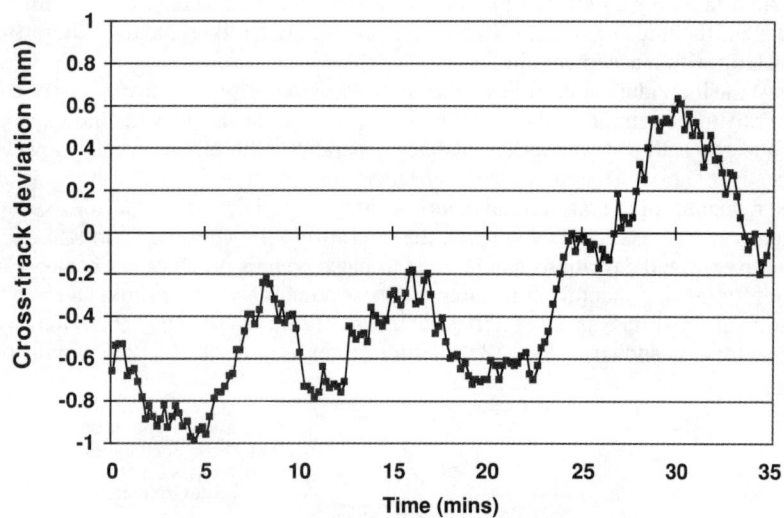

Fig. 15   Coss-track deviation example (GA aircraft).

---

*Personal communication with Mike Paglione, FAA Technical Center, Atlantic City, NJ, Jan. 2000.

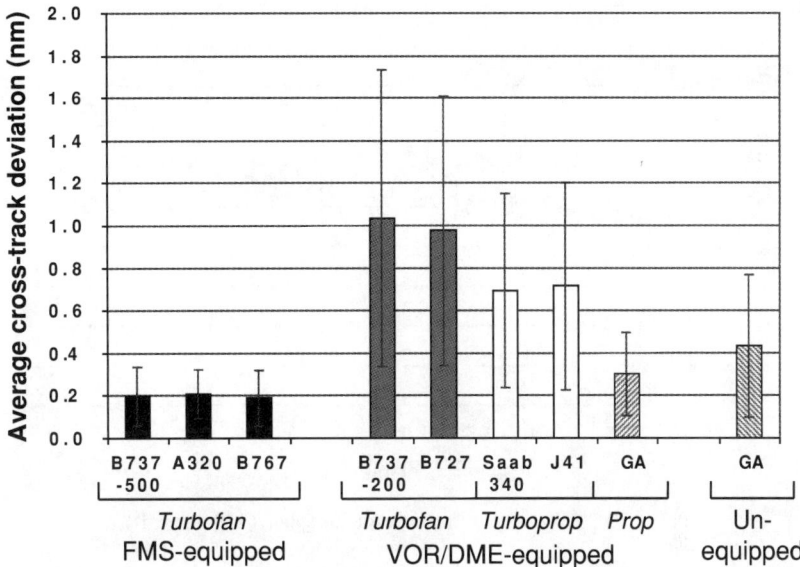

Fig. 16 Summary cross-track conformance capability by aircraft type and equipage (1 standard deviation error bars shown, minimum of 5 h of data per type).

at what point the controller would determine that this aircraft is not conforming to the assumed intent and take corrective action. The analysis carried out suggests that aircraft within similar propulsion (e.g., turbofan, turboprop, etc.) and navigation system equipage [e.g., flight management system (FMS), Very high frequency Omni-directional Range/Distance Measuring Equipment (VOR/DME), unequipped] categories have remarkably consistent lateral conformance capabilities, as shown in Fig. 16. Each bar was based on a minimum of 5 h of radar track data. It is postulated that the controller develops an internal model of the tracking behavior of different aircraft constructed through experience of watching their behaviors over extended periods of time. Hence, this could be a basis for setting internal thresholds on expected behaviors of different types (and equipages) of aircraft to determine when nonconformance occurs. The controller is therefore likely to give larger thresholds (i.e., more track-conformance flexibility) to an unequipped GA aircraft than to an autoflight-equipped aircraft capable of much better tracking performance, such as that exhibited by the A320 in Fig. 17.

When a controller determines that an aircraft is not conforming to the assumed intent, then the uncertainty regarding the aircraft's intent would increase significantly, as shown in Fig. 18. Because of the intent uncertainty, the controller would need to significantly increase the separation buffer for the aircraft and route other aircraft out of its way until consistent intent is reestablished and the separation plan reinstated. A similar situation exists when aircraft lose radio contact and ATC must increase the separation buffer around it, presuming it will either fly the original flight plan or divert to an alternate.

In addition to the importance of aircraft type and equipage, similar arguments can also be made for how the conformance thresholds may differ between different flight phases. Much tighter thresholds should be established in situations

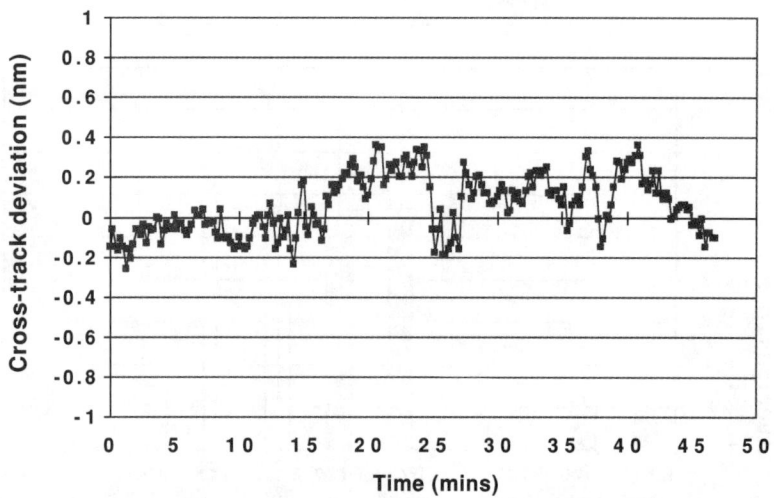

Fig. 17 Cross-track deviation example (A320 aircraft).

in which intent is well defined (e.g., on the final approach path) or in which behavior is more observable compared to when intent is more ambiguous or behavior is less observable. With tighter thresholds on expected behavior, an aircraft will not have to deviate very far off its expected course before nonconformance is detectable. Therefore, deviations from intent should be detected sooner and hence aircraft can be allowed closer together while still maintaining a certain level of safety. In this way, the relationships between surveillance of traditional

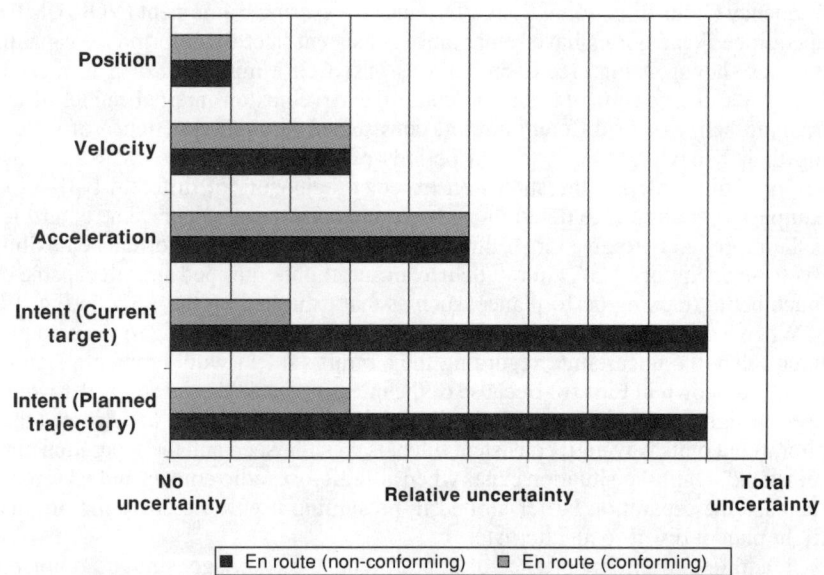

Fig. 18 En route environment state uncertainty (nonconforming aircraft).

dynamic states, intent conformance monitoring, and separation minima can be observed.

IX. Conclusions

Many proposed strategies for increasing airspace capacity call for a reduction in aircraft separation minima. Before they can be implemented, methods of modeling procedural or technical changes are essential to ensure that safety is maintained. The surveillance state vector approach combines the traditional dynamic states of position, velocity, and acceleration with higher-order intent states representing the future behavior of the aircraft. The intent states are of fundamental importance to the separation assurance task. Investigating the benefits of formal surveillance and communication of intent in future air traffic control environments (such as may be enabled by ADS-B and controller-pilot datalink technologies)[17] is therefore an important area of research in the context of separation minima reductions.

The conformance monitoring task of the controller involves combining state surveillance uncertainties and intent with the aircraft separation function. By examining the impact that new procedures and technologies may have on the state uncertainties, knowledge of intent and the conformance monitoring function can also indicate their impact on separation safety. Hence this approach provides a basis for exploring new technologies and procedures. Further work is required to formally investigate these relationships under future operating environments. Although much of the discussion has assumed an environment in which ground controllers are primarily responsible for aircraft separation, a similar approach can also be used in an aircraft self-separation environment in which the cockpit crew take on more responsibility for keeping their aircraft separated from other traffic.

Acknowledgments

This work was supported by the Federal Aviation Administration (FAA)/NASA Joint University Program for Air Transportation. The authors gratefully acknowledge the significant contribution to this work by Mike Paglione at the FAA Technical Center, and also the discussions with Len Tobias and Dallas Denery at NASA Ames Research Center and Hayley Davison at Massachusetts Institute of Technology. Tom Reynolds would also like to thank the British-American Foundation and the Fulbright Commission for their support.

References

[1] "Aeronautics and Space Transportation Technology: Three Pillars for Success," Office of Aeronautics and Space Transportation Technology, Alliance Development Office, NASA, Washington, DC, 1997 (available through NASA website).

[2] "Air Traffic Control," Federal Aviation Administration, Order 7110.65M, Feb. 2000.

[3] "An Authoritative Guide to Oceanic Operations," Federal Aviation Administration, FAA Advisory Circular, AC No. 91-70, Sept. 1994.

[4]"North Atlantic MNPS Airspace Operations Manual," 8th ed., North Atlantic Systems Planning Group/U.K. National Air Traffic Services, Ltd., London, April 1999 (available through website).

[5]"Final Report of RTCA Task Force 3 Free Flight Implementation," RTCA, Inc., Washington, DC, Oct. 1995.

[6]Shank, E. M., and Hollister, K. M., "Precision Runway Monitor," *MIT Lincoln Laboratory Journal,* Vol. 7, No. 2, 1994, pp. 329–353.

[7]Perry, R. B., Hinton, D. A., and Stuever, R. A., "NASA Wake Vortex Research for Aircraft Spacing," AIAA Paper 97-0057, Jan. 1997.

[8]Rockman, M. J., "A Review of the Current Radar Separation Minima and Some Thoughts on Reducing Them," MITRE Corp., MITRE WN 94W000099, Washington, DC, July 1994.

[9]Thompson, S. D., "Terminal Area Separation Standards: Historical Development, Current Standards and Processes for Change," MIT Lincoln Laboratory Rept. ATC-258, Cambridge, MA, Jan. 1997, p. 16.

[10]Ballin, M. G., and Erzberger, H., "An Analysis of Landing Rates and Separations at the Dallas/Fort Worth International Airport," NASA TM-110397, July 1996.

[11]Braff, R., Powell, J. D., and Dorfler, J., "Applications of the GPS to Air Traffic Control," *Global Positioning System: Theory and Applications Volume II,* edited by B. W. Parkinson and J. J. Spilker Jr., Vol. 164, Progress in Astronautics and Aeronautics, AIAA, Reston, VA, 1996, Chap. 12, pp. 327–374.

[12]Brookner, E., "Tracking and Kalman Filtering Made Easy," Wiley, New York, 1998.

[13]Skolnik, M. I., "Introduction to Radar Systems," 2nd ed., McGraw-Hill, New York, 1980, pp. 409–411.

[14]"Airport Surveillance Radar: Civil Aeronautics Administration Specification," Dept. of Commerce, CAA-R-864a, 1 January 1956, p. 10.

[15]"Maintenance of Airport Surveillance Radar (ASR) facilities," Federal Aviation Administration, Washington, DC, SMP 6310.1, 23 July 1963, p. 15.

[16]Cole, E. L., Hodges, M. J., Oliver, R. G., and Sullivan, A. C., "Novel Accuracy and Resolution Algorithms for the Third Generation MTD," *Proceedings of the IEEE National Radar Conference,* Los Angeles, CA, March 12–13, 1986, pp. 41–47.

[17]Turney, K. A., and Syblon, W., "Airspace Efficiency and Technological Advances," *Handbook of Airline Operations,* 1st ed., edited by G. F. Butler and M. R. Keller, McGraw-Hill, New York, 2000, Chap. 35, pp. 517–531.

Section VII: Cognitive Workload Analysis and the Changing Role of the Air Traffic Controller

The four chapters presented in this section deal with the evaluation of potential future roles of the air traffic controller. The section includes the evaluation of decision support tools (DSTs) that use the power of computers to augment the air traffic controllers cognitive processes. Emphasis will be placed on cognitive workload, one of the primary metrics that is used in such an evaluation.

In the mid to late 1990s, the concern with increases in controller workload and with aviation system capacity constraints led to increased research and early introduction of controller decision support tools. These tools change the way that controllers do their job and interact with one another and thus can have a significant impact on controller workload and performance. The introduction, in the 1970s and 1980s, of new avionics systems, including sophisticated flight management systems intended to assist pilots and to increase the safety of flight, provided us with some valuable lessons. New systems were not often subject to the necessary human factors analysis and usually introduced independently of one another, without a clear understanding of the combined impact of all of the new systems on pilot performance. The result was a cockpit that, at times, increased rather than decreased workload, and had the potential for compromising safety. This experience pointed to the need to understand how human operators and the systems they use and control interact when changes are introduced and to understand the effect the new roles and systems have on the human operator. The chapters in this section report on various aspects of research into the changing role of controllers dictated by advances in air traffic management.

The first chapter reports on a human factors study conducted during an operational evaluation of a DST for final approach spacing. The evaluation considered controller-reported levels of acceptance and usability, effects on coordination and communication between controllers, and impact on overall workload. This was one of the first human factors evaluation during field testing of a new DST and provides a number of "lessons learned." The second chapter provides a comprehensive discussion of controller workload measurement. A set of measures are proposed and evaluated for workload and the related areas of task load, complexity, and performance. The third and fourth chapters in this section explore the shifting role of the air traffic controller. In the third chapter a concept involving a number of DSTs (including trajectory negotiation and a multisector planner) and a shift in the responsibilities between planning and tactical controller are evaluated. In this research, human factors issues were evaluated using both fast-time and human-in-the-loop simulations. The fourth chapter reports on the analytic evaluation of several alternative concepts for eliminating some of the sector-related procedural barriers to true strategic planning.

Chapter 35. Passive Final Approach Spacing Tool Human Factors Operational Assessment, Katharine K. Lee, NASA Ames Research Center; and Beverly D. Sanford, Cadence Design Systems, Inc., 1998.

Chapter 36. Evaluating Taskload Measures Derived from Routinely Recorded Air Traffic Control Data, Carol A. Manning, Cynthia M. Fox, and Elaine Pfleiderer, FAA Civil Aeromedical Institute; and Scott H. Mills, SBC Technology Resources, Inc., 2000.

Chapter 37. Controller Roles—Time to Change, Robert Graham, Alan Marsden, Isabelle Pichancourt, and Franck Dowling, EEC, 2000.

Chapter 38. Trajectory Orientation: Technology-Enabled Concept Requiring Shift in Controller Roles and Responsibilities, Kenneth J. Leiden, Micro Analysis and Design; and Steven M. Green, NASA, 2000.

Chapter 35

Passive Final Approach Spacing Tool Human Factors Operational Assessment

Katharine K. Lee*
NASA Ames Research Center, Moffett Field, California
and
Beverly D. Sanford[†]
Cadence Design Systems, Inc., San Jose, California

I. Introduction

THE Center-Terminal Radar Approach Control (TRACON) Automation System (CTAS) is software being developed at NASA Ames Research Center, Moffett Field, California, in conjunction with the Federal Aviation Administration (FAA). CTAS provides decision-making assistance to air traffic controllers in the current terminal and en route air traffic environments, through optimizing arrival traffic flow and generating advisories.[1] Several elements of the CTAS tool suite, the Traffic Management Advisor (TMA), the Descent Advisor (DA), and the Passive Final Approach Spacing Tool (pFAST), have all undergone thousands of hours of simulation testing and, in the past several years, have been the focus of extensive field evaluations in Dallas/Fort Worth, Texas, and Denver, Colorado. The focus of this chapter is the human factors results from the field testing of the terminal area tool, pFAST. Information regarding the development and testing of TMA and DA can be found in Refs. 2–9. Passive FAST is currently being deployed at selected terminal area facilities in the National Airspace System (NAS) as part of the FAA's Free Flight Phase 1 Program.

Passive FAST integrates runway assignment and sequence number advisories into the full datablock (FDB) of the TRACON arrival controllers' radar displays. The "passive" designation indicates the split of the full functionality of FAST[10]; pFAST provides runway and sequence number advisories. The "active" phase of

---

Copyright © 2001 by the American Institute of Aeronautics and Astronautics, Inc. No copyright is asserted in the United States under Title 17, U.S. Code. The U.S. Government has a royalty-free license to exercise all rights under the copyright claimed herein for Governmental purposes. All other rights are reserved by the copyright owner.
*Human Factors Engineer, Terminal Area ATM Research Branch, Aviation Systems Division.
[†]Human Factors Engineer, Research and Development Division.

FAST will additionally display heading, speed, and turn advisories.[11] Active FAST is currently under development at NASA Ames Research Center.[12,13]

The engineering specifications, methodology, and results of the pFAST field evaluation are reported in Refs. 14–16. Overall, an increase in throughput and runway balancing efficiency was shown, coupled with improvements in surface operations.[14] However, in addition to such operational benefits, a successful air traffic control decision-support tool must offer direct benefits to the air traffic controllers and air traffic facility who will be using it.[17] Therefore, an important part of the evaluation and assessment of CTAS is to understand the impact upon the controller. This emphasis contributes to the distinctively human-centered design of CTAS,[1] and is examined under the framework of human factors. The CTAS tool development process has successfully coupled engineering and human factors efforts.

The emphasis in this chapter is on the coordination and communication data from the 1996 operational field assessment of pFAST. References 14, 18, and 19 describe the overall human factors assessment in greater detail; preliminary results have appeared in Refs. 14 and 19.

II. Methods

Prior to the operational testing of pFAST at the Dallas/Fort Worth (DFW) TRACON, several years of development and simulation were performed with a team of controllers from DFW who were well trained on the use of pFAST.[10] This pFAST assessment team, consisting of eight controllers and one area supervisor, provided input into software functionality and the refinement of the human factors questionnaires and data collection methods. The assessment team helped to ensure that the questionnaires that would be used in the operational assessment were understandable and meaningful and that the proposed methods would not be intrusive to live operations. All of the human factors data in the operational assessment were collected from the assessment team, with the exception of two substitute controllers who participated when there was a staffing shortage. The substitute controllers were chosen by the assessment team and were briefed on the operation of pFAST prior to their participation in the assessment.

The specific framework of the human factors operational assessment of pFAST was built on previous human factors evaluations of TMA and DA,[19,20] as well as the assessment of the Computer Oriented Metering Planning and Advisory System (COMPAS), a decision-support tool developed by Deutsche Forschungsanstalt für Luft-und Raumfahrt (DLR) for German air traffic control.[21] The general approach included developing an understanding of 1) the existing operational environment, 2) the tasks for which the controllers, area supervisors, and traffic management coordinators (TMCs) are responsible, and 3) the constraints of conducting a test in an operational environment. This approach required significant interaction between the researchers and controllers, and allowed both groups to define the operational tasks and the testing objectives, while respecting the boundaries and needs of both parties during testing activities. In addition, these interactions contributed to refinement of data collection procedures and interpretation of results.

The human factors assessment examined the usability, suitability, and acceptance[22] of the pFAST advisories. Usability issues reflect the physical characteristics of the equipment and displays; in the case of pFAST, usability questions were posed about keyboard and slewball use and ability to detect the advisories themselves. Suitability issues reflect how pFAST is incorporated into the controller's tasks and involved questions of overall workload as well as coordination and communication. Acceptance is influenced by both usability and suitability, as well as larger issues such as job satisfaction. The human factors operational assessment data, described here, addresses acceptance, usability, and suitability issues based on controller coordination observations during the operational test periods and questionnaire responses.

During the six months of operational testing of pFAST, controllers used pFAST advisories during 25 arrival rush periods across 7 different rush times. Baseline observations (when pFAST was not being used) were collected during 12 rush periods. Engineering data, such as airport throughput, in-trail separation on final approach, and adherence to the pFAST sequence and runway advisories, were collected; these findings are described in Refs. 14–16. The engineering team was stationed in a room adjacent to the operational TRACON. In this separate area, the engineering data were collected and stored, and the overall system was monitored during operational use of pFAST. During the assessment, the human factors engineers conducted their data collection activities on the operational floor, and recorded observations and limited their interaction with the controllers, except to answer questions about pFAST.

The DFW airport operates primarily in either north flow (with traffic arriving and departing toward the north) or south flow (traffic arriving and departing toward the south), with south flow the predominant airport configuration. Six of the 25 test rushes were in north flow. All of the data presented in this chapter reflect a three-arrival runway configuration; the operational assessment took place prior to the addition of a fourth arrival runway at DFW in October 1996.

The pFAST advisories, which consist of runway assignments and sequence numbers, are incorporated into the FDBs of the arrival aircraft on the existing full digital automated radar terminal system (ARTS) displays (FDADs), utilized by the TRACON arrival controllers. A few additional keyboard entries were required to input runway changes and accept runway advisories when they differed from default runway assignments. No other physical manipulation of the equipment was required when using pFAST.

The display of the pFAST advisories is described in Fig. 1. The second line of the pFAST FDB shows time-shared information: in one mode, the default runway assignment and the aircraft type are shown, and in a second mode, the aircraft's altitude and speed are shown. On the third line, the aircraft's sequence number to the pFAST-allocated runway is shown, together with the pFAST runway advisory. A runway advisory is displayed on the third line of the FDB only if it differs from the default runway assignment (shown on the second line).

In the example shown in Fig. 1, the pFAST runway advisory is to 17L, but the default runway assignment is 17C. Until the controller acknowledges the pFAST runway advisory (either accepting it or rejecting it through a keyboard entry), the 17L advisory continues to be displayed in the third line of the FDB. If pFAST's runway advisory did not differ from the default runway assignment, there would be no runway information in the third line of the FDB.

The sequence number displayed in the third line is for the pFAST-advised runway. If the controller chooses not to direct the aircraft to the pFAST-suggested runway, another entry can be made to indicate the controller's runway assignment, and the sequence number would update accordingly.

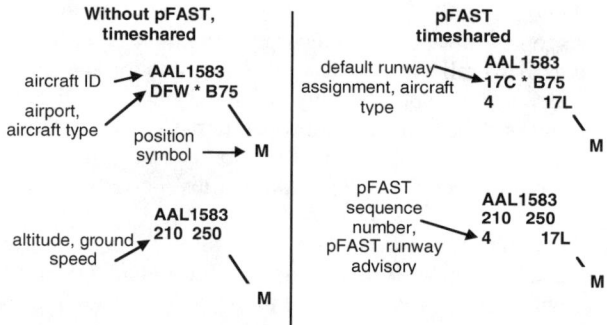

Fig. 1 Full datablocks without, and with, pFAST advisory information.

A. Questionnaire Data

After each test rush, controllers were asked to answer questionnaires regarding overall workload [using a scale based on the NASA Task Load Index (TLX)[23]], the contributors to their workload, and acceptability [using the controller acceptance rating scale (CARS)[10]]. Approximately once per every three rushes, controllers were given an in-depth survey including questions regarding controlling strategy, perceived coordination, and perceptions about the traffic feed from the center.

B. Controller Observation Data

During both baseline and pFAST test conditions, observations were recorded by two human factors engineers at two positions along the arrival wall: one between the two parallel finals and one on the busy side of the rush (typically this was the east side). Figure 2 illustrates the locations of the controller and observer positions.

West side operations were located on the left of the arrival wall, and east side operations were located on the right. The two feeder positions [feeder west (FW) and feeder east (FE)] were assisted by handoff positions (designated by "h" preceding

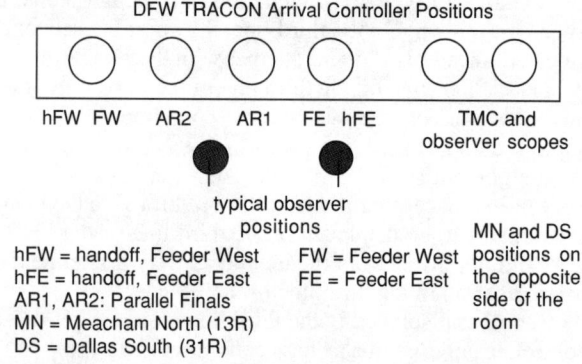

Fig. 2 Arrival controller and observer positions.

the feeder name in Fig. 2). Feeder controllers are responsible for merging the multiple streams of traffic that arrive from the center and vectoring these streams (which may be separated by altitude as well as arrival fix) into single streams toward the runways. In the DFW airspace configuration during the operational assessment, the FW controller was responsible for merging traffic arriving over both west arrival fixes, and the FE controller was responsible for merging traffic arriving over both east arrival fixes.

Feeder controllers hand off traffic to the final controllers, who are responsible for controlling the traffic to their final approach courses. AR2 and AR1 were the final controllers who were each responsible for working one of the two parallel runways. Either the Meacham north (MN) or the Dallas south (DS) position was responsible for the diagonal runways, 13R and 31R, respectively. The MN and DS positions were not co-located along the arrival wall, and observations were not collected from these positions.

Basic characteristics of each observed rush were noted by the human factors engineers, including airport configuration, weather conditions, changes to staffing, and coordination between the area supervisor and the TMCs, and between the area supervisor/TMCs and the tower and the center. Specific attention was paid to coordination between the arrival controllers and, when possible, coordination between arrival controllers and the center. Coordination was defined as an instance of any verbal or nonverbal contact that was related to controlling traffic. The observations from the human factors engineers were merged into a single transcript for each observed rush. Any observations that were incomplete, or unrelated to the traffic situation, were not included in the analysis.

Codes were assigned to each observed coordination event. The codes fell into nine general categories: runway, sequence, TRACON situation, aircraft status, point-outs/handoffs, weather, traffic management issues, communication issues, and equipment problems. Each of the nine major categories were further subdivided into two to six subcategories. The 33 subcategories are described in Ref. 18.

III. Results

The results in the following are described according to the general framework of acceptance, usability, and suitability. More extensive discussion of the results can be found in Ref. 18.

A. Acceptance

The user acceptance of the system was determined through direct ratings using the CARS. The CARS (see Refs. 10 and 19) was developed with the direct participation of the pFAST assessment team, who helped to define the various descriptors tied to the different rating levels, as well as the definitions behind the ratings of adequate vs desired performance.

The overall acceptance rating across all the test rushes was 7.8 [Standard Deviation (SD) = 1.1], on a scale of 1 to 10, where 1 is the least desirable rating and 10 is the most desirable rating. The overall rating, rounded to 8, is associated with the following description of the system: "Mildly unpleasant deficiencies. System is acceptable and minimal compensation is needed to meet desired performance." Extensive comments were collected with the CARS ratings and reflected the controllers' concerns about the accuracy of the sequence advisories. These comments related to overtakes and general disagreement with some sequences. However, the engineering data show very positive results for *adherence* to both runway and sequence advisories during the operational test.[15] It is possible that the controllers felt that the advisories needed to be "perfect," and thus their ratings may reflect their tendency to characterize a less-than-perfect test rush as problematic.

A controller's definition of perfect advisories was likely to match her/his view of the traffic situation; this does not account for pFAST's knowledge of traffic outside of the controller's perception. In contrast, a perfect rush in terms of the flow efficiency (measured by the pFAST researchers) was one in which delay was minimized. Some disagreement between controller preferences and pFAST calculations is inevitable but is considered in the evaluation of acceptance.

B. Usability

Usability was examined through questionnaires, with questions focusing on information presentation and equipment interaction. The controller ratings indicated that additional inputs required did not significantly increase workload. At best, the runway advisories were acceptable enough to require few corrections, or at worst, did not impact controller workload significantly when changes were required.

When changes to the runway advisories were required, the greatest concern that the controllers voiced had to do with the associated update delay on their displays. The delay was not rated as excessive and was reported to contribute minimally to their workload.

Coordination and communication were also examined from a usability perspective to see how pFAST affected the controller communication frequency. The controllers rated the amount of communication that they had with the aircraft under their control. On average, the controllers reported talking to each aircraft between 2 and 5 times. The reported average over all of the controllers was 3.8 times (SD = .80). None of the controllers reported having to talk to any aircraft more frequently due to the pFAST advisories.

The controllers also rated the level of coordination required (with other controllers and facilities) when using pFAST. They reported that the required level of coordination did not exceed what they normally experienced. These results indicate that pFAST is not creating additional interactions either with other controllers or with the aircraft.

C. Suitability

Suitability was assessed from questionnaires and observations. The workload impact and the effect on controller tasks as well as communication and coordination are described in the following.

1. Workload

Workload ratings were collected to examine overall workload, as well as to examine specific elements of the controllers' tasks. The controllers did not report a dramatic increase or decrease in their overall workload when using pFAST. Controllers rated the runway and sequence advisories as contributing no more than "somewhat" to their overall workload. The amount of effort required to use the pFAST advisories was rated as about the same as the controllers were accustomed

Table 1 Five most frequently observed coordination categories

Category	Description
pFAST/ARTS-related issues	pFAST-required keyboard entries, as well as display issues related to pFAST pFAST being turned on or off, or ARTS-related display problems
Point-outs	Aircraft requiring special handling Aircraft crossing through airspace to which they were not normally assigned Approval requests (APREQs) from departure airports within the TRACON Utilizing another controller's airspace, but retaining control of the aircraft Often non-verbal
Handoff issues	Asking for handoffs Frequency changes Ownership
Runway assignments	What the runway assignments were Changes to runway assignments
Aircraft altitude changes	Expedited descents Coordinating or inquiring about aircraft altitude

to working. They also reported that overall pFAST had no effect on their ability to control traffic in their sectors. Given the demonstrated throughput benefits, this "non-effect" can be seen as a positive result, demonstrating that pFAST did not detract from operations.

2. Coordination and Communication

Coordination (between controllers) helps ensure safe aircraft handling.[17] The five most frequently observed categories of coordination (regardless of test condition) were pFAST/ARTS-related issues, point-outs, handoff issues, runway assignments, and aircraft altitude changes. These categories are described in detail in Table 1.

Although the controllers did not report any significant increase in controller-to-aircraft or controller-to-controller coordination, changes in coordination were observed between baseline and pFAST conditions. The means and standard deviations of the baseline data compared to the pFAST test data are shown in Fig. 3. Runway assignments, sequences, and spacing were discussed with significantly greater frequency under the pFAST conditions. This is somewhat expected, as the new information provided to the controller, as well as the testing environment itself, would likely promote discussion about the advisories. Increased discussion regarding status checking was also found under pFAST conditions relative to baseline, but this may be an artifact of the operational assessment itself. It is likely that the testing environment prompted the assessment team to increase their monitoring and awareness of operations to identify problems.

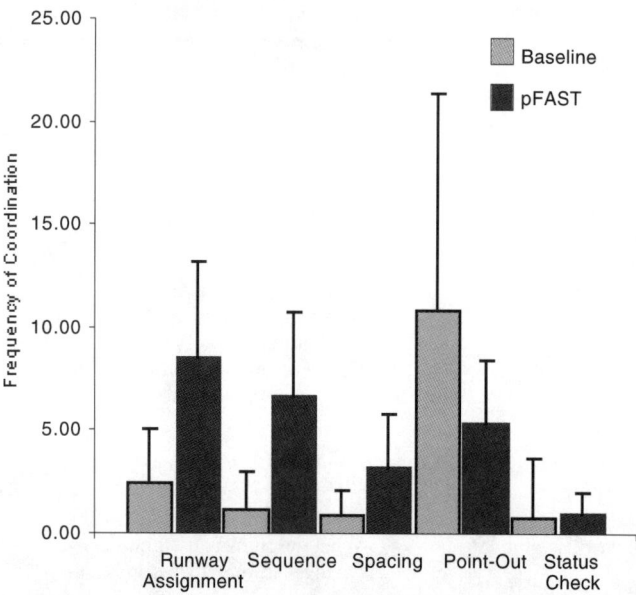

Fig. 3 Baseline vs pFAST coordination comparison: mean frequency of coordination with standard deviation bars.

The point-outs category was the only one that demonstrated a significant decrease under pFAST conditions relative to baseline. Point-outs are defined as coordination with another position and utilizing another controller's airspace, but retaining communication and control. Twice as many point-outs occurred under baseline conditions as occurred under pFAST conditions.

Reducing the number of point-outs could allow controllers to spend more time separating and monitoring aircraft, rather than being concerned with coordination.* It could also allow controllers to coordinate regarding other aspects of the traffic control process; perhaps more advance planning could be accomplished given more time to evaluate the traffic situation, therefore resulting in controllers using each others' airspace less than they would have to otherwise. Alternatively, point-outs could be reduced out of necessity as there was increased discussion regarding the advisories. However, if this were the case, the controllers should have indicated difficulties with coordination. In contrast, the controllers did not report any difficulties with the amount of coordination that they experienced and did not report that the amount of coordination required was increased by the use of the pFAST advisories. The controllers reported that point-outs themselves did not contribute, on average, more than minimally to their overall workload.

Tables 2 and 3 present the mean frequency (and standard deviation) of instances of coordination per rush for the five most frequent categories of coordination in the baseline and test conditions. There were three categories of coordination whose frequencies were common to both baseline and test conditions: altitude changes,

---

*Personal communication with M. Prichard, FAA, DFWTRACON, 1997.

Table 2 Most common coordination categories under baseline conditions

Category	Mean frequency (SD)
Point-outs	10.9 (10.5)
Altitude changes[a]	4.9 (3.5)
Handoffs[a]	4.2 (3.5)
Heading changes	4.1 (2.9)
Runway assignment[a]	2.5 (2.6)
Weather[b]	2.5 (4.1)

[a] Categories common to baseline and test conditions.
[b] Weather conditions were more uniform under pFAST testing than during baseline.

runway assignments, and handoffs. All three of these categories were discussed with greater frequency in the test condition than in the baseline condition.

As shown in Tables 2 and 3, the overall frequency of discussion under pFAST conditions is higher and is more evenly distributed among the top five categories. In the baseline condition, one category, point-outs, occurs twice as frequently as the other top categories of coordination. It is likely that under baseline conditions, the controllers coordinate only about what is unusual, as procedures exist to accomplish routine coordination. The increase in coordination across the major categories in the pFAST condition may indicate the controllers' need to confirm with one another the changes that they noted in their operations. As pFAST use increases, the general increase in coordination could diminish over time as the controllers become more accustomed to a new way of operating. Regardless, the additional coordination was not reported to negatively impact overall controller workload.

IV. Lessons Learned

A. Constraints of Field Testing

The human factors results gathered from the pFAST operational assessment need to be interpreted in the light of the constraints inherent in field testing. Some of the constraints were not anticipated prior to the pFAST evaluation and, as a result, the human factors data are "noisy." Consequently, care must be exercised in extrapolating the results to other decision-support tools or ATC environments.

Table 3 Most common coordination categories under pFAST test conditions

Category	Mean frequency (SD)
Runway assignment[a]	8.6 (4.7)
ARTS problems	7.5 (5.4)
Handoffs[a]	7.2 (4.2)
Sequence	6.8 (4.1)
Altitude changes[a]	5.7 (4.5)

[a] Categories common to baseline and test conditions.

Some of the problems that were faced from testing in an operational ATC facility would have been eliminated had the testing been confined to more controllable settings. However, through the development process of pFAST, it became clear to both researchers and controllers that an operational evaluation was necessary. High-fidelity simulations, both at NASA Ames Research Center and the FAA's William J. Hughes Technical Center, in Atlantic City, New Jersey, could not adequately approximate the operational setting for fully evaluating all of pFAST's functionality. The experience from the pFAST field assessment is instructive for future testing of ATC decision-support tools and for identifying some of the problems faced in testing in an operational environment.

During the planning of the operational assessment, the researchers determined that data collection in the field, especially over several months, would be subject to numerous restrictions. Engineering data collection was less problematic than the human factors data collection, as the engineering data were obtained in an automated fashion. In contrast, to collect the human factors data, observations from live operations were required. Controllers, tasked with completing their usual workload, were also asked to provide feedback about pFAST. The collection of the human factors data in the midst of actual operations meant that boundaries were set on data collection activities; all test personnel clearly understood that operational demands took priority over any type of evaluation activity. Weather, training requirements, or other facility demands on space or time could require that the human factors data collection activities be curtailed. In addition, more questionnaire data were originally proposed than what was eventually collected because of concerns that questionnaire results could have become affected by controller fatigue.

Other challenges in the operational assessment affected both engineering and human factors data. There was no opportunity to exercise experimental control over air traffic conditions, such as the airport configurations and airport acceptance rates, which would have simplified the data analysis. The test periods themselves were limited by facility concerns about the traffic, or conditions of severe or unpredictable weather.

There were also no guarantees about the staffing of the controller positions; although the majority of the rushes were staffed by the assessment team controllers, some substitute controllers participated when assessment team members were not available. Although the assessment team carefully chose the substitute controllers, and the substitute controllers were fully briefed on pFAST, variations in controller perception are introduced into the data.

The pFAST researchers established the requirement that controllers evaluating pFAST be as familiar as possible with the development process and the philosophy behind pFAST. Consequently, the number of actual participants in the pFAST evaluation was restricted relative to the overall facility population, and the facility at-large was not well acquainted with pFAST. Situations arose during the testing phase in which the facility experienced problems with their own systems that were attributed to pFAST even when pFAST was not being used. This is a collateral problem with field testing that was encountered to some degree in other CTAS field evaluations. Although it can be disconcerting to have to prepare to defend a decision-support tool against events totally unrelated to actual testing activities, researchers should be aware that unsubstantiated rumors about system performance can be damaging to the credibility of their decision-support tool.

Data collection was also affected by lack of staffing. For example, the human factors team was not able to directly observe or measure the impact of pFAST on the center, although there were some anecdotal reports that pFAST operations helped to reduce holding in one area.

Finally, the problems that arise in research in general, such as unexpected loss of data and the unavailability of data, also occurred. Extending the testing period would have enabled the researchers to collect more data samples per configuration and rush period, but time and resource constraints prevented this from occurring.

There are obviously tremendous benefits to field testing. In addition to the definite engineering advantages, including exercising pFAST under real-time, real-traffic loads, operational testing also provided valuable insight into human factors issues, such as training, procedural changes, job satisfaction, and coordination and communication effects. The effects on these areas are not easily captured or anticipated through simulations.

Field exposure also provides significant benefits to the researchers who are able to see how the software integrates with existing operations. Direct interaction with the end-users also makes researchers more aware of the users' specific requirements in the target environment. The dialogue between the researchers and the end-users is enhanced through this interaction, thus facilitating coordination required for ongoing development. By being exposed to the development process, the users are able to educate the rest of the facility on the functionality of the decision-support tools. Thus, misconceptions that inevitably arise when such decision-support tools are introduced can be better minimized.

B. Recommendations

The pFAST assessment highlighted a number of procedures that should be required in the human factors data collection and analysis of ATC decision-support tools. Ways to improve the collection and analysis of human factors data in future operational assessments are also identified:

1) Extensive involvement by the target facility (or target user-group) is essential. The feedback from the assessment team helped to define the boundaries of the data collection, as well as to clarify the goals of the evaluation. The assessment team input was also important for making the data collection materials (surveys and debriefing questions) understandable and usable. The facility involvement in this process should be expanded to inform a larger group of target users about the development process, and to help mitigate misconceptions about the new decision-support tool.

2) Flexibility in data collection is needed; the researchers need to be able to accept some noise in the data and plan for the eventuality of missing and lost data. In doing so, it may be necessary to reduce a planned test matrix to try to capture more samples but with a limited scope.

3) The number of observers could be increased or rotated to different observation points to obtain a more complete picture of coordination, as long as this is balanced against disrupting operations.

4) The data should be evaluated partway through the test to streamline the data collection. This would help to weed out questions in questionnaires that reveal few meaningful responses and highlight other topics that need to be investigated more closely. Such an approach might help focus the data analysis, as well as reduce the fatigue involved with responding to long questionnaires.

The human factors evaluation of pFAST emphasizes the need for a testing framework for human factors assessments, describing how data should be gathered and interpreted for multiple, interdependent users of ATC decision-support tools. Some of the findings from this operational assessment can contribute to such a testing framework.

V. Concluding Remarks

The human factors results are as important as the engineering results to the overall evaluation of pFAST. The engineering data show benefits of runway balancing and throughput. The human factors data describe the outcome of these benefits on the controllers themselves. Because of the heightened throughput and more efficient runway balancing during the pFAST operational evaluation, it would not have been surprising if controllers reported increased workload. Also, the nature of the information being provided (runway assignments and sequences to the runways) might have led to increased controller workload through increased coordination and communication. The human factors data instead bear out a different conclusion: despite the increased number of aircraft controlled during the field test, the controllers did not report any significant increase in overall workload. Furthermore, through direct ratings, the controllers indicated that using pFAST was acceptable in an operational setting. The controllers did not report having to increase their coordination activities between themselves, or between the TRACON and the center due to the use of pFAST. They also did not report having to talk to aircraft more frequently to achieve the pFAST-assigned runways or sequences.

The operational assessment at DFW TRACON gave the pFAST researchers a unique opportunity to see their ATC decision-support tool tested under the rigorous traffic conditions of one of the busiest ATC facilities in the United States. The lessons learned from the human factors assessment contribute to defining human factors testing guidelines for field evaluations.

The success of pFAST, as demonstrated by the operational assessment, is due to the long history of controller involvement and input in the design and testing of pFAST. Controllers need to understand new systems in order to effectively utilize and integrate them into their existing knowledge and experience.[17] The development of pFAST, from concept definition through operational testing, employed a strategy of closely coupling the researchers and the controllers working within the boundaries of an existing, and highly specialized, setting. The human factors involvement in the development process contributed to identifying controller needs and determining if those needs were being met. This design approach resulted in the trust of the air traffic controllers and their willingness to test pFAST operationally. Without controller understanding and support of the system, benefits might never have been able to be identified.

The attention that has been paid to the human factors issues has helped to define CTAS and ensure that it will meet controller needs. The human factors findings from the pFAST operational evaluation help to validate the processes that guided pFAST (and CTAS) development and demonstrate how benefits are achieved not only in terms of overall airport throughput and efficiency, but in terms of what impact may be experienced by the controller. The positive human factors findings increases the confidence in the operational deployment of pFAST by ensuring that key issues from the controllers' perspective have been examined.

References

[1] Erzberger, H., Davis, T. J., and Green, S. M., "Design of Center-TRACON Automation System," *Proceedings of the 56th Symposium on Machine Intelligence in Air Traffic Management*, AGARD, Berlin, 1993, pp. 52-1–52-14.

[2] Denery, D. G., and Erzberger, H., "The Center-TRACON Automation System: Simulation and Field Testing," NASA TM 110366, Aug. 1995.

[3] Erzberger, H., and Nedell, W., "Design of Automated System for Management of Arrival Traffic," NASA TM 102201, June 1989.

[4] Harwood, K., and Sanford, B., "Denver TMA Assessment," NASA CR-4554, Oct. 1993.

[5] Hoang, T., and Swenson, H. N., "The Challenges of Field Testing the Traffic Management Advisor in an Operational Air Traffic Control Facility," AIAA Paper 97-3734, Aug. 1997.

[6] Swenson, H. N., Hoang, T., Engelland, S., Vincent, D., Sanders, T., Sanford, B., and Heere, K., "Design and Operational Evaluation of the Traffic Management Advisor at the Dallas/Fort Worth Air Route Traffic Control Center," *Proceedings of the 1st USA/Europe Air Traffic Management Research & Development Seminar*, FAA and EUROCONTROL, June 1997.

[7] Green, S. M., and Vivona, R. A., "Field Evaluation of Descent Advisor Trajectory Prediction Accuracy," AIAA Paper 96-3764, July 1996.

[8] Green, S. M., Vivona, R. A. and Sanford, B., "Descent Advisor Preliminary Field Test," AIAA Paper 95-3368, Aug. 1995.

[9] Sanford, B. D., Smith, N. M., Lee, K. K., and Green, S. M., "Decision-Aiding Automation for the En Route Controller: A Human Factors Field Evaluation," Presented at the 10th International Symposium on Aviation Psychology, The Ohio State Univ., Columbus, OH, May 1999.

[10] Lee, K. K., and Davis, T. J., "The Development of the Final Approach Spacing Tool (FAST): A Cooperative Controller-Engineer Design Approach," *Control Engineering Practice*, Vol. 4, No. 8, Aug. 1996, pp. 1161–1168.

[11] Davis, T. J., Erzberger, H., Green, S. M., and Nedell, W., "Design and Evaluation of an Air Traffic Control Final Approach Spacing Tool," *Journal of Guidance, Control, and Dynamics*, Vol. 14, No. 4, 1991, pp. 848–854.

[12] Quinn, C., and Robinson, J. E., III, "A Human Factors Evaluation of Active Final Approach Spacing Tool Concepts," Presented at the 3rd USA/Europe Air Traffic Management R&D Seminar, FAA and EUROCONTROL, June 2000.

[13] Robinson, J. E., III, and Isaacson, D. R., "A Concurrent Sequencing, and Deconfliction Algorithm for Terminal Area Air Traffic Control," AIAA Paper 2000-4473, Aug. 2000.

[14] Davis, T. J., Isaacson, D. R., Robinson, J. E., III, den Braven, W., Lee, K. K., and Sanford, B., edited by Papageorgiou, M. and Pouliezos, A. "Operational Test Results of the Final Approach Spacing Tool," *Proceedings of the IFAC Eighth Symposium on Transportation Systems '97*, Technical Univ. of Crete, Chania, Greece, 1997, pp. 190–196.

[15] Isaacson, D. R., Davis, T. J., and Robinson, J. E., III, "Knowledge-Based Runway Assignment for Arrival Aircraft in the Terminal Area," AIAA Paper 97-3543, Aug. 1997.

[16] Robinson, J. E., III, Davis, T. J., and Isaacson, D. R., "Fuzzy Reasoning-Based Sequencing of Arrival Aircraft in the Terminal Area," AIAA Paper 97-3542, Aug. 1997.

[17] Hopkin, V. D., "Air Traffic Control," *Human Factors in Aviation*, edited by E. L. Wiener and D. C. Nagel, Academic Press, San Diego, CA, 1988, pp. 639–663.

[18] Lee, K. K., and Sanford, B. D, "Human Factors Assessment: The Passive Final Approach Spacing Tool (pFAST) Operational Evaluation," NASA TM 208750, Oct. 1998.

[19]Harwood, K., Sanford, B. D., and Lee, K. K., "Developing ATC Automation in the Field: It Pays to Get Your Hands Dirty," *ATC Quarterly Journal*, Vol. 6, No. 1, 1998, pp. 45–70.

[20]Harwood, K., and Sanford, B., "Evaluation in Context: ATC Automation in the Field," *Human Factors Certification of Advanced Aviation Technologies*, edited by J. Wise, V. D. Hopkin, and P. Stager, Embry-Riddle Aeronautical Univ. Press, Daytona Beach, FL, 1994, pp. 247–262.

[21]Volckers, U., Brokof, U., Dippe, D., and Schubert, M., "Contributions of DLR to Air Traffic Capacity Enhancement Within a Terminal Area," AGARD 56th Conference of GCP, AGARD, Berlin, 1993, pp. 10-1–10-11.

[22]Harwood, K., "Defining Human-Centered System Issues for Verifying and Validating Air Traffic Control Systems," *Verification and Validation of Complex and Integrated Human Machine Systems*, edited by J. Wise, V. D. Hopkin, and P. Stager, Springer-Verlag, Berlin, 1993, pp. 115–129.

[23]Hart, S. G., and Staveland, L. E., "Development of a NASA-TLX (Task Load Index): Results of Empirical and Theoretical Research," *Human Mental Workload*, edited by P. A. Hancock and N. Meshkati, North-Holland, Amsterdam, 1988, pp. 139–183.

Chapter 36

Evaluating Taskload Measures Derived from Routinely Recorded Air Traffic Control Data

Carol A. Manning*
Federal Aviation Administration Civil Aeromedical Institute,
Oklahoma City, Oklahoma

Scott H. Mills[†]
SBC Technology Resources, Inc., Austin, Texas

Cynthia M. Fox[‡] and Elaine Pfleiderer[‡]
Federal Aviation Administration Civil Aeromedical Institute,
Oklahoma City, Oklahoma
and
Henry Mogilka[§]
Federal Aviation Administration Training Academy, Oklahoma City, Oklahoma

I. Introduction

IT IS necessary to measure workload, taskload, complexity, and performance in air traffic control (ATC) to evaluate the effects of new systems and procedures on individual air traffic controllers and on the ATC system as a whole.[1] The effects of using different display designs or alternative procedures on controllers' workload and performance must be assessed before they are implemented. When new ATC systems are introduced in field facilities, it is necessary to document their effects on individual and system performance, both soon after implementation and later, after controllers have become accustomed to using them. Computing measures of taskload and performance on a system level, while accounting for sector complexity, may also contribute to better prediction of overloads at specific sectors.

This material is declared a work of the U.S. Government and is not subject to copyright protection in the United States.
*Engineering Research Psychologist.
[†]Senior Member of Technical Staff.
[‡]Research Assistant.
[§]Supervisory Air Traffic Control Instructor.

II. Defining Controller Workload, Taskload, Sector Complexity, and Performance

Although many methods have been used to measure ATC workload, taskload, sector complexity, and controller performance, definitions of the terms are not widely agreed on. In general, mental workload typically refers to the physical and mental effort an individual exerts to perform a task. In this sense, ATC workload may be differentiated from taskload in that taskload refers to air traffic events to which the controller is exposed, whereas mental workload describes the effort expended by the controller to manage those events.

Sector complexity describes the static and dynamic characteristics of the air traffic environment that combine with the taskload to produce a given level of controller mental workload.[2] In that sense, complexity can mediate the relationship between taskload and mental workload. Federal Aviation Administration (FAA) *Air Traffic Control*[3] states, "The primary purpose of the ATC system is to prevent a collision between aircraft operating in the system and to organize and expedite the flow of traffic." Thus, measurement of controller performance involves determining the effectiveness with which an individual controller's activities accomplished these goals.

A. Measures of ATC Mental Workload, Taskload, Sector Complexity, and Controller Performance

Many methods have been developed to measure mental workload, taskload, sector complexity, and controller performance in air traffic control (see Hadley et al.[4] for a database containing 162 measures). The dynamic nature of ATC (encompassing both movements of an individual aircraft and constant changes in relative positions of groups of aircraft) makes it necessary to take the passage of time into consideration when measuring these constructs. Even when time is considered, it is even more difficult to measure controller performance and mental workload than it is to measure taskload and sector complexity. The reason is that taskload can be measured by counting recorded ATC events, and sector complexity can be measured by recording observable sector characteristics and other factors about the ATC situation. Mental workload and controller performance, on the other hand, include factors that cannot be easily observed, and thus are not easy to measure. For example, controllers constantly review aircraft positions, directions, and speeds, and mentally project aircraft positions, but they may only occasionally take observable actions. It is possible to count or otherwise evaluate certain observable activities, such as making keyboard entries and marking or moving flight progress strips. However, the relationship between these activities (taskload) and the amount of cognitive effort expended (mental workload) or the effectiveness of the results (controller performance) is unclear. Even actions that appear to be easily interpretable (e.g., commission of operational errors resulting in losses of separation) may not be very meaningful because they occur so infrequently as to be of little value in assessing individual performance.

This section discusses some of the methods that have been used to measure mental workload, taskload, sector complexity, and controller performance. The advantages and disadvantages of using these methods will also be discussed.

B. Mental Workload Measures

Mental workload, controllers' subjective reactions to the taskload they experienced, is hypothesized to include components that cannot be explained by taskload alone. Measures of mental workload in ATC may be obtained either during a simulated scenario or after its completion. For example, the NASA Task Load Index[5] (TLX) is obtained from controllers after they complete a scenario. Controllers provide separate ratings for each of six scales: mental demand, physical demand, temporal demand, effort, frustration, and performance.

In contrast, the air traffic workload input technique[6] (ATWIT) measures mental workload in real time. The ATWIT presents auditory and visual cues (a tone and illumination, respectively) that prompt a controller to press one of seven buttons within a specified amount of time to indicate the amount of mental workload experienced at that moment. The workload assessment keypad (WAK) device records each mental workload rating as well as the time it took to respond to the prompt.

The primary advantage of using a real-time mental workload measure is that the respondent can report the experience soon after it occurs. However, the process of providing a real-time rating may increase the controller's perception of his or her mental workload or, worse yet, may interfere with the performance of certain tasks. On the other hand, a workload rating obtained after a scenario is complete may be overly influenced by earlier or more recent events (i.e., primacy or recency effects), or the controller may forget to consider certain events altogether.

C. Taskload Measures

Several measures describing controller taskload have been derived from recordings of either operational National Airspace System (NAS) activities or simulation data. For example, Buckley et al.[7] developed a set of computer-derived measures obtained during ATC simulations. They identified four factors that summarized the measures: conflict, occupancy, communications, and delay. Galushka et al.[8] used counts of controller activities [along with over-the-shoulder (OTS) subjective performance ratings] to assess en route air traffic controller baseline performance during a simulation study.

Using data extracted from the log and track files generated by the Data Analysis and Reduction Tool[9] (DART), Mills[10] developed an extensive set of computer-derived taskload measures. Performance and Objective Workload Evaluation Research (POWER) software measures information about controlled aircraft, handoffs, number of altitude changes, number of controller data entries and data entry errors, and variations in aircraft headings, speeds, and altitudes. Mills[11] described these measures in more detail.

D. Sector Complexity Measures

Several measures of sector complexity have also been developed. These typically include both physical characteristics of a sector and factors specific to the air traffic situation. For example, Grossberg[2] identified three groups of factors (control adjustments such as merging, spacing, and speed changes; climbing and descending flight paths; and the mix of aircraft types) that contributed to the complexity of operations in different sectors. Mogford et al.[12] identified 15 complexity factors using multidimensional scaling techniques. The complexity construct has also been found useful in research. Rodgers et al.[13] found a significant multiple correlation between the overall rate of operational errors at Atlanta Center and the 15 complexity factors of Mogford et al.[12]

If sector information is available, it should be relatively easy to measure these factors. The number of factors necessary to describe sector complexity is not clear (though the constructs proposed should be closely related). Nevertheless, it appears that the complexity construct may provide information beyond what is available from the taskload construct.

E. Controller Performance Measures

One of the challenges associated with measuring controller performance is evaluating the different approaches controllers use to control traffic. Most approaches used by a controller to maintain aircraft separation and a smooth flow of air traffic would be considered acceptable. However, such individuality of technique makes it difficult to evaluate the effectiveness of an individual controller's actions to move a set of aircraft through a sector.

To accommodate these differences in technique, subject matter expert (SME) observations are the most frequently used methods for measuring ATC performance. Several processes have been developed to record SME observations. The behavioral summary scales[14] (BSS) were developed as a criterion measure against which the air traffic selection and training (AT-SAT) selection battery[15] could be validated. The BSS scales included 10 distinct performance categories and measured typical rather than maximum performance, that is, how well controllers performed consistently over time rather than how well they could perform under peak traffic conditions.

Several other procedures have been developed to evaluate maximum performance (during high-fidelity simulations). For example, Bruskiewicz et al.[16] developed two other procedures for measuring controller performance that were used in a high-fidelity simulation study conducted to evaluate the AT-SAT performance measures. These were the OTS rating form and the behavior and event checklist (BEC). The OTS rating form, used to evaluate controller performance across broad dimensions, was based in part on the BSS. The BEC was used to record specific mistakes made during the simulation exercises.

The advantage of using SME observations as a basis for evaluating controller performance is that SMEs (especially instructors involved in controller training) possess detailed knowledge about the job and, thus, can evaluate aspects of controllers' behavior beyond what can be obtained from counting events. They are also very accustomed to observing the actions of other controllers.

However, several problems may be associated with SME observations. First, determining appropriate performance ratings and identifying mistakes require considerable interpretation on the part of the observer. To assure the reliability of these subjective ratings and error counts, extensive SME training and practice sessions are required. It is also not always possible to obtain SME observations because often an insufficient number of controllers is available to participate in these activities.

III. Purpose of Study

Our challenge was to develop measures describing aspects of ATC activity that are objective, reliable, valid, and easy to obtain. Measures (such as SME observations or mental workload ratings) that may require extensive training to provide but have more apparent validity than taskload measures are frequently not available, while recorded ATC data usually are. On the other hand, taskload measures may not be adequate if the numbers derived from recorded ATC data do not sufficiently account for the subjective aspects of mental workload or controller performance.

The goal of this study was to determine whether the POWER taskload measures derived from routinely recorded ATC data could sufficiently describe mental workload, sector complexity, and controller performance. Although a set of POWER measures has been computed, as yet no empirical evidence is available to indicate whether or not these numbers actually measure the constructs they were intended to measure. In particular, we hypothesized that some POWER measures may be related to measures of controller workload and/or performance for individual controllers (see Table 1). Likewise, some POWER measures may also reflect ATC complexity.

If the POWER measures are found to relate to measures of sector complexity, mental workload, or controller performance, it may be possible to use them in situations in which it would not otherwise be possible to evaluate these variables (when SMEs are unavailable or controllers were unable to provide workload evaluations). For example, a validated set of POWER measures could provide information that would allow post-implementation evaluation of the operational effects of new ATC systems.

IV. Method

A. Participants

Participants were 16 en route air traffic control instructors from the FAA Academy in Oklahoma City, Oklahoma. All were previously fully qualified controllers at en route air route traffic control centers (ARTCCs). Two participants previously controlled traffic at some of the sectors represented in the traffic samples, though none worked all sectors included in the study.

B. Materials

1. Traffic Samples

System analysis report (SAR) and voice communication tapes were obtained for 12 traffic samples obtained from four sectors in the Kansas City ARTCC. The traffic samples consisted of routine operations and contained no accidents or incidents.

The SAR data used for the traffic samples were extracted by the DART program and the National Track Analysis Program[17] (NTAP). Resulting files were processed both by Systematic Air Traffic Operations Research Initiative[18] (SATORI) and POWER[10] software. SATORI synchronizes information from DART and NTAP files with tapes containing the radar (R) controller's voice communications, using the time code common to both data sources. POWER uses a subset of the DART files to compute measures of sector and controller activity.

Table 1 Expected relationships for POWER measures

Power measure	Complexity	Performance	Workload
Total number of aircraft controlled	X		X
Maximum aircraft simultaneously controlled	X		X
Average time aircraft under control	X	X	X
Average heading variation	X	X	X
Average speed variation	X	X	X
Average altitude variation	X	X	X
Total number of altitude changes	X	X	X
Total number of handoffs (HO)	X		X
Total number of handoffs accepted			X
Average time to accept handoff		X	X
Total number of handoffs initiated			X
Average time until initiated HOs are accepted			X
Total number of data entries			X
Total number of data entry errors			X
Number of radar controller data entries		X	X
Number of radar controller data entry errors		X	X
Number of route displays		X	X
Number of radar controller pointouts		X	X
Number of data block offsets		X	X
Total number of conflict alerts		X	X
Number of conflict alert suppression entries		X	
Number of distance reference indicators requested		X	X
Number of distance reference indicators deleted		X	X
Number of track reroutes			X
Number of strip requests			X

Three traffic samples were recreated for each of the four sectors. One traffic sample (used for training) was 8 min long. The two remaining experimental traffic samples were both 20 min long.

2. *Sector Training Materials*

Computerized training sessions were developed that described characteristics and procedures applicable to each sector. Participants also examined copies of sector maps on which important information was highlighted. These maps and a copy of the sector binder (containing additional sector information) were available for the participants to review while they watched the traffic samples. Participants also had access to flight plan information (derived from flight strip messages) for each aircraft controlled by the sector during the traffic sample.

3. Mental Workload and Controller Performance Measures

Participants provided three types of mental workload measures (the ATWIT, the NASA TLX, and an estimate of the traffic sample's activity level) and two types of controller performance measures (an OTS form and a BEC) for each traffic sample they observed. ATWIT ratings were elicited every 4 min during each traffic sample using the WAK. Although controllers typically rate their own mental workload, in this study the participants used the WAK to rate the amount of mental workload they thought the R controller experienced in reaction to the taskload that occurred during the traffic sample.

Participants also completed the NASA TLX after each traffic sample. Again, instead of rating their own mental workload, participants used the NASA TLX to rate their perception of the mental workload experienced by the R controller. The TLX ratings were entered using a computerized form. The activity level rating was provided for each traffic sample using a 5-point scale ranging from "Not at all busy" to "Very busy."

Participants rated controller performance using a revised version of the OTS form originally developed for the AT-SAT high-fidelity validation study.[15] The OTS form was revised because participants had access to only the R controllers' voice communications and thus were unable to evaluate all the events that occurred at the sector during the traffic sample. Participants also used the BEC to record errors made by the R controller during the traffic sample.

4. Sector Complexity

The sector complexity measures used in this study were based on the 15 complexity factors of Mogford et al.[12] These factors were combined into three variables. Static complexity included variables that remained constant over the course of a traffic sample. These were numbers of adjacent sectors, transfer control points, sequencing functions, military operations, major airports, Very high frequency Omni-directional Range Tactical air navigation System (VORTACS), airway and jetway intersections, miles of airways, shelves, and airspace size. This information was derived from letters of agreement for each sector and from Kansas City ARTCC's adaptation control environmental system (ACES) map files.

Dynamic complexity included variables that would be expected to vary during a traffic sample. These were numbers of pilot/controller transmissions, interphone communications, maximum number of Hs and Ls displayed (indicating high and low weather activity), amount of climbing/descending traffic, percentages of jets and visual flight rules (VFR) aircraft, number of military aircraft, percentages of arrivals/departures for St. Louis Lambert airport, clearances issued for traffic, altitude and speed restrictions issued, heading changes or vectors issued, conversations about holding, conversations about weather, and a variable reflecting traffic volume. This information was derived from the traffic samples. An overall complexity variable was also computed by averaging standardized scores for all variables included in the static and dynamic complexity measures.

C. Procedure

Participants reviewed a description of the purpose and methods for the experiment, completed consent and biographical information forms, then viewed instructions for completing the workload and performance measures. For each of the four sectors, participants 1) reviewed training materials, 2) observed one 8-min training traffic sample, and 3) observed two 20-min experimental traffic samples. To ensure continuity, all traffic samples for a sector were shown together as a block. The order in which the four blocks of traffic samples were observed was counterbalanced, as was the order in which the two experimental traffic samples were presented within each block.

As each traffic sample progressed, participants recorded any mistakes they observed on the BEC. The ATWIT aural signal occurred every 4 min. Participants responded by entering a number between 1 and 7 on the WAK keypad. At the end of the traffic sample, participants completed the NASA TLX, summed the errors they had marked on the BEC, then completed the OTS rating form. Finally, they rated the activity level for that traffic sample.

Reviewing the training materials and observing the three traffic samples for each sector required about 1.5 h. After observing the traffic samples for all four sectors, participants answered questions about their experiences during the observation process.

V. Results

To simplify the analysis, the original 24 controller performance and mental workload variables were combined into seven categories (see Manning et al.[19] for a description). The five controller performance categories included three principal components derived from the BEC that described the types of errors made, 1) inactivity, 2) disorganization, and 3) inefficient but safe, the overall OTS rating, and the TLX performance scale. The two mental workload categories were 1) the ATWIT and traffic sample activity level measures, along with the mental, physical, temporal, and effort TLX scales comprising a component called "Activity"; and 2) the TLX frustration scale, along with the mental, physical, temporal, and effort TLX scales comprising a component called "Frustration."

Values for the seven mental workload and controller performance categories (averaged across raters) were computed for each traffic sample. The static, dynamic, and overall complexity factors and the associated POWER measures were computed for the eight experimental scenarios. These data were matched with the seven performance and workload measures, and descriptive statistics were computed (see Table 2).

Some of the POWER measures occurred fairly often over the 20-min period. Other data entries [e.g., point-outs, data block offsets, distance reference indicators (DRIs), otherwise known as J-rings, track reroutes, and strip requests] occurred very infrequently. In fact, many data entries did not occur at all during most traffic samples, resulting in near-zero means and standard deviations greater than the corresponding means. The complexity measures, which were standardized, and the controller performance and mental workload measures, derived from orthogonally rotated principal components, had mean values of zero, but the standard deviations indicate their relative reliability.

Tables 3–5 show correlations of the remaining POWER measures with the sector complexity, controller performance, and mental workload measures. Correlations significant at the .05 level or lower are starred. Since the number of traffic samples analyzed was so small ($N=8$) and the number of correlations computed was so large ($N=338$), it is likely that some of the statistically significant correlations could have occurred due to chance. However, this result is less likely if a POWER

Table 2 Descriptive statistics for POWER, sector complexity, controller performance, and mental workload measures ($N = 8$)

Measures	Mean	SD
Power measures		
Total number of aircraft controlled	15.25	5.23
Maximum aircraft controlled simultaneously	6.88	2.47
Average time aircraft under control	389.75	97.62
Average heading variation	11.64	3.02
Average speed variation	1.28	.68
Average altitude variation	.84	.51
Total number of altitude changes	12.13	5.38
Total number of handoffs (HO)	19.50	7.37
Total number of handoffs accepted	5.88	3.98
Average time to accept handoff	39.22	19.54
Total number of handoffs initiated	10.0	4.0
Average time until initiated HOs are accepted	50.47	27.26
Number of radar controller data entries	56.75	22.70
Number of radar controller data entry errors	1.13	1.36
Number of data controller data entries	9.63	5.73
Number of data controller data entry errors	.38	.52
Number of route displays	2.00	2.27
Number of radar controller pointouts	0.38	0.74
Number of data controller pointouts	0.38	1.06
Number of data block offsets	0.75	0.89
Total number of conflict alerts displayed	0.50	0.53
Number of conflict alert suppression entries	0.13	0.35
Number of distance reference indicators requested	0.25	0.46
Number of distance reference indicators deleted	0.13	0.35
Number of track reroutes	0.38	0.74
Number of strip requests	0.13	0.35
Sector complexity measures		
Static complexity	0.0	2.31
Dynamic complexity	0.0	5.57
Overall complexity	0.0	7.62
Controller performance measures		
Average OTS rating	3.79	0.24
TLX performance	44.53	4.89
Inactivity component	0.0	0.47
Disorganization component	0.0	0.13
Inefficient but safe component	0.0	0.33
Mental workload measures		
Activity component	0.0	0.59
Frustration component	0.0	0.52

measure was correlated with more than one measure of a construct or if several similar POWER measures were correlated with the same construct. On the other hand, many of the constructs are independent (because they are component scores produced by a principal components analysis with varimax rotation). Thus, a lack of relationship of a POWER measure with multiple components is not unexpected.

A. Relationship with Sector Complexity

Table 3 shows the relationship of POWER measures with the three sector complexity measures. Higher static complexity (based on sector characteristics) was related to higher average speed variation, fewer R controller point-outs, and fewer data block offsets. Higher dynamic complexity (based on situational

Table 3 Correlations of POWER measures with measures of sector complexity ($N = 8$)

Power measure	Static complexity	Dynamic complexity	Overall complexity
Total number of aircraft controlled	−.52	.16	−.20
Maximum aircraft controlled simultaneously	−.39	.66	.25
Average time aircraft under control	−.14	.72[a]	.44
Average heading variation	.58	−.40	.06
Average speed variation	.72[a]	.18	.57
Average altitude variation	.38	.18	.36
Total number of altitude changes	.25	.37	.42
Total number of handoffs (HO)	−.51	−.13	−.41
Total number of handoffs accepted	−.61	.57	.05
Average time to accept handoff	−.41	−.57	−.67
Total number of handoffs initiated	−.52	−.17	−.44
Average time until initiated HOs are accepted	.62	.08	.44
Number of radar controller data entries	−.29	.48	.17
Number of radar controller data entry errors	.12	.48	.43
Number of data controller data entries	−.20	−.17	−.25
Number of data controller data entry errors	.32	−.34	−.06
Number of route displays	−.35	−.02	−.22
Number of radar controller pointouts	−.78[a]	.33	−.23
Number of data controller pointouts	−.66	.00	−.40
Number of data block offsets	−.81[a]	.36	−.23
Total number of conflict alerts displayed	.20	.49	.48
Number of conflict alert suppression entries	.62	.34	.63
Number of distance reference indicators requested	−.29	.40	.11
Number of distance reference indicators deleted	−.05	−.17	−.15
Number of track reroutes	.57	.24	.52
Number of strip requests	.46	.27	.48

[a] $p < .05$.

characteristics) was related to longer times that aircraft were under control. Higher overall complexity (combining the components of both static and dynamic complexity) had no significant correlations with any of the POWER measures.

B. Relationship with Controller Performance

Table 4 shows the relationship of POWER measures with measures of controller performance. Higher average OTS ratings were related to more conflict alerts displayed. Higher TLX performance scores (indicating lower performance) were related to more R and data (D) controller point-outs made, and more data block offsets made. It was also related to lower average heading variation.

Higher scores on the BEC inactivity scale were related to lower average heading variation, more handoffs accepted, and more R and D controller point-outs. Higher scores on the BEC disorganization scale were related to higher altitude variation and more DRIs deleted. Higher scores on the BEC inefficient but safe scale were related to more aircraft controlled simultaneously, more handoffs accepted, more R and D controller point-outs, and more data block offsets.

C. Relationship with Mental Workload

Table 5 shows the relationship of POWER measures with measures of mental workload. Higher scores on the activity scale were related to more R controller data entries and more conflict alerts displayed. Higher scores on the frustration scale were related to more aircraft controlled, more handoffs accepted, more R and D controller point-outs, and more data block offsets.

VI. Conclusions

Twenty-four measures of mental workload and controller performance were collected from 16 SME observers to assess the relationships with a set of taskload measures derived from routinely recorded ATC data. A reduced set of these measures was derived, and a correlational analysis was performed. However, given the small number of observations, the results should be interpreted with caution. Nevertheless, some interesting relationships between controller and sector activities and the constructs of sector complexity, controller performance, and mental workload were evident.

The interpretation of the relationships between POWER measures and sector complexity may have been related to the structure and functions of the sectors used in the study. For example, the POWER measures that were related to static complexity (a variable based on sector characteristics), such as higher speed variation (suggesting more speed changes were issued), and making fewer data block offsets and R controller point-outs during traffic samples, may be influenced by the way the sector boundaries were configured. Moreover, other POWER measures may have been related to dynamic complexity (based on events that occurred during the traffic sample) because of the functions of the sectors. For example, having aircraft under control for a longer time is related to sector size, busyness, and purpose (i.e., arrival, departure, overflight sectors).

The relationships between the POWER measures and measures of controller performance are not as easy to interpret. Specifically, higher performance ratings (on the reverse-scaled TLX performance scale) and higher inactivity were related

Table 4 Correlations of POWER measures with measures of controller performance ($N = 8$)

Power measure	Controller performance measure				
	Average OTS rating	TLX performance	BEC inactivity	BEC disorganization	BEC inefficient but safe
Total number of aircraft controlled	.36	.23	.46	−.43	.62
Maximum aircraft controlled simultaneously	.43	.26	.56	−.30	.78[a]
Average time aircraft under control	−.05	.29	.47	.40	.60
Average heading variation	.45	−.91[a]	−.71[a]	−.11	−.57
Average speed variation	−.11	−.38	−.29	.51	−.12
Average altitude variation	−.09	−.24	−.25	.72[a]	−.18
Total number of altitude changes	.14	−.26	.04	.51	.38
Total number of handoffs (HO)	.42	.09	.25	−.42	.35
Total number of handoffs accepted	−.02	.65	.79[a]	−.01	.91[a]
Average time to accept handoff	−.01	−.05	−.12	−.20	−.22
Total number of handoffs initiated	.42	.11	.24	−.44	.29
Average time until initiated HOs are accepted	−.01	−.18	−.42	−.06	−.38
Number of radar controller data entries	.66	−.08	.21	−.35	.44
Number of radar controller data entry errors	−.00	.25	.08	−.42	.02
Number of data controller data entries	.07	.20	−.13	.18	.01
Number of data controller data entry errors	.44	−.40	−.51	−.69	−.53
Number of route displays	.33	.18	−.16	.20	−.04
Number of radar controller pointouts	−.34	.80[a]	.95[a]	−.15	.89[a]
Number of data controller pointouts	−.56	.74[a]	.88[a]	−.10	.79[a]
Number of data block offsets	−.04	.75[a]	.70	−.01	.72[a]
Total number of conflict alerts displayed	.83[a]	−.39	−.07	−.24	.24
Number of conflict alert suppression entries	.26	−.58	−.10	.16	.21
Number of distance reference indicators requested	.05	.30	−.01	.51	.08
Number of distance reference indicators deleted	−.35	.19	−.24	.79[a]	−.21
Number of track reroutes	.08	−.46	−.21	.52	.10
Number of strip requests	−.01	.00	−.09	−.47	−.17

[a] $p < .05$.

Table 5 Correlations of POWER measures with mental workload measures ($N = 8$)

	Mental workload measure	
Power measure	Activity	Frustration
Total number of aircraft controlled	.56	.73[a]
Maximum aircraft controlled simultaneously	.69	.64
Average time aircraft under control	.34	.28
Average heading variation	.31	−.43
Average speed variation	−.04	−.30
Average altitude variation	.14	−.44
Total number of altitude changes	.53	.29
Total number of handoffs (HO)	.58	.51
Total number of handoffs accepted	.33	.81[a]
Average time to accept handoff	−.08	.15
Total number of handoffs initiated	.57	.41
Average time until initiated HOs are accepted	−.45	−.40
Number of radar controller data entries	.83[a]	.38
Number of radar controller data entry errors	−.44	−.07
Number of data controller data entries	.05	.23
Number of data controller data entry errors	−.08	−.36
Number of route displays	.40	.01
Number of radar controller pointouts	−.04	.91[a]
Number of data controller pointouts	−.28	.90[a]
Number of data block offsets	.24	.72[a]
Total number of conflict alerts displayed	.96[a]	.03
Number of conflict alert suppression entries	.47	.06
Number of distance reference indicators requested	.16	.00
Number of distance reference indicators deleted	−.26	−.11
Number of track reroutes	.32	.00
Number of strip requests	−.50	−.30

[a] $p < .05$.

to lower average heading variation (suggesting that controllers who received poorer performance ratings turned aircraft less often), more handoffs accepted, more R and D controller point-outs, and more data block offsets. Likewise, accepting more handoffs, making more point-outs, and more data block offsets were also related to higher inefficiency. Perhaps the controllers making these entries were more engaged in housekeeping activities than handling the traffic efficiently and effectively. Higher altitude variation (suggesting more altitude changes were made) and more DRIs deleted were related to higher disorganization scores. Finally, more conflict alerts displayed were related to higher average OTS ratings. This last result suggests that the OTS performance ratings made by the SME observers may have been partially based on the workload the participant perceived to occur during the traffic sample.

Mental workload seems to be related to aircraft and controller activities. The two workload components used in this study (activity and frustration) appear to measure different aspects of mental workload, in part because they were typically correlated with different POWER measures. The activity component was related to R controller data entries and conflict alerts displayed (which was also significantly correlated with the average OTS rating). Data entries and conflict alerts are an indicator of how busy the controller is. On the other hand, the frustration component was related to total numbers of aircraft controlled, total handoffs accepted, R and D controller point-outs, and data block offsets made. This component seems to be related to the extent to which higher aircraft activity requires additional controller effort (such as pointing out an aircraft to another sector or moving data blocks to be able to continue to see aircraft information).

Several POWER measures appear to be unrelated to sector complexity, controller performance, and mental workload. These included average time until initiated handoffs are accepted (hypothesized to be related to mental workload), number of R controller data entry errors (hypothesized to be related to controller performance), number of D controller data entries (hypothesized to be related to mental workload), number of route displays (hypothesized to be related to controller performance), number of DRIs requested (hypothesized to be related to both controller performance and mental workload), number of track reroutes (hypothesized to be related to mental workload), and number of strip requests (hypothesized to be related to mental workload). Although some of the hypothesized relationships in fact may not exist, some of these variables occurred very infrequently during traffic samples used in this study, and so it may be inappropriate to conclude at this time that they are unrelated to any of the constructs.

Although this exploratory study provided important information about the POWER measures, additional research is needed to better understand the relationships observed here. It may be possible to compare controller performance and mental workload ratings collected during simulation studies with POWER measures obtained for those traffic samples to obtain additional evidence about the validity of the measures. It will also be necessary to analyze larger blocks of POWER data to examine the statistical characteristics of the measures and perhaps identify a smaller set of POWER measures that accounts for differences in sector complexity, controller performance, and mental workload. When the properties and limitations of these measures are better understood, they may then be used to calculate baseline measures for the current National Airspace System and may eventually be used to assess the effects of implementing new ATC systems.

Acknowledgments

This research could not have been conducted and reported without the support of the following individuals and groups: Dennis Rester and Greg Ruffin (computer support), Rick Day and Dale Engel (provided access to the SAR data), Clarence Jones (coordinated obtaining the SAR data), the FAA Academy instructors who participated in the study, and those internal and external reviewers who provided valuable comments regarding this document.

References

[1] Wickens, C. D., Mavor, A. S., Parasuraman, R., and McGee, J. P. (eds), *The Future of Air Traffic Control: Human Operators and Automation,* National Academy Press, Washington, DC, 1998, pp. 216–217.

[2] Grossberg, M., "Relation of Airspace Complexity to Operational Errors," *Quarterly Report of the Federal Aviation Administration's Office of Air Traffic Evaluation and Analysis,* April 1989.

[3] *Air Traffic Control,* Federal Aviation Administration, Order 7110.65M, Feb. 2000.

[4] Hadley, G. A., Guttman, J. A., and Stringer, P. G., "Air Traffic Control Specialist Performance Measurement Database," William J. Hughes Technical Center, DOT/FAA/CT-TN99/17, June 1999.

[5] Hart, S. G., and Staveland, L. E., "Development of NASA-TLX (Task Load Index): Results of Empirical and Theoretical Research," *Human Mental Workload,* edited by P. A. Hancock and N. Meshkati, North-Holland, Amsterdam, 1988, pp. 139–183.

[6] Stein, E. S., "Air Traffic Controller Workload: An Examination of Workload Probe," Federal Aviation Administration Technical Center, DOT/FAA/CT-TN84/24, 1985.

[7] Buckley, E. P., DeBaryshe, B. D., Hitchner, N., and Kohn, P., "Methods and Measurements in Real-Time Air Traffic Control System Simulation," Dept. of Transportation/Federal Aviation Administration Technical Center, DOT/FAA/CT-83/26, 1983.

[8] Galushka, J., Frederick, J., Mogford, R., and Krois, P., "Plan View Display Baseline Research Report," Federal Aviation Administration Technical Center, DOT/FAA/CT-TN95/45, Sept. 1995.

[9] *Multiple Virtual Storage (MVS); User's Manual; Data Analysis and Reduction Tool (DART),* Federal Aviation Administration, NASP-9247-PO2, 1993.

[10] Mills, S. H., Manning, C. A., and Pfleiderer, E. M., "Computing En Route Baseline Measures with POWER," Poster presented at Tenth International Symposium on Aviation Psychology, Columbus, OH, May 1999.

[11] Mills, S. H., "Performance and Objective Workload Evaluation Research (POWER)," technical report in preparation, 2001.

[12] Mogford, R. H., Murphy, E. D., Roske-Hofstrand, R. J., Yastrop, G., and Guttman, J. A., "Research Techniques for Documenting Cognitive Processes in Air Traffic Control: Sector Complexity and Decision Making," Federal Aviation Administration Technical Center, DOT/FAA/CT-TN94/3, June 1994.

[13] Rodgers, M. D., Mogford, R. H., and Mogford, L. S, "The Relationship of Sector Characteristics to Operational Errors," Federal Aviation Administration Office of Aviation Medicine, DOT/FAA/AM-98-14, 1998.

[14] Borman, W. C., Hedge, J. W., Hanson, M. A., Bruskiewicz, K. T., Mogilka, H., Manning, C., Bunch, L. B., and Horgen, K. E., "Development of Criterion Measures of Air Traffic Controller Performance," *Documentation of Validity for the AT-SAT Computerized Test Battery,* Contract No. DTFA01-95-C-00052, Caliber Associates, 1999.

[15] *Documentation of Validity for the AT-SAT Computerized Test Battery,* Caliber Associates, Contract No. DTFA01-95-C-00052, Fairfax, VA, 1999.

[16] Bruskiewicz, K. T., Hedge, J. W., Manning, C. A., and Mogilka, H. J., "The Development of a High Fidelity Performance Measure for Air Traffic Controllers," *Measuring Air Traffic Controller Performance in a High-Fidelity Simulation,* edited by C. A.

Manning, DOT/ FAA/AM-00/2, Federal Aviation Administration Office of Aviation Medicine, Jan. 2000.

[17]*Multiple Virtual Storage (MVS); Subprogram Design Document; National Track Analysis Program (NTAP),* Federal Aviation Administration, NASP-9114-H04, 1991.

[18]Rodgers, M. D., and Duke, D. A., "SATORI: 'Situation Assessment Through Recreation of Incidents," *Journal of Air Traffic Control,* Vol. 3, No. 54, 1993, pp. 10–14.

[19]Manning, C. A., Mills, S. H., Pfleiderer, E., Fox, C., and Mogilka, H. J., "Investigating the Validity of Performance and Objective Workload Evaluation Research (POWER)," Federal Aviation Administration Office of Aviation Medicine, technical report in press, 2001.

Chapter 37

Controller Roles—Time to Change

Robert Graham,* Alan Marsden,† Isabelle Pichancourt,‡
and Franck Dowling§
EUROCONTROL Experimental Centre, Brétigny-sur-Orge, France

I. Introduction

THIS chapter presents concepts and initial simulation results put forward by controller tools and transition trials[1] (C3T) a study project whose aim is to define, develop, and evaluate controller roles, tasks, and working methods suited to the introduction of (ATM) decision support tools (DST) and data link communication. The project is in support of the EUROCONTROL ATM2000+ strategy[2] and will apply both model and real-time simulation techniques.

To address the need for change, C3T has identified a number of operational improvements linked to technology expectations including:

1) Task sharing, with the planning controller (PC) organizing tactical controller (TC) activity;

2) PC empowerment (which means facilitating PC definition of tactical problem resolutions), using predefined alternate clearances (PAC), system assisted coordination (SYSCO), and data link (CPDLC) to resolve problems and deliver clearances early;

3) Multisector planning (MSP), reducing complexity and density, balancing traffic load and planning traffic flows over several sectors;

4) Delegation, from controller to pilot for specific separation assurance;

5) Trajectory negotiation permits contracting agreed constraints for which there is a high expectation of achievement.

This chapter discusses results arising from a model-based and a real-time simulation of a number of planning controller empowerment scenarios.

Copyright © 2001 by EUROCONTROL. Published by the American Institute of Aeronautics and Astronautics, Inc., with permission.
*Programme Manager-ERIS.
†System Analyst.
‡Human Factor Specialist.
§Fast-time Simulation Specialist.

II. Controller Tools and Transition Trials—C3T

C3T aims to identify benefits accrued to airspace users, service providers, and controllers from the proposed concepts and to define a pragmatic implementation of its findings. The need for change is clear; capacity walls have been reached in numerous sectors in core European airspace and traditional methods of providing extra capacity are almost exhausted (sector split, more controllers, improved procedures, etc.).

An opportunity exists for operational staff to embrace technology and forge it to their own requirements. The motivation for C3T study is based on the expectation that:

1) Task sharing will balance workload in the sector through validation and prioritization of problems by the planner for the tactical controller;

2) Cost in workload to the controller (especially the tactical) and in economics to the airline will be reduced through early problem resolution;

3) Delegation to the pilot of separation assurance will increase airspace capacity through reduced controller workload;

4) Data link communication (CPDLC) will facilitate advance delivery of problem resolutions by planning controllers, and enable trajectory negotiation.

We empower the PC to define problem resolutions through use of conflict probing (CP), system supported coordination (SYSCO[3]), and advanced human machine interaction (HMI). Resolutions are proposed to the TC of the offering sector via SYSCO or in the form of an alternate clearance (PAC) to their own sector TC for execution or rejection in real time.

The C3T project employs the results of a number of research projects whose global aim has been the development of modified control practices and decision support tools. Major contributors to C3T include EATCHIP III,[4] ERATO,[5] PHARE,[6] FACTS,[7] URET,[8] and CTAS,[9] and these are briefly introduced in the following.

A. EATCHIP III

EATCHIP III has proposed a series of real-time simulations to evaluate system supported and civil/military coordination[10] and "added functions" i.e., monitoring aids MONA,[11] medium term conflict detection (MTCD),[12] safety nets (SNET),[13] and air/ground DATALINK.[14]

B. ERATO

ERATO is a decision support tool whose concept of operation is based on cognitive model of the principal tasks of the controller—situation awareness development, detection, and resolution of conflicts. ERATO aims at assisting the controller in the decision-making process associated with these tasks, principally by presenting the most pertinent information in a timely manner to the controller, who remains the decision making authority.

C. PHARE

The Program for Harmonized Air traffic management Research in EUROCONTROL (PHARE) provided for the development of a future ATM concept that supported the introduction of MSP and air/ground integration.[15,16] This role included reduction of traffic complexity and sector load balancing.

D. FACTS

The U.K. Future Area Control Tools Support (FACTS) Program is developing and evaluating a series of controller support tools using trajectory prediction and medium-term conflict prediction. The controller tools evaluated include a fully electronic coordination capability and planning and tactical controller tools. These have been well received by the controllers participating in simulation trials.

E. URET

User Request Evaluation Tool (URET) includes a conflict probe facility that checks flight plan trajectories for strategic conflicts and a trial planning function that allows the controller to check a desired flight plan amendment for potential conflicts before a clearance is issued. The system classifies conflicts according to their occurrence probability and provides a notification time helping the controller to better organize tasks and optimize time management.

F. CTAS

Center Terminal Radar Approach Control (TRACON) Automation System (CTAS) has been developed to support conflict detection and resolution in the form of a conflict prediction and trial planning tool for field test evaluation. This includes a conflict detection function that estimates conflict probability and a trial planning function that allows the controller to check the efficiency of a desired resolution.

III. C3T Study Concepts

A number of operational improvements are necessary to move the PC from today's air traffic control (ATC) to an empowered PC using data link technology to transmit preplanned clearances to the pilot, and ultimately the introduction of an MSP role covering several sectors. The C3T improvements evaluated for PC and TC in the model-based simulation are:

1) Task sharing. The concept for task sharing is that the PC evaluates and resolves entry and exit problems and limited in-sector resolution proposals (cleared flight level and direct route). The TC role is similar to today. The task sharing technology is the EATCHIP III$^+$ tool set, i.e., SNET, CP (including filtering function and prioritization of problem according to the resolution time), MONA, and SYSCO. This is expected in 2007 with a 20% capacity increase.

2) Empowerment. The concept for empowerment is that the PC uses tools to identify and resolve problems excluding problems requiring TC intervention (e.g., radar vectoring). The PC predefined alternate (deferred) clearance (PAC) is distributed through SYSCO or proposed directly to the TC through the HMI. The TC may reject, change, or execute the PAC based on judgment of the current situation.

The TC role is directed to validation of the PAC and to complex problem resolution more easily determined by the human being. The PAC is linked to CPDLC for delivery at aircraft sector log-on. The technology is the EATCHIP III tool set with trajectory editor (TED) and CPDLC. This is expected in 2010–2015 with a 40% capacity increase.

IV. Model-Based Study—Hypothesis

The model simulations had to evaluate the potential for reduction in controller workload and improved capacity through increased task sharing between the sector TC and PC. The task of defining problem resolutions (excluding tactical radar control) is delegated to the PC. This study assumed that CP, MONA, SYSCO, TED, and CPDLC for delivery of predefined clearance proposals would be available.

Improved task sharing can be achieved between sector team members through proposed problem resolutions and organizing TC intervention in time, resulting in increased capacity at an acceptable cost through productivity gains. The use of SYSCO and CPDLC with the empowered PC role will increase sector capacity without a significant increase in sector team workload, ensuring asynchronous communication between ground and ground, and between air and ground.

V. Model Based Study—Scenarios

Both PC/TC task sharing and empowered PC with CPDLC were evaluated, which, with the baseline, provided three study scenarios, referred to as organizations (see Fig. 1). The baseline operational parameters employed within the fast-time model were derived from a EUROCONTROL study of Polish airspace validated by Polish ATC staff. A traffic sample equating to a 24-h period in 1997 was augmented by 100% for the purposes of this study. The performance of the Polish airspace and procedures under this future traffic volume was used as the baseline.

A. Organization 0—Baseline

Two controllers (PC and TC) staff each sector. The ATC environment assumed radar, paper strip displays with R/T, and telephone used for clearance delivery and coordination.

B. Organization 1—Task Sharing

The sector configuration remained the same, however, controller tasks were amended to model task sharing between the TC and PC. This organization assumed that both conflict detection and clearance monitoring were the responsibility of the system(via CP and MONA).

Certain conflict types were deemed as being the responsibility of the PC for resolution proposal, including solutions involving direct route and planned level changes. SYSCO was assumed for coordinating changes with the appropriate sectors for TC delivery. Within the model, a reduced TC cost (task execution time) was simulated as a result of the simplification of the conflict resolution task

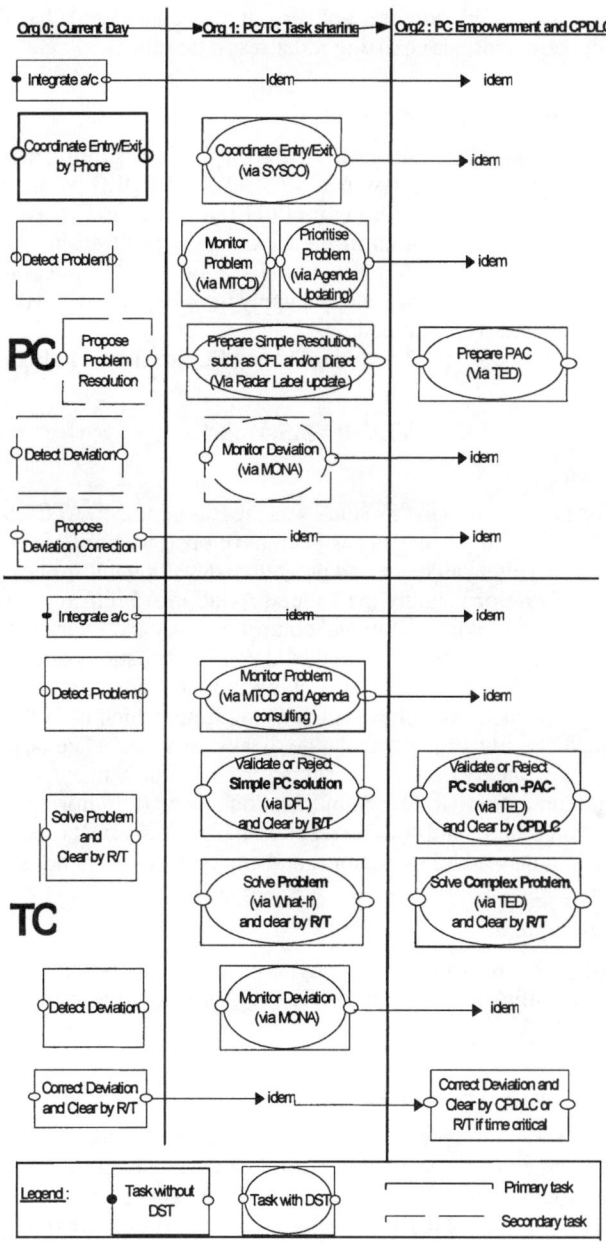

Fig. 1 PC/TC roles and tasks.

to that of clearance validation and delivery. It was considered that the PC cost would remain consistent with existing radar resolution tasks.

C. Organization 2—Empowerment

The controller tasks were reallocated so that only radar vectoring tasks remained with the TC. This task sharing assumed SYSCO and CPDLC would be available and that a graphical trajectory-editing tool (TED) would also be available to the PC. It was assumed that the output from this tool would be translated into either a coordination message to the sector $N-1$ or $N+1$, or a CPDLC message posted to the sector TC. The TC would validate and apply the clearance when the concerned aircraft logged on and was assumed.

The evolution of the roles of both the TC and PC are shown in Fig. 1.

VI. RAMS Model

A. Operation

The fast-time model used in this study was the Re-organized ATC Mathematical Simulator (RAMS).[17] This model has been developed within EUROCONTROL and is used extensively in airspace and procedure development studies both within European administrations and by the Federal Aviation Administration (FAA).

RAMS operates by taking individual controller tasks and their execution times and applying them to resolution of predicted conflict situations and also standard ATC procedures such as coordination, assume and transfer of control. Task execution times are translated into an overall working time, which is further translated into a workload classification of moderate, heavy, or severe. The tasks simulated for this study are categorized as: coordination, flight data management, routine radio/telecommunication (R/T) communication, radar (monitoring contract adherence), conflict search, and conflict resolution. Random effects such as weather, system failures, nonstandard operations, and emergencies were not simulated.

B. Traffic Sample

The Poland fast-time simulation study from which the baseline scenario was derived used a traffic sample from 1997. For this study, the traffic sample was augmented by 100%.

C. Sector Capacity Assessment

The capacity estimation provided by RAMS is defined as the maximum number of flights that can enter a specified control area in a defined period while maintaining an acceptable level of controller workload. This maximum is the heavy load threshold (HLT). The HLT value selected for radar controller position is 70%, which corresponds to 42 min measured working time in 1 h, leaving 18 min for other tasks not defined within the model.

VII. Model-Based Study—Preliminary Results

A. Organization 0—Baseline

Figure 2 indicates the workload (% time performing specific tasks) over the busiest three-hour period within the baseline organization for the chosen traffic sample

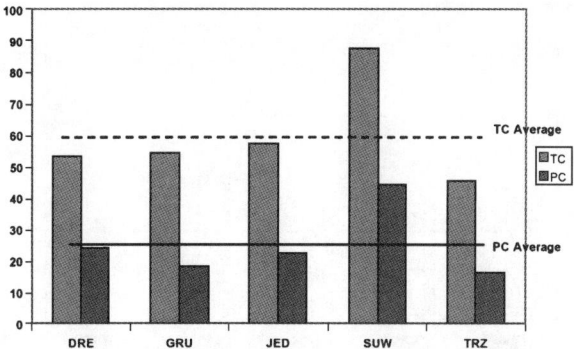

Fig. 2 Workload in baseline organization.

(1997, 100% augmented). Four of the sectors used in the study are upper sectors (FL320 and above), whereas sector SUW extended from ground to unlimited.

Traditionally within RAMS studies, a workload over a three-hour period in excess of 40% is deemed heavy and in excess of 50% is referred to as severe. Figure 2 highlights the extremes of workload for the TC and PC of each sector.

B. Organization 1—Task Sharing

Within this organization the task times associated with telephone coordination (PC task) were reduced to reflect the availability of SYSCO. The physical tasks associated with flight data management (strip manipulation) were deleted, and the task times were made consistent with the task of manipulating, reading, and assimilating electronic information. Routine R/T communication task execution times were modified to reflect the availability of shared information between air and ground systems.

Tasks associated with radar surveillance were not significantly modified because the controller is still required to build and maintain the same level of situational awareness as with the baseline scenario. The system does assist the controller in areas such as clearance monitoring.

Conflict search tasks remained unchanged although additional tasks were assigned, including conflict resolution, evaluation, and proposition (PC) and evaluation and application of the solution (TC). Figure 3 shows the change in workload associated with task sharing.

C. Organization 2—Empowerment

The task times associated with coordination and flight data management remained consistent with those used in task sharing. Figure 4 shows the change in workload associated with early problem resolution (empowerment) by the PC.

Other task changes included:

1) Communication. Times were modified to reflect the availability of CPDLC for transmission of the PC proposed PAC.

2) PAC definition. An additional task was assigned to the PC.

3) Radar surveillance. Times were reduced slightly to reflect the availability of TED and MONA, and improved aircraft conformance.

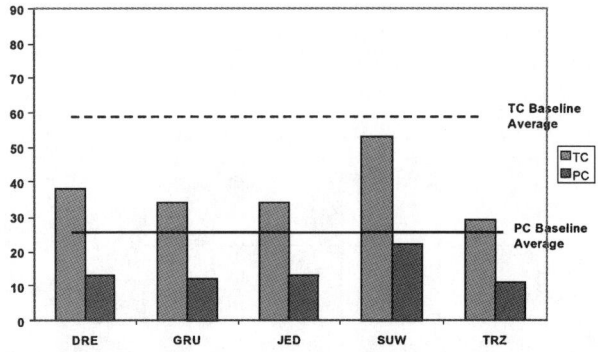

Fig. 3 Workload in organization 1.

4) Conflict resolution. Times were slightly reduced for the PC because the availability of TED with feed-back should reduce the time taken to construct conflict-free trajectories.

D. Sector Capacity Analysis

The analysis of sector capacity was performed for each of the sectors and each of the study organizations. Figure 5 shows that, in the baseline scenario, the capacity is indicated to be approximately 37 aircraft per hour in this sector and under task sharing the capacity would potentially rise to a value in excess of 50 aircraft per hour.

For the PC empowerment scenario, the workload figures were not sufficiently high to produce a reliable capacity estimate.

VIII. Real-Time Simulation—Hypotheses

The EUROCONTROL MTCD real-time simulation (EATCHIP 3a.bis)[18] was used by C3T to understand initial ideas concerning task sharing. This simulation

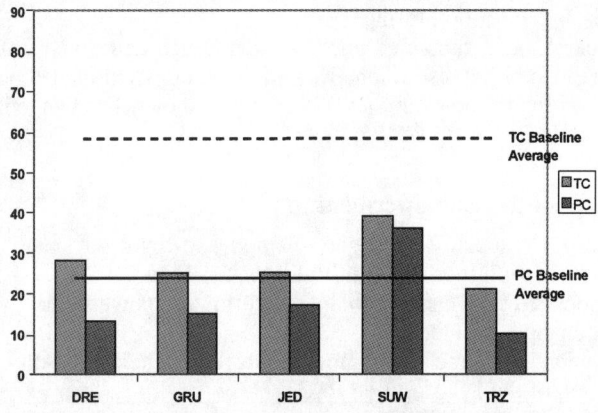

Fig. 4 Workload in organization 2.

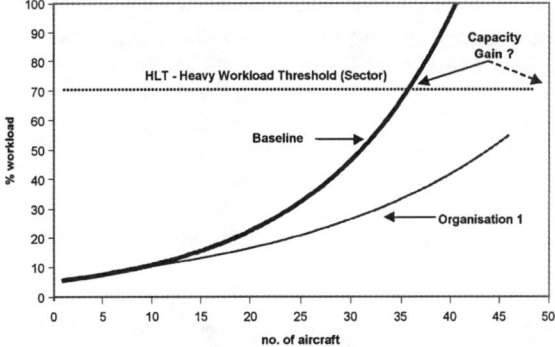

Fig. 5   Capacity estimates in baseline and organization 1.

used the EATCHIP tool set, which includes SNET, MONA, MTCD (CP), and a limited version of SYSCO supporting only flight-level coordination.

The primary objectives were to evaluate controller confidence towards MTCD and the impact of MTCD on controller roles, tasks, and working methods. The concept of operation was based on a layered filtering of conflict information as follows:

I. MTCD detects problems;
II. MONA detects noncompliance by aircraft of clearance (deviation) and advises on imminent actions, e.g., top of descent, frequency change, etc.,
III. The PC validates the conflict (is it real?);
  A. If no, PC low lights it;
  B. If yes, can PC resolve it?
    1) If yes, PC uses SYSCO to propose solution (or telephone);
    2) If no, is there urgency/severity? (e.g., less than 8 n miles)
      a. If yes, PC, puts a warning for TC
      b. If no, PC transfers earlier/delegates to TC.
IV. MTCD provides an on-line configurable prewarning (e.g., 5 min);
V. The TC uses the conflict list (including delegated and late detected conflicts) to prioritize work, but the role and working method is generally unchanged;
VI. The PC is not tasked to specifically monitor the conflict resolution and/or clearance or critical delay for sake of safety;
VII. STCA provides last warning option (e.g., 90 s);
VIII. The PC may verbally warn the TC about a conflict or an aircraft to be treated with priority and may provide assistance to the TC by updating the system on TC request.

IX.   Real-Time Simulation—Preliminary Results

Unfortunately, at this early stage of simulation analysis, the results discussion is mainly based on the subjective data provided by controllers through questionnaires and debriefing, and from human factors observation. Subjective data on perceived workload by each controller [instantaneous subjective analysis (ISA) every 2 min of an exercise] and objective data on traffic load and controller data input have been recorded.

The simulation hypotheses that the use of SYSCO and DST would reduce the PC workload and impact positively on TC workload were not confirmed. However, all participants agreed that the DST reduced PC workload.

It was considered that TC workload was not reduced, however, controllers worked scenarios with a 20% traffic increase without a perceived change in workload. This was confirmed by the ISA analysis.

Controllers were frustrated by the fact that SYSCO only permitted boundary level change coordination. They found the facility of system assisted coordination to be very good and wanted this to include direct and speed to reduce the telephone coordination.

C3T hypotheses concerning task sharing were confirmed. Controllers said that the EATCHIP tool set enabled them to more readily plan and resolve sector entry conflicts. Planning controllers perceived this to be time saving in conflict detection, providing more time to assist and monitor the TC.

Participants agreed that DST supported the PC to plan in-sector and sector exit conflict resolutions in advance.

Initial analysis shows that the EATCHIP tools support PC intervention more easily when traffic is stable than evolving; in these situations, the PC preferred to delegate problem resolution to the TC.

When a conflict was urgent or severe, PC put a warning on the conflict to attract the TC's attention. However, the simulation did not permit the PC to organize the TC work apart from the PC to TC delegation of problems, which were prioritized by the system according to time and severity. Such an eventuality would permit the PC to reduce the TC time pressure provoked by the warning, by prioritizing the TC intervention time.

The use of a conflict pre-warning displayed in the radar label was considered to be very useful. This was generally configured to provide a warning less than 5 min before the start of a problem, reminding the TC to plan a resolution.

The PC monitored in-sector conflicts and verbally coordinated solutions with the TC, when workload permitted.

Participant's confidence with the DST evolved with time. This resulted in a request to view conflicts in the next sector. The objective was to improve service by ensuring that coordination would be accepted and by anticipating problems within an area of interest comprising current and next sector.

Results provided an insight to the task of traffic picture building. This requires the controller to be aware of the current traffic situation, future situation, and future workload. Some controllers used the sector inbound lists and conflict lists as future workload indicators.

Nevertheless, some limitations remain:

1) The simulation showed that there is a need for a filter, which checks conflicts before displaying them to the controller. This would include borderline separations in which variations in speed provoke intermittent conflict prediction and display.

2) Controllers were concerned about trajectory prediction for aircraft in evolution and did not have much confidence in system indications for such traffic.

3) It was felt that simple rules such as climb as soon as possible and descend as late as possible are no longer sufficient. The use of a trajectory editor may improve the trajectory matching with the planned or actual clearances issued.

It should be noted that controllers did not feel challenged by the traffic during the simulation exercises despite a traffic increase of 20% between high and low samples.

X. Conclusions

The motivation for C3T is based on the expectations that:

1) Task sharing will balance workload in the sector through validation and prioritization of problems by the planner for the tactical controller;

2) Cost in workload to the controller (especially the tactical) and in economics to the airline will be reduced through early problem resolution;

3) Delegation to the pilot of separation assurance will increase airspace capacity through reduced controller workload;

4) Data link communication (CPDLC) will facilitate advance delivery of problem resolutions by planning controllers, and enable trajectory negotiation.

Airline economics were not part of these studies.

The model-based simulation study concentrated on 1, 2, and 3 and was conducted to explore a number of hypotheses relating to the potential for reduced controller workload resulting from task sharing and PC empowerment. The EATCHIP III real-time simulation was conducted to validate the MTCD concept and investigate DST impact on working methods, tasks, and roles; for C3T this included aspects of 1 and 2.

The results are considered to be encouraging for the introduction of task sharing and planning controller empowerment.

Concerning the hypothesis, *"Improved task sharing can be achieved between sector team members...,"* the results in fast-time simulation have demonstrated that the introduction of task sharing and PC empowerment may lead to reduced sector team workload with associated capacity benefits. The TC workload is seen to reduce in each of the organizations as PC involvement in conflict resolution increases. The PC workload is seen to reduce in task sharing, as a result of the introduction of SYSCO and modification of flight data management tasks, although it increases in PC empowerment as a result of the delegation of conflict resolution tasks. The PC workload remains at an acceptable level. There is a more balanced distribution of workload between the TC and PC.

Concerning the hypothesis, *"The use of SYSCO and CPDLC with the empowered PC role will increase sector capacity...,"* the model capacity analysis performed demonstrated potential for capacity increase within the task-sharing scenario in excess of 35%. This figure is higher than initial expectations and now needs to be validated in real-time simulation where more reliable information concerning workload and capacity issues will be gathered. (Note that in EATCHIP 3a-bis real-time simulation, controllers worked scenarios with a 20% traffic increase without a perceived change in workload.)[18]

This study has used an existing fast-time scenario as its baseline to provide a high level of confidence in the individual ATC tasks and their duration as well as the airspace and traffic samples. The development of the task timings for the task sharing and PC empowerment scenarios was performed using expert judgment as to the likely effect on tasks as a result of modified procedures and equipment availability. The results are sensitive to these timings.

In C3T real-time simulation, the next concept and scenario to be investigated will be:

1) PC with the ability to edit the trajectory and to test if these alternatives are conflict free.

2) With a full SYSCO version, PC will solve any entry conflict. This PC empowerment should be accepted without any doubt.

3) On the other hand, the ability to prepare and implement in-sector conflict solution merits an in-depth validation process.

The concept of PC empowerment opens up issues such as training and teamwork, which will need to be addressed when considering the human factors issues associated with the roles of the controller in the future in the face of increased automation. C3T will simulate a PC empowerment scenario related to the EUROCONTROL Conflict Resolution Assistant Project (CORA). It is hoped that this will be an iterative loop of real-time and fast-time studies in which more detailed task timing information from real-time study will be fed back into fast-time.

References

[1] Graham, R., "C3T Management Plan," EUROCONTROL, April 1999.

[2] "Air Traffic Management Strategy for 2000+," Vol. 2 EUROCONTROL, Oct. 1998.

[3] "Standard for On Line Data Interchange," EUROCONTROL, DPS.ET1.ST06.STD-01-00, 1997.

[4] Brain, C. J., "EATCHIP III Evaluation and Demonstration Controller Notes," EUROCONTROL, Dec. 1997.

[5] Leroux, M., "ERATO Une aide au contrôleur aérien," (ERATO an Aid to the Air Traffic Controller), *Control Magazine*, 1997.

[6] Marsden, A., "PHARE EEC PD/3 Final Report," PHARE, Doc. No. 99-70-01, Oct. 1998.

[7] Russell, S., Springall, L., and McLoughlin, D., "Future Area Control Computer Assistance Tools," FACTS1B Trial, Rept. 9957.

[8] Brudnicki, D. J., and McFarland, A. L., "User Request Evaluation Tool (URET) Conflict Probe Performance and Benefits Assessment," MITRE Corp., DTFA01-93-C-00001, 1997.

[9] McNally, B. D., Bach, R. E., and Chan, W., "Field Test Evaluation of the CTAS Conflict Prediction and Trial Planning Capability," American Institute of Aeronautics and Astronautics Guidance, Navigation, and Control Conference, Boston, MA, Aug. 10–12, 1998.

[10] Dorbes, A., "EATCHIP III Evaluation and Demonstration PHASE 2 Experiment 1; Civil Military co-ordination," EUROCONTROL, No. 318, 1997.

[11] "EATCHIP OR, Monitoring Aids," OPR.ET1.ST04.1000ORD-01.00, EUROCONTROL, 1995.

[12] "EATCHIP OR, Medium Term Conflict Detection," OPR.ET1.DEL04.MTCD, EUROCONTROL, 1995.

[13] "EATCHIP OR, Safety Nets," OPR.ST04.DEL04.2, EUROCONTROL, 1995.

[14] EATCHIP 3. And COM-ATN-EOLIA project—Experiment 3B: A/G DATALINK Final Report. EEC Rept. No. 340.

[15] "PD/3 Operational Scenarios," EUROCONTROL, PHARE/NLR/PD3-1.1.3.2.2/OSD1;2.22.

[16] Potential Application of Multi Sector Planning (MSP), Testing Operational Scenarios for Concepts in ATM (TOSCA), Reference: TOSCA/DERA/WPR/6/11.

[17] "RAMS User Manual," Model Development Group, EUROCONTROL, 1996.

[18] MTCD Concept of Operation, EATCHIP III Evaluation and Demonstration PHASE 3A–BIS, EEC NOTE 15/99, Project ODP-7-EI.

Trajectory Orientation: Technology-Enabled Concept Requiring Shift in Controller Roles and Responsibilities

Kenneth J. Leiden*
Micro Analysis and Design, Boulder, Colorado
and
Steven M. Green,[†]
NASA Ames Research Center, Moffett Field, California

I. Introduction

IN SUPPORT of Free Flight, many new tools and technologies are being developed to improve the efficiency of the National Airspace System (NAS) through evolutionary enhancements. Automation enhancements to current practices will offer immediate benefit. However, the greatest potential for improvement will come from new practices and procedures enabled by new decision support tool (DST) technologies. A case in point is the En Route Descent Advisor (EDA), a Center Terminal Radar Approach Control (TRACON) Automation System (CTAS) DST under development at the NASA Ames Research Center.[1] EDA assists controllers with the separation and flow-rate conformance (i.e., time-based metering and/or distance-based spacing) of air traffic in en route airspace. Although utilization of conflict probe and metering tools within today's sector-oriented operational paradigm will provide some free flight benefits, significantly greater benefits would be realized by a shift to a trajectory-oriented operational paradigm. This is the goal for developing EDA.

Trajectory orientation is a procedural concept that enables en route controllers to plan and coordinate trajectories across sector boundaries while efficiently maintaining separation and conforming to flow-rate constraints (i.e., time-based metering and/or distance-based spacing). This concept requires a new set of en route controller roles, responsibilities, and procedures that will enable controllers to

---

Copyright © 2001 by the American Institute of Aeronautics and Astronautics, Inc. No copyright is asserted in the United States under Title 17, U.S. Code. The U.S. Government has a royalty-free license to exercise all rights under the copyright claimed herein for Governmental purposes. All other rights are reserved by the copyright owner.
*Senior Systems Engineer. Member AIAA.
[†]Manager, En route Systems. Member AIAA.

efficiently plan across sectors. The role of new DST technologies is to enable trajectory oriented operations, which in turn will be a large step toward realizing the full economic and workload benefits envisioned by free flight proponents.

The content of this chapter focuses on presenting the trajectory orientation concept, identifying issues in today's operations that inhibit it, and suggesting solutions to enable it. In the following section, EDA is discussed sufficiently to highlight DST capabilities necessary to support trajectory orientation. This is followed by a detailed overview of the trajectory orientation concept. Section IV outlines the objectives of this research and the operations assessment approach to meet those objectives. Next, Section V discusses three topics. The first topic addresses procedures and techniques in today's operations that inhibit a trajectory orientation in addition to proposing viable solutions. The second topic discusses the advantages of tactical conflict detection. The third topic compares different controller operational concepts in their ability of orchestrating a trajectory orientation.

II. Background

EDA or EDA-like DSTs are the primary enabling technologies that make trajectory orientation possible. EDA will enable controllers to more easily accommodate user-preferred trajectories while efficiently assuring traffic separation and conformance with flow-rate restrictions. EDA will accurately detect separation and flow-rate conformance problems up to 20 min into the future (generally across 1–2 sectors). The CTAS trajectory-prediction accuracy that supports such advisories has undergone extensive field testing and validation.[2,3] The controller is also provided with resolution advisories that are nominally problem free over a 20-min time horizon [i.e., conflict free and in conformance with air traffic control (ATC) constraints such as required time of arrival, spacing restrictions, and crossing restrictions.] [The key to EDA is the integration of flow-rate conformance and conflict detection and resolution (CDR) advisories. Integration not only reduces conflict-probe false-alarm and missed-alert rates when needed most (under high-density delay conditions), it leads to more efficient traffic plans that are nominally conflict free.] Trial planning capability allows controllers to direct EDA advisories according to their own operational preferences. A significant economic and workload benefit will be enabled by EDA's capability to develop path-independent flow-rate conformance advisories. Instead of forcing flow-constrained flights in trail to establish spacing, EDA allows controllers to delay merges and minimize deviations from user-preferred trajectories. This approach reduces the concentration of metered flights in any one sector, thus distributing the workload across sectors and away from the final merge point.

EDA and the trajectory orientation concept mesh well with the NASA advanced air transportation technologies goal of developing longer term technologies that will support user flexibility and distributed air-ground (DAG) traffic management.[4] EDA will provide controllers with the decision support needed to manage a free flight environment characterized by 1) significant reduction in procedural restrictions, 2) significant increase in dynamically imposed flow restrictions (to mitigate capacity overloads), and 3) significant increase in dynamic flight replanning by the user.

The biggest challenge is transitioning flights to/from high-density terminal areas. Economically, it does not make sense for the user/ATC community to heavily invest for en route savings only to lose those benefits upon transition to a congested airport. Many concepts, ranging from trajectory negotiation to free maneuvering, have been proposed to maximize user flexibility in en route airspace. In any case, the EDA/trajectory orientation combination may be viewed as an enabling step to transition free flight aircraft smoothly and efficiently to and from the terminal area.

III. Trajectory Orientation Concept

Trajectory orientation is a concept, developed at NASA Ames Research Center,[1] that offers an alternative to today's sector-oriented ATC operations. Trajectory orientation requires a fundamental shift in thinking about intersector coordination.

Today's sector-oriented operations are characterized by controller emphasis on actions to protect their sector's internal airspace. The primary focus is on the planning and tactical separation of aircraft within their sector. This planning also includes consideration for constraints, such as crossing restrictions, both within the sector and within close proximity to the sector boundary (to facilitate a hand-off to the next sector). The hand-off process is used to ensure that incoming flights are at least tactically separated. However, there is little visibility or control over the conformance of incoming flights to flow-rate restrictions. The sector closest to flow-restricted airspace not only has the greatest concentration of impacted flights, but also has the greatest potential responsibility for conformance. Sector-oriented operations generally involve just enough cooperation between adjacent sectors to permit a hand–off, but not enough to achieve an efficient flow of traffic.

Trajectory orientation, on the other hand, focuses on efficient flight planning that nominally conforms to all ATC constraints within a time horizon (e.g., 15–20 min) independent of airspace boundaries. In addition to separation, this approach emphasizes the upstream strategic planning of actions to conform to flow-rate restrictions in downstream sectors. The result is a distribution of workload away from the flow-impacted airspace. Instead of controllers operating relatively independently, with the main focus on protecting their sector's internal airspace, the controllers would work cooperatively across sectors and depend on each other for well-planned, conflict-free flow of traffic.

Trajectory orientation will require new roles, responsibilities, and procedures for en route controllers that could potentially be quite different from today's operations. Trajectory orientation is the ATC counterpart to the orientation of a pilot in operating his/her aircraft. Pilot actions not only consider their current state and tactical challenges (e.g., weather and traffic avoidance), but also their strategic goal to complete the trajectory (i.e., they maintain a continuously updated trajectory plan for completion of the flight).

Two example cases are presented next to illustrate two specific differences between sector-oriented and trajectory-oriented operations. Figure 1 depicts a sector orientation for the first case. In this example, the aircraft in sectors 1 and 2 are compliant with all constraints within their respective sectors as well as any sector 3 hand-off constraints. However, to solve a downstream capacity problem, traffic management requires a 20 n mile spacing at the sector 3/4 boundary for aircraft A, B, and C. This restriction corresponds to an approximately 20-min time horizon from their current positions in sectors 1 and 2. In today's environment, the delay maneuvers for spacing conformance would most likely occur in sector 3, the

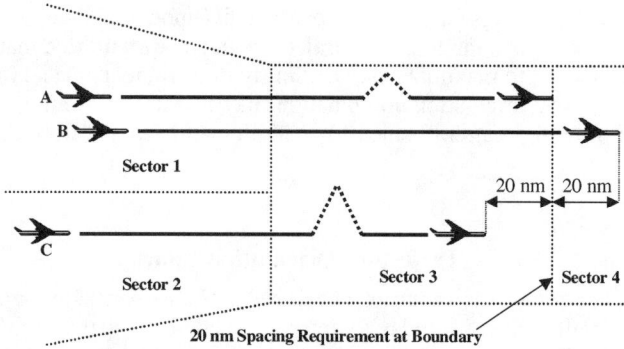

Fig. 1 Sector orientation example 1.

downstream sector. This is illustrated in Fig. 1 by the vector deviations to aircraft A and C within sector 3 airspace.

Figure 2 illustrates the trajectory-oriented version of the first example. In this case, the delay maneuvers to meet the spacing requirements would occur in the upstream sectors. The longer time horizon allows the upstream sectors to use speed control to achieve most, if not all, of the spacing requirement. Additional action by the downstream controller (sector 3) is needed only to adjust for unplanned disturbances (i.e., actions required by exception rather than the rule).

Figure 3 depicts a sector orientation for the second case. For this example, traffic management still requires a 20 n mile spacing at the sector 4 boundary. However, they also "pass back" a spacing restriction of 40 miles in trail at the sector 1/3 and 2/3 boundaries to assist the sector 3 controller with absorbing the required delay for the final spacing at the sector 3/4 boundary. In this example, the sector 1 controller must delay aircraft A 35 n miles to achieve the 40 n mile spacing at the sector 1/3 boundary. However, this is an unnecessary delay because there are no aircraft in sector 2 that must be spaced between aircraft A and B. In fact, aircraft C is the only aircraft in sector 2 during this time period, and because it is ahead of the aircraft in sector 1, aircraft C requires no delay at all. If the controller in sector 1

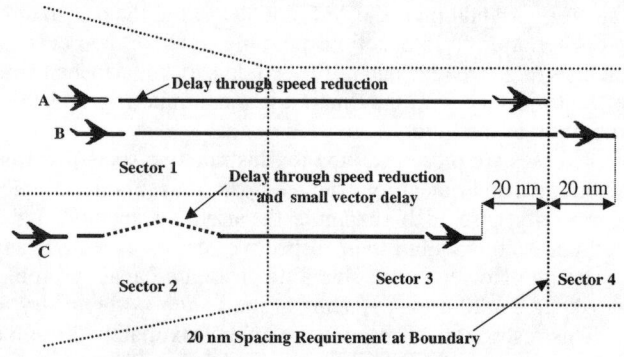

Fig. 2 Trajectory orientation example 1.

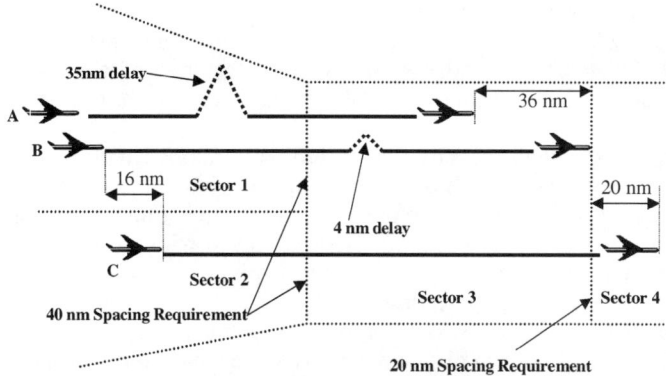

Fig. 3 Sector orientation example 2.

were aware of this, he/she could have delayed aircraft A 15 n miles instead of 35 n miles. As such, aircraft A crosses the Sector 3/4 boundary with 16 n miles of excessive spacing.

In Fig. 4, DST technology enables upstream controllers to determine the accurate relative spacing between aircraft in adjacent sectors due to downstream constraints. This allows the controllers in sectors 1 and 2 to effectively coordinate downstream spacing while maximizing the independent operations of both sectors.[5] Because of this capability, traffic management would not need to artificially pass back spacing requirements to the upstream sectors. This prevents inefficient gaps or missed slots that prevent airspace from being used to its true capacity.

Although not explicitly depicted in Fig. 3, the pass back procedure commonly used by traffic management highlights another problem. Pass back procedures rarely result in a seamless transition of merging streams for the downstream controller. Using the sector geometry and traffic management constraints depicted in Fig. 3, a seamless transition requires a projected spacing of 20 n miles in the downstream sector (i.e., sector 3) between aircraft currently positioned in sector 1 and sector 2 in addition to 40 n miles between aircraft in the same

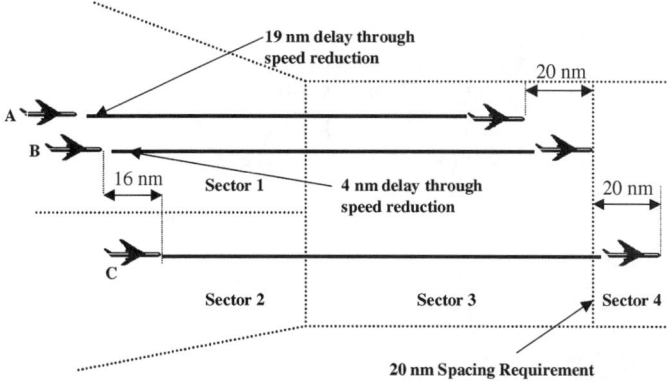

Fig. 4 Trajectory orientation example 2.

sector. In other words, even if the aircraft in a given sector are each appropriately spaced relative to other aircraft in that sector, anything other than a projected 20 n mile "relative" spacing between aircraft in sector 1 and aircraft in sector 2 would require the sector 3 controller to delay one of the streams to achieve the final spacing.

The problems with the pass back procedures as described in the preceding paragraph can be solved with a transition toward a time-based metering environment. This is because flow conformance of aircraft due to metering is measured relative to time rather than to other aircraft. However, even with meter-fix delay times displayed on an upstream sector's plan view display, the tactical and gross nature of today's techniques and procedures leaves the upstream controller ill equipped to plan the actions necessary to accurately absorb the required delay. The downstream controller must then plan new actions to meet the required metering times. Ironically, the accuracy of the metering is often invalidated by the imprecision of today's delay tactics.

In summary, the strategic nature of trajectory orientation offers several advantages when supported by DST technology. The efficiency of flow-rate conformance (delay absorption) is increased in three ways. First, the strategic planning reduces the need for tactical corrections (interruptions) because each maneuver action is calculated to nominally result in conformance. Second, as depicted in example 1, this approach enables greater use of speed control by increasing the time horizon for conformance. Third, as depicted in example 2, excessive spacing between aircraft can be significantly reduced because DST functionality provides relative spacing information between aircraft in adjacent sectors. With respect to workload, as depicted in both examples, trajectory orientation results in a more even distribution of workload from the downstream sector (where traffic is converging) to sectors further upstream.

Although the examples presented are useful in describing a trajectory orientation, they only focus on the aspects related to intersector coordination. The other key aspect relates to the complementary subject of intrasector coordination. The operational roles and responsibilities of individual controller positions [e.g., radar (R-side) and radar associate (D-side)] define the building blocks from which intersector procedures may be created. Evaluating these roles and responsibilities is the primary objective of this research. The results of this evaluation are discussed in Section V.C.

IV. Research Approach

The top-level objectives of this research were:

1) Determine the core issues in today's en route operations that inhibit a trajectory orientation.

2) Identify potential technology and procedural solutions that address those issues.

3) Evaluate specific candidate concepts to determine which have the highest potential of achieving a trajectory orientation.

An assessment of today's en route operations was performed to meet the goals of this study. The assessment included a literature review of current en route operations.[6-10] In addition, future free flight concepts[11-18] were researched because of their potential impact on current operations. Controller participation from the Denver and Cleveland air route traffic control centers (ARTCC) was supported by the regional offices of the Federal Aviation Administration (FAA) and the National Air Traffic Controllers Association. Denver center was chosen because it has been a research site for several DST development efforts at NASA Ames Research Center and was within close proximity of the principal researcher. Cleveland center was chosen because of the complexity of the air traffic it manages on a daily basis. Located between Chicago and the northeast corridor, it is the busiest center in the United States, with over 2.6 million operations in 1999.[19] Each sector within Cleveland center must handle arrival, departure, and over-flight traffic on a regular basis.

A total of 15 controllers and traffic management personnel were briefed on EDA functionality and the trajectory orientation concept. A formal set of questions was utilized to determine the extent and nature of how controllers perform strategic planning for separation and flow-rate conformance. The controllers were interviewed for four categories of information:

1) Specific examples of scenarios, events, and procedures for which they felt that DST technology may not be sufficiently accurate to predict aircraft trajectories.

2) Specific examples that would inhibit a trajectory orientation.

3) Requirements for decision-support capability and usability that would enable them to perform their jobs most efficiently under a trajectory orientation.

4) Operational advantages and disadvantages of the set of candidate concepts for controller roles and responsibilities.

V. Results and Discussion

A. Operations Assessment

The single most important issue identified from the operations assessment is the inability of controllers to perform accurate strategic planning (i.e., time horizon of 15–20 min) that includes the intentions of both the upstream and downstream controllers as well as the intentions of the aircraft. This section decomposes strategic planning issues into smaller issues that are more easily understood and are thus more likely to have solutions. The focus of this section is to identify and address issues related to strategic detection and resolution of conflicts and flow-rate conformance problems. More detailed results of this research are included in Ref. 20.

One important point from the controller interviews is the fundamental fact that controllers have been trained to act and think tactically, not strategically. Emerging DST capabilities have demonstrated, under limited conditions in the field, the ability to enable more strategic planning by controllers. However, simply making DSTs available to controllers would not necessarily result in strategic planning because the controller's mindset and procedures are still based on a tactical culture and environment that dates back several decades.

In addition, controllers are reluctant to strategically resolve flow-rate conformance and conflict problems (a fundamental requirement of the trajectory orientation concept). This reluctance is due, in large part, to the general uncertainty, and lack of predictability, they expect over a strategic time horizon. To clarify these issues further, results from the controller assessment are summarized in the following sections. Eight core issues were identified as obstacles that prevent or inhibit controllers from performing effective strategic planning. These issues are ranked here in terms of their impact on enabling a trajectory orientation:

1) Controllers are not responsible for resolving conflicts or meeting flow-rate constraints of other sectors.

2) Strategic resolutions may be insufficient in resolving conflicts or meeting flow-rate constraints.

3) Intersector resolutions may interfere with an adjacent controller's plans.

4) Strategic resolutions may lead to conflicts with other aircraft because of inadequate situation awareness.

5) Strategic resolutions have a lower priority compared to other controller tasks.

6) Conflicts may resolve themselves because they are actually false alarms.

7) Conflicts may resolve themselves because of unpredictable events.

8) Strategic resolutions may lead to conflicts or flow-rate conformance problems with other aircraft because of simultaneous and conflicting actions by adjacent controllers.

The following sections describe these core issues in greater detail and present potential solutions.

1. *Controllers are not responsible for resolving conflicts or meeting flow-rate constraints of other sectors.*

The current ATC system clearly assigns responsibility and control authority to individual sectors. Although there are exceptions, generally speaking, controllers are responsible only for resolving conflicts that occur in their own sectors. Similarly, controllers are responsible for meeting flow-rate constraints at their respective exit boundaries or metering fixes. The advantage to these methods is that in the case of an operational error (e.g., violation of the minimum separation rule), the fault is readily determined. The disadvantage is that there is no impetus for controllers to collaborate on trajectory oriented intersector planning. Without such intersector planning, achieving a trajectory orientation is not possible.

Potential Solutions: Unlike many of the other issues discussed in the following sections, the solution for this issue requires changes to many aspects of today's ATC operations. New tools and procedures must evolve that give controllers confidence that trajectory-oriented planning is beneficial to all sectors. Only when all eight issues pertaining to strategic planning are addressed will the right conditions exist for proactive, widespread participation in trajectory-oriented, intersector planning.

2. *Strategic resolutions may be insufficient in resolving conflicts or meeting flow-rate constraints.*

Inadequate strategic resolutions would most likely be caused by controllers using manual "rule of thumb" approaches that are too gross for the given scenario. Such approaches do not accurately account for variations in wind and true airspeed, or trajectory and conflict geometry. The controller's experience and skill becomes an important factor in calculating a resolution advisory that is sufficient, but not excessive. The challenge becomes more difficult for strategic time horizons because position uncertainties tend to grow linearly with time.

Potential Solutions: A solution for inadequate resolutions due to conflict geometry and wind variation is DST automatic resolution and/or trial-planning capability that addresses both separation and flow-rate conformance requirements, with consideration for trajectory-prediction uncertainty.

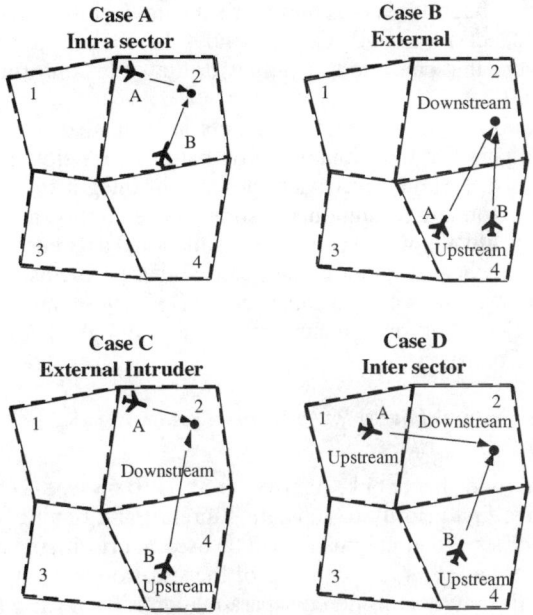

Fig. 5 Conflict scenarios.

3. *The intersector resolution may interfere with an adjacent controller's flow-rate conformance plans.*

As an example, if two aircraft in an upstream sector (see external case in Fig. 5) will conflict 5 min into the downstream sector, the upstream controller is required to resolve the conflict before the aircraft is handed off. When spacing constraints are required in the downstream sector, the conflict resolution might interfere with the aircraft spacing plans of the downstream controller. He/she would then need to resolve the new problem and issue an additional clearance to the aircraft. One method currently employed by controllers to avoid this situation is to hand off the aircraft early so that the downstream controller can resolve both the conflict and the spacing conformance. The drawback to this method is that the controllers are dealing with the traffic tactically rather than strategically, which is particularly inefficient for delay (metering) situations.

Potential Solutions: The solution is strategic (upstream) planning of resolution maneuvers aided by DST functionality that accounts for downstream flow-rate conformance. The technology must support a common situational awareness across sectors to ensure that plans and actions are complementary.

4. *Strategic resolutions may lead to conflicts with other aircraft because of inadequate situation awareness.*

A primary reason that strategic resolutions lead to conflicts with other aircraft is inadequate situation awareness across sectors. Controller situation awareness can be negatively affected by lack of data, high workload, complacency, or lack of vigilance. Regardless of the cause of inadequate situation awareness, or which sector it occurs in, the result is the controller's mental picture of the airspace does not accurately reflect all aircraft. Consequently, the controller may determine a resolution advisory that can lead to conflicts or flow-rate conformance problems with those aircraft.

Potential Solutions: The problem for conflicts due to inadequate situation awareness can be mitigated by DST functionality that provides situational awareness cues on several levels. At one level, such cues may be integrated as part of the trial planning and/or automatic resolution advisories for separation and flow-rate conformance. This would augment a controller's situational awareness by alerting the controller to cases where resolution plans cause other problems. At a more basic level, situational awareness can be enhanced by DST cues that call a controller's attention to situations requiring greater scrutiny (e.g., a flight that is not correlated with its predicted path).

5. *Strategic resolutions have a lower priority compared to other controller tasks.*

The FAA controller handbook, referred to as 7110.65 (see Ref. 7), states on page 2-1-1: "Give first priority to separating aircraft and issuing safety alerts as required in this order. Good judgment shall be used in prioritizing all other provisions of this order based on requirements of the situation at hand." No controller interviewed for this study considered separating aircraft, based on a conflict detection time horizon of 15–20 min, a "first priority." Obviously, controllers would deal with tactical conflicts before the strategic conflicts/flow-rate conformance problems because the safety of the aircraft is more imminent. In the event that there are no tactical conflicts, the controller in most situations would be inclined to perform low priority tasks, such as housekeeping, over strategic resolution. As one controller stated, "Twenty minutes is an eternity to a controller," but 20 min is also the preferred time horizon for efficient flow-rate conformance. Ironically, the lack of strategic planning today results in a higher tactical workload that, in turn, reduces the opportunity to perform strategic planning.

Potential Solutions: The solution requires a fundamental change to the environment that controllers have been trained to support. It also implies a shift in controller roles and responsibilities. The circumstances presented to controllers in any given situation must have adequate solutions, via new tools and procedures, to give them confidence that by acting and thinking strategically, they are improving the overall traffic flow and are not increasing their workload.

6. *Conflicts may resolve themselves because they are actually false alarms.*

Prediction errors occur because of uncertainty in actual ground speed, altitude rate, and radar track when a controller projects each aircraft's trajectory. The controller may falsely predict a conflict situation that, if left alone, would have resolved itself.

Potential Solutions: The solution for prediction errors and the resulting false alarms is through automated four-dimensional trajectory predictor algorithms, such as those residing in the CTAS EDA[2,3] and the User Request Evaluation Tool (URET).[21,22] These algorithms have been studied for several years and are well suited for addressing this particular problem. Effectiveness can be improved by including functionality to accurately reflect or model the intentions of the pilot and/or adjacent controllers.

7. *Conflicts may resolve themselves because of unpredictable events.*

The longer the conflict detection time horizon, the higher the probability that something unpredictable or unintended will occur that results in the conflict resolving itself. For example, the pilot of one of the conflicting aircraft may request an altitude and/or speed change (e.g., due to turbulence) or a heading change (e.g., due to a weather cell in its path) prior to the conflict becoming tactical (i.e., within the time horizon of today's radar controller). In these cases, granting the pilot request resolves the conflict.

Potential Solutions: Because this issue pertains primarily to separation conflicts, rather than flow-rate conformance problems, one solution is for the controllers to delay a conflict resolution until the probability is high for the conflict to occur. This is discussed in detail in Section V.B.

8. *Strategic resolutions may lead to conflicts or flow-rate conformance problems with other aircraft because of simultaneous and conflicting actions by adjacent controllers.*

This case is rare, but it was the cause for a near-miss between two aircraft. Simultaneous trajectory changes to aircraft in two separate, but adjacent, sectors can lead to what otherwise would have been a preventable conflict. In this case, the controller has adequate knowledge of aircraft in adjacent sectors to determine a strategy to resolve a conflict in his sector that would not interfere with aircraft in the adjacent sectors. However, the controller does not know the actions being performed simultaneously by a controller in an adjacent sector unless one of the controllers takes the initiative to coordinate with the other. The simultaneous actions have the potential to negate the intended effect, resulting in another conflict or flow-rate conformance problem. Following normal sector-to-sector communication procedures would prevent this case from occurring, but the fact that it does occur is a cause for concern.

Potential Solutions: Although a rare problem, the solution for conflicts and flow-rate conformance problems due to simultaneous actions in today's operations is to emphasize correct procedures related to sector-to-sector communication. In a future DST environment that supports trial planning and automatic resolution, this problem can and must be avoided because controller trust in the DST is at stake. As one controller stated, "Trust is hard to gain, but easy to lose." The option that appears to be most favorable is cross-referencing. Cross-referencing tests any newly created trial plan, whether controller derived or computer derived, against all active flight plans as well as all pending trial plans that correspond to other sectors. If the newly created trial plan conflicts with any of the other plans, then the controller is notified of the discrepancy and must decide on a new course of action.

B. Tactical Detection vs Strategic Detection

Early in this research, the assumption was made that strategic detection (15–20-min time horizon) of conflicts and flow-rate conformance problems would generally result in the most optimal/efficient resolutions. However, after the analysis of the strategic planning issues in the preceding section, it became apparent that the inability of controllers to perform accurate strategic planning was due to their concerns about strategic detection and resolution of conflicts (specifically as presented in problems 5–7) rather than the strategic detection and resolution of flow-rate conformance problems. Two points are discussed in the following paragraphs that suggest a trajectory orientation is still achievable despite a strategy that permits tactical detection of conflicts.

The first point is that, even with accurate DSTs, wind uncertainties over a 20-min time horizon can still result in detection errors along the flight path that are significantly large relative to the 5 n mile separation criteria.[23] Depending on the conflict geometry, this can result in false alarms or missed alerts that needlessly distract the controller. However, those same detection errors are much smaller relative to typical traffic management spacing requirements of 10–40 n miles. Consider the case in which two merging aircraft are currently predicted to be spaced 5 n miles apart but the requirement is for a 20 n mile spacing. Even with an uncertainty of +/−3 n miles in a DST advisory, the upstream controller can nominally plan to absorb all the delay, leaving the downstream controller with the responsibility for correcting any unacceptable deviations that develop. With the nominal conformance plan, the downstream controller only needs to intervene by exception, rather than by the rule. When required, such exceptional actions would only require fine-tuning compared to the original delay-absorption plan.

The second point is that waiting to resolve a separation conflict tactically is not nearly as inefficient as waiting to resolve a flow-rate conformance problem tactically. The maximum amount an aircraft needs to be maneuvered to resolve a conflict would be slightly greater than the separation criteria (e.g., 5 n miles in radar-controlled en route airspace). In comparison, delays for flow-rate restrictions (e.g., arrival metering) can typically exceed 4 min (approximately equivalent to 20–30 n miles of flight). A longer time horizon is required for efficient flow-rate conformance than for conflict resolution.

The purpose of mentioning these two points is to suggest that a trajectory orientation could still be achieved by detecting conflicts on a tactical time horizon rather than a strategic horizon. In contrast, detecting flow-rate conformance problems on a tactical time horizon would clearly inhibit a trajectory orientation—strategic detection is mandatory. Lastly, tactical detection of conflicts would reduce the number of false alarms and missed alerts because the reduced time horizon limits the growth of trajectory-prediction uncertainties such as wind and pilot/controller intent.

One final point concerning tactical vs strategic detection of conflicts needs to be clarified. When considering time horizons, it is important to distinguish between problem detection and problem resolution. Regardless of whether a conflict is detected/alerted on a tactical or strategic time horizon, if it involves a flow-restricted flight, the resolution should be strategic in nature (i.e., in conformance with flow-rate restriction and nominally conflict free to the meter fix). The point is that, if it is necessary to replan a flight to resolve a problem, automation-assisted resolutions should help the controller avoid new problems in the foreseeable future. For example, if a flight must be replanned for a metering delay, the replan should be nominally conflict free to the meter fix. Alternatively, if a metered flight falls into conflict while in conformance, the conflict-resolution action should be nominally in conformance with the metering restriction. In summary, if the controller must throw a stone, they might as well use the decision support technology to aim the stone to hit two birds. This hybrid concept allows the best aspects related to problem detection and resolution to be combined.

C. Evaluation of Candidate Controller Roles, Responsibilities, and Procedures

Seven candidate concepts that define controller roles, responsibilities, and procedures were evaluated to determine which concept(s) would be most effective in achieving a trajectory orientation in a new operational paradigm. This set of concepts was formulated in an attempt to reflect the various options under consideration in the United States and Europe.

The study evaluated each candidate concept against a set of generic conflict and flow-rate conformance problems. These problems were further expanded to consider the generic range of possible aircraft/airspace combinations that were originally illustrated in Ref. 24 and are depicted here in Fig. 5. A major element of the evaluation was a qualitative analysis of the amount of controller coordination required by each concept. It was assumed for this evaluation that electronic flight strips or some other replacement to paper flight strips would be available.

The concepts are explained briefly in the following sections, but full details are included in Ref. 20. Candidates 5 and 7 are discussed in greater detail because these concepts were down-selected for further study. The findings are organized and summarized in Table 1.

1. Candidate 1 URET-Like Procedures

This concept was inspired by combining Free Flight Phase 1 URET procedures[21,22] with EDA-like DST capabilities. To achieve a trajectory orientation, the downstream D-side uses a DST for strategic conflict detection and flow-rate conformance. The DST is configured to notify only the D-side position in the sector where the spacing violation and/or conflict is predicted to occur. The downstream D-side then cues the upstream D-side controller(s) who confers with his/her R-side partners to agree on a resolution strategy for one or both aircraft. Upon agreement, the upstream R-side(s) issues the clearance to the aircraft. In the intersector case (case D of Fig. 5), the D-side would potentially need to coordinate with upstream controllers in sectors 1 and 4. In the external case, the D-side would need to coordinate only with sector 4. In the external intruder case, the D-side would need to coordinate with his/her R-side and/or the sector 4 controller.

2. Candidate 2 EUROCONTROL PHARE

This candidate concept represents trajectory negotiation roles defined for the Program for Harmonized Air Traffic Management Research (PHARE) in EUROCONTROL demonstration 3 (PD/3).[25,26] In this concept, the downstream D-side strategically plans the upstream using a DST as configured in candidate 1. The D-side issues the resolution advisories to one or both aircraft via controller-pilot data link communication (CPDLC) rather than voice communication. The advisory involves planned changes to the flight path that become effective at the sector boundary to minimize coordination with the R-side. This is explained more fully in the candidate 5 concept.

3. Candidate 3 Upstream D-Side

In this concept, as well as for the candidate 4 and 7 concepts, the upstream sector has the responsibility to resolve an aircraft's downstream conflict/spacing/metering problems. The upstream D-side strategically plans the downstream. The DST is configured so that predicted problems are displayed to the sector that currently

Table 1 Candidate concept comparison

Candidate concept	1 URET	2[b] EUROCONTROL PHARE	3	4	5[b] EUROCONTROL MSP	6 NASA AC	7[b]
Position responsible for planning strategic resolution	Down stream D-side	Down stream D-side	Upstream D-side	Upstream R-side	MSP	AC	Either upstream position
Position responsible for coordination	Down stream D-side	Down stream D-side	Upstream D-side	Upstream D-side	—	AC	Upstream D-side
Position responsible for issuing strategic clearance and monitoring for compliance	Upstream R-side	Down stream D-side	Upstream R-side	Upstream R-side	MSP	Upstream R-side	Either upstream Position
Strategic clearance becomes effective where/when?	Upon issue	Sector boundary	Upon issue	Upon issue	Sector boundary	Upon issue	Upon issue
Strategic planning coordination required between R-side and D-side position?[a]							
Intrasector	Yes	Yes	Yes	No	—	—	No
External	No	No	Yes	No	—	—	No
External intruder	Yes	Yes	Yes	Yes	—	—	No
Intersector	No	No	Yes	Yes	—	—	No
Number of sectors that require coordination[a] (sectors adjacent to the sector responsible for strategic planning							
Intrasector	0	0	0	0	0	1	0
External	1	0	0	0	0	1	0
External intruder	1	0	1	1	0	2	1
Intersector	2	0	1	1	0	2	1

[a]For the purposes of comparison, the assumption is made that both aircraft will be issued resolution advisories. In many cases, the controller may choose to focus on one aircraft, which would reduce the coordination that is indicated by this table.
[b]Assumes CPDLC is available.

has jurisdiction for the aircraft involved in a conformance problem and/or conflict. For example in Fig. 5, an intersector conflict and/or spacing problem would be displayed to the upstream D-side positions in sectors 1 and 4. The D-side positions would need to coordinate to determine if one or both aircraft need maneuvering.

4. Candidate 4 Upstream R-Side

This concept is similar to candidate 3, but the upstream R-side, rather than the upstream D-side, is responsible for strategic planning of the downstream. Similar to the Automated En Route ATC (AERA 2) concept,[14] DST advisories are integrated and blended into the R-side's primary traffic display. With some modifications to the configuration of the DST, this concept has further potential for reducing intersector coordination. For intersector conflicts/spacing problems, the DST could be configured so that, of the two upstream sectors that own the aircraft in conflict, only one sector is notified initially of the conflict by the DST. The notification logic in the DST could be based on objective measures of controller workload or, alternatively, which aircraft in the conflict pair can be maneuvered more fuel efficiently.

5. Candidate 5 EUROCONTROL MSP

This candidate is based in part on the EUROCONTROL concept that proposes a new multi-sector planner (MSP) position.[25,26] The EUROCONTROL MSP concept combines controller functions with traffic management functions. However, for the purposes of this study, the MSP candidate concept represents only one aspect of the full EUROCONTROL MSP concept (i.e., that aspect related to the planning and contracting of user-ATM negotiated trajectories). The reference is made here to give credit for inspiring the intersector coordination aspect evaluated in this study.

Each MSP monitors a group of sectors within a center. The number of MSPs per center will depend on traffic density and other criteria to be determined. The MSP is responsible for strategic planning of aircraft within his/her defined airspace. The MSP issues clearances based on advisories from the DST via CPDLC that become effective at the boundary of the next sector. The MSP is responsible for monitoring of compliance.

a. Advantages. This concept requires minimal intersector coordination if any. By limiting flight plan changes to ones that are initiated after the next sector's boundary, this concept does not require the MSP to coordinate with the current sector to issue a clearance. From the perspective of the next downstream sector, the MSP change simply appears as the current flight plan when the flight comes under that sector's control. The MSP effectively "inserts" the flight plan update between the two sectors. In this way, the MSP position is autonomous, which will permit him/her to focus specifically on achieving a trajectory orientation.

b. Disadvantages. On the surface, this concept appears to be the best candidate for achieving a trajectory orientation. However, there are several issues that were identified in the assessment worth discussing. To begin with, there is a risk that the effectiveness of the MSP position at the busiest centers would be limited during peak periods of traffic—a time when trajectory orientation is most needed. The risk is related to the number of sectors the MSP must serve. This issue may be answered through controller-in-the-loop simulation. Reducing the number of sectors per MSP position to improve efficiency could result in diminishing returns when compared to the other sector-based candidate concepts.

Second, because the resolution becomes effective at the sector boundary, it can become obsolete if the upstream R-side issues a tactical clearance to the aircraft after the MSP has issued the strategic clearance via CPDLC. This causes a problem because the DST resolution is based on the assumption that the aircraft would follow the original flight plan until the sector boundary.

In addition, it is necessary that the MSP work seamlessly with the controllers/sectors in his/her jurisdiction. Otherwise, controllers would be very resistant to what they might view as outside interference with their basic roles and responsibilities. A strong understanding of the operations and traffic flow of all sectors in his/her domain is necessary to avoid impeding the actions of the controllers in those sectors. The controllers expressed the opinion that the MSP position would require a controller who is highly skilled and well respected amongst his/her peers. Otherwise, it is unlikely that the concept would be effective in achieving a trajectory orientation.

The MSP also would have authority to issue clearances, but whether he/she should be responsible for an operational error (i.e., violation of the 5 n mile standard) needs to be determined. For example, the MSP might fail to adequately monitor for compliance of a strategic clearance that results in a tactical operational error. Who is responsible for the error, the MSP or R-side controller who "owns" the aircraft at the time of the operational error? Operational acceptance of this concept requires answers to these questions.

6. *Candidate 6 NASA Airspace Coordinator*

This candidate represents the NASA Airspace Tool concept[24] of creating a new position called the Airspace Coordinator (AC). (For the purpose of maintaining consistency between the other candidate concepts, the assumption is made that the DST available to the AC has EDA-like functionality rather than functionality envisioned in the original Airspace Tool concept.) The AC sees the airspace of many sectors within a center. The AC, assisted with a DST, is able to provide more intelligent solutions for efficient air traffic management than a single controller assisted by DST capability at the sector. This concept has many similarities to candidate 5. The only significant difference between the two concepts is the method for issuing the strategic clearances. In this concept, the AC resolves the conflict with the aid of the DST, but he/she must coordinate with the R-side, via the display system, for agreement. Upon agreement, the R-side issues the clearance and is responsible for monitoring for compliance. As such, unlike candidate 5, the R-side is also clearly responsible for any operational errors that may occur.

7. *Candidate 7 Upstream Team*

This concept was proposed based on feedback from controllers at the Denver center and is essentially a combination of upstream D-side and upstream R-side concepts. These two concepts shared a common characteristic favored by the controllers, namely that the upstream sector resolves downstream problems. This minimizes intersector coordination compared to some of the other concepts and would allow controllers to be more focused on strategic planning.

The controllers disliked the aspect of the upstream D-side characteristic that only the D-side controller would have access to EDA-like decision support. From a workstation perspective, they thought it would be most efficient to have both R-side and D-side positions supported by the decision support capabilities. Certainly this would be more convenient for the R-side if he/she was the only controller working a sector during slower periods of air traffic. On the other hand, the primary drawback of the upstream R-side candidate, based on controller feedback, was its heavy dependence on the R-side position to support strategic planning tasks during busy periods (a time when the R-side is already experiencing high workload). This dependence may inhibit a trajectory orientation during periods when it is needed most.

In the upstream team concept, both the R-side and D-side are supported by EDA-like capability. The R-side would manage all tactical conflicts and, as the team leader, delegate strategic problems to the D-side depending on workload and other circumstances. If the R-side were too busy with tactical situations, the D-side would work alone on strategic planning, otherwise the strategic planning would be shared between both positions. Until CPDLC becomes available, the R-side must concur with the D-side resolution. Prior to the availability of CPDLC, the R-side would be responsible for issuing clearances to implement the strategic plans. With CPDLC, the R-side would have the option to delegate clearance communications to the D-side position as appropriate. This approach maximizes a controller team's flexibility to manage their traffic and workload. If the sector team includes a new controller to be checked out, the R-side team leader could require concurrence with D-side resolutions prior to D-side issuance of clearances. This provides a method for the more experienced controllers to supervise and mentor the less experienced with minimal risk (analogous to what occurs in a flight deck between a senior captain and a junior first officer). With or without CPDLC, the controller who resolves the conflict is responsible for monitoring the aircraft for compliance (e.g., if the R-side issues the clearance, the R-side must monitor for compliance).

This concept can reduce intersector coordination if the supporting DST technology is configured to distribute problem alerts/advisories to the appropriate sectors.

a. Advantages. By having EDA-like DSTs available to both controllers, this team concept appears to be the most effective of all the concepts for consistently supporting strategic planning. As stated before, strategic planning is the single most important criteria for achieving a trajectory orientation. The team concept allows for a balancing of workload between the R-side and D-side positions. If the R-side is not too busy with tactical conflicts, both controllers can work on aircraft conflicts further out on the time horizon, possibly to 20 min out. In contrast, if the R-side were busy with tactical conflicts, the D-side would perform all the strategic planning, but perhaps only work on conflicts with time horizons of 10–15 min out. This concept has a natural ebb and flow that should work well to smooth out the conflicts for air traffic patterns that have their own peaks and troughs.

Another advantage to this concept is its robust nature to the elemental changes that the ATC system will experience during the evolution to Free Flight. For example, it is fully effective with or without CPDLC. In contrast, the MSP concept requires CPDLC. In this concept, aircraft can be strategically planned whether or not they are equipped with CPDLC. Furthermore, in comparison to the MSP concept, the upstream team may plan resolutions that include immediate action rather than being restricted to flight plan changes that are initiated after the boundary to the next sector.

b. Disadvantages. One disadvantage to this concept is the need to provide EDA-like DST capabilities for both controller positions at each sector. The most significant disadvantage, however, is the risk associated with implementing upstream-team-based procedures. The operational viability of this concept rests on the dependence between sectors to receive traffic flows that are nominally planned to be in conformance with ATC constraints. Like posts supporting a picket fence, each post must carry its weight. Each downstream sector is, in turn, an upstream sector to someone else. The added workload to plan nominal conformance upstream translates into a lower workload in the next sector. Assuming that the net workload remained constant, but was redistributed, the airspace would benefit from a more predictable and robust flow of traffic. In any case, most if not all sectors must adopt the practice to realize the net benefit.

D. Rationale for Candidate Down-Selection

The two candidate concepts that were selected for further NASA research were candidates 5 and 7, the MSP and upstream team concepts, respectively. Candidate 1, URET-like procedures, and candidate 6, the airspace coordinator, were not selected because of the amount of inter/intrasector coordination required between controllers. Candidate 3 (upstream D-side) and candidate 4 (upstream R-side) were not selected, in favor of candidate 7 (upstream team), because candidate 7 combined the best features of both with few of the disadvantages of either. A detailed description of the advantages and disadvantages of each candidate concept is presented in Ref. 20.

VI. Conclusions

Trajectory orientation is a concept that, coupled with advanced en route DST capabilities, enables controllers to facilitate fuel-efficient, conflict-free trajectories across several sectors of airspace while conforming to flow-rate constraints. An operations assessment determined core issues in today's en-route operations that inhibit a trajectory orientation. In addition, seven concepts for new controller roles, responsibilities, and procedures were evaluated for their potential in achieving a trajectory orientation. Two concepts, one inspired by the EUROCONTROL MSP concept and one based on the upstream team concept, were determined to be the most likely candidates for achieving a trajectory orientation. Further research will focus on developing detailed controller procedures and requirements for supporting DST capabilities to facilitate trajectory-oriented ATC operations. In addition, a controller-in-the-loop simulation will be developed for procedural and DST concept validation.

References

[1]Green, S., "NASA AATT Milestone 5.10," NASA Advanced Air Transportation Technologies Project Office, NASA Ames Research Center, Dec. 1999.

[2]Green, S., Williams, D., and Grace, M., "Flight Test Results: CTAS and FMS Cruise/Descent Trajectory Prediction Accuracy," *ATM-2000, 3rd USA/Europe Air Traffic Management R&D Seminar*, Italian Agency for Air Navigation Services (ENAV), Paper 84, June 2000.

[3]Green, S., Vivona, B., Grace, M., and Fang, T. C., "Field Evaluation of Descent Advisor Trajectory Prediction Accuracy for En Route Clearance Advisories," AIAA *Guidance, Navigation, and Control Conference* 98-4479, CNO26576610, AIAA, Reston, VA, 1998, pp. 1668–1685.

[4] "Concept Definition for Distributed Air/Ground Traffic Management (DAG/TM)," Version 1.0, Advanced Air Transportation Technologies Project, NASA, Sept. 1999.

[5] Green, S. M., "En Route Spacing Tool: Efficient Conflict-Free Spacing to Flow-Restricted Airspace," *ATM-2000, 3rd USA/Europe Air Traffic Management R&D Seminar*, Italian Agency for Air Navigation Services (ENAV), Paper 30, June 2000.

[6] Rodgers, M. D. and Drechsler, G. K., "Conversion of the CTA, Incorporated, En Route Operations Concepts Database into a Formal Sentence Outline Job Task Taxonomy," Civil Aeromedical Institute, Federal Aviation Administration, DOT/FAA/AM-93/1, Jan. 1993.

[7] *Air Traffic Control*, Federal Aviation Administration, 7110.65M, 24 Feb. 2000.

[8] Ammerman, H. L., Bergen, L. J., Davies, D. K., Hostetler, C. M., Inman, E. E., and Jones, G. W., "FAA Air Traffic Control Operations Concepts Volume VI: ARTCC-Host En route Controllers," Federal Aviation Administration, DOT/FAA/AP/87-01, 1987.

[9] Nolan, M. S., *Fundamentals of Air Traffic Control*, Wadsworth, Inc., Belmont, CA, 1994.

[10] Smolensky, M. W., and Stein, E. S. (eds.), *Human Factors in Air Traffic Control*, Academic Press, San Diego, CA, 1998.

[11] Wickens, C. D., Mavor, A. S., Parasuraman, R., and McGee, J. P. (eds.), *Flight to the Future Human Factors in Air Traffic Control*, National Academy Press, Washington, 1997.

[12] Wickens, C. D., Mavor, A. S., Parasuraman, R., and McGee, J. P. (eds.), *The Future of Air Traffic Control Human Operators and Automation*, National Academy Press, Washington, DC, 1998.

[13] Kingsbury, J. A, "Air Traffic Control Automation, an AERA for This Century," MITRE Corp., MP86W28, McLean, VA, Oct. 1986.

[14] Celio, J. C., "Controller Perspective of AERA 2," MITRE Corp., MP-88W00015 Rev. 1, Feb. 1990.

[15] Arthur, W. C., Fernow, J. P., and Taber, N. J., "Functional Description of AERA 2 Automated Problem Resolution, Volume 1 Problem Management," MITRE Corp., MTR-85W223-01, McLean, VA, Jan. 1986.

[16] Fernow, J. P., Frolow, I., Pool, D. A., Shively, C. A., Taber, N. J., Walker, G. R. and Weidner, J. F., "Functional Description of AERA 2 Automated Problem Resolution, Volume 2 Aircraft Conflicts," MITRE Corp., MTR-85W223-02, McLean, VA, Jan. 1986.

[17] Dubofsky, D. F., Pool, D. A., and Walker, G. R., "Functional Description of AERA 2 Automated Problem Resolution, Volume 3 Airspace Conflicts," MITRE Corp., MTR-85W223-03, McLean, VA, Jan. 1986.

[18] Dubofsky, D. F. and Reierson, J. D., "Functional Description of AERA 2 Automated Problem Resolution, Volume 4 Flow Restriction Violations," MITRE Corp., MTR-85W223-04, McLean, VA, Jan. 1986.

[19] "Administrator's Fact Book," Federal Aviation Administration, U.S. Department of Transportation, APF-100, Dec. 1999.

[20] Leiden, K., and Green, S., "Research Task Order 34B Final Report," NASA Advanced Air Transportation Technologies Project Office, NASA Ames Research Center, May 2000.

[21] Kerns, K., and McFarland, A. L., "Conflict Probe Operational Evaluation and Benefits Assessment," MITRE Corp., MP98W239, McLean, VA, Nov. 1999.

[22] Brudnicki, D. J., and McFarland A. L., "User Request Evaluation Tool (URET) Interfacility Conflict Probe Performance Assessment," MITRE Corp., MP98W204, McLean, VA, Dec. 1998.

[23] Cole, R., Green, S., Jardin, M., Schwartz, B., and Benjamin, S., "Wind Prediction Accuracy for Air Traffic Management Decision Support Tools," *ATM-2000, 3rd USA/Europe Air*

Traffic Management R&D Seminar, Italian Agency for Air Navigation Services (ENAV), Paper 110, June 2000.

[24] Vivona, R. A., Ballin, M. G., Green, S. M., Bach, R. E., and McNally, B. D., "A System Concept for Facilitating User Preferences in En Route Airspace," NASA TM 4763, Nov. 1996.

[25] Nicolaon, J. P., *Multisector Planning Controller*, EUROCONTROL Experimental Centre, Bretigny-sur-Orge, France, (no date indicated).

[26] Meckiff, C., Chone, R., and Nicolaon, J. P., "The Tactical Load Smoother for Multi-sector Planning," EUROCONTROL Experimental Centre, BP 91222, Bretigny-sur-Orge, France, 1999.

Section VIII: Emerging Issues in Aircraft Self-Separation

One of the basic elements of Free Flight is aircraft "self-separation." Objectives include reduction in controller workload, support for more aircraft autonomy in flight, and enhancements to approach and landing in instrument meteorological conditions. There is considerable promise for capacity improvements if one can use aircraft situation displays that would allow pilots to conduct the same kind of self-spacing on instrument approaches that are used during visual approaches.

A number of issues need to be addressed by researchers in support of the development of self-separation schemes. These include 1) the role and responsibility of the pilot and of the controller; 2) information available to the pilot and controller; 3) nature of conflict detection and resolution algorithms; 4) knowledge of intent and amount of cooperation in conflict resolution; and 5) metrics for evaluation rules. The four chapters in this book are reports on some of the research into these issues.

Chapters 39 and 40 deal with approaches to cooperative airborne separation assurance. The authors of Chapter 39 propose and evaluate an interesting agent-based artificial intelligence approach to dealing with the conflict resolution problem. They use both safety and economic criteria in selecting conflict avoidance maneuvers. The authors of Chapter 40 explore some of the cost, safety, and efficiency tradeoffs for different degrees of sophistication/complexity of maneuver coordination rules. They propose an interesting mathematically defined, rule efficiency measure and also report on a computer model-based evaluation of different rule structures.

The authors of Chapter 41 look at the general problem of determining the conflict potential of a set of planned or predicted aircraft trajectories. They propose a novel probabilistic approach for solving this problem and, using a simulation, compare it to three other, commonly used approaches. The metrics to compare the approaches are flexibility of usage and imposed restrictions on aircraft behavior.

Chapter 42 reports on an actual operational evaluation of some of the technologies (i.e., Global Positioning System (GPS) and Automatic Dependent Surveillance-Broadcast mode (ADS-B)) that underlie airborne separation assurance and the integration of these technologies into enhanced pilot displays. The trials focused on ADS-B benefits for enhanced visual acquisition and enhanced visual approaches. This operational demonstration represents an ambitious partnership between government and industry to find ways of accelerating the implementation of new technology.

Chapter 39. Cooperative Optimal Conflict Airborne Separation Assurance in Free Flight Airspace, Colin Goodchild, Miguel A. Vilaplana, and Stefano Elefante, University of Glasgow, 2000.

Chapter 40. Operational Efficiency of Maneuver Coordination Rules for Airborne Separation Assurance System, R. Schild, and J. K. Kuchar, MIT, 2000.

Chapter 41. Probabilistic Approaches Toward Conflict Prediction, G. J. Bakker, H. J. Kremer, and H. A. P. Blom, NLR, 2000.

Chapter 42. Safe Flight 21: 1999 Operational Evaluation of ADS-B Applications, James J. Cieplak, Edward Hahn, and Baltazar O. Olmos, MITRE/Corporation, 2000.

Chapter 39

Cooperative Optimal Airborne Separation Assurance in Free Flight Airspace

Colin Goodchild,* Miguel A. Vilaplana,[†]
and Stefano Elefante[†]
University of Glasgow, Glasgow, United Kingdom

I. Introduction

THE goal of air traffic management (ATM) according to the International Civil Aviation Organization's (ICAO) committee on future air navigation systems (FANS) is: *"To enable aircraft operators to meet their planned time of departure and arrival and adhere to their preferred flight profiles with minimum constraints without compromising agreed levels of safety."*[1]

The key technologies required for the implementation of the ATM system are communications, navigation, and surveillance (CNS). Advances in these technologies and developments involving their integration into systems architectures provide the information framework for a global ATM.

One approach to realizing the CNS/ATM objective for the en route phase of aircraft operations is the concept of Free Flight, proposed by the Federal Aviation Administration (FAA). The Free Flight concept establishes a regime of instrument flight rules (IFR) airspace where aircraft are allowed to fly user-preferred routes with the task of separation assurance delegated to the aircraft. The motivation for introducing Free Flight is based on evidence[2] that it will bring significant cost savings to the airlines in addition to improving the efficiency and capacity of the current restricted airspace system.

EUROCONTROL proposes to implement Free Flight operations in certain regions of airspace that will be promulgated by the airspace planning and management services on a daily basis according to the expected traffic demand.[3] The user-preferred route scheme will also be available for suitably equipped aircraft in managed airspace, where a ground-based control center will be responsible

---

Copyright © 2001 by the American Institute of Aeronautics and Astronautics, Inc. All rights reserved.
*Senior Lecturer in ATM and Avionics, ATM Research Group, Department of Aerospace Engineering.
[†]Research Assistant, ATM Research Group, Department of Aerospace Engineering.

for separation assurance. However, the ability to fly operator-preferred routes is of questionable economic benefit if aircraft have to make frequent maneuvers to avoid proximate traffic conflicts. This has motivated a number of recent studies on various aspects of conflict avoidance in Free Flight.

One approach that has been followed in several studies of conflict avoidance is based on hybrid control techniques and is reported in Refs. 4–6. The method is a scheme to verify the safety of resolution maneuvers in different conflict scenarios. Another study proposes a probabilistic approach to conflict avoidance. In this work, presented in Ref. 7, a technique that employs an airborne alerting logic based on the probability of conflict is developed for encounters with one intruder. The EUROCONTROL Free-Route Experimental Encounter Resolution-1 (FREER-1) study[8] analyzes the basic requirements for autonomous airborne separation assurance in Free Flight and introduces an interactive human-centered resolution scheme. A set of extended flight rules (EFR) is defined to provide coordinated conflict resolution by assigning a priority to each aircraft involved in a conflict.

An approach employing optimal control theory is reported in Ref. 9. With approach, optimal control techniques are used to develop an algorithm for the resolution of conflicts involving two aircraft. This algorithm is based on the maximization of the distance between two aircraft at the point of their closest approach. An advisory system based on this algorithm is introduced in the form of a set of maneuver charts that are to be used as "rules of the air." Another study, reported in Ref. 10, uses an economic model to analyze the cost-benefits of different resolution maneuvers for long look-ahead time strategic conflicts.

Algorithms for resolving three-dimensional conflicts involving multiple aircraft are presented in Ref. 11. These algorithms are based on trajectory optimization methods and provide resolution actions that minimize a certain cost function.

When considering conflict scenarios in Free Flight, the approach proposed in this chapter assumes that aircraft will be equipped with the airborne segment of the planned CNS system. In this system, accurate navigation information will be provided by the Global Navigation Satellite System (GNSS), and an information link with the proximate aircraft will be established through automatic dependent surveillance-broadcast (ADS-B) and data link communications. By using these advanced CNS technologies, the requirement for look-ahead time in the conflict avoidance systems for Free Flight is estimated to be between 10 and 30 min.[12] This enhanced look-ahead time capability, together with accurate trajectory prediction and the possibility of data exchanges between aircraft, will permit cooperative conflict avoidance strategies that incorporate optimization criteria that consider the interests of all the conflicting aircraft.

Regarding the preceding assumptions, this chapter makes two main contributions to airborne separation assurance in free flight airspace:

1) the definition of a formal framework that supports cooperative airborne separation assurance in free flight airspace.

2) the development of a dynamic programming algorithm to provide a weighted-cost sharing resolution to multiple aircraft conflicts.

The proposed framework supports the methodology for a multiple aircraft, strategic airborne separation assurance system in which aircraft share the costs involved in the conflict resolution. This methodology is based on the theory of multiagent systems in distributed artificial intelligence (DAI).[13–16] A set of proximate aircraft operating in free flight airspace is cast as the multiagent system. These aircraft are modeled as intelligent agents having joint responsibility[16] to establish a defined joint goal[16] of separation assurance. This joint goal is achieved through a joint solution or common conflict resolution plan. Upon detection of a predicted conflict by one or more aircraft, a team of the conflicting aircraft is formed with the purpose of resolving the conflict. A set of conventions[15] establishes the foundation for a communication protocol that allows the cooperation and negotiation of resolution plans.[17–19] A dynamic programming algorithm enables the agent aircraft conflict scenario to compute resolution actions consisting of a set of airspeed controls for each aircraft involved in the conflict. These plans are implemented by the team members according to the preceding framework. The proposed conflict avoidance algorithm is based on the concept of motion planning in robotics.[20–22]

All of the techniques described in this chapter would be implemented as a facility to enhance the pilot situation awareness and aid the human decision-making process. However, these ideas could form the basis of fully autonomous aircraft operations in the future.

II. Operational Methodology

A. Multiagent System Model

The operational methodology for the proposed airborne separation assurance system for aircraft in free flight airspace is based on the theory of multiagent systems within the broader scope of DAI.[16] Proximate aircraft flying in free flight are modeled as autonomous and intelligent entities (agents) that constitute a multiagent system. They fly according to their self-interests (user-preferred routes) but are willing to cooperate to avoid conflicts. Conflicting aircraft associate in teams to cooperate in separation assurance. Team members adopt the avoidance of the predicted conflict as a joint goal[16] and commit to the execution of a conflict avoidance plan (joint solution).[16]

A set of rules called conventions known to the agents regulates the team formation process according to the model of joint responsibility[16] and establishes the foundation for a cooperation protocol.

The attributes of the aircraft forming a separation assurance team are:
1) The aircraft share minimum separation conflict avoidance as the common goal (G).
2) Each aircraft in the team has a defined priority for its free flight operations, which is communicated to the other aircraft in the team.
3) All aircraft in the team are aware of the free flight priorities of the other team members.

B. Protocol for Cooperative Conflict Avoidance

Let A represent a group or set of proximate aircraft flying in free flight airspace. Each aircraft in this set is monitoring the tracks and communications of the other aircraft in the set to determine the possibility of the violation of separation minima.

When an aircraft $a_0 \in A$ detects a possible violation of separation minima involving itself and one or more other aircraft in A, the aircraft a_0 defines the subset $A_C \subset A$. The subset A_C consists of the aircraft in A that are on courses that are predicted to conflict with the single aircraft a_0. The aircraft a_0 that is detecting the conflicts, broadcasts a message to the aircraft in the set A_C. The purpose of this message is to form a team of aircraft A_T that will cooperate to establish a common resolution plan. In this team a_0 assumes the role of team organizer. Each aircraft belonging to the subset A_C considers its defined set of conventions to assess the request from a_0. When the assessment is complete, each aircraft replies to a_0 indicating whether it will join the team or continue with its current intentions. Thus a team A_T also included in the set A is organized by a_0. It follows that the membership of A_T consists of a_0 and the members of A_C that are willing to cooperate in the resolution of the conflict.

Once the team has been formed, a_0 designs a common resolution plan P and transmits this plan to the other team members. The plan P consists of a set of actions that each aircraft in A_T must execute to maintain safe separation. This plan provides a strategy to solve the predicted conflicts in a coordinated manner, while simultaneously taking account of the costs of the resolution for all the members of the team. When the members of the team formed by a_0 have received the plan P designed by a_0, they assess the plan against their conventions and their current situation. Each aircraft in the team A_T communicates its imminent intentions to the team organizer a_0. If a team member a_i decides to drop its commitment to the execution of the plan P, it is assumed it does so because it is unable to execute the actions assign to it in the plan P devised by the team organizer a_0. This may occur for the following reasons:

1) The team organizer a_0 has incomplete knowledge of a_i's current situation and suggests a plan that a_i cannot execute because it has suffered an emergency and/or an equipment failure.

2) The aircraft a_i is aware of another resolution action that is more effective to its own situation and that it can apply to itself to maintain safe separation from the other aircraft in A_T. This resolution action can be made only if it does not interfere with the actions that the plan P assigns to the other team members.

In the case of 1), the team organizer replans with a new team that excludes a_i and considers it as an intruder that becomes a constrain in the new plan. In the case of 2), either the original plan is executed by the remaining team members or a significantly more efficient plan is devised by the organizer. Figure 1 illustrates the structure of the set of conventions that can be implemented by the individual aircraft agent a_i.

It is considered that the airborne communication, navigation, and surveillance technologies being developed for the future ATM system will be capable of supporting the airborne separation assurance scheme outlined in the preceding.

III. Planning Algorithm

This section presents a dynamic programming algorithm that enables a team organizer to design cooperative strategic resolution plans for two-dimensional conflicts involving multiple aircraft. These plans consist of a set of feasible speed control actions for each of the team members and provide weighted-cost sharing

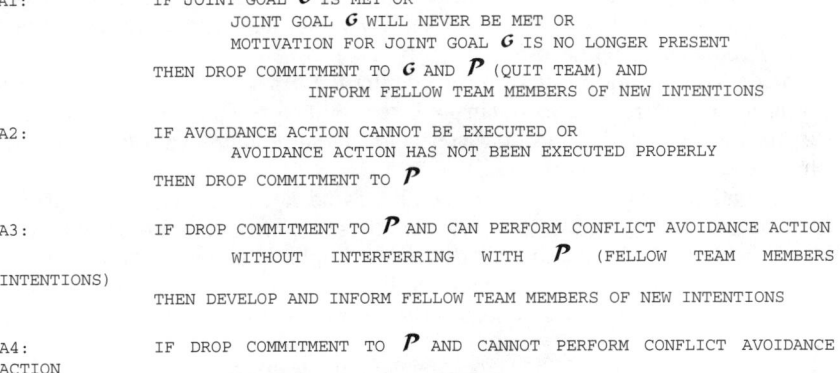

Fig. 1 High-level conventions for joint responsibility.

solutions suitable to be implemented according to the operational methodology described.

A team organizer a_0 uses its knowledge of its current flight plan and the proximate aircraft's intentions to predict possible conflicts. The aircraft a_0 forms a team A_T and designs a common conflict avoidance plan. As the designer of this plan, a_0 acts as a central planner[20] and searches for the speed control actions to enable the members of the team A_T to track cost-effective conflict-free trajectories along their intended paths.

For simplicity, it is supposed that there are neither navigation nor conflict prediction errors. The conflicts considered involve aircraft in free flight airspace flying along straight-line tracks at their optimal cruise speed.

The use of speed control actions as a strategic conflict resolution technique in free flight is justified by the fact that conflicts can be solved considering cost savings without changing the aircraft's preferred paths. Speed control conflict resolution can be implemented in a cost-effective manner and considering passengers' comfort if the conflict is detected at least 10 min prior to the time of closest approach.[10] Moreover, because the aircraft's intended paths remain unchanged, this technique presents the additional advantage of a low probability of creation of new conflicts with the proximate aircraft as a result of the execution of the resolution actions.

A. Mathematical Foundation

The conflict avoidance planning algorithm presented in this chapter is based on concepts from motion planning in robotics,[20–22] game theory,[23] optimal control,[24] and dynamic programming.[25,26] Conflict avoidance is modeled as a multistage decision-making process consisting of k stages and involving n decision makers. A scalar-valued stage-additive functional

$$L^i = \sum_{s=1}^{k} l_s^i \tag{1}$$

is defined for each team member. The functional L^i is called the loss functional of the ith decision maker and l_s^i represents an accumulative cost. The loss functionals are defined to guide the decision makers in the search for a sequence of actions that provides optimal conflict avoidance. Therefore, loss functionals encompass both safety and economic costs.

Safety costs are introduced by considering the times to the points of closest approach for the aircraft involved in predicted conflicts.[27] The economic costs that loss functionals take into account are those that accumulate with time, such as flight time, fuel consumption, etc.

The mathematical framework for conflict avoidance presented in this chapter introduces a hybrid architecture in which discrete-time analysis (decision making) is combined with continuous dynamics (implementation of the decisions). This hybrid architecture together with a discretization of the state and decision spaces produces feasible actions that can be implemented by the flight management system.

B. Multiobjective Optimization

Designing a separation assurance plan consists of searching for a sequence of decisions (speed control actions) for each decision maker (team member) that enables it to track a conflict-free trajectory while optimizing its loss functional.

Because cooperative conflict is the goal, the issue becomes a multiobjective minimization[28] of the loss functionals, which is subject to the constraints associated with aircraft dynamics and separation minima. The aircraft dynamic models used by the algorithm are based on the EUROCONTROL database called the Base of Aircraft Data (BADA).[29]

To obtain solutions that minimize the losses in an equitable manner for all the team members, the weighting-objectives method[28] is applied. Therefore, the problem is changed to a scalar minimization of a global functional of the form:

$$L = \sum_{i=1}^{n} w_i \cdot L^i = \sum_{i=1}^{n} w_i \cdot \left(\sum_{s=1}^{k} l_s^i \right) \tag{2}$$

where the scalars $w_i \geq 0$ are the weighting coefficients. Optimal strategies obtained by minimizing Eq. (2) are Pareto nondominated solutions[28] of the multiobjective optimization problem. Therefore, the global solutions are optimal in the sense that the value of any individual loss functional cannot be reduced without increasing the value of at least one of the other individual loss functionals.

The default values for the weighting coefficients in the planning algorithm are $w_i = (1/n)$ for each of the n members of the team. These values can be changed to represent the relative importance of the loss functionals according to the aircraft priorities.

The development of analytical continuous-time solutions to the problem of the minimization of Eq. (2) would require detailed analysis of the specific models and geometry of the conflicts. However, the introduction of a multistage decision-making process and the discretization of the decision and state spaces make the application of dynamic programming possible for the resolution of the problem.[23,25,26] Dynamic programming provides the means for the numerical computation of solutions for a certain discretization of both the decision and state spaces. Therefore, a planning algorithm based on dynamic programming can be adapted to other discretizations.

In the algorithm presented in this chapter, dynamic programming is used to find sequences of speed control actions (decisions) for all the aircraft in the team. These control actions are designed to enable all the aircraft in the team to avoid the predicted conflicts while at the same time minimizing the global loss functional, Eq. (2). This loss functional provides a tradeoff between the self-interests of each aircraft and a cost sharing between team members and assures safe separation.

Dynamic programming is carried out in a forward stepping manner applying the Theorem of Optimality for discrete and deterministic multistage decision processes.[23] Hence, optimal sequences of decisions are found through an iterative minimization process:

$$L^*_{s+1}(x_{s+1}) = \min \left(L^*_s + \sum_{i=1}^{n} w_i \cdot l^i_s \right) \quad (3)$$

where L^*_s is a minimum of the global loss computed from Eq. (2) up to stage s. Equation (3) is minimized at every stage until the conflict is avoided. Constraints associated with separation minima and the aircraft dynamics must be satisfied at each stage and during the intervals of transition between stages.

IV. Computed Example

An example of the application of the operational methodology and the planning algorithm presented in this chapter is shown in the following. A scenario is considered in which three aircraft, A1, A2, and A3, are flying along straight tracks at the same altitude in free flight airspace and at their selected speeds. They are assumed to be equipped with GNSS and ADS-B or some other type of data link for interaircraft communications. In addition, each aircraft is equipped with the cooperative separation assurance system presented in the preceding. A range of 120 n miles for ADS-B and data link communications is assumed.[8] A conflict alert is declared if a separation of less than 5 n miles between any two aircraft is detected along the aircraft's intended trajectories.

The initial configuration is displayed in Fig. 2. Figure 3 shows the evolution of the distances between the aircraft if their current speeds are maintained. Three predicted violations of the separation minima are displayed. At the initial configuration aircraft A1 predicts three violations of separation minima shown in Fig. 3. A1

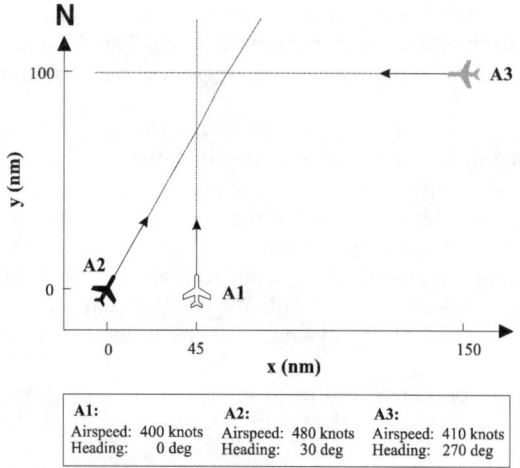

Fig. 2   Initial configuration for three aircraft conflict.

initiates the formation of a team with A2 and A3 to avoid the conflict cooperatively. A1 can communicate with A2 and A3, but A2 and A3 cannot communicate with each other.

Different solutions to this conflict scenario can be obtained depending on 1) whether A2 and A3 join the team, and 2) the economic criteria considered in the loss functionals. Three possible solutions are presented in the following sections.

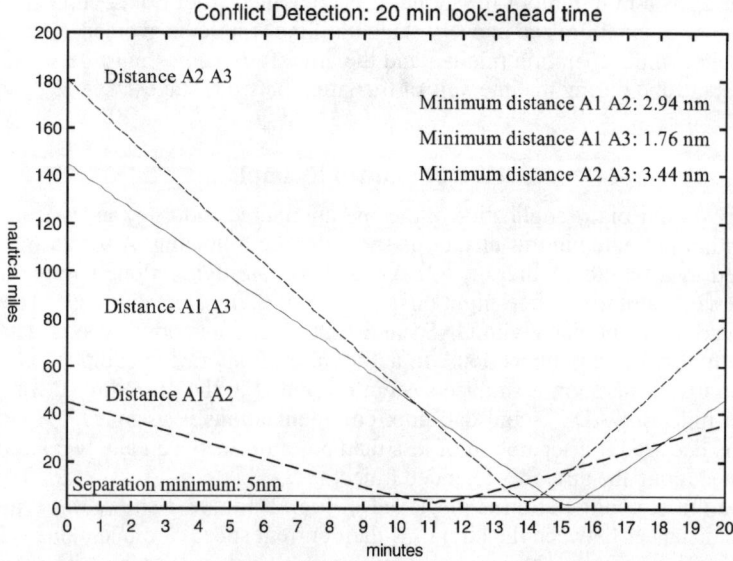

Fig. 3   Prediction of possible conflicts by A1 at initial configuration.

A. Solution 1: Team Solution, Separation Assurance with Minimum Maneuvering Costs

If both A2 and A3 are willing to cooperate in the conflict avoidance and therefore join the team, A1 applies its planning algorithm to search for a global solution to the conflicts that minimizes a given loss functional. This solution consists of speed control actions for the three aircraft involved in the conflict.

The global loss functional minimized by A1 encompasses safety as well as economic criteria for the three aircraft involved in the conflicts. The economic criteria considered in this case are the time of operation at nonoptimal speed and the additional costs due to accelerations and decelerations. Therefore, the solution assures separation with the minimum number of maneuvers and at the same time keeps optimal speeds for as long as possible. Deviation from the four-dimensional intended trajectory has not been considered as an economic cost in this case.

The three weighting factors in the global loss functional are set to the same value to distribute equitably the costs among the three aircraft involved in the conflicts. The results of this case are shown in Figs. 4 and 5 and in Table 1.

B. Solution 2: Team Solution, Minimization of Four-Dimensional Losses

Taking again the scenario described for solution 1, consider a situation in which A1 is required to compute a global loss functional that takes account of the team members performing separation assurance actions that are causing them to deviate from their individual optimal four-dimensional trajectories.

The minimization of this new global loss functional, carried out by A1, provides all three aircraft with a new set of speed control actions that guarantees coordinated separation assurance. Providing the same value is assigned to the three weighting

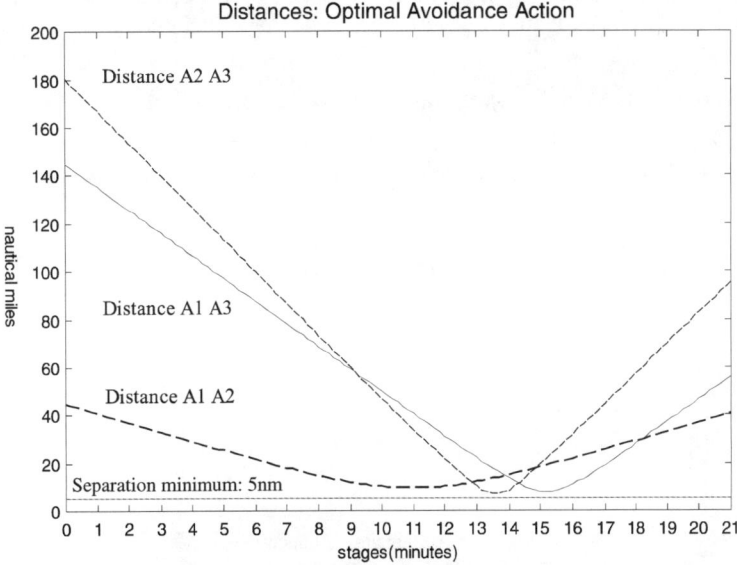

Fig. 4 Distances for conflict avoidance action (case 1).

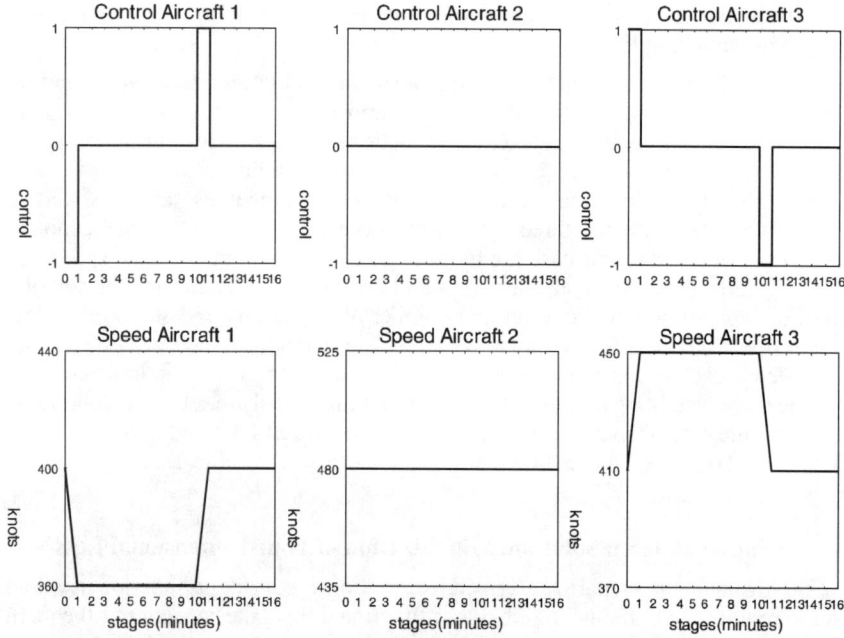

Fig. 5 Conflict avoidance speed control actions (solution 1).

factors in the global loss functional, this new set of control actions minimizes both the deviations from the optimal intended four-dimensional trajectories and the economic criteria considered in solution 1.

This team solution distributes the costs equitably between the three members. The results of this case are shown in Figs. 6 and 7 and in Table 2.

C. Solution 3: No Team Solution (A3 Does Not Join the Team)

The case is now considered in which the scenario of solution 2 is modified with A3 not willing to cooperate in the conflict resolution plan. It is assumed that A3

Table 1   Conflict avoidance solution 1

Minimum distance A1 A2: 9.6 n miles
Minimum distance A1 A3: 7.7 n miles
Minimum distance A2 A3: 7.1 n miles
$t_{4D} - t_{avoid}{}^a$:
  A1: 1.0 min
  A2: 0 min
  A3: −0.98 min

---

[a]Delay from the optimal four-dimensional time at the point of the trajectory in which the conflicts are considered as avoided.

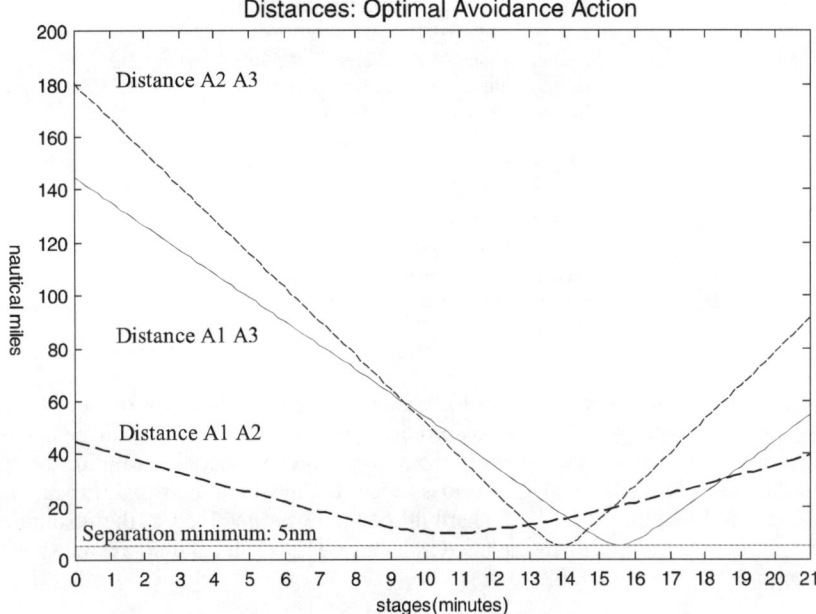

Fig. 6 Distances for conflict avoidance action (solution 2).

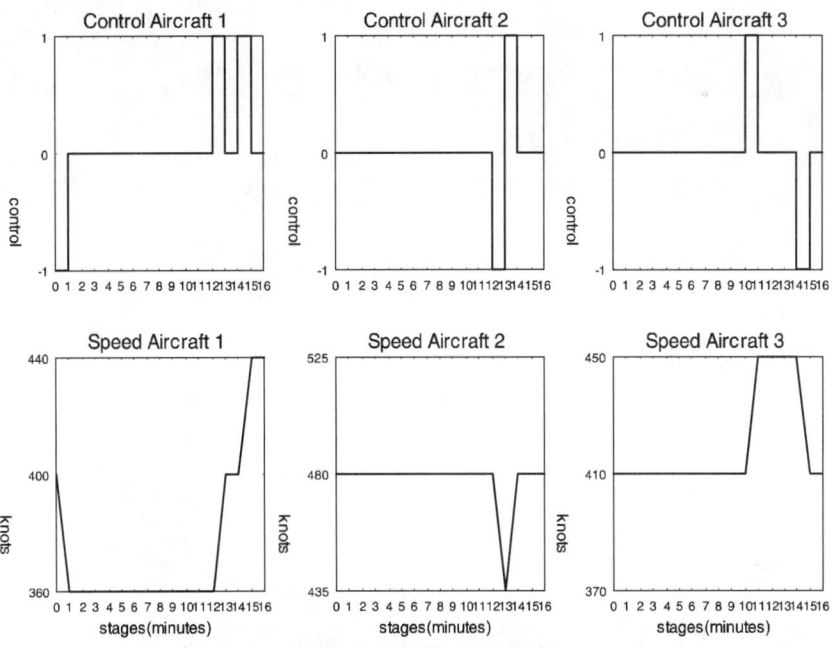

Fig. 7 Conflict avoidance speed control actions (solution 2).

Table 2 Conflict avoidance solution 2

Minimum distance A1 A2: 9.85 n miles
Minimum distance A1 A3: 5.37 n miles
Minimum distance A2 A3: 5.1 n miles
$t_{4D} - t_{avoid}{}^a$:
A1: 1.05 min
A2: 0.09 min
A3: -0.39 min

[a] Delay from the optimal four-dimensional time at point of the conflicts trajectory in which the are considered as avoided.

intends to maintain its current trajectory. Thus the team in this case is reduced to the two aircraft A1 and A2. The solution presented here provides an avoidance plan for A1 and A2 that accounts for the optimization criteria considered in solution 2. The aircraft A1 minimizes a new global loss functional in which the weighting factors are adjusted so that A1 and A2 share all of the costs involved in the resolution leaving, A3 to continue on its intended trajectory. The results of this case are shown in Figs. 8 and 9 and in Table 3.

V. Conclusions

A new framework based on DAI has been proposed to implement autonomous airborne separation assurance in free flight airspace. This framework supports

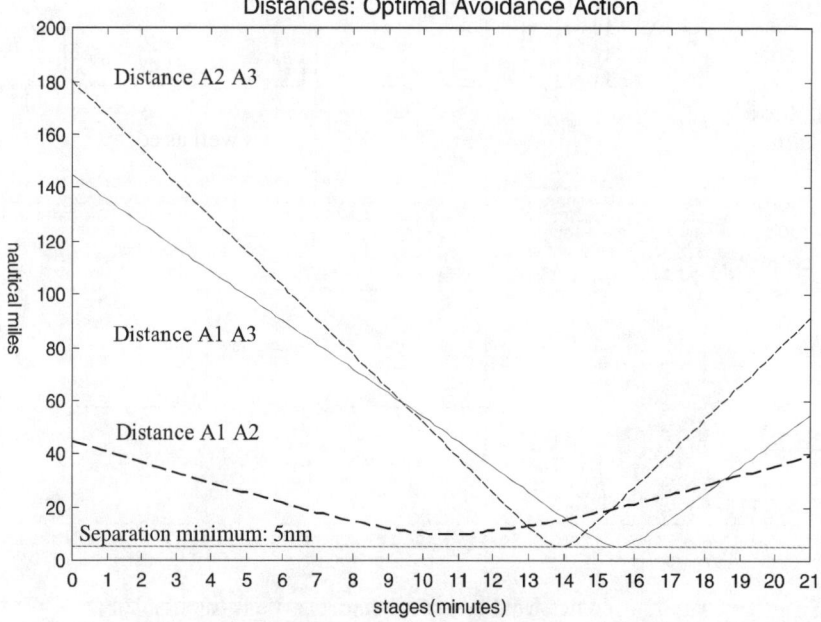

Fig. 8 Distances for conflict avoidance action (solution 3)

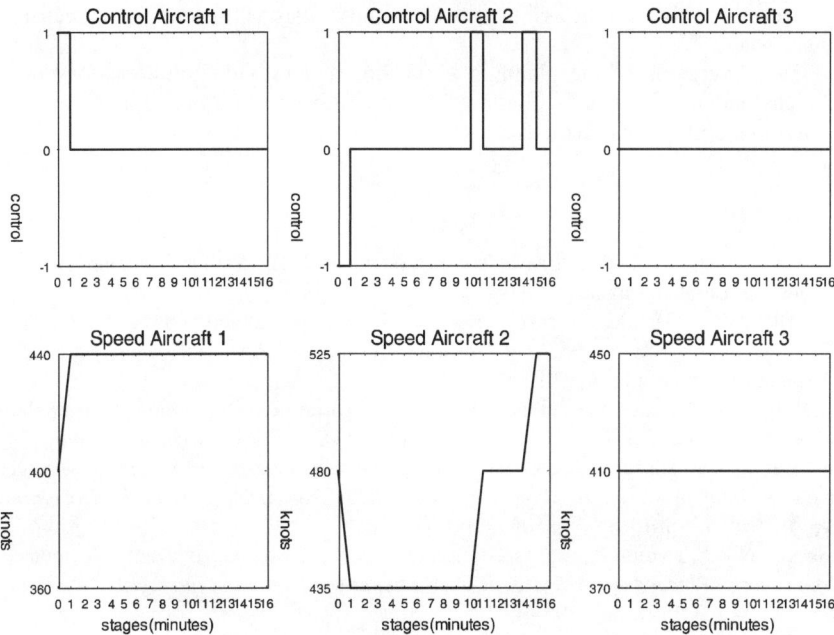

Fig. 9 Conflict avoidance speed control actions (solution 3).

cooperation and costs optimization in airborne strategic conflict avoidance. Within the framework, conflicting aircraft form teams to cooperatively resolve conflicts. An operational protocol for team formation has been presented. The detailed definition of this protocol will depend on the future technical requirements of interaircraft data communication systems (ADS-B and data links).

A dynamic programming algorithm has been developed for airborne centralized conflict avoidance planning. The algorithm provides avoidance strategies that minimize a global loss functional that considers safety as well as economic costs.

The conflict avoidance maneuvers considered in the current version of the algorithm are speed control actions. Therefore, the existence of a global optimal avoidance strategy is not guaranteed and depends on the loss functional that is being minimized. However, solutions can be found in more complex cases

Table 3 Conflict avoidance solution 3

Minimum distance A1 A2: 11.05 n miles
Minimum distance A1 A3: 8.43 n miles
Minimum distance A2 A3: 6.85 n miles
$t_{4D} - t_{avoid}{}^a$:
  A1: -1.55 min
  A2: 0.80 min
  A3: 0 min

[a] Delay from the optimal four-dimensional time at point of the trajectory in which the conflicts are considered as avoided.

by increasing the weight assigned to the safety costs within the individual loss functionals.

Future versions of the planning algorithm will provide solutions for three-dimensional conflicts and will allow for cooperative resolution that includes both horizontal and vertical maneuvers.

References

[1] Galotti, V., "The Future Air Navigation System (FANS)," Avebury Aviation, Ashgate Publishing Limited, England, 1997.

[2] Allen, D. L., Haraldsdottir, A., Lawler, R. W., Pirotte, K., and Schwab, R. W., "The Economic Evaluation of CNS/ATM Transition," CNS/ATM Projects, Boeing Commercial Airplane Group, Seattle, April 1998.

[3] "Operational Concept Document (OCD)," Ed. 1.1, European Organization for the Safety of Air Navigation (EUROCONTROL), Bretigny-sur-Orge, France, 4 January 1999.

[4] Tomlin, C., Pappas, G. J., and Sastry, S., "Conflict Resolution for Air Traffic Management: A Study in Multiagent Hybrid Systems," *IEEE Transactions on Automatic Control*, Vol. 43, No. 4, April 1998, pp. 509–521.

[5] Kosecka, J., Tomlin, C., Pappas, G. J., and Sastry, S., "Generation of Conflict Resolution Maneuvers," *Proceedings of the IEEE International Conference on Intelligent Robots and Systems*, Vol. 3, 1997, pp. 1598–1603.

[6] Tomlin, C., Pappas, G. J., and Sastry, S., "Noncooperative Conflict Resolution," *Proceedings of the 36th Conference on Decision and Control*, Institute of Electrical and Electronics Engineers, Vol. 2, Dec. 1997, pp. 1816–1821.

[7] Yang, L. C., and Kuchar, J. K., "Prototype Conflict Alerting System for Free Flight," *Journal of Guidance, Control, and Dynamics*, Vol. 20, No. 4, July–Aug. 1997, pp. 768–773.

[8] Duong, V. N., Hoffman, E., and Nicolaon, J.-P., "Initial Results of Investigation into Autonomous Aircraft Concept (FREER-1)," *paper from 1st USA/Europe ATM R&D Seminar*, June 1997.

[9] Krozel, J., Mueller, T., and Hunter, G., "Free Flight Conflict Detection and Resolution Analysis," *AIAA Guidance, Navigation, and Control Conference*, Paper No. AIAA-96-3763, AIAA, Reston, VA, July 1996.

[10] Krozel, J., and Peters, M., "Strategic Conflict Detection and Resolution for Free Flight," *Proceedings of the 36th Conference on Decision and Control*, Institute of Electrical and Electronics Engineers, Dec. 1997, pp. 1822–1829.

[11] Menon, P. K., and Sweriduk, G. D., "Optimal Strategies for Free-Flight Air Traffic Conflict Resolution," *Journal of Guidance, Control, and Dynamics*, Vol. 22, No. 2, March–April 1999, pp. 202–211.

[12] Warren, A., "A Methodology and Initial Results Specifying the Requirements for Free Flight Transitions," Boeing Commercial Group *paper from 1st USA/Europe ATM R&D Seminar*, June 1997.

[13] Mandiau, R., and Piechowiak, S., "Conflict Solving into the Multi-Agent Distributed Planning," *Universite de Valenciennes*, France, 1998.

[14] Woodlridge, M., "The Logical Modelling of Computational Multi-Agent Systems," Ph.D. Thesis, Univ. of Manchester Inst. of Science and Technology, Manchester, Oct. 1992.

[15] Singh, M. P., Rao, A. S., and Georgeff, M. P., "Formal Methods in DAI: Logic-Based Representation and Reasoning," *Multi-Agent Systems: A Modern Approach to Distributed Artificial Intelligence*, edited by Gerhard Weiss, MIT Press, Cambridge, MA, 1999, Chap. 8, pp. 331–376.

[16]Jennings, N., *Cooperation in Industrial Multi-Agent Systems*, World Scientific Publishing Co. Pte. Ltd., Singapore, 1994.

[17]Davis, R., and Smith, R. G., "Negotiation as a Metaphor for Distributed Problem Solving," *Artificial Intelligence*, Vol. 20, No. 1, 1983, pp. 63–109.

[18]Wangermann, J. P., and Stengel, R. F., "Distributed Optimization and Principled Negotiation for Advanced Air Traffic Management," *Proceedings of the 1996 IEEE International Symposium on Intelligent Control*, Dearborn, MI, Sept. 1996, pp. 156–161.

[19]Wangermann, J. P., and Stengel, R. F., "Optimization and Coordination of Multiagent Systems Using Principled Negotiation," *Journal of Guidance, Control, and Dynamics*, Vol. 22, No. 1, Jan.–Feb. 1999, pp. 43–50.

[20]LaValle, S. M., and Hutchinson, S. A., "Optimal Motion Planning for Multiple Robots Having Independent Goals," *IEEE Transactions on Robotics and Automation*, Vol. 14, No. 6, Dec. 1998, pp. 912–925.

[21]LaValle, S. M., "A Game-Theoretic Framework for Robot Motion Planning," Ph.D. Thesis, Univ. of Illinois, Urbana-Champaign, IL, 1995.

[22]Mediavilla, M., Fraile, J. C., Peran, J. R., and Dodds, G. I., "Optimization of Collision Free Trajectories in Multi-Robot Systems," *Proceedings of the 1998 IEEE International Conference on Robotics & Automation* Leuven, Belgium, May 1998, pp. 2910–2915.

[23]Kauffmann, A., *Graphs, Dynamic Programming, and Finite Games*, Academic Press, New York, 1967.

[24]Lewis, F. L., and Syrmos, V. L., *Optimal Control*, 2nd ed., Wiley, New York, 1995.

[25]Bellman, R., *Dynamic Programming*, Princeton Univ. Press, Princeton, NJ, 1957.

[26]Gluss, B., *An Elementary Introduction to Dynamic Programming. A State Equation Approach*, Allyn and Bacon, Boston, 1972.

[27]Fulton, N. L., "Airspace Design: Towards a Rigorous Specification of Conflict Complexity Based on Computational Geometry," *The Aeronautical Journal*, Vol. 103, No. 1020, Feb. 1999, pp. 75–84.

[28]Osyczka, A., *Multicriterion Optimization in Engineering with FORTRAN Programs*, Ellis Horwood, Chichester, England, 1984.

[29]Baulleret, P., "User Manual for the Base of Aircraft Data (BADA)–Revision 3.1," EEC Note No 25/98, EUROCONTROL, Bretigny-sur-Orge, France, Nov. 1998.

Chapter 40

Operational Efficiency of Maneuver Coordination Rules for Airborne Separation Assurance System

R. Schild* and J. K. Kuchar[†]
Massachusetts Institute of Technology, Cambridge, Massachusetts

I. Introduction

THE air traffic conflict detection and resolution process consists of several tasks to ensure separation or avoid collisions depending on the scope of the system (see Fig. 1).[1] The first task is to check the environment for potential conflicts with the help of state and possibly velocity vector information. Based on the information available, future positions can then be estimated.

Conflict detection is based on the estimation of future vehicle positions through the application of predefined metrics on the situation to decide whether or not a conflict is present. This metric may include a sole parameter (e.g., distance) or may be a combination of several parameters (e.g., distance, time, and maneuvering cost). After the detection of a conflict, a conflict resolution phase requires appropriate maneuver action and information distribution to all aircraft involved in the conflict.

Under certain operational conditions, rule systems such as visual flight rules (VFR)[2] in aviation or collision avoidance rules for shipping[3] are used for the resolution maneuver coordination. These rules define priorities to vehicles involved in the conflict and suggest a corresponding resolution maneuver. Explicit maneuver coordination is possible but not necessarily required. Envisioned self-separation environments such as free flight[4] will likely need some form of resolution maneuver coordination.

In a decentralized system, each agent has to coordinate decisions with other agents involved in the conflict. Predictability of decisions and resolution maneuvers is essential to maintain an orderly flow of traffic and to prevent collisions. Rule systems for maneuver coordination (priority determination and resolution maneuver) would facilitate this predictability.

Copyright © 2001 by the Massachusetts Institute of Technology. Published by the American Institute of Aeronautics and Astronautics, Inc., with permission.
*Research Affiliate, Department of Aeronautics and Astronautics.
[†]Assistant Professor, Department of Aeronautics and Astronautics. Senior Member AIAA.

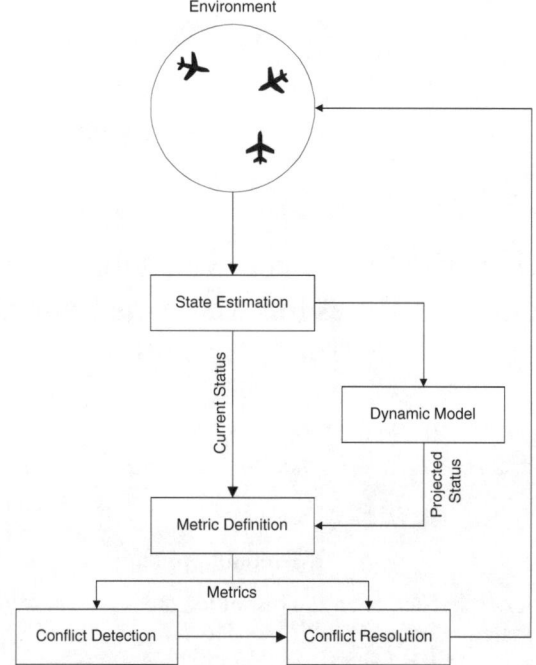

Fig. 1 Generic conflict detection and resolution process.

The complexity of maneuver coordination rule systems depends on the operational environment and its characteristics. In today's existing rule-based systems, quite simple rule sets are used and work well (e.g., VFR). Responsibility for decisions rely with the human operator (for legal reasons), who has to understand the underlying principles. The use of too complex rule systems holds the potential for human misinterpretation, while they might not necessarily increase operational system performance.

Additionally, the increase of rule complexity follows the law of diminishing returns, meaning that the additional gain through more input information and higher rule complexity is getting smaller. Finally, increasing rule complexity may make it more difficult for human operators to understand the basis behind conflict decisions, potentially resulting in nonconformance and distrust of the system.

II. Evaluation of Rule Systems

New system designs are generally tested on whether or not specific predefined criteria, e.g., operational needs, are met. Careful attention has to be placed on the definition of these criteria. In the case of evaluating different designs of maneuver coordination rules, criteria such as "*route efficiency, time efficiency, fuel efficiency and other practical aspects related to displaying and executing the resolutions*" can be used.[5]

For the particular evaluation scheme, several processes may be considered:

1) A system evaluation can be done through a team of proven experts. Most often the system operators are involved. In the evaluation process of the extended

flight rules (EFR) proposed by EUROCONTROL,[6] experts from fields such as piloting, air traffic control (ATC) operation, and aircraft dynamics and equipment have been involved.

2) A computer based simulation (i.e., a fast-time simulation with no human operator involved) would be a more complex but more objective evaluation method. It is certainly less biased to well-established beliefs within the operator or expert community and offers the advantage of being able to objectively run through far more scenarios and options in significantly less time. Further, if necessary, advanced methods to optimize the design can be applied. However, the analytical criteria used in a simulation must reflect operational needs as closely as possible, and it is often difficult to formulate those criteria (i.e., if cognitive aspects are critical). Additionally, it is difficult to model human behavior in fast-time simulations beyond a simple level. Often it is assumed that humans will act exactly as the rules dictate, when in reality there may be cases in which this is not so.

3) Finally, a human-in-the-loop computer-based simulation can be a means to allow cost efficient but quite realistic evaluation (compared to flight trials with real aircraft) of even complex cognitive aspects of designs. However, the number of scenarios subject to test are limited due to time constraints related to the duration of a human-in-the-loop simulation run and the availability of suitable human test subjects.

III. Rule Design

A. Information Needed for Maneuver Coordination

For maneuver coordination based on rules, essentially three different kinds of information can be used: 1) position or state information, 2) velocity vector information, and 3) additional (intent) information, e.g., emergency status, target altitude, next waypoint. All of this information is envisioned available in an automatic dependent surveillance broadcast (ADS-B) message. The available information pieces can be used solely or in combination. Table 1 lists examples of maneuver coordination rules used in today's operational environment or those proposed in the literature with the corresponding information used in the priority determination process. The column labeled "Evaluated" indicates

Table 1 Information used for maneuver coordination with examples

Information	Example	Evaluated
State	Visual flight rules (VFR)	Yes
	ATLAS autonomous flight rules[6]	Yes
	Instrument flight rules (IFR)/VFR flight level rule[2]	Yes
	NLR's altitude for direction of flight (AFDOF)[5]	(partly)
State and velocity vector	Distance to closest point of approach—EFR rules used in FREER[6]	
State, velocity vector, and flight phase	Maneuverability/availability—EFR rules used in FREER[6]	Yes
	Phase-of-flight priority (PFP)—NLR free flight trials[5]	

whether this information was used for priority determination rules in this chapter.

The most basic information about conflicting aircraft is their positions (either absolute or relative). Priority determination in simple coordination rules such as VFR are entirely based on state information (i.e., the other vehicle's relative position).

The velocity vector could also be used to determine priority based on the vertical rates of the aircraft involved. Considering that aircraft fly at constant altitudes in cruise, a vertically moving aircraft could be assigned lower priority than level cruising aircraft. This would require only the knowledge of the vertical rate.

Further state information and the velocity vector can be used to determine priority. The proposed EFR rules use a reliable estimation of the distance to the closest point of approach (CPA) based on current states and the explicit knowledge of the velocity vector of the aircraft involved in the conflict. In case both aircraft are in the same (sub)flight phase, the aircraft farther away from CPA, i.e., the faster aircraft, will have to give way to the slower aircraft.

Even more additional information (priority categories such as emergencies or ambulance flights), possibly in conjunction with state and/or the velocity vector, can be used to determine the flight phase of particular aircraft and to assign priority between conflicting aircraft based on it.

B. Resolution Dimensions

Initiating a resolution maneuver requires at least one aircraft to change its flight path, i.e., to change its velocity vector. Three directions are possible:

1) The speed of the aircraft (length of the velocity vector) can be increased or decreased.

2) A turn can be initiated, i.e., a change of the velocity vector in the horizontal plane.

3) The vertical rate can be changed, i.e., change of the velocity vector in the vertical plane.

Changes of the velocity vector are possible as a single-state maneuver (e.g., change of the speed only) or in combinations (e.g., speed and heading). Combinations of maneuvers can be performed simultaneously or in sequence. Further, maneuver strength can be shared among conflicting aircraft or not (e.g., Eby's[5] potential field model used in NLR free-flight trials) (see Table 2).

Table 2 lists some potential resolution options possible and examples if known. As in Table 1, the column labeled "Evaluated" indicates whether the resolution option was evaluated in the computer simulation discussed here.

IV. Rule Evaluation Criteria

System evaluation, besides many other aspects, requires careful selection of the proper evaluation criteria. These have to be chosen according to the operational requirements and constraints of the environment in which the system will be used.

For aircraft separation purposes, the most important operational constraint for maneuver coordination rules is definitely that proper application ensures safe separation of all aircraft at all times. This constraint can be considered a hard constraint, disqualifying those rules not meeting it.

Table 2 Conflict resolution options with examples

Order	Example	Evaluated
Shared resolution; only one change of the velocity vector	NLR repulsive force principle (with sole maneuver)[5] NLR extended VFR overtaking rules (EVOR)[5] ATLAS head-on encounter[6]	Yes (partly)
Shared resolution; sequential changes of velocity vector (e.g., level off followed by a turn)	—	Yes (partly)
Not shared resolution; only one change of the velocity vector	Visual flight rules[2]	Yes
Not shared resolution; sequential changes of velocity vector (e.g., level off followed by a turn)	—	Yes (heading and altitude)

Between rules meeting the constraint of ensuring safe separation, however, differences in operational efficiency of different rules should be taken into account. In an effort to evaluate rule sets, a cost function was used.[7] Besides other parameters, the cost function is based on cost for additional fuel and time, required by resolution maneuvers. However, one disadvantage of this approach is the requirement of detailed information about different parameters, e.g., cost indexes and fuel flow depending on the aircraft mass, which might vary between users and equipment. Another approach using only cross-track distance from the initial planned route instead of fuel and time reflects the operational efficiencies in an appropriate way and is far easier to model. The latter approach was chosen for this study.

As a third evaluation criterion, the number of maneuvers necessary to achieve safe separation was chosen. Free Flight Studies involving pilots showed that pilots tend to dislike the necessity of several resolution maneuvers to achieve safe separation.[8] Pilots are requested to keep the number of resolution maneuvers low. Therefore, the number of maneuvers resulting from the application of coordination and resolution rules was inserted into the evaluation function.

In summary, three criteria are used in this study: 1) safe separation, 2) cross-track distance from the initial planned route, and 3) number of resolution maneuvers. The criteria are mathematically combined, giving the following evaluation value (EV) function:

$$EV = \frac{k}{Offset \cdot Maneuver^2 + Penalty}$$

Offset (in kilometers) is the average of the maximum cross-track distances from the initial planned route required by a (combination of) resolution maneuver(s).

Maneuver is defined as the average number of changes of the velocity vector to resolve a conflict. The square of maneuver is used to emphasize a low number of maneuvers. A *Penalty* of 100 is added if the required separation is not met. The penalty is needed for the genetic algorithm to prevent reproduction of rule combinations not meeting separation minima. The parameter k was chosen to be 10 to provide convenient magnitudes for *EV*, though any value could be used.

The EV function defines operational efficiency and is used to compare different rule combinations. The higher the EV value, the higher the achieved operational efficiency.

V. Rule Evaluation

For rule evaluation a fast-time flight simulation was developed. The simulation integrates: 1) equations of motion allowing the calculation of flight paths for point mass aircraft; 2) aircraft performance data of Airbus A340 aircraft based on EUROCONTROL's Base of Aircraft Data (BADA)[9]; 3) a database defining different meta-rules for priority determination and conflict resolution; 4) a database with a discrete number of conflict situations; and 5) the calculation of the evaluation value based on the preceding specified criteria.

A. Rule Definition

Different meta-rules were defined and stored in a database. Meta-rules are based on existing rule systems (e.g., VFR), suggested systems found in the literature, or logical combinations of priority determination and resolution maneuver options as described in the preceding.

A meta-rule consists of different rules that are combined forming a decision tree. Several rule inputs are connected to a rule output (decision or action). Additionally meta-rule parameters such as sectors (see Fig. 2) or a maneuver strength (bank angle or vertical rate) are variable and are subject to optimization in the evaluation process.

The simulation is linked to an optimization system using genetic algorithms.[10] The EV function is used as the objective function for the genetic algorithm, optimizing parameters of meta-rule combinations for comparability. In the meta-rules, parameters such as the angles β_1, β_2, and β_3, which describe sectors around the aircraft in the horizontal plane (as used in VFR), are variable (see Fig. 2). Varying parameters within a meta-rule result in different operational efficiencies.

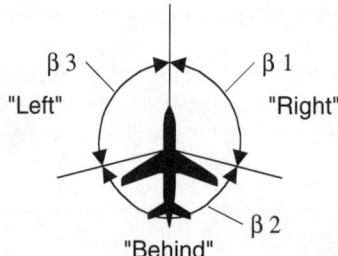

Fig. 2 Parameters β_1, β_2, and β_3 for sector definition in a meta-rule.

Table 3   Meta-rules used in the evaluation process

Meta rules	Explanation
Priority determination	
VFR	Models the priority decision part of existing VFR rules. The sector geometry (β_1, β_2, and β_3) is constant.
State	Like VFR, but the sector geometry is used for priority determination. The sector geometry (β_1, β_2, and β_3) is variable.
State and vector	Like state, but aircraft having a vertical rate unequal zero have lower priority than aircraft with a vertical rate equal zero.
State, vector, and flight phase	Like state and vector, but the velocity vector is used to differentiate between climb, cruise, and descend. Cruise has the highest priority. Climb has a higher priority than a descent.
Conflict resolution	
Maneuver lateral	Resolution maneuvers are performed in the lateral plane only. Two meta-rules are used. Turn in the direction to pass behind the conflicting aircraft. Always turn to the right (corresponding resolution maneuver to VFR).
Maneuver lateral and level off	Sequential changes of the speed vector in the lateral and vertical plane. Changes in vertical plane are in the form of a temporary level-off or a decrease of the vertical rate. Five meta-rules are in this group.
Maneuver lateral and vertical	Sequential changes of the speed vector in the lateral and vertical plane are possible. Possible velocity vector changes in the vertical plane include a level-off, a decrease of the vertical rate, or a change of the flight level in case of cruising. Seven meta-rules are in this group.

For evaluation purposes different combinations of meta-rules were tried. Each combination should yield the best evaluation values possible for a predefined, discrete number of conflict situations. Thus, optimization of meta-rule parameters is necessary. Genetic algorithms were found to be suitable for this task.

Eleven meta-rules, four for priority determination and seven for conflict resolution, were defined and evaluated. Tables 1 and 2 show which rule designs were used. Table 3 provides explanations about the meta-rules used. The seven conflict resolution meta-rules are organized, according to the necessary changes of the velocity vector, in three groups. In the simulations the resolution meta-rules in the specific groups were combined with the meta-rules for priority determination, and parameters were optimized using genetic algorithms.

B. Situations

The rules were evaluated using 19 different conflict situations varying in geometry (horizontally: head-on, crossing, and slow-closure encounters; vertically: combinations of cruise, climb, and descent), and flight parameters such as velocity and rate of climb/descend are used. Each scenario was weighted by its likelihood of occurrence, based on data provided by the EUROCONTROL airborne collision avoidance system (ACAS) evaluation report[11] and a Boeing report about en route free flight simulations in the Cleveland airspace.[12]

Modeling situations by continuously varying relative closure angles, vertical rates, and turns on a stochastical basis is possible and such trials were conducted. These trials showed no significant evaluation value difference compared to using a discrete number of situations weighted with a likelihood of occurrence. Computational effort, however, is much higher in the former case.

The 19 different conflict situations were used for three different separation environments varying in horizontal and vertical separation minima: 1) 5 n miles horizontal and 2000 ft vertical, 2) 5 n miles horizontal and 1000 ft vertical, and 3) 3 n miles horizontal and 1000 ft vertical. In the simulations it was assumed that crews would initiate a resolution maneuver following rules between 10 min and 5 min before reaching the closest point of approach (CPA) in the conflict. The resolution initiation thus was modeled for 600, 500, 400, and 300 s before reaching the CPA. The resulting evaluation values were then averaged. For simplification purposes at this stage, the simulations used only pairs of conflicting aircraft.

C. Results

In Figs. 3–5, the rule evaluation results for the three separation environments are plotted. Along the x axis, the priority determination rules used in the study are shown. The points (combined with lines) mark the different combinations of priority determination and resolution maneuver rule options. The EV of the simulation is plotted along the y axis. Higher EV indicates better operational rule efficiency according to the criteria specified in the preceding.

The solid line shows different priority rule efficiencies if only lateral changes of the flight paths are applied for resolution. The broken line shows priority rule

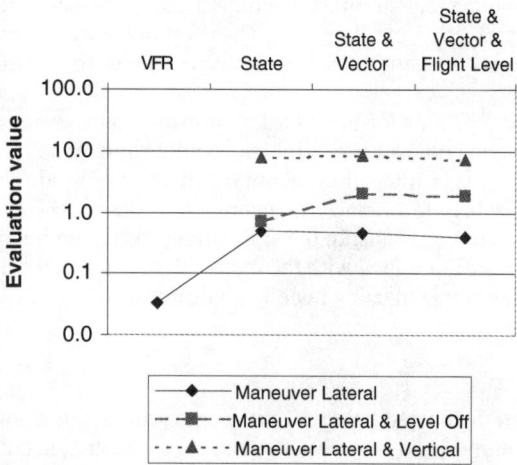

Fig. 3 Evaluation results for a separation requirement of 5 n miles laterally and 2000 ft vertically.

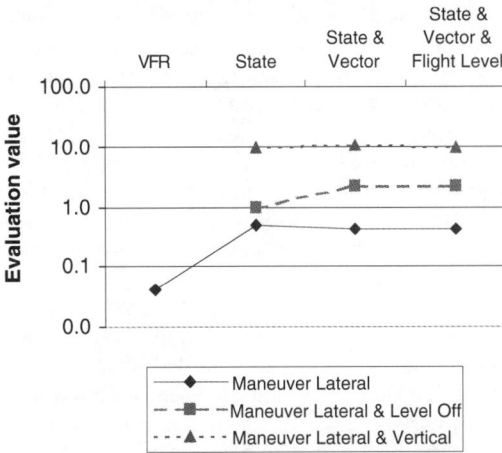

Fig. 4 Evaluation results for a separation requirement of 5 n miles laterally and 1000 ft vertically.

efficiencies if lateral and vertical direction changes are applied solely or in sequence. The simulation showed that for aircraft that have to give way and are either in a climb or descent, modifying the vertical rate is an efficient option to resolve the conflict.

The dotted line shows priority rule efficiencies if meta-rules include lateral maneuvers, level-off, and flight-level changes. Adapting the vertical rate is the best option for aircraft in climb or descent. For aircraft in cruise, because of the

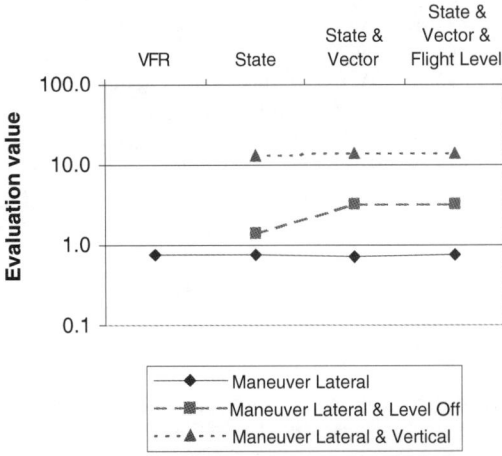

Fig. 5 Evaluation results for a separation requirement of 3 n miles laterally and 1000 ft vertically.

lower vertical separation required, changing the altitude results in a higher EV value than a heading change.

Resolution maneuver rules show a general trend to use vertical direction changes (i.e., reduced rate of climb/descent or changing altitude), which can be explained through the large separation standard ratio between horizontal and vertical and to the form of the evaluation function that was used. Without considering passenger comfort, engine wear, or other aspects, changes of the flight path in the vertical direction is more efficient than lateral flight path changes.

Visual flight rules are used as a baseline in all three scenarios. However, the application of VFR violated separation requirements of 5 n miles/2000 ft and 5 n miles/1000 ft in cases where the resolution maneuver was initiated at a smaller time to closest point of approach limit (e.g., 300 s) in slow-closure encounter situations. The separation violation, because of the penalty, results in a very low EV value (see Figs. 3 and 4). With the lower required separation of 3 n miles/1000 ft, the VFR evaluation yields a more comparable EV value (see Fig. 5).

Figures 3–5 show that the use of additional information in the priority determination process does not necessarily lead to better results in terms of efficiency. The only major increase in efficiency due to rule complexity across all separation standards was in the maneuver lateral and level-off line (broken line). An increase of the EV value occurs if the velocity vector is used for priority determination and maneuvers in the vertical direction are allowed. Using additional information such as the differentiation between flight phases does not result in an increase of rule efficiency for the particular situations modeled.

The average number of maneuvers used to resolve conflicts was generally close to 1. The simulations showed a slight but consistent increase of the average number of maneuvers if rules suggested resolution maneuvers in the vertical direction. The overall evaluation value, however, is getting higher. The reason once again lies in the different distances for horizontal and vertical separation.

VI. Human Factors and Rules

Despite the use of a rule structure, the human decision makers in the cockpit and on the ground will form decisions based on their own internal logic. When a human disagrees with rule-based logic, the outcome may range from a delay in action to a chronic loss of trust in the rules. Mismatches between human decisions and the rule structure may occur in three general areas:

1) Observable information. The human may not be able to observe the same information that is used by the rule base, and vice versa. For example, the human will likely not be able to estimate range, bearing, or altitude of another aircraft as accurately or rapidly as an automated rule-based system. Accordingly, the human may disagree with the automation as far as the proper course of action due to this mismatch in the perceived values of the inputs to the decision. Similarly, the human generally has access to more information regarding an encounter situation than is used in the rule base. This may include voice communications that indicate the intentions of another aircraft, weather, etc. Then, a rule-based decision may be clearly unacceptable due to a conflict with this other information.

2) Differences in decision-making logic. Humans are very adept at making complex judgments that may be traceable to some form of rule structure, yet there are also many elements of human decision making that cannot be represented

formally by a set of rules. Because humans do not naturally follow a rigid series of rules when making a decision, they may also have difficulty fully understanding or predicting the behavior of air traffic in a complex rule environment.

3) Differences in the degrees of freedom of action. A rule base is limited in the scope of actions or outputs that it can provide. This may involve restrictions to only lateral maneuvers, for example. The human may consider other options, though the human may also be limited as far as the complexity of action is concerned. A complex rule-based suggestion may not be as well accepted as a simple, single heading vector, for example, solely due to the desire of the human to limit complexity.

The result of these potential areas of conflict is that human performance will likely place constraints on the design of a rule base used for traffic management. Just because a certain rule base is shown to enhance traffic separation in ideal simulations does not necessarily mean that the same rule base would be acceptable to the pilot or controller. Thus, the results presented in this chapter also need to be considered in the light of how they may impact interactions between the humans and the rule structure. To be the most effective, the rules should provide for a flexible, rather than restrictive, structure that can adapt to the full range of situations that may unfold during operation.

VII. Conclusions

The results of the rule evaluation using a fast-time simulation show that higher rule complexity, assuming a proper design, can lead to additional gains in safety and maneuver efficiency. In the simulations, a genetic algorithm is used to optimize rule parameters. Additional study is required to determine whether the rule generated by the genetic algorithm are indeed optimal, or if other solutions may be viable. Furthermore, the proper form of an evaluation function is open to debate and would likely change from the one used here to better reflect actual operating costs, safety levels, and human preferences.

The additional gain in rule efficiency through higher rule complexity appears to be driven mainly by the allowance of vertical maneuvers, which, because of separation standards, are more efficient than lateral maneuvers.

The results also suggest that the most efficient use of rules occurs when the information used in the rules is matched with the information used in the resolution maneuvers. For example, there is little if any benefit to adding the vertical vector in the rules if only horizontal maneuvers are allowed for resolution. Similarly, there is a significant benefit to including the vertical vector when using vertical resolution maneuvers.

Explicit human interaction in the form of human-in-the-loop simulation was not investigated. However, it is important to note that human acceptance of rules to solve traffic conflicts is influenced by the complexity of rules and their flexibility of application related to the operational environment.

As air traffic management transitions to a more flexible and efficient mode of operation, it will be necessary to maintain some degree of structure and predictability of traffic through the use of rules. These rules play a role similar to a centralized controller by ensuring coordination and consistency between aircraft actions. The proper selection of rules will be critical to balance complexity, effectiveness, flexibility, equipment requirements, failure robustness, and human acceptance. The study described here represents an initial evaluation of only one component of this important area, and further research by the community is recommended.

References

[1]Kuchar, J. K., and Yang, L. C., "Survey of Conflict Detection and Resolution Methods," AIAA-97-3732, *AIAA Guidance, Navigation, and Control Conference*, AIAA, Reston, VA, 1997.

[2]Mensen, H., *Moderne Flugsicherung*, Springer-Verlag, Berlin, 1993.

[3]Paul, W., *Seeverkehrsrecht*, DSV Verlag, Hamburg, 1995.

[4]"Final Report of RTCA Task Force 3: Free Flight Implementation," RTCA Task Force, 26 Oct. 1995.

[5]Van Gent, R. N. H. W., Hoekstra, J. M., and Ruigrok, R. C. J., "Free Flight with Airborne Separation Assurance," National Aerospace Laboratory NLR report, Amsterdam, 1997.

[6]Duong, V., Hoffman, E., Floc'hic, L., Nicolaon, J. P., and Bossu, A., *Extended Rules-of-the-Air to Apply to the Resolution of Encounters in Autonomous Airborne Separation*, EUROCONTROL Experimental Centre, Bretigny-Sur-Orge, France, 1996.

[7]Schild, R., "Rule Optimization for Airborne Aircraft Separation," Ph.D. Thesis, Vienna Technical Univ., Vienna, 1998.

[8]Lozito, S., McGann, A., Mackintosh, M. A., Cashion, P., "Free Flight and Self-Separation from the Flight Deck Perspective," EUROCONTROL/FAA ATM Seminar 1997, Paris, 1997.

[9]Bos, A., *User Manual for the Base of Aircraft Data (BADA), Revision 2.6*, EUROCONTROL, Note No. 23/97, Sept. 1997.

[10]*Evolver User Reference Handbook*, Ver. 4.0, Palisade Corp., 1998.

[11]Hager, G., "European ACAS Operational Evaluation—Final Report," EUROCONTROL, Rept. No. 316, July 1997.

[12]Warren, A., "A Methodology and Initial Results Specifying Requirements for Free Flight Transitions," EUROCONTROL/FAA ATM Seminar 1997, Paris, 1997.

Chapter 41

Probabilistic Approaches Toward Conflict Prediction

G. J. Bakker,* H. J. Kremer,[†] and H. A. P. Blom[‡]
*National Aerospace Laboratory NLR,
Amsterdam, The Netherlands*

I. Introduction

IN THIS chapter, the conflict prediction part of conflict probing will be considered. We will consider four approaches concerning conflict prediction. The first approach is the classical geometrical approach; the second approach is the probabilistic approach described in Refs. 1 and 2; the third approach is a variation of the second approach, and the fourth approach is a novel probabilistic approach, which is based on collision risk formulae.[3]

The objective of all conflict prediction approaches is to evaluate a set of planned or predicted trajectories for their conflict potential and to supply other air traffic management (ATM) subsystems with the conflict information. In this chapter the focus will be on the detection of conflicts between predicted aircraft trajectories in ATM.

When predicting aircraft trajectories, the prediction uncertainty increases with the prediction period. This is caused by the fact that prediction errors accumulate over time. It is assumed that the trajectories that predict the future aircraft behavior are four-dimensional trajectories. A four-dimensional trajectory is defined by predicted three-dimensional positions and corresponding predicted times that are given for all points on that trajectory. These four-dimensional trajectory predictions are evaluated for their conflict potential.

This chapter will compare the mentioned conflict prediction approaches with which conflict potential is evaluated with pairs of predicted four-dimensional trajectories. It is a continuation of Ref. 4. The chapter is organized as follows.

Copyright © 2001 by NLR. Published by the American Institute of Aeronautics and Astronautics, Inc., with permission.
*Air Traffic Management researcher, Air Transport Division, Air Traffic Management Department.
[†] Air Traffic Management researcher, Air Transport Division; currently at KPN Research.
[‡] Group Leader Air Traffic Management Modelling, Air Transport Division, Air Traffic Management Department.

First, the classical geometric conflict prediction approach will be considered. Some limitations will be highlighted that create the reason why we will study probabilistic conflict prediction approaches. The first probabilistic conflict prediction approach that will be considered is based on conflict probability.[1,2] This approach will be reviewed briefly. The second probabilistic approach is based on overlap probability and is introduced as a variation of the first probabilistic approach. The third probabilistic approach that will be considered is based on collision risk formulae.[3] This approach will be explained briefly.

Issues like flexibility of usage and restrictions on aircraft behavior of the four approaches will be discussed, and conclusions will be drawn.

II. Conflict Prediction Approaches

A. Geometric Conflict Prediction Approaches

The classical geometric conflict prediction approach that is performed with a pair of predicted four-dimensional trajectories will be considered. Input for the geometric conflict prediction is the predicted four-dimensional trajectory. The uncertainty of the predicted four-dimensional trajectory is translated into areas around the predicted trajectory. Let us refer to these areas as protection zones. The protection zones are such that at any time in the future, the probability that an aircraft is inside its protection zone is larger than some threshold. The size and shape of the protection zones may vary with time. The protection zones for the horizontal plane and for the vertical plane are defined independently. Horizontal and vertical distances between protection zones should be such that they are safe. Two aircraft are said to be in geometric conflict when the distance between the protection zones of those aircraft becomes smaller than the minimum allowed distance between them [e.g., defined by the International Civil Aviation Oraganization (ICAO)]. Information like the duration (e.g., time interval in which two aircraft are in geometric conflict) and minimum distance between the protection zones can be generated (e.g., Refs. 5 and 6).

B. Limitations of Geometric Approaches

Let us start by considering various causes that result in aircraft deviating from their predicted four-dimensional trajectories. These causes exist in all parts of ATM, some examples are 1) wind modeling and prediction errors, and 2) tracking, navigation, and control errors.

Large wind modeling and prediction errors can result in aircraft that deviate from their predicted trajectory. The same result applies for large tracking, navigation, and control errors. Conflict prediction methods predict aircraft deviations from their predicted trajectory, and on the basis of this prediction, conflict potential is evaluated. Geometric conflict prediction approaches translate the mentioned prediction uncertainties in areas around the predicted aircraft positions (protection zones). The main limitation of this geometric approach to conflict prediction is its tendency to be overly conservative in handling uncertainties in aircraft behavior. For example, climbing or descending aircraft are given a lot of moving space. To improve conflict prediction, uncertainties should be handled less conservatively than geometric approaches handle them. However, uncertainties should still be handled conservatively enough to keep the sky safe.

The key attribution of this chapter is that the mentioned limitation of geometric approaches towards conflict prediction can be overcome by an appropriate probabilistic approach. Furthermore, using probabilistic conflict prediction, more information about conflicts or encounters can be provided (e.g., probabilities, collision risks), which can be exploited for an improved quality of the decision whether there is a conflict or not. (Thus, one might expect the number of false and missed conflicts to be reduced.) Therefore, there is a clear reason to study probabilistic conflict prediction approaches.

In this chapter three probabilistic approaches are discussed. The first probabilistic approach is the conflict probability approach.[1,2] The second probabilistic approach is a variation of the first probabilistic approach and is based on overlap probability (also based on the method described in Refs. 1 and 2). The third probabilistic approach is based on collision risk formulae.[3] There are some basic differences between the probabilistic approaches. These differences will become clear when the approaches are described.

C. Conflict Probability Approach

The authors of Refs. 1 and 2 have developed a method to evaluate conflict probabilities. The approach is initially developed to predict conflicts in the horizontal plane only. In their approach a conflict is defined as a situation in which the separation between aircraft falls below a certain separation threshold. Evaluation of conflict potential is done based on the evaluated conflict probabilities.

In Refs. 1 and 2 the conflict prediction is focused on free flight. The future deviations of the aircraft from the expected four-dimensional trajectories are predicted by probability density functions. They realized that in free flight the further you predict a trajectory in the future, the less certain these predictions are. Note that this does not need to be the case in the four-dimensional ATM philosophy, in which aircraft are kept within some boundaries around their planned four-dimensional trajectory.

In the case of free flight, the decision whether aircraft will approach each other too closely is seen as a tradeoff between efficiency and certainty. To optimize this tradeoff, the authors of Refs. 1 and 2 developed a method to describe the certainty. The approach aims to predict the probability that the separation between two aircraft falls below a certain separation threshold (e.g., ICAO separation standards). This probability is called conflict probability. The goal is to keep the conflict probability below some acceptable level. To evaluate the conflict probability, they assume that it is realistic to model the deviations of the aircraft from their expected trajectories by Gaussian density functions. Using the direction of the relative velocity at time of minimum predicted separation, i.e., as the minimal horizontal distance between the expected trajectories, the probability density function of the relative position at that time is obtained. An analytical expression is obtained to estimate the conflict probability. For a more extended treatment, the reader is referred to Refs. 1 or 2.

D. Overlap Probability Approach

So far, the approach described in Refs. 1 and 2 is used to predict the probability that the separation between two aircraft falls below a threshold that is determined by (e.g., ICAO) separation standards; this probability is called conflict probability. If, however, for this threshold a value such as the size of an aircraft is used, then the same approach yields the overlap probability. Thus, overlap probability follows from a variation of the approach developed in Refs. 1 and 2; with the threshold reduced to the size of an aircraft, the overlap probability reflects the probability that the aircraft physical volumes overlap.

E. Collision Risk Approach

In our novel probabilistic approach, the conflict potential is evaluated through collision risk formulae,[3] which are a generalized version of Reich's collision risk approach[7] adopted by ICAO. The generalizations have been developed because the Reich model applies under rather restrictive assumptions only.

The resulting collision risk equals the probability of collision between two aircraft. The steps that have to be taken in the novel approach are as follows. First, the joint probability density functions of the positions and velocities of individual aircraft are predicted, and then the joint probability density function of the relative position and velocity of an aircraft pair is evaluated. Then the collision risk for the aircraft pair is evaluated using the generalized Reich collision risk equations. This novel collision risk approach will be briefly elaborated next.

III. Collision Risk Modeling

A. Generalized Reich Collision Risk Model

In this section we briefly discuss the generalized Reich collision risk model without going too much into the mathematical details. For a detailed description we refer to Refs. 3 and 8.

Let the stochastic process $\{S_t^i\}$ represent the position of the center of aircraft i, and let $\{v_t^i\}$ represent its velocity.

Next, with s_t^i and s_t^j representing the positions of the centers of aircraft pair (i,j), the relative position is represented by the process $s_t \triangleq s_t^i - s_t^j$, and the relative velocity is represented by the process $v_t \triangleq v_t^i - v_t^j$.

Now we define an in-crossing of a certain area D around the origin as follows. The relative position s_t enters D at time t, if

$$s_{t-\Delta} \in D^c \text{ and } s_t \in D \quad \text{for } \Delta \downarrow 0$$

where D^c is an open set in R^3 and equals the complement of D. Each entering of D by the relative position s_t is called an in-crossing.

The in-crossing rate is defined as the expected number of in-crossings at time t per unit time and is denoted by $\varphi(t)$. In Ref. 3, the in-crossing rate is defined as

$$\varphi(t) \triangleq \lim_{\Delta \downarrow 0} \frac{P\{s_t \in D, s_{t-\Delta} \in D^c\}}{\Delta} \tag{1}$$

We can express the collision risk between aircraft (probability of an in-crossing) for a time period $[t_1, t_2]$, denoted by $P_{ic}(t_1, t_2)$, as follows[3]:

$$P_{ic}(t_1, t_2) = \int_{t_1}^{t_2} \varphi(t)\, dt \qquad (2)$$

In Ref. 3, a characterization of the in-crossing rate $\varphi(t)$ has been derived under very general conditions. This model is called the generalized Reich collision risk model, in which it is assumed that the process $\{s_t, v_t\}$ admits a density function $P_{s_t, v_t}(.)$. For numerical evaluation of $\varphi(t)$, there is a need to characterize the probability density function $P_{s_t, v_t}(s, v)$ for the relative position s_t and the relative velocity v_t. Characterizing this probability density thus is an important part of the collision risk prediction problem.

B. Gaussian Case

To be able to compare the collision risk approach with the other approaches, we will assume that the position and velocity of each individual aircraft is Gaussian distributed with some mean and covariance. Using the well-known fact that a linear combination of Gaussian variables is also Gaussian, it is clear that the relative position and velocity are also Gaussian distributed.

Using the Gaussian probability density function of relative position and velocity, the in-crossing rate (1) can then be evaluated.

Next, the collision risk approach, the conflict probability approach, the overlap probability approach, and the classical geometrical approach will be compared by applying them to Gaussian ATM examples.

IV. Comparison of Approaches

First of all, it should be noted that the collision risk approach deals with the problem of conflict (collision) prediction in a three-dimensional sense: horizontal and vertical movements are incorporated, also when they are not independent of each other. This implies a significant improvement over geometric approaches in which the horizontal and vertical distances between protection zones are monitored independently and the conflict prediction approach in Refs. 1 and 2, which tends to define the probability of a horizontal conflict or overlap independently from the probability of a vertical conflict or overlap.

Next, the conflict prediction approaches are compared with each other by applying them to a two-dimensional example that was already described in Ref. 1. In the described ATM example, aircraft move in the horizontal plane only.

A. Situation and Modeled Uncertainties

In the examples, some parameters that define the situation can be distinguished. Which parameters and how they were taken is explained in the following. The exact values of the appropriate parameters are given in the sections in which the examples are discussed.

The probability density functions of the positions of the aircraft at a certain time are characterized by the predicted positions and their uncertainties in the across-track and along-track direction (the uncertainties are assumed to be Gaussian distributions, and so they are characterized by the standard deviations). The deviations in along-track and across-track direction are assumed independent of each other.

The positions of both aircraft are predicted in time. The expected magnitudes of the ground speeds are assumed constant for both aircraft. The predicted across-track uncertainty in position (standard deviation) is constant for both aircraft. The predicted along-track uncertainty in position (standard deviation) is 0 for both aircraft at current time and increases linearly in time (given by a growth rate). The routes that can be formed by connecting the predicted aircraft positions are straight lines in the horizontal planes that cross each other, except for a path angle of 0 deg, in which case the aircraft are predicted to fly on parallel routes.

The situation described in the preceding is visualized in Fig. 1. All conflict prediction approaches will be applied to this situation. In the simulations we have evaluated 1) the predicted minimum distance between protection zones around the aircraft (classical geometric approach, e.g., Refs. 5 and 6); 2) the conflict probability (see Refs. 1–3) the overlap probability (version of Refs. 1 and 2 with threshold at 50 m); and 4) the collision risk following our novel approach.

The threshold on which the geometric approach and the conflict probability approach defined in Refs. 1 and 2 are based is 5 n miles (5 n miles is the currently used ICAO separation standard for en route airspace). The threshold used for evaluating the overlap probability is set to 50 m. The novel probabilistic approach needs extra input parameters, the across-track standard deviation of the velocity,

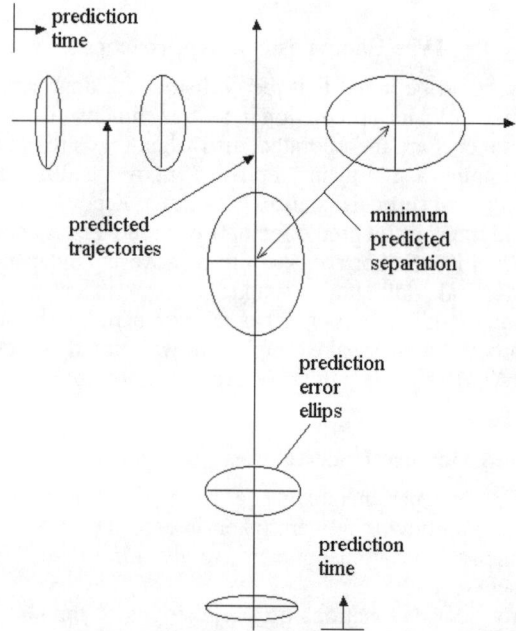

Fig. 1   ATM example in the horizontal plane.

the along-track standard deviation of the velocity, and the size of the boxes that represent the aircraft. For these parameters, some reasonable values were used: standard deviation of the velocity is 2% of the ground speed in either direction and independent of each other. The length and width of the box enclosing one aircraft are 50 m. The collision risk is evaluated for the time interval that starts 5 min before the aircraft reach their minimum predicted separation until 5 min after they have reached their minimum predicted separation. In the geometric approach, the size of the protection zone is defined as a box whose length is equal to the along-track standard deviation of position and whose width is equal to the across-track standard deviation of position. The length of the box lies in the predicted velocity direction.

The evaluation of the minimum predicted distances between the protection zones, the conflict probability, the overlap probability, and the collision risk can be done for various sets of simulation parameters. Performances of the conflict prediction approaches in various situations are compared by varying the following simulation parameters: 1) minimum predicted separation, 2) path crossing angle, 3) predicted ground speed of the aircraft, 4) time before minimum predicted separation, 5) growth rate of the along-track standard deviation of position, and 6) across-track standard deviation of position.

The results of some examples will be shown.

B. Example 1

In this example, from Ref. 1, the minimum predicted separation between the aircraft is 6 n miles. The path angle between the predicted aircraft routes is 90 deg. The predicted ground speed magnitude of both aircraft is 480 kn. The time before minimum predicted separation is varied from 40 min to 1 min. The growth rate of the along-track standard deviation of position is 15 kn for both aircraft, and the across-track standard deviation of position is 1 n mile and constant for both aircraft.

The result of varying the time before minimum predicted separation is that the along-track standard deviation of position at time of minimum predicted separation is varied from 10 n miles to 0.25 n mile.

In the geometric approach, the minimum predicted distances between the protection zones are evaluated. If a 'geometric' conflict is detected, the probability of a conflict is 1; otherwise it is 0. The geometric approach was used with a 1-sigma value for the assumed area of aircraft; the length and width of the area are equal to the along-track and across-track standard deviation, respectively. Figure 2 shows the results of the geometric and conflict probability approach. Figure 3 shows the results of the conflict probability and the collision risk approach. In Fig. 3, all curves are normalized (to fit within a linear scale figure).

To make the difference between the probabilistic approaches more clear, we use a logarithmic scale to plot the results of the example see Fig. 4 for all three probabilistic approaches.

C. Example 2

In this example, also from Ref. 1, we change the minimum predicted separation between the aircraft from 6 n miles to 4 n miles in the set of simulation parameters

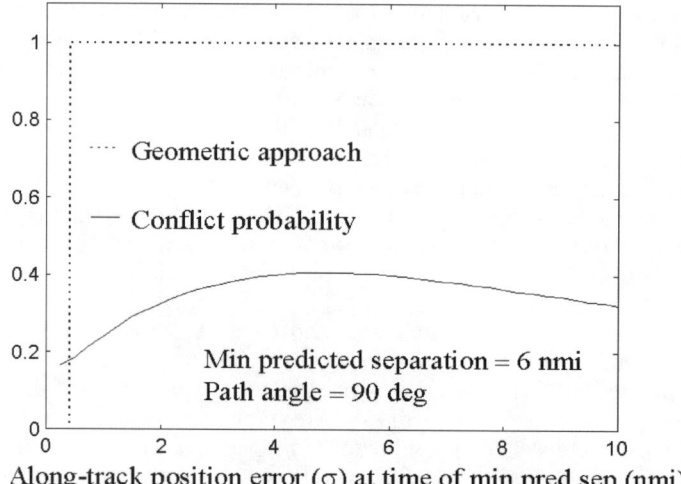

Fig. 2 Geometric approach and conflict probability in example 1.

for example 1. In Fig. 5, conflict probability, overlap probability, and collision risk are plotted using a log scale.

D. Example 3

In this example, we further compare collision risk and overlap probability. The minimum predicted separation is 6 n miles. The path angle is varied from 0 to 360 deg. The ground speed of one aircraft is 420 kn, and the ground speed of the other aircraft is 480 kn. In all situations, the faster aircraft crosses behind the

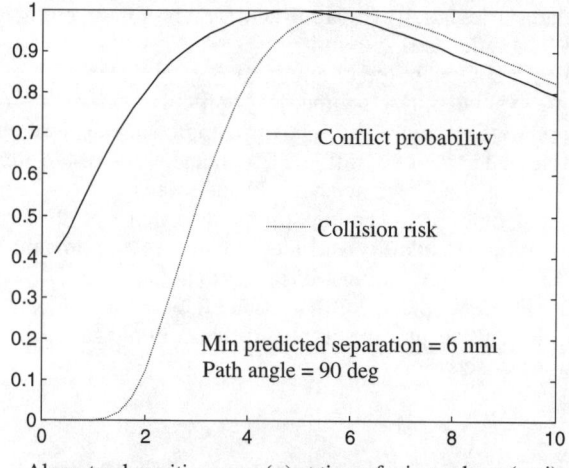

Fig. 3 Conflict probability and collision risk in example 1 (normalized).

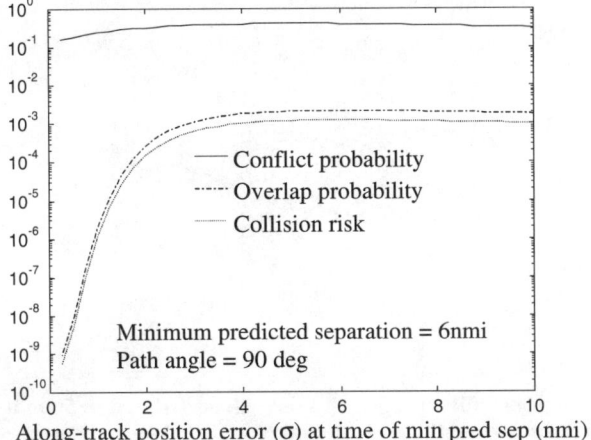

Fig. 4 Conflict probability, overlap probability (with threshhold reduced to 50 m), and collision risk in example 1 (log scale).

slower aircraft (except for path angle 0, when the routes are parallel). The time before minimum predicted separation is varied from 20 min to 1 min. The growth rate of the along-track standard deviation of position is 10 kn for both aircraft. The across-track standard deviation of position is 1 n mile and constant for both aircraft. The overlap probability (threshold reduced to 50 m) and collision risk are evaluated.

The results are given in Figs. 6–9. In the three-dimensional figures, the horizontal axes represent the time to minimum predicted separation (minutes) and the path angle. The position of one aircraft at the time of minimum predicted separation is translated to (0,0) in the horizontal plane. All points on a circle in the horizontal plane represent the same time that this aircraft needs to fly from

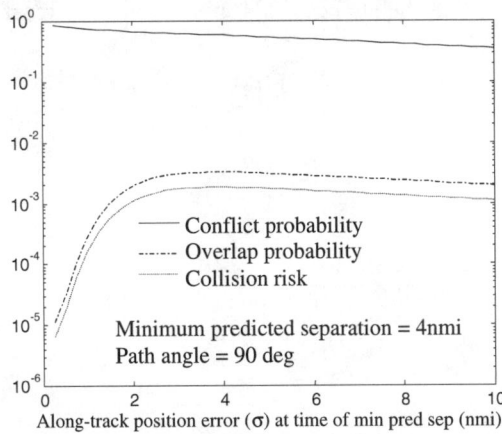

Fig. 5 Conflict probability, overlap probability (with threshhold reduced to 50 m), and collision risk in example 2 (log scale).

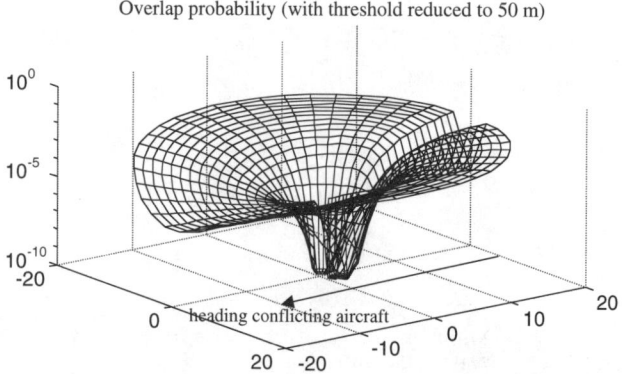

Fig. 6 Overlap probability (z axis) with threshhold reduced to 50 m is represented for various path angles and times to minimum predicted separation (horizontal axis).

its current position to its position at the time of minimum predicted separation (0,0). Thus each point in the horizontal plane represents a possible position where one aircraft currently is. The heading of the other aircraft (conflicting aircraft) is given in the figures. The vertical axis represents the overlap probability with respect to the collision risk. In the two-dimensional figures, the axes are the same as the horizontal axes of the three-dimensional figures. Possible current positions of one aircraft relative to its position at the time of minimum predicted separation are colored according to the value of the overlap probability with respect to collision risk (the coloring scale is shown in the figures).

Figures 6 and 8 do not give a very clear view of the differences between overlap probability and collision risk. Figures 7 and 9, however, do show a clear difference between overlap probability and collision risk, especially when the aircraft are close to the position where the predicted separation reaches its minimum. Therefore, in Fig. 10 the overlap probability and collision risk are evaluated for situations in which the aircraft are 4 min before they reach their minimum

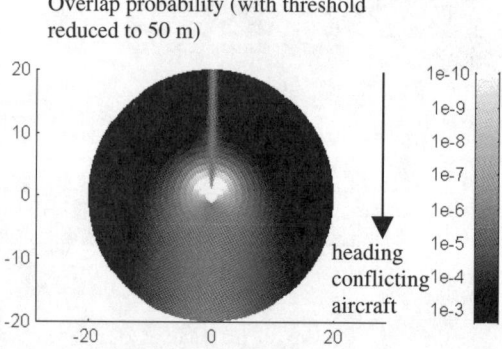

Fig. 7 Overlap probability (with threshhold reduced to 50 m) is represented by colors for various path angles and times to minimum predicted separation.

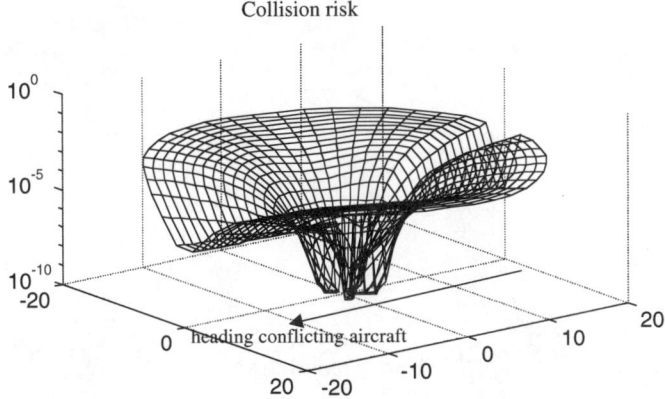

Fig. 8 Collision risk (z axis) is represented for various path angles and times to minimum predicted separation (horizontal axis).

predicted separation. The path angles are varied from 0 to 360 deg. This figure shows a significant difference between overlap probability and collision risk.

E. Example 4

The situation simulated in this example is the same as was simulated for example 3, except for the fact that the faster aircraft now crosses before the slower aircraft instead of behind the slower aircraft.

The overlap probability (with threshold reduced to 50 m) and collision risk are evaluated for path angles between 0 and 180 deg and the time before minimum predicted separation is 4 min. Figure 11 shows the overlap probabilities and the collision risk results. The overlap probabilities are the same for the situations in examples 3 and 4. The collision risk results for examples 3 and 4 differ.

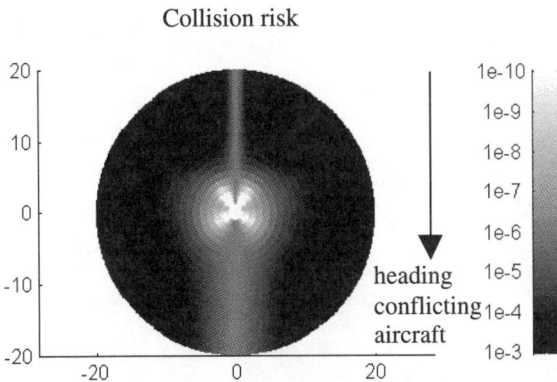

Fig. 9 Collision risk is represented by colors for various path angles and times to minimum predicted separation.

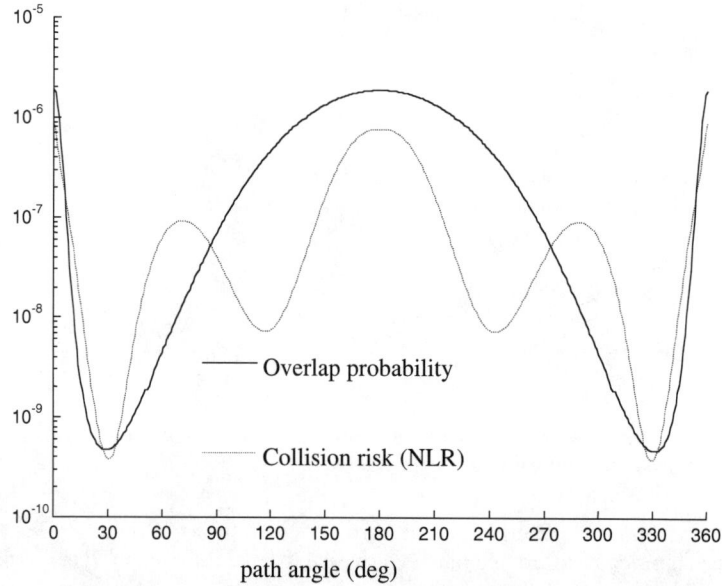

Fig. 10 Overlap probability (with threshhold reduced to 50 m) and collision risk for various path angles and 4 min before time of minimum predicted separation of 6 n miles.

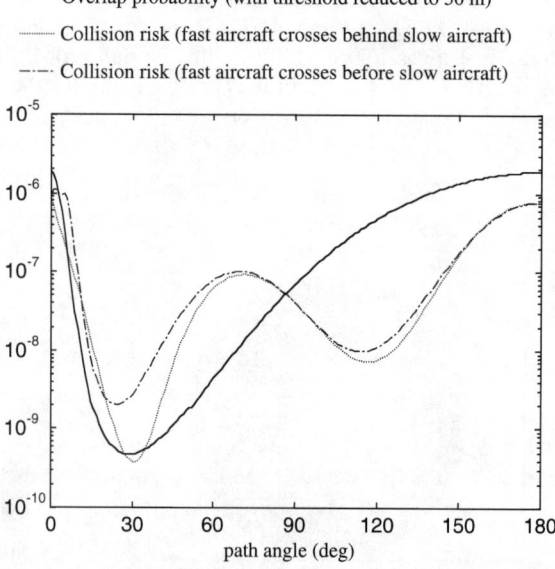

Fig. 11 Difference between situations in which a faster aircraft crosses before or after a slower aircraft. Evaluated overlap probability (with threshold reduced to 50 m) and collision risk are shown.

V. Discussion of Results

A. Flexibility of Usage

From the results of example 1, we can conclude that even with a protection zone that is represented by an uncertainty area of 1-sigma (70% containment), the geometric approach does not show flexibility in its usage. It can be seen from Figs. 2 and 3 that the probabilistic approaches show more flexibility to changes in along-track standard deviation of position. Thus let us take a closer look at the probabilistic approaches.

In Fig. 4, it can be seen that for large uncertainties in the along-track position, the conflict probability, overlap probability, and the collision risk are approximately equally sensitive to changes in the uncertainty. Thus for tools that concentrate on situations in which large uncertainties are common, all probabilistic approaches can be used. A good example in which large uncertainties are common is a flow management tool. However, if the uncertainties in along-track position become smaller, collision risk and overlap probability are much more sensitive to changes in along-track uncertainty than conflict probability. When the uncertainty in the along-track position decreases in Fig. 4, the collision risk and overlap probability values decrease very fast to very small values, whereas conflict probability values decrease very slow. Thus in this example, when the aircraft get closer to the point of minimum predicted separation, the more flexible collision risk and overlap probability become the conflict probability. For small uncertainties, it is easier to separate safe situations from unsafe situations by using collision risk or overlap probability than conflict probability.

In Refs. 1 and 2, the authors already concluded that conflict probabilities for minimum predicted separations below 5 n miles have a different shape than for minimum predicted separations above 5 n miles. If the minimum predicted separation is larger than 5 n miles, the shape of the conflict probabilities is as plotted in Fig. 3; first with increasing along-track position uncertainty from 0, the conflict probability increases from 0 to a maximum, and then it decreases again. If the minimum predicted separation is smaller than 5 n miles, with increasing along-track position uncertainty from 0, the conflict probability decreases from its maximum monotonically toward 0. For collision risk and overlap probability, such a distinction is not necessary, as can be seen from Figs. 4 and 5. For a minimum predicted separation of 4 and 6 n miles, with decreasing along-track position uncertainties, the collision risk and overlap probability slowly increase to a certain maximum and then decrease to very small values. As a result, we conclude that conflict probabilities can give no information with regard to the possible modifications of ICAO separation standards, whereas collision risk and overlap probability can.

B. Tradeoff Between Velocity Magnitude and Period of Encounter

Figures 4 and 5 show that in the simulated situations the overlap probability and collision risk have similar shapes. Thus in these situations they show similar flexibility in their usage. This can be explained as follows.

The overlap probability (and conflict probability) is evaluated based on a random indication of relative position and velocity. The magnitude of the relative velocity does not have any effect on the overlap probability (and conflict probability), only the direction of the relative velocity does. The implication of using a random indication for evaluation of the conflict and overlap probability is that the period of the encounter or possible conflict is not taken into account and thus has no effect on the result. Uncertainty in relative velocity is also not taken into account.

Collision risk is evaluated from in-crossing rates integrated over time. At any moment in time, the magnitude, direction, and uncertainty of the relative velocity are used for evaluation of the in-crossing rate at that time. Thus the magnitude, direction, and uncertainty of the relative velocity are all incorporated in the collision risk. The implication of integrating the in-crossing rates over time is also that the period of encounter or possible conflict is incorporated in the collision risk.

In general, the magnitude of the velocity and the period of encounter or potential conflict will have opposite effects on the collision risk and no effect on overlap probability (and conflict probability).

This can be explained as follows. The faster the aircraft fly, the shorter the encounter or period of potential conflict will be. The larger magnitude of the relative velocity will enlarge the in-crossing rates during the period of encounter or possible conflict. The consequence of the smaller time period is a potential reduction of collision risk, whereas the larger in-crossing rates create a potential increase of collision risk. Collision risk will show a tradeoff between these effects; overlap probability (and conflict probability) will not.

From the results given in Figs. 4 and 5, it can be concluded that in these situations, the tradeoff is such that the two consequences balance each other out. This, however, is not always the case.

From the results of example 3 (Figs. 6–10), it is straightforward that the overlap probability and collision risk are symmetric with respect to 0 path angle, i.e., overlap probability/collision risk in case path angle is β, is the same as in case path angle is $-\beta$. To obtain a complete picture of the situation, the results for all path angles between 0 and 360 deg are given in the figures.

The overlap probability with threshold reduced to 50 m and collision risk are evaluated for various path angles and times to minimum predicted separation. The shapes of the light/dark areas in Figs. 7 and 9 show a difference. From this difference, it can be concluded that the opposite effects of the magnitude of the velocity and the period of encounter or potential conflict are not always in balance. This is further explained by considering the overlap probability and collision risk on the 4-min circle (all positions on this circle represent a possible position of one aircraft 4 min before its position at time of minimum predicted separation).

The results are shown in Fig. 10. In this figure, the focus is on the possible situations 4 min before time of minimum predicted separation. Overlap probabilities show that the worst situations are represented by the 0 (or 360) and 180 deg path angle. The best situation is represented by the 30 deg path angle. Based on collision risk, the worst situations are reached for path angles of 0 (or 360) and 180 deg. The best situation is represented by the 30 (or 330) deg path angle. The path angles for which the collision risk really differs from the overlap probability (up to a factor 15) are the path angles between 30 and 180 deg (or 330 and 180 deg). Around 70 deg path angle a local maximum in collision risk (local minimum in safety) is achieved and around 120 deg a local minimum in collision risk appears. These results can be explained by a tradeoff between the magnitude of the relative velocity and the period of encounter or potential conflict.

The overlap probability is not capable of taking the described opposite effects into account. Collision risk, however, does take these effects into account. Example 4 shows the tradeoff when a faster aircraft crosses before a slower aircraft instead of crossing behind the slower aircraft, which was simulated in example 3. Figure 11 shows that overlap probability is the same for both cases, but collision risk differs significantly for path angles around 30 deg. Thus collision risk can distinguish between the simulated situations and overlap probability cannot.

C. Imposed Restrictions on Aircraft Behavior

Most conflict prediction approaches assume some restrictions on aircraft behavior. In the geometric approach, the more dynamic the aircraft behavior, the more difficult it is to define an appropriate deterministic protection zone around the aircraft and the more difficult it is to evaluate distances between protection zones. Therefore, geometric conflict prediction approaches tend to be complex in the case of dynamic aircraft behavior. To reduce complexity, most geometric conflict prediction approaches assume that aircraft fly in straight lines.

The probabilistic approach in Refs. 1 and 2 yields a search for the moment of minimum predicted separation. It is assumed that the aircraft velocities and prediction errors are constant during the encounter or period of potential conflict. The conflict probability and overlap probability are derived from a random indication of the aircraft positions and velocities together with their uncertainty corresponding to the moment of minimum predicted separation. Therefore, dynamic aircraft behavior may cause incorrect conflict or overlap probabilities.

The novel probabilistic conflict prediction approach is based on collision risk. Collision risk is evaluated from in-crossing rates integrated over time. At any moment in time, predicted aircraft positions and velocities together with their uncertainties are used for evaluation of the in-crossing rate at that time. The in-crossing rates are evaluated for the whole encounter or period of potential conflict. Collision risk is derived from these in-crossing rates, thus incorporating all dynamics.

D. Advanced Application: Dynamic Spacing

The conflict prediction approaches were compared by considering restrictions on aircraft behavior, flexibility of usage, and conservatism. Now an advanced application of the conflict prediction approaches will be discussed: dynamic spacing.

If the meteorological conditions change, the ATM system should be able to absorb this information and to translate it into use. If we focus on conflict prediction, in bad weather it may be needed to increase aircraft separations (spacing). One possible way to realize this is to change the separation threshold to a value that everybody agrees on and use conflict prediction approaches that make use of this separation threshold.[1,2]

Let us refer to methods that dynamically change the separation threshold according to changes in (meteorological) conditions, as dynamic spacing methods. A procedure could be that the right people judge (meteorological) conditions and select a certain separation threshold, based on their experience.

If dynamic spacing methods are developed and used in line with geometric or the conflict probability approach in Refs. 1 and 2, (meteorological) conditions should be translated into separation thresholds that apply for all aircraft. This way an ATM system can be created in which the capabilities of highly equipped (expensive) aircraft are not fully used.

The overlap probability approach (variation of the method of Paielli and Erzberger) and the novel probabilistic approach use a probabilistic separation threshold. If the probability density function of position and velocity are given dependent on the (meteorological) conditions, conflict potentials will be predicted dependent on the (meteorological) conditions. In this approach, the dynamic spacing method yields that (meteorological) conditions are translated into probability density functions of position and velocity.

If dynamic spacing methods are developed in line with the novel probabilistic conflict prediction approach (or the overlap probability approach), every aircraft will be judged on its capability to navigate in current conditions. Using this probabilistic approach, the spacing between two highly equipped aircraft may be smaller than the spacing between aircraft with less equipment on board, thus making full use of all aircraft capabilities.

VI. Conclusions

In this chapter, an overview is given of four conflict prediction approaches, the classical geometric approach, the conflict probability approach,[1,2] the overlap probability approach (a variant of the approach in Refs. 1 and 2), and a novel probabilistic approach. The objective of all conflict prediction approaches is to evaluate a set of planned or predicted trajectories for their conflict potential and to supply other ATM subsystems with the conflict information.

The reason for studying probabilistic conflict prediction approaches is that the classical geometric approach tends to be overly conservative in handling uncertainties in aircraft behavior. In the probabilistic conflict prediction models, modeling of the trajectory uncertainties causes the predictions to be less conservative. The conservatism that is seen as a limitation in the geometric approach can be overcome by an appropriate probabilistic approach.

The first two approaches are briefly reviewed and the overlap probability approach is introduced. Overlap probabilities are evaluated with the threshold reduced to the size of an aircraft. The novel probabilistic approach is explained in more detail. The approaches are compared for various qualities. The results of the comparisons are summarized in the following.

In the studied examples, only two-dimensional straight predicted flight paths were simulated. The reason for simulating straight flight paths lies in the imposed restrictions on aircraft behavior. Dynamic aircraft behavior would cause large complexity in the classical geometric conflict prediction approach and may cause incorrect predictions in the conflict probability and overlap probability evaluated according to the approach of Paielli and Erzberger.[1,2] However, the novel probabilistic conflict prediction approach, based on collision risk formulae, incorporates all aircraft behavior.

The conflict prediction approaches were compared for flexibility of usage. Flexibility was judged for the amount of impact the input trajectory uncertainties have on the output of the prediction. The more sensitive the output of the prediction is with respect to changes in the input, the better the approach can distinguish safe from unsafe situations. In this respect, the classical geometric approach showed the worst flexibility in its usage. The conflict probabilities proved to be much less sensitive to changes in the probability density functions than overlap probability and collision risk. Overlap probability and collision risk showed a lot of sensitivity to changes in the probability density functions, especially for small uncertainties in position. The latter makes overlap probability and collision risk extremely valuable in environments in which small uncertainties in position are common (e.g., four-dimensional ATM, short-term conflict prediction).

In some situations, the overlap probability and collision risk showed similar flexibility. However, the flexibility of the overlap probability and collision risk was not similar in all situations. This was explained by the tradeoff between the period of encounter or potential conflict and the magnitude of the relative velocity. The overlap probability evaluated according to the method in Refs. 1 and 2 does not take the magnitude of the velocity and the period of encounter or possible conflict into account. The evaluated collision risk incorporates all aircraft behavior, and so magnitude of velocity and period of encounter or potential conflict are also taken into account (and the tradeoff between them). An ATM example was simulated in which this tradeoff made a difference. Evaluated collision risks indicated that for some path angles it would be safer for a fast aircraft to cross behind a slower aircraft than crossing before the slower aircraft. Overlap probabilities could not distinguish between these situations.

For an ATM system to make full use of (meteorological) conditions information, dynamic spacing methods are necessary. Briefly, this means that known (meteorological) conditions are translated in an amount of space that is necessary to separate aircraft so that they are safe. If the classical geometric approach or the conflict probability approach in Refs. 1 and 2 is used, dynamic spacing methods need to be developed that translate (meteorological) conditions in separation thresholds. This means that all aircraft are treated equally, which induces no full use of (expensive) aircraft equipment. If the overlap probability or the novel probabilistic approach is used, dynamic spacing means that models need to be developed that represent aircraft behavior in all (meteorological) conditions. These approaches have the option of taking the quality of the equipment of individual aircraft into account, thus making full use of the aircraft equipment.

In the qualities described, the novel probabilistic approach proves to be the most promising and enables other advanced applications such as the incorporation of the probability density functions for all possible (meteorological) conditions, and the incorporation of collision risk prediction capability into the ATM design.

References

[1] Paielli, R. A., and Erzberger, H., "Conflict Probability Estimation for Free Flight," *Journal of Guidance, Control, and Dynamics*, Vol. 20, No. 3, May–June 1997, pp. 588–596.

[2] Erzberger, H., Paielli, R. A., Isaacson, D. R., and Eshow, M. M., "Conflict Detection and Resolution in the Presence of Prediction Error," *1st USA/Europe Air Traffic Management R&D Seminar*, FAA/EUROCONTROL, 1997.

[3] Bakker, G. J., and Blom, H. A. P., "Air Traffic Collision Risk Modelling," *32nd IEEE Conference on Decision and Control*, Vol. 2, Institute of Electrical and Electronics Engineers, New York, 1993, pp. 1464–1469.

[4] Kremer, H. J., Bakker, G. J., and Blom, H. A. P., "Geometric Versus Probabilistic Conflict Probing," *1st U.S.A./Europe Air Traffic Management R&D Seminar*, FAA/EUROCONTROL, 1997.

[5]Kremer, H. J., Vertegaal, W. C., and Jansen, R. B. H. J., "PHARE Advanced Tools Conflict Probe," National Aerospace Laboratory NLR, NLR Rept. DOC 98-70-18 (Vol. 3 of 10), Amsterdam, 1998.

[6]Vink, A., de Jong, C. J. M., and Beers, J. N. P., "Medium Term Conflict Detection," *1st USA/Europe Air Traffic Management R&D Seminar*, FAA/EUROCONTROL, 1997.

[7]Reich, P. G., "A Theory of Safe Separation Standards for Air Traffic Control," Royal Aircraft Establishment, 64041, Malvern, UK, 1964.

[8]Bakker, G. J., Kremer, H. J., and Blom, H. A. P., "Geometric and Probabilistic Approaches Towards Conflict Prediction in ATM," (submitted for publication).

Chapter 42

Safe Flight 21: 1999 Operational Evaluation of Automatic Dependent Surveillance Broadcast Applications

James J. Cieplak,* Edward Hahn,[†] and Baltazar O. Olmos[‡]
MITRE Corporation, McLean, Virginia

I. Introduction

SAFE Flight 21 is a cooperative government/industry effort to develop enhanced capabilities for free flight based on evolving communications, navigation, and surveillance (CNS) technologies. Safe Flight 21 will demonstrate the in-cockpit display of traffic, weather, and terrain information for pilots and will provide improved information for controllers. The new technologies on which this program is based include the Global Positioning System (GPS), automatic dependent surveillance broadcast (ADS-B), flight information services broadcast (FIS-B), traffic information service broadcast (TIS-B), and their integration with enhanced pilot and controller information displays. Safe Flight 21 will evaluate the safety, service, and procedure improvements these technologies make possible. The primary objective of the Safe Flight 21 program is to enable and expedite decisions by stakeholders on implementing nine operational enhancements identified by this forum. The program will do this by working with industry to demonstrate and evaluate these enhancements. Prior to committing the Federal Aviation Administration (FAA) and the users to a full-scale implementation of these enhancements, there needs to be consensus among the FAA and industry on the feasibility and business case for the enhancements. On 10 July 1999, a year's worth of work in developing ADS-B technology and procedures was culminated in an operational evaluation in Wilmington, Ohio. That evaluation brought together 24 aircraft comprised of general aviation, commercial, military, and government aircraft. Specific scenarios were flown that day to allow data to be taken to support the eventual implementation of this technology.

Copyright © 2001 by the MITRE Corporation. Published by the American Institute of Aeronautics and Astronautics, Inc., with permission.
*Principal Engineer, Center for Advanced Aviation System Development. Senior Member AIAA.
[†]Principal Engineer, Center for Advanced Aviation System Development.
[‡]Lead Engineer, Center for Advanced Aviation System Development.

This chapter will focus on the results from the Ohio Valley evaluation, including the benefits of enhanced visual acquisition and enhanced visual approaches using ADS-B and the data that will be used to support the ADS-B link decision. The information in this chapter is based on the Phase I Operational Evaluation Final Report.[1]

II. Operational Evaluation 1999

The primary operational objectives of the Cargo Airline Association/Safe Flight 21 Operational Evaluation (OpEval) for 1999 were to 1) demonstrate ADS-B technology, 2) evaluate specific air-air and air-ground applications, and 3) develop a wide support base within the aviation community for the advancement of ADS-B implementation. These objectives were to be met by a series of high- and low-altitude flight tests consisting of multiple aircraft types, avionics platforms, and a government/industry ground station configuration.[2]

Other goals included the demonstration of a prototype version of Lockheed-Martin's MicroEARTS ground system to give the controller the benefit of improved surveillance via ADS-B, the demonstration of the compatibility of ADS-B with traffic alert and collision avoidance systems (TCAS), and an attempt to stimulate industry toward the production of certified air and ground systems.

In brief, the following was accomplished in the OpEval on 10 July 1999:

1) Twenty-four aircraft and 1 ground vehicle were equipped with avionics from UPS Aviation Technologies, AlliedSignal, Rockwell-Collins, Honeywell, and BF Goodrich. The various aircraft included 12 cargo airline, 4 avionics industry test aircraft, 3 FAA test aircraft, 3 General Aviation/Universities, 1 NASA research aircraft, and 1 U.S. Navy flight test aircraft.

2) There were 600+ h of flight time logged leading up to and including 10 July 1999.

3) Three ADS-B ground stations were operational (McLean, Virginia; Wilmington, Ohio; Louisville, Kentucky).

4) There was integration of ADS-B and radar surveillance data on controller-display equipment by a consortium including Lockheed-Martin, Sensis Corporation, and Harris Corporation.

The roles of the MITRE Corporation/Center for Advanced Aviation System Development (MITRE/CAASD) in the planning and execution of OpEval were extensive. First, MITRE/CAASD is the primary author of the Safe Flight 21 Master Plan, which lays out the multiyear schedule and priorities for evaluating the Safe Flight 21 applications. The CAASD Integration and Interaction Simulation Laboratory (I-Lab) was used during the spring of 1999 to develop the final air traffic control (ATC) and pilot procedures that were flown during the 10 July test. CAASD also developed and coordinated the overall flight profiles and schedules that were flown. The ground broadcast servers (GBSs) were developed and installed by MITRE, and were manned by CAASD personnel during the integration flight tests and the actual OpEval. Finally, CAASD personnel monitored all 24 aircraft and updated the flight schedule during the OpEval. All of these activities were performed in close coordination with the Cargo Airline Association (CAA), FAA headquarters, NASA, Radio Technical Commission for Aeronautics (RTCA), and union and staff controllers from Dayton Terminal Radar Approach Control (TRACON), Indianapolis Air Route Traffic Control Center (ARTCC), and Wilmington Tower.

Flight activities during OpEval were prioritized to focus on near-term cockpit display of traffic information (CDTI) applications that may provide benefits in a relatively short time period with few, if any, changes to current ATC procedures. The scenarios were designed to assess crew performance, operational procedures/benefits, and data link technical performance. The CDTI applications were prioritized in the following manner (from highest to lowest priority): 1) formal evaluation of enhanced visual acquisition for "see and avoid," 2) formal evaluation of enhanced visual approaches, 3) demonstration of airport surface situation awareness, 4) demonstration of enhanced in-trail (or lead) climbs/descents, 5) demonstration of station keeping, 6) demonstration of departure spacing, and 7) demonstration of final approach spacing.

A. System Description—Airborne Elements

The CDTI, depicted in Fig. 1, is a flight-deck display that presents relative position of other traffic in the vicinity with respect to own aircraft. In addition to aircraft position, other information, such as navigational aids and obstructions, may be displayed. Traffic information for the CDTI may be obtained from various sources, including ADS-B, Traffic Information Service (TIS), or an on-board TCAS. To display traffic information, the CDTI may use a dedicated display device or a shared multifunction display (MFD). Even though a visual, graphical presentation of the traffic on a heads-down display will be the most common in the near term, other types of presentation (e.g., aural, graphical, heads up) are also possible.

The Link and Display Processor Unit (LDPU), manufactured by UPS Aviation Technologies of Salem, Oregon, receives traffic surveillance data from each of the three candidate ADS-B data links: mode select (mode S) transponder (1090 MHz), universal access transceiver (UAT) (966 MHz), and very high frequency (VHF)

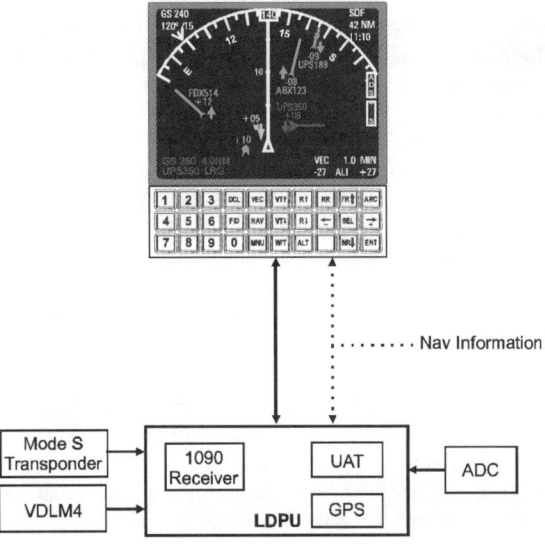

Fig. 1 LDPU and CDTI display.

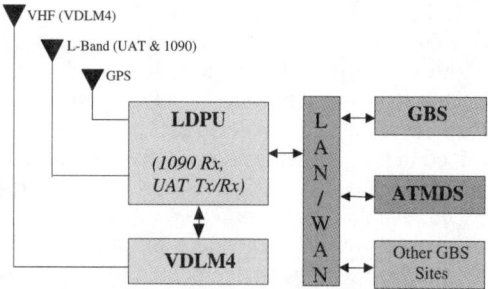

Fig. 2 Safe Flight 21 ADS-B ground station general architecture.

data link mode 4 (108–136 MHz range). It then transfers the data to the CDTI for display in the cockpit. A block diagram of the LDPU can be found in Fig. 1. The LDPU contains a 1090 MHz receiver, a UAT transceiver, and an embedded GPS card for position information. Each aircraft is equipped with a mode S transponder that is used for transmitting the ADS-B "extended squitter" on 1090 MHz and receiving TIS information on 1030 MHz. The VDLM4 radio is also separate from the LDPU.

B. System Description—Ground Elements

A ground system including receivers of the three data link candidates and an air traffic management system was located at Wilmington to receive and record ADS-B data. The ground system also demonstrated that existing ground technologies can accommodate ADS-B information. The ground system consisted of an SF21 ADS-B ground station and an air traffic management demonstration system (ATMDS).

The SF21 ADS-B ground station (Fig. 2) is a proof-of-concept ADS-B implementation integrated and installed by the MITRE Corporation. It received broadcasts from all three candidate links and logged and output the data for processing by the ATMDS. During OpEval, two SF21 ADS-B ground stations were utilized, one at Wilmington Airpark (ILN) and another at Louisville International Airport (SDF).

The CAA invited industry participation in the OpEval. The OpEval industry team was led by Lockheed Martin Air Traffic Management (LMATM) and included Harris Information Systems Division (ISD) of Melbourne, Florida, and Sensis Corporation of Syracuse, New York. The ATMDS received ADS-B data from the SF21 ADS-B ground station and from two additional 1090 MHz receivers supplied by Harris and Sensis. The system also received terminal radar data from the Airport Surveillance Radar—Model 9 and Monopulse Secondary Surveillance Radar (ASR-9/MSSR) on site at Wilmington and fused the two data types for display.

The ATMDS was intended to support the OpEval and to: 1) demonstrate that existing ground system technologies can accommodate ADS-B, 2) provide a basis for planning evaluation of ADS-B operational concepts and procedures, 3) demonstrate the benefits of ADS-B to air traffic management, 4) demonstrate ability to use ADS-B in transitional environment, and 5) provide early visibility into requirements for implementation of ADS-B into the National Airspace System

(NAS). It should be noted that FAA controllers provided air traffic services during the evaluation from FAA Dayton TRACON using existing ATC systems that do not have ADS-B capability. In parallel the ATM demonstration system processed ADS-B data for observers, who included air traffic controllers. Feedback from observers is included in lessons learned from demonstrations.

C. Flight Profiles

The OpEval consisted of three major flight periods, a morning, mid-day, and afternoon flight period. The CAA aircraft primarily flew during the morning and afternoon flight periods. Flight profiles consisted of multiple approaches at the Wilmington airport and also a high-altitude en route segment.

III. Method of Test

A. Operational Concepts

One product of the OpEval was the validation of the CDTI operational concepts as developed by RTCA SC-186. This section documents the feedback from OpEval into the RTCA CDTI application description document,[3] which was used to identify the near-term CDTI applications that were evaluated or demonstrated during OpEval. OpEval experience was provided as input to the document at a Working Group 1 (WG1) meeting in October 1999. The following is a brief description of that input for the main applications under consideration.

1. Enhanced Visual Acquisition

The CDTI enhanced visual acquisition application is a capability that aids pilots in visually acquiring other proximate traffic as well as increasing their traffic awareness. The feasibility of this concept was validated in the OpEval flight environment with multiple equipped aircraft of various types.

The CDTI assists the pilot with this see-and-avoid visual scan by providing a display of traffic. From the pilot's point of view, this capability should be considered as a complement to the traffic advisory service provided by ATC. CDTI can also enhance the pilot's ability to visually acquire traffic called out by ATC. The CDTI is currently intended only to assist in visual acquisition of other aircraft in visual meteorological conditions (VMC). It does not relieve the pilot of responsibility to see and avoid other aircraft. Currently, there are no aircraft evasive maneuvers recommended, authorized, or provided for as a sole result of the CDTI or CDTI alerts.

One finding of OpEval that was further discussed at SC-186 WG1 is the benefit of having an aural/visual alert to draw the pilot's attention to the CDTI when there is visual acquisition traffic of interest. This could increase the utility/benefit of the application. However, also noted at OpEval and further discussed is the need for carefully designed alerting logic that minimizes false or nuisance alerts. WG1 felt alerts were not a minimum requirement for this application, but is adding it as an option in the operational concept.

2. Enhanced Visual Approaches

The OpEval also validated the feasibility of this application and provided feedback to refining the operational concept. The normal conduct of visual approaches was enhanced with the use of the CDTI (see Section IV.A). In addition, the concept feasibility of allowing the pilots to respond with traffic call sign was validated; however, concerns were brought up on specific phraseology, voice communication clutter/misinterpretation, and the controller's potential use to call out traffic by flight identification (ID). This last point is in the RTCA CDTI operational concept, but was not formally implemented during this OpEval. These issues will need to be addressed during further procedure development. Allowing the pilots to close up spacing once cleared for a visual approach was not in the original CDTI operational concept for enhanced visual approaches, but it was part of the operational concept for final approach spacing. Given the pilot's ability to perform this task, it has now been added to the SC-186 concept for enhanced visual approaches, with specific references to monitoring and closing up spacing during visual approaches.

B. Operational Procedure Development

OpEval procedures were specifically designed to replicate commercial air carrier "line" operations to the maximum extent possible. Likewise, no waivers to existing procedures in FAA Order 7110.65L[4] were requested. All separation criteria and communication procedures were per current operational guidelines. These restrictions were established to ensure a more line-oriented evaluation of the ADS-B/CDTI system. In addition, the results can more easily be extrapolated into current line operations if the test environment was representative of present-day NAS operations.

Although OpEval was specifically designed to minimize changes to current ATC procedures, there was, nonetheless, a need to provide flight crews with guidance on procedures for the various CDTI applications. In addition, a departure from normal pilot responses to an ATC traffic call-out was developed that allowed flight crews to use the aircraft call sign. Both of these items are discussed in the following.

1. Procedure Development and Flight Crew Maneuver Cards

The process for developing the OpEval procedures began by reviewing the RTCA SC-186's CDTI application description document.[3] This document helped to establish the required procedures and tasks that a pilot must be able to perform to use the CDTI effectively. (The CDTI tasks performed by the flight crews were additional to and augmented normal flight tasks, e.g., flight crew must visually acquire aircraft, but can reference the CDTI to assist in the visual search.)

Upon considering those tasks, and in conjunction with input from the I-Lab simulations, a set of flight maneuver cards was developed. The purpose of the flight maneuver cards was to assist flight crews during training and briefing, and as an in-flight reference for OpEval procedures, tasks, and techniques.

2. Flight ID Phraseology

During the planning of OpEval, concerns were expressed by both pilots and controllers regarding the potential for confusion with respect to one flight crew being able to use the call sign of another aircraft. From the pilot's perspective, they are highly attuned to their own call sign, and might assume that any transmission including their call sign was for them, probably adding to their workload, and to frequency congestion and controller workload as they resolved the confusion. From the controller's perspective, they might not be sure if the call was from the call sign aircraft or from another aircraft about the identified aircraft, again increasing frequency congestion and workload while the confusion was resolved. Augmenting these concerns were reports of those situations actually occurring during some of the I-Lab simulations. As a result of the simulation work, communications procedures were developed for OpEval to minimize call sign confusion.

C. Collected Data

The objectives of the CDTI human factors analysis, which was primarily conducted by personnel affiliated with the NASA Ames Research Center, were as follows: 1) evaluate CAA-specified applications, 2) evaluate flight crew and controller aspects of workload and traffic situational awareness, 3) evaluate potential traffic management and procedure implications, and 4) evaluate effects of demonstration/evaluation on resulting data. The human factors data collection plan identified observer records, flight crew and controller feedback, GBS track data, and ATC facility radar track and voice communications tapes as data sources. It further identified the following avenues for soliciting and obtaining flight crew and controller feedback: 1) during simulation studies at MITRE, 2) in-flight data collection with flight crews, 3) controller workload ratings and questionnaires, and 4) post-flight debriefing of the OpEval flight crews and controllers.

Based on the planned flight schedule detailed in the Test and Evaluation Master Plan (TEMP),[2] there were 96 scheduled observable trials planned consisting of 32 baseline and 64 CDTI. Crews flew as within-subjects in cells that were counterbalanced for runway, flying pilot, order effects, baseline, CDTI, and for two (a.m. and p.m.) visibility factors.

Table 1 provides a description of the data resources and the applicable performance measures that were collected and summarized in this report.

1. Flight Crew Human Factors Observers

NASA selected, trained, and scheduled observers who observed the flight crews' interaction with and without the CDTI during the OpEval flights.

The observers took notes on the observer forms, administered the questionnaire, and debriefed the flight crews after the test scenarios were completed. The observer

Table 1 OpEval performance measures

Dependent measure	Data resource
Visual acquisition time	ATC voice tapes, observer reports
Flight crew and ATC workload	ATC voice tapes, questionnaires
Aircraft spacing, terminal, and en-route	Dayton TRACON radar track data, MITRE GBS
CDTI feature preferences	Questionnaire and debrief
ATC responses to CDTI	Questionnaire and debrief
Flight crew response to CDTI	Questionnaire and debrief
Flight crew traffic awareness	Observer records, questionnaire, and debrief
Flight ID phraseology	ATC voice tapes

records were used to support the collection of 1) response time to each traffic call, 2) assessment of CDTI use during visual traffic acquisition, and 3) assessment of the impact of CDTI on normal cockpit duties.

2. Flight Crew HF Post-Event Questionnaire

The flight crew questionnaire was designed to elicit a variety of responses from flight crews on a number of CDTI related issues. In the results we report on questions relevant to each application.

Pilot opinion ratings were gathered during the post-flight debriefing, after each crew completed their post-flight duty requirements. The questionnaires were completed prior to a structured interview and the combined activity lasted about one hour. Each question was designed to elicit specific information from the flight crews on the usefulness of individual features and functions of the CDTI, and the impact of the CDTI on specific flight-related tasks. Items in the questionnaire were scaled from 1 to 3 or from 1 to 5 to support a Likert scale analysis. Additionally, a selection of "Not Applicable/Did Not Use" was an available choice.

3. Air Traffic Control Observers and Post-Event Questionnaires

A separate controller was assigned to each runway, with responsibility for aircraft up to 5000 ft. A separate radio frequency was used for each runway. In accordance with standard operating procedures, there was a relief shift provided by a third controller partway through the morning and afternoon sessions. A fourth controller and the operations manager performed high-level coordination of the OpEval aircraft.

Two human factors specialists collected data at the Dayton TRACON during the ADS-B OpEval exercise, focusing on the low-altitude flights. As each set of aircraft completed a full cycle around the traffic pattern (i.e., all aircraft in the pattern had executed a missed approach or landed), a specialist asked the controller to estimate his or her workload during that cycle on an 8-point scale. The scale ranged from 1 = very low workload, through 4–5 = moderate workload, to 8 = very high workload. They were also able to add comments about why their workload was at its current level. After the morning low-altitude session, and then again following the afternoon session, the three controllers who had worked the traffic completed a three-part questionnaire.

After the questionnaires were completed independently by each controller, the human factors specialists conducted an oral debriefing of the three controllers who worked the traffic and the controller who worked as a coordinator. The four were debriefed as a group. The debriefings provided an opportunity for the specialists to ask questions about specific events that occurred during the session and for the controllers to discuss their experiences.

4. Air Traffic Control Voice Tapes

Five hours of audiotaped pilot/ATC communications were provided by the Dayton TRACON facility. The two-channel tapes included all calls made on the recorded radio frequency on one tape channel and the universal time coordinated (UTC) time code on the other channel expressed in terms of date, hour, minute, and whole second.

5. Aircraft Track Data

Post-processed computer recordings of aircraft track data were obtained from Indianapolis ARTCC and Dayton TRACON radar tapes, and from ADS-B surveillance data. These track data were used to calculate statistics showing both the accuracy and variance of CDTI-based spacing, in both ARTCC and TRACON airspace. Track time-tags also permitted the correlation of aircraft position with concurrent time-tagged events recorded in other data. Inferences from these statistics pertained to the same objectives for which the other indices were obtained, namely insight into effects of CDTI on workload, and traffic throughput, implications for future procedures, and effects of test conditions.

IV. Results

A. Human Factors

This section presents a synopsis of the results presented in the Phase I—Operational Evaluation Final Report[1] and their implications for CDTI applications. OpEval has provided a unique opportunity for investigators to evaluate CDTI in an operational flight environment. A considerable amount of flying time was achieved, allowing for the collection of a large quantity of objective performance data. There are clear trends in the performance data showing operational benefits of the CDTI. However, due to operational flight issues (weather, maintenance, etc.), a balanced data collection protocol could not be maintained, and therefore most data were not subjected to standard statistical tests. In addition to the objective data, both controllers and flight crews were very willing to share their experience and opinions after the event, resulting in a wealth of subjective opinion data. Analysis of the performance and subjective data has provided valuable insight into the benefits and issues surrounding CDTI.

Collectively, a review of both the performance and subjective data revealed no "showstoppers" that would indicate serious obstacles toward the implementation of CDTI for the applications evaluated and demonstrated at OpEval. Comments from both flight crews and controllers were generally positive, although the performance data did not always support their positive opinions. Overall, flight crews agreed that the CDTI aided visual acquisition, visual approaches, station keeping, in-trail climbs/descents, and high-altitude departure flights. [A majority of crewmembers agreed that overall the CDTI System was an aid to 1) high-altitude departure flight (mean 1.6, $p < 0.05$, $df = 3$); 2) station keeping (mean rating 1.2, $p < 0.05$, $df = 3$); 3) in-trail or lead climbs and descents (mean 1.4, $p < 0.05$, $df = 3$); 4) visual approach (mean 1.4, $p < 0.05$, $df = 24$); and 5) visual acquisition (mean 1.4, $p < 0.05$, $df = 23$.] Crews ratings were mixed on the use of CDTI as an aid to surface awareness (mean rating 2.88, $p > 0.05$, $df = 24$). In addition, controller opinions suggested that the CDTI aided the enhanced visual acquisition and enhanced visual approach applications, and they were generally positive about the use of CDTI.

We should note that controllers did express some concerns over potential CDTI effects in an operational environment (e.g., flight crews having too much discretion over how closely to follow traffic). It is, however, important to differentiate between concerns over what might happen and what actually happened during OpEval. Most of their concerns were not apparent in the OpEval data, but we cannot state whether or not these concerns would be substantiated under different operational conditions.

1. General Issues

Flight crews identified three issues that need to be considered as we proceed with the design and use of CDTI to support visual traffic acquisition and other ADS-B applications. Crews identified display integration, clutter, and heads-down time as issues that need to be addressed in future CDTI implementations.

a. Display location and integration. Overall, flight crews reported using the CDTI effectively during the approach phase of flight, although some flight crews reported that display integration was an issue. Specifically, flight crews reported that the location of the display outside the primary visual scan made it difficult to integrate it into the normal scan, and that this location may have caused additional intracockpit communication. Intuitively, integrating the CDTI with the navigation display in a glass cockpit aircraft should improve CDTI usability and reduce heads-down time; however, this remains to be demonstrated. Regardless, flight crews reported the present CDTI implementation to be effective as an aid to visual acquisition, either with or without an ATC traffic call, and that maintaining awareness of multiple traffic targets was less difficult with the CDTI. This would suggest that the CAA's initial CDTI implementation, on a stand-alone display, was adequate for both the enhanced visual acquisition and enhanced visual approach applications.

Some crews identified issues with the location of the CDTI. It was difficult for the second officer in B-727 aircraft to see and use the CDTI, which was located forward of the throttles, and he or she could not reach it without leaving his or her seat. The DC-9 First Officers also had less access to the CDTI, which was located nearer to the captain's side. The impact of the placement of the display is dependent on flight crew procedures for operation of the CDTI. Overall, the display location required flight crews to develop alternatives to their usual cockpit scan to include the CDTI and make use of the information being presented.

b. Display clutter. In general, display clutter was reported to be manageable during airborne operations, even in the relatively densely populated low-flight scenarios in which aircraft were conducting visual approaches. Display clutter was, however, especially evident during airport surface operations, in which a large number of targets were located in close proximity. This combined with large target and data tag size, as well as lack of a surface map, contributed to a number of adverse remarks about the CDTI's usability on the airport surface. Because the CDTI was not designed for use on the airport surface, these adverse comments are not surprising. A surface map will be added to the CDTI and evaluated in future operational evaluations. This may introduce other issues that will be addressed during that evaluation.

c. Heads-down time. Many flight crews commented on the increase in heads-down time while using the CDTI, while at the same time suggesting it was an effective aid to visual acquisition and visual approaches, both currently out-the-window tasks. One possible explanation for this reported increase in heads-down time is that flight crews were relatively inexperienced with the CDTI and received only moderate instruction in its operation. There was, however, some evidence that flight crews' confidence in and efficiency with the CDTI improved over the course of the day (shown in Fig. 3). This would suggest that the increased heads-down time may be mitigated with training and experience.

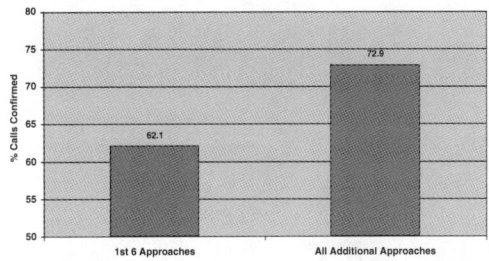

Fig. 3 Percentage of calls confirmed during p.m. visual approaches.

Previous research on visual target acquisition during low visibility surface operations suggests that, for visual search, the increase in CDTI heads-down time may, in fact, have a positive impact on visual acquisition performance[5]. That is, the time spent heads down may translate to a more effective visual scan in the out-the-window view. However, the determination of the extent to which any improved efficiency for visual acquisition would be offset by increased heads-down time would require a controlled experiment.

2. Enhanced Visual Acquisition

Crew comments on the use of the CDTI for enhanced visual acquisition were positive. Flight crews reported that the CDTI was very useful as an aid to visual traffic acquisition during marginal visual meteorological conditions (VMC) and in VMC. Flight crews also reported that the CDTI improved the efficiency of the visual acquisition task, and that they found the workload associated with the use of the CDTI acceptable. In addition, they reported that the CDTI was useful as an aid to maintaining awareness of multiple targets and to reacquire previously acquired traffic when needed.

Figure 4 shows that in the absence of an ATC traffic call, flight crews acquired traffic 76% of the time using the CDTI, either before, after, or without acquiring the traffic out the window (OTW). Only 24% of the time was traffic acquired without the aid of the CDTI. These data strongly support the flight crews' assertions that the CDTI was an aid to visual acquisition. After an ATC traffic call (Fig. 5), almost half the responses are OTW visual acquisitions only, but CDTI was still used for acquisition in the remaining responses, including 33% of the total responses where traffic is first acquired on the CDTI. This suggests that, although many responses to an ATC call are traditional OTW visual acquisitions, the CDTI is still a significant aid in that process.

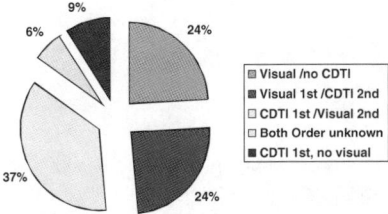

Fig. 4 Distribution of traffic acquisitions without an ATC traffic call—total number of acquisitions confirmed and coded $n = 33$.

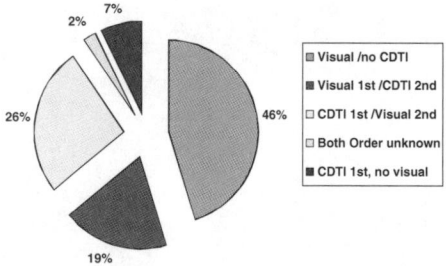

Fig. 5 Distribution of traffic acquisitions after ATC traffic call—total number of acquisitions confirmed and coded $n = 42$.

Controllers also reported that the CDTI enhanced visual acquisition in ways beneficial to ATC. However, much of the time they reported being unable to determine when CDTI was in use. They thought CDTI was in use all of the time after the first baseline approach, when in fact it was not. Nonetheless, they suggested that the CDTI increased situational awareness for both pilots and controllers when traffic was called out.

The controllers stated that CDTI allowed them to call traffic earlier than normal. For instance, one controller reported that on two successive departures he was able to call traffic to follow on the crosswind leg, and the subject aircraft followed without incident. This would appear to enable controllers to better manage and reduce their workload by reducing time. However, this statement is not supported by the data presented in Fig. 6, which does not indicate greater call out distances for CDTI aircraft. Additionally, no evidence indicated that CDTI reduced the duration or the number of messages that comprised the ATC communication sets, and so we were not able to confirm the controllers' perceived reduction in workload.

a. Visual acquisition time. In light of findings from previous simulation studies,[6,7] we expected that visual acquisition time would be reduced with the aid of the CDTI. Two independent ways of measuring visual acquisition time were developed. The first was the flight-deck observer measuring the time between an

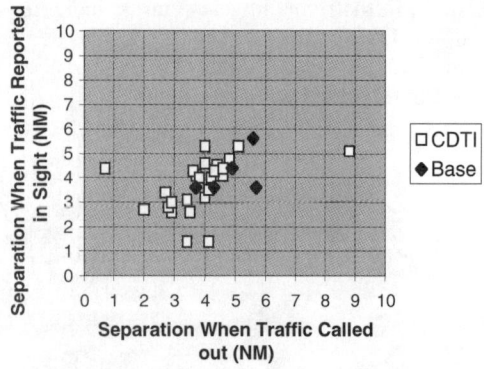

Fig. 6 Visual acquisition distances.

ATC call and the first indication by any member of the flight crew that they had acquired the traffic. This indication could take the form of a reply to ATC, or an intracockpit communication or gesture. A second measure came from analysis of the ATC communication voice tapes and was the time interval between ATC calling traffic and the flight crew responding to ATC. Analysis of the data shows that neither metric indicated a significant difference in acquisition time.

For the acquisition times measured by the observer, the average is about 20 s both with and without the CDTI. ATC communication tape responses were analyzed both for all responses, and then for only those responses indicating "traffic in sight." The traffic in sight responses show no significant difference between baseline and CDTI conditions, although there is a slight trend toward reduced response time with the CDTI. Also, there is a trend toward longer "looking" or "not in sight" response times with the CDTI than without. This may indicate crews with the CDTI used it as part of their search before replying to ATC, as is also indicated by the observer data reported in Fig. 4. The ATC communication tape data set for all responses show that CDTI aircraft had a higher frequency of responses less than 3 s (Fig. 7). These are practically immediate responses, expected to reflect the pilots' OTW perception at the time of the ATC call, either traffic in sight (i.e., previously acquired), looking, or not in sight. More crews with the CDTI chose to give an immediate reply, rather than looking for the traffic before replying. This may imply some additional level of comfort due to the improved situational awareness afforded by the CDTI, so that the first reaction is to reply to the ATC call before looking for or reconfirming the traffic.

Analysis of visual acquisition time is confounded by the fact that each ATC traffic call is for a unique separation distance and encounter geometry. One would expect faster acquisition performance for closer aircraft, and for aircraft called out near the 12 o'clock position. Acquisition time proved to be correlated to separation distance at the time of the ATC call, for the combined CDTI and baseline data. Inspection of the data in Fig. 6 shows that most traffic was called out at distances ranging from 2 to 6 miles. It is clear that there are too many uncontrolled variables, including range, geometry, visibility conditions, and flight crew workload, all of which could affect acquisition time, for us to evaluate the true effect of the CDTI on visual acquisition performance.

The data showed an apparent increase in use of the CDTI as the day progressed (Fig. 3). Traffic was acquired using the CDTI more frequently during the later afternoon approaches, and less frequently visual only. This suggests flight crews

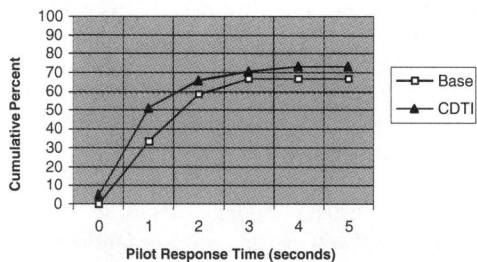

Fig. 7 Cumulative distribution of time for all pilot responses to ATC traffic call-outs (within 5 s).

were learning to integrate the use of CDTI into their usual procedures, and that some training may be required to develop optimal patterns of use.

b. Pilot-initiated visual acquisition reports. There were eight occasions when flight crews initiated a traffic call. At least some appeared to represent implicit requests to follow the traffic reported in sight. This was a concern expressed by all three controllers, because responding to pilot-initiated requests in terminal airspace can potentially increase controller workload and might impact their plans for handling the traffic. Of the eight calls, four came from one aircraft, and four were in the late afternoon, suggesting that possibly these calls are not representative of the frequency of pilot-initiated requests that would be experienced under normal operational conditions.

3. Enhanced Visual Approaches

Generally, both flight crews and controllers commented that the CDTI was a positive aid for visual approaches, in that it aided overall traffic awareness and in closing to a comfortable and appropriate final approach spacing. Flight crews perceived the workload for gauging the distance behind the aircraft ahead to be acceptable, although heads-down time was reported to increase.

a. Spacing from lead aircraft. Analysis of the spacing data obtained from radar tracks and ADS-B for baseline and CDTI approaches shows a clear trend toward spacing reduction with the CDTI, although this difference was found to be only marginally significant. Specifically, analysis of the radar track data indicates an approximate 6% probability that the 1.4 mile mean reduction in the median spacing with the CDTI was due to chance and would not be repeated if the same flight scenarios were flown again. Again, the operational conditions and limited data sample may have contributed to this trend not reaching significance. The mean approach times with and without CDTI show a 15% reduction (72.5 to 65.8 s) with the CDTI. Although not conclusive, these data strongly suggest the CDTI is aiding the efficiency of visual approaches.

The test scenario may have affected spacing, because each flight crew was aware that the aircraft ahead would go around and not land. This could induce a spacing comfort level well below what would be comfortable were the lead aircraft to land. During normal VMC operations, an aircraft cleared for a visual approach behind another aircraft will maintain spacing sufficient to ensure the lead aircraft has cleared the runway before it crosses the threshold, allowing for the lead aircraft to land long and/or delay its turn off. This distance is usually not less than about 1.5 to 2 miles, but for other reasons (e.g., wake vortex considerations or the spacing that existed when the trailing aircraft was cleared for the visual approach) may be much larger. The approach spacing data indicate that aircraft did not get closer than 2 miles in almost all cases, suggesting that flights did not get close enough with the CDTI that they would have risked having to go around because the lead aircraft had not cleared the runway. However, it is still possible that flight crews were more comfortable with closer spacing on approach than they would have been if the aircraft were landing instead of going around. A controller commented that flight crews closed up less when coming in for a landing than when performing low approaches. Analysis of the spacing data to evaluate whether this was in fact true was not attempted because there were insufficient numbers of approach events that ended in a landing.

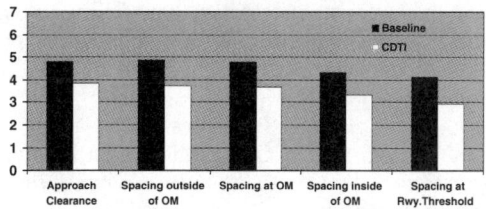

Fig. 8 Baseline and CDTI spacing for aircraft cleared to follow lead aircraft.

CDTI usage appeared to increase over the course of the day, and so one might presume that spacing might also be reduced as pilots became more familiar with the CDTI (Fig. 3). However, no trend toward greater spacing reduction over the course of the afternoon was observed.

ADS-B and radar data show that aircraft spacing was closer between CDTI aircraft when the aircraft were cleared for the approach (Fig. 8). The exact reason for this is unclear. CDTI aircraft, being closer when turning on to the final approach course, had less time to reduce spacing, and less spacing distance to reduce in that time. Even with the reduced closure opportunity, CDTI aircraft closed by about the same percentage as baseline aircraft, possibly suggesting the CDTI had a beneficial effect on spacing over the flight pattern. There were no statistically or operationally significant differences in spacing variability between CDTI and baseline aircraft.

Figure 9 shows the distributions of all approach event spacing distances. The CDTI distribution shows evidence of a positive skew, whereas the base distribution assumed a bimodal shape.

b. Flight ID phraseology potential pilot/controller confusion. Confusion over hearing their call sign repeated by another fight crew was not mentioned by any flight crew or controller during the debrief sessions. Examinations of complete call-sign use, nonstandard communication practices, call-sign location in pilot replies to traffic call-outs, and confusions in pilot replies, failed to find statistically robust evidence of problems that could be attributed to CDTI. Procedures to utter the own ship call sign first, and traffic call sign in the middle or end of the message were generally followed, although a few instances of pilots and controllers confusing traffic and own ship call signs were nonetheless recorded. No aircraft whose call sign was used in a traffic reference mistakenly responded. There was one instance when ATC used a traffic call sign when communicating with a plane that did not have access to CDTI. This illustrates the need for ATC to know whether aircraft have CDTI available (not the case at OpEval). Although communications

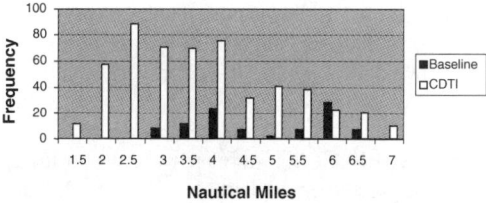

Fig. 9 Frequency distribution of spacing measurements for the p.m. session.

problems occurred in the low to moderate workload OpEval environment at a rate comparable to normal TRACON operations, communications problems and their antecedents in pilot and controller use of flight ID should continue to be studied, especially under higher workload conditions.

    c. Aid to positive identification. One problem that is often reported during visual approach operations is that of flight crews misidentifying the traffic called out by ATC.[8] As a result, a primary objective of the enhanced visual approach application is to minimize these instances of misidentification (described in the application description as "Aid to Positive Identification"). A review of the data shows three or four occasions when a flight crew's use of call sign alerted the controller to the fact that the wrong aircraft had been identified, thus indicating that the CDTI is, in fact, acting as an aid to positive identification. This suggests a major CDTI benefit in providing early warning to ATC if an incorrect aircraft has been identified. This information will allow the controller to correct the situation well before it can become a potential conflict. This benefit was reflected in both pilot and controller comments.

    d. ATC communications workload. Overall, controller workload was perceived as low to moderate during OpEval. This was largely because a controller was assigned to each runway, resulting in an average of less than three aircraft on frequency, much lower than many terminal operations. A statistically significant increase in the number of transmissions when aircraft were using the CDTI was found. This is possibly attributable to the higher throughput for aircraft using CDTI, resulting in less time for the communications to occur. The communications rate decreased slightly with increasing exposure to CDTI. Messages were slightly longer and more complex with the CDTI, although these differences did not reach statistical significance. Increases in complexity and duration may have been due to the addition of traffic call signs in pilot messages. Lack of sufficient baseline data inhibits further analysis of this trend. It is possible that the apparent increase in communications frequency and complexity with the CDTI was due to flight crews and controllers attempting to maximize CDTI usage. Although no definite implications can be drawn about CDTI's effects on communications workload, if CDTI increases throughput, it may also increase communications workload.

    e. Use of CDTI in IMC and MVMC. Instrument landing system (ILS) approaches were flown during the morning low-flight scenarios. Aircraft were vectored onto the final approach course outside the outer marker, resulting in longer final approaches than were flown for the afternoon visual scenarios. The data show a distinct trend toward reduced separation with the CDTI (Fig. 10), but the results did not meet the normal criteria for statistical significance. Specifically, there is an 11% probability that the 1.1 mile measured reduction in median spacing with the CDTI is merely due to chance, and would not be repeated if the same flight scenarios were to be flown again. Again, the observed spacing difference between baseline and CDTI aircraft existed before the aircraft were vectored onto the final approach course.

    Crews commented that the CDTI increased their confidence in maintaining awareness of traffic that passed in and out of clouds. Because crews are generally searching the visual scene for traffic in both inclement meteorological conditions (IMC) and VMC, this type of system should aid them in this task.

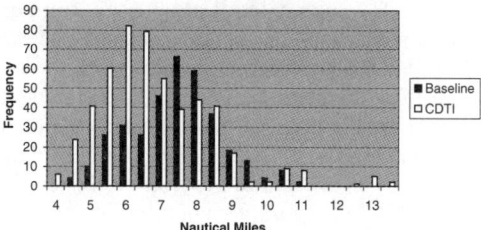

Fig. 10   Frequency distribution of spacing measurements for the a.m. session.

Overall, the value of CDTI for traffic situational awareness was reported to be very positive.

B. Assessment of Maturity of Concepts and Technology

Following the RTCA "Development and Implementation Planning Guide for ADS-B Applications,"[9] the following paragraphs summarize the implementation feasibility and the technology availability of the applications investigated during OpEval.

The OpEval afforded the look into the feasibility of several of the ADS-B applications. The OpEval Coordination Group makes the following conclusions about the ADS-B applications evaluated/demonstrated at OpEval:

1) Evaluate enhanced visual acquisition for see and avoid. As one of the main applications being evaluated at OpEval, this application is seen as very feasible. The operational concept for this application is mature and has been approved by the RTCA SC-186. The operational concept and CDTI requirements have been evaluated within the OpEval.

2) Evaluate enhanced visual approaches. The operational concept for this application is mature and has been approved by the RTCA SC-186. The majority of this application and its associated CDTI requirements have been evaluated at OpEval. The remaining item from the operational concept that has not been demonstrated is the controller use of call sign. However, this procedure will be evaluated in FY00 as part of the Safe Flight 21 program.

V.   Conclusions

The CAA's initial CDTI implementation was evaluated with respect to several near-term applications. For the enhanced visual acquisition evaluation, the mature operational concept and associated CDTI requirements were tested. Both pilots and controllers felt the CDTI augmented the visual acquisition task and improved pilot awareness of surrounding traffic. No significant problems with respect to the operational approval of this application were revealed. One potential issue that was raised, which needs to be addressed as part of future CDTI training programs, is that of flight crews initiating unwarranted requests from ATC. For the enhanced visual approach evaluation, the CDTI requirements as outlined in the operational concept were tested. The results have revealed significant performance benefits in the form of enhanced spacing awareness and a reduction in the misidentification of aircraft called out by ATC. As with enhanced visual acquisition, both pilots and controllers felt the CDTI augmented flight crew performance during visual approaches, and no significant problems with respect to the operational approval of this application were revealed. The remaining applications were demonstrated, and the findings will support future operational concept development and evaluations.

Although the airborne applications were the primary focus of the operational evaluation, the ground system demonstrated the potential to support air-ground ADS-B applications. The ADS-B Minimum Aviation System Performance Standards (MASPS) (RTCA DO-242) identifies a number of potential near-term ADS-B air traffic services surveillance applications. Many of these are reflected in the Safe Flight 21 program's near-term plans.

Acknowledgment

The contents of this material reflect the views of the authors and/or the Director of the Center for Advanced Aviation System Development. Neither the Federal Aviation Administration nor the Department of Transportation makes any warranty or guarantee, or promise, expressed or implied, concerning the content or accuracy of the views expressed herein.

References

[1] Operational Evaluation Coordination Group (OCG), *Phase I Operational Evaluation Final Report*, FAA, 2000.

[2] Operational Evaluation Coordination Group (OCG), *Test and Evaluation Master Plan (TEMP), Phase I–Operational Evaluation*, FAA, 1999.

[3] RTCA Inc., *Application Description for Initial Cockpit Display of Traffic Information (CDTI) Applications,* Do-259, Washington, DC, 2000.

[4] Federal Aviation Administration, *Air Traffic Control Handbook, Order 7110.65L,* Department of Transportation, FAA, Washington, DC, 1999.

[5] Battiste, V., Down, M. and McCann, R. S., "Advanced Taxi Map Display Design for Low-Visibility Operations," *Proceedings of the 40th Annual Meeting of the Human Factors and Ergonomics Society,* 1996.

[6] Wickens, C. D., Helleburge, J., and Xu, X., "Maneuver Choice in Free Flight," *Proceedings of the World Aviation Congress (99WAC-146),* Society of Automotive Engineers, Inc., 1999.

[7] Olmos, B. O., Mundra, A. D., Cieplak, J. J., Domino, D. A., and Stassen, H. P., "Evaluation of Near-term Applications for ADS-B/CDTI Implementation," *Proceedings of SAE/AIAA World Aviation Congress,* Society for Automotive Engineers, Inc., 1998.

[8] Stassen, H. P., "Enhancing the Safety and Efficiency of Runway Approaches Using a Cockpit Display of Traffic Information," The MITRE Corporation/Center for Advanced Aviation System Development, McLean VA, 1998.

[9] RTCA Inc., "Development and Implementation Planning Guide for Automatic Dependent Surveillance-Broadcast (ADS-B) Applications," DO-249, Washington, DC, 1999.

Chapter 43

Conclusions and Observations

I. Introduction

A VERY large body of data and analysis has been presented in this book. This is the first attempt the editors are aware of to develop a comprehensive theory of air transportation systems engineering. The model (proposed by Haraldsdottir in Chapter 2) of an air traffic management (ATM) system as a sequence of nested/inter-related feedback control loops provides deep insight into how ATM systems operate. The remaining chapters in Section I show that, although the U.S. and European systems use essentially the same equipment, they employ them somewhat differently. There are as many differences as there are similarities in the two world-dominant ATM systems.

II. U.S. Air Traffic Management System

The U.S. system attempts to maximize throughput capacity assuming that airports have unlimited and uncontrolled capacity and operate under ideal meteorological conditions. Most U.S. airports do not employ slot controls. In addition, ATM controllers under visual meteorological conditions do not tightly control the spacing of approaching aircraft. This leads to high capacity operations under ideal conditions but creates a more fragile system that leads to significant, unpredictable delays and flight cancellations due to inherently unpredictable weather events or airport restriction conditions.

Since 1978, deregulation of the privately owned airlines has led to a privately organized and competitively operated hub-and-spoke transportation network. The hub-and-spoke system requires flight schedules to be banked into tightly grouped arrival and departure times that produce large queueing delays at the hub airports as they approach maximum operational capacity fractions. These queueing delays insert schedule disruptions into the entire transportation network because of the non-linear propagation of these delays throughout the entire system.

The philosophical nature of the U.S. government is toward a minimum regulation structure, and thus most technical efforts are currently aimed at maximizing hub airport throughput at high operational load factors with a minimum of regulation. This philosophy is epitomized by the goal of the United States to evolve to a Free Flight paradigm. Taken literally, a true free flight, minimum restriction ATM system will theoretically lead to large systemic queueing delays at high capacity fractions. Most professionals more properly think of Free Flight as the elimination of unnecessary restrictions.

III. European Air Traffic Management System

Although both the United States and Europe use International Civil Aviation Organization (ICAO) approved communications, navigation, and surveillance (CNS) and ATM equipment and procedures, they employ them quite differently. Europe is much more disposed to transportation systems that are predictable and run on time, all the time. Therefore, the ATM system is more highly regulated, and airport slot controls are the rule rather than the exception. The existence of slot controls and an increasingly powerful central flow control system operated by EUROCONTROL in Brussels eases the burden on airports. European airports are lightly loaded by U.S. standards. Arriving aircraft into European airports are always spaced at inclement meteorological conditions (IMC) or positive control separation distances, thus greatly decreasing visual meteorological conditions (VMC) capacity. The high value placed on transportation system predictability leads the Europeans to operate the system under IMC standards as a matter of course. In addition, European airspace design is far from optimum due to the fact that each individual country operates its low-altitude airspace. Non-optimal airspace design leads to en route sector saturation as the prime cause of ATM delays. These delays are more or less allocated to major airports uniformly in the interest of international equity even though the European airports are operating at relatively low capacity fractions. The European research focus is thus on en route controller workload, which is deemed to be the largest source of delay in the system.

IV. Public-Private Nature of Air Transportation

The true technical nature of the partnership between the public and the private sector in the United States has not received much attention in the past. Increasingly, the airline-operated air operation center (AOC) is becoming as important in the overall operational efficiency as the air route traffic control center (ARTCC) and the central flow management control (CFMC). Little has been published about the operation of these centers, and Section II presents an important insight into both the operations of an AOC as well as the economic consequences of delays and disruptions on the commercial air carriers.

One of the prime innovations in the modernization of the U.S. air traffic control (ATC) system is the introduction of collaborative decision making (CDM) between the Federal Aviation Administration (FAA) control centers and the airline AOCs. To date, the primary purpose of CDM has been to intervene during periods of adverse weather to attempt to minimize disruptions to the overall network flow. Unfortunately, much of the potential of CDM for capacity increase has not yet been realized. Chapter 15 by Campbell et al. illustrates that the National Airspace System (NAS) is a complex adaptive system with many players optimizing different object functions. CDM can be used to increase airline profitability as well as passenger throughput and timeliness.

The nested feedback control loop model described by Haraldsdottir in Chapter 2 illustrates the underlying causes for this disappointing performance. For the United States, weather prediction is the primary input to planned ground delay programs (GDPs) that are negotiated with FAA centers and airline AOC's. Transcontinental flights (of approximately 5 h flight duration) would like to have GDP information at least one hour prior to aircraft scheduled departure time. Thus, the feedback control loop has a time constant of roughly 6 h. Unfortunately, convective weather predictions are not very accurate (especially to the scale of airport terminal areas) 6 h in advance. Giving more authority to CFMC for aggressive GDPs is like turning up the gain on an amplifier with a low signal-to-noise-ratio input signal. Control theory tells us that this can lead to system instability, which we have observed in the summers of 1999 and 2000 in the United States. For short-duration flights, the weather forecasts are much more accurate and the CDM/GDP program shows much better performance. Proper use of "reservoirs" of planes in the vicinity of airports helps dampen out the unpredictability.

Europeans use their CFMC in a much more predicable way, perhaps to greater effect. Weather patterns in Europe are significantly different from those encountered in the United States. Whereas the United States struggles with largely unpredictable spring and summer thunderstorms and winter snowstorms, Europe experiences large periods of relatively predictable low clouds and steady rain. The European CFMC facility uses its strategic flight information to anticipate uncoordinated high-altitude sector loading congestion. It uses this information to issue GDP holds to the system airports on almost a daily and predictable basis to compensate for inadequate airline network scheduling coordination.

V. Safety Is Much Discussed But Little Analyzed

Despite all of the discussion by governmental agencies, the modeling and understanding of ATM system underlying safety principles are grossly inadequate. Section VI is the most comprehensive discussion of this subject in a single volume that the editors are aware of. The majority of this effort has come from one organization, the National Aerospace Laboratory NLR in the Netherlands, under the leadership of Henk Blom. Four chapters are devoted to describing the philosophy and giving examples of the Traffic Organization and Perturbation Analyzer (TOPAZ) modeling approach. The TOPAZ approach develops a system of stochastic differential equations to model the air transportation system in specific phases of flight. A considerable amount of analysis is required to model each specific phase of flight for each technical system and envisioned operational procedure. Once the applicable probability density functions have been estimated and the Petri nets have been constructed, probabilities of aircraft collision are computed and compared to the ICAO safety goal of 5×10^{-9} hull losses per aircraft flight hour.

Not only is the analysis a daunting task, but the ability for many government officials to understand the analysis is questionable. Much more work needs to be done in this area. Both the underlying mathematics and model validation need attention by air transportation certification officials. A theoretically less rigorous, but more intuitive, approach is to conduct an operational hazard safety analysis, similar to the approach traditionally taken by the FAA for new aircraft design certification. Both of these approaches are a significant departure for ATM service providers, but the trend is clearly in the direction of better ATM system safety design in the future. Safety is the combination of a much broader analysis conducted during all aspects of system design and development. Upon deployment, safety must get cultural focus during initial system operation.

Air transportation system safety and capacity are clearly related functions. One such relationship is hypothesized in Chapter 1 by one of the editors (Donohue). Although most ATM design engineers understand that this relationship must exist, government transportation officials rarely discuss it. This lack of attention is continuing to lead to poor policy decisions and must be improved in the future.

VI. Air Traffic Controller—Pilot Cognitive Workload Substitution Function

Sections VII and VIII discuss the emerging debate on how the aircraft separation function should be accomplished in the future. Increasingly, the limitations of air transportation capacity are not going to be dictated by the beam width or scan rate of a ground-based radar. Global Positioning System accuracy and digital data links are providing aircraft state vector and intent information that allow exceptionally good 30-min position predictions for flight management system (FMS) equipped aircraft. The ability for computerized decision support systems (DSS) such as the User Request Evaluation Tool (URET) and the Final Approach Spacing Tool (FAST) to utilize this information has been demonstrated in both the United States and Europe in the last several years.

These DSS can be used either to increase ATM controller productivity or to decrease controller workload. To date, they have been used only to decrease controller workload. Both in the United States and in Europe, controller union contracts set productivity levels (i.e., number of aircraft per sector per controller per hour). The advent of the new DSS tools and automatic dependent surveillance broadcast (ADS-B) allow, at least in principle, some of the current controller workload to be transferred to the flight deck and some to be transferred to the DSS. The international research community is increasingly coming to the conclusion that tactical aircraft separation authority and responsibility must be transferred to the flight deck. The overall responsibility for central flow and sequencing will remain a collaborative decision between ATM and AOCs. In the future, as economic incentives begin to take on a more important role, these collaborative decisions may take on more of the form of auctions.

VII. Final Comments

We have tried to provide the reader with the best and most informed thinking by some of the top research personnel in the world in this book. The subject is a very complicated one, and much more work needs to be done. Research is continuing (primarily in the United States and in Europe) to ensure that the primary mode of international transportation does not fail us early in the 21st century. The consequences of such a failure could lead to world economic recession and a breakdown in international relationships. This would come at a time when international economic and institutional barriers are in a rapid state of change. Loss of dependable, affordable international transportation options could turn the world into new and uncharted waters. As the old chart makers used to say, sailing in these waters, "There be dragons!"

Subject Index

accelerated departure procedures (ADP), 344
acceleration uncertainty, 573
accident risk, 464–468, 503–504
Active Final Approach Spacing Tool
  (AFAST), 270, 279
Advanced Air Transportation Program
  (AATT), 177
agent-based modeling (ABM), 229
Air Carrier Cost-Benefit Model (ACCBM),
  88–97
  air carrier operating costs, 93–96
  benefit categories
    payload opportunity, 92
    schedule recovery, 92
    utilization opportunity, 91–92
  life-cycle cost module, 92–93
air carrier investment model (ACIM), 81–83
air navigation systems (ANS), 29
Air Route Surveillance Radar (ARSR),
  571–572, 576
air route traffic control center (ARTCC), 177,
  179, 182, 188, 209, 220, 249, 270,
  424, 449, 603
air traffic control (ATC), 1, 165–167, 179–182,
  184–187, 387, 482, 547, 557, 629
  ATC effect, 507
  ATC effort, 507
air traffic control center (ATCC), 38–39
Air Traffic Control Systems Command Center
  (ATCSCC), 117, 208, 220, 239, 245,
  290, 422
Air Traffic Controller (ATCo), 474, 488,
  490–491, 494
  tactical ATCo model, 494–495
  reduced tactical ATCo model, 499
air traffic flow management (ATFM) program,
  52–53, 193, 409
air traffic management (ATM), 4–6, 49, 103,
  110–114, 161, 191–194, 205, 253,
  367
  operations concepts, 7, 9–11
  flow management
    local traffic flow management, 17
    national flow management, 52
  work system, 11–12
air traffic management demonstration system
  (ATMDS), 698
air traffic workload input technique (ATWIT),
  601, 605
airborne collision avoidance system (ACAS),
  552, 556–560, 672
airborne conflict probability, 543

airborne separation, 533, 537, 556, 649
airborne separation assurance system (ASAS),
  462, 536, 547–556
  ASOR, 558
  hazard classification matrix, 550
  hazard likelihood analysis, 555
aircraft separation minima, 564, 574
airline operations center (AOC), 2, 127–133,
  163, 179, 182–187, 209, 249
  queueing network model of the AOC,
  130–133
airline sequencing model (ASM), 339
Airline Service Quality Performance (ASQP)
  database, 106–109, 121–125, 132,
  136, 307, 414, 430
airport acceptance rate (AAR), 248, 266, 292,
  301
airport capacity curve, 295
airport runway configuration, 309
Airport Surveillance Radar (ASR), 571–572,
  576
American Airlines System Operations Control
  (AALSOC), 180–181
annual ATM interruption benefits, 375
Annual Delay Cost Savings, 278
ARINC Communications Addressing and
  Reporting System (ACARS), 3, 215
arrival flow, 298
Arrival Fuel Cost, 373
Arrival/Departure Optimization Model,
  293–294
  priority demand, 292
ATC control loops, 569
ATC mental workload, 600–601, 605, 607, 611
ATC taskload, 600–601
ATC/Air Carrier Information Exchange,
  180–185
ATM decision support tools (ATM DSTs),
  269, 367, 615
ATM interaction architecture
  strategic, 209
  tactical, 208
    airline/ATM (AA), 208, 221–222
    controller/controller cross-facility (CCX),
      208, 217–221
    controller/controller interfacility (CCI),
      208, 215, 216
    controller/pilot (CP), 208, 209–213
    pilot/airline (PA), 208, 214, 215
ATM interruptions model, 368–369, 374
ATM safety
  collision risk analysis, 482
  hazard analysis, 482

SUBJECT INDEX

automated radar terminal system (ARTS), 587, 591
automatic dependent surveillance (ADS), 23
automatic dependent surveillance broadcast (ADS-B), 547, 650, 667, 695
average flight delay, 84–86
aviation system
  hub-and-spoke service, 31
  point-to-point service, 31
Aviation System Analysis Capability (ASAC), 87

Bank, 256

capacity and delays, 52, 54, 62, 71
CAS (cruise speed), 371–372
CDMNet, 241, 243
CENA (French research center for civil aviation), 438–439, 559
  CENA PD/3, 438–440, 444
Center for Advanced Aviation Systems Development (CAASD), 103–104, 115, 449, 696
Center-TRACON Automation System (CTAS), 177–188, 254, 256, 269, 306–309, 368, 422, 585, 617, 636
  Collaborative Arrival Planning (CAP) project, 179–187
  traffic management advisor (TMA)
    P-GUI, 178–179, 181
    T-GUI, 178–179, 181
central flow control management (CFCM), 2, *see also* CFMC
central flow management control (CFMC), 714–715, *see also* CFMC
central flow management unit (CFMU), 12, 28, 34, 46, 49, 52, 164–165, 198
Chain of impact from ATM changes to fleet finances, 110–111
CNS/ATM, 2, 16–18, 78, 269, 272, 276–285, 461
CNS/ATM Transition Logic Digram, 19–24
cockpit display of traffic information (CDTI), 697–712
Collaborative Arrival Planning (CAP), 179–188
collaborative decision making (CDM), 2, 161, 174, 191, 200, 228–235, 239, 245–249, 290, 352, 422
collaborative routing (CR) function, 422
collaborative routing coordination tool (CRCT), 249
collision risk, 556, 680–691
collision risk tree, 473
communication, navigation, and surveillance (CNS), 269, 649
complex adaptive system (CAS), 229
configuration planner, 350–351

conflict avoidance planning algorithm, 654, 658, 661
conflict detection, 665–666
conflict prediction, 677–678
conflict probability, 679, 684–685
conformance monitoring, 577–580
Consolidated Operations and Delay Analysis System (CODAS) database, 64, 310, 350, 430
constrained forecast of demand, 81
controlled times of departure (CTD), 290
controller acceptance rating scale (CARS), 588–589
controller performance, 600, 602, 605, 607, 610
controller tools and transition trials (C3T), 615–617
Controller-Pilot Data Link Communication (CPDLC), 205, 615–618, 625, 639, 643
cooperating intelligent agents (CIA), 201
cross-track deviation, 578–579

daily entry rate (DER), 41
decision support tool (DST), 627, 632, 636, 638, 641
declared airport capacity, 31, 33–34
declared sector capacity, 41
demand forecast, 79
departure flow, 298
departure planner (DP), 343, 347, 361
Descent Advisor (DA), 585
digital communications, 3
direct routing, 386, 394, 538
dispatcher, 134–140
distributed artificial intelligence (DAI), 651
dynamically colored Petri net (DCPN), 469–473

En Route Descent Advisor (EDA), 270, 368, 371, 376, 627–628, 636
enhanced traffic management system (ETMS), 243, 244, 350, 430
ERATO, 616
estimated on time (E ON), 182
estimated time of arrival (ETA), 170, 254, 256–260
estimated time of departure (ETD), 171–172
EUROCONTROL, 1, 27, 49–50, 475, 649
evaluation value (EV), 669, 672
expected further clearance (EFC) time, 184
extended flight rules (EFR), 650, 667–668

FCFS scheduling algorithms, 254, 258–259, 262
Federal Aviation Administration (FAA), 1, 27, 115, 290–292, 397, 714

SUBJECT INDEX 719

flexible airspace modeling environment (FLAME), 387
flight management system (FMS), 22, 269, 273, 579
flight planning, 12–13, 136–139
flight schedule monitor (FSM), 239, 254, 290, 301
flow rate, 634–635
free flight, 451, 460, 461, 533, 649
free flight conflict resolution algorithm, 539–542
Free Flight Phase 1 (FFP1), 368, 421, 425
Free-Route Experimental Encounter Resolution (FREER), 667
Future Area Control Tools Support (FACTS) Program, 617

Generalized Reich Collision Risk Model, 680
geometric conflict prediction approach, 678
global conflict probability, 543
gradient algorithm, 148–149, 154
ground delay program (GDP), 229–235, 240, 242, 245–246, 248, 290–292, 295
Ground Delay Program Enhancement (GDP-E), 239
ground operations model, 322

human air traffic controller, 385
human performance/reliability model, 483
 contextual control mode model, 485–487
 human error modeling, 487–488
  human error identification (HEI), 487
  human error probability (HEP), 487
 multiple resource model, 483–485
Hypothetical pilot-controller workload substitution curve, 5–6
Hypothetical safety-capacity substitution curves, 5

inclement meteorological condition (IMC), 28, 31, 277
information distribution process, 162
information gaps, 166–173
information sharing, 205–206
instrument flight rules (IFR), 29, 273, 295, 313
instrument landing system (ILS), 575
integrated predictive error (IPE), 241–242
Intelligent Agent-Based Model for Policy Analysis of Collaborative TFM (IMPACT), 227–228, 231–233
intent, 569–570
International Civil Aviation Organization (ICAO), 1, 3, 55, 61, 200, 209, 464, 519, 547, 680

Joint Air Navigation Experiments (JANE), 191

limitations on growth
 airport capacity, 57
 environment, 58
Low Visibility Landing and Surface Operations (LVLASO), 88, 97–102

Macroscopic Capacity Model (MCM), 62–71
maneuver coordination, 665, 667
medium term conflict detection (MTCD), 616, 622–625
meta-rules, 670–671
metering conformance cost model, 373–375
miles-in-trail (MIT) restrictions, 355–356
multi-sector planner (MSP), 641–642

National Airspace System (NAS), 78, 115, 127, 269, 290, 397, 400
 NAS long-range performance objectives, 9
 NAS preliminary design process, 9, 10
no-action, 229–230

operational environment definition (OED), 548
Operational Evaluation (OpEval), 696, 698–699, 701
operational hazard analysis (OHA), 549, 554
operational safety assessment (OSA) methodology, 548
operator, 128–129
 aircraft routers, 128
 ATC coordinators, 128
 crew schedulers, 128
 system operations controllers, 128
origin-destination (OD), 145–146, 151
original estimated time of departure (OETD), 243–244
over-the-shoulder (OTS), 601, 605, 606, 611
overlap probability, 680, 685–686

passive Final Approach Spacing Tool (pFAST), 110, 177, 424, 430, 585–595
PC/TC roles, 619
Performance and Objective Workload Evaluation Research (POWER), 601, 603–611
planning controller (PC), 615, 617
position shift (PS), 261
position uncertainty, 571–572
precision runway monitor (PRM), 576
pricing model of ATM, 144–147
priority scheduling, 255, 260, 265
probabilistic discrete choice models, 146
 multinomial Logit model, 147
 multinomial Probit model, 147

Program for Harmonized Air Traffic
  Management Research in
  EUROCONTROL (PHARE),
  437–438, 616, 639

Radio Technical Commission for Aeronautics
  (RTCA), 462, 533, 548
radio telephony (RT) communication, 474,
  490, 620
rate control index (RCI), 247
reduced vertical separation minima (RVSM),
  38, 563
Reorganized ATC Mathematical Simulator
  (RAMS), 465, 620
revenue passenger miles (RPM), 81, 94
risk assessment spiral development cycle,
  514
risk curve, 505
RNP1, 474
rule evaluation, 668, 670
runway capacity, 72, 306

safe separation, 519
safety and capacity substitution curve, 4–5
schedule disruptions, 115–117
scheduled time of arrival (STA), 178, 184,
  369, 376
sector complexity, 600–601, 605, 607–608
sector controller, 14, 15
sector orientation, 630–632
sector planning, 14, *see also* sector controller
sector vector model, 567–568
  general state vector approach, 567
  surveilled acceleration vector, 568
  surveilled position vector, 568
  surveilled velocity vector, 568
self-separation, 647, 665
separation assurance budget, 565
  minimum separation requirement, 565
  personal safety buffer, 565
  procedural safety buffer, 565
  protection zone, 565
  surveillance uncertainty, 565
separation standards, 2, 15
short-term conflict alert (STCA), 474–476, 482
simulated annealing algorithm, 149–150, 155
sliding hourly entry rate (SHER), 41
standard instrument departures (SID), 369, 399
standard terminal arrival routes (STAR), 348,
  399, 569
stochastic differential equation (SDE),
  469–470
strategic collaborative arrival planning
  (S-CAP), 186
surface movement advisor (SMA), 424
system assisted coordination (SYSCO),
  615–625
system operations controller (SOC), 128, 131
system-wide excess costs, 106

Task Load Index (TLX), 601, 605–606
Terminal Radar Approach Control
  (TRACON), 34, 177, 216, 219, 256,
  274–276, 306, 430, 585, 703
Total Airport and Airspace Modeller (TAAM),
  411, 465
total system throughput, 84
Traffic Alert and Collision Avoidance System
  (TCAS), 211–213, 536, 543–544,
  552, 696
traffic density, 389
traffic flow management (TFM), 18–19,
  227–229, 290, 301
traffic growth, 50, 535–536
  European traffic growth, 61
traffic information service (TIS), 547
Traffic Management Advisor (TMA), 18–19,
  46, 177, 205, 218–219, 254, 270,
  368, 375, 424, 438, 585
traffic management coordinator (TMC),
  208–209, 217, 586
traffic management phases, 44
Traffic Organization and Perturbation
  Analyzer (TOPAZ), 461, 466, 715
  TOPAZ accident risk assessment cycle, 468
traffic spacing and performance factors,
  15–16
Trajectory Accuracy and Traffic Spacing
  Model, 273–274
trajectory interaction, 387
trajectory orientation, 628–629, 631, 638

uncertainty in daily flow plan, 12–14
unconstrained forecast of demand, 80
User Request Evaluation Tool (URET), 216,
  401, 403–405, 422, 449, 617, 636,
  639, 716
  computer human interface (CHI), 452
  restriction relaxation, 453–455
  URET FFP1 sites, 450

velocity uncertainty, 573
virtual queue manager (VQM), 344, 348,
  352–353
visual acquisition time, 706
visual flight rules (VFR), 273, 674
visual meteorological condition (VMC), 28,
  31, 277

wake vortex risk assessment, 516
  COLLIR, 516
  CRITER, 516
  SIMULATOR, 514
  TAXIR, 516
  WAVIR, 516
window control, 332–336
workload assessment keypad (WAK),
  601, 605

PROGRESS IN ASTRONAUTICS AND AERONAUTICS SERIES VOLUMES

*1. **Solid Propellant Rocket Research (1960)**
Martin Summerfield
Princeton University

*2. **Liquid Rockets and Propellants (1960)**
Loren E. Bollinger
Ohio State University
Martin Goldsmith
The Rand Corp.
Alexis W. Lemmon Jr.
Battelle Memorial Institute

*3. **Energy Conversion for Space Power (1961)**
Nathan W. Snyder
Institute for Defense Analyses

*4. **Space Power Systems (1961)**
Nathan W. Snyder
Institute for Defense Analyses

*5. **Electrostatic Propulsion (1961)**
David B. Langmuir
Space Technology Laboratories, Inc.
Ernst Stuhlinger
NASA George C. Marshall Space Flight Center
J. M. Sellen Jr.
Space Technology Laboratories, Inc.

*6. **Detonation and Two-Phase Flow (1962)**
S. S. Penner
California Institute of Technology
F. A. Williams
Harvard University

*7. **Hypersonic Flow Research (1962)**
Frederick R. Riddell
AVCO Corp.

*8. **Guidance and Control (1962)**
Robert E. Roberson
Consultant
James S. Farrior
Lockheed Missiles and Space Co.

*9. **Electric Propulsion Development (1963)**
Ernst Stuhlinger
NASA George C. Marshall Space Flight Center

*10. **Technology of Lunar Exploration (1963)**
Clifford I. Cumming
Harold R. Lawrence
Jet Propulsion Laboratory

*11. **Power Systems for Space Flight (1963)**
Morris A. Zipkin
Russell N. Edwards
General Electric Co.

*12. **Ionization in High-Temperature Gases (1963)**
Kurt E. Shuler, Editor
National Bureau of Standards
John B. Fenn, Associate Editor
Princeton University

*13. **Guidance and Control-II (1964)**
Robert C. Langford
General Precision Inc.
Charles J. Mundo
Institute of Naval Studies

*14. **Celestial Mechanics and Astrodynamics (1964)**
Victor G. Szebehely
Yale University Observatory

*15. **Heterogeneous Combustion (1964)**
Hans G. Wolfhard
Institute for Defense Analyses
Irvin Glassman
Princeton University
Leon Green Jr.
Air Force Systems Command

*16. **Space Power Systems Engineering (1966)**
George C. Szego
Institute for Defense Analyses
J. Edward Taylor
TRW Inc.

*17. **Methods in Astrodynamics and Celestial Mechanics (1966)**
Raynor L. Duncombe
U.S. Naval Observatory
Victor G. Szebehely
Yale University Observatory

*18. **Thermophysics and Temperature Control of Spacecraft and Entry Vehicles (1966)**
Gerhard B. Heller
NASA George C. Marshall Space Flight Center

*19. **Communication Satellite Systems Technology (1966)**
Richard B. Marsten
Radio Corporation of America

---

*Out of print.

*20. **Thermophysics of Spacecraft and Planetary Bodies: Radiation Properties of Solids and the Electromagnetic Radiation Environment in Space (1967)**
Gerhard B. Heller
NASA George C. Marshall Space Flight Center

*21. **Thermal Design Principles of Spacecraft and Entry Bodies (1969)**
Jerry T. Bevans
TRW Systems

*22. **Stratospheric Circulation (1969)**
Willis L. Webb
Atmospheric Sciences Laboratory, White Sands, and University of Texas at El Paso

*23. **Thermophysics: Applications to Thermal Design of Spacecraft (1970)**
Jerry T. Bevans
TRW Systems

*24. **Heat Transfer and Spacecraft Thermal Control (1971)**
John W. Lucas
Jet Propulsion Laboratory

25. **Communication Satellites for the 70's: Technology (1971)**
Nathaniel E. Feldman
The Rand Corp.
Charles M. Kelly
The Aerospace Corp.

26. **Communication Satellites for the 70's: Systems (1971)**
Nathaniel E. Feldman
The Rand Corp.
Charles M. Kelly
The Aerospace Corp.

27. **Thermospheric Circulation (1972)**
Willis L. Webb
Atmospheric Sciences Laboratory, White Sands, and University of Texas at El Paso

28. **Thermal Characteristics of the Moon (1972)**
John W. Lucas
Jet Propulsion Laboratory

*29. **Fundamentals of Spacecraft Thermal Design (1972)**
John W. Lucas
Jet Propulsion Laboratory

*30. **Solar Activity Observations and Predictions (1972)**
Patrick S. McIntosh
Murray Dryer
Environmental Research Laboratories, National Oceanic and Atmospheric Administration

*31. **Thermal Control and Radiation (1973)**
Chang-Lin Tien
University of California at Berkeley

*32. **Communications Satellite Systems (1974)**
P. L. Bargellini
COMSAT Laboratories

*33. **Communications Satellite Technology (1974)**
P. L. Bargellini
COMSAT Laboratories

*34. **Instrumentation for Airbreathing Propulsion (1974)**
Allen E. Fuhs
Naval Postgraduate School
Marshall Kingery *Arnold Engineering Development Center*

*35. **Thermophysics and Spacecraft Thermal Control (1974)**
Robert G. Hering
University of Iowa

36. **Thermal Pollution Analysis (1975)**
Joseph A. Schetz
Virginia Polytechnic Institute
ISBN 0-915928-00-0

*37. **Aeroacoustics: Jet and Combustion Noise; Duct Acoustics (1975)**
Henry T. Nagamatsu, Editor
General Electric Research and Development Center
Jack V. O'Keefe, Associate Editor
The Boeing Co.
Ira R. Schwartz, Associate Editor
NASA Ames Research Center
ISBN 0-915928-01-9

*38. **Aeroacoustics: Fan, STOL, and Boundary Layer Noise; Sonic Boom; Aeroacoustics Instrumentation (1975)**
Henry T. Nagamatsu, Editor
General Electric Research and Development Center
Jack V. O'Keefe, Associate Editor
The Boeing Co.
Ira R. Schwartz, Associate Editor
NASA Ames Research Center
ISBN 0-915928-02-7

*39. **Heat Transfer with Thermal Control Applications (1975)**
M. Michael Yovanovich
University of Waterloo
ISBN 0-915928-03-5

---

*Out of print.

*40. **Aerodynamics of Base Combustion (1976)**
S. N. B. Murthy, Editor
J. R. Osborn,
Associate Editor
Purdue University
A. W. Barrows
J. R. Ward,
Associate Editors
Ballistics Research Laboratories
ISBN 0-915928-04-3

*41. **Communications Satellite Developments: Systems (1976)**
Gilbert E. LaVean
Defense Communications Agency
William G. Schmidt
CML Satellite Corp.
ISBN 0-915928-05-1

*42. **Communications Satellite Developments: Technology (1976)**
William G. Schmidt
CML Satellite Corp.
Gilbert E. LaVean
Defense Communications Agency
ISBN 0-915928-06-X

*43. **Aeroacoustics: Jet Noise, Combustion and Core Engine Noise (1976)**
Ira R. Schwartz, Editor
NASA Ames Research Center
Henry T. Nagamatsu,
Associate Editor
General Electric Research and Development Center
Warren C. Strahle,
Associate Editor
Georgia Institute of Technology
ISBN 0-915928-07-8

*44. **Aeroacoustics: Fan Noise and Control; Duct Acoustics; Rotor Noise (1976)**
Ira R. Schwartz, Editor
NASA Ames Research Center
Henry T. Nagamatsu,
Associate Editor
General Electric Research and Development Center
Warren C. Strahle,
Associate Editor
Georgia Institute of Technology
ISBN 0-915928-08-6

*45. **Aeroacoustics: STOL Noise; Airframe and Airfoil Noise (1976)**
Ira R. Schwartz, Editor
NASA Ames Research Center
Henry T. Nagamatsu,
Associate Editor
General Electric Research and Development Center
Warren C. Strahle,
Associate Editor
Georgia Institute of Technology
ISBN 0-915928-09-4

*46. **Aeroacoustics: Acoustic Wave Propagation; Aircraft Noise Prediction; Aeroacoustic Instrumentation (1976)**
Ira R. Schwartz, Editor
NASA Ames Research Center
Henry T. Nagamatsu,
Associate Editor
General Electric Research and Development Center
Warren C. Strahle,
Associate Editor
Georgia Institute of Technology
ISBN 0-915928-10-8

*47. **Spacecraft Charging by Magnetospheric Plasmas (1976)**
Alan Rosen
TRW Inc.
ISBN 0-915928-11-6

*48. **Scientific Investigations on the Skylab Satellite (1976)**
Marion I. Kent
Ernst Stuhlinger
NASA George C. Marshall Space Flight Center
Shi-Tsan Wu
University of Alabama
ISBN 0-915928-12-4

*49. **Radiative Transfer and Thermal Control (1976)**
Allie M. Smith
ARO Inc.
ISBN 0-915928-13-2

*50. **Exploration of the Outer Solar System (1976)**
Eugene W. Greenstadt
TRW Inc.
Murray Dryer
National Oceanic and Atmospheric Administration
Devrie S. Intriligator
University of Southern California
ISBN 0-915928-14-0

*51. **Rarefied Gas Dynamics, Parts I and II (two volumes) (1977)**
J. Leith Potter
ARO Inc.
ISBN 0-915928-15-9

*52. **Materials Sciences in Space with Application to Space Processing (1977)**
Leo Steg
General Electric Co.
ISBN 0-915928-16-7

---

*Out of print.

*53. Experimental
Diagnostics in Gas Phase
Combustion Systems
(1977)
Ben T. Zinn, Editor
*Georgia Institute of
Technology*
Craig T. Bowman,
Associate Editor
Stanford University
Daniel L. Hartley,
Associate Editor
Sandia Laboratories
Edward W. Price,
Associate Editor
*Georgia Institute of
Technology*
James G. Skifstad,
Associate Editor
Purdue University
ISBN 0-915928-18-3

*54. Satellite
Communication: Future
Systems (1977)
David Jarett
TRW Inc.
ISBN 0-915928-18-3

*55. Satellite
Communications:
Advanced Technologies
(1977)
David Jarett
TRW Inc.
ISBN 0-915928-19-1

*56. Thermophysics of
Spacecraft and Outer
Planet Entry Probes
(1977)
Allie M. Smith
ARO Inc.
ISBN 0-915928-20-5

*57. Space-Based
Manufacturing from
Nonterrestrial Materials
(1977)
Gerald K. O'Neill, Editor
Brian O'Leary,
Assistant Editor
Princeton University
ISBN 0-915928-21-3

*58. Turbulent
Combustion (1978)
Lawrence A. Kennedy
*State University of
New York at Buffalo*
ISBN 0-915928-22-1

*59. Aerodynamic
Heating and Thermal
Protection Systems (1978)
Leroy S. Fletcher
University of Virginia
ISBN 0-915928-23-X

*60. Heat Transfer and
Thermal Control Systems
(1978)
Leroy S. Fletcher
University of Virginia
ISBN 0-915928-24-8

*61. Radiation Energy
Conversion in Space
(1978)
Kenneth W. Billman
NASA Ames Research Center
ISBN 0-915928-26-4

*62. Alternative
Hydrocarbon Fuels:
Combustion and Chemical
Kinetics (1978)
Craig T. Bowman
Stanford University
Jorgen Birkeland
Department of Energy
ISBN 0-915928-25-6

*63. Experimental
Diagnostics in Combustion
of Solids (1978)
Thomas L. Boggs
Naval Weapons Center
Ben T. Zinn
*Georgia Institute of
Technology*
ISBN 0-915928-28-0

*64. Outer Planet Entry
Heating and Thermal
Protection (1979)
Raymond Viskanta
Purdue University
ISBN 0-915928-29-9

*65. Thermophysics and
Thermal Control (1979)
Raymond Viskanta
Purdue University
ISBN 0-915928-30-2

*66. Interior Ballistics of
Guns (1979)
Herman Krier
*University of Illinois at
Urbana–Champaign*
Martin Summerfield
New York University
ISBN 0-915928-32-9

*67. Remote Sensing of
Earth from Space: Role of
"Smart Sensors" (1979)
Roger A. Breckenridge
*NASA Langley Research
Center*
ISBN 0-915928-33-7

*68. Injection and Mixing
in Turbulent Flow (1980)
Joseph A. Schetz
*Virginia Polytechnic
Institute and State
University*
ISBN 0-915928-35-3

*69. Entry Heating and
Thermal Protection (1980)
Walter B. Olstad
NASA Headquarters
ISBN 0-915928-38-8

*70. Heat Transfer,
Thermal Control, and
Heat Pipes (1980)
Walter B. Olstad
NASA Headquarters
ISBN 0-915928-39-6

*71. Space Systems and
Their Interactions with
Earth's Space
Environment (1980)
Henry B. Garrett
Charles P. Pike
Hanscom Air Force Base
ISBN 0-915928-41-8

---

*Out of print.

*72. Viscous Flow Drag
Reduction (1980)
Gary R. Hough
*Vought Advanced Technology
Center*
ISBN 0-915928-44-2

*73. Combustion
Experiments in a Zero-
Gravity Laboratory
(1981)
Thomas H. Cochran
*NASA Lewis Research
Center*
ISBN 0-915928-48-5

*74. Rarefied Gas
Dynamics, Parts I and II
(two volumes) (1981)
Sam S. Fisher
University of Virginia
ISBN 0-915928-51-5

*75. Gasdynamics of
Detonations and
Explosions (1981)
J. R. Bowen
*University of Wisconsin
at Madison*
N. Manson
Universite de Poitiers
A. K. Oppenheim
*University of California
at Berkeley*
R. I. Soloukhin
*Institute of Heat and Mass
Transfer, BSSR Academy
of Sciences*
ISBN 0-915928-46-9

*76. Combustion in
Reactive Systems (1981)
J. R. Bowen
*University of Wisconsin
at Madison*
N. Manson
Universite de Poitiers
A. K. Oppenheim
*University of California
at Berkeley*
R. I. Soloukhin
*Institute of Heat and
Mass Transfer, BSSR
Academy of Sciences*
ISBN 0-915928-47-7

*77. Aerothermodynamics
and Planetary Entry
(1981)
A. L. Crosbie
University of Missouri-Rolla
ISBN 0-915928-52-3

*78. Heat Transfer and
Thermal Control (1981)
A. L. Crosbie
University of Missouri-Rolla
ISBN 0-915928-53-1

*79. Electric Propulsion
and Its Applications to
Space Missions (1981)
Robert C. Finke
*NASA Lewis Research
Center*
ISBN 0-915928-55-8

*80. Aero-Optical
Phenomena (1982)
Keith G. Gilbert
Leonard J. Otten
*Air Force Weapons
Laboratory*
ISBN 0-915928-60-4

*81. Transonic
Aerodynamics (1982)
David Nixon
*Nielsen Engineering
& Research, Inc.*
ISBN 0-915928-65-5

*82. Thermophysics of
Atmospheric Entry (1982)
T. E. Horton
University of Mississippi
ISBN 0-915928-66-3

*83. Spacecraft Radiative
Transfer and Temperature
Control (1982)
T. E. Horton
University of Mississippi
ISBN 0-915928-67-1

*84. Liquid-Metal
Flows and
Magnetohydrodynamics
(1983)
H. Branover
*Ben-Gurion University
of the Negev*
P. S. Lykoudis
Purdue University
A. Yakhot
*Ben-Gurion University
of the Negev*
ISBN 0-915928-70-1

*85. Entry Vehicle
Heating and Thermal
Protection Systems: Space
Shuttle, Solar Starprobe,
Jupiter Galileo Probe (1983)
Paul E. Bauer
*McDonnell Douglas
Astronautics Co.*
Howard E. Collicott
The Boeing Co.
ISBN 0-915928-74-4

*86. Spacecraft Thermal
Control, Design, and
Operation (1983)
Howard E. Collicott
The Boeing Co.
Paul E. Bauer
*McDonnell Douglas
Astronautics Co.*
ISBN 0-915928-75-2

*87. Shock Waves,
Explosions, and
Detonations (1983)
J. R. Bowen
University of Washington
N. Manson
Universite de Poitiers
A. K. Oppenheim
*University of California
at Berkeley*
R. I. Soloukhin
*Institute of Heat and Mass
Transfer, BSSR Academy
of Sciences*
ISBN 0-915928-76-0

---

*Out of print.

*88. Flames, Lasers, and
Reactive Systems (1983)
J. R. Bowen
University of Washington
N. Manson
Universite de Poitiers
A. K. Oppenheim
*University of California at
Berkeley*
R. I. Soloukhin
*Institute of Heat and Mass
Transfer, BSSR Academy of
Sciences*
ISBN 0-915928-77-9

*89. Orbit-Raising and
Maneuvering Propulsion:
Research Status and
Needs (1984)
Leonard H. Caveny
*Air Force Office of
Scientific Research*
ISBN 0-915928-82-5

*90. Fundamentals
of Solid-Propellant
Combustion (1984)
Kenneth K. Kuo
Pennsylvania State University
Martin Summerfield
*Princeton Combustion
Research Laboratories, Inc.*
ISBN 0-915928-84-1

91. Spacecraft
Contamination: Sources
and Prevention (1984)
J. A. Roux
University of Mississippi
T. D. McCay
*NASA Marshall Space
Flight Center*
ISBN 0-915928-85-X

92. Combustion Diagnostics by Nonintrusive
Methods (1984)
T. D. McCay
*NASA Marshall Space
Flight Center*
J. A. Roux
University of Mississippi
ISBN 0-915928-86-8

93. The INTELSAT Global
Satellite System (1984)
Joel Alper
COMSAT Corp.
Joseph Pelton
INTELSAT
ISBN 0-915928-90-6

94. Dynamics of Shock
Waves, Explosions, and
Detonations (1984)
J. R. Bowen
University of Washington
N. Manson
Universite de Poitiers
A. K. Oppenheim
*University of California
at Berkeley*
R. I. Soloukhin
*Institute of Heat and Mass
Transfer, BSSR Academy
of Sciences*
ISBN 0-915928-91-4

95. Dynamics of Flames and
Reactive Systems (1984)
J. R. Bowen
University of Washington
N. Manson
Universite de Poitiers
A. K. Oppenheim
*University of California
at Berkeley*
R. I. Soloukhin
*Institute of Heat and Mass
Transfer, BSSR Academy
of Sciences*
ISBN 0-915928-92-2

96. Thermal Design of
Aeroassisted Orbital
Transfer Vehicles (1985)
H. F. Nelson
University of Missouri-Rolla
ISBN 0-915928-94-9

97. Monitoring Earth's
Ocean, Land, and
Atmosphere from Space—
Sensors, Systems, and
Applications (1985)
Abraham Schnapf
*Aerospace Systems
Engineering*
ISBN 0-915928-98-1

98. Thrust and Drag: Its
Prediction and
Verification (1985)
Eugene E. Covert
*Massachusetts Institute
of Technology*
C. R. James
Vought Corp.
William F. Kimzey
*Sverdrup Technology
AEDC Group*
George K. Richey
U.S. Air Force
Eugene C. Rooney
*U.S. Navy Department
of Defense*
ISBN 0-930403-00-2

99. Space Stations and Space
Platforms— Concepts,
Design, Infrastructure, and
Uses (1985)
Ivan Bekey
Daniel Herman
NASA Headquarters
ISBN 0-930403-01-0

100. Single- and Multi-
Phase Flows in an
Electromagnetic Field:
Energy, Metallurgical, and
Solar Applications (1985)
Herman Branover
*Ben-Gurion University
of the Negev*
Paul S. Lykoudis
Purdue University
Michael Mond
*Ben-Gurion University
of the Negev*
ISBN 0-930403-04-5

101. MHD Energy
Conversion: Physiotechnical
Problems (1986)
V. A. Kirillin
A. E. Sheyndlin
Soviet Academy of Sciences
ISBN 0-930403-05-3

---

*Out of print.

102. Numerical Methods for Engine-Airframe Integration (1986)
S. N. B. Murthy
Purdue University
Gerald C. Paynter
Boeing Airplane Co.
ISBN 0-930403-09-6

103. Thermophysical Aspects of Re-Entry Flows (1986)
James N. Moss
NASA Langley Research Center
Carl D. Scott
NASA Johnson Space Center
ISBN 0-930430-10-X

*__104. Tactical Missile Aerodynamics (1986)__
M. J. Hemsch
PRC Kentron, Inc.
J. N. Nielson
NASA Ames Research Center
ISBN 0-930403-13-4

105. Dynamics of Reactive Systems Part I: Flames and Configurations; Part II: Modeling and Heterogeneous Combustion (1986)
J. R. Bowen
University of Washington
J.-C. Leyer
Universite de Poitiers
R. I. Soloukhin
Institute of Heat and Mass Transfer, BSSR Academy of Sciences
ISBN 0-930403-14-2

106. Dynamics of Explosions (1986)
J. R. Bowen
University of Washington
J.-C. Leyer
Universite de Poitiers
R. I. Soloukhin
Institute of Heat and Mass Transfer, BSSR Academy of Sciences
ISBN 0-930403-15-0

*__107. Spacecraft Dielectric Material Properties and Spacecraft Charging (1986)__
A. R. Frederickson
U.S. Air Force Rome Air Development Center
D. B. Cotts
SRI International
J.A. Wall
U.S. Air Force Rome Air Development Center
F. L. Bouquet
Jet Propulsion Laboratory, California Institute of Technology
ISBN 0-930403-17-7

*__108. Opportunities for Academic Research in a Low-Gravity Environment (1986)__
George A. Hazelrigg
National Science Foundation
Joseph M. Reynolds
Louisiana State University
ISBN 0-930403-18-5

109. Gun Propulsion Technology (1988)
Ludwig Stiefel
U.S. Army Armament Research, Development and Engineering Center
ISBN 0-930403-20-7

110. Commercial Opportunities in Space (1988)
F. Shahrokhi
K. E. Harwell
University of Tennessee Space Institute
C. C. Chao
National Cheng Kung University
ISBN 0-930403-39-8

111. Liquid-Metal Flows: Magnetohydrodynamics and Application (1988)
Herman Branover
Michael Mond
Yeshajabu Unger
Ben-Gurion University of the Negev
ISBN 0-930403-43-6

112. Current Trends in Turbulence Research (1988)
Herman Branover
Micheal Mond
Yeshajahu Unger
Ben-Gurion University of the Negev
ISBN 0-930403-44-4

113. Dynamics of Reactive Systems Part I: Flames; Part II: Heterogeneous Combustion and Applications (1988)
A. L. Kuhl
R&D Associates
J. R. Bowen
University of Washington
J.-C. Leyer
Universite de Poitiers
A. Borisov
USSR Academy of Sciences
ISBN 0-930403-46-0

114. Dynamics of Explosions (1988)
A. L. Kuhl
R&D Associates
J. R. Bowen
University of Washington
J.-C. Leyer
Universite de Poitiers
A. Borisov
USSR Academy of Sciences
ISBN 0-930403-47-9

115. Machine Intelligence and Autonomy for Aerospace (1988)
E. Heer
Heer Associates, Inc.
H. Lum
NASA Ames Research Center
ISBN 0-930403-48-7

---

*Out of print.

116. Rarefied Gas Dynamics: Space Related Studies (1989)
E. P. Muntz
University of Southern California
D. P. Weaver
U.S. Air Force Astronautics Laboratory (AFSC)
D. H. Campbell
University of Dayton Research Institute
ISBN 0-930403-53-3

117. Rarefied Gas Dynamics: Physical Phenomena (1989)
E. P. Muntz
University of Southern California
D. P. Weaver
U.S. Air Force Astronautics Laboratory (AFSC)
D. H. Campbell
University of Dayton Research Institute
ISBN 0-930403-54-1

118. Rarefied Gas Dynamics: Theoretical and Computational Techniques (1989)
E. P. Muntz
University of Southern California
D. P. Weaver
U.S. Air Force Astronautics Laboratory (AFSC)
D. H. Campbell
University of Dayton Research Institute
ISBN 0-930403-55-X

119. Test and Evaluation of the Tactical Missile (1989)
Emil J. Eichblatt Jr.
Pacific Missile Test Center
ISBN 0-930403-56-8

120. Unsteady Transonic Aerodynamics (1989)
David Nixon
Nielsen Engineering & Research, Inc.
ISBN 0-930403-52-5

121. Orbital Debris from Upper-Stage Breakup (1989)
Joseph P. Loftus Jr.
NASA Johnson Space Center
ISBN 0-930403-58-4

122. Thermal–Hydraulics for Space Power, Propulsion and Thermal Management System Design (1990)
William J. Krotiuk
General Electric Co.
ISBN 0-930403-64-9

123. Viscous Drag Reduction in Boundary Layers (1990)
Dennis M. Bushnell
Jerry N. Hefner
NASA Langley Research Center
ISBN 0-930403-66-5

*__124. Tactical and Strategic Missile Guidance (1990)__
Paul Zarchan
Charles Stark Draper Laboratory, Inc.
ISBN 0-930403-68-1

125. Applied Computational Aerodynamics (1990)
P. A. Henne
Douglas Aircraft Company
ISBN 0-930403-69-X

126. Space Commercialization: Launch Vehicles and Programs (1990)
F. Shahrokhi
University of Tennessee Space Institute
J. S. Greenberg
Princeton Synergetics Inc.
T. Al-Saud
Ministry of Defense and Aviation Kingdom of Saudi Arabia
ISBN 0-930403-75-4

127. Space Commercialization: Platforms and Processing (1990)
F. Shahrokhi
University of Tennessee Space Institute
G. Hazelrigg
National Science Foundation
R. Bayuzick
Vanderbilt University
ISBN 0-930403-76-2

128. Space Commercialization: Satellite Technology (1990)
F. Shahrokhi
University of Tennessee Space Institute
N. Jasentuliyana
United Nations
N. Tarabzouni
King Abulaziz City for Science and Technology
ISBN 0-930403-77-0

*__129. Mechanics and Control of Large Flexible Structures (1990)__
John L. Junkins
Texas A&M University
ISBN 0-930403-73-8

130. Low-Gravity Fluid Dynamics and Transport Phenomena (1990)
Jean N. Koster
Robert L. Sani
University of Colorado at Boulder
ISBN 0-930403-74-6

131. Dynamics of Deflagrations and Reactive Systems: Flames (1991)
A. L. Kuhl
Lawrence Livermore National Laboratory
J.-C. Leyer
Universite de Poitiers
A. A. Borisov
USSR Academy of Sciences
W. A. Sirignano
University of California
ISBN 0-930403-95-9

---

*Out of print.

132. Dynamics of Deflagrations and Reactive Systems: Heterogeneous Combustion (1991)
A. L. Kuhl
Lawrence Livermore National Laboratory
J.-C. Leyer
Universite de Poitiers
A. A. Borisov
USSR Academy of Sciences
W. A. Sirignano
University of California
ISBN 0-930403-96-7

133. Dynamics of Detonations and Explosions: Detonations (1991)
A. L. Kuhl
Lawrence Livermore National Laboratory
J.-C. Leyer
Universite de Poitiers
A. A. Borisov
USSR Academy of Sciences
W. A. Sirignano
University of California
ISBN 0-930403-97-5

134. Dynamics of Detonations and Explosions: Explosion Phenomena (1991)
A. L. Kuhl
Lawrence Livermore National Laboratory
J.-C. Leyer
Universite de Poitiers
A. A. Borisov
USSR Academy of Sciences
W. A. Sirignano
University of California
ISBN 0-930403-98-3

135. Numerical Approaches to Combustion Modeling (1991)
Elaine S. Oran
Jay P. Boris
Naval Research Laboratory
ISBN 1-56347-004-7

136. Aerospace Software Engineering (1991)
Christine Anderson
U.S. Air Force Wright Laboratory
Merlin Dorfman
Lockheed Missiles & Space Company, Inc.
ISBN 1-56347-005-0

137. High Speed Flight Propulsion Systems (1991)
S. N. B. Murthy
Purdue University
E. T. Curran
Wright Laboratory
ISBN 1-56347-011-X

138. Propagation of Intensive Laser Radiation in Clouds (1992)
O. A. Volkovitsky
Yu. S. Sedenov
L. P. Semenov
Institute of Experimental Meteorology
ISBN 1-56347-020-9

139. Gun Muzzle Blast and Flash (1992)
Günter Klingenberg
Fraunhofer-Institut für Kurzzeitdynamik, Ernst-Mach-Institut
Joseph M. Heimerl
U.S. Army Ballistic Research Laboratory
ISBN 1-56347-012-8

140. Thermal Structures and Materials for High-Speed Flight (1992)
Earl. A. Thornton
University of Virginia
ISBN 1-56347-017-9

141. Tactical Missile Aerodynamics: General Topics (1992)
Michael J. Hemsch
Lockheed Engineering & Sciences Company
ISBN 1-56347-015-2

142. Tactical Missile Aerodynamics: Prediction Methodology (1992)
Michael R. Mendenhall
Nielsen Engineering & Research, Inc.
ISBN 1-56347-016-0

143. Nonsteady Burning and Combustion Stability of Solid Propellants (1992)
Luigi De Luca
Politecnico di Milano
Edward W. Price
Georgia Institute of Technology
Martin Summerfield
Princeton Combustion Research Laboratories, Inc.
ISBN 1-56347-014-4

144. Space Economics (1992)
Joel S. Greenberg
Princeton Synergetics, Inc.
Henry R. Hertzfeld
HRH Associates
ISBN 1-56347-042-X

145. Mars: Past, Present, and Future (1992)
E. Brian Pritchard
NASA Langley Research Center
ISBN 1-56347-043-8

146. Computational Nonlinear Mechanics in Aerospace Engineering (1992)
Satya N. Atluri
Georgia Institute of Technology
ISBN 1-56347-044-6

147. Modern Engineering for Design of Liquid-Propellant Rocket Engines (1992)
Dieter K. Huzel
David H. Huang
Rocketdyne Division of Rockwell International
ISBN 1-56347-013-6

148. Metallurgical Technologies, Energy Conversion, and Magnetohydrodynamic Flows (1993)
Herman Branover
Yeshajahu Unger
Ben-Gurion University of the Negev
ISBN 1-56347-019-5

149. Advances in Turbulence Studies (1993)
Herman Branover
Yeshajahu Unger
Ben-Gurion University of the Negev
ISBN 1-56347-018-7

150. Structural Optimization: Status and Promise (1993)
Manohar P. Kamat
Georgia Institute of Technology
ISBN 1-56347-056-X

151. Dynamics of Gaseous Combustion (1993)
A. L. Kuhl
Lawrence Livermore National Laboratory
J.-C. Leyer
Universite de Poitiers
A. A. Borisov
USSR Academy of Sciences
W. A. Sirignano
University of California
ISBN 1-56347-060-8

152. Dynamics of Heterogeneous Gaseous Combustion and Reacting Systems (1993)
A. L. Kuhl
Lawrence Livermore National Laboratory
J.-C. Leyer
Universite de Poitiers
A. A. Borisov
USSR Academy of Sciences
W. A. Sirignano
University of California
ISBN 1-56347-058-6

153. Dynamic Aspects of Detonations (1993)
A. L. Kuhl
Lawrence Livermore National Laboratory
J.-C. Leyer
Universite de Poitiers
A. A. Borisov
USSR Academy of Sciences
W. A. Sirignano
University of California
ISBN 1-56347-057-8

154. Dynamic Aspects of Explosion Phenomena (1993)
A. L. Kuhl
Lawrence Livermore National Laboratory
J.-C. Leyer
Universite de Poitiers
A. A. Borisov
USSR Academy of Sciences
W.A. Sirignano
University of California
ISBN 1-56347-059-4

155. Tactical Missile Warheads (1993)
Joseph Carleone
Aerojet General Corporation
ISBN 1-56347-067-5

156. Toward a Science of Command, Control, and Communications (1993)
Carl R. Jones
Naval Postgraduate School
ISBN 1-56347-068-3

*****157. Tactical and Strategic Missile Guidance Second Edition (1994)**
Paul Zarchan
Charles Stark Draper Laboratory, Inc.
ISBN 1-56347-077-2

158. Rarefied Gas Dynamics: Experimental Techniques and Physical Systems (1994)
Bernie D. Shizgal
University of British Columbia
David P. Weaver
Phillips Laboratory
ISBN 1-56347-079-9

159. Rarefied Gas Dynamics: Theory and Simulations (1994)
Bernie D. Shizgal
University of British Columbia
David P. Weaver
Phillips Laboratory
ISBN 1-56347-080-2

160. Rarefied Gas Dynamics: Space Sciences and Engineering (1994)
Bernie D. Shizgal
University of British Columbia
David P. Weaver
Phillips Laboratory
ISBN 1-56347-081-0

161. Teleoperation and Robotics in Space (1994)
Steven B. Skaar
University of Notre Dame
Carl F. Ruoff
Jet Propulsion Laboratory, California Institute of Technology
ISBN 1-56347-095-0

162. Progress in Turbulence Research (1994)
Herman Branover
Yeshajahu Unger
Ben-Gurion University of the Negev
ISBN 1-56347-099-3

163. Global Positioning System: Theory and Applications, Volume I (1996)
Bradford W. Parkinson
Stanford University
James J. Spilker Jr.
Stanford Telecom
Penina Axelrad, Associate Editor
University of Colorado
Per Enge, Associate Editor
Stanford University
ISBN 1-56347-107-8

---

*Out of print.

164. Global Positioning System: Theory and Applications, Volume II (1996)
Bradford W. Parkinson
Stanford University
James J. Spilker Jr.
Stanford Telecom
Penina Axelrad, Associate Editor
University of Colorado
Per Enge, Associate Editor
Stanford University
ISBN 1-56347-106-X

165. Developments in High-Speed Vehicle Propulsion Systems (1996)
S. N. B. Murthy
Purdue University
E. T. Curran
Wright Laboratory
ISBN 1-56347-176-0

166. Recent Advances in Spray Combustion: Spray Atomization and Drop Burning Phenomena, Volume I (1996)
Kenneth K. Kuo
Pennsylvania State University
ISBN 1-56347-175-2

167. Fusion Energy in Space Propulsion (1995)
Terry Kammash
University of Michigan
ISBN 1-56347-184-1

168. Aerospace Thermal Structures and Materials for a New Era (1995)
Earl A. Thornton
University of Virginia
ISBN 1-56347-182-5

169. Liquid Rocket Engine Combustion Instability (1995)
Vigor Yang
William E. Anderson
Pennsylvania State University
ISBN 1-56347-183-3

170. Tactical Missile Propulsion (1996)
G. E. Jensen
United Technologies Corporation
David W. Netzer
Naval Postgraduate School
ISBN 1-56347-118-3

171. Recent Advances in Spray Combustion: Spray Combustion Measurements and Model Simulation, Volume II (1996)
Kenneth K. Kuo
Pennsylvania State University
ISBN 1-56347-181-7

172. Future Aeronautical and Space Systems (1997)
Ahmed K. Noor
NASA Langley Research Center
Samuel L. Venneri
NASA Headquarters
ISBN 1-56347-188-4

173. Advances in Combustion Science: In Honor of Ya. B. Zel'dovich (1997)
William A. Sirignano
University of California
Alexander G. Merzhanov
Russian Academy of Sciences
Luigi De Luca
Politecnico di Milano
ISBN 1-56347-178-7

174. Fundamentals of High Accuracy Inertial Navigation (1997)
Averil B. Chatfield
ISBN 1-56347-243-0

175. Liquid Propellant Gun Technology (1997)
Günter Klingenberg
Fraunhofer-Institut für Kurzzeitdynamik, Ernst-Mach-Institut
John D. Knapton
Walter F. Morrison
Gloria P. Wren
U.S. Army Research Laboratory
ISBN 1-56347-196-5

176. Tactical and Strategic Missile Guidance Third Edition (1998)
Paul Zarchan
Charles Stark Draper Laboratory, Inc.
ISBN 1-56347-279-1

177. Orbital and Celestial Mechanics (1998)
John P. Vinti
Gim J. Der, Editor
TRW
Nino L. Bonavito, Editor
NASA Goddard Space Flight Center
ISBN 1-56347-256-2

178. Some Engineering Applications in Random Vibrations and Random Structures (1998)
Giorz Maymon
RAFAEL
ISBN 1-56347-258-9

179. Conventional Warhead Systems Physics and Engineering Design (1998)
Richard M. Lloyd
Raytheon Systems Company
ISBN 1-56347-255-4

180. Advances in Missile Guidance Theory (1998)
Joseph Z. Ben-Asher
Isaac Yaesh
Israel Military Industries—Advanced Systems Division
ISBN 1-56347-275-9

181. Satellite Thermal Control for Systems Engineers (1998)
Robert D. Karam
ISBN 1-56347-276-7

182. Progress in Fluid Flow Research: Turbulence and Applied MHD (1998)
Yeshajahu Unger
Herman Branover
Ben-Gurion University of the Negev
ISBN 1-56347-284-8

183. Aviation Weather Surveillance Systems (1999)
Pravas R. Mahapatra
Indian Institute of Science
ISBN 1-56347-340-2

184. Flight Control Systems (2000)
Rodger W. Pratt
Loughborough University
ISBN 1-56347-404-2

185. Solid Propellant Chemistry, Combustion, and Motor Interior Ballistics (2000)
Vigor Yang
Pennsylvania State University
Thomas B. Brill
University of Delaware
Wu-Zhen Ren
China Ordnance Society
ISBN 1-56347-442-5

186. Approximate Methods for Weapons Aerodynamics (2000)
Frank G. Moore
ISBN 1-56347-399-2

187. Micropropulsion for Small Spacecraft (2000)
Michael M. Micci
Pennsylvania State University
Andrew D. Ketsdever
Air Force Research Laboratory, Edwards Air Force Base
ISBN 1-56347-448-4

188. Structures Technology for Future Aerospace Systems (2000)
Ahmed K. Noor
NASA Langley Research Center
ISBN 1-56347-384-4

189. Scramjet Propulsion (2000)
E. T. Curran
Department of the Air Force
S. N. B. Murthy
Purdue University
ISBN 1-56347-322-4

190. Fundamentals of Kalman Filtering: A Practical Approach (2000)
Paul Zarchan
Howard Musoff
Charles Stark Draper Laboratory, Inc.
ISBN 1-56347-455-7

191. Gossamer Spacecraft: Membrane and Inflatable Structures Technology for Space Applications (2001)
Christopher H. M. Jenkins
South Dakota School of Mines
ISBN 1-56347-403-4

192. Theater Ballistic Missile Defense (2001)
Ben-Zion Naveh
Azriel Lorber
WALES Ltd.
ISBN 1-56347-385-2

193. Air Transportation Systems Engineering (2001)
George L. Donohue
George Mason University
Andres G. Zellweger
Embry Riddle Aeronautical University
ISBN 1-56347-474-3

194. Physics of Direct Hit and Near Miss Warhead Technology (2001)
Richard M. Lloyd
Raytheon Electronics Systems
ISBN 1-56347-473-5